# INTRODUCTION TO CANCER METASTASIS

This is Volume 1 in the

CANCER METASTASIS series
Series Editor: Aamir Ahmad

# INTRODUCTION TO CANCER METASTASIS

*Edited by*

AAMIR AHMAD
*Mitchell Cancer Institute*
*University of South Alabama*
*Mobile, AL, United States*

ELSEVIER

AMSTERDAM • BOSTON • HEIDELBERG • LONDON
NEW YORK • OXFORD • PARIS • SAN DIEGO
SAN FRANCISCO • SINGAPORE • SYDNEY • TOKYO
Academic Press is an imprint of Elsevier

Academic Press is an imprint of Elsevier
125 London Wall, London EC2Y 5AS, United Kingdom
525 B Street, Suite 1800, San Diego, CA 92101-4495, United States
50 Hampshire Street, 5th Floor, Cambridge, MA 02139, United States
The Boulevard, Langford Lane, Kidlington, Oxford OX5 1GB, United Kingdom



**Library of Congress Cataloging-in-Publication Data**
A catalog record for this book is available from the Library of Congress

**British Library Cataloguing-in-Publication Data**
A catalogue record for this book is available from the British Library

ISBN: 978-0-12-804003-4

For information on all Academic Press publications
visit our website at https://www.elsevier.com/



*Publisher:* Mica Haley
*Acquisition Editor:* Rafael Teixeira
*Editorial Project Manager:* Pat Gonzalez
*Production Project Manager:* Lucía Pérez
*Designer:* Matthew Limbert

Typeset by TNQ Books and Journals

# Contents

**List of Contributors** ix
**Preface** xi

# I
# CANCER METASTASIS FROM PRIMARY ORGANS

## 1. Cancer Metastasis: An Introduction
H. ZUBAIR AND A. AHMAD

1. Introduction 3
2. Disruption of Cell Adhesion: Initiation of Metastasis 3
3. Matrix Metalloproteinases in Cancer Metastasis 5
4. EMT in Cancer Metastasis 6
5. Organ-Specific Metastasis 7
6. microRNAs and Cancer Stem Cells in Cancer Metastasis 8
7. Conclusions and Perspectives 8

References 9

## 2. Breast Cancer Metastasis
W. TSUJI AND J.A. PLOCK

Abbreviations 13
1. Introduction 13
2. Biology 14
3. Metastatic Dissemination 18
4. Prognostic Factors After Recurrence 20
5. Treatment for Metastatic Breast Cancer 20
6. Conclusions 28

References 28

## 3. Prostate Cancer Metastasis
C. NICHOLSON, I. VELA AND E.D. WILLIAMS

1. Epidemiology and Overview of Metastatic Prostate Cancer 33
2. Cellular Biology of Prostate Cancer Metastases 35
3. The Metastatic Cascade 41
4. Conclusions 53

References 53

## 4. Lung Cancer Metastasis
P. CHANVORACHOTE AND P. CHUNHACHA

1. General Information 61
2. Mechanism of Metastasis 62
3. Molecular Mechanism in the Regulation of Lung Cancer Metastasis 63
4. Epithelial to Mesenchymal Transition in Lung Cancer 67
5. Cancer Stem Cells 69
6. Concluding Remarks 71

Acknowledgments 71
References 71

## 5. Cervical Cancer Metastasis
S.W. AZIZ AND M.H. AZIZ

Abbreviations 77
1. Introduction 77
2. Risk Factors 78
3. Pathogenesis 80
4. Histopathology 81
5. Routes of Dissemination 81
6. Major Molecular Factors 83
7. Conclusion 88

References 88

6. Colorectal Cancer Metastasis
M.A. AZIZ

1. Introduction 95
2. Molecular Aspects of Colorectal Cancer Metastasis 96
3. Main Sites of Colorectal Cancer Metastasis 107
4. Therapeutic Options for Metastatic Colorectal Cancer 109
5. Challenges and Probable Solutions 110
6. Conclusions 112
References 112

7. Metastatic Pancreatic Cancer: Current State and Future Directions
P. MARTINELLI AND E. LONARDO

1. Introduction 117
2. Management of Pancreatic Ductal Adenocarcinoma 119
3. Molecular Mechanisms Driving Metastasis 122
4. In Vivo Experimental Models 127
5. Open Question: Onset and Dynamics of Metastases 129
6. Conclusions 130
References 131

8. Gastric Cancer Metastasis
K. DEBIEC AND J. WYDMANSKI

1. Epidemiology and Screening Programs 137
2. Metastatic GC—Clinical Aspects 139
3. Diagnosis of Metastatic GC With a Particular Focus on 18-F-FDG PET/CT 141
4. Prognostic and Predictive Factors 152
5. Biomarkers 153
6. Treatment Methods of Metastatic GC 154
7. Conclusion 155
Acknowledgments 156
References 156

9. Hepatocellular Carcinoma: Metastatic Disease
S. CHANDRADAS, C.B. SKILLIN AND C.T. FRENETTE

1. Introduction 163
2. Epidemiology 163
3. Diagnosis 167
4. Prognosis and Staging 170
5. Treatment 171
6. Conclusions 174
References 174

10. Metastatic Bladder Cancer
J.L. CHIN, K.M. SIDDIQUI AND K.-C. TRAN

1. Introduction 177
2. Epidemiology 177
3. Staging 179
4. Grading 182
5. Staging Tools 182
6. Emerging Biomarkers 184
7. Prognostic Factors 185
8. Clinical Presentation of Metastatic Bladder Disease 186
9. Management of Clinically Localized Bladder Cancer, Positive Pelvic Nodes and Metastatic Disease 186
10. Clinical Prognosticators 188
11. Management of Metastatic Bladder Cancer 189
12. New Agents 191
13. Immunotherapy 191
14. Role of Surgery in Metastatic Bladder Cancer 192
15. Conclusions 193
References 194

11. Metastatic Cutaneous Squamous Cell Carcinomas
I. GARCIA-DIEZ AND A. TOLL

1. Introduction 199
2. Defining the High-Risk Variant 200
3. Staging Systems 211
4. Treating High-Risk Cutaneous Squamous Cell Carcinoma 212
5. Sentinel Node Biopsy 213
6. Treatment of Locally Advanced or Metastatic Cutaneous Squamous Cell Carcinoma 213
7. Concluding Remarks 214
References 215

12. Osteosarcoma Metastasis—Prognostic Factors and Treatment Strategies
S. IWATA

1. Introduction 223
2. Clinical Features of Metastasis From Osteosarcoma 223

3. Treatment for Metastasis From Osteosarcoma 227
4. Conclusion 229
References 229

# II
# CANCER METASTASIS TO DISTANT ORGANS

## 13. Lymph Node Metastasis

S.D. NATHANSON, K. ROSSO, D. CHITALE AND M. BURKE

1. Introduction 235
2. History 237
3. Mechanisms—How Tumors Metastasize to Lymph Nodes 238
4. Clinical Aspects of Lymph Node Metastasis 244
5. Pathology 250
6. Conclusions & Summary 255
References 255

## 14. Molecular Involvement of the Bone Marrow Microenvironment in Bone Metastasis

S.F. MILER, C.Y. THOMAS AND Y. SHIOZAWA

1. Introduction 263
2. Types of Bone Metastasis 265
3. Pathophysiology of Bone Metastasis 267
4. Treatment Options for Bone Metastasis 269
5. Discussion/Conclusion 271
Acknowledgments 273
References 273

## 15. An Overview on Hepatic Metastasis

F. UGGERI, E. PINOTTI, M. GARANCINI AND F. ROMANO

1. Introduction 277
2. Hepatic Physiology and Cancer Biology 278
3. Hepatic Functional and Surgical Anatomy 280
4. Clinical Presentation and Diagnosis 281
5. Treatment of Hepatic Metastases 284
6. Conclusions 293
References 293

## 16. Pulmonary Metastasis

M. GONZALEZ, T. KRUEGER AND J.Y. PERENTES

1. Development of Pulmonary Metastases 297
2. Rationale for Local Treatment of Pulmonary Metastases 297
3. Diagnosis of Pulmonary Metastases 299
4. Preoperative Exams 300
5. Surgical Resection 302
6. Radiological Ablation 304
7. Radiotherapy 305
8. Lymph Node Involvement 305
9. Resection of Recurrent Metastases 306
10. Lung Metastasis in Given Tumor Types 307
11. Conclusion 310
References 311

## 17. Brain Metastasis: Basic Biology, Clinical Management, and Insight From Experimental Model Systems

D.H. MURRELL, F. PERERA, A.F. CHAMBERS AND P.J. FOSTER

1. Introduction 317
2. Biology 318
3. Clinical Management 321
4. Experimental Model Systems 327
5. Preclinical Findings 329
6. Conclusion 331
References 331

# III
# EMERGING TARGETS FOR CANCER METASTASIS

## 18. Developmental Pathways: Emerging Therapeutic Targets for Cancer Metastasis

A.H. NWABO KAMDJE, L. VECCHIO, P. TAKAM KAMGA, P.F. SEKE ETET, J.M. MULLER, G. BASSI, M. KRAMPERA

1. Introduction 337
2. Prometastatic Changes and Developmental Pathways 338
3. Notch Signaling and Metastasis 341

4. Antimetastatic Potential of Wnt and Hedgehog Targeting 343
5. Concluding Remarks 346
References 347

19. Emerging Therapeutic Targets for Cancer Metastasis: From the Perspective of Embryo Implantation
D.-Q. LI AND Z.-M. SHAO

1. Introduction 353
2. The Similarity Between Embryo Implantation and Tumor Metastasis 355
3. Trophoblast Invasion Inhibitory Factors Derived From the Maternal Endometrium During Embryo Implantation and Their Implications in Cancer Metastasis 358
4. Conclusions and Perspectives 365
Acknowledgments 366
References 366

20. Wnt Signaling as a Therapeutic Target in Cancer and Metastasis
R. MORGAN, R. ANKRAH, S. EL-TANANI, P.M. LOADMAN, L. PATTTERSON, P.S. RUDLAND AND M. EL-TANANI

1. Introduction 375
2. Wnt Signaling and Cancer 378
3. The Role of Wnt Signaling in Epithelial–Mesenchymal Transition and Cancer Stem Cells 379
4. Therapeutic Strategies Targeting Wnt Signaling 381
5. Protein Knockdown Strategies 387
6. Viral-Based Therapies 388
7. Conclusion 390
References 390

**Index 395**

# List of Contributors

**A. Ahmad** University of South Alabama, Mobile, AL, United States

**R. Ankrah** University of Bradford, Bradford, United Kingdom

**M.A. Aziz** King Saud Bin Abdulaziz University for Health Sciences, Riyadh, Kingdom of Saudi Arabia

**M.H. Aziz** East Tennessee State University, Johnson City, TN, United States

**S.W. Aziz** IPC Healthcare, Inc., Johnson City, TN, United States

**G. Bassi** University of Verona, Verona, Italy

**M. Burke** Henry Ford Health System, Detroit, MI, United States

**A.F. Chambers** Western University, London, ON, Canada; London Health Sciences Centre, London, ON, Canada

**S. Chandradas** Scripps Clinic/Green Hospital, La Jolla, CA, United States

**P. Chanvorachote** Chulalongkorn University, Bangkok, Thailand

**J.L. Chin** Western University, London, ON, Canada

**D. Chitale** Henry Ford Health System, Detroit, MI, United States

**P. Chunhacha** Chulalongkorn University, Bangkok, Thailand

**K. Debiec** Maria Sklodowska-Curie Memorial Cancer Center and Institute of Oncology, Gliwice, Poland

**M. El-Tanani** University of Bradford, Bradford, United Kingdom

**S. El-Tanani** University of Bradford, Bradford, United Kingdom

**P.J. Foster** Robarts Research Institute, London, ON, Canada; Western University, London, ON, Canada

**C.T. Frenette** Scripps Clinic/Green Hospital, La Jolla, CA, United States

**M. Garancini** San Gerardo Hospital, Monza, Italy

**I. Garcia-Diez** Hospital del Mar. Parc de Salut Mar, Barcelona, Spain

**M. Gonzalez** Centre Hospitalier Universitaire Vaudois, Lausanne, Switzerland

**S. Iwata** Chiba Cancer Center, Chiba, Japan

**M. Krampera** University of Verona, Verona, Italy

**T. Krueger** Centre Hospitalier Universitaire Vaudois, Lausanne, Switzerland

**D.-Q. Li** Fudan University, Shanghai, China

**P.M. Loadman** University of Bradford, Bradford, United Kingdom

**E. Lonardo** Institute for Research in Biomedicine, Barcelona, Spain; Institute of Genetics and Biophysics 'A. Buzzati-Traverso', Naples, Italy

**P. Martinelli** Medical University Vienna, Austria

**S.F. Miler** Wake Forest School of Medicine, Winston-Salem, NC, United States

**R. Morgan** University of Bradford, Bradford, United Kingdom

**J.M. Muller** University of Poitiers, Poitiers, France

**D.H. Murrell** Robarts Research Institute, London, ON, Canada; Western University, London, ON, Canada

**S.D. Nathanson** Henry Ford Health System, Detroit, MI, United States

**C. Nicholson** Queensland University of Technology, Woolloongabba, QLD, Australia; Queensland Health, Woolloongabba, QLD, Australia

**A.H. Nwabo Kamdje** University of Ngaoundere, Ngaoundere, Cameroon

**L. Pattterson** University of Bradford, Bradford, United Kingdom

**J.Y. Perentes** Centre Hospitalier Universitaire Vaudois, Lausanne, Switzerland

**F. Perera** London Health Sciences Centre, London, ON, Canada

**E. Pinotti** University of Milano-Bicocca, Monza, Italy

**J.A. Plock** University Hospital Zurich, Zurich, Switzerland

**F. Romano** University of Milano-Bicocca, Monza, Italy

**K. Rosso** Henry Ford Health System, Detroit, MI, United States

**P.S. Rudland** University of Liverpool, Liverpool, United Kingdom

**P.F. Seke Etet** Qassim University, Buraydah, Saudi Arabia

**Z.-M. Shao** Fudan University, Shanghai, China

**Y. Shiozawa** Wake Forest School of Medicine, Winston-Salem, NC, United States

**K.M. Siddiqui** Western University, London, ON, Canada

**C.B. Skillin** Scripps Clinic/Green Hospital, La Jolla, CA, United States

**P. Takam Kamga** University of Verona, Verona, Italy

**C.Y. Thomas** Wake Forest School of Medicine, Winston-Salem, NC, United States

**A. Toll** Hospital del Mar. Parc de Salut Mar, Barcelona, Spain

**K.-C. Tran** Western University, London, ON, Canada

**W. Tsuji** Shiga Medical Center for Adults, Shiga, Japan

**F. Uggeri** University of Milano-Bicocca, Monza, Italy

**L. Vecchio** Qassim University, Buraydah, Saudi Arabia

**I. Vela** Queensland University of Technology, Woolloongabba, QLD, Australia; Queensland Health, Woolloongabba, QLD, Australia

**E.D. Williams** Queensland University of Technology, Woolloongabba, QLD, Australia

**J. Wydmanski** Maria Sklodowska-Curie Memorial Cancer Center and Institute of Oncology, Gliwice, Poland

**H. Zubair** University of South Alabama, Mobile, AL, United States

# Preface

We are well aware that cancer is a very difficult disease to manage and treat. The challenges that it presents are manifold—from the many different types and subtypes of cancer that we know to the inadequate knowledge we have with regards to its initiation and progression. Research in the last few decades has helped make us understand the disease a little better but the knowledge is far from conclusive. The one single factor that makes the clinical management of cancer patients so challenging is the process of "metastasis." Cancers confined to primary organs are generally well managed but once they metastasize, they become extremely difficult to treat. With these challenges in mind, it is my distinct pleasure and privilege to present this first volume *Introduction to Cancer Metastasis*, which is part of the series *Cancer Metastasis*.

Although the series *Cancer Metastasis* will explore the various events/hallmarks that are responsible for onset or progress of metastasis, this first volume in the series is an attempt to introduce and summarize the problem of cancer metastasis. Through the chapters on individual primary cancers and their metastasis and those on common metastatic sites, this volume comprehensively informs readers on the broader knowledge so far on cancer metastasis. Since most of the organ-specific information for cancer metastasis is widespread, readers will appreciate all the relevant information in one place, with the most updated information. This volume is a solid platform for the other volumes to follow in the series where we will be analyzing, in greater detail, the information on mechanisms and management of cancer metastasis.

For the convenience of readers, I have organized this volume into three distinct sections. Section 1 (*Cancer Metastasis from Primary Organs*) is the largest section that is composed largely of chapters focused on individual human cancers. The only exception is Chapter 1, which introduces the problem of metastasis and is not organ specific. This introductory chapter summarizes the general information we have on the subject, along with the different factors that are thought to contribute to the overall process of metastasis. This chapter is followed by primary site-specific chapters. Chapters 2–12 focus on the metastases from primary breast cancer, prostate cancer, lung cancer, cervical cancer, colorectal cancer, pancreatic cancer, gastric cancer, hepatocellular cancer, bladder cancer, squamous-cell cancer, and osteosarcoma, respectively. These chapters provide updated information on the current knowledge on the metastases from these primary organs, with a focus on the various factors that are implicated in respective metastases along with a discussion on the most common sites these individual cancers metastasize to.

The second section in the volume (*Cancer Metastasis to Distant Organs*) focuses on the most common secondary and the tertiary sites/organs that the primary cancers metastasize to. Although almost all the organs have been reported to be invaded by the metastasizing cells, there are some organs that remain the preferential sites for metastasis. Chapters 13–17 discuss these common metastatic sites viz. lymph nodes, bones, liver, lungs, and brain. As indicated above, metastasis to distant organs is the primary reason for cancer-associated mortality.

These chapters in the second section detail the challenges of treating metastatic cancers, however, the overall focus of these chapters is to showcase the promising leads we have obtained through the concerted scientific investigations in the last several years. There have been some very positive developments, both in the diagnosis of metastases and the treatment strategies, and these topics remain the point of discussion in this section.

The third and the final section (*Emerging Targets for Cancer Metastasis*) has three chapters focused entirely on the progress made in recent years for the targeted therapy of cancer metastases. These chapters, Chapters 18–20, focus mostly on the developmental pathways that are being actively investigated for their involvement in cancer metastasis. Without any bias, these chapters present the most consensus information on the role of several key signaling pathways in metastasis before building a case for the selective targeting of these pathways to control/eliminate distant metastases. It is pertinent to mention that a few primary cancers or the metastatic site-specific therapies have been discussed in individual chapters in the first two sections, as most appropriate.

Again, it has been nothing short of absolute pleasure working with the world-renowned scientists who are well regarded in their own individual fields, and compiling this volume for the benefit of cancer researchers worldwide. This volume would not have been possible without the time and dedication of all these contributors. Of course, all of the hard work put in by numerous researchers, cited here or not, leading up to the current information, needs to be acknowledged as well. The staff at the Elsevier Inc. has been nothing short of phenomenal, ready to help and support in every possible way.

I hope that this volume, and the series, serves as an excellent resource for the young inquisitive brains just entering the field, in addition to providing an easy to access latest information to the more seasoned researchers on the topics covered.

**Aamir Ahmad**
*Mitchell Cancer Institute*
*University of South Alabama*
*Mobile, AL, United States*

PART I

# CANCER METASTASIS FROM PRIMARY ORGANS

CHAPTER

1

# Cancer Metastasis: An Introduction

*H. Zubair, A. Ahmad*

**University of South Alabama, Mobile, AL, United States**

## 1. INTRODUCTION

Cancer remains a leading cause of death worldwide, resulting in about 8.2 million deaths in 2012 (Stewart and Wild, 2014). Metastases of cancer account for a vast majority of morbidity and mortality of cancer patients and is associated with about 90% of all cancer-associated deaths (Mehlen and Puisieux, 2006). Cancer metastasis is defined as the formation of new tumors (secondary and tertiary tumor nests) in tissues and organs away from the primary site of tumor origin. Although significant strides have been made in understanding the mechanism of metastasis, vital information on the process of metastasis, the factors responsible for the cancer spread and establishment at distal secondary locations are still relatively unknown (Hunter et al., 2008). Nearly half of all cancer patients present clinically detectable metastatic disease (Martin et al., 2000), whereas a larger number of clinically diagnosed cancer patients also have micrometastasis that is beyond detection limit of techniques currently employed (Winter et al., 2015).

Cancer is the result of the accumulation of multiple genetic aberrations and epigenetic modifications which can be considered as the basis of metastasis as well (Vogelstein and Kinzler, 2004; Jaenisch and Bird, 2003). Metastasis requires a careful choreography of chain-of-events to be completed for successful colonization; which otherwise can lead to the elimination of emigrating cells at any stage of metastasis (van et al., 2011). The major events involved in the "metastatic cascade" include loss of cell–cell adhesion, detachment from the extracellular matrix and invasion in to the basement membrane, intravasation to blood stream and survival in circulation, identification of organ suitable for secondary growth, extravasation from circulation and establishment of distant colony followed by angiogenesis to overcome dormancy (Wan et al., 2013) (Fig. 1.1). Identification of underlying mechanism(s) during these crucial events can lead to design of novel targeted therapies that can limit cancer invasion and result in better management and treatment of cancer.

## 2. DISRUPTION OF CELL ADHESION: INITIATION OF METASTASIS

Initiation of the metastatic cascade in cancers derived from the epithelium primarily requires the detachment of cells from the tumor mass

*Introduction to Cancer Metastasis*
http://dx.doi.org/10.1016/B978-0-12-804003-4.00001-3

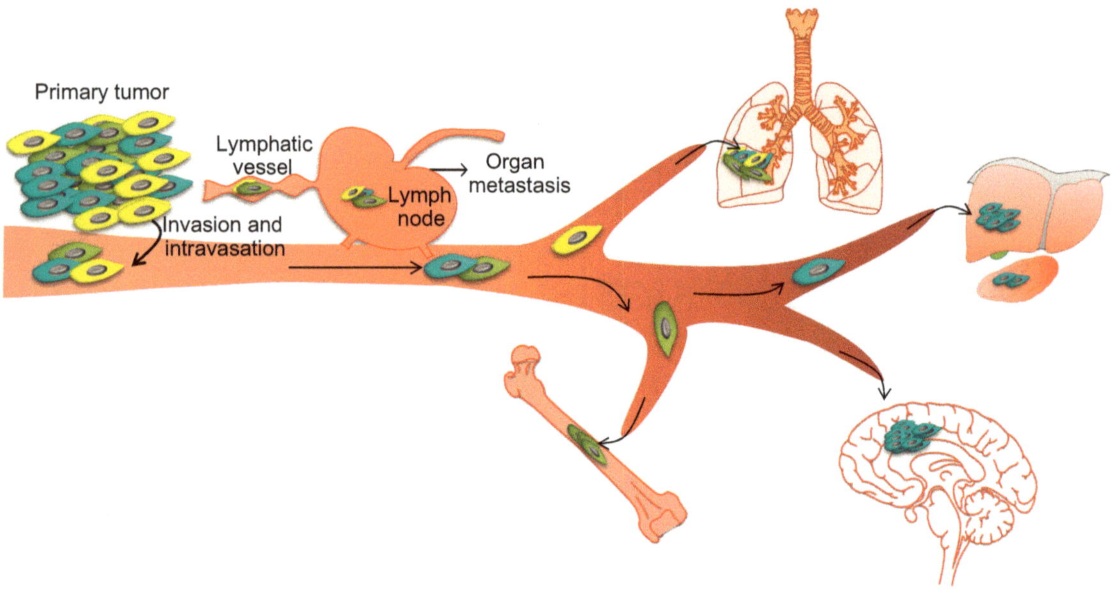


FIGURE 1.1 **An overview of the metastatic cascade.** In metastasis, tumor cells from the primary site initiate loss of cell–cell adhesion and detach from the extracellular matrix to invade the basement membrane from where they intravasate to blood stream. The tumor cells survive in circulation and they seek an organ suitable for their secondary growth, and extravasate from circulation to establish distant colony. Alternatively, these cells may propagate to the lymph nodes from which they can move to distant lymph nodes and other organs.

with changes in the extracellular matrix and the surrounding basement membrane, relieving the cells of cell–cell interactions (Palmer et al., 2011; Bersini et al., 2014). Adhesiveness between cells is maintained by adherens junctions (AJs), tight junctions (TJs), and gap junctions (GPs).

## 2.1 Adherens Junctions

The AJs are the source of the strongest adhesion in the epithelium, followed by TJs and GJs that provide a modest adhesion (Coopman and Djiane, 2016). AJs are cell–cell interacting microdomains (Giepmans and van Ijzendoorn, 2009). Cadherin protein family represents the widely studied AJs and E-cadherin is one of the very well-characterized family member. E-cadherin, apart from maintaining cell–cell adhesion, has also been demonstrated to maintain the levels of cell cycle regulator $p27^{kip1}$(Massarelli et al., 2005). Loss of E-cadherin in cancer cells leads to a simultaneous loss of these cell cycle inhibitors leading to disruption of contact inhibition (Motti et al., 2005). Moreover, the cadherins also have catenin-binding domains that interact with and bind to β-catenin or γ-catenin (plakogobin), which forms a link with the actin cytoskeleton (Conacci-Sorrell et al., 2002). Thus, in relation to cancer, E-cadherins have been delineated as potent tumor suppressors due to their essential role in maintaining cell–cell interactions. E-cadherin's downregulation is also one of the initial step in initiation of epithelial to mesenchymal transition (EMT). However, soluble E-cadherin has been demonstrated to activate multiple signaling pathways enhancing tumor progression (Hu et al., 2016). Furthermore, somatic and germline E-cadherin mutations have been observed in a number of cancers (Berx et al., 1998a,b). Lobular breast cancers, among other cancer types, have been observed to harbor somatic mutations that correlate with tumor progression toward highly invasive and metastatic phenotype of the cancer. Some cases of familial gastric carcinoma have been reported to present germline mutations (Black et al., 2014). However, most tumor cases demonstrate E-cadherin reduction by transcriptional silencing through direct targeting of the promoter sequence (Venza et al., 2016; Han et al., 2016; Liu et al., 2016). Thus, there seems to be overwhelming evidence supporting metastasis-suppressive action of E-cadherin, and the loss or downregulation of E-cadherin is one of the early event in cancer metastasis.

## 2.2 Tight Junctions

TJ complexes are located at the apical membranes of epithelial and endothelial cells and are subdivided into integral membrane proteins, junctional adhesion molecules, the claudin family of proteins, and the cytoplasmic adaptor proteins (Balda and Matter, 2016). These complexes line the plasma membranes between adjacent cells and create an intercellular barrier, occluding the extracellular matrix (Brucher and Jamall, 2014) and are often considered one of the earliest barrier that the cells must overcome to metastasize (Singh and Dhawan, 2015). Studies indicate that in order for a cell to metastasize, similar changes in the tumor cell as well as the neighboring epithelial cells occur that involve the modulation of relevant TJ proteins and result in the loss of cell–cell association causing uncontrolled growth, loss of adhesion, and the degradation of the basement membrane (Forster, 2008). The role of claudin family of tight junction proteins in pathophysiological conditions including cancers has only recently been started to be understood (Ding et al., 2013). However, alterations in claudin expression have been observed to be regulated differently in different cancer types. For example, in pancreatic cancers, claudin-1 expression and distribution have been associated with the activation of the mitogen-activated protein kinases (MAPK) signaling (Kondo et al., 2008), whereas claudin-7 is observed to be decreased in head and neck cancers, metastatic breast cancer, and invasive ductal adenocarcinomas (Singh et al., 2010). In contrast, claudin-4 and -3 are frequently upregulated in various cancers such as prostate (Romanov et al., 2014; Bartholow et al., 2011), ovarian (Liu et al., 2014; Boylan et al., 2011), breast (Ma et al., 2015; Todd et al., 2015), and pancreatic ductal adenocarcinoma (Borka, 2009). Although it has been speculated that mutations in claudins are causative of tumor formation, there are no claudin-specific sequencing data to support this idea. However, hypermethylation of the promoter region of claudin has been observed as the mechanism of claudin-gene silencing in tumors (Maryan et al., 2015; Roll et al., 2013; Riaz et al., 2013; Agarwal et al., 2009; Nakayama et al., 2008; Osanai et al., 2007).

# 3. MATRIX METALLOPROTEINASES IN CANCER METASTASIS

In the initiation as well as progress of metastasis, some of the well-characterized steps involve matrix metalloproteinases (MMPs)-mediated remodeling of the extracellular matrix (ECM) wherein the MMPs induce proteolytic degradation of the ECM altering cell–cell and cell–matrix interactions thereby enhancing migration, angiogenesis, and establishment of the secondary foci at the metastatic site (Thakur and Bedogni, 2016; Candido et al., 2016; Jiang et al., 2015; Al-Alem and Curry, 2015; Deryugina and Quigley, 2015; Kessenbrock et al., 2015). The enhanced activity and overexpression of these MMPs are a result of the concerted effect of both tumor cells and surrounding stromal cells that are activated in a paracrine manner by the tumor cells. Stimulation of the stromal cells or host cells such as fibroblasts is achieved through the tumor-secreted interleukins, growth factors, and extracellular MMP inducers (Ito et al., 1995). The function of the MMPs is largely determined by the nature of the cells that secrete them, for example, MMP-9 secretion by neutrophils results in the activation of the proteinase more readily (Kessenbrock et al., 2010; Fernandez et al., 2005). Similarly, the treatment of stromal cells with transforming growth factor-β (TGF-β) enhances the secretion of pro-MMP-9 protein, whereas TGF-β treatment of prostate and breast cancer cells induces the secretion of both MMP-2 and MMP-9 (Dong et al., 2001). Similar tumor-stromal crosstalk in breast tumors has demonstrated the secretion of MMP-9 by breast fibroblasts upon activation by tumor-secreted tumor necrosis factor alpha (TNF-α) and TGF-β (Stuelten et al., 2005). Furthermore, evidence in the recent past has demonstrated that MMP-family members have different effects on the development and progression of different types of cancers (Gialeli et al., 2011). Different studies have identified the expression of different MMP family members (e.g., MMP-1, -3, -7, -9, -12, and -14) in a variety of solid tumors and non-Hodgkin's lymphomas, suggesting a direct correlation between tumor progression and MMP-expression in various human cancers (Kessenbrock et al., 2015; Gialeli et al., 2011; Gong et al., 2014). Even the preliminary in vivo studies using natural tissue inhibitor of metalloproteinase (TIMP) and synthetic MMP inhibitors to suppress tumor invasion and metastasis in mouse tumor models generated significant interest in the use of these inhibitors in the clinic.

However, not all MMP proteins exhibit the same function in the progression of all types of cancer. The expression of MMP-8 has been correlated with improved survival of patients of squamous cell carcinoma of the tongue (Korpi et al., 2008). Furthermore, in breast cancer, the expression of MMP-8 decreases the metastatic potential of breast cancer cells (Gutierrez-Fernandez et al., 2008). It thus appears that MMP family members may have distinct, sometimes even conflicting, roles in metastasis progression. This is an interesting paradox and clearly more studies are needed to completely understand the complex functionality of MMPs. Studies have also identified other functions of MMPs. For example, cleavage of heparin-binding epidermal growth factor-like growth factor by MMP3 leads to the release of proangiogenic and active growth factor (Suzuki et al., 1997). MMP-1, -2, and -3 can cleave insulin-like growth factor binding protein-3 and other surface growth factor receptors (Fowlkes et al., 1994). They can alter cell cycle progression and affect genomic stability by altering cell adhesion (Bourboulia and Stetler-Stevenson, 2010). MMPs can also induce programmed-cell death in cells, thus enhancing the selection of anoikis resistant cells and promoting tumor or inhibiting tumor growth (Reunanen and Kahari, 2000).

## 4. EMT IN CANCER METASTASIS

The metastatic ability of cancer cells depends a lot on their motility. Motility and migration of cells are crucial for normal orgnanogenesis, wound healing, and inflammation; changes in these signaling pathways can lead to the pathological processes of tumor cell invasion and metastasis. EMT requires the loss of cell adhesion and polarization of the epithelial morphology with characteristic loss of epithelial junctional proteins such as E-cadherin, claudins, etc., as discussed above, and the concomitant increase in the expression of mesenchymal markers and proteins, for example, N-cadherin, fibronectin, vimentin, Zeb1, Zeb2, and cytoskeletal reorganization. It is no wonder that the EMT features of typical developmental phase are recapitulated during the oncogenic EMT (Samatov et al., 2013), but with own unique signature and characteristics, that are the subject of a lot of research studies on cancer. At times this even leads to the generation of hybrid phenotypes that arise as a result of having both epithelial and mesenchymal cell properties (Jolly et al., 2015).

To execute EMT, cells invoke transcriptional alterations of a number of genes critical for EMT and these are considered as the core EMT regulators. The first group of genes upregulated are the transcription factors Snail1 and Snail2, members of the zinc finger family, both of which can directly bind the E-boxes of E-cadherin promoter, repressing its expression (Medici et al., 2008). Zeb1 and Zeb2, zinc finger E-box binding homeobox family proteins, are the second group of regulators that also repress E-cadherin transcription (Wong et al., 2014) via the microRNA (miRNA-200) family expression (Korpal et al., 2008; Ahmad et al., 2011). Interestingly, both Snail and Zeb transcription factors are also able to modulate other junctional proteins such as claudins and ZO-1 (Singh et al., 2011). The third group involves Twist1, Twist2, and E12/E47, basic helix–loop–helix family of transcription factors, which can induce the EMT alone or in cooperation with other transcription factors of other groups (Ansieau et al., 2008). For example, Twist1 induces the expression of Snail transcription factor to repress E-cadherin expression (Lamouille et al., 2014; Yu et al., 2013) and can also activate tumor invasion independently (Mikheeva et al., 2010). Interestingly, the induction of EMT-related genes has not yet been identified to be in association with genetic alteration, instead, the EMT induction in tumor cells is considered to be in response to a combination of extracellular signals present in the tumor microenvironment.

The most notable tumor microenvironment inducer of EMT is the TGF-β pathway (Xu et al., 2009). TGF-β induces the upregulation of all of the EMT core transcription factors, vis-à-vis Zeb1/2, Snail1/2, and Twist1/2[23]. Sma- and Mad-related protein (SMAD) family of transcription factors mediates the EMT modulation by TGF-β (Lamouille et al., 2014), as verified by genetic manipulation of these signaling molecules. However, TGF-β/SMAD signaling induction of EMT is significantly dependent on cooperation with several other signaling pathways, such as activation of Ras kinase cascade and the induction of wingless-type MMTV integration site family member/β-catenin/lymphoid enhancer binding factor 1 (LEF-1) signaling pathway (Cordray and Satterwhite, 2005; Ahmad et al., 2012). TGF-β in the tumor microenvironment is induced by the residential stromal fibroblast cells and may be secreted in response to tumor-induced signals (Taylor et al., 2011). In addition to growth factor signaling, tumor microenvironment has been shown to release inflammatory cytokines, such as TNF-α, C-X-C motif chemokine ligand 12 (CXCL12), IL-6, etc. These cytokines activate the p65-NF-κB (nuclear factor kappa B) signaling inducing the expression of Twist1, which induces EMT (Li et al., 2012; Wu and Zhou, 2010). In addition these cytokines can also induce signal transducer and activator of transcription 3 (STAT3) signaling via activation of Janus kinase (JAK) kinases to induce the expression of Twist1 and vimentin and suppression of E-cadherin (Huang et al., 2016; Cheng et al., 2008). TNF-α has also been observed to enhance the stability of Snail1 through NF-κB activation (Wu and Zhou, 2010). Hypoxic tumor microenvironment activates the hypoxia-inducible factor-1 (HIF-1) transcription factor and augments the expression of Snail1 and Twist1 inducing EMT (Zhang et al., 2013). The induction of EMT, therefore, is intricately connected to the metastatic process and has been proposed as a valid target for therapeutic intervention (Ginnebaugh et al., 2014).

## 5. ORGAN-SPECIFIC METASTASIS

In early stages, the tumor cells are believed to be confined to the primary site within the surrounding stromal tissue and extracellular matrix. However, with the progression of disease, these tumor cells undergo genetic and even epigenetic changes (Ahmad et al., 2014a; Valastyan and Weinberg, 2011), becoming more aggressive, and breach the surrounding cells to invade the local tissue or disseminate via the vascular, lymphatic, or transcoelomic routes (Fig. 1.1). The vascular route of metastases is one of the most studied and better understood. Cancer cells are thought to invade the basement membrane and invade in to nearby blood vessels, thus gaining access to circulation and can, therefore, be transported to distal organs. These cells then extravasate through the blood vessels and invade the tissue of the secondary site to establish distal metastatic nests. Tumor cells, at times, invade in to the lymphatic system of the host whereby they are transported to the regional and distant lymph nodes and then to secondary sites (Wong and Hynes, 2006). Unfortunately, there are no current preoperative investigations to predict the involvement of lymph nodes, although targeted biopsies have a better chance of predicting positive lymph node but only in cases where the entire lymph node has been taken over by tumor cells (AlNoury et al., 2015). Moreover, tumor recurrence in lymph nodes has been observed in patients where it was thought that the tumor was completely resected (Wong and Hynes, 2006), thus suggesting that lymph nodes may be important sites for the occult disease. The third route of spreading of metastatic tumors is in the body cavity and is broadly referred to as transcoelomic metastasis; these usually arise from the cancers of the pancreas, colon, ovary, gastrointestine, and cervix (Bacac and Stamenkovic, 2008).

In the "seed and soil" theory more than a century ago, a predisposal of certain body sites to home metastatic cells and generate secondary tumor nests was proposed (Paget, 1989). Emerging evidence has further strengthened this concept and it has been observed that many cancers have an increased propensity to metastasize to certain sites in the body (Jiang et al., 2015). Bone metastasis is frequently observed in breast and prostate cancers; liver and lung metastases are mostly observed in the cancers of the gastrointestinal region (Mundy, 2002; Suva et al., 2011). Lung is the first site to encounter tumor cells that are present in circulation as venous drainage passes directly through the lungs (Martin et al., 2000). In fact the most prominent sites for metastases are the lung, liver, brain, and bone; however, low metastasis is observed in the spleen, whereas heart rarely hosts metastatic nests. The factors responsible for these predispositions are largely unknown and are currently of significant interest.

# 6. MICRORNAS AND CANCER STEM CELLS IN CANCER METASTASIS

In addition to all classical and novel factors and signaling pathways discussed earlier, miRNAs and the cancer stem cells (CSCs) have attracted a lot of attention in recent years for their involvement in cancer metastasis. With the central hypothesis that miRNAs are involved in the initiation (Ma et al., 2007) as well as progression (Tavazoie et al., 2008) of cancer metastasis, they are being investigated as putative biomarkers (Mishra, 2014; Heneghan et al., 2010) and even as novel therapeutic targets in metastatic cancers (Heneghan et al., 2010; Hur, 2015; Nicoloso et al., 2009). They are also being investigated for organ-specific cancer metastasis (Ahmad et al., 2014b; Asangani et al., 2008). Similar to miRNAs, CSCs have also been connected to metastatic process (Hermann et al., 2007). It is not surprising that there seems to be a connection between miRNAs and the CSCs (Croce and Calin, 2005). In fact, there is evidence for a vicious connection between miRNAs, EMT, and the CSCs (Wang et al., 2010; Taube et al., 2013), suggesting a close association between all these factors leading to an efficient coordination of events determining the metastatic outcome. A meaningful discussion on the exact role of miRNAs or the CSCs in cancer metastases is beyond the scope of this short review. There is a wealth of published literature on the topic. More importantly, despite the numerous reports, our knowledge on the subject still seems to be evolving. Focused and in-depth articles on the role of miRNAs and CSCs in cancer metastasis are scheduled to be published later as part of this series.

# 7. CONCLUSIONS AND PERSPECTIVES

In conclusion, metastasis to distant organs remains a major cause of cancer-related deaths. Some progress has been made in recent years in identifying the multitude of factors that contribute to cancer metastases. However, we are still not any close to finding a targeted therapy to stop, control, or eliminate these metastases. One reason for this could be the high adaptability of cancer cells. As suggested by numerous reports, targeted therapies often result in activation of alternate pathways, rendering the targeted therapy completely ineffective. This is further complicated by the realization that cancers from a single primary site metastasize to different secondary and tertiary sites, and there seems to be some role of specific molecular signatures in guiding metastatic cancer cells to their eventual destination. It can be envisioned that even an effective targeted therapy might be able to influence metastasis to just a single site, and the molecular signatures for metastases to other sites might still be functional. Also although EMT seems to be important for initiation and progression of metastasis, a reverse process, mesenchymal to epithelial transition, is believed to be functional when the metastatic cancer cells finally reach their destination site. This ensures their "homing" and growth at the new site. Thus, there has been a debate that therapies targeting EMT might be hastening the process of "homing" at distant sites for the cancer cells that have already escaped from the primary site. These are some of the unanswered questions that need to be addressed. A lot of progress has been made in recent years, which has evidently improved our understanding of the whole metastatic process, but clearly, a lot more needs to be accomplished.

## References

Agarwal, R., Mori, Y., Cheng, Y., Jin, Z., Olaru, A.V., et al., 2009. Silencing of claudin-11 is associated with increased invasiveness of gastric cancer cells. PLoS One 4, e8002.

Ahmad, A., Aboukameel, A., Kong, D., Wang, Z., Sethi, S., et al., 2011. Phosphoglucose isomerase/autocrine motility factor mediates epithelial-mesenchymal transition regulated by miR-200 in breast cancer cells. Cancer Res. 71, 3400–3409.

Ahmad, A., Sarkar, S.H., Bitar, B., Ali, S., Aboukameel, A., et al., 2012. Garcinol regulates EMT and Wnt signaling pathways in vitro and in vivo, leading to anticancer activity against breast cancer cells. Mol. Cancer Ther. 11, 2193–2201.

Ahmad, A., Li, Y., Bao, B., Kong, D., Sarkar, F.H., 2014a. Epigenetic regulation of miRNA-cancer stem cells nexus by nutraceuticals. Mol. Nutr. Food Res. 58, 79–86.

Ahmad, A., Sethi, S., Chen, W., Ali-Fehmi, R., Mittal, S., Sarkar, F.H., 2014b. Up-regulation of microRNA-10b is associated with the development of breast cancer brain metastasis. Am. J. Transl. Res. 6, 384–390.

Al-Alem, L., Curry Jr., T.E., 2015. Ovarian cancer: involvement of the matrix metalloproteinases. Reproduction 150, R55–R64.

AlNoury, M.K., Almuhayawi, S.M., Alghamdi, K.B., Al-Noury, K.I., 2015. Preoperative imaging modalities to predict the risk of regional nodal recurrence in well-differentiated thyroid cancers. Int. Arch. Otorhinolaryngol. 19, 116–120.

Ansieau, S., Bastid, J., Doreau, A., Morel, A.P., Bouchet, B.P., et al., 2008. Induction of EMT by twist proteins as a collateral effect of tumor-promoting inactivation of premature senescence. Cancer Cell 14, 79–89.

Asangani, I.A., Rasheed, S.A., Nikolova, D.A., Leupold, J.H., Colburn, N.H., et al., 2008. MicroRNA-21 (miR-21) post-transcriptionally downregulates tumor suppressor Pdcd4 and stimulates invasion, intravasation and metastasis in colorectal cancer. Oncogene 27, 2128–2136.

Bacac, M., Stamenkovic, I., 2008. Metastatic cancer cell. Annu. Rev. Pathol. 3, 221–247. http://dx.doi.org/10.1146/annurev.pathmechdis.3.121806.151523.

Balda, M.S., Matter, K., 2016. Tight junctions as regulators of tissue remodelling. Curr. Opin. Cell Biol. 42, 94–101. http://dx.doi.org/10.1016/j.ceb.2016.05.006.

Bartholow, T.L., Chandran, U.R., Becich, M.J., Parwani, A.V., 2011. Immunohistochemical profiles of claudin-3 in primary and metastatic prostatic adenocarcinoma. Diagn. Pathol. 6 (12), 12–16. http://dx.doi.org/10.1186/1746-1596-6-12.

Bersini, S., Jeon, J.S., Moretti, M., Kamm, R.D., 2014. In vitro models of the metastatic cascade: from local invasion to extravasation. Drug Discov. Today 19, 735–742.

Berx, G., Becker, K.F., Hofler, H., van, R.F., 1998a. Mutations of the human E-cadherin (CDH1) gene. Hum. Mutat. 12, 226–237.

Berx, G., Nollet, F., van, R.F., 1998b. Dysregulation of the E-cadherin/catenin complex by irreversible mutations in human carcinomas. Cell Adhes. Commun. 6, 171–184.

Black, M.D., Kaneshiro, R., Lai, J.I., Shimizu, D.M., 2014. Hereditary diffuse gastric cancer associated with E-cadherin germline mutation: a case report. Hawaii J. Med. Public Health 73, 204–207.

Borka, K., 2009. Claudin expression in different pancreatic cancers and its significance in differential diagnostics. Magy. Onkol. 53, 273–278.

Bourboulia, D., Stetler-Stevenson, W.G., 2010. Matrix metalloproteinases (MMPs) and tissue inhibitors of metalloproteinases (TIMPs): positive and negative regulators in tumor cell adhesion. Semin. Cancer Biol. 20, 161–168.

Boylan, K.L., Misemer, B., De Rycke, M.S., Andersen, J.D., Harrington, K.M., et al., 2011. Claudin 4 is differentially expressed between ovarian cancer subtypes and plays a role in spheroid formation. Int. J. Mol. Sci. 12, 1334–1358.

Brucher, B.L., Jamall, I.S., 2014. Cell-cell communication in the tumor microenvironment, carcinogenesis, and anticancer treatment. Cell Physiol. Biochem. 34, 213–243.

Candido, S., Abrams, S.L., Steelman, L.S., Lertpiriyapong, K., Fitzgerald, T.L., et al., 2016. Roles of NGAL and MMP-9 in the tumor microenvironment and sensitivity to targeted therapy. Biochim. Biophys. Acta 1863, 438–448.

Cheng, G.Z., Zhang, W.Z., Sun, M., Wang, Q., Coppola, D., et al., 2008. Twist is transcriptionally induced by activation of STAT3 and mediates STAT3 oncogenic function. J. Biol. Chem. 283, 14665–14673.

Conacci-Sorrell, M., Zhurinsky, J., Ben-Ze'ev, A., 2002. The cadherin-catenin adhesion system in signaling and cancer. J. Clin. Invest. 109, 987–991.

Coopman, P., Djiane, A., 2016. Adherens junction and E-cadherin complex regulation by epithelial polarity. Cell Mol. Life Sci. 73, 3535–3553.

Cordray, P., Satterwhite, D.J., 2005. TGF-beta induces novel Lef-1 splice variants through a Smad-independent signaling pathway. Dev. Dyn. 232, 969–978.

Croce, C.M., Calin, G.A., 2005. miRNAs, cancer, and stem cell division. Cell 122, 6–7.

Deryugina, E.I., Quigley, J.P., 2015. Tumor angiogenesis: MMP-mediated induction of intravasation- and metastasis-sustaining neovasculature. Matrix Biol. 44–46, 94–112. http://dx.doi.org/10.1016/j.matbio.2015.04.004 (Epub 2015 Apr 22.).

Ding, L., Lu, Z., Lu, Q., Chen, Y.H., 2013. The claudin family of proteins in human malignancy: a clinical perspective. Cancer Manag. Res. 5, 367–375. http://dx.doi.org/10.2147/CMAR.S38294.

Dong, Z., Nemeth, J.A., Cher, M.L., Palmer, K.C., Bright, R.C., Fridman, R., 2001. Differential regulation of matrix metalloproteinase-9, tissue inhibitor of metalloproteinase-1 (TIMP-1) and TIMP-2 expression in co-cultures of prostate cancer and stromal cells. Int. J. Cancer 93, 507–515.

Fernandez, C.A., Yan, L., Louis, G., Yang, J., Kutok, J.L., Moses, M.A., 2005. The matrix metalloproteinase-9/neutrophil gelatinase-associated lipocalin complex plays a role in breast tumor growth and is present in the urine of breast cancer patients. Clin. Cancer Res. 11, 5390–5395.

Forster, C., 2008. Tight junctions and the modulation of barrier function in disease. Histochem. Cell Biol. 130, 55–70.

Fowlkes, J.L., Enghild, J.J., Suzuki, K., Nagase, H., 1994. Matrix metalloproteinases degrade insulin-like growth factor-binding protein-3 in dermal fibroblast cultures. J. Biol. Chem. 269, 25742–25746.

Gialeli, C., Theocharis, A.D., Karamanos, N.K., 2011. Roles of matrix metalloproteinases in cancer progression and their pharmacological targeting. FEBS J. 278, 16–27.

Giepmans, B.N., van Ijzendoorn, S.C., 2009. Epithelial cell-cell junctions and plasma membrane domains. Biochim. Biophys. Acta 1788, 820–831.

Ginnebaugh, K.R., Ahmad, A., Sarkar, F.H., 2014. The therapeutic potential of targeting the epithelial-mesenchymal transition in cancer. Expert Opin. Ther. Targets 18, 731–745.

Gong, Y., Chippada-Venkata, U.D., Oh, W.K., 2014. Roles of matrix metalloproteinases and their natural inhibitors in prostate cancer progression. Cancers. (Basel) 6, 1298–1327.

Gutierrez-Fernandez, A., Fueyo, A., Folgueras, A.R., Garabaya, C., Pennington, C.J., et al., 2008. Matrix metalloproteinase-8 functions as a metastasis suppressor through modulation of tumor cell adhesion and invasion. Cancer Res. 68, 2755–2763.

Han, T., Jiao, F., Hu, H., Yuan, C., Wang, L., et al., 2016. EZH2 promotes cell migration and invasion but not alters cell proliferation by suppressing E-cadherin, partly through association with MALAT-1 in pancreatic cancer. Oncotarget 7, 11194–11207.

Heneghan, H.M., Miller, N., Kerin, M.J., 2010. MiRNAs as biomarkers and therapeutic targets in cancer. Curr. Opin. Pharmacol. 10, 543–550.

Hermann, P.C., Huber, S.L., Herrler, T., Aicher, A., Ellwart, J.W., et al., 2007. Distinct populations of cancer stem cells determine tumor growth and metastatic activity in human pancreatic cancer. Cell Stem Cell 1, 313–323.

Hu, Q.P., Kuang, J.Y., Yang, Q.K., Bian, X.W., Yu, S.C., 2016. Beyond a tumor suppressor: soluble E-cadherin promotes the progression of cancer. Int. J. Cancer 138, 2804–2812.

Huang, Q., Han, J., Fan, J., Duan, L., Guo, M., et al., 2016. IL-17 induces EMT via Stat3 in lung adenocarcinoma. Am. J. Cancer Res. 6, 440–451.

Hunter, K.W., Crawford, N.P., Alsarraj, J., 2008. Mechanisms of metastasis. Breast Cancer Res. 10 (Suppl. 1), S2. http://dx.doi.org/10.1186/bcr1988.

Hur, K., 2015. MicroRNAs: promising biomarkers for diagnosis and therapeutic targets in human colorectal cancer metastasis. BMB Rep. 48, 217–222.

Ito, A., Nakajima, S., Sasaguri, Y., Nagase, H., Mori, Y., 1995. Co-culture of human breast adenocarcinoma MCF-7 cells and human dermal fibroblasts enhances the production of matrix metalloproteinases 1, 2 and 3 in fibroblasts. Br. J. Cancer 71, 1039–1045.

Jaenisch, R., Bird, A., 2003. Epigenetic regulation of gene expression: how the genome integrates intrinsic and environmental signals. Nat. Genet. (33 Suppl.), 245–254.

Jiang, W.G., Sanders, A.J., Katoh, M., Ungefroren, H., Gieseler, F., et al., 2015. Tissue invasion and metastasis: molecular, biological and clinical perspectives. Semin. Cancer Biol. (35 Suppl.), S244–S275. http://dx.doi.org/10.1016/j.semcancer.2015.03.008 (Epub 2015 Apr 10.).

Jolly, M.K., Boareto, M., Huang, B., Jia, D., Lu, M., et al., 2015. Implications of the hybrid epithelial/mesenchymal phenotype in metastasis. Front. Oncol. 5, 155. http://dx.doi.org/10.3389/fonc.2015.00155 (eCollection 2015.).

Kessenbrock, K., Plaks, V., Werb, Z., 2010. Matrix metalloproteinases: regulators of the tumor microenvironment. Cell 141, 52–67.

Kessenbrock, K., Wang, C.Y., Werb, Z., 2015. Matrix metalloproteinases in stem cell regulation and cancer. Matrix Biol. 44–46, 184–190. http://dx.doi.org/10.1016/j.matbio.2015.01.022 (Epub 2015 Feb 3.).

Kondo, J., Sato, F., Kusumi, T., Liu, Y., Motonari, O., et al., 2008. Claudin-1 expression is induced by tumor necrosis factor-alpha in human pancreatic cancer cells. Int. J. Mol. Med. 22, 645–649.

Korpal, M., Lee, E.S., Hu, G., Kang, Y., 2008. The miR-200 family inhibits epithelial-mesenchymal transition and cancer cell migration by direct targeting of E-cadherin transcriptional repressors ZEB1 and ZEB2. J. Biol. Chem. 283, 14910–14914.

Korpi, J.T., Kervinen, V., Maklin, H., Vaananen, A., Lahtinen, M., et al., 2008. Collagenase-2 (matrix metalloproteinase-8) plays a protective role in tongue cancer. Br. J. Cancer 98, 766–775.

Lamouille, S., Xu, J., Derynck, R., 2014. Molecular mechanisms of epithelial-mesenchymal transition. Nat. Rev. Mol. Cell Biol. 15, 178–196.

Li, C.W., Xia, W., Huo, L., Lim, S.O., Wu, Y., et al., 2012. Epithelial-mesenchymal transition induced by TNF-alpha requires NF-kappaB-mediated transcriptional upregulation of Twist1. Cancer Res. 72, 1290–1300.

Liu, L., Gou, M., Yi, T., Bai, Y., Wei, Y., Zhao, X., 2014. Antitumor effects of heparin-polyethyleneimine nanogels delivering claudin-3-targeted short hairpin RNA combined with low-dose cisplatin on ovarian cancer. Oncol. Rep. 31, 1623–1628.

Liu, L., Xu, Z., Zhong, L., Wang, H., Jiang, S., et al., 2016. Enhancer of zeste homolog 2 (EZH2) promotes tumour cell migration and invasion via epigenetic repression of E-cadherin in renal cell carcinoma. BJU Int. 117, 351–362.

Ma, L., Teruya-Feldstein, J., Weinberg, R.A., 2007. Tumour invasion and metastasis initiated by microRNA-10b in breast cancer. Nature 449, 682–688.

Ma, X., Miao, H., Jing, B., Pan, Q., Zhang, H., et al., 2015. Claudin-4 controls the proliferation, apoptosis, migration and in vivo growth of MCF-7 breast cancer cells. Oncol. Rep. 34, 681–690.

Martin, T.A., Ye, L., Sanders, A.J., Lane, J., Jiang, W.G., 2000. Cancer invasion and metastasis: molecular and cellular perspective. In: Madame Curie Bioscience Database [Internet]. Landes Bioscience, Austin (TX), 2000–2013.

Maryan, N., Statkiewicz, M., Mikula, M., Goryca, K., Paziewska, A., et al., 2015. Regulation of the expression of claudin 23 by the enhancer of zeste 2 polycomb group protein in colorectal cancer. Mol. Med. Rep. 12, 728–736.

Massarelli, E., Brown, E., Tran, N.K., Liu, D.D., Izzo, J.G., et al., 2005. Loss of E-cadherin and p27 expression is associated with head and neck squamous tumorigenesis. Cancer 103, 952–959.

Medici, D., Hay, E.D., Olsen, B.R., 2008. Snail and slug promote epithelial-mesenchymal transition through beta-catenin-T-cell factor-4-dependent expression of transforming growth factor-beta3. Mol. Biol. Cell 19, 4875–4887.

Mehlen, P., Puisieux, A., 2006. Metastasis: a question of life or death. Nat. Rev. Cancer 6, 449–458.

Mikheeva, S.A., Mikheev, A.M., Petit, A., Beyer, R., Oxford, R.G., et al., 2010. TWIST1 promotes invasion through mesenchymal change in human glioblastoma. Mol. Cancer 9, 194. http://dx.doi.org/10.1186/1476-4598-9-194.

Mishra, P.J., 2014. MicroRNAs as promising biomarkers in cancer diagnostics. Biomark Res. 2, 19.

Motti, M.L., Califano, D., Baldassarre, G., Celetti, A., Merolla, F., et al., 2005. Reduced E-cadherin expression contributes to the loss of p27kip1-mediated mechanism of contact inhibition in thyroid anaplastic carcinomas. Carcinogenesis 26, 1021–1034.

Mundy, G.R., 2002. Metastasis to bone: causes, consequences and therapeutic opportunities. Nat. Rev. Cancer 2, 584–593.

Nakayama, F., Semba, S., Usami, Y., Chiba, H., Sawada, N., Yokozaki, H., 2008. Hypermethylation-modulated down-regulation of claudin-7 expression promotes the progression of colorectal carcinoma. Pathobiology 75, 177–185.

Nicoloso, M.S., Spizzo, R., Shimizu, M., Rossi, S., Calin, G.A., 2009. MicroRNAs–the micro steering wheel of tumour metastases. Nat. Rev. Cancer 9, 293–302.

Osanai, M., Murata, M., Chiba, H., Kojima, T., Sawada, N., 2007. Epigenetic silencing of claudin-6 promotes anchorage-independent growth of breast carcinoma cells. Cancer Sci. 98, 1557–1562.

Paget, S., 1989. The distribution of secondary growths in cancer of the breast. 1889. Cancer Metastasis Rev. 8, 98–101.

Palmer, T.D., Ashby, W.J., Lewis, J.D., Zijlstra, A., 2011. Targeting tumor cell motility to prevent metastasis. Adv. Drug Deliv. Rev. 63, 568–581.

Reunanen, N., Kahari, V.M., 2000. Matrix metalloproteinases in cancer cell invasion. In: Madame Curie Bioscience Database [Internet]. Landes Bioscience, Austin (TX), 2000–2013.

Riaz, M., van Jaarsveld, M.T., Hollestelle, A., Prager-van der Smissen, W.J., Heine, A.A., et al., 2013. miRNA expression profiling of 51 human breast cancer cell lines reveals subtype and driver mutation-specific miRNAs. Breast Cancer Res. 15, R33.

Roll, J.D., Rivenbark, A.G., Sandhu, R., Parker, J.S., Jones, W.D., et al., 2013. Dysregulation of the epigenome in triple-negative breast cancers: basal-like and claudin-low breast cancers express aberrant DNA hypermethylation. Exp. Mol. Pathol. 95, 276–287.

Romanov, V., Whyard, T.C., Waltzer, W.C., Gabig, T.G., 2014. A claudin 3 and claudin 4-targeted *Clostridium perfringens* protoxin is selectively cytotoxic to PSA-producing prostate cancer cells. Cancer Lett. 351, 260–264.

Samatov, T.R., Tonevitsky, A.G., Schumacher, U., 2013. Epithelial-mesenchymal transition: focus on metastatic cascade, alternative splicing, non-coding RNAs and modulating compounds. Mol. Cancer 12, 107–112.

Singh, A.B., Dhawan, P., 2015. Claudins and cancer: fall of the soldiers entrusted to protect the gate and keep the barrier intact. Semin. Cell Dev. Biol. 42, 58–65. http://dx.doi.org/10.1016/j.semcdb.2015.05.001 (Epub 2015 May 27.).

Singh, A.B., Sharma, A., Dhawan, P., 2010. Claudin family of proteins and cancer: an overview. J. Oncol. 2010, 541957. http://dx.doi.org/10.1155/2010/541957 (Epub 2010 Jul 8.).

Singh, A.B., Sharma, A., Smith, J.J., Krishnan, M., Chen, X., et al., 2011. Claudin-1 up-regulates the repressor ZEB-1 to inhibit E-cadherin expression in colon cancer cells. Gastroenterology 141, 2140–2153.

Stewart, B., Wild, C.P., 2014. World Cancer Report 2014. International Agency for Research on Cancer, Lyon.

Stuelten, C.H., DaCosta, B.S., Arany, P.R., Karpova, T.S., Stetler-Stevenson, W.G., Roberts, A.B., 2005. Breast cancer cells induce stromal fibroblasts to express MMP-9 via secretion of TNF-alpha and TGF-beta. J. Cell Sci. 118, 2143–2153.

Suva, L.J., Washam, C., Nicholas, R.W., Griffin, R.J., 2011. Bone metastasis: mechanisms and therapeutic opportunities. Nat. Rev. Endocrinol. 7, 208–218.

Suzuki, M., Raab, G., Moses, M.A., Fernandez, C.A., Klagsbrun, M., 1997. Matrix metalloproteinase-3 releases active heparin-binding EGF-like growth factor by cleavage at a specific juxtamembrane site. J. Biol. Chem. 272, 31730–31737.

Taube, J.H., Malouf, G.G., Lu, E., Sphyris, N., Vijay, V., et al., 2013. Epigenetic silencing of microRNA-203 is required for EMT and cancer stem cell properties. Sci. Rep. 3, 2687.

Tavazoie, S.F., Alarcon, C., Oskarsson, T., Padua, D., Wang, Q., et al., 2008. Endogenous human microRNAs that suppress breast cancer metastasis. Nature 451, 147–152.

Taylor, M.A., Lee, Y.H., Schiemann, W.P., 2011. Role of TGF-beta and the tumor microenvironment during mammary tumorigenesis. Gene Expr. 15, 117–132.

Thakur, V., Bedogni, B., 2016. The membrane tethered matrix metalloproteinase MT1-MMP at the forefront of melanoma cell invasion and metastasis. Pharmacol. Res. 111, 17–22. http://dx.doi.org/10.1016/j.phrs.2016.05.019.

Todd, M.C., Petty, H.M., King, J.M., Piana Marshall, B.N., Sheller, R.A., Cuevas, M.E., 2015. Overexpression and delocalization of claudin-3 protein in MCF-7 and MDA-MB-415 breast cancer cell lines. Oncol. Lett. 10, 156–162.

Valastyan, S., Weinberg, R.A., 2011. Tumor metastasis: molecular insights and evolving paradigms. Cell 147, 275–292.

van, Z.F., Krupitza, G., Mikulits, W., 2011. Initial steps of metastasis: cell invasion and endothelial transmigration. Mutat. Res. 728, 23–34.

Venza, M., Visalli, M., Catalano, T., Biondo, C., Beninati, C., et al., 2016. DNA methylation-induced E-cadherin silencing is correlated with the clinicopathological features of melanoma. Oncol. Rep. 35, 2451–2460.

Vogelstein, B., Kinzler, K.W., 2004. Cancer genes and the pathways they control. Nat. Med. 10, 789–799.

Wan, L., Pantel, K., Kang, Y., 2013. Tumor metastasis: moving new biological insights into the clinic. Nat. Med. 19, 1450–1464.

Wang, Z., Li, Y., Ahmad, A., Azmi, A.S., Kong, D., et al., 2010. Targeting miRNAs involved in cancer stem cell and EMT regulation: an emerging concept in overcoming drug resistance. Drug Resist. Updat. 13, 109–118.

Winter, M., Gibson, R., Ruszkiewicz, A., Thompson, S.K., Thierry, B., 2015. Beyond conventional pathology: towards preoperative and intraoperative lymph node staging. Int. J. Cancer 136, 743–751.

Wong, S.Y., Hynes, R.O., 2006. Lymphatic or hematogenous dissemination: how does a metastatic tumor cell decide? Cell Cycle 5, 812–817.

Wong, T.S., Gao, W., Chan, J.Y., 2014. Transcription regulation of E-cadherin by zinc finger E-box binding homeobox proteins in solid tumors. Biomed. Res. Int. 2014, 921564. http://dx.doi.org/10.1155/2014/921564 (Epub 2014 Aug 13.).

Wu, Y., Zhou, B.P., 2010. TNF-alpha/NF-kappaB/Snail pathway in cancer cell migration and invasion. Br. J. Cancer 102, 639–644.

Xu, J., Lamouille, S., Derynck, R., 2009. TGF-beta-induced epithelial to mesenchymal transition. Cell Res. 19, 156–172.

Yu, W., Zhang, Y., Ruest, L.B., Svoboda, K.K., 2013. Analysis of Snail1 function and regulation by Twist1 in palatal fusion. Front Physiol. 4, 12. http://dx.doi.org/10.3389/fphys.2013.00012 (eCollection 2013.).

Zhang, L., Huang, G., Li, X., Zhang, Y., Jiang, Y., et al., 2013. Hypoxia induces epithelial-mesenchymal transition via activation of SNAI1 by hypoxia-inducible factor -1alpha in hepatocellular carcinoma. BMC Cancer 13, 108. http://dx.doi.org/10.1186/1471-2407-13-108.

CHAPTER

2

# Breast Cancer Metastasis

W. Tsuji[1], J.A. Plock[2]

[1]Shiga Medical Center for Adults, Shiga, Japan; [2]University Hospital Zurich, Zurich, Switzerland

## Abbreviations

**5-FU** 5-fluorouracil
**ABC** Advanced breast cancer
**AI** Aromatase inhibitor
**CEP** Circulating endothelial cell progenitor
**CI** Confidence interval
**CSCs** Cancer stem cells
**CTCs** Circulating tumor cells
**DCIS** Ductal carcinoma in situ
**DFI** Disease free interval
**DNA** Deoxyribonucleic acid
**DTCs** Disseminated tumor cells
**ER** Estrogen receptor
**FDA** Food and drug administration
**GPA** Graded prognostic assessment
**HER** Human epidermal growth factor receptor
**HR** Hazard ratio
**LHRH** Luteinizing hormone releasing hormone
**MBC** Metastatic breast cancer
**MTD** Maximum tolerable dose
**ORR** Overall response rate
**OS** Overall survival
**PFS** Progression free survival
**PgR** Progesterone receptor
**QOL** Quality of life
**RANK** Receptor activator of nuclear factor-kappa B
**RANKL** Receptor activator of nuclear factor-kappa B ligand
**SERD** Selective estrogen downregulator
**SERM** Selective estrogen receptor modulator
**SRE** Skeletal related events
**SRS** Stereotactic radiosurgery
**SRT** Stereotactic radiotherapy
**TTP** Time to progression
**VEGF** Vascular endothelial growth factor
**WBRT** Whole brain radiotherapy

## 1. INTRODUCTION

Breast cancer is the most prevalent malignant disease among women worldwide (Coates et al., 2015). Remarkable progress has been made regarding breast cancer diagnosis and treatment. Recently, gene-based predictive tools such as Oncotype DX or MammaPrint have become commercially available, and adjuvant chemotherapy or endocrine therapy is now applied based on the individual's recurrence risk (Paik et al., 2004; van't Veer et al., 2002). However, 20–30% of patients progress on to metastatic disease after the initial treatment (EBCTCG, 2005). The median overall survival of metastatic breast cancer (MBC) patients is still only 2–3 years, although the range is wide. Some patients may experience metastatic relapse within months, whereas others go several years or even decades without distant metastasis (Aguirre-Ghiso, 2007; Uhr and Pantel, 2011). On the other hand, approximately 4–6% of women are reported to present with metastatic disease at the time of the initial diagnosis (Cardoso et al., 2011). MBC is rarely curable, and patients have to live with the disease. The main purpose of MBC

*Introduction to Cancer Metastasis*
http://dx.doi.org/10.1016/B978-0-12-804003-4.00002-5

treatment is prolongation of overall survival (OS) and maintaining patients' quality of life.

## 2. BIOLOGY

### 2.1 Breast Cancer Proliferation Signals

There are two types of breast cancer proliferation signals; hormone receptor pathway via estrogen receptor (ER) and transmembrane receptor pathway representing human epidermal growth factor receptor 2 (HER2).

Estrogen is well recognized to play an important role in the development and progression of breast cancer. In the classical genomic pathway, estrogen-bound ER directly binds to the estrogen response element (ERE) present in the target gene promoters (Fig. 2.1). In the nongenomic pathway, ER can regulate transcription without directly binding to DNA. ER acts as a coactivator via interaction with other transcription factors, which regulates the gene transcriptions at their specific DNA sites. Membrane ER represents only a small fraction of total ER and is considered to be the same protein as nuclear ER. Membrane ER has been reported as being associated with many growth factor receptors such as IGF-1R, epidermal growth factor receptor (EGFR) (HER1), and HER2 [avian erythroblastosis oncogene B 2 (ERBB2)]. The estrogen-bound membrane ER rapidly activates several signals in a cell-specific manner including calcium currents, cyclic adenosine monophosphate, inositol phosphate, G proteins, Src, and Shc. This leads to activation of downstream kinases such as mitogen-activated protein kinase (MAPK) and phosphatidylinositol 3-kinase (PI3K)/Akt.

HER2 (ERBB2) is one of the HER families (EGFR or HER1, HER2, HER3, and HER4), and a growth factor that exists in the cell membrane (Fig. 2.2). HER receptors exist as both monomers and dimers, either homo- or heterodimers. The receptor consists of an extracellular domain, with four subdomains including two cysteine-rich domains, a transmembrane domain, and an intracellular domain consisting of a juxtamembrane region, a tyrosine kinase domain, and carboxyl tail harboring autophosphorylation site. Ligands binding

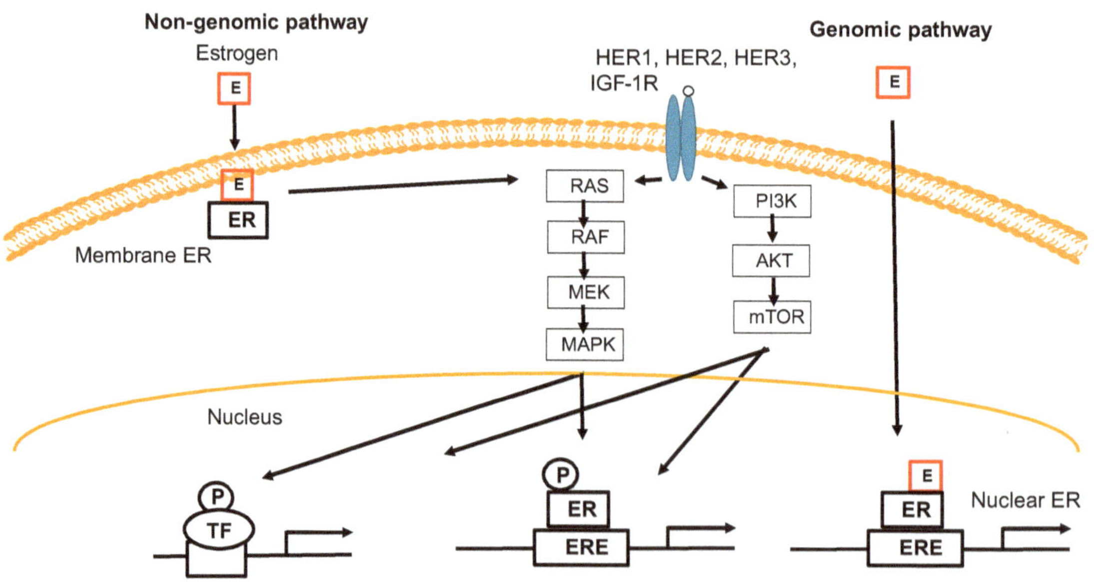


FIGURE 2.1 ER signaling. *E*, estrogen; *ER*, estrogen receptor; *ERE*, estrogen response element; *HER*, human epidermal growth factor receptor; *MAPK*, mitogen-activated protein kinase; *P*, protein; *PI3K*, phosphatidylinositol 3-kinase; *mTOR*, mammalian target of rapamycin; *TF*, transcription factor.

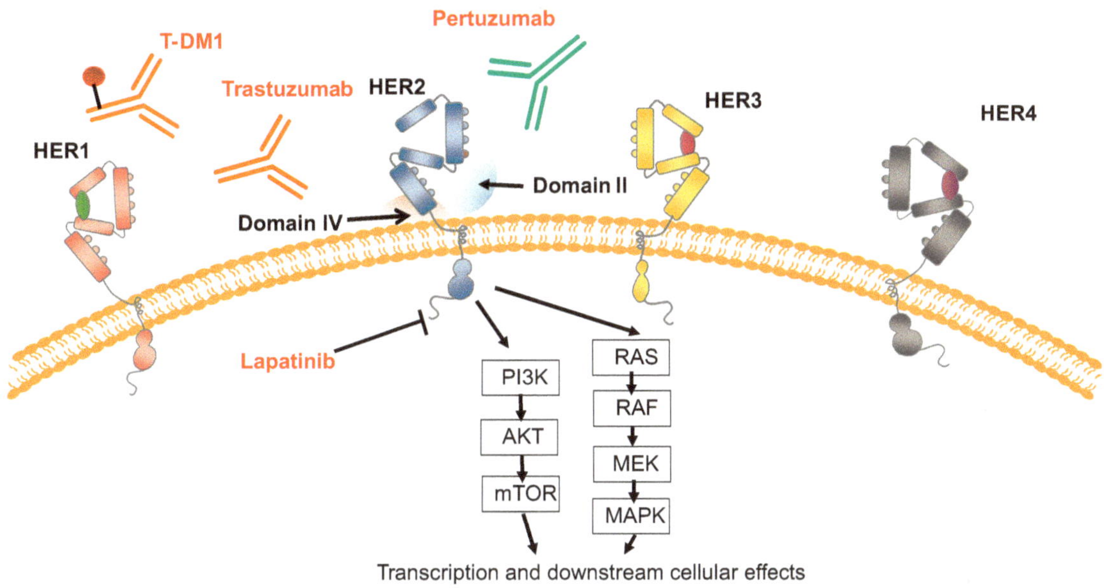


FIGURE 2.2 HER family. *HER*, human epidermal growth factor receptor; *MAPK*, mitogen-activated protein kinase; *mTOR*, mammalian target of rapamycin; *PI3K*, phosphatidylinositol 3-kinase; *T-DM1*, Trastuzumab emtansine. *Adapted from Cho, H.S., et al. 2003. Nature 421, 756–760; Fendly, B.M., et al., 1990. Cancer Res 50, 1550–1558; Franklin, M.C., et al., 2004. Insights into ErbB signaling from the structure of the ErbB2-pertuzumab complex, Cancer Cell 5, 317–328; Nahta, R., et al., 2004. Cancer Res 64, 2343–2346; Scheuer, W., et al., 2009. Cancer Res 69, 9330–9336.*

to HER1, HER3, or HER4 induce rapid receptor dimerization, with a marked preference for HER2 as a dimer partner. Moreover, HER2-containing heterodimers generate intracellular signals that are significantly stronger than signals emanating from other HER combinations. A range of growth factors serve as ligands, but no specific ligands have been identified in association with the HER2 receptor. When HER2 is overexpressed, multiple HER2 heterodimers are formed and cell signaling becomes stronger, resulting in enhanced responsiveness to growth factors and malignant growth. This is the reason why HER2 overexpression is a predictor of poor prognosis in breast cancers.

## 2.2 Breast Cancer Microenvironment

Metastastatic development occurs via complex interactions between cancer cells, microenvironment, and host. Metastases progress through the progressive acquisition of characteristics that allow malignant cells, originating from the mammary epithelium, to disseminate and colonize secondary sites. Once cancer cells acquire invasion ability, they migrate entering the blood or lymphatic system.

The concept of colonization of secondary sites by cancer cells was developed from the "seed and soil" hypothesis suggested by Paget in 1989. Here, the interactions between cancer cells and the microenvironment of secondary homing sites are emphasized (Paget, 1989).

Hormone-dependent breast cancer proliferation is supplied by stromal cells, which exist around breast tumor cells. The incidence of breast cancer is high even in postmenopausal women when the ovaries have ceased to produce estrogen. In postmenopausal breast cancer patients, adrenally derived androgen is converted into estrogen by aromatase that exists in stromal fat tissue. Estrogen levels are higher in the breast cancer tissue than normal breast tissue. Furthermore, aromatase expression levels in breast cancer tissues are significantly higher than those in benign breast lesions. These facts verify the reciprocal relationship between breast cancer cells and stromal cells. Aromatase inhibitor effectiveness signifies that this cycle is vital for breast cancer cell survival and proliferation.

Stromal fibroblasts play an important role in the progression of breast cancers. Cancer cells actively recruit stromal cells, such as fibroblasts, inflammatory cells, and endovascular cells, into the tumor to create a supportive microenvironment for their own growth. Stromal fibroblasts secrete various growth factors, chemokines, and proteases, which facilitate angiogenesis, tumor growth, invasion, and metastases.

Recently, adipocytes and adipose progenitor cells among stromal cells were reported to promote breast cancer cells, an observation made based on the fact that obesity is associated with a greater frequency and progression of several types of cancer including breast cancer. Obesity-related destruction of the energy homeostasis results in inflammation and alterations of adipokine signaling and may foster cancer initiation and progression. Leptin produced by adipocytes and adipose progenitor cells promote breast cancer cell survival and progression. It has been suggested that adiponectin produced by mature adipocytes suppress breast cancer development (Vona-Davis and Rose, 2007). Adipose progenitor cells secrete multiple growth factors such as HGF, VEGF, and basic fibroblast growth factor; these growth factors are known to promote cancer cells. Additionally, adipose progenitor cells are reported to have an immunosuppressive effect (Plock et al., 2015), which may benefit breast cancer invasion, survival, and proliferation.

## 2.3 Disseminated Tumor Cells and Circulating Tumor Cells

Only invasive breast cancer types are considered to shed isolated tumor cells into the bloodstream and infiltrate lymph nodes. However, a couple of studies have shown that tumor cell dissemination may occur before stromal invasion (Banys et al., 2012; Sanger et al., 2011).

The presence of disseminated tumor cells (DTCs) in bone marrow is a common phenomenon, observed in up to 40% of patients at primary diagnosis of breast cancer. Pooled analysis of bone marrow samples from more than 4700 patients revealed that the presence of micrometastases in bone marrow is correlated with poor progression free survival (PFS) and OS (Braun et al., 2005). Bone marrow puncture is needed to detect DTCs, whereas circulating tumor cells (CTCs), which are identical to DTCs in peripheral blood, are obtained from the peripheral blood. Numerous studies report that the number of CTCs is useful as a biomarker and prognostic factor (Cristofanilli et al., 2004, 2005). Cristofanilli et al. (2004) investigated patients with MBC before starting a new line of treatment, and were able to show that patients with more than 0.67 CTCs/mL of whole blood had significantly shorter PFS and OS.

Circulating DNA fragments carrying tumor-specific sequence alterations are found in the cell-free fraction in blood, representing a variable and generally small fraction of the total circulating DNA. Recently, circulating tumor DNA was found to be informative, specific, and a highly sensitive biomarker in patients with MBC (Dawson et al., 2013).

Breast cancer specific biomarkers are needed to monitor the disease progression and therapeutic effect. As a biomarker, cancer antigen 15-3 (CA 15-3) is a breast cancer specific tumor marker that is clinically useful because it can be detected in the peripheral blood with a sensitivity of about 60–70% (Harris et al., 2007). Thus, CTC (or circulating tumor DNA) has been recently noted as a biomarker for MBC patients.

## 2.4 Breast Cancer Dormancy

Breast cancer dormancy is the term describing a prolonged quiescent state in which cancer cells are present, but disease progression is not clinically apparent. DTCs in bone marrow could explain the early onset of MBC. In a significant population of breast cancer patients, distant metastases emerge after years or even decades of latency. How DTCs are kept quiescent and what awakens them are fundamental questions in tumor biology.

There are multiple hypotheses on the orchestration of DTC dormancy. One is that growth restriction of micrometastases could be due to their inability to recruit extra blood vessels required for further growth (Holmgren et al., 1995). Another is related to an effective immune response keeping the cancer population stable at low levels (Koebel et al., 2007). Also the cell-cycle of DTCs in G0-1 remains latent, because the new microenvironment lacks the cues to reverse the epithelial–mesenchymal transition (Radisky and LaBarge, 2008). DTCs have been found in the blood of breast cancer patients who had no symptoms of recurrence for decades. Interestingly DTCs are nevertheless thought to have a short life-span (Meng et al., 2004). This evokes a hypothesis indicating that long-term dormancy in breast cancer is maintained by micrometastases, that contain proliferating cells, rather than by solitary quiescent cells.

## 2.5 Cancer Stem Cells

The ability of a tumor to metastasize appears to be aggravated by the inherent property of a subset of metastatic cancer stem cells (CSCs), as well as by changes induced by local microenvironmental factors. CSC hypothesis postulates that tumors are maintained by a self-renewing CSC population that is capable of differentiating into non-self-renewing cell populations that constitute the bulk of the tumor. Breast CSCs have been characterized as $CD44^{+}CD24^{low/-}$ by Al-Hajj et al., in 2003 (Al-Hajj and Clarke, 2004). They showed that $CD44^{+}CD24^{low/-}$ cell population maintained high tumorigenicity by using a mouse xenograft model. In $CD44^{+}CD24^{low/-}$ cell population, TGF-beta, Wnt, and Hedgehog signals are activated. Furthermore, gene expressions related to chemotaxis, angiogenesis, and invasion are upregulated in this population. Tumor cells are a highly heterogenous population, and CSCs represent only a minor subpopulation (Shackleton et al., 2009). An estimated 1–10% of tumor cells are considered to be CSCs. Besides the cell surface marker, CSCs are also identified by their function. Side population is one of the CSC isolation methods. CSCs cannot be stained by Hoechst33342 because they have adenosine triphosphate-binding cassette transporters such as MDR1 (multidrug resistance one) and excrete foreign contaminations. Floating mammosphere-forming assay is widely used to identify the CSC population. Mammospheres originate from a small population of cancer cells with stem cell features that are able to grow in a suspension culture. These CSCs are resistant against any therapy, including chemotherapy and radiotherapy, and can thus evade systemic treatment. The CSC percentage increased in the residual tumor after neoadjuvant chemotherapy (Li et al., 2008). Thus, recurrence may occur from remnant CSCs after neoadjuvant therapy. CSC-targeted therapy as a cure could be a future route for efficacious cancer treatment by implementing it in, for example, gene-targeted therapy against Wnt, Notch, Hedgehog, or B lymphoma Mo-MLV insertion region 1 homolog, which plays an important role in CSC self-renewal. Niches for CSCs have been well characterized in the bone marrow (Scadden, 2006), and their application can be further extended to target breast cancer treatment.

## 2.6 Intrinsic Subtypes

In 2001, Sorlie et al. classified breast cancers based on variations in gene expression patterns derived from complementary DNA (cDNA) microarrays using the intrinsic gene set 456 cDNA clones. By cluster analysis, 78 tumor specimens were divided into six subtypes based on differences in gene expression: basal-like, ERBB2+, normal breast-like, luminal subtype A, luminal subtype B, and luminal subtype C.

TABLE 2.1 Breast Cancer Subtypes

| | ER | PgR | HER2 | Treatment |
|---|---|---|---|---|
| Luminal | + | ± | − | Endocrine therapy |
| Luminal HER2 | + | ± | + | Endocrine therapy ± anti-HER2 therapy ± chemotherapy |
| HER2 | − | − | + | Anti-HER2 therapy ± chemotherapy |
| Triple negative | − | − | − | Chemotherapy |

*ER*, estrogen receptor; *HER*, human epidermal growth factor receptor; *PgR*, progesterone receptor.

Furthermore, the Kaplan–Mayer curves based on the subclasses showed a highly significant difference in OS between the subtypes, with the basal-like and ERBB2+ (HER2+) subtypes associated with the shortest survival times. Similar results were obtained on relapse-free survival.

These days it is well known that intrinsic subtypes affect the clinical course and prognosis of both primary breast cancer and MBC (Perou et al., 2000; Lobbezoo et al., 2013). Clinically, ER, progesterone receptor (PgR), HER2, and proliferation marker (Ki67) are assessed by immunohistochemistry rather than gene expression (Table 2.1). Treatments are planned based on the subtypes. Luminal type breast cancer has a better prognosis, and endocrine therapy is effective. Luminal HER2 type breast cancer may need chemotherapy in addition to trastuzumab and endocrine therapy. HER2 type breast cancer previously had a graver prognosis, however, this has improved since trastuzumab was approved. However, basal-like breast cancer has the worst prognosis and chemotherapy is the only treatment. The breast cancer subtype of metastases can also switch and develop differently from the primary tumor. Thus, biopsies from any metastatic sites are recommended.

## 3. METASTATIC DISSEMINATION

Breast cancer imaging is recommended after primary diagnosis and treatment to detect local or regional recurrence. X-ray, computed tomography (CT) or magnetic resonance imaging scan, and bone scintigraphy are potential modalities. Frequent distant metastatic organs are bone (40–75%), lung (5–15%), pleura (5–15%), liver (3–10%), and brain (less than 5%).

### 3.1 Clinical Symptoms

Various symptoms occur depending on the metastatic site or general symptoms such as pain, fatigue, depression, and sleeping disorder. Many MBC patients experience major concerns, including fear of dying, declining quality of life (QOL), treatment side effects, concerns about the ability to take care of their families, and concerns about palliative care (Bower, 2008). MBC patients with bone metastases may suffer from pain, hypercalcemia, pathologic fractures, and loss of mobility. Progressed disease with lung metastases may also be associated with pain, dyspnea, hemoptysis, and coughing, whereas liver metastases may be associated with jaundice, loss of appetite, and abdominal distention from ascites.

### 3.2 Locoregional Recurrence

Diagnosis of locoregional recurrence should be done with cytology or histology. Subcutaneous nodules or swollen axillary lymph nodes are highly suspect and need clarification. Some populations of local recurrence are still curable if resection is possible. Recent publications have shown some benefit after complete resection of a first locoregional isolated recurrence followed by a systemic approach, such as endocrine therapy and chemotherapy (Aebi et al., 2014; Borner et al., 1994). The chemotherapy for isolated locoregional recurrence of breast cancer study demonstrated that the use of chemotherapy after surgery resulted in a significant reduction of systemic recurrence [hazard ratio (HR) = 0.59, $P = 0.046$] (Aebi et al., 2014). If new nodules arise after resection, then re-resection is recommended for such lesions if systemic metastases are absent.

## 3.3 Bone Metastasis

Approximately 65–75% of patients with MBC develop bone metastases, which result in osteolysis and skeletal-related events (SREs) including pathologic fracture, bone pain, and spinal cord compression (Coleman, 2001). Breast cancer cells preferentially metastasize to the skeleton because of the fertile environment provided by bone that is enriched further through interactions between cancer cells and some of the cells within the bone, including osteoblasts and osteoclasts. These interactions stimulate bone resorption, thereby increasing the availability of tumor-promoting factors, which in turn results in cancer cell proliferation (Roodman, 2004).

Some bone metastases occur long after the first treatment. Bone metastasis tends to be symptom free in the beginning. However, with disease progression, SREs can be associated with severe pain, impaired mobility, and reduced QOL, as well as being associated with increased mortality (Sathiakumar et al., 2012). According to a Danish cohort study, 4% of all breast cancer patients were diagnosed with bone metastasis either at the time of their initial breast cancer diagnosis or during follow-up (median follow up, 3.5 years) (Jensen et al., 2011). More than 40% of the patient population had an SRE such as pathological fracture, spinal cord compression, or pain requiring palliative therapy.

## 3.4 Lung Metastasis, Pleural Metastasis

Lung metastasis is relatively common among MBC patients. A partial explanation for this relatively high prevalence may be that our entire cardiac output circulates through the lung-capillary network. Most commonly, lung metastatic lesions initially develop at the level of small pulmonary arterioles, where they must either burst through, or otherwise breach, both the tight endothelial junctions of lung blood vessels and the underlying basal membrane. Once in the lung parenchyma, metastatic cells must survive and grow in this unique microenvironment, which contains a highly organized extracellular matrix and specialized cell types for the purpose of respiration (Gupta and Massague, 2006). Lung metastases themselves are asymptomatic for a long time, but patients may develop dyspnea, cough, and fatigue over time with an increasing tumor load.

## 3.5 Liver Metastasis

The liver is a well-vascularized organ that is supplied by both the portal and systemic circulatory systems. Breast cancer cells can metastasize to the liver via the artery after filtering through the lung. This is the difference of liver metastases originating from the colon or rectum. Colon or rectal cancers metastasize through the portal vein, and the liver is the first organ to metastasize. Breast cancer patients with liver metastases can cause another organ to metastasize such as bone or lung. Liver metastases are asymptomatic in the beginning, but become life threatening as they grow.

## 3.6 Brain Metastasis

Brain metastases are diagnosed in less than 5% of MBC patients, however, brain metastases have been identified in 16–30% of patients with breast cancer at autopsy (Tsukada et al., 1983). Brain metastases usually occur in the late phase of MBC, and are associated with poor prognosis. The 1- and 2-year survival rates are 20% and <2%, respectively (DiStefano et al., 1979). Screening breast cancer patients for brain metastasis is not currently recommended, because no data are available showing that effective screening prolongs OS and prevents serious symptoms. Patients have various symptoms, such as headache, thirst, nausea, or neck pain, depending on the part of the brain where the tumor occurs. Risk factors for brain metastasis from breast cancer are young age, negative ER status, poor differentiation, HER2 overexpression, and basal type breast cancer (Souchon et al., 2010). Recently, a trend toward increased incidence of brain metastasis has been noted, this is because MBC patients survive longer and can thus be treated more effectively during the early stages of MBC.

### 3.7 Peritoneal Dissemination

Peritoneal dissemination is rare and difficult to detect. The main symptoms are abdominal fullness, constipation, ascites, or inguinal herniation. Lobular carcinoma is reported to have higher rates of peritoneal metastases compared to ductal carcinoma (Harris et al., 1984). Intra-abdominal metastases tend to involve the serosal surfaces and retroperitoneum, or ovaries. During surgery, exploration sometimes reveals thickening of the serosal, mesenteric, and retroperitoneal tissues, with no discrete mass. Ureteral obstruction can develop as a consequence of retroperitoneal metastases. In other cases, there is a diffuse spread to the uterus and ovaries with ovarian enlargement creating the features of Krukenberg's tumor. CT scanning has proven especially useful for the detection of abdominal metastases of invasive lobular carcinoma.

## 4. PROGNOSTIC FACTORS AFTER RECURRENCE

Demicheli et al. (2008) showed that the hazard rate for any recurrence displayed a bimodal pattern with a first surge peaking at about 24 months and a second peak occurring at almost 60 months. The same trend was observed when the whole recurrence risk was split into the risk of local recurrence and the risk of distant metastasis. They concluded that their findings support the view that the disease course after surgical removal of the primary tumor follows a common pathway with well-defined steps and that the recurrence risk pattern results from inherent features of the metastasis development process, which are apparently attributed to tumor cells. Additionally, a Japanese group found that the hazard of recurrence within 2 years after treatment was higher for ER-negative patients than ER-positive patients. The hazard rate beyond 3 years was higher for ER-positive patients than ER-negative patients (Kiba et al., 2008).

In general, prognosis after breast cancer recurrence is poor. MBC patients rarely make a complete recovery after chemotherapy, and a long survival time even after complete response cannot be anticipated (Decker et al., 1979; Rahman et al., 1999).

Disease-free interval (DFI) is a prognostic factor for MBC patients (Alanko et al., 1985; Insa et al., 1999). Patients with short DFI tend to have a shorter OS and survival after recurrence. Short DFI means a rapidly progressing disease. As mentioned earlier, ER-negative breast cancer tends to be fast growing.

A multiple cohort study showed that ER positivity is the most reliable prognostic factor not only for primary breast cancer but also for MBC (Alanko et al., 1985; Saphner et al., 1996; Hess et al., 2003).

## 5. TREATMENT FOR METASTATIC BREAST CANCER

The primary goals of MBC treatment include maintaining patients' QOL and extension of survival (Aranda et al., 2005; Chung and Carlson, 2003). QOL is subjective and difficult to assess for clinicians. European Organization for Research and Treatment of Cancer Quality of Life Questionnaire and Functional Assessment of Cancer Therapy are questionnaires developed for this specific purpose. Although MBC is a largely incurable stage of the disease, there are multiple options to effectively achieve these goals. Systemic therapy is therefore well established for MBC patients according to Hortobagyi's decision tree or National Cancer Comprehensive Network guidelines. New drugs have been developed and multiple treatment strategies are available. In addition, an International Consensus Guidelines Conference on Advanced Breast Cancer (ABC) was held for the first time in 2011. Since then, ABC has been held biannually and a series of guidelines were proposed (Cardoso et al., 2014). Consequently, survival has been chronologically extended with improved and standardized strategies (Chia et al., 2007; Giordano et al., 2004; Tsuji et al., 2014).

## 5.1 Endocrine Therapy

Approximately 70–80% of breast cancer patients show hormone receptor expression, defined by ER- and/or PgR-positivity (Huang et al., 2005). Estrogen stimulates not only normal breast epithelial cells, but also ER-positive breast cancer cells. Endocrine therapy now interferes with estrogen stimulation of breast cancer and is the standard of care in the management of hormone receptor-positive MBC patients. Compared to chemotherapy, endocrine therapy has less adverse side effects, thus a majority of MBC patients favor endocrine therapy as the treatment of choice. Multiple endocrine therapies are available nowadays (Table 2.2).

### 5.1.1 *Selective Estrogen Receptor Modulators*

The orally administrated selective estrogen receptor modulators (SERMs) act as hormone receptor antagonists and compete with estrogen for available estrogen binding sites to halt the cancer cell proliferation. For decades, SERMs such as tamoxifen or toremifene were the most widely used endocrine therapies for HR-positive breast cancer. SERMs are considered to be first-line endocrine therapy options for premenopausal HR-positive MBC patients. In premenopausal MBC patients, a luteinizing hormone releasing hormone (LHRH) analog for drug-induced ovarian suppression should be considered to deplete estrogen levels either prior to administration of hormonal therapy or subsequent to first-line endocrine therapy with tamoxifen (Cardoso et al., 2011).

### 5.1.2 *Aromatase Inhibitors*

In postmenopausal women low levels of androgen are converted to estrogen by the enzyme aromatase, which exists in many tissues including breast cancer (Buzdar et al., 2002). Aromatase inhibitors (AIs) inhibit the synthesis of estrogen. Two classes of AIs are currently available: nonsteroidal AIs, such as anastrozole and letrozole, and steroidal AIs, such as exemestane. Nonsteroidal AIs compete with the endogenous ligands androstendione and testosterone for the active binding site of aromatase, where they form a strong but reversible bond, blocking both ligands and oxygen from aromatase. The steroidal AIs bind irreversibly to the active site through their metabolites, causing permanent inactivation of aromatase, even after elimination of the drug from the circulation (Buzdar et al., 2002). All three drugs have been proven to be superior compared to tamoxifen in a phase III randomized study regarding time to progression (TTP) (Bonneterre et al., 2001; Mouridsen et al., 2001; Paridaens et al., 2008). AIs are considered the first option for the treatment of postmenopausal HR-positive MBC patients.

### 5.1.3 *Selective Estrogen Receptor Downregulators*

ER downregulators bind to the ER and block the estrogen-binding site. Selective estrogen receptor downreguators (SERDs) even completely inactivate and destroy ERs. Fulvestrant (Faslodex, 17-beta-estradiol) is an ER antagonist

TABLE 2.2 Endocrine Therapy

| Category | Drugs | Menopausal Status | Route of Administration |
|---|---|---|---|
| SERM | Tamoxifen, Toremifene | Pre and postmenopausal | Oral |
| AI (non-steroidal) | Anastrozole, Letrozole | Postmenopausal | Oral |
| AI (steroidal) | Exemestane | Postmenopausal | Oral |
| SERD | Fulvestrant | Postmenopausal | Intramuscular injection |
| LHRH analog | Goserelin acetate, Leuprorelin acetate | Premenopausal | Subcutaneous injection |

*AI*, aromatase inhibitor; *LHRH*, luteinizing hormone releasing hormone; *SERD*, selective estrogen downregulator; *SERM*, selective estrogen receptor modulator.

without agonistic properties that downregulate cellular levels of ER in a dose-dependent manner (Howell et al., 2000). Fulvestrant exhibits lack of cross-reactivity with tamoxifen. Consequently, patients on fulvestrant with progressed disease may maintain sensitivity to treatments with other endocrine therapies (Robertson et al., 2005).

Fulvestrant is recommended for postmenopausal MBC patients only. However, fulvestrant can be administered to premenopausal women, who have received ovarian ablation. Di Leo et al. conducted a double-blind phase III study headed "Comparison of Faslodex in Recurrent or Metastatic Breast Cancer" and reported that a 500 mg dose of fulvestrant reduced the risk of death by 20% as compared to half the dosage. Additionally, higher dosages of fulvestrant did not increase the toxicity (Di Leo et al., 2010).

## 5.2 Chemotherapy

Chemotherapy is administered for hormone receptor negative cases or patients with resistance to endocrine therapy. The main argument for recommendation of chemotherapy is the sequential use of single agents. Combined chemotherapy is reserved for situations of visceral crisis, rapid progress, or highly symptomatic disease.

### 5.2.1 *Anthracyclines*

Anthracyclines are isolated from bacteria or produced semisynthetically. They are antibiotics with cytostatic activity. Anthracyclines lead to inhibition of DNA replication and transcription. The cell cycle is blocked during the S phase and mitosis, as a result of topoisomerase II enzyme inhibition and intercalation in DNA. Epirubicin and doxorubicin are both known anthracyclines.

In phase III trials for doxorubicin, first-line remission rates of 36% were reported, as well as a median time to therapy failure of 5.8 months, a median PFS of 7.8 months, and a median OS time of roughly 20 months (Harris et al., 2002; O'Brien et al., 2004). However, major adverse events to be considered are hematological toxicity, alopecia, and cardiotoxicity.

### 5.2.2 *Taxanes*

Taxanes are cytostatic agents that occur naturally in certain types of yew tree, however, nowadays they are produced semisynthetically. Their mechanism of action is targeting the cytoskeleton. They lead to an enhanced polymerization of microtubules and then prevent disaggregation of the spindle apparatus through binding to a tubulin subunit. Consequently, the cell cycle is blocked in the G2 phase or the M phase, respectively. Paclitaxel or docetaxel are taxanes in clinical use.

Nab-paclitaxel is a suspension of nanoparticles from human serum albumin to which the paclitaxel molecules are bound. Albumin serves as a physiological carrier molecule for hydrophobic substances. In a phase III trial, there was no significant survival benefit under nab-paclitaxel therapy for the entire population. Yet, there was a survival benefit for patients from the second line of treatment onwards (Gradishar et al., 2005).

### 5.2.3 *Vinorelbine*

The vinca alkaloid vinorelbine is a spindle inhibitor. In contrast to the taxanes, it inhibits the polymerization of tubulin and blocks mitosis in the G2-M phase. MBC patients who received vinorelbine as a monotherapy after pretreatment with anthracycline and taxane achieved an OS response rate of 26%, a median PFS of 4 months, and a median OS of 16.4 months. Neutropenia and anemia are typical severe side effects of vinorelbine (Martin et al., 2007; Ejlertsen et al., 2004).

Vinorelbine is also approved in combination with trastuzumab and has yielded equal or superior results to both paclitaxel and docetaxel, when combined with trastuzumab in the HER2 positive MBC in the HERTANA (Andersson et al., 2011) and TRAVIOTA trials (Burstein et al., 2007).

### 5.2.4 *Capecitabine*

Capecitabine is a prodrug that is transformed intracellularly to the actual active substance, 5-fluorouracil (5-FU). 5-FU possesses structural similarity to uracil and is incorporated into the RNA instead of uracil. It inhibits thymidylate synthetase, a key enzyme of pyrimidine synthesis, which occurs both in healthy tissue and in higher concentrations in tumor tissue. Capecitabine thus leads to an inhibition of cell growth especially in cells with higher replication rates. After anthracycline pretreatment, the combined therapy with docetaxel and capecitabine, in comparison to monotherapy with docetaxel, was demonstrated to have significant improvements in the overall response rate (42% vs. 30%, $P=0.006$), prolongation of time to progression (median TTP 6.1 vs. 4.2 months, HR = 0.652, $P=0.0001$), and overall survival (median OS 14.5 vs. 11.5 months, HR = 0.775, $P=0.0126$) (O'Shaughnessy et al., 2002).

Monotherapy with capecitabine results in an overall response rate of up to 24.3%, a median PFS of up to 5.2 months, and a median OS of 22.4 months after treatment with anthracycline and taxane (Pallis et al., 2012).

Capecitabine is an oral drug that maintains patients' QOL, however, diarrhea and hand foot syndrome are typical severe side effects (O'Shaughnessy et al., 2002; Pallis et al., 2012).

### 5.2.5 *Eribulin*

Eribulin belongs to the class of halichondrines that has been isolated from a marine sponge, and has been available since 2011. Eribulin inhibits the growth of microtubules but does not inhibit their polymerization. The cell cycle is blocked in the G2-M phase. In a phase III study, eribulin has provided an OS benefit (median OS 13.1 vs. 10.6 months, HR = 0.61, $P=0.041$) in heavily pretreated patients (up to five lines of treatment) (Cortes et al., 2011) and has had similar results for PFS and OS compared to capecitabine after pretreatment with an anthracycline and taxane (O'Shaughnessy et al., 2002; Cortes et al., 2011). Major adverse events are neutropenia, fatigue, and peripheral neuropathy.

### 5.2.6 *Platinum Salts*

Cisplatin and carboplatin are platinum salts that act by interlinking DNA single and double strands. On account of their DNA-damaging mechanism, platinum compounds are expected to be particularly active in tumors deficient of DNA damage repair capacity. Available data are mostly derived from the neoadjuvant setting, and less from the MBC setting. The ABC2 (Advanced Breast Cancer 2) panel supported the inclusion of platinum-containing regimens in the treatment of BRCA1/2 mutant patients pretreated with anthracyclines and taxanes and who demonstrated to be endocrine-resistant (Cardoso et al., 2014).

## 5.3 Metronomic Chemotherapy

Karbel et al. proposed metronomic chemotherapy, which is characterized by relatively low dosages and more frequent administration of the chemotherapeutic agent. This typically results in fewer side effects and better QOL (Kerbel and Kamen, 2004). It is thought that the metronomic administration of chemotherapy inhibits tumor growth by antiangiogenic effects directly targeting proliferating endothelial cells in the tumor vasculature and circulating endothelial progenitor cells (CEPs). The maximum tolerable dose (MTD) of chemotherapy has induced robust CEP mobilization in the past causing tumors to rapidly become resistant (Bertolini et al., 2003). Metronomical administration of cyclophosphamide has been reported to maintain low levels of viable circulating endothelial cells for long periods by avoiding mobilization of CEPs (Bertolini et al., 2003). Because of the different target, metronomic chemotherapy may therefore be effective in tumors resistant to chemotherapy with MTD (Munoz et al., 2006; Browder et al., 2000).

## 5.4 Molecular Targeted Therapy

Approximately 20% of breast cancers show overexpression of HER2, a tyrosine kinase transmembrane receptor. The HER family consists of HER1 (EGFR), HER2, HER3, and HER4 (Fig. 2.2). Each receptor is composed of an extracellular domain and an intracellular domain with tyrosine kinase activity. HER2 has no ligand, and HER3 does not have tyrosine kinase activity (Lee-Hoeflich et al., 2008). HER2 overexpression is associated with a more aggressive phenotype and a poor prognosis. The American Society of Clinical Oncology (ASCO) clinical practice guideline recommends HER2-targeted therapy for HER2-positive MBC patients, except for those with clinical congestive heart failure or significantly compromised left ventricular ejection fraction (Giordano et al., 2014). Trastuzumab, pertuzumab, T-DM1, and lapatinib are antibodies targeting the HER2 receptor.

### *5.4.1 Trastuzumab*

Trastuzumab is a humanized monoclonal antibody that binds to the subdomain IV of the HER2 extracellular domain (Molina et al., 2001) (Fig. 2.2). In 1987, the development of trastuzumab (Herceptin) brought prolonged PFS and OS (Slamon et al., 1987). In 2001, Slamon et al. revealed that the addition of trastuzumab on paclitaxel was associated with longer TTP (7.4 months vs. 4.6 months, $P<0.001$), a higher overall response rate (50% vs. 32%, $P<0.001$), a longer duration of response (9.1 months vs. 6.1 months, $P<0.001$), a lower rate of death at 1 year (22% vs. 33%, $P=0.01$), and a 20% reduction in the risk of mortality. Based on this trial, trastuzumab was approved in combination with paclitaxel as a first choice of treatment for HER2-positive MBC.

### *5.4.2 Lapatinib*

Lapatinib is an orally administered small molecule that is a potent and reversible inhibitor of the ATP-binding site at intracellular tyrosine kinase domains of the EGFR (HER1) and HER2 receptor (Fig. 2.2). Lapatinib is approved for use in combination with capecitabine for the treatment of patients with HER2 overexpressing locally advanced or MBC patients who have received prior therapy, including trastuzumab, anthracycline, or taxanes. In a phase III study, 399 patients were assigned to receive lapatinib 1250 mg/m$^2$/day plus capecitabine 2000 mg/m$^2$/day or single agent capecitabine 2500 mg/m$^2$/day for 14 days (Cameron et al., 2008; Geyer et al., 2006). The combination of lapatinib with capecitabine significantly prolonged the time to progression in comparison to capecitabine alone (HR=0.57, $P<0.001$), but did not show any difference in OS (HR=0.87, $P=0.206$) (Cameron et al., 2008; Geyer et al., 2006). Hand foot syndrome is one of the major side effects when combined with capecitabine. Other possible side effects are diarrhea, oral ulcer, rush, liver dysfunction, and interstitial pneumonia.

### *5.4.3 T-DM1*

Trastuzumab emtansine (T-DM1) is an antibody active substance conjugate of trastuzumab and the cytotoxin DM1. The linker molecule is [*N*-maleimidomethyl]cyclohexane-1-carboxylate (MCC). T-DM1 and trastuzumab have similar binding affinities for HER2. After T-DM1 binds HER2 on the cell surface, the HER2–T-DM1 complex enters the cell via receptor-mediated endocytosis, resulting in the release of an active metabolite, lysine–MCC-DM1. Internalized endocytic vesicles form early endosomes. The load of early endosomes can be recycled back into the cell membrane. Intracellular lysine–MCC DM1 inhibits microtubule assembly, causing mitotic arrest, apoptosis, mitotic catastrophe, and disrupted intracellular trafficking.

In the phase III EMILIA trial, 991 patients with HER2-positive MBC, who had previously been treated with trastuzumab and a taxane, were randomly assigned to T-DM1 or lapatinib/capecitabine combination. The T-DM1 group showed significant prolongation of the median PFS [9.6 months vs. 6.4 months, 95% confidence interval (CI): 0.55–0.77, $P<0.001$], median OS (30.9 months vs. 25.1 months, HR for death from any cause: 0.68, 95% CI: 0.55–0.85, $P<0.001$), and ORR (overall response rate) (43.6% vs. 30.8%, $P<0.001$) with a lower incidence of grade 3 toxicity (57% vs. 41%) (Verma et al., 2012). Based on these results, the FDA approved T-DM1 as a second-line agent in HER2-positive MBC. Three-weekly 3.6 mg/kg of T-DM1 is given intravenously. Major side effects are thrombocytopenia, anemia, neutropenia, and diarrhea.

### 5.4.4 *Pertuzumab*

Pertuzumab is a humanized monoclonal antibody that binds to a different epitope of HER2 extracellular domain (subdomain II) than trastuzumab (Franklin et al., 2004) (Fig. 2.2). Pertuzumab prevents HER2/HER3 dimerization, which is considered to be the most potent signaling pair driving cell proliferation in HER2-positive cancer. In the phase III Clinical Evaluation of Pertuzumab and Trastuzumab study the combination of pertuzumab, trastuzumab, and docetaxel demonstrated prolongation of PFS to 6.1 months (HR = 0.62, $P<0.001$) without cardiac toxicity in the first-line setting (Baselga et al., 2012). Pertuzumab is given every 3 weeks with trastuzumab and docetaxel until disease progress or unmanageable toxicity occurs. Six cycles of docetaxel as a minimum are recommended. Possible side effects of pertuzumab are infusion reaction, left ventricular dysfunction, and anaphylaxis.

### 5.4.5 *Bevacizumab*

Bevacizumab is a recombinant humanized monoclonal IgG-antibody against vascular endothelial growth factor (VEGF) A. The docking of VEGF on the VEGF receptors of endothelial cells is prevented to inhibit endothelial proliferation and angiogenesis. Accordingly, the mechanism is assumed to involve a deficient supply of oxygen and nutrients to the tumor resulting in subsequent growth inhibition. Three phase III trials on the first-line therapy were performed: the E2100 trial for the combination of bevacizumab with weekly paclitaxel (Gray et al., 2009), the RIBBON-1 trial for the combination of bevacizumab with capecitabine (Robert et al., 2011), and the Avastin and docetaxel trial for the combination of bevacizumab with docetaxel (Miles et al., 2010). However, none of these revealed any advantage in OS when bevacizumab was used in the combination with chemotherapy. In 2008, bevacizumab in combination with paclitaxel was granted accelerated approval by the FDA for the treatment of patients who had not received chemotherapy for metastatic HER2-negative breast cancer (Miller et al., 2007). However, in 2010, the FDA withdrew bevacizumab for treatment of MBC, because safety and efficacy had not been demonstrated. Possible side effects of bevacizumab are hypertension, kidney dysfunction, gastrointestinal tract perforation, and infusion reaction.

#### 5.4.6 *Everolimus*

Everolimus is a selective inhibitor of the mammalian target of rapamycin (mTOR), a serine threonine kinase that participates in regulation of the cell cycle, angiogenesis and glycolysis, the activities of which are upregulated in many human tumors. The phase III breast cancer trials of oral everolimus II trial recruited hormone receptor positive MBC patients, who had made progress under previous therapy with a nonsteroidal AI. This study identified significant PFS prolongation with a combination of evelorimus and exemestane in comparison to exemestane alone (median PFS: 6.9 vs. 2.8 months, HR = 0.43, $P < 0.001$) (Baselga et al., 2012). However, this was not associated with significant OS prolongation (HR = 0.89, $P = 0.14$) (Piccart et al., 2014). Typical side effects are oral ulcer, interstitial pneumonia, and compromised disease. Indications for everolimus must be made on a case-by-case basis, after discussion with a well-informed patient, and administered by physicians experienced in managing severe adverse effects (Cardoso et al., 2014).

### 5.5 Bisphosphonate

Bisphosphonates have profound effects on osteoclasts, and are potent inhibitors of osteoclastic bone resorption (Fleisch, 2002). Treatment with bisphosphonates resulted in a significant decrease of morbidity associated with bone metastases. Four bisphosphonates, namely clodronate, pamidronate, ibandronate, and zoledronic acid, have been widely used to treat MBC patients with bone metastases. In the phase III study comparing zoledronic acid and pamidronate, patients receiving zoledronic acid had a significant 21% reduced risk of developing an SRE (Rosen et al., 2004). To date, zoledronic acid is known to be the most potent bisphosphonate to reduce the risk of SREs. Standard doses of zoledronic acid have been consistently reported to induce selective stimulation of gamma delta T cells, which exert an additional beneficial antitumor effect in vivo (Santini et al., 2009). Generally, 4 mg of zoledronic acid is given intravenously every 3–4 weeks. Major side effects of zoledronic acid are hypocalcemia, osteonecrosis, and acute renal failure.

### 5.6 RANK-Ligand (RANKL)

RANK (Receptor Activator of Nuclear Factor-kappa B) is a surface receptor mainly expressed on mature osteoclasts and their progenitors (Boyle et al., 2003). Its primary function is to induce osteoclastogenesis and control calcium metabolism (Li et al., 2000). The RANK–RANK ligand (RANKL) pathway plays a key role in the metastasis growth and bone destruction cycle. RANKL is a polypeptide that belongs to type II transmembrane proteins. RANKL is found on the surface of osteoblasts and bone stromal cells, but may also be found in soluble form within the osseous microenvironment (Suda et al., 1999). RANKL binds to its receptor on preosteoclasts and mature osteoclasts. Inhibition of RANKL, using the human monoclonal antibody denosumab, has demonstrated efficacy in delaying SRE associated with bone metastases in a phase III trial (Stopeck et al., 2010). Denosumab also significantly decreased levels of bone turnover (Stopeck et al., 2010). Denosumab is given by subcutaneous injection every 4 weeks. Possible side effects are joint osteonecrosis, renal adverse events, and hypocalcemia.

### 5.7 Radiotherapy

Radiotherapy is indicated for MBC patients as an outpatient procedure with low incidence of side effects. It is difficult to eradicate metastatic cancer but radiotherapy enables shrinking tumor masses and reducing pain. MBC patients receive radiotherapy for palliation, which aims to maintain QOL.

Radiotherapy is recommended for brain metastases. Whole brain radiotherapy (WBRT), stereotactic radiosurgery (SRS), and stereotactic radiotherapy (SRT) are valid options. Graded prognostic assessment (GPA) is utilized for evaluation on an individual basis. GPA is a prognostic index that takes into account age, Karnofsky performance status, the number of brain metastases, and extracranial metastases. Additionally, the breast cancer-specific GPA score is calculated with breast cancer subtypes. For MBC patients with a better GPA score, SRT is recommended. MBC patients with multiple brain metastases generally have a poorer prognosis, and WBRT is recommended (Chang et al., 2009). WBRT resulted in neurological improvement in 70% of patients with brain metastases and the efficacy was maintained for a longer period of time (Gaspar et al., 2010).

For bone metastasis with SRE, radiotherapy is also recommended. External beam radiotherapy provides significant palliation of painful bone metastases in 50–80% of patients, with up to one-third of patients achieving complete relief (Chow et al., 2007). External beam radiotherapy is most effective when the disease is localized, allowing the highest dose to be delivered to the lesion. The mechanism of pain relief remains unknown, but is assumed caused by a reduction of tumor size and cancer cell viability. Numbers of prospective randomized clinical trials have shown that 30 Gy in 10 fractions, 24 Gy in six fractions, 20 Gy in five fractions, or 8 Gy in a single fraction provide excellent pain control and minimal side effects (Lutz et al., 2011). Several evidence-based guidelines have been published by radiation oncology groups (Lutz et al., 2011; Wu et al., 2004). For patients who still suffer from painful bone metastases after radiotherapy, repetitive treatment is effective in up to 58% of patients. Re-irradiation should be considered 4 weeks after primary radiotherapy (Huisman et al., 2012).

## 5.8 Surgery

Since OS has improved with advances in MBC treatment, surgery for primary breast cancer with synchronous metastases has been reevaluated (Ali and Le Scodan, 2011). In some populations the QOL of MBC patients could be improved.

Surgery is effective for locoregional recurrence when provided in tandem with systemic therapy. In a randomized phase III study physicians assigned 162 patients postoperatively to either chemotherapy or no chemotherapy. This study showed that the use of chemotherapy after complete resection resulted in a significant reduction in systemic recurrence (HR = 0.59, 95% CI = 0.35, $P$ = 0.046) (Aebi et al., 2014). There was a significant improvement in survival for the ER-negative subgroup.

Surgery is also effective for patients with oligo metastases. Oligo metastasis is a subset of MBC with a limited number and sites of metastasis, and constitutes up to 20% of the MBC population. The oligo metastatic state implies that a few metastases (usually ≤5) exist before tumor cells acquire widespread metastatic potential (Weichselbaum and Hellman, 2011). Theoretically, removing the primary oligo MBC results in elimination of a potential source of further metastatic seeding, restoration of immune competence, and reduction of chemoresistance by reducing the number of clones (Pagani et al., 2010).

Surgical treatment should be carefully considered for each patient.

## 5.9 Palliative Care

Various symptoms occur depending on the site of metastasis. Common problems in MBC patients are pain, depression, fatigue, and sleeping disorder. Integrated palliative care has become more important as patients tend to live with metastatic disease for longer periods of time (Chia et al., 2007; Giordano et al., 2004; Tsuji et al., 2014). Palliative care focuses on symptom management and has been shown to improve survival in metastatic cancer patients in general (Temel et al., 2010). Palliative care given concurrently with systemic therapies should be considered. Pharmacologic and nonpharmacologic interventions such as rehabilitation referral or cognitive behavioral therapy are effective.

Pain is the most prevalent and distressing symptom in MBC patients. Pain is categorized as somatic pain, visceral pain, and neuropathic pain. Initially, a systematic pain assessment should be performed: type of pain, timing, duration, and effective drugs. The World Health Organization has developed a three-step cancer pain relief regime in adults.

In a comparative analysis, there were no significant differences in the prevalence of psychiatric disorders between 303 early breast cancer patients and 200 MBC patients (Kissane et al., 2004). Among MBC patients, 31% had mood disorders, 24.5% had minor depression, 6.5% had major depression, and 6% had anxiety disorders (Kissane et al., 2004). Antidepressant or anxiolytics can be effective for these patients. However, paroxetine, fluoxetine, and bupropion being cytochrome P450 2D6 inhibitors interfere with tamoxifen efficacy.

## 6. CONCLUSIONS

There has been substantial progress regarding the basic science, diagnosis, identification of prognostic factors, and treatments of MBC recently. Currently, MBC has become not only curable but also treatable. As drug development and treatment options have improved over the years, survival rates after first recurrence are improving accordingly.

## References


Aebi, S., et al., 2014. Chemotherapy for isolated locoregional recurrence of breast cancer (CALOR): a randomised trial. Lancet Oncol. 15, 156–163.

Aguirre-Ghiso, J.A., 2007. Models, mechanisms and clinical evidence for cancer dormancy. Nat. Rev. Cancer 7, 834–846.

Al-Hajj, M., Clarke, M.F., 2004. Self-renewal and solid tumor stem cells. Oncogene 23, 7274–7282.

Alanko, A., Heinonen, E., Scheinin, T., Tolppanen, E.M., Vihko, R., 1985. Significance of estrogen and progesterone receptors, disease-free interval, and site of first metastasis on survival of breast cancer patients. Cancer 56, 1696–1700.

Ali, D., Le Scodan, R., 2011. Treatment of the primary tumor in breast cancer patients with synchronous metastases. Ann. Oncol. 22, 9–16.

Andersson, M., et al., 2011. Phase III randomized study comparing docetaxel plus trastuzumab with vinorelbine plus trastuzumab as first-line therapy of metastatic or locally advanced human epidermal growth factor receptor 2-positive breast cancer: the HERNATA study. J. Clin. Oncol. 29, 264–271.

Aranda, S., et al., 2005. Mapping the quality of life and unmet needs of urban women with metastatic breast cancer. Eur. J. Cancer Care 14, 211–222.

Banys, M., et al., 2012. Hematogenous and lymphatic tumor cell dissemination may be detected in patients diagnosed with ductal carcinoma in situ of the breast. Breast Cancer Res. Treat. 131, 801–808.

Baselga, J., et al., 2012a. Pertuzumab plus trastuzumab plus docetaxel for metastatic breast cancer. N. Engl. J. Med. 366, 109–119.

Baselga, J., et al., 2012b. Everolimus in postmenopausal hormone-receptor-positive advanced breast cancer. N. Engl. J. Med. 366, 520–529.

Bertolini, F., et al., 2003. Maximum tolerable dose and low-dose metronomic chemotherapy have opposite effects on the mobilization and viability of circulating endothelial progenitor cells. Cancer Res. 63, 4342–4346.

Bonneterre, J., et al., 2001. Anastrozole is superior to tamoxifen as first-line therapy in hormone receptor positive advanced breast carcinoma. Cancer 92, 2247–2258.

Borner, M., et al., 1994. First isolated locoregional recurrence following mastectomy for breast cancer: results of a phase III multicenter study comparing systemic treatment with observation after excision and radiation. Swiss Group for Clinical Cancer Research. J. Clin. Oncol. 12, 2071–2077.

Bower, J.E., 2008. Behavioral symptoms in patients with breast cancer and survivors. J. Clin. Oncol. 26, 768–777.

Boyle, W.J., Simonet, W.S., Lacey, D.L., 2003. Osteoclast differentiation and activation. Nature 423, 337–342.

Braun, S., et al., 2005. A pooled analysis of bone marrow micrometastasis in breast cancer. N. Engl. J. Med. 353, 793–802.

Browder, T., et al., 2000. Antiangiogenic scheduling of chemotherapy improves efficacy against experimental drug-resistant cancer. Cancer Res. 60, 1878–1886.

Burstein, H.J., et al., 2007. Trastuzumab plus vinorelbine or taxane chemotherapy for HER2-overexpressing metastatic breast cancer: the trastuzumab and vinorelbine or taxane study. Cancer 110, 965–972.

Buzdar, A.U., Robertson, J.F., Eiermann, W., Nabholtz, J.M., 2002. An overview of the pharmacology and pharmacokinetics of the newer generation aromatase inhibitors anastrozole, letrozole, and exemestane. Cancer 95, 2006–2016.

Cameron, D., et al., 2008. A phase III randomized comparison of lapatinib plus capecitabine versus capecitabine alone in women with advanced breast cancer that has progressed on trastuzumab: updated efficacy and biomarker analyses. Breast Cancer Res. Treat. 112, 533–543.

Cardoso, F., et al., 2014. ESO-ESMO 2nd international consensus guidelines for advanced breast cancer (ABC2). Breast 23, 489–502.

Cardoso, F., Fallowfield, L., Costa, A., Castiglione, M., Senkus, E., 2011. Locally recurrent or metastatic breast cancer: ESMO Clinical Practice Guidelines for diagnosis, treatment and follow-up. Ann. Oncol. 22 (Suppl. 6), vi25–vi30.

Chang, E.L., et al., 2009. Neurocognition in patients with brain metastases treated with radiosurgery or radiosurgery plus whole-brain irradiation: a randomised controlled trial. Lancet Oncol. 10, 1037–1044.

Chia, S.K., et al., 2007. The impact of new chemotherapeutic and hormone agents on survival in a population-based cohort of women with metastatic breast cancer. Cancer 110, 973–979.

Chow, E., Harris, K., Fan, G., Tsao, M., Sze, W.M., 2007. Palliative radiotherapy trials for bone metastases: a systematic review. J. Clin. Oncol. 25, 1423–1436.

Chung, C.T., Carlson, R.W., 2003. Goals and objectives in the management of metastatic breast cancer. Oncologist 8, 514–520.

Coates, A.S., et al., 2015. Tailoring therapies-improving the management of early breast cancer: St Gallen international expert consensus on the primary therapy of early breast Cancer 2015. Ann. Oncol. 26, 1533–1546.

Coleman, R.E., 2001. Metastatic bone disease: clinical features, pathophysiology and treatment strategies. Cancer Treat Rev. 27, 165–176.

Cortes, J., et al., 2011. Eribulin monotherapy versus treatment of physician's choice in patients with metastatic breast cancer (EMBRACE): a phase 3 open-label randomised study. Lancet 377, 914–923.

Cristofanilli, M., et al., 2004. Circulating tumor cells, disease progression, and survival in metastatic breast cancer. N. Engl. J. Med. 351, 781–791.

Cristofanilli, M., et al., 2005. Circulating tumor cells: a novel prognostic factor for newly diagnosed metastatic breast cancer. J. Clin. Oncol. 23, 1420–1430.

Dawson, S.J., et al., 2013. Analysis of circulating tumor DNA to monitor metastatic breast cancer. N. Engl. J. Med. 368, 1199–1209.

Decker, D.A., et al., 1979. Complete responders to chemotherapy in metastatic breast cancer. Characterization and analysis. JAMA 242, 2075–2079.

Demicheli, R., Biganzoli, E., Boracchi, P., Greco, M., Retsky, M.W., 2008. Recurrence dynamics does not depend on the recurrence site. Breast Cancer Res. 10, R83.

Di Leo, A., et al., 2010. Results of the CONFIRM phase III trial comparing fulvestrant 250mg with fulvestrant 500mg in postmenopausal women with estrogen receptor-positive advanced breast cancer. J. Clin. Oncol. 28, 4594–4600.

DiStefano, A., Yong Yap, Y., Hortobagyi, G.N., Blumenschein, G.R., 1979. The natural history of breast cancer patients with brain metastases. Cancer 44, 1913–1918.

EBCTCG, 2005. Effects of chemotherapy and hormonal therapy for early breast cancer on recurrence and 15-year survival: an overview of the randomised trials. Lancet 365, 1687–1717.

Ejlertsen, B., et al., 2004. Phase III study of intravenous vinorelbine in combination with epirubicin versus epirubicin alone in patients with advanced breast cancer: a Scandinavian Breast Group Trial (SBG9403). J. Clin. Oncol. 22, 2313–2320.

Fleisch, H., 2002. Development of bisphosphonates. Breast Cancer Res. 4, 30–34.

Franklin, M.C., et al., 2004. Insights into ErbB signaling from the structure of the ErbB2-pertuzumab complex. Cancer Cell 5, 317–328.

Gaspar, L.E., et al., 2010. The role of whole brain radiation therapy in the management of newly diagnosed brain metastases: a systematic review and evidence-based clinical practice guideline. J. Neurooncol 96, 17–32.

Geyer, C.E., et al., 2006. Lapatinib plus capecitabine for HER2-positive advanced breast cancer. N. Engl. J. Med. 355, 2733–2743.

Giordano, S.H., et al., 2004. Is breast cancer survival improving? Cancer 100, 44–52.

Giordano, S.H., et al., 2014. Systemic therapy for patients with advanced human epidermal growth factor receptor 2-positive breast cancer: American Society of Clinical Oncology clinical practice guideline. J. Clin. Oncol. 32, 2078–2099.

Gradishar, W.J., et al., 2005. Phase III trial of nanoparticle albumin-bound paclitaxel compared with polyethylated castor oil-based paclitaxel in women with breast cancer. J. Clin. Oncol. 23, 7794–7803.

Gray, R., Bhattacharya, S., Bowden, C., Miller, K., Comis, R.L., 2009. Independent review of E2100: a phase III trial of bevacizumab plus paclitaxel versus paclitaxel in women with metastatic breast cancer. J. Clin. Oncol. 27, 4966–4972.

Gupta, G.P., Massague, J., 2006. Cancer metastasis: building a framework. Cell 127, 679–695.

Harris, M., et al., 1984. A comparison of the metastatic pattern of infiltrating lobular carcinoma and infiltrating duct carcinoma of the breast. Br. J. Cancer 50, 23–30.

Harris, L., et al., 2002. Liposome-encapsulated doxorubicin compared with conventional doxorubicin in a randomized multicenter trial as first-line therapy of metastatic breast carcinoma. Cancer 94, 25–36.

Harris, L., et al., 2007. American Society of Clinical Oncology 2007 update of recommendations for the use of tumor markers in breast cancer. J. Clin. Oncol. 25, 5287–5312.

Hess, K.R., Pusztai, L., Buzdar, A.U., Hortobagyi, G.N., 2003. Estrogen receptors and distinct patterns of breast cancer relapse. Breast Cancer Res. Treat. 78, 105–118.

Holmgren, L., O'Reilly, M.S., Folkman, J., 1995. Dormancy of micrometastases: balanced proliferation and apoptosis in the presence of angiogenesis suppression. Nat. Med. 1, 149–153.

Howell, A., Osborne, C.K., Morris, C., Wakeling, A.E., 2000. ICI 182,780 (Faslodex): development of a novel, "pure" antiestrogen. Cancer 89, 817–825.

Huang, H.J., et al., 2005. Association between tumour characteristics and HER-2/neu by immunohistochemistry in 1362 women with primary operable breast cancer. J. Clin. Pathol. 58, 611–616.

Huisman, M., et al., 2012. Effectiveness of reirradiation for painful bone metastases: a systematic review and meta-analysis. Int. J. Radiat. Oncol. Biol. Phys. 84, 8–14.

Insa, A., et al., 1999. Prognostic factors predicting survival from first recurrence in patients with metastatic breast cancer: analysis of 439 patients. Breast Cancer Res. Treat. 56, 67–78.

Jensen, A.O., et al., 2011. Incidence of bone metastases and skeletal-related events in breast cancer patients: a population-based cohort study in Denmark. BMC Cancer 11, 29.

Kerbel, R.S., Kamen, B.A., 2004. The anti-angiogenic basis of metronomic chemotherapy. Nat. Rev. Cancer 4, 423–436.

Kiba, T., et al., 2008. The reversal of recurrence hazard rate between ER positive and negative breast cancer patients with axillary lymph node dissection (pathological stage I-III) 3 years after surgery. BMC Cancer 8, 323.

Kissane, D.W., et al., 2004. Psychiatric disorder in women with early stage and advanced breast cancer: a comparative analysis. Aust. N.Z. J. Psychiatry 38, 320–326.

Koebel, C.M., et al., 2007. Adaptive immunity maintains occult cancer in an equilibrium state. Nature 450, 903–907.

Lee-Hoeflich, S.T., et al., 2008. A central role for HER3 in HER2-amplified breast cancer: implications for targeted therapy. Cancer Res. 68, 5878–5887.

Li, J., et al., 2000. RANK is the intrinsic hematopoietic cell surface receptor that controls osteoclastogenesis and regulation of bone mass and calcium metabolism. Proc. Natl. Acad. Sci. USA 97, 1566–1571.

Li, X., et al., 2008. Intrinsic resistance of tumorigenic breast cancer cells to chemotherapy. J. Natl. Cancer Inst. 100, 672–679.

Lobbezoo, D.J., et al., 2013. Prognosis of metastatic breast cancer subtypes: the hormone receptor/HER2-positive subtype is associated with the most favorable outcome. Breast Cancer Res. Treat. 141, 507–514.

Lutz, S., et al., 2011. Palliative radiotherapy for bone metastases: an ASTRO evidence-based guideline. Int. J. Radiat. Oncol. Biol. Phys. 79, 965–976.

Martin, M., et al., 2007. Gemcitabine plus vinorelbine versus vinorelbine monotherapy in patients with metastatic breast cancer previously treated with anthracyclines and taxanes: final results of the phase III Spanish Breast Cancer Research Group (GEICAM) trial. Lancet Oncol. 8, 219–225.

Meng, S., et al., 2004. Circulating tumor cells in patients with breast cancer dormancy. Clin. Cancer Res. 10, 8152–8162.

Miles, D.W., et al., 2010. Phase III study of bevacizumab plus docetaxel compared with placebo plus docetaxel for the first-line treatment of human epidermal growth factor receptor 2-negative metastatic breast cancer. J. Clin. Oncol. 28, 3239–3247.

Miller, K., et al., 2007. Paclitaxel plus bevacizumab versus paclitaxel alone for metastatic breast cancer. N. Engl. J. Med. 357, 2666–2676.

Molina, M.A., et al., 2001. Trastuzumab (herceptin), a humanized anti-Her2 receptor monoclonal antibody, inhibits basal and activated Her2 ectodomain cleavage in breast cancer cells. Cancer Res. 61, 4744–4749.

Mouridsen, H., et al., 2001. Superior efficacy of letrozole versus tamoxifen as first-line therapy for postmenopausal women with advanced breast cancer: results of a phase III study of the International Letrozole Breast Cancer Group. J. Clin. Oncol. 19, 2596–2606.

Munoz, R., et al., 2006. Highly efficacious nontoxic preclinical treatment for advanced metastatic breast cancer using combination oral UFT-cyclophosphamide metronomic chemotherapy. Cancer Res. 66, 3386–3391.

O'Brien, M.E., et al., 2004. Reduced cardiotoxicity and comparable efficacy in a phase III trial of pegylated liposomal doxorubicin HCl (CAELYX/Doxil) versus conventional doxorubicin for first-line treatment of metastatic breast cancer. Ann. oncol. 15, 440–449.

O'Shaughnessy, J., et al., 2002. Superior survival with capecitabine plus docetaxel combination therapy in anthracycline-pretreated patients with advanced breast cancer: phase III trial results. J. Clin. Oncol. 20, 2812–2823.

Pagani, O., et al., 2010. International guidelines for management of metastatic breast cancer: can metastatic breast cancer be cured? J. Natl. Cancer Inst. 102, 456–463.

Paget, S., 1989. The distribution of secondary growths in cancer of the breast. 1889. Cancer Metastasis Rev. 8, 98–101.

Paik, S., et al., 2004. A multigene assay to predict recurrence of tamoxifen-treated, node-negative breast cancer. N. Engl. J. Med. 351, 2817–2826.

Pallis, A.G., et al., 2012. A multicenter randomized phase III trial of vinorelbine/gemcitabine doublet versus capecitabine monotherapy in anthracycline- and taxane-pretreated women with metastatic breast cancer. Ann. Oncol. 23, 1164–1169.

Paridaens, R.J., et al., 2008. Phase III study comparing exemestane with tamoxifen as first-line hormonal treatment of metastatic breast cancer in postmenopausal women: the European Organisation for Research and Treatment of Cancer Breast Cancer Cooperative Group. J. Clin. Oncol. 26, 4883–4890.

Perou, C.M., et al., 2000. Molecular portraits of human breast tumours. Nature 406, 747–752.

Piccart, M., et al., 2014. Everolimus plus exemestane for hormone-receptor-positive, human epidermal growth factor receptor-2-negative advanced breast cancer: overall survival results from BOLERO-2dagger. Ann. Oncol. 25, 2357–2362.

Plock, J.A., et al., 2015. Adipose- and bone marrow-derived mesenchymal stem cells prolong graft survival in vascularized composite allotransplantation. Transplantation 99, 1765–1773.

Radisky, D.C., LaBarge, M.A., 2008. Epithelial-mesenchymal transition and the stem cell phenotype. Cell Stem Cell 2, 511–512.

Rahman, Z.U., et al., 1999. Results and long term follow-up for 1581 patients with metastatic breast carcinoma treated with standard dose doxorubicin-containing chemotherapy: a reference. Cancer 85, 104–111.

Robert, N.J., et al., 2011. RIBBON-1: randomized, double-blind, placebo-controlled, phase III trial of chemotherapy with or without bevacizumab for first-line treatment of human epidermal growth factor receptor 2-negative, locally recurrent or metastatic breast cancer. J. Clin. Oncol. 29, 1252–1260.

Robertson, J.F., et al., 2005. Sensitivity to further endocrine therapy is retained following progression on first-line fulvestrant. Breast Cancer Res. Treat. 92, 169–174.

Roodman, G.D., 2004. Mechanisms of bone metastasis. N. Engl. J. Med. 350, 1655–1664.

Rosen, L.S., et al., 2004. Zoledronic acid is superior to pamidronate for the treatment of bone metastases in breast carcinoma patients with at least one osteolytic lesion. Cancer 100, 36–43.

Sanger, N., et al., 2011. Disseminated tumor cells in the bone marrow of patients with ductal carcinoma in situ. Int. J. Cancer 129, 2522–2526.

Santini, D., et al., 2009. In vivo effects of zoledronic acid on peripheral gammadelta T lymphocytes in early breast cancer patients. Cancer Immunol. Immunother 58, 31–38.

Saphner, T., Tormey, D.C., Gray, R., 1996. Annual hazard rates of recurrence for breast cancer after primary therapy. J. Clin. Oncol. 14, 2738–2746.

Sathiakumar, N., et al., 2012. Mortality following bone metastasis and skeletal-related events among women with breast cancer: a population-based analysis of U.S. Medicare beneficiaries, 1999–2006. Breast Cancer Res. Treat. 131, 231–238.

Scadden, D.T., 2006. The stem-cell niche as an entity of action. Nature 441, 1075–1079.

Shackleton, M., Quintana, E., Fearon, E.R., Morrison, S.J., 2009. Heterogeneity in cancer: cancer stem cells versus clonal evolution. Cell 138, 822–829.

Slamon, D.J., et al., 1987. Human breast cancer: correlation of relapse and survival with amplification of the HER-2/neu oncogene. Science 235, 177–182.

Slamon, D.J., et al., 2001. Use of chemotherapy plus a monoclonal antibody against HER2 for metastatic breast cancer that overexpresses HER2. N. Engl. J. Med. 344, 783–792.

Sorlie, T., et al., 2001. Gene expression patterns of breast carcinomas distinguish tumor subclasses with clinical implications. Proc. Natl. Acad. Sci. USA 98, 10869–10874.

Souchon, R., et al., 2010. Clinical recommendations of DEGRO and AGO on preferred standard palliative radiotherapy of bone and cerebral metastases, metastatic spinal cord compression, and leptomeningeal carcinomatosis in breast cancer. Breast Care 5, 401–407.

Stopeck, A.T., et al., 2010. Denosumab compared with zoledronic acid for the treatment of bone metastases in patients with advanced breast cancer: a randomized, double-blind study. J. Clin. Oncol. 28, 5132–5139.

Suda, T., et al., 1999. Modulation of osteoclast differentiation and function by the new members of the tumor necrosis factor receptor and ligand families. Endocr. Rev. 20, 345–357.

Temel, J.S., et al., 2010. Early palliative care for patients with metastatic non-small-cell lung cancer. N. Engl. J. Med. 363, 733–742.

Tsuji, W., Teramukai, S., Ueno, M., Toi, M., Inamoto, T., 2014. Prognostic factors for survival after first recurrence in breast cancer: a retrospective analysis of 252 recurrent cases at a single institution. Breast Cancer 21, 86–95.

Tsukada, Y., Fouad, A., Pickren, J.W., Lane, W.W., 1983. Central nervous system metastasis from breast carcinoma. Autopsy study. Cancer 52, 2349–2354.

Uhr, J.W., Pantel, K., 2011. Controversies in clinical cancer dormancy. Proc. Natl. Acad. Sci. USA 108, 12396–12400.

van't Veer, L.J., et al., 2002. Gene expression profiling predicts clinical outcome of breast cancer. Nature 415, 530–536.

Verma, S., et al., 2012. Trastuzumab emtansine for HER2-positive advanced breast cancer. N. Engl. J. Med. 367, 1783–1791.

Vona-Davis, L., Rose, D.P., 2007. Adipokines as endocrine, paracrine, and autocrine factors in breast cancer risk and progression. Endocr. Relat. Cancer 14, 189–206.

Weichselbaum, R.R., Hellman, S., 2011. Oligometastases revisited. Nat. Rev. Clin. Oncol. 8, 378–382.

Wu, J.S., et al., 2004. Radiotherapy fractionation for the palliation of uncomplicated painful bone metastases – an evidence-based practice guideline. BMC Cancer 4, 71.

CHAPTER

# 3

# Prostate Cancer Metastasis

*C. Nicholson*[1,2], *I. Vela*[1,2], *E.D. Williams*[1]

[1]Queensland University of Technology, Woolloongabba, QLD, Australia; [2]Queensland Health, Woolloongabba, QLD, Australia

## 1. EPIDEMIOLOGY AND OVERVIEW OF METASTATIC PROSTATE CANCER

Global estimates show that, in 2012, prostate cancer was diagnosed in 1.1 million men and caused the death of 307,000 men, making it the second most common cancer and the fifth leading cause of cancer-related mortality in males worldwide (Ferlay et al., 2015).

Due to regional diversity in genetic and environmental influences together with variations in prostate-specific antigen (PSA) screening practices, the incidence rates between localities vary more than 25-fold (Ferlay et al., 2015). The highest rates are reported in Australia/New Zealand and North America (age-standardized incidence rates of 111.6 and 97.2 per 100,000, respectively) as well as Western and Northern Europe (Ferlay et al., 2015). Moderately high incidences occur in less developed regions such as the Caribbean (79.8 per 100,000), South Africa (61.8 per 100,000), and South America (60.1 per 100,000). The lowest incidence rates are reported in Asia (4.5–10.5 per 100,000) (Ferlay et al., 2015). Globally, prostate cancer is the fifth leading cause of death from cancer in males (Ferlay et al., 2015).

The natural history of prostate cancer typically follows an indolent course for the first 10–20 years following diagnosis of early, prostate-confined disease (Popiolek et al., 2013). Although intent-to-cure treatments such as radiation therapy and surgery are applied successfully in the majority of localized cancer cases, metastatic disease remains synonymous with incurable disease. Of newly diagnosed cases, a significant proportion (10–20%) will have metastatic disease detected on initial presentation. A further 20–30% of men progress to metastatic prostate cancer after failed therapy for "localized" or locally advanced disease (Felici et al., 2012).

During the initial metastatic stages, genotypic, phenotypic, and microenvironmental changes result in malignant cells detaching from the primary tumor and circulating through hematogenous and lymphatic routes to establish themselves in more distant sites. A classic feature of advanced prostate cancer is the formation of osteoblastic bony metastases; the overwhelmingly dominant site in cases of overt metastases diagnosed at first presentation (Fig. 3.1). At autopsy, histological evidence of bony involvement can be found in over 90% of cases (Bubendorf et al., 2000). Prostate cancer

*Introduction to Cancer Metastasis*
http://dx.doi.org/10.1016/B978-0-12-804003-4.00003-7

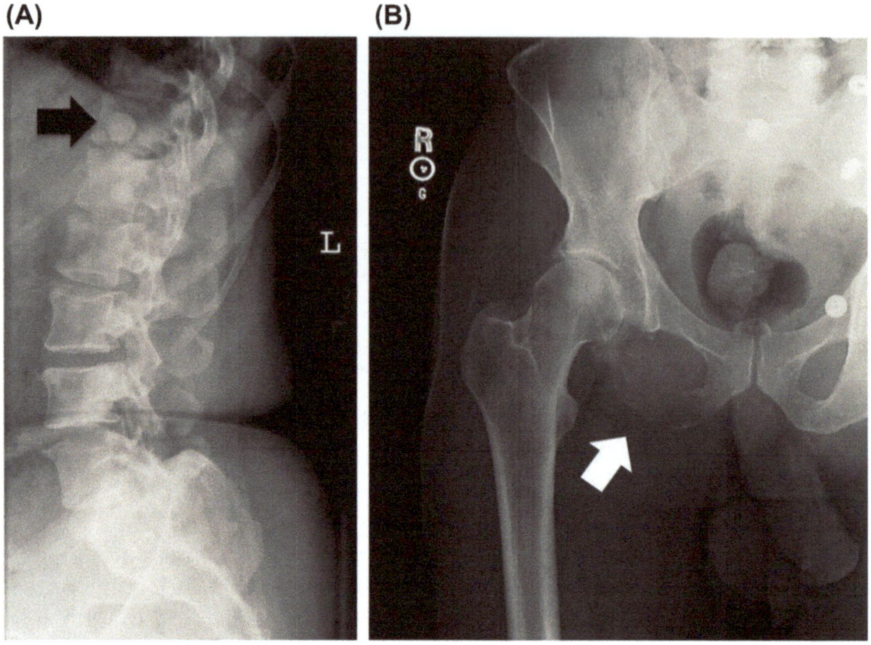


FIGURE 3.1 (A) Typical appearance of prostate cancer osteoblastic bone metastasis on plain X-ray (*black arrow*) compared to (B) the typical appearance of an osteolytic metastasis (*white arrow*), which is the more common metastatic response in bone to most other cancers (in this particular case due to lung cancer).

metastases are also found in other locations such as the liver, lungs, and brain, although almost universally these are in association with bony involvement (Bubendorf et al., 2000; Gandaglia et al., 2015).

In prostate cancer, the classic definition of metastatic cancer is accompanied by the challenging scenario of patients who develop PSA recurrence after definitive therapy as their only evidence of potential metastatic disease. This phenomenon is termed biochemical recurrence (BCR).

In its early stages, metastatic prostate cancer, similar to benign prostate tissue, is composed primarily of cells heavily reliant on androgens such as testosterone for survival and growth. Hence, despite its terminal nature, the initial response to first-line androgen ablation therapy results in significant remission in over 80% of men (Rashid and Chaudhary, 2004). This phase of the disease is termed castrate-sensitive disease. Unfortunately, this hormone-responsive period only persists for a variable and limited duration before selective mutations and adaptive responses inevitably result in progression to castration-resistant prostate cancer (CRPC). Despite castrate levels of testosterone, the median time to progression to CRPC has been approximated at 12–18 months and median survival is approximately 1–3 years (Ross et al., 2008; Kirby et al., 2011).

The development of disseminated disease is the overwhelming cause of morbidity and mortality related to prostate cancer. Mortality rates tend to be higher in black populations (Caribbean 29 per 100,000 and sub-Saharan Africa 19–24 per 100,000), intermediate in the Americas and Oceania, and low in Asian populations (2.9 per 100,000) (Ferlay et al., 2015).

Where cure is not yet attainable, maintaining quality of life (QOL) becomes the primary objective. Given the predilection of metastatic prostate cancer for bone, much of the associated morbidity can be attributed subjectively to bony pain and objectively to the frequency of skeletal-related events (SREs), which include pathologic fractures, spinal cord compression, or need for surgery or radiation therapy to bony metastases.

Therapeutic options for metastatic prostate cancer include first- and second-line hormonal treatments (androgen targeted therapies), chemotherapy, bone-targeted therapy, radiation therapy and surgery, delayed disease progression, increase cancer specific survival, improved QOL, and reduced disease-related symptoms. Advancements in technology and disease knowledge are altering the treatment paradigm resulting in novel treatments and approaches using existing therapies as well as fostering concepts previously foreign to prostate cancer, for example, genetic counseling and precision medicine.

## 2. CELLULAR BIOLOGY OF PROSTATE CANCER METASTASES

In the 70 years since the seminal work of Huggins and Hodges illustrated the initial reliance of metastatic prostate cancer on androgens, the scientific community has made significant advances in identifying the cellular and subcellular determinants of progressive metastatic prostate cancer and how this knowledge might be translated to improved clinical outcomes for patients.

Prostate cancer is a heterogeneous entity with no singular causative genetic alteration identified in its initiation or progression but rather a constellation of alterations spanning multiple processes from fundamental germline mutations to pathways involved in cellular proliferation, survival, and DNA repair. It is important to recognize that involved pathways do not function in isolation and significant crosstalk occurs.

A discussion of associated mutations and pathways consistently expressed in advanced prostate cancer follows to provide a foundation on which the cellular biology of metastatic disease and the rationale for its treatment can be comprehended, but it is by no means a complete account of all mutations discovered in an extensive and growing body of literature on the subject.

### 2.1 The Androgen Receptor Signaling Pathway

The androgen receptor (AR) signaling pathway has been shown to be involved in up to 60% of metastatic prostate cancer specimens (Robinson et al., 2015).

Androgens, such as testosterone and 5α-dihydrotestosterone (DHT), are the principle ligands for the AR. Testosterone is produced by the Leydig cells of the testes under the influence of luteinizing hormone (LH) from the pituitary gland, which is itself regulated by gonadotropin releasing hormone (GnRH) secreted by the hypothalamus. Androgens are also produced to a lesser degree in the adrenal glands and peripheral tissues including the prostate. In the prostate, free testosterone is converted intracellularly by the 5-α-reductase enzyme to the more potent DHT, which has a two- to ten-fold higher affinity for binding the AR (Titus et al., 2005).

Physiologically, the AR pathway plays a major role in exerting the effects of androgens to initiate the differentiation and development of male gender characteristics. The AR coding gene is located on chromosomes X q11–q12. The AR acts as a transcription factor and consists of four major functional domains: the N-terminal domain (NTD), the C-terminal ligand binding domain (LBD), the DNA-binding domain (DBD), and the hinge region (Lu et al., 2006; Gelmann, 2002). The NTD contains the activation function 1 (AF1) domain and confers constitutional activity to the AR in the absence of the LBD (Lu et al., 2006; Gelmann, 2002). The activation function 2 (AF2) domain in the LBD is ligand dependent and also serves as the binding site of cytoplasmic chaperones, which maintains the AR in a favorable conformational state for ligand binding in their absence (Lu et al., 2006; Gelmann, 2002). Both AF1 and AF2 are susceptible to modulation by coregulatory proteins resulting in upregulation (coactivators) or downregulation (corepressors) of their activity (Cano et al., 2013; Bennett et al., 2010). The hinge region possesses a nuclear localization signal, which is important in AR translocation to the nucleus and the DBD stabilizes the AR–DNA complex through binding to promoter and enhancer regions of AR-regulated genes (Cano et al., 2013; Bennett et al., 2010). When androgens bind to the LBD, cytoplasmic chaperones such as heat shock protein 90 become dissociated and nuclear importin factors attach to the AR resulting in its translocation to the nucleus (Cano et al., 2013; Bennett et al., 2010). Within the nucleus, the AR binds to the promoter regions of target genes and recruits various coregulators to induce transcription and modulation of downstream cellular responses (Cano et al., 2013; Bennett et al., 2010).

Androgen deprivation therapy (ADT) triggers cell death or cell-cycle arrest in prostate cancer cells with analyses showing that AR is a critical regulator of the growth 1 (G1) synthesis (S) transition, the cell cycle boundary beyond which cells are committed to cell division (Balk and Knudsen, 2008). Androgen depletion in prostate epithelial benign or castrate-sensitive cancer cells results in early G1 phase arrest, with the loss of expression of cyclin proteins D1 and D3, which normally bind and activate the G1 cyclin-dependent kinases (CDK) (Balk and Knudsen, 2008). Subsequent attenuation of CDK4 results in reduced formation of CDK4/6–cyclin complexes and hypophosphorylation (activation) of the retinoblastoma (RB) tumor suppressor protein consequently inhibiting cell cycle progression (Balk and Knudsen, 2008). RB activation results in the assembly of transcriptional repressor complexes on promoters of target genes so transcription cannot be initiated (Balk and Knudsen, 2008).

Following a period of regression on ADT, an inevitable progression to CRPC evolves. Castration resistance can be broadly defined as the expansion of preexisting disease, biochemical progression, or the appearance of new metastases despite complete androgen blockade (Attar et al., 2009). It was thought initially that these tumors had developed independence from the AR-signaling pathway, however, a growing body of evidence supports the theory that they continue to be dependent on AR pathway activation through multiple escape mechanisms (Attar et al., 2009).

One key mechanism is overexpression of AR secondary to AR gene amplification, which is detected in up to 80% of tumors from patients who develop CRPC following ADT (Linja et al., 2001; Koivisto et al., 1997; Ford et al., 2003). In contrast, untreated prostate cancers rarely contain AR amplifications (Linja et al., 2001; Koivisto et al., 1997; Ford et al., 2003). Alternative mechanisms for increased AR expression in the absence of gene amplification include increased transcription or stabilization of mRNA. The net outcome of any of these processes is a lowered threshold of androgen sensitivity in tumor cells (Attar et al., 2009).

Invoking alternative pathways of androgen production can also help maintain AR activity despite a substantial reduction in testosterone to castrate levels through medical or surgical castration interventions. Certainly, the intratumoral testosterone levels in CRPC patients have been shown in several studies to be equivalent or higher compared to those in non-CRPC patients (Titus et al., 2005; Mohler et al., 2004). The multitude of sources for alternative androgen production includes intratumoral *de novo* synthesis through upregulation of androgen-producing enzymes, cholesterol, and adrenal androgen precursor conversion pathways in CRPC tissue through altered expression of steroidogenic enzymes and downregulation of androgen-inactivating enzymes (Holzbeierlein et al., 2004; Stanbrough et al., 2006; Montgomery et al., 2008; Jernberg et al., 2013).

AR point mutations, observed in 10–30% of patients treated with first-generation antiandrogens, have been shown to result in AR activation through promoting ligand promiscuity (Han et al., 2001; Eisermann et al., 2013; Ware et al., 2014). Commonly described AR mutations in CRPC involve the LBD or NTD and increased the promiscuity of the AR to allow other hormones, steroids, and even antiandrogens to activate the AR (Han et al., 2001; Eisermann et al., 2013; Ware et al., 2014). The W741C AR mutation discovered in bicalutamide-treated patients and the T878A mutation in flutamide-treated patients conferred a growth advantage to cancer cells implying their roles in converting antagonists to agonists (Yoshida et al., 2005). Other mutations involving coregulatory proteins can also affect the ligand specificity of the AR (Wang et al., 2011; Chmelar et al., 2007; Debes et al., 2003).

AR splice variants (so-called due to alternative splicing of the LBD with preservation of the NTD and DBD) have recently been recognized with multiple variants (e.g., ARV7, ARV567ES) facilitating constitutively activated, ligand-independent AR signaling in CRPC (Sun et al., 2010; Watson et al., 2010; Dehm et al., 2008; Antonarakis et al., 2014). AR-V7 is the most prominent variant detected in CRPC and has shown to confer resistance to suppression by AR-directed treatments such as enzalutamide and abiraterone in patient-derived circulating tumor cells (CTCs) (Antonarakis et al., 2014).

Although the AR-signaling pathway plays a central role in prostate cancer progression, many other signaling pathways vitally contribute through independent and interdependent signaling.

## 2.2 The Phosphatidyl Inositol 3-Kinase/Akt-Signaling Pathway

Dysregulation of the phosphatidylinositol 3-kinase (PI3K)/Akt pathway has been implicated in multiple malignancies including prostate cancer through its role as a critical modulator of signal transduction connecting a wide variety of membrane receptors to diverse cellular functions (Courtney et al., 2010; Liu et al., 2009; Dienstmann et al., 2014; Liang and Slingerland, 2003; Cantley, 2002). Stimuli that induce phosphorylation of the plasma membrane phospholipid phosphatidylinositol (4,5) biphosphate (PIP2) to phosphatidylinositol (3,4,5) triphosphate (PIP3) by PI3K include receptor tyrosine kinases, integrins, oncogenic proteins, cytokine receptors, and G-protein coupled receptors (Courtney et al., 2010; Liu et al., 2009; Dienstmann et al., 2014; Liang and Slingerland, 2003; Cantley, 2002; Guo and Giancotti, 2004). PIP3 accumulation provokes recruitment and activation of the serine/threonine-specific protein kinase Akt, which translocates to the nucleus where it phosphorylates and influences the activity of several transcriptional regulators (Courtney et al., 2010; Liu et al., 2009; Dienstmann et al., 2014; Liang and Slingerland, 2003; Cantley, 2002). Akt regulates cellular growth by phosphorylating and relieving the inhibition of the tumor suppressor TSC2 (tuberous sclerosis) complex on the mammalian target of rapamycin (mTOR1) signaling pathway in response to growth factors (Courtney et al., 2010; Liu et al., 2009; Dienstmann et al., 2014; Liang and Slingerland, 2003; Cantley, 2002). Furthermore, Akt regulates cellular proliferation via phosphorylation of the two G1-check point CDK inhibitors, p21 and p27, resulting in their cytoplasmic localization and inability to interact with their nucleic targets (Liang and Slingerland, 2003). Akt mediates cell survival through direct inhibition of proapoptotic proteins such as BAD (Bcl-2-associated death promoter) and inhibition of proapoptotic signals generated by transcription pioneers such as FOXO (Forkhead Box O family of transcription factors) (Liang and Slingerland, 2003; Zhang et al., 2011; Catz and Johnson, 2003). Cell growth inhibition or apoptosis by FOXO occurs via increased expression of death receptor ligands, for example, TNF-related apoptosis inducing ligand, increasing levels of CKD inhibitors, as well as stimulating proapoptotic Bcl2 family members (Liang and Slingerland, 2003; Zhang et al., 2011; Catz and Johnson, 2003).

Data suggest that 30–50% of all prostate cancer cases involve the PI3K/Akt/mTOR upregulation largely due to mutation of the phosphatase and tensin homolog (PTEN) gene resulting in a loss of its normal role (dephosphorylating PIP3 to PIP2) as a negative regulator of Akt (Liang and Slingerland, 2003; Morgan et al., 2009). PTEN mutation is more common in metastatic prostate cancers compared to localized tumors (63% vs. 30%) (Wang et al., 2003). Reduced PTEN expression in advanced prostate cancer results in highly activated Akt, which has been shown to reduce the transcription of key regulators of cell cycle progression such as p27, possibly through inactivation of FOXO1 (Zhang et al., 2011; Wang et al., 2003; Abate-Shen and Shen, 2000). p27 (also known as CDKN1B), the CDKI located on chromosome 12p12–3, displays reduced levels in aggressive prostate cancers associated with poor prognosis (Abate-Shen and Shen, 2000). Loss of the p27 expression results in unregulated prostate cancer cell growth and is observed in 23% of localized tumors, 30% of regional lymph nodes metastases, and 47% of distant metastases (Abate-Shen and Shen, 2000; Kibel et al., 2000).

Akt directly phosphorylates BAD, a key modulator of the apoptotic process, resulting in its inactivation and inability to interact with antiapoptotic members of the Bcl-2 family of proteins (Zhang et al., 2011; Catz and Johnson, 2003). The Bcl-2 protein is not highly expressed in normal prostate tissue but is frequently expressed in prostate cancer, principally in advanced hormone-refractory cases (Catz and Johnson, 2003). Its upregulation promotes tumor cell survival and prostate cancer progression through inhibition of apoptotic pathways (Zhang et al., 2011; Catz and Johnson, 2003).

PI3K/Akt/mTOR alterations occur at the genomic and transcriptional level in most cases of advanced prostate cancer and result in stimulation of mTOR signaling and activation of various transcription genes and translation initiation machinery (Liu et al., 2009; Dienstmann et al., 2014; Liang and Slingerland, 2003; Cantley, 2002). Increased translation of mRNAs coding for positive regulators of cell cycle progression such as cyclin D1 (overexpressed in prostate cancer xenografts) and proto-oncogene c-Myc, with downregulation of cell cycle inhibitors are downstream effects of mTOR activation (Liu et al., 2009; Dienstmann et al., 2014; Liang and Slingerland, 2003; Cantley, 2002). c-Myc is a proto-oncogene that belongs to the family of transcription factors called the basic helix–loop–helix–leucine zipper family (Wang et al., 2003). Constitutional overexpression of c-Myc promotes cellular transformation, cell cycle progression, and apoptosis. C-Myc amplification has been observed in 30% of metastatic lesions and its upregulation significantly correlated with higher Gleason score and poor prognosis in advanced disease (Kibel et al., 2000; Sato et al., 1999; Nupponen et al., 1998).

## 2.3 Cell Cycle Pathways

In addition to the pathways already described, components of the complex modulatory system comprised RB proteins, cyclins, CDKs, and CDK inhibitors involved in cell cycle progression are often mutated in prostate cancer (Efeyan and Serrano, 2007; Navone et al., 1993; Macleod, 2010; Sharma et al., 2010; Zhang et al., 2006; Hguygen et al., 2000). Defective cell cycle control check points as a consequence of tumor suppressor gene mutation result in chromosomal volatility and loss of safeguard senescence mechanisms (Efeyan and Serrano, 2007; Navone et al., 1993; Macleod, 2010; Sharma et al., 2010; Zhang et al., 2006; Hguygen et al., 2000). p53 is a tumor suppressor gene commonly mutated in most human cancer types due to its pivotal role promoting apoptosis in cells with damaged DNA attempting to enter the S phase of the cell cycle

(Efeyan and Serrano, 2007; Navone et al., 1993). Although p53 is not frequently mutated in primary prostate cancer (incidence of 10–20%), mutations are found in 42% of advanced disease cases including bony metastases and castration-resistant disease (Bookstein et al., 1993; Navone et al., 1999; Voeller et al., 1994). Its abnormal expression has been correlated with higher grade, stage, and clinical progression (Hughes et al., 2005). In light of the higher occurrence of p53 abnormalities in advanced prostate cancer it is interesting to note that in vitro cellular models using lines characterized by complete loss of p53 function are frequently associated with decreased radiosensitivity, however, cell lines which express functional p53 protein undergo cellular senescence as the dominant form of death following radiotherapy (Lehmann et al., 2007). This suggests that in the contemporaneous setting of early detection of prostate cancer secondary to PSA screening (where p53 mutations are rare), terminal growth arrest may be the primary modality of cell death postradiotherapy, correlating with the clinical observation that complete regression of prostate cancer treated with radiation therapy can take over one year (Lehmann et al., 2007).

As described previously, RB is a key regulator of the G1 phase of apoptosis and the cell cycle (Macleod, 2010). Mutations and heterozygous loss of the RB protein have been reported in both localized and advanced prostate cancers with one study showing a high prevalence of 85% in small cell prostatic carcinomas (Sharma et al., 2010; Tan et al., 2014). The CDKN2A gene is frequently inactivated in prostate cancer and codes for two proteins, p16 and p14, which act as CDK inhibitors and negatively regulate cell cycle progression (Zhang et al., 2006; Hguygen et al., 2000; Jarrard et al., 1997). p16 prevents activation of CDK4/cyclin D kinase complex through sequestration of CDK4 resulting in activated hypophosphorylated RB protein accumulation and cell cycle arrest (Zhang et al., 2006; Hguygen et al., 2000). p14 activates the tumor suppressor p53 (Zhang et al., 2006; Hguygen et al., 2000).

## 2.4 DNA Repair Pathways

The PARP (poly ADP-ribose polymerase) family of enzymes modifies a subset of nuclear proteins by poly(ADP-ribose) polymerization and plays a critical role in DNA damage repair and transcription (Karanika et al., 2015; Deshmukh and Qui, 2015). Inhibition of major member PARP1 is selectively lethal to tumors harboring homologous recombination-mediated DNA repair defects through a pharmacological synergism called genetic synthetic lethality (Karanika et al., 2015). In homeostasis, homologous recombination repairs double-stranded DNA breaks through reconfiguration of homologous DNA sequences (Karanika et al., 2015; Deshmukh and Qui, 2015). PARP inhibitors have shown sustainable responses in ovarian cancer patients carrying the homologous recombination-deficient BRCA1/BRCA2 (breast cancer, early onset genes) tumor suppressor genes (Kaufman et al., 2015). Preliminary results of the Phase II TOPARP trial suggest a similar benefit and might be seen in prostate cancer patients with these genetic mutations (Mateo et al., 2015). PARP1 has a second significant effect on chromatin as a transcriptional coregulator modifying chromatin structure and transcription factor activity (Schiewer et al., 2012). PARP enzymatic activity is critical for AR occupancy of chromatin and for AR transcription activity in castrate-sensitive and resistant models (Schiewer et al., 2012). These observations in conjunction with sequencing studies identifying DNA repair gene aberrations in up to a third of sporadic CRPC have led to the evaluation of PARP inhibitors as a strategy for simultaneously impeding AR activity and manipulating prostate cancer susceptibility to gene-induced cytotoxicity (Robinson et al., 2015; Mateo et al., 2015; Schiewer et al., 2012; Beltran et al., 2013).

## 2.5 Wnt Pathways

The Wnt family is a group of glycoproteins with a homeostatic role in embryogenesis and in adults affecting cellular proliferation, differentiation, and epithelial–mesenchymal interactions (Kypta and Waxman, 2012). Wnt proteins exert their effects through both canonical and noncanonical pathways that are differentiated by their ability to stabilize the coregulatory protein β-catenin (Kypta and Waxman, 2012). Nucleic accumulation of β-catenin promotes gene transcription through transcription factors such as T-cell factor (TCF) and lymphoid enhancer factor (LEF) families, as well as AR-dependent transcription (Kypta and Waxman, 2012). β-Catenin exists in a cytosolic complex with axin, adenomatous polyposis coli, and glycogen synthase kinase 3B (GSK3B), thus forming a β-catenin degradation complex (Kypta and Waxman, 2012; Clevers, 2006). In the absence of Wnt-induced phosphorylation and inactivation of GSK3B, β-catenin is itself phosphorylated triggering ubiquitination, signaling for its further degradation by proteasomes (Kypta and Waxman, 2012; Clevers, 2006). Wnt binding to transmembrane frizzled receptors promotes disheveled protein dephosphorylation of the axin component that reduces β-catenin degradation complexes with accrual of free β-catenin (Kypta and Waxman, 2012; Clevers, 2006). The β-catenin/LEF/TCF complex transcriptionally activates a variety of Wnt target genes such as c-Myc, vascular endothelial growth factor (to facilitate angiogenesis), and matrix metalloproteinases (MMPs) (to facilitate invasion) among many others (Kypta and Waxman, 2012; Clevers, 2006). β-Catenin may also function as an upregulator of AR activity, as a target gene for activated TCF/LEF (Kypta and Waxman, 2012; Karantanos et al., 2013).

Wnt family members have been extensively studied in prostate cancer progression and a variety of the Wnt ligand themselves are frequently upregulated in prostate cancer (Zhu et al., 2004; Chen et al., 2004; Uysal-Onganer et al., 2010). One example is the protein Wnt-11, demonstrated to have increased expression in CRPC and correlates with higher PSA levels and potential progression to the androgen-resistant state (Zhu et al., 2004; Uysal-Onganer et al., 2010). The exact mechanism of action of Wnt-11 in prostate cancer is not fully delineated; however, some preliminary studies suggest it may play a role in neuroendocrine differentiation (Zhu et al., 2004; Uysal-Onganer et al., 2010). Alterations of β-catenin expression have also been implicated in prostate cancer tumorigenesis in vitro (Chen et al., 2004; Voeller et al., 1998; Horvath et al., 2005). Mutated β-catenin proteins, which inhibit binding to the degradation complex in prostatectomy samples, have also been reported (Voeller et al., 1998; Horvath et al., 2005). Loss of the PTEN gene results in activation of downstream pathway factors such as Akt, which inactivates a component of the β-catenin degradation complex, representing another mechanism for increased levels of β-catenin. Downstream transcriptional targets of the β-catenin signaling pathway such as c-Myc and cyclin-D1 facilitate tumor cell growth (Thakur and Mishra, 2013).

## 2.6 Coregulator Factors

Dysregulated expression of coregulator and pioneer factors has a significant effect on the regulation of AR by facilitating the recruitment of transcription factors and chromatin remodeling (Sung and Cheung, 2014). For example, upregulation of the p160 steroid receptor coactivator (SRC) family, particularly SRC1 and SRC3, has known roles in mediating steroid receptor transcription in AR activity as well as prostate cancer progression (Sung and Cheung, 2014; Chandrasekar et al., 2015). Speckle-type POZ protein (SPOP) mutations are found in up to 15% of prostate cancers and result in a failure to repress SRC3 activity (Geng et al., 2013; Barbieri et al., 2012). In a prospective study analyzing 150 metastatic prostate cancer samples using whole exome and transcriptome sequencing, many of the key genes reported to be altered are known modulators of AR signaling including AR, SPOP, and FOXA1 (Robinson et al., 2015). FOXA1, a Forkhead Box pioneer factor, assists normal AR activity by promoting the open chromatin state and initiation of specific transcriptional programs (Zhang et al., 2011; Barbieri et al., 2012).

Comprehensive profiling of the genetic basis of human metastatic prostate cancer highlights the high number of actionable mutations translatable to potentially more rewarding targets for future therapeutic interventions.

## 3. THE METASTATIC CASCADE

The process of metastases follows through a sequence of steps known as the metastatic cascade, which are comparable among most solid malignancies. These steps include neovascularization, lymphangiogenesis, loss of cell adhesion, altered microenvironment (at primary tumor site and metastatic sites), altered immunity, degradation of the local stroma, intravasation, extravasation, survival of CTCs in transit, and homing of cells to specific organs (Gupta and Massague, 2006). The process of metastasis is remarkably inefficient as demonstrated by in vivo studies showing a lack of formation of metastatic foci following injection of prostate cancer cells into the circulation in mice (Rosol et al., 2003). The metastatic processes involved will be reviewed here with specific reference to prostate cancer.

To support its growth and metastases, a primary tumor must establish its own neovasculature from the preexisting blood supply and recruit vascular progenitors through induction of an angiogenic phenotype resulting in the secretion of proangiogenic factors, which include vascular endothelial growth factor (VEGF) and fibroblast growth factor (FGF) from tumor cells and the surrounding extracellular matrix (ECM) (Huss et al., 2001; Jendraschak and Sage, 1996). At least one study correlated higher intratumoral microvessel densities with advancing tumor stage in prostatic adenocarcinoma (Bostwick et al., 1996).

Lymphangiogenesis is an extension of embryological blood vessel development and therefore the family of growth factors stimulating lymphangiogenesis and neoangiogenesis are the same. VEGF-A is the most potent angiogenic factor, although VEGF-C and VEGF-D are potent lymphangiogenic factors whose proliferative and migratory actions are mediated by the tyrosine kinase VEGF receptors-2 and -3 (VEGFR2 and VEGFR3) located in lymphatic endothelial cells (Datta et al., 2010). Positive staining for VEGFR3 in the lymphatic endothelial cells has been reported in a subset of prostate cancer samples, along with VEGF-C and VEGF-D positive staining in all prostate cancer specimens (Zeng et al., 2004). In this study, VEGFR3-positive vessels correlated with the presence of lymph node metastases, Gleason score, and extracapsular extension (Zeng et al., 2004). Other studies have shown a positive relationship between overexpression of VEGF-C and the formation of regional lymph nodes metastases in prostate carcinoma (Jennbacken et al., 2005).

Switching to a migratory phenotype with loss of normal epithelial cell connections is a vital early step in the metastatic cascade. This is instigated by decreased expression of transmembrane proteins called cell adhesion molecules (CAMs) by malignant cells allowing homotypic or heterotypic cellular interactions (Gupta and Massague, 2006). As prostate cancer cells become progressively more aggressive, E-cadherin is commonly lost, with its downregulation seen in prostate cancer cell lines LNCaP and PC-3 (Bussemakers et al., 2000). Furthermore, a study comparing expression of E-cadherin mRNA and proteins in 14 paired primary prostate specimens and bone metastases in treatment-naive patients showed that although E-cadherin mRNA was present in

13 of the 14 primary tumors and 11 of the bone metastases, protein expression was downregulated in all bone metastases, but only five of the primary cancers compared to benign prostatic epithelial cells (Bryden et al., 2002).

Another group of CAMs called integrins are involved in invasion of the local stromal site and at the metastatic site (Datta et al., 2010). Integrins mostly mediate cell–matrix interactions with components comprising the basement membrane and interstitial stroma (e.g., fibronection, vitronectin) (Goel et al., 2008). For example, LNCaP cells have been demonstrated to use primarily α6-β4-integrin for attachment, adhesion, and invasion on laminin, vitronectin, and osteopontin (Edlund et al., 2001). α6-β4-integrins are upregulated in the LNCaP subline C4-2 and have been associated with an increase in metastatic behavior and enhanced cell migration (Edlund et al., 2001).

To degrade the constituents of the basement membrane (e.g., laminin) and surrounding interstitial connective tissue (e.g., collagen fibers, proteoglycans), prostate cancer cells secrete various proteinases including MMPs, serine proteinases, cysteine proteinase, and aspartyl proteinases (Datta et al., 2010; Ganguly et al., 2014). MMPs are proenzymes activated extracellularly, and produced by prostate cancer cells and surrounding fibroblasts (Ganguly et al., 2014). Several studies have demonstrated higher levels of MMPs in prostate cancer compared to normal prostate tissues (e.g., MMP-1 in metastatic prostate cancer cell lines PC-3 and DU145 cells) (Ganguly et al., 2014; Pulukuri, 2008). Serine proteinases [e.g., hepsin, trypsin, urokinase type plasminogen activator (uPA)] are also overexpressed in prostate cancer, in particular hepsin has been associated with lymph node metastases (Ganguly et al., 2014; Li et al., 2009). uPA converts inactive plasminogen to the more potent protease, plasmin, which degrades ECM and activates other latent proteases and is expressed in higher quantities in prostate cancer bone metastases compared to primary tumors (Dong et al., 2008). Increasing ECM–laminin degradation results in a more invasive phenotype (Datta et al., 2010). Cysteine proteinases are comprised of several cathepsins, including cathepsin B, which degrade several components of the ECM and have been implicated in prostate cancer metastases (Arya et al., 2006; Miyake et al., 2007).

The ability of CTCs to shed from the primary tumor or a metastatic deposit, circulate, and reestablish tumors at a different site suggests acquisition of an alternate phenotype to facilitate hematogenous spread (Datta et al., 2010; Hu et al., 2013). Epithelial–mesenchymal transition (EMT) is a bidirectional process whereby adherent epithelial cells gain migratory and invasive properties and vice versa and is thought to be important in the pathogenesis of CTCs, which express mesenchymal markers (Datta et al., 2010; Hu et al., 2013). The tumorigenic potential of CTCs is representative of the nature of prostate cancer generally, that is heterogeneous, with Carvalho et al. illustrating that patient-derived CTCs implanted in mice were unable to elicit tumors suggesting that prostate cancer metastasis through hematogenous CTC deposition is reliant on a diminutive subset of aggressive cells (Carvalho et al., 2013). CTC enumeration has been validated as a prognostic tool in prostate cancer and more recently, the molecular analysis has been used to characterize CTCs (Fig. 3.2; Hu et al., 2013).

After invading through the basement membrane and into the vasculature or lymphatics, tumor cell survival requires strategies of immune evasion and protection against the turbulent environment of the circulatory system (Arya et al., 2006). The immune system requires the presence of immunogenic tumor antigens effectively processed by antigen presenting cells to elicit cytotoxic T lymphocyte (CTL)-mediated killing (Arya et al., 2006; Blades et al., 1995). Tumors, including prostate cancers, circumvent this by downregulating their Class I human leucocyte antigen expression (Blades et al., 1995). Additionally, prostate cancers have been shown

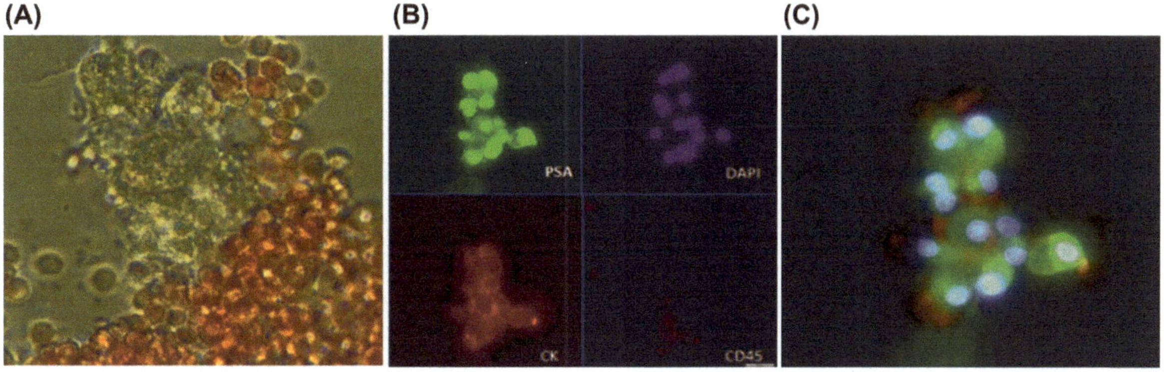


FIGURE 3.2 (A) Bright-field image of circulating tumor cell (CTC) cluster (40×). (B) Immunofluorescence (IF) analysis of CTCs showing staining for DNA (DAPI; nucleus), prostate-specific antigen (PSA), and epithelial cytokeratins indicating prostate origin. Negative staining for hematopoietic cell marker CD45 indicates that these cells are not of white blood cell origin. (C) Merged IF image of CTC cluster.

to influence the intracellular processing of presented antigens by producing altered peptide ligands, which interfere with tumor recognition by antigen-specific CTL (Arya et al., 2006). For protection against the shear forces of the circulatory system, prostate cancer cells may travel clumped with other tumor cells and/or platelets as part of a clot through integrin-related attachments and recruitment of platelets through activation of the protease-activated receptor 1 (PAR-1), which increases platelet adhesion (Datta et al., 2010; Arya et al., 2006). There is modest evidence for PAR-1 activation also inferring a potential metastatic advantage as it is more highly expressed in cells derived from bone metastases (Chay et al., 2002).

The seeding of a particular metastatic site by prostate cancer cells, resulting in the characteristic spread of the disease, is likely to involve a combination of theories such as Paget's "seed and soil" theory, which suggests that certain tumors or "seeds" have an affinity for certain metastatic sites representing the "soil" and compatibility between the two determines if a "seed" will form tumors at a specific location (Datta et al., 2010; Fidler, 2003). Success in establishing metastases is thought to rely on the "correct" combination of a variety of metastatic genotypes to aid the cells survival in particular environments along the metastatic cascade (Datta et al., 2010). Other plausible contributories include Ewing's mechanical theory of characteristic metastatic spread resulting from circulatory and lymphatic flow patterns (Fidler, 2003). In line with this hypothesis, a higher proportion of lumbar vertebral metastases in prostate cancer has been associated with the presence of Batson's plexus, a portal-like venous system between the prostate and lower lumbar vertebrae (Bubendorf et al., 2000; Fidler, 2003; Maccauro et al., 2011). CTCs are also distributed via complementary routes of hematogenous spread such as the inferior vena cava (IVC), carrying blood to the lungs before systemic distribution (Maccauro et al., 2011). This could account for the recurrent finding across several studies of an inverse correlation between lung and skeletal metastases representing the two possible hematogenous routes (Bubendorf et al., 2000; Maccauro et al., 2011). The lymphatic system also converges with the circulatory system either through lymphatic blood vessel shunts within Batson's plexus or via the thoracic duct (Maccauro et al., 2011). The subanalysis involving 1589 autopsies of men with prostate cancer showed an increased incidence in bony metastases compared to other malignancies and noted that the highest frequency of metastases was in the lower lumbar spine, and higher along the vertebrae, followed by the ribs, long bones, and skull suggestive of an upward metastatic spread along spinal veins (Bubendorf et al., 2000). It was also uncommon for less frequently affected bones or visceral organs to be affected without involvement of the more frequently affected bones (Bubendorf et al., 2000). The more recent "homing" theory suggests chemoattractant molecules produced by the target organ resulting in intentional migration of tumor cells toward specific locations (Datta et al., 2010; Cooper and Pienta, 2000). This has been observed in vivo when prostate carcinoma cell lines injected into severe combined immunodeficiency (SCID) mice adjacent to human bone migrate toward the bone, indicating the production of organ-specific chemoattractant factors (Jacob et al., 1999). Epidermal growth factor (EGF), expressed by prostate cancer bone and lymph node metastases, enhanced in vitro chemomigration in bladder cancer cell line TSU-Pr1 (which was originally described as a prostate cancer cell line), although antihuman ECF receptor monoclonal antibodies inhibited this effect (Cooper and Pienta, 2000). A plethora of other growth factors and stromal components have been shown to be chemoattractants in vitro for cancer cells, including insulin-like growth factors (IGF) and transforming growth factor (TGF) (Cooper and Pienta, 2000). A group of cytokines called chemokines are thought to play a role in organ-specific metastases through their upregulated expression in malignancy and role as regulatory chemotactic proteins. The chemokine CXCL12 is shown in vitro to be concentrated in organs commonly colonized by metastatic deposits, such as bone and liver, compared to organs that are not (e.g., eye, tongue) and interacts with chemokine receptors such as CXRCR4 on prostate cancer cells (Vindrieux et al., 2009; Weilbaecher et al., 2011).

After embedding in the capillary beds of distal organs, metastatic cells attach to the endothelium and extravasate into the local environment through a "dock and lock" mechanism (Datta et al., 2010). In vitro studies have shown that prostate cancer cell lines adhere preferentially to human bone marrow endothelial cells as opposed to umbilical vein, aortic, or dermal endothelial cells (Cooper and Pienta, 2000). CAMs involved in the "docking" phase, such as vascular CAM, confer adhesion of tumor cells to the endothelium, whereas the "locking" mechanism attains a more secure attachment through the complex collaboration of integrins (Cooper and Pienta, 2000).

A discussion of prostate cancer cell interactions with the secondary microenvironments of specific organs necessitates a review of the spectrum of involved metastatic sites separated here into bone, lymph node, and visceral metastases.

## 3.1 Bone Metastases

Bone metastases represent the majority of metastatic disease in prostate cancer occurring in 65–75% of men with advanced disease and over 90% at autopsy, and large population-based cohort studies showing 5-year survival rates of only 3% in those diagnosed with bony metastases (Bubendorf et al., 2000; Nørgaard et al., 2010; Coleman, 2001). Skeletal involvement tends to occur in a predictable manner appearing first in the axial skeleton and then in the appendicular skeleton (Bubendorf et al., 2000).

Knowledge of normal bone physiology is essential to understanding the effects of prostate cancer bony metastases since they induce an imbalance of the normal physiological processes. Bone homeostasis is the result of a balance between bone formation and resorption. Bone formation is mediated by osteoblasts, whereas bone degradation (osteolysis) is mediated by osteoclasts (Roodman, 2004; Keller and Brown, 2004). The best known modulator of this remodeling process is the receptor activator of nuclear factor kappa B (RANK)/RANK ligand (RANKL)/OPG system, which involves a transmembrane receptor on osteoclast precursor cells called RANK, which bind to the RANKL presenting on osteoblasts and bone marrow stromal cells (Roodman, 2004; Keller and Brown, 2004). The binding results in promotion of osteoclast maturation, activation, and survival (Roodman, 2004; Keller and Brown, 2004). Principal growth factors released during osteolysis, including TGF-β and IGF2, trigger paracrine stimulation of osteoblasts to repair the bone at the site of osteolysis (Ganguly et al., 2014; Keller and Brown, 2004). This process is to be further altered by the presence of osteoprotegerin (OPG), a soluble alternative receptor for RANKL produced by mature osteoblasts and stromal cells, which inhibits osteoclastogenesis (Roodman, 2004; Keller and Brown, 2004).

One of the critical preliminary steps in establishing bone metastases is thought to be osteolysis. This prepares a niche allowing prostate cancer cells to seed the bone in a supportive microenvironment rich in growth factors such as IGF1, TGF-β1, bone morphogenic proteins (BMPs), basic FGF, interleukin (IL)-6, endothelin-1, and parathyroid hormone-related protein (PTHrP), and promote their survival (Keller and Brown, 2004). This osteolytic phase is orchestrated by the prostate cancer cells through chemotactic attraction and stimulation of osteoclast precursor cells to mature at seeding sites (Keller and Brown, 2004). Prostate cancer cells also secrete several potential osteoclast-inducing humoral factors, including TGF-β, BMPs, macrophage colony-stimulating factor, PTHrP, urokinase-type plasminogen activator (uPA), IL-1, and IL-6 (Ganguly et al., 2014; Keller and Brown, 2004).

Following osteolysis, the tumor cells penetrate the bone matrix altering the microenvironment to induce the actions of osteoblasts (Golshayan and Antonarakis, 2013).

In contrast to other solid malignancies (e.g., breast and lung) with a predilection for bones, which tend to be characterized by osteolytic lesions, prostate cancer bony lesions initially display osteolytic features but later become predominantly osteoblastic radiographically (Fig. 3.1) (Roodman, 2004; Keller and Brown, 2004). Clinically, increased systemic indices for both osteolysis (increased cross-linked N-telopeptide of type-1 collagen and deoxypyridinoline) and osteoblastic activity (increased bone specific alkaline phosphatase) are observed in men with prostate cancer-related skeletal metastases (Fizazi et al., 2015; Takeuchi et al., 1996). The bony microenvironment, bone, and tumor cells all function as autocrine and paracrine effectors resulting in tumor cell survival, stimulation of osteoblastic activity with increased bone deposition, and augmented OPG expression by prostate cancer cells to reduce local osteoclastogenesis (Keller and Brown, 2004). The increased bioavailability of growth factors such as IGF-1 is key to the survival of prostate cancer cells in the bony environment and is induced by the cancer cells through synthesis of uPA, which attach to its receptor on the surface of osteoblasts, activating proteolytic activity at sites neighboring the osteoblasts (Koutsilieris and Polychronakos, 1992). This action results in locally increased proteolysis either directly by uPA or indirectly by uPA-mediated generation of plasmin with ensuing activation of MMPs (Koutsilieris and Polychronakos, 1992). uPA-mediated hydrolysis of IGF-binding proteins releases IGF-1, resulting in its increased local availability for supporting growth of both osteoblasts and prostate cancer cells (Koutsilieris and Polychronakos, 1992). Growth factors and cytokines released by tumor cells can have direct mitogenic effects on osteoblasts and can also stimulate production of RANKL by osteoblasts with resultant increased binding and stimulation of RANK with augmented differentiation of pro-osteoclasts to mature osteoclasts (Coleman, 2001; Roodman, 2004). Production of key components of the bone matrix (e.g., osteocalcin, bone sialoprotein, osteopontin, and type-1 collagen) deposited through osteoblastic activity is regulated by the osteoblastic-specific transcription factor, core binding factor-1 (or Runx-2), and is suboptimal in CaP with reduced strength (Roodman, 2004; Steeg, 2006). Tumor cells functionally contribute to the process of bone matrix formation by acquiring an osteoblastic phenotype (osteomimicry) demonstrated in vitro by the production of osteocalcin and bone sialoprotein in prostate cancer cell lines (Huang et al., 2005).

## 3.2 Lymph Node Metastases

In prostate cancer, regional lymph node metastasis is an early metastatic event. Approximately 35% of radical prostatectomy (RP) patients with lymph node positive cancer will have distant metastases within 5 years (Touijer et al., 2014). Patterns of dissemination from primary prostatic tumor to lymph nodes demonstrate a predictable distribution (Nathanson, 2003). Regional lymphatic drainage of the prostate plays an important part in the location of lymph node metastases as most initially occur in the pelvic lymph nodes suggesting a potential contribution from peri- or intratumoral lymphatic vessels of the primary tumor to seeding of distant sites (Nathanson, 2003). The significance of intratumoral lymphatics in prostate cancer has been controversial with conflicting studies on the correlation between increased lymphatic vessel density (LVD) and the occurrence of lymph node metastases (Zeng et al., 2004; Cheng et al., 2008). Preclinical models have demonstrated that selectively inhibiting intratumoral lymphangiogenesis or using specific tumor models, which do not result in intratumoral lymphangiogenesis, had a limited impact on the development of lymph mode metastases (Wong et al., 2005).

Lymph node microenvironment factors have not been comprehensively investigated, however, gene expression comparisons of prostate cancer cells derived from primary sites of orthotopic implantation in mice versus those derived from resulting lymph node metastases revealed the most commonly upregulated metastasis-related genes included ECM degrading enzymes, adhesion molecules, transcription factors, and cell surface receptors (Wong et al., 2005). These factors have a known role in the metastatic cascade and may also encourage or facilitate metastasis to lymph nodes. Additionally as noted previously, VEGF-C and VEGF-D represent the major lymphangiogenic cytokines and have been putatively correlated with lymph node metastasis in several in vivo preclinical models (Nathanson, 2003).

## 3.3 Visceral Metastases

In a large autopsy series of over 1500 patients with advanced prostate cancer, metastases to visceral organs were present in the liver and lungs in 25% and 46% of cases, respectively (Bubendorf et al., 2000). The presence of visceral metastases is an independent, negative prognostic factor in men with bone metastases (Pond et al., 2014). A retrospective study of the prognostic impact of the site of visceral metastasis in data from the TAX 327 phase III trial on overall survival (OS) revealed men with liver metastases with or without other metastases had the shortest median OS (10 months) compared to men with lung metastases with or without bone or nodal metastases (14.4 months) (Pond et al., 2014). The best overall median OS was seen with lymph node-only disease (26.7 months), followed by those with bone-only metastases (19 months) and bone-plus-node disease (15.7 months) (Pond et al., 2014).

Despite its conspicuous presence in advanced disease and significantly worse prognosis for men with prostate cancer, only a handful of studies to date have directly examined the cellular biology or microenvironment factors, which may contribute to the visceral disease. Understanding differences between tumor cells of various metastatic locales may illustrate targetable genes or pathways specific to preventing or treating the disease. A few small studies have shown the potential for alternate dysregulation of antiapoptotic pathways, proangiogenic factors and transcriptional regulators in visceral compared to bony metastatic samples (Jin et al., 2013; Akfirat et al., 2013; Morrissey et al., 2008). Further investigation of these processes may also inform the likelihood of response to treatments as well as the choice of treatment.

Pulmonary metastases are rarely a significant event in prostate cancer in terms of symptomatology; however, it can be associated with common respiratory symptoms such as dyspnea and hemoptysis (Fabozzi et al., 1995; Tohfe et al., 2008). The reported incidence of clinically apparent lung metastases is variable from 5% up to 30%, whereas the autopsy prevalence among prostate cancer patients is reported at a rate of approximately 45% (Bubendorf et al., 2000; Fabozzi et al., 1995). Generally, isolated lung metastases are rare, rather they are more frequently encountered in the context of multiorgan involvement (Bubendorf et al., 2000; Fabozzi et al., 1995). Two basic radiologic patterns predominate with prostate cancer pulmonary metastases; a lymphangitic pattern is the most common and presumed to be secondary to infiltration of the pulmonary lymphatics and less commonly as discrete nodular densities (Fabozzi et al., 1995).

Intracranial metastases are likewise an uncommon event in prostate carcinoma (0.2–6%) with most cases seen in association with disseminated disease and the majority of the literature derived from large autopsy series and occasional case reports (Gómez et al., 2009; Remont-Lukats et al., 2003). Symptoms related to increased intracranial pressure, such as headache and cognitive impairment, were the most frequently reported symptoms (Remont-Lukats et al., 2003). In the MD Anderson institutional analysis of 16,280 prostate cancer patients, 103 patients with intraparenchymal brain metastases were diagnosed with 78 cases being supratentorial, 22 infratentorial, and 3 in both locations (Remont-Lukats et al., 2003). The most frequently associated histology was adenocarcinoma, however, rarer histologic subtypes such as small cell were also found at higher frequencies compared to other sites of metastasis in the body (Remont-Lukats et al., 2003). The mean survival of men with intracranial metastasis is estimated at 1–3 months in untreated patients, 3.5–6 months in those treated with radiotherapy, and up to 9 months where surgery was also performed (Gómez et al., 2009; Remont-Lukats et al., 2003; McCutcheon et al., 1999). Overall the published literature suggests a possible trend toward improved survival outcomes in appropriately selected, surgically treated patients with brain metastasis. Dural metastases are reported in up to 19.5% of prostate cancer cases in autopsy studies and are seen most commonly in advanced disease (Bubendorf et al., 2000). Cranioaxial spread is likely hematogenous via the IVC or Batsons' plexus (Weiner et al., 2013). Symptomatic disease presentations and treatment options are similar to those of intraparenchymal brain metastases (Weiner et al., 2013).

Liver metastases are likewise reported as a rare and late occurrence (~5%) associated with other systemic disease after therapy with hormone and/or chemotherapy (Singha et al., 2015; Pouessel et al., 2007). They may be detected incidentally during a routine follow-up imaging or present with specific symptoms. The most frequently reported symptoms associated with prostate cancer hepatic metastases were upper abdominal pain, jaundice, and abdominal distension (Wang et al., 2014). Pouessel et al. demonstrated a median OS time of 6 months postdiagnosis of hepatic metastases (Pouessel et al., 2007).

## 3.4 Clinical Considerations in Prostate Cancer Metastasis

### 3.4.1 *Disease Recurrence and Risk Assessment*

Serial serum PSA measurements are routinely employed posttherapy for presumed prostate-confined disease for early detection of disease recurrence, leading to the identification of individuals with PSA-only BCR that have no other evidence of locoregional recurrence or distant metastases. In a review of over 2809 men treated for localized-disease, 31% developed BCR with a mean time to relapse of 2.9 years (median of 2.4 years) post-RP (Roberts et al., 2001). The significance of BCR varies considerably and not all cases warrant immediate treatment as the disease can behave indolently. As demonstrated by one series of 450 men monitored following RP with the development BCR, 134 developed metastases (29.8%), with a median metastasis-free survival duration of 10 years (Antonarakis et al., 2012).

The diagnosis of BCR differs according to the initial treatment was undertaken. Successful surgical management with RP involves removal of all prostatic tissue, hence detectable serum PSA (by standard immunoassay) postoperatively is indicative of residual prostatic tissue in the body, presumably undetected locoregional or systemic disease. The American Urological Association guidelines define BCR post-RP as a serum PSA ≥0.2 ng/mL on two separate occasions (Paller et al., 2013). The diagnosis of BCR postradiation therapy is more challenging as residual benign prostate tissue remains posttreatment and therefore the PSA is unlikely to fall to undetectable levels. The currently accepted definition is derived from the recommendations of the American Society for Radiation Oncology panel in 2005 as a PSA rise of ≥2 ng/mL above the PSA nadir, succeeding the original definition of BCR as following three consecutive PSA rises from the nadir (Paller et al., 2013). The decline in serum PSA postradiation therapy is gradual and can take 18 months or longer to reach the nadir (Crook et al., 1998).

A number of factors that predict the development of BCR are also those shown to predict the development of aggressive metastatic disease and including T stage, pretreatment PSA level, Gleason score, and adverse pathological features on prostatectomy specimen (T stage, seminal vesicle involvement, surgical margin, and nodal status) (Swanson and Basler, 2010). Posttreatment factors including interval to PSA failure and rapid PSA-doubling time (DT) were also associated with increased rates of distant metastases (Swanson and Basler, 2010; Pound et al., 1999). One series illustrated that those with a PSA-DT of <3 months had a 20-fold increased risk of dying of prostate cancer compared to those with a PSA-DT of ≥3 months (D'Amico et al., 2003).

The major predicting factor of survival in metastatic disease is the site of metastatic involvement illustrated by a recent metaanalysis that presented data from approximately 4000 men treated with docetaxel for metastatic CRPC (mCRPC) (Halabi et al., 2014). OS progressively declined with involvement of lymph nodes, bone, lung, and liver (approximately 27, 20, 17, and 12 months, respectively) (Halabi et al., 2014). Higher enumeration of systemic CTCs in prostate cancer patients has also been associated with shorter OS in several trials (Danila et al., 2007; Goldkorn et al., 2014). In one phase III trial, CTCs were assayed in 238 CRPC patients randomly assigned to treatment with docetaxel or docetaxel plus endothelial receptor antagonist atrasentan (Goldkorn et al., 2014). Median OS was significantly prolonged in those with <5 CTCs per 7.5 mL compared to those with ≥5 CTCs per 7.5 mLat baseline (Goldkorn et al., 2014).

The disordered bone homeostasis seen frequently in men with bony metastases is associated with disturbed markers of bone metabolism potentially providing prognostic or treatment progress data. A phase III trial in mCRPC patients assayed two markers of bone resorption (N-telopeptide, pyridinoline) and two markers of bone formation (C-terminal collagen propeptide, bone alkaline phosphatase). Across all markers, a higher serum level at baseline line was correlated with a shorter OS time (Lara et al., 2014).

## 3.5 Overview of Treatment for Prostate Cancer Metastases

Contemporaneous research has resulted in the development of numerous active treatment modalities for metastatic disease, with the goals of prolonged survival, reduced complications, and preserving QOL. Therapies for patients with BCR are the same as those used for patients with overt metastatic disease; however, the potentially protracted clinical course following BCR dictates the timing of treatment initiation.

Although ADT is palliative, initially it objectively normalizes serum PSA in up to 90% of patients and subjectively improves metastasis-related symptoms (Harris et al., 2009). The duration of ADT response in metastatic prostate cancer is highly variable, with the majority of patients eventually experiencing disease progression. The initial reliance of castrate-sensitive metastatic prostate cancer on androgens for growth provides the rationale for androgen ablation accomplished by either bilateral orchiectomy or medical castration. Surgical castration provides irreversible and immediate reduction in serum testosterone levels through loss of testosterone producing Leydig cells (Harris et al., 2009). Medical castration is reversible and exerts its effects on testosterone via the hypothalamic pituitary axis by continuous treatment with GnRH agonist (such as leuprolide, goserelin, or triptorelin) with consequent reduction in the synthesis of testicular androgens within 3–4 weeks of administration (Harris et al., 2009; Shore et al., 2013). ADT aims to reduce serum testosterone to levels seen postbilateral orchidectomy, historically correlating to a level of <1.7 nmol/L (50 ng/dL) (Harris et al., 2009). A transient surge in LH and consequently serum testosterone levels following initial administration of a GnRH agonist is often observed as downregulation of the GnRH receptors in response to continuous GnRH stimulation occurs after 1 week (Harris et al., 2009). This "flare" phenomenon can cause a transient worsening of disease symptoms, which can be a concern in the setting of impending spinal cord compression or bladder outlet obstruction, and therefore an antiandrogen (e.g., flutamide or bicalutamide) is often administered for 2–4 weeks in conjunction with the initiation of a GnRH agonist to avert its occurrence (Harris et al., 2009; Shore et al., 2013). First-generation antiandrogens (e.g., flutamide, bicalutamide, nilutamide) competitively bind ARs preventing attachment of testosterone and DHT thus its use with GnRH agonists result in combined androgen blockade (Harris et al., 2009). Alternatively GnRH antagonists, such as degarelix, which inhibit GnRH receptors on pituitary gondaotropin-producing cells, can be administered instead of GnRH agonists, avoiding the testosterone "flare" caused by the sudden release of LH (Harris et al., 2009; Shore et al., 2013; Klotz et al., 2008). Degarelix sustainably suppressed serum testosterone within 3 days of administration in 96% of participants in a randomized controlled trial (RCT) comparing it to the GnRH agonist leuprolide (Klotz et al., 2008). Side effects experienced with ADT use can be attributed to castrate levels of serum testosterone and include increased adiposity, cardiovascular and metabolic abnormalities, reduced muscle mass and bone mineral density, hot flashes, gynecomastia, fatigue, altered cognition, and sexual dysfunction (Harris et al., 2009; Shore et al., 2013). Intermittent administration of ADT has been proposed as a strategy to abate the impact of these side effects on QOL by withholding active therapy once patients experience a treatment response and reinstituting it when there is evidence of disease progression (Harris et al., 2009).

Evidence of disease progression such as a rising PSA, new or progressive metastases, while on ADT is considered to indicate castration-resistant disease. No standardized definition of CRPC is widely accepted especially in the setting of PSA-only progression and is largely reliant on the treating physician, with many using a defined rising serum PSA of >2 ng/mL over 1 month, whereas others initiate discussion of additional treatment when the PSA doubles (Hong and Kim, 2014). As discussed previously, the presence of CRPC does not imply total independence from AR-stimulated growth and second-line antiandrogens may be effective.

Second-line hormonal treatments abiraterone and enzalutamide have demonstrated significantly improved OS in both CRPC patients who were chemotherapy naive or postdocetaxel, compared with placebo in phase III trials (Omlin et al., 2014; Fizazi et al., 2012; Ryan et al., 2013; Scher et al., 2012; Beer et al., 2014). Abiraterone is an orally administered irreversible inhibitor of CYP17 gene products, resulting in reduced intratumoral and adrenal-related synthesis of androgens (Tan et al., 2014). MDV3100 or enzalutamide is also orally administered and attaches to the AR exerting multilevel effects by inhibiting nuclear translocation, coactivator recruitment, and binding to target DNA ARs (Tan et al., 2014; Richards et al., 2012; Golshayan and Antonarakis, 2013).

Previous strategies for second-line hormonal manipulation involved the addition of first-generation antiandrogens to patients with progression despite GnRH antagonism or surgical orchidectomy, withdrawal of antiandrogen therapy from men with progression despite taking antiandrogens (may result in androgen withdrawal syndrome, an unexplained decline in serum PSA observed in ~20% of cases managed with complete androgen blockade following the withdrawal of an antiandrogen), administration of ketoconazole or steroids to suppress adrenal androgen production, and estrogens or progesterones to further suppress GnRH (Lorente et al., 2015). These strategies may still offer modest beneficial effects for select patients who progress on novel second-generation antiandrogens in whom more toxic therapies are not suitable.

Immune-directed therapies also have utility in prostate cancer, as the cancer cells are associated with a number of tumor-specific antigens, which may be targeted including prostatic acid phosphatase (PAP), PSA, and prostate-specific membrane antigen (PSMA) (Hammerstrom et al., 2011). The dendritic cell vaccine, Sipuleucel-T is patient specific and derived from mononuclear cells obtained by leukapheresis, which are exposed to an immune-stimulating recombinant protein of PAP and human granulocyte macrophage colony-stimulating factor (Hammerstrom et al., 2011; Kantoff et al., 2010). Sipuleucel-T prolonged OS compared with placebo in men with minimally symptomatic, metastatic prostate cancer, however, it did not significantly increase progression-free survival or affect the serum PSA (Kantoff et al., 2010).

The taxane chemotherapeutic agents such as docetaxal and cabazitaxel demonstrate efficacy in men with CRPC (Berthold et al., 2008; De Bono et al., 2010). In the TAX 327 phase III trial docetaxel administered in combination with daily prednisone prolonged OS compared to an alternative chemotherapy agent mitoxantrone plus prednisone (Berthold et al., 2008). On those findings, docetaxel plus prednisone is currently the standard initial regimen for CRPC when chemotherapy is indicated. Cabazitaxel is a taxane derivative shown in a phase III trial to significantly increase OS compared with mitoxantrone plus prednisone in CRPC patients progressing on docetaxel (De Bono et al., 2010).

Sequential therapy with hormonal therapies and follow-up treatment with docetaxel for progressive castrate-resistant disease remains the standard of care across most settings. However, recent studies have shown a combination of ADT plus docetaxel in men with castrate-sensitive prostate cancer and high volume disease may offer clinically meaningful improvements (James et al., 2015a). One arm of the large STAMPEDE trial randomly assigning men to either ADT plus docetaxel chemotherapy or ADT alone and recruited 1184 men with metastatic or high-risk disease (James et al., 2015a,b). Population features included distant metastases in 61% of patients, 15% with node positive disease, and 24% high-risk localized-disease on enrolment (James et al., 2015a,b). Initial results presented at the 2015 American Society of Clinical Oncology meeting showed the primary endpoint of OS was significantly increased in those treated with ADT plus docetaxel compared with ADT alone (median of 77 vs. 67 months) as well as progression-free survival (James et al., 2015b). In a subanalysis based on the presence or absence of overt metastases, OS was improved with the addition of docetaxel in those with metastatic disease (median survival of 65 vs. 43 months) (James et al., 2015b). These findings confirm the results of the previous CHAARTED trial, which enrolled 790 men with hormone-sensitive metastatic prostate cancer and randomized them to docetaxel + ADT or ADT alone. The ADT + chemotherapy arm demonstrated a significant increase in the primary endpoint of OS (57.6 months vs. 44 months, $p < .0001$) compared to ADT alone (Sweeney et al., 2015).

Infrequently patients with mCRPC may have low serum PSA values, and this is a likely indicator of more aggressive de-differentiated disease or alternative histopathological features (e.g., neuroendocrine) (Sella et al., 2000). These cases are more likely to result in visceral and osteolytic metastases rather than osteoblastic bone lesions (Sella et al., 2000). There have been no RCTs to direct management of such cases, however, limited evidence suggests neuroendocrine (small cell) features are relatively sensitive to chemotherapy regimens utilized for lung small-cell cancers (e.g., platinum/etoposide combinations), with response rates of over 50% (at least initially) (Sella et al., 2000). Poorly differentiated adenocarcinoma follows a clinically aggressive path with no clear benefit demonstrated to using one chemotherapeutic strategy over the others.

## 3.6 Treatment of Symptomatic Metastatic Disease

For the treatment of symptomatic metastases, systemic first-line ADT should be promptly initiated to palliate symptoms and to reduce the likelihood of complications such as urinary obstruction, pathological fractures, and spinal cord compression.

For the specific management of symptomatic bony metastases external beam radiation therapy is standard treatment for those with focally symptomatic bony lesions not adequately palliated by appropriate analgesics. Overall it has been shown to elicit a subjective benefit in 80% of patients and complete resolution of symptoms in 25% (Di Lorenzo et al., 2003).

Prior to the discovery of alpha-particle emitting radiopharmaceuticals, beta-particle emitting radioisotopes such as strontium-89 were employed for pain management related to osteoblastic metastases, particularly if no relief was achieved postmaximal ERBT (Lewington et al., 1991). Usurped by radium-223 (an alpha-particle radiopharmaceutical which delivers a targeted and highly localized dose of radiation to cancer cells), it has demonstrated improved OS and time to first symptomatic SRE in the phase III ALSYMPCA trial in patients with symptomatic, bone-only metastases (Parker et al., 2013).

Bisphosphonates such as zolendronic acid inhibit bone resorption via inhibition of osteoclast activity and have been shown through RCTs to provide a moderate effect in the provision of pain relief and to reduce the frequency of SREs in patients with CRPC bony metastases (Saad et al., 2002). Denosumab, which works through binding to the critical osteoclastogenesis ligand RANK, also reduces SRE rates in CRPC metastatic to bone (Fizazi et al., 2011).

Surgical intervention for bone lesions is generally reserved for cases of pathologic fractures or spinal cord compression.

Pelvic symptoms of advanced prostate cancer also impact significantly on QOL including lower urinary tract symptoms, abdominopelvic pain, hematuria, and occasionally obstructive rectal symptoms. In small prospective studies, systemic therapy followed by EBRT in refractory cases has shown potential benefit in symptom palliation (Cameron et al., 2015). Urological interventions may be useful in the palliative management of lower urinary tract symptoms, symptomatic ureteric obstruction, or refractory hematuria.

## 3.7 Future Areas of Development

Prostate cancer management is in an exciting phase of expansion after a prolonged period of quiescence where improved understanding of genetic and mechanistic cellular biology in conjunction with new technologies has resulted in the development of novel treatments and examination procedures. A brief discussion of some of these developments is discussed here.

## 3.8 Liquid Biopsy and Precision Medicine

The interest in CTC enumeration as a noninvasive indicator of disease burden over time has given way to interest in CTC molecular characterization to guide individualized therapeutic strategies. Despite the genotypically heterogeneous nature of metastatic prostate cancer with its extensive array of molecular drivers, treatment regimens remain relatively standardized. CTC analysis to elicit the individual genetic aberrations of a patients' tumor over time can provide clinically valuable information regarding treatment susceptibility or resistance ultimately leading to personalized and precise therapeutic strategies. This evolving application was illustrated recently in a study undertaken in men with mCRPC where the presence of CTCs demonstrating the specific AR splice variant AR-V7 was highly predictive of clinical abiraterone acetate and enzalutamide resistance such as an absence of PSA responses to either agent, and reduced progression-free survival and OS compared with men without detectable CTC AR-V7 (Antonarakis et al., 2014).

## 3.9 Imaging Modalities

Novel molecular imaging agents such as PSMA (a type II transmembrane protein overexpressed in prostate cancer) in conjunction with positron emission tomography, computed tomography, and magnetic resonance imaging with increased sensitivity and specificity compared to conventional imaging are consolidating their role in diagnosing and guiding metastatic prostate cancer management (Fig. 3.3) (Afshar-Oromieh et al., 2014).

## 3.10 Investigational Therapies

ADT is highly effective initial therapy for metastatic prostate cancer but its effects are not sustainable; therefore, additional approaches

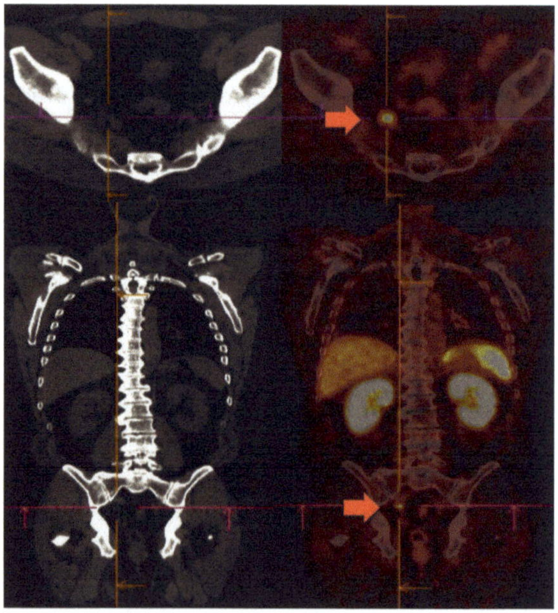

FIGURE 3.3 Comparative computed tomography (CT; LHS) and prostate-specific membrane antigen (PSMA) positron emission tomography (PET) CT (RHS) scans demonstrating a positive lymph node prostate cancer metastasis on PSMA PET/CT (*arrows*), which did not meet the diagnostic criteria on the standard CT scan.

are required to control disease and prolong the survival. An increased understanding of the biology of prostate cancer is leading to the development of a multitude of new therapies that can target specific abnormalities in prostate cancer.

Apalutamide or ARN-509 is a novel second-generation antiandrogen, which competitively inhibits the AR to hinder its nuclear translocation, AR-regulated gene transcription, and DNA binding (Clegg et al., 2012). In vivo it showed greater CRPC tumor suppression in murine xenograft models than enzalutamide and at lower comparative doses (Leibowitz-Amit and Joshua, 2012; Kim and Ryan, 2012; Schweizer and Antonarakis, 2012). The results of the Phase II clinical trial are still pending publication; however, Phase I clinical trials demonstrated a >50% PSA response rates in >60% of patients with mCRPC (Rathkopf et al., 2013).

The concept of "theranostics" (a combination of therapeutics and diagnostics) is gaining widespread interest in metastatic prostate cancer with the investigation of Lutetium-177 labeled PSMA (Lu-PSMA). Several Phase I and II trials are ongoing around the world investigating this new therapeutic option. Initial results to date are scant but eagerly awaited.

## 4. CONCLUSIONS

Metastatic prostate cancer is currently incurable, progress has been made in disease control, prolonging survival and improving QOL, although research is still required to identify new therapeutic targets and strategies to delay the evolution of the lethal phenotype. Recent advances in imaging have enabled the earlier detection of metastatic deposits and provide a platform to investigate whether interventions targeting low volume disease can further improve outcomes for men with prostate cancer.

## References

Abate-Shen, C., Shen, M.M., 2000. Molecular genetics of prostate cancer. Genes Dev. 14 (19), 2410–2434.

Afshar-Oromieh, A., Haberkorn, U., Schlemmer, H.P., et al., 2014. Comparison of PET/CT and PET/MRI hybrid systems using a 68Ga-labelled PSMA ligand for the diagnosis of recurrent prostate cancer: initial experience. Eur. J. Nucl. Med. Mol. Imaging 41 (5), 887–897.

Akfirat, C., Zhang, X., Ventura, A., et al., 2013. Tumour cell survival mechanisms in lethal metastatic prostate cancer differ between bone and soft tissue metastases. J. Pathol. 230 (3), 291–297.

Antonarakis, E.S., Feng, Z., Trock, B.J., Humphreys, E.B., 2012. The natural history of metastatic progression in men with prostate-specific antigen recurrence after radical prostatectomy: long-term follow-up. BJU Int. 109 (1), 32–39.

Antonarakis, E.S., Chang Xue, L., Wang, H., et al., 2014. AR-V7 and resistance to enzalutamide and abiraterone in prostate cancer. N. Engl. J. Med. 371 (11), 1028–1038.

Arya, M., Bott, S., Shergill, I., 2006. The metastatic cascade in prostate cancer. Surg. Oncol. 15 (3), 117–128.

Attar, R., Takimoto, C., Gottardis, M., 2009. Castration-resistant prostate cancer: locking up the molecular escape routes. Clin. Cancer Res. 15 (10), 3251–3252.

Balk, S., Knudsen, K., 2008. AR, the cell cycle, and prostate cancer. Nucl. Recept. Signal 6, e001.

Barbieri, C.E., Baca, S.C., Lawrence, M.S., Demicheis, F., et al., 2012. Exome sequencing identifies recurrent sPOP, FOXA1 and MED12 mutations in prostate cancer. Nat. Genet. 44 (6), 685–689.

Beer, T.M., Armstrong, A.J., Rathkopf, D.E., et al., 2014. Enzalutamide in metastatic prostate cancer before chemotherapy. N. Engl. J. Med. 371 (5), 424–433.

Beltran, H., Yelensky, R., Frampton, G.M., et al., 2013. Targeted next-generation sequencing of advanced prostate cancer identifies potential therapeutic targets and disease heterogeneity. Eur. Urol. 63 (5), 920–926.

Bennett, N.C., Gardiner, R.A., Hooper, J.D., et al., 2010. Molecular cell biology of androgen receptor signalling. Int. J. Biochem. Cell Biol. 42 (6), 813–827.

Berthold, D.R., Pond, G.R., Soban, F., de Wit, R., Eisenberger, M., Tannock, I.F., 2008. Docetaxel plus prednisone or mitoxantrone plus prednisone for advanced prostate cancer: updated survival in the TAX 327 study. J. Clin. Oncol. 26 (2), 242–245.

Blades, R.A., Keating, P.J., McWilliam, L.J., George, N.J., Stern, P.L., 1995. Loss of HLA class I expression in prostate cancer: implications for immunotherapy. Urology 46 (5), 681–686.

Bookstein, R., MacGrogan, D., Hilsenbeck, S.G., Sharkey, F., Allred, D.C., 1993. p53 is mutated in a subset of advanced-stage prostate cancers. Cancer Res. 53 (14), 3369–3373.

Bostwick, D.G., Wheeler, T.M., Blute, M., et al., 1996. Optimised microvessel density analysis improves prediction of cancer stage from prostate needle biopsies. Urology 48 (1), 47–57.

Bryden, A.A., Hoyland, J.A., Freemont, A.J., et al., 2002. E-cadherin and beta-cadherin are down-regulated in prostatic bone metastases. BJU Int. 89 (4), 400–403.

Bubendorf, L., Schopfer, A., Wagner, U., Sauter, G., Moch, H., Willi, N., Gasser, T.C., Mihatsch, M., 2000. Metastatic patterns of prostate cancer: an autopsy study of 1,589 patients. Hum. Pathol. 31, 578–583.

Bussemakers, M.J., Van Bokhoven, A., Tomita, K., Jansen, C., Schalken, J.A., 2000. Complex cadherin expression in human prostate cancer cells. Int. J. Cancer 85 (3), 446–450.

Cameron, M.G., Kersten, C., Vistad, I., et al., 2015. Palliative pelvic radiotherapy for symptomatic incurable prostate cancer – a prospective multicenter study. Radiother. Oncol. 115 (3), 314–320.

Cano, L.Q., Lavery, D.N., Bevan, C.L., 2013. Mini-review: foldosome regulation of androgen receptor action in prostate cancer. Mol. Cell Endocrinol. 369 (1–2), 52–62.

Cantley, L.C., 2002. The phosphoinositide 3-kinase pathway. Science 296 (5573), 1655–1657.

Carvalho, F.L., Simons, B.W., Antonarakis, E.S., et al., 2013. Tumorigenic potential of circulating prostate tumor cells. Oncotarget 4 (3), 413–421.

Catz, S.D., Johnson, J.L., 2003. BCL-2 in prostate cancer: a mini review. Apoptosis 8 (1), 29–37.

Chandrasekar, T., Yang, J., Gao, A., Evans, C., 2015. Mechanism of resistance in castration resistant prostate cancer (CRPC). Transl. Androl. Urol. 4 (3), 365–380.

Chay, C.H., Cooper, C.R., Gendernalik, J.D., et al., 2002. A functional thrombin receptor (PAR1) is expressed on bone-derived prostate cancer cell lines. Urology 60 (5), 760–765.

Chen, G., Shukeir, N., Potti, A., et al., 2004. Up-regulation of wnt-1 and B-catenin production in patients with advanced metastatic prostate cancer. Cancer 101 (6), 1345–1356.

Cheng, L., Bishop, E., Zhou, H., et al., 2008. Lymphatic vessel density in radical prostatectomy specimens. Hum. Pathol. 39 (4), 610–615.

Chmelar, R., Buchanan, G., Need, E.F., Tilley, W., Greenberg, N.M., 2007. Androgen receptor coregulators and their involvement in the development and progression of prostate cancer. Int. J. Cancer 120 (4), 719–733.

Clegg, N.J., Wongvipat, J., Joseph, J.D., et al., 2012. ARN-509: a novel antiandrogen for prostate cancer treatment. Cancer Res. 72 (6), 1494–1503.

Clevers, H., 2006. Wnt/β-catenin signalling in development and disease. Cell 127 (3), 469–480.

Coleman, R.E., 2001. Metastatic bone disease: clinical features, pathophysiology and treatment strategies. Cancer Treat. Rev. 27 (3), 165–176.

Cooper, C.R., Pienta, K.J., 2000. Cell adhesion and chemotaxis in prostate cancer metastasis to bone: a minireview. Prostate Cancer Prostatic Dis. 3 (1), 6–12.

Courtney, K.D., Corcoran, R.B., Engelman, J.A., 2010. The PI3K pathway as drug target in human cancer. J. Clin. Oncol. 28 (6), 1075–1083.

Crook, J.M., Choan, E., Perry, G.A., Robertson, S., Esche, B.A., 1998. Serum prostate-specific antigen profile following radiotherapy for prostate cancer: implications for patterns of failure and definition of cure. Urology 51 (4), 566–572.

D'Amico, A.V., Moul, J.W., Carroll, P.R., Sun, L., Lubeck, D., Chen, M.H., 2003. Surrogate end point for prostate cancer–specific mortality after radical prostatectomy or radiation therapy. J. Natl. Cancer Inst. 95 (18), 1376–1383.

Danila, D.C., Heller, G., Gignac, G.A., et al., 2007. Circulating tumor cell number and prognosis in progressive castration-resistant prostate cancer. Clin. Cancer Res. 13 (23), 7053–7058.

Datta, K., Muders, M., Zhang, H., Tindall, D.J., 2010. Mechanism of lymph node metastasis in prostate cancer. Future Oncol. 6 (5), 823–836.

De Bono, J.S., Oudard, S., Ozguroglu, M., et al., 2010. Prednisone plus cabazitaxel or mitoxantrone for metastatic castration-resistant prostate cancer progressing after docetaxel treatment: a randomised open-label trial. Lancet 376 (9747), 1147–1154.

Debes, J.D., Sebo, T.J., Lohse, C.M., Murphy, L.M., Haugen, D.A., Tindall, D.J., 2003. p300 in prostate cancer proliferation and progression. Cancer Res. 63 (22), 7638–7640.

Dehm, S.M., Schmidt, L.J., Heemers, H.V., Vessella, R.L., Tindall, D.J., 2008. Cancer Splicing of a novel androgen receptor exon generates a constitutively active androgen receptor that mediates prostate cancer therapy resistance. Cancer Res. 68 (13), 5469–5477.

Deshmukh, D., Qui, Y., 2015. Role of PARP-1 in prostate cancer. Am. J. Clin. Exp. Urol. 3 (1), 1–12.

Di Lorenzo, G., Autorino, R., Ciardiello, F., et al., 2003. External beam radiotherapy in bone metastatic prostate cancer: impact on patients' pain relief and quality of life. Oncol. Rep. 10 (2), 399–404.

Dienstmann, R., Rodon, J., Serra, V., Tabernero, J., 2014. Picking the point of inhibition: a comparative review of PI3K/AKT/mTOR pathway inhibitors. Mol. Cancer Ther. 13 (5), 1021–1031.

Dong, Z., Saliganan, A., Meng, H., et al., 2008. Prostate cancer cell-derived urokinase-type plasminogen activator contributes to intraosseous tumor growth and bone turnover. Neoplasia 10 (5), 439–449.

Edlund, M., Miyamoto, T., Sikes, R.A., et al., 2001. Integrin expression and usage by prostate cancer cell lines on laminin substrate. Cell Growth Differ. 12 (2), 99–107.

Efeyan, A., Serrano, M., 2007. p53: guardian of the genome and policeman of the oncogenes. Cell Cycle 6 (9), 1006–1010.

Eisermann, K., Wang, D., Jing, Y., Pascal, L.E., Wang, Z., 2013. Androgen receptor gene mutation, rearrangement, polymorphism. Transl. Androl. Urol. 2 (3), 137–147.

Fabozzi, S.J., Schellhammer, P.F., el-Mahdi, A.M., 1995. Pulmonary metastases from prostate cancer. Cancer 75 (11), 2706–2709.

Felici, A., Pino, M.S., Carlini, P., 2012. A changing landscape in castration-resistant prostate cancer treatment. Front Endocrinol. 3, 85.

Ferlay, J., Soerjomataram, I., Dikshit, R., Eser, S., Mathers, C., Rebelo, M., Parkin, D.M., Forman, D., Bray, F., 2015. Cancer incidence and mortality worldwide: sources, methods and major patterns in GLOBOCAN 2012. Int. J. Cancer 136 (5), E359–E386.

Fidler, I.J., 2003. The pathogenesis of cancer metastasis: the 'seed and soil' hypothesis revisited. Nat. Rev. Cancer 3 (6), 453–458.

Fizazi, K., Carducci, M., Smith, M., et al., 2011. Denosumab versus zoledronic acid for treatment of bone metastases in men with castration-resistant prostate cancer: a randomised, double-blind study. Lancet 377 (9768), 813–822.

Fizazi, K., Scher, H.I., Molina, A., et al., 2012. Abiraterone acetate for treatment of metastatic castration-resistant prostate cancer: final overall survival analysis of the COU-AA-301 randomised, double-blind, placebo-controlled phase 3 study. Lancet Oncol. 13 (10), 983–992.

Fizazi, K., Massard, C., Smith, M., et al., 2015. Bone-related parameters are the main prognostic factors for overall survival in men with bone metastases from castration-resistant prostate cancer. Eur. Urol. 68 (1), 42–50.

Ford, O.H., Gregory, C.W., Kim, D., et al., 2003. Androgen receptor gene amplification and protein expression in recurrent prostate cancer. J. Urol. 170 (5), 1817–1821.

Gandaglia, G., Karakiewicz, P.I., Briganti, A., et al., 2015. Impact of the site of metastases on survival in patients with metastatic prostate cancer. Eur. Urol. 68 (2), 325–334.

Ganguly, S., Li, X., Miranti, C., 2014. The host environment influences prostate cancer invasion, systemic spread, bone colonization, and osteoblastic metastasis. Front Oncol. 4 (364), 1–16.

Gelmann, E.P., 2002. Molecular biology of the androgen receptor. J. Clin. Oncol. 20 (13), 3001–3015.

Geng, C., He, B., Xu, L., Barbieri, C.E., et al., 2013. Prostate cancer-associated mutations in speckle-type POZ protein (SPOP) regulate steroid receptor coactivator 3 protein turnover. Proc. Natl. Acad. Sci. U.S.A. 110 (17), 6997–7002.

Goel, H., Li, J., Kogan, S., Languino, L., 2008. Integrins in prostate cancer progression. Endocr. Relat. Cancer 15 (3), 657–664.

Goldkorn, A., Ely, B., Quinn, D.I., et al., 2014. Circulating tumor Cell counts are prognostic of overall survival in SWOG S0421: a phase III trial of docetaxel with or without atrasentan for metastatic castration-resistant prostate cancer. J. Clin. Oncol. 32 (11), 1136–1142.

Golshayan, A.R., Antonarakis, E.S., 2013. Enzalutamide: an evidence-based review of its use in the treatment of prostate cancer. Core Evid. 8, 27–35.

Gómez, J.E.C., Herranz, U.A., Castro, A.C., Caamafio, A.G., Mateos, L.A.L., Vázquez, C.P., López, R.L., 2009. Brain metastases from prostate adenocarcinoma. Clin. Trans. Oncol. 11 (1), 63–64.

Guo, W., Giancotti, F.G., 2004. Integrin signalling during tumour progression. Nat. Rev. Mol. Cell Biol. 5 (10), 816–826.

Gupta, G.P., Massague, J., 2006. Cancer metastasis: building a framework. Cell 127, 679–695.

Halabi, S., Kelly, W.K., Zhou, H., et al., 2014. The site of visceral metastases (mets) to predict overall survival (OS) in castration-resistant prostate cancer (CRCP) patients (pts). J. Clin. Oncol. 32 (5s) (abstr 5002).

Hammerstrom, A.E., Cauley, D.H., Atkinson, B.J., Sharma, P., 2011. Cancer immunotherapy: sipuleucel-T and beyond. Pharmacotherapy 31 (8), 813–828.

Han, G., Foster, B.A., Mistry, S., et al., 2001. Hormone status selects for spontaneous somatic androgen receptor variants that demonstrate specific ligand and cofactor dependent activities in autochthonous prostate cancer. J. Biol. Chem. 276 (14), 11204–11213.

Harris, W.P., Mostaghel, E.A., Nelson, P.S., Montgomery, B., 2009. Androgen deprivation therapy: progress in understanding mechanisms of resistance and optimizing androgen depletion. Nat. Clin. Pract. Urol. 6 (2), 76–85.

Hguygen, T.T., Nguyen, C.T., Gonzales, F.A., Nichols, P.W., et al., 2000. Analysis of cyclin-dependent kinase inhibitor expression and methylation patterns in human prostate cancers. Prostate 43 (3), 233–242.

Holzbeierlein, J., Lal, P., LaTulippe, E., et al., 2004. Gene expression analysis of human prostate carcinoma during hormonal therapy identifies androgen-responsive genes and mechanisms of therapy resistance. Am. J. Pathol. 164 (1), 217–227.

Hong, J.H., Kim, I.Y., 2014. Nonmetastatic castration-resistant prostate cancer. Korean J. Urol. 55 (3), 153–160.

Horvath, L.G., Henshall, S.M., Lee, C.S., et al., 2005. Lower levels of nuclear beta-catenin predicts for a poorer prognosis in localised prostate cancer. Int. J. Cancer 113 (3), 415–422.

Hu, B., Rochefort, H., Golkorn, A., 2013. Circulating tumor cells in prostate cancer. Cancers 5 (4), 1676–1680.

Huang, W.C., Xie, Z., Konaka, H., Sodek, J., et al., 2005. Human osteocalcin and bone sialoprotein mediating osteomimicry of prostate cancer cells: role of cAMP-dependent protein kinase A signalling pathway. Cancer Res. 65 (6), 2303–2313.

Hughes, C., Murphy, A., Martin, C., Sheils, O., O'Leary, J., 2005. Molecular pathology of prostate cancer. J. Clin. Pathol. 58 (7), 673–684.

Huss, W.J., Hanrahan, C.F., Barrios, R.J., Simons, J.W., Greenberg, N.M., 2001. Angiogenesis and prostate cancer: identification of a molecular progression switch. Cancer Res. 61 (6), 2736.

Jacob, K., Webber, M., Benayahu, D., Kleinman, H.K., 1999. Osteonectin promotes prostate cancer cell migration and invasion: a possible mechanism for metastasis to bone. Cancer Res. 59 (17), 4453–4457.

James, N.D., Spears, M.R., Clarke, N.W., et al., 2015a. Survival with newly diagnosed metastatic prostate cancer in the "docetaxel era": data from 917 patients in the control arm of the STAMPEDE trial (MRC PR08, CRUK/06/019). Eur. Urol. 67 (6), 1028–1038.

James, N.D., Sydes, M.R., Mason, M.D., et al., 2015b. Docetaxel and/or zoledronic acid for hormone-naïve prostate cancer: first overall survival results from STAMPEDE. J. Clin. Oncol. 33 (suppl.) (abstr 5001).

Jarrard, D.F., Bova, G.S., Ewing, C.M., et al., 1997. Deletional, mutational, and methylation analyses of CDKN2 (p16/MTS1) in primary and metastatic prostate cancer. Genes Chromosomes Cancer 19 (2), 90–96.

Jendraschak, E., Sage, E.H., 1996. Regulation of angiogenesis by SPARC and angiostatin: implications for tumor cell biology. Semin. Cancer Biol. 7, 139–146.

Jennbacken, K., Vallbo, C., Wang, W., Damber, J.E., 2005. Expression of vascular endothelial growth factor C (VEGF-C) and VEGF receptor-3 in human prostate cancer is associated with regional lymph node metastasis. Prostate 65 (2), 110–116.

Jernberg, E., Thysell, E., Bovinder Ylitalo, E., et al., 2013. Characterization of prostate cancer bone metastases according to expression levels of steroidogenic enzymes and androgen receptor splice variants. PLoS One 8 (11), e77407.

Jin, R., Steriling, J., Edwards, J., DeGraff, D., et al., 2013. Activation of NF-Kappa B signalling promotes growth of prostate cancer cells in bone. PLoS One 8 (4), e60983.

Kantoff, P.W., Higano, C.S., Shore, N.D., et al., 2010. Sipuleucel-T immunotherapy for castration-resistant prostate cancer. N. Engl. J. Med. 363 (5), 411–422.

Karanika, S., Karantanos, T., Li, L., Corn, P.G., Thompson, T.C., 2015. DNA damage response and prostate cancer: defects, regulation and therapeutic implications. Oncogene 34 (22), 2815–2822.

Karantanos, T., Corn, P.G., Thompson, T.C., 2013. Prostate cancer progression after androgen deprivation therapy: mechanisms of castrate resistance and novel therapeutic approaches. Oncogene 32 (49), 5501–5511.

Kaufman, B., Shapira-Frommer, R., Schmutzler, R.K., Audeh, M.W., et al., 2015. Olaparib monotherapy in patients with advanced cancer and a germline mutation BRCA1/2 mutation. J. Clin. Oncol 33 (3), 244–250.

Keller, E.T., Brown, J., 2004. Prostate cancer bone metastases promote both osteolytic and osteoblastic activity. J. Cell Biochem. 91 (4), 718–729.

Kibel, A.S., Faith, D.A., Bova, G.S., Isaacs, W.B., 2000. Loss of heterozygosity at 12P12-13 in primary and metastatic prostate adenocarcinoma. J. Urol. 164 (1), 192–196.

Kim, W., Ryan, C.J., 2012. Androgen receptor directed therapies in castration-resistant metastatic prostate cancer. Curr. Treat Options Oncol. 13 (2), 189–200.

Kirby, M., Hirst, C., Crawford, E.D., 2011. Characterising the castration-resistant prostate cancer population: a systematic review. Int. J. Clin. Pract. 65 (11), 1180–1192.

Klotz, L., Boccon-Gibod, L., Shore, N., et al., 2008. The efficacy and safety of degarelix: a 12-month, comparative, randomized, open-label, parallel-group phase III study in prostate cancer patients. BJU Int. 102 (11), 1531–1538.

Koivisto, P., Kononen, J., Palmberg, C., et al., 1997. Androgen receptor gene amplification: a possible molecular mechanism for failure of androgen deprivation therapy in prostate cancer. Cancer Res. 57 (2), 314–319.

Koutsilieris, M., Polychronakos, C., 1992. Proteinolytic activity against IGF-binding proteins involved in the paracrine interactions between prostate adenocarcinoma cells and osteoblasts. Anticancer Res. 12 (3), 905–910.

Kypta, R.M., Waxman, J., 2012. Wnt/β-catenin signalling in prostate cancer. Nat. Rev. Urol. 9 (8), 418–428.

Lara, P.N., Ely, B., Quinn, D.I., et al., 2014. Serum biomarkers of bone metabolism in castration-resistant prostate cancer patients with skeletal metastases: results from SWOG 0421. J. Natl. Cancer Inst. 106 (4) dju013.

Lehmann, B.D., McCubrey, J.A., Jefferson, H.S., Paine, M.S., et al., 2007. A dominant role for p53-dependent cellular senescence in radiosensitization of human prostate cancer cells. Cell Cycle 6 (5), 595–605.

Leibowitz-Amit, R., Joshua, A.M., 2012. Targeting the androgen receptor in the management of castration-resistant prostate cancer: rationale, progress, and future directions. Curr. Oncol. 19 (Suppl. 3), S22–S31.

Lewington, V.J., McEwan, A.J., Ackery, D.M., et al., 1991. A prospective, randomised double-blind crossover study to examine the efficacy of strontium-89 in pain palliation in patients with advanced prostate cancer metastatic to bone. Eur. J. Cancer Clin. Oncol. 27 (8), 954–958.

Li, W., Wang, B., Moran, P., et al., 2009. Pegylated kunitz domain inhibitor suppresses hepsin-mediated invasive tumor growth and metastasis. Cancer Res. 69 (21), 8395–8402.

Liang, J., Slingerland, J.M., 2003. Multiple roles of the PI3k/PKB (Akt) pathway in cell cycle progression. Cell Cycle 2 (4), 336–342.

Linja, M.J., Savinainen, K.J., Saramäki, O.R., Tammela, T.L., Vessella, R.L., Visakorpi, T., 2001. Amplification and overexpression of androgen receptor gene in hormone-refractory prostate cancer. Cancer Res. 61 (9), 3550–3555.

Liu, P., Cheng, H., Roberts, T.M., Zhao, J.J., 2009. Targeting the phosphoinositide 3-kinase (PI3K) pathway in cancer. Nat. Rev. Drug Discov. 8 (8), 627–644.

Lorente, D., Mateo, J., Zafeiriou, Z., et al., 2015. Switching and withdrawing hormonal agents for castration-resistant prostate cancer. Nat. Rev. Urol. 12 (1), 37–47.

Lu, N.Z., Wardell, S.E., Burnstein, K.L., et al., 2006. International Union of Pharmacology. LXV. The pharmacology and classification of the nuclear receptor superfamily: glucocorticoid, mineralocorticoid, progesterone, and androgen receptors. Pharmacol. Rev. 58 (4), 782–797.

Maccauro, G., Spinelli, M.S., Mauro, S., Perisano, C., Graci, C., Rosa, M.A., 2011. Physiopathology of spine metastasis. Int. J. Surg. Oncol. 2011, 107969.

Macleod, K.F., 2010. The RB tumour suppressor: a gatekeeper to hormone independence in prostate cancer? J. Clin. Invest. 120 (12), 4179–4182.

Mateo, J., Sandhu, S., Miranda, S., Carreira, S., et al., 2015. DNA repair defects and olaparib in metastatic prostate cancer. New Engl. J. Med. 373 (18), 1697–1708.

McCutcheon, I.E., Eng, D.Y., Logothetis, C.J., 1999. Brain metastasis from prostate carcinoma: antemortem recognition and outcome after treatment. Cancer 86 (11), 2301–2311.

Miyake, H., Hara, I., Eto, H., 2007. Serum level of cathepsin B and its density in men with prostate cancer as novel markers of disease progression. Anticancer Res. 24 (4), 2573–2577.

Mohler, J.L., Gregory, C.W., Ford, O.H., et al., 2004. The androgen axis in recurrent prostate cancer. ClinCancer Res. 10 (2), 440–448.

Montgomery, R.B., Mostaghel, E.A., Vessella, R., et al., 2008. Maintenance of intratumoral androgens in metastatic prostate cancer: a mechanism for castration-resistant tumour growth. Cancer Res. 68 (11), 4447–4454.

Morgan, T.M., Koreckij, T.D., Corey, E., 2009. Targeted therapy for advanced prostate cancer: inhibition of the PI3K/Akt/mTOR pathway. Curr. Cancer Drug Targets 9 (2), 237–249.

Morrissey, C., True, L.D., Roudier, M.P., et al., 2008. Differential expression of angiogenesis associated genes in prostate cancer bone, liver and lymph node metastases. Clin. Exp. Metastasis 25 (4), 377–388.

Nathanson, S., 2003. Insights into the mechanisms of lymph node metastasis. Cancer 98 (2), 413–423.

Navone, N.M., Troncoso, P., Pisters, L.L., et al., 1993. p53 protein accumulation and gene mutation in progression of human prostate carcinoma. J. Natl. Cancer Inst. 85 (20), 1657–1669.

Navone, N.M., Labate, M.E., Troncoso, P., et al., 1999. p53 mutations in prostate cancer bone metastases suggest that selected p53 mutants in the primary site define foci with metastatic potential. J. Urol. 161 (1), 304–308.

Nørgaard, M., Jensen, A.Ø., Jacobsen, J.B., Cetin, K., Fryzek, J.P., Sørensen, H.T., 2010. Skeletal related events, bone metastasis and survival of prostate cancer: a population based cohort study in Denmark (1999 to 2007). J. Urol. 184 (1), 162–167.

Nupponen, N.N., Kakkola, L., Koivisto, P., Visakorpi, T., 1998. Genetic alterations in hormone-refractory recurrent prostate carcinomas. Am. J. Pathol. 153 (1), 141–148.

Omlin, A., Pezaro, C., Sommer, S., 2014. Sequential use of novel therapeutics in advanced prostate cancer following docetaxel chemotherapy. Ther. Adv. Urol. 6 (1), 3–14.

Paller, C.J., Antonarakis, E.S., Eisenberger, M.A., Carducci, M.A., 2013. Management of patients with biochemical recurrence after local therapy for prostate cancer. Hematol. Oncol. Clin. North Am. 27 (6), 1205–1219.

Parker, C., Nilsson, S., Heinrich, D., et al., 2013. Alpha emitter radium-223 and survival in metastatic prostate cancer. N. Engl. J. Med. 369 (3), 213–223.

Pond, G.R., Sonpavde, G., de Wit, R., Eisenberger, M.A., Tannock, I.F., Armstrong, A.J., 2014. The prognostic importance of metastatic site in men with metastatic castration-resistant prostate cancer. Eur. Urol. 65 (1), 3–6.

Popiolek, M., Rider, J.R., Andren, O., Andersson, S.O., Holmberg, L., Adami, H.O., Johansson, J.E., 2013. Natural history of early, localized prostate cancer: a final report from three decades of follow-up. Eur. Urol. 63 (3), 428–435.

Pouessel, D., Gallet, B., Bibeau, F., Avances, C., et al., 2007. Liver metastases in prostate carcinoma: clinical characteristics and outcome. BJU Int. 99 (4), 807–811.

Pound, C.R., Partin, A.W., Eisenberger, M.A., Chan, D.W., Pearson, J.D., Walsh, P.C., 1999. Natural history of progression after PSA elevation following radical prostatectomy. JAMA 281 (17), 1591–1597.

Pulukuri, S., Rao, J., 2008. Matrix metalloproteinase-1 promotes prostate tumour growth and metastasis. Int. J. Oncol. 32 (4), 757–765.

Rashid, M.H., Chaudhary, U.B., 2004. Intermittent androgen deprivation therapy for prostate cancer. Oncologist 9 (3), 295–301.

Rathkopf, D.E., Morris, M.J., Fox, J.J., et al., 2013. Phase I study of ARN-509, a novel antiandrogen, in the treatment of castration-resistant prostate cancer. J. Clin. Oncol. 31 (28), 3525–3530.

Remont-Lukats, I.W., Bobustuc, G., Lagos, G.K., Lolas, K., Kyritsis, A.P., Puduvalli, V.K., 2003. Brain metastases from prostate carcinoma: the M. D. Anderson Cancer Centre experience. Cancer 98 (2), 363–368.

Richards, J., Lim, A.C., Hay, C.W., et al., 2012. Interactions of abiraterone, eplerenone, and prednisolone with wild-type and mutant androgen receptor: a rationale for increasing abiraterone exposure or combining with MDV3100. Cancer Res. 72 (9), 2176–2182.

Roberts, S.G., Blute, M.L., Bergstralh, E.J., Slezak, J.M., Zincke, H., 2001. PSA doubling time as a predictor of clinical progression after biochemical failure following radical prostatectomy for prostate cancer. Mayo Clin. Proc. 76 (6), 576–581.

Robinson, D., Van Allen, E.M., Wu, Y., et al., 2015. Integrative clinical genomics of advanced prostate cancer. Cell 161 (5), 1215–1228.

Roodman, G.D., 2004. Mechanisms of bone metastasis. N. Engl. J. Med. 350 (16), 1655–1664.

Rosol, T.J., Tannehill-Gregg, S.H., LeRoy, B.E., et al., 2003. Animal models of bone metastases. Cancer 97 (3 Suppl.), 748–757.

Ross, R.W., Xie, W., Regan, M.M., et al., 2008. Efficacy of androgen deprivation therapy (ADT) in patients with advanced prostate cancer: association between Gleason score, prostate specific antigen level, and prior ADT exposure with duration of ADT effect. Cancer 112 (6), 1247–1253.

Ryan, C.J., Smith, M.R., de Bono, J.S., et al., 2013. Abiraterone in metastatic prostate cancer without previous chemotherapy. New Engl. J. Med. 368 (2), 138–148.

Saad, F., Gleason, D.M., Murray, R., et al., 2002. A randomized, placebo-controlled trial of zoledronic acid in patients with hormone-refractory metastatic prostate carcinoma. J. Natl. Cancer Inst. 94 (19), 1458–1468.

Sato, K., Qian, J., Slezak, J.M., Liber, M.M., 1999. Clinical significance of alterations of chromosome 8 in high-grade, advanced, nonmetastatic prostate carcinoma. J. Natl. Cancer Inst. 91 (18), 1574–1580.

Scher, H., Fizazi, K., Saad, F., et al., 2012. Increased survival with enzalutamide in prostate cancer after chemotherapy. N. Engl. J. Med. 367 (13), 1187–1197.

Schiewer, M.J., Goodwin, J.F., Han, S., et al., 2012. Dual roles of PARP-1 promote cancer growth and progression. Cancer Discov. 2 (12), 1134–1149.

Schweizer, M.T., Antonarakis, E.S., 2012. Abiraterone and other novel androgen-directed strategies for the treatment of prostate cancer: a new era of hormonal therapies is born. Ther. Adv. Urol. 4 (4), 167–178.

Sella, A., Konichezky, M., Flex, D., Sulkes, A., Baniel, J., 2000. Low PSA metastatic androgen- independent prostate cancer. Eur. Urol. 38 (3), 250–254.

Sharma, A., Yeow, W., Ertel, A., et al., 2010. The retinoblastoma tumour suppressor controls androgen signalling and human prostate cancer progression. J. Clin. Invest. 120 (12), 4478–4492.

Shore, N.D., Abrahamsson, P.A., Anderson, J., Crawford, E.D., Lange, P., 2013. New considerations for ADT in advanced prostate cancer and the emerging role of GnRH antagonists. Prostate Cancer Prostatic Dis. 16 (1), 7–15.

Singha, A., Cheedellaa, N., Shakil, S., Gulmi, F., et al., 2015. Liver metastases in prostate carcinoma represent a relatively aggressive subtype refractory to hormonal therapy and short-duration response to docetaxel monotherapy. World J. Oncol. 6 (1), 265–269.

Stanbrough, M., Bubley, G.J., Ross, K., et al., 2006. Increased expression of genes converting adrenal androgens to testosterone in androgen-independent prostate cancer. Cancer Res. 66 (5), 2815–2825.

Steeg, P., 2006. Tumor metastasis: mechanistic insights and clinical challenges. Nat. Med. 12 (8), 895–904.

Sun, S., Sprenger, C.C., Vessella, R.L., et al., 2010. Castration resistance in human prostate cancer is conferred by a frequently occurring androgen receptor splice variant. J. Clin. Invest. 120 (8), 2715–2730.

Sung, Y., Cheung, E., 2014. Androgen receptor co-regulatory networks in castration resistant prostate cancer. Endocr. Relat. Cancer 21 (1), R1–R11.

Swanson, G.P., Basler, J.W., 2010. Prognostic factors for failure after prostatectomy. J. Cancer 2, 1–19.

Sweeney, C.J., Chen, Y.H., Carducci, M., et al., 2015. Chemohormonal therapy in metastatic hormone-sensitive prostate cancer. N. Engl. J. Med. 373 (8), 737–746.

Takeuchi, S., Arai, K., Saitoh, H., Yoshida, K., Miura, M., 1996. Urinary pyridinoline and deoxypyrridinoline as potential markers of bone metastasis in patient with prostate cancer. J. Urol. 156 (5), 1691–1695.

Tan, H., Sood, A., Rahimi, H.A., et al., 2014. Rb loss is characteristic of prostatic small cell neuroendocrine carcinoma. Clin. Cancer Res. 20 (4), 890–903.

Tan, P.S., Haaland, B., Montero, A.J., Kyriakopoulos, C.E., Lopes, G., 2014. Hormonal therapeutics enzalutamide and abiraterone acetate in the treatment of metastatic castration-resistant prostate cancer (mCRPC) post-docetaxel—an indirect comparison. Clin. Med. Insights Oncol. 8, 29–36.

Thakur, R., Mishra, D.P., 2013. Pharmacological modulation of beta-catenin and its applications in cancer therapy. J. Cell Mol. Med. 17 (4), 449–456.

Titus, M.A., Schell, M.J., Lih, F.B., et al., 2005. Testosterone and dihydrotestosterone tissue levels in recurrent prostate cancer. Clin. Cancer Res. 11 (13), 4653–4657.

Tohfe, M., Baki, S., Saliba, W., et al., 2008. Metastatic prostate adenocarcinoma presenting with pulmonary symptoms: a case report and review of the literature. Cases J. 1 (1), 316.

Touijer, A., Mazzola, C.R., Sjoberg, D.D., Scardino, P.T., Eastham, J.A., 2014. Long-term outcomes of patient with lymph node metastasis treated with radical prostatectomy without adjuvant androgen-deprivation therapy. Eur. Urol. 65 (1), 20–25.

Uysal-Onganer, P., Kawano, Y., Caro, M., et al., 2010. Wnt-11 promotes neuroendocrine-like differentiation, survival and migration of prostate cancer cells. Mol. Cancer 9, 55.

Vindrieux, D., Escobar, P., Lazennec, G., 2009. Emerging roles of chemokines in prostate cancer. Endocr. Relat. Cancer 16 (3), 663–673.

Voeller, H.J., Sugars, L.Y., Pretlow, T., Gelmann, E.P., 1994. p53 oncogene mutations in human prostate cancer specimens. J. Urol. 151 (2), 492–495.

Voeller, H.J., Truica, C.I., Gelmann, E.P., 1998. Beta-catenin mutations in human prostate cancer. Cancer Res. 58 (12), 2520–2523.

Wang, S., Gao, J., Lei, Q., Rozengurt, N., et al., 2003. Prostate-specific deletion of the murine Pten tumour suppressor gene leads to metastatic prostate cancer. Cancer Cell 4 (3), 209–221.

Wang, Y., Li, J.Q., Shao, C., et al., 2011. Androgen receptor coregulators NOCR1, TIF2, and ARA70 may account for the hydroxyflutamide insensitivity of prostate cancer cells. Ir. J. Med. Sci. 180 (4), 865–872.

Wang, H., Li, B., Zhang, P., Yao, Y., Chang, J., 2014. Clinical characteristics and prognostic factors of prostate cancer with liver metastases. Tumor Biol. 35 (1), 595–601.

Ware, K.E., Garcia-Blanco, M.A., Armstrong, A.J., Dehm, S.M., 2014. Biologic and clinical significance of androgen receptor variants in castration resistant prostate cancer. Endocr. Relat. Cancer 21 (4), T87–T103.

Watson, P.A., Chen, Y.F., Balbas, M.D., et al., 2010. Constitutively active androgen receptor splice variants expressed in castration-resistant prostate cancer require full-length androgen receptor. Proc. Natl. Acad. Sci. U.S.A. 107 (39), 16759–16765.

Weilbaecher, K.N., Guise, T.A., McCauley, L.K., 2011. Cancer to bone: a fatal attraction. Nat. Rev. Cancer 11 (6), 411–425.

Weiner, A.B., Cortes-Mateus, S., De Luis, E., Duran, I., 2013. Dural metastases in advanced prostate cancer: a case report and review of the literature. Curr. Urol. 7 (3), 166–168.

Wong, S., Haack, H., Crowley, D., Barry, M., et al., 2005. Tumour-secreted vascular endothelial growth factor-C is necessary for prostate cancer lymphangiogenesis but lymphangiogenesis is unnecessary for lymph node metastasis. Cancer Res. 65 (21), 9789–9798.

Yoshida, T., Kinoshita, H., Segawa, T., Nakamura, E., Inoue, T., Shimizu, Y., et al., 2005. Antiandrogen bicalutamide promotes tumour growth in a novel androgen-dependent prostate cancer xenograft model derived from a bicalutamide-treated patient. Cancer Res. 65 (21), 9611–9616.

Zeng, Y., Opeskin, K., Baldwin, M.E., et al., 2004. Expression of vascular endothelial growth factor receptor-3 by lymphatic endothelial cells is associated with lymph node metastasis in prostate cancer. Clin. Cancer Res. 10 (15), 5137–5144.

Zhang, Z., Rosen, D.G., Yao, J.L., Huang, J., Liu, J., 2006. Expression of p14ARF, p15INK4b, p16INK4a, and DCR2 increases during prostate cancer progression. Mod. Pathol. 19 (10), 1339–1343.

Zhang, X., Tang, N., Hadden, T., Rishi, A., 2011. Akt, FoxO and regulation of apoptosis. Biochim. Biophys. Acta 1813 (11), 1978–1986.

Zhu, H., Mazor, M., Kawano, Y., Walker, M.M., et al., 2004. Analysis of wnt gene expression in prostate cancer: mutual inhibition of Wnt11 and the androgen receptor. Cancer Res. 64 (21), 7918–7926.

CHAPTER

# 4

# Lung Cancer Metastasis

*P. Chanvorachote, P. Chunhacha*
**Chulalongkorn University, Bangkok, Thailand**

## 1. GENERAL INFORMATION

Throughout the world, lung cancer has been shown to be the leading cause of cancer-related death with the estimated number of 1.5 million deaths each year (Jemal et al., 2011; Youlden et al., 2008; Siegel et al., 2016). An evidence shows that the dominant cause of death in lung cancer involves metastasis (Mehlen and Puisieux, 2006), with the widely accepted concept that tumor metastasis is a major obstacle of success in the treatment of lung cancer. Metastatic lung cancer is a cancer that originates within lung and spreads from its original site to nearby lymph nodes and other sites in the body. The percentage of metastasis-related death in lung cancer is as high as 90% of all lung cancer mortality (Mehlen and Puisieux, 2006), highlighting the importance of metastasis in this cancer.

Generally, lung cancer is classified into two pathological types: non-small-cell lung cancer (NSCLC) and small-cell lung cancer (SCLC). The NSCLC and SCLC account for approximately 80–85% and 20% of all diagnosed lung cancers, respectively. The most frequent forms of NSCLC are adenocarcinoma (30–50% of NSCLC) and squamous cell carcinoma (30% of NSCLC). For NSCLC, the early lymphatic metastasis can be detectable in stage II. Stages III and IV of NSCLC involve the spread of the tumor cells to lymph nodes in the middle of the chest and distant sites, respectively. For SCLC, it was found to have a higher rate in both cancer growth and metastasis in comparison to that of NSCLC (Heidemann et al., 2014). As it has a higher rate of tumor spreading, SCLC is classified to the limited stage with the cancer found in one side of the lung together with the metastasis to nearby lymph nodes and the extensive stage with the tumor cells spread to distant regions (Heidemann et al., 2014). In addition, the late detection of lung cancer can be critical in determining the rate of death. As most of the lung cancer is found at late stages, this makes the cancer impossible to be completely treated by surgery or radiation. In fact, approximately 70% of diagnosed lung cancer cases is associated with local lymph node metastasis or distant spreading of the cancer (Hirsch and Hansen, 2001). The prognoses of disease and the 5-year survival strictly depend on the stage. In NSCLC, the 5-year survival of NSCLC with stage I (no metastasis) is 70% in comparison to only 2% of that of stage IV (metastatic lung cancer) (Goldstraw et al., 2011). SCLC is more aggressive and is associated with at least lymphatic metastasis in almost all cases; the 5-year survival of this disease is not more than 10% (Heidemann et al., 2014).

*Introduction to Cancer Metastasis*
http://dx.doi.org/10.1016/B978-0-12-804003-4.00004-9

## 2. MECHANISM OF METASTASIS

The appearance of the metastatic cancer cells is generally similar to that of primary tumor cells; however, metastatic lung cancer cells have several different molecular features. Although the metastatic lung cancer cells preserve the profile of certain proteins and chromosome alterations, the metastatic cells have some specific features that enhance their ability to metastasize. Lung is an essential organ, which is highly vascularized and rich in the supply of lymphatic vessels providing comfortable routes for cancer cells to spread to nearby lymph nodes as well as to other parts of the body.

Lung cancer originates from the normal cells that are exposed genotoxic substances and/or chronic inflammatory substances that confer abnormal genetic expression contributing to neoplastic transformation, nonstop cell proliferation and differentiation, and eventually tumor formation. Cancer cells then generate new blood vessels to supply sufficient nutrient and oxygen to themselves, which is called neovascularization. However, the overgrowth of such cancer cells does not have enough blood supply, causing cancer cells to move from the originate site. Metastasis of cancer belongs to multiple processes. In general, lung cancer cell metastasis involves the following steps, as found in most solid tumors (Fig. 4.1).

- **Detachment of lung cancer cells from the extracellular matrix (ECM), local invasion, and migration**: Cancer cells dissociate from their adjacent cells and basement membrane, migrate, as well as invade nearby tissues. The cells produce proteolytic enzymes to degrade the ECM that provides the route of invasion.
- **Intravasation into blood or lymphatic systems**: Cancer cells attach to the endothelial wall, invade, and move through the walls of blood or lymph vessels.
- **Survive in circulation**: Cancer cells circulate or move in the blood or lymphatic systems to other sites.

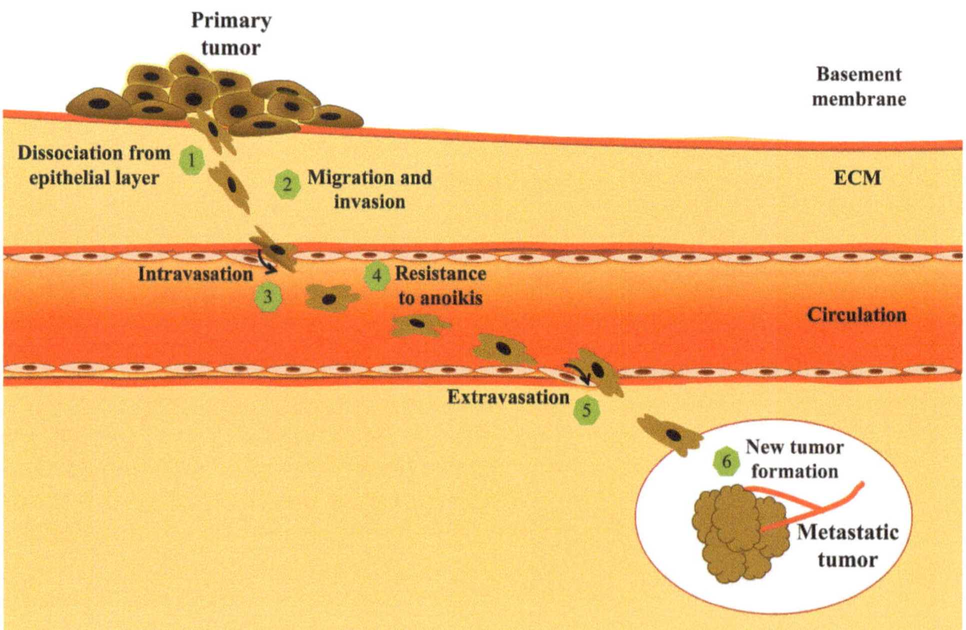


FIGURE 4.1 **Schematic overview of lung cancer metastasis.** Metastasis of lung cancer cells is initiated by the dissociation of the cells from their original tumor. The cells then migrate and invade to reach circulations where they can transfer to distant sites of the body, extravasate, and establish new tumor. *ECM*, extracellular matrix.

- **Arrest and extravasation at metastatic site**: Cancer cells adhere to the capillary walls at the site of metastasis. They penetrate or invade the walls of the capillaries and migrate to the new location.
- **Formation of metastatic tumor**: Cancer cells proliferate to form small tumors.

During metastasis, the cancer cells must overcome several hindrances. The success of metastasis depends on several factors for which either the cancer or the noncancerous cells create the microenvironment. Such noncancerous cells include immune cells, present at the primary tumor, in lymph and blood, and at distant metastatic sites. Not all cancerous cells possess the ability to escape from all types of immune attacks. In addition, abilities of the cells to survive in anchorage-independent condition and tumorigenic potential have been extensively shown to be critical for metastasis in lung cancer.

As the metastasis is a complex multistep process, metastatic cells require special properties to overcome hindrances and, most importantly, survive in detached condition, invade, and generate the new tumors. Sufficient recent studies have suggested the important impact of cellular phenotypes of epithelial to mesenchymal transition (EMT) and cancer stem cell (CSC) as key features rendering cancer cell dissemination. Although the information as well as mechanistic regulations involving in EMT and CSC of lung cancer is still largely unknown, they have garnered increasing attentions in lung cancer research.

## 3. MOLECULAR MECHANISM IN THE REGULATION OF LUNG CANCER METASTASIS

During the early phase of metastasis, metastatic cells will detach from their membrane as well as adjacent cells. Such a process makes the cells lack of integrin-mediated survival signals (Boudreau and Jones, 1999) and lead to apoptosis mechanism termed "anoikis." Anoikis is activated in most of the adherent cells during the insufficiency or loss of cell to cell and/or cell to ECM adhesions (Frisch and Francis, 1994). Since anoikis is a process to ensure the appropriate adherent site of the cells, it is an important primary mechanism in eliminating the detached cancer cells when they travel in blood and lymphatic systems. The deregulation or disruption of executive anoikis, therefore, allows the survival of cells in circulation, readhesion to new matrices, and growth as secondary tumors.

Anoikis of cancer cells is achieved through a complex cascade of events; however, in general, the cellular balance between survival and apoptotic signals is disrupted and the proapoptotic molecules overcome and precede their machinery resulting in apoptosis. According to genotypic information, the common driver for oncogenesis in lung cancer involves the mutations that activate epidermal growth factor receptor (EGFR), Kirsten rat sarcoma (KRAS), MET proto-oncogene, and proto-oncogene B-Raf (Heidemann et al., 2014). Thus, these findings lead to the development of targeted therapies. When compared with the standard chemotherapeutic treatments, such new targeted therapies show sufficient improvement in progression-free survival with less toxicity (Johnson et al., 2014); however, the overall survivals are still not much improved (Fukuoka et al., 2011; Shaw et al., 2013). The EGFR is highly expressed in many malignancies including NSCLC (Grandis and Sok, 2004; Merrick et al., 2006). In tumors, the EGFR can activate several cellular signals including mitogen-activated protein kinase (MAPK), signal transducer and activator of transcription (STAT), phosphatidylinositol 3′-kinase (PI3K)/protein kinase B (AKT), proto-oncogene factors, such as fos, jun, and myc, resulting in the augmentation of tumor growth, survival, and progression (Grandis and Sok, 2004). In terms of anoikis, its activity to sustain the survival pathway of proto-oncogene tyrosine-protein kinase Src–extracellular signal-regulated kinase (Src–ERK) cascade through its intracellular domain tyrosine kinase has been shown to protect lung adenocarcinoma cells from detachment-induced apoptosis (Wei et al., 2004). Src family kinases (SFKs) are the other group of nonreceptor tyrosine kinases involved in the regulation of various signaling pathways that promotes cell survival and motility. Overexpression or hyperactivity of SFKs is frequently found in many types of human epithelial tumors, including NSCLCs (Masaki et al., 2003; Mazurenko et al., 1991, 1992). One member of SFKs that has been found in a tight regulation with EGFR is c-Src. Enhanced c-Src activation is mostly found in tumors harboring activated EGFR, and inhibition of c-Src can reverse the malignant properties of cells overexpressing EGFR (Karni et al., 1999). In lung cancer, SFKs are activated in NSCLC biopsy samples and promote the survival of NSCLC cells in EGFR-dependent fashion. Moreover, src inhibition was shown to abrogate lung cancer cell migration and invasion, suggesting that src is involved in malignant phenotype of lung cancer cells (Leung et al., 2009). Src-mediated pathways mainly contribute to tumor survival and metastatic potential by employing STAT-3 and focal adhesion kinase (FAK). Src-kinase activity enhances STAT-3 DNA binding activity and promotes lung cancer survival (Song et al., 2003).

The WZ4002, a third-generation EGFR inhibitor, was shown to sensitize EGFR-mutant lung adenocarcinoma cells to anoikis, supporting the significant effect of EGFR on anoikis regulation in lung cancer (Sakuma et al., 2012). Interestingly, such a study provides information indicating that WZ4002 overcomes anoikis resistance in EGFR-mutant lung adenocarcinomas more efficiently than Src inhibitors, suggesting the existence of other strong survival signals downstream of EGFR. Accordingly, EGFR was also shown to function via heterodimerization with human epidermal growth factor receptor 2 (HER2) and this heterodimerization is more potent than EGFR homodimerization (Graus-Porta et al., 1997; Olayioye et al., 2000; Tzahar et al., 1996). The overexpression of HER2 is also observed in NSCLC and other tumors; but the expression of HER2 is less than that of EGFR (Hirsch et al., 2002). The tumor cells that express both EGFR and HER2 will have highly aggressive growth pattern; therefore, the expression of both receptors may have a prognosis value in prediction of aggressiveness in NSCLC (Hirsch et al., 2009). Novel inhibitor targeting both EGFR and HER2 has been evaluated in the phase I clinical trial for lung cancer treatment with the hope to improve the outcome of the lung cancer management (Zhang et al., 2014). AST1306 is an irreversible potent inhibitor of EGFR (erbB1), HER2 (erbB2), and HER4 (erbB4). Although its main function is the anticancer activity in advanced solid tumor, an evidence of HER2-targeted monoclonal antibody, trastuzumab, for the treatment of metastatic breast cancer has supported the potential of this drug for the treatment of metastatic lung cancer.

Cellular movement in normal physiological such as tissue morphogenesis or pathological such as tumor metastasis heavily relies on the function of integrins. It has been shown that tobacco alkaloid, nicotine, stimulates lung cancer cell growth by increasing fibronectin (FN) mRNA and protein synthesis (Zheng et al., 2007). The major FN receptor is the integrin $\alpha5\beta1$ and the inhibition of this integrin abrogates the mitogenic effect of nicotine in lung cancer cells. The evidence of integrin $\alpha5\beta1$ in the promotion of lung cancer metastasis is the study that provides the correlation of increased integrin $\alpha5\beta1$ and lymph node metastasis of NSCLC (Adachi et al., 2000; Han et al., 2003). Moreover, the 5-year survival rate of node-negative NSCLC patients bearing the high expression of integrin $\alpha5\beta1$ is significantly less than that of the patients having normal integrin $\alpha5\beta1$ expression (Adachi et al., 2000; Han et al., 2003). This highlights the role of integrin $\alpha5\beta1$ in the metastatic potential of NSCLC patients.

The role of integrin $\alpha5\beta1$ in the migratory behavior of NSCLC has been further characterized in the study of Tuomi et al. (2009). It has been shown that integrin α5β1 preferentially interacts with the tight junction protein zonula occludens-1 (ZO-1) at the lamellae of migrating cells and this requires protein kinase C epsilon (PKCε). Moreover, the silencing of α5 integrin reduces the migration and invasion of lung cancer cells. Overall, a spatially restricted α5 integrin–ZO-1 interaction regulated by PKCε triggers persistent migration in lung cancer cells. In some instance, the cooperative signaling between integrins and growth factor receptors is also important for the regulation of lung cancer cell migration and invasion. It was found that integrin α5β1 cooperates with the EGFR in enhancing lung cancer cell migration (Caswell et al., 2008). In the presence of αvβ3 integrin, integrin α5β1 has a modest interaction with Rab coupling protein (RCP) due to αvβ3 integrin–RCP interaction. However, in the presence of osteopontin (OPN), which is the soluble ligand of αvβ3 integrin, the interaction of αvβ3 integrin–OPN increases the levels of free RCP to interact with integrin α5β1. These situations promote the association of integrin α5β1 with RCP and EGFR, enhance EGFR and integrin α5β1 recycling, and finally increase EGFR autophosphorylation and activation of the proinvasive PKB/Akt signaling. Moreover, OPN activates Akt pathway through the direct interaction with αvβ3 integrin and increases the phosphorylation of FAK, p85 subunit of PI3K, serine 437 of AKT, and ERK (Fong et al., 2009). Nevertheless, whether this signaling pathway is dependently regulated by the interaction of integrin α5β1 and EGFR still requires further characterization. Recently, there is evidence suggesting that β1 integrin is required for A549 cell migration via the EGFR signaling (Morello et al., 2011). Altogether, cooperation between integrins and EGFR affects many aspects of lung cancer progression, especially in cell migration and invasion.

The αvβ6 integrin is also an efficient FN receptor apart from integrin α5β1. The positive correlation between the expression level of αvβ6 integrin and poor prognostic factor for the survival of NSCLC patients emphasizes the role of αvβ6 integrin in promoting adhesion, migration, and invasion in the FN-rich ECM around the tumor mass (Elayadi et al., 2007). However, αvβ6 integrin has a distinct mechanism involving its ability to promote the release of ECM-associated transforming growth factor β (TGF-β) by interacting with small latent complex in the ECM to induce conformational modification and to cultivate the release of the sequestered TGF-β (Sheppard, 2004; Wipff et al., 2007). The release of TGF-β, a well-characterized inducer of EMT, results in enhanced cell migration and invasion.

Other molecules have been shown to implicate in the metastasis of lung cancer cells including αB-crystallin (Malin et al., 2015), FAK, and Cav-1. αB-crystallin was shown to be induced in the lung cancer cells after cell detachment and to cause anoikis resistance (Malin et al., 2015). The FAK activity is implicated in the tumor growth, motility, and progression. Regarding anoikis, FAK was shown in lung cancer cell line A549 to render survival in detached condition via PI3K/AKT and MAPK-ERK pathways (Liu et al., 2008). Besides, the cooperation of Src and FAK pathways was shown to promote anchorage-dependent and anchorage-independent cell growth in many types of tumors including NSCLC (Laird et al., 2003). In terms of cell motility regulation, FAK mediates migration in lung cancer cells through FAK/PI3K/AKT pathway. Meng et al. demonstrated that Src, ERK1/2, PI3K, and Akt are activated in response to FAK stimulation by FN engagement with integrins (Meng et al., 2009). The inhibition of PI3K with its inhibitors, wortmannin or LY294002, abrogates FAK-stimulated cell migration (Reiske et al., 1999). Thus, this situation suggests that PI3K is required for FAK-mediated cell migration. In addition, Src homology 2 domain of Grb7 is required for the interaction with PI3K and promotes migration via the FAK-dependent mechanism. Overall, the cooperation of FAK/PI3K/Grb7 is required for the regulation of cancer cell migration. Not only is FAK a principal regulator of cell migration, but it also cooperates with integrin to attenuate the immune system that may attack cancer cells. In lung cancer, the phosphorylation of tyrosine residues of FAK is essential in β1 integrin-induced reduction of intercellular adhesion molecule-1, resulting in the reduction of adhesion between lung cancer cells to T cells, allowing them to escape from the immunological surveillance (Yasuda et al., 2001).

Likewise, Cav-1, a 21- to 24-kDa protein component of the plasma membrane microdomains termed caveolae, has been shown to regulate motility, growth in anchorage-independent manner, and anoikis resistance in lung cancer cells (Chanvorachote et al., 2015). The Cav-1 was shown to bind with antiapoptotic Mcl-1 protein and prevent it from degradation via the ubiquitin–proteasome pathway (Chunhacha et al., 2012). The sustained level of Mcl-1 thus prevents the cancer cells from anoikis. Besides, another anoikis-inhibiting role of Cav-1 was shown in breast cancer cells due to the upregulation of insulin-like growth factor-I receptors and signaling (Ravid et al., 2005). The small interfering RNA-mediated downregulation of Cav-1 was shown to cause stable arrest of proliferation in metastatic lung cancer cells (Pancotti et al., 2012). From the clinical observation aspects, the expression of Cav-1 in lung adenocarcinoma was shown to be largely absent in the primary tumors; however, the expression was significantly augmented in the disease with nodal metastasis and in the lymph nodes metastases (Ho et al., 2002; Sloan et al., 2009). High level of Cav-1 was shown to be associated with poor prognosis in lung cancer (Moon et al., 2005; Yoo et al., 2003). In addition, advanced NSCLC patients receive gemcitabine-based chemotherapy, Cav-1 is linked with drug resistance and poor prognosis (Ho et al., 2008). In terms of Cav-1, many studies have highlighted the role of Cav-1 in NSCLC, which has been extensively reviewed elsewhere (Chunhacha and Chanvorachote, 2012).

ECM is the main scaffold that provides a structural support to the tissue. During embryonic development, this scaffolding function of ECM is particularly important for guiding the cells for migration. Not only does ECM provide physical interaction that supports cell structure and their survival, but ECM also provides a substrate for cell anchorage through the cell adhesion receptors especially integrins. This interaction between ECM and integrins is vital for cellular growth, differentiation, and apoptosis. Matrix metalloproteinases (MMPs) are families of 21 multidomain, multifunctional proteins mainly degrading ECM components to help facilitate cancer cell migration. Regulation of MMP function occurs at multiple levels. MMP mRNA expression is under cell type–dependent control and the individual MMPs expression are associated with specific inflammatory, connective tissue, or epithelial cell types. At basal level, MMP transcripts are expressed at low levels but dramatically increase with some circumstances such as inflammation or cancer. MMPs are mostly stored in inflammatory cell granules as well as secreted and found anchored on the cell membrane, and oftentimes they present as a tethering form with the other proteins on the cell surface or within the ECM. In lung cancer, increase in MMP-9 expression has been shown to present in metastatic lung cancer both in the tumor and in the serum (Hoikkala et al., 2006; Ylisirnio et al., 2000). The 92-kDa MMP-9 is the member of type IV collagenase that plays a critical role during lung cancer invasion and metastasis (Hrabec et al., 2001). The regulation of MMP expression is the complex process; however, there are significant proportions of NSCLC tumor that coexpress MMP-9 and EGFR. The presence of these two markers is associated with poor prognosis (Cox et al., 2000). There is evidence that MMP-9 is regulated by EGFR at both transcriptional and activity (Cowden Dahl et al., 2008), suggesting that the cooperation of MMP-9 and EGFR is implicated in tumorigenicity. Since MMP-9 seems to be involved in the metastatic potential of lung cancer, the researchers are interested in

MMP-9 as a target for cancer gene therapy. One study that used a replication-deficient adenovirus capability of expressing antisense urokinase-type plasminogen activator receptor (uPAR) and antisense MMP-9 transcripts to downregulate uPAR and MMP-9 in vitro seems to reduce invasive and metastatic potential of lung cancer cells significantly (Rao et al., 2005). Altogether, MMP-9 might be one of the good molecular targets for preventing lung cancer metastasis.

The influence of cancer microenvironment factors has been shown to affect the aggressive as well as metastatic behaviors of lung cancer. Chemokines are small secreted protein that exhibits the pleiotropic role in promoting tumor metastasis as well as leukocytes maturation, intravasation, and eventually homing at the specific tissue sites (Strieter, 2001). Chemokines bind to their seven-transmembrane G-protein-coupled receptors and induce the migration toward the chemokine gradient in the process called chemotaxis. The structure of chemokines can be divided into the two main chemokine subfamilies based on the arrangement of N-terminal cysteine residues in which there is an amino acid between them (CXC), or just adjacent (CC). Among several members of the chemokine receptor family, CXCR4 has received much attention because this receptor facilitates the entry of T-tropic (X4) HIV-1 strain into CD4+ T cells (Bleul et al., 1996). After this discovery, the growth in CXCR4 research has brought to light several subsequent findings including the work from Muller et al. (2001). Muller and coworkers demonstrated that CXCR4 and CCR7 are highly expressed in human breast cancer cells and also play a crucial role in breast tumors and metastases. They suggested that the signaling from the interactions of CXCR4 and its ligand stromal cell-derived factor-1 (SDF-1) triggers actin polymerization and pseudopodia formation, which both contribute to invasive characteristics. However, neutralizing these interactions with antibodies successfully inhibits metastasis capabilities of breast cancer cells to lymph nodes and lung.

In lung cancer, CXCR4 is also expressed in tumor specimens resected from patients as well as in NSCLC cell lines (Phillips et al., 2003). This study also demonstrated that SDF-1 is highly expressed in the preferred destined metastasis sites rather than it does occur at the primary site. The role of CXCR4–SDF-1 in promoting metastatic potential in NSCLC patients has been further characterized by Su et al. (2005). Highly expressed CXCR4 tumors tend to have more clinical metastasis in comparison to low expression tumors. CXCR4 specific antibodies abrogate the CXCR4 role in promoting migration, invasion, and adhesion in vitro. The expression of SDF-1 is also well correlated with NSCLC progression; high level of SDF-1 expression was found in the majority of NSCLC tissue section taken from stage IA–IIB patients with more advance disease stage (Wald et al., 2006). Recently, CXCR4 has been implicated in the maintenance of lung cancer stemness. The researchers found that the CXCR4 level was increased in gefitinib-resistant A549 cells; this cell population has a high potential of self-renewal activity as well as shows an increase in tumorigenicity (Jung et al., 2013). Overall, SDF-1–CXCR4 axis appears to be a canonical pathway in NSCLC pathogenesis, acting at both primary and metastatic sites as a chemoattractant and growth-promoting factor for lung cancer (Fig. 4.2).

## 4. EPITHELIAL TO MESENCHYMAL TRANSITION IN LUNG CANCER

Evidence has shown that EMT is a process in which epithelial cells undergo phenotypic change to mesenchymal cells. Accumulating studies demonstrated that EMT is involved in the metastasis and treatment resistance, and associated with the progression of many type of tumors including lung. Cancer cells require having the phenotypic switching from epithelial to mesenchymal to migrate and invade, and later survive in the circulation. Such cells that contain

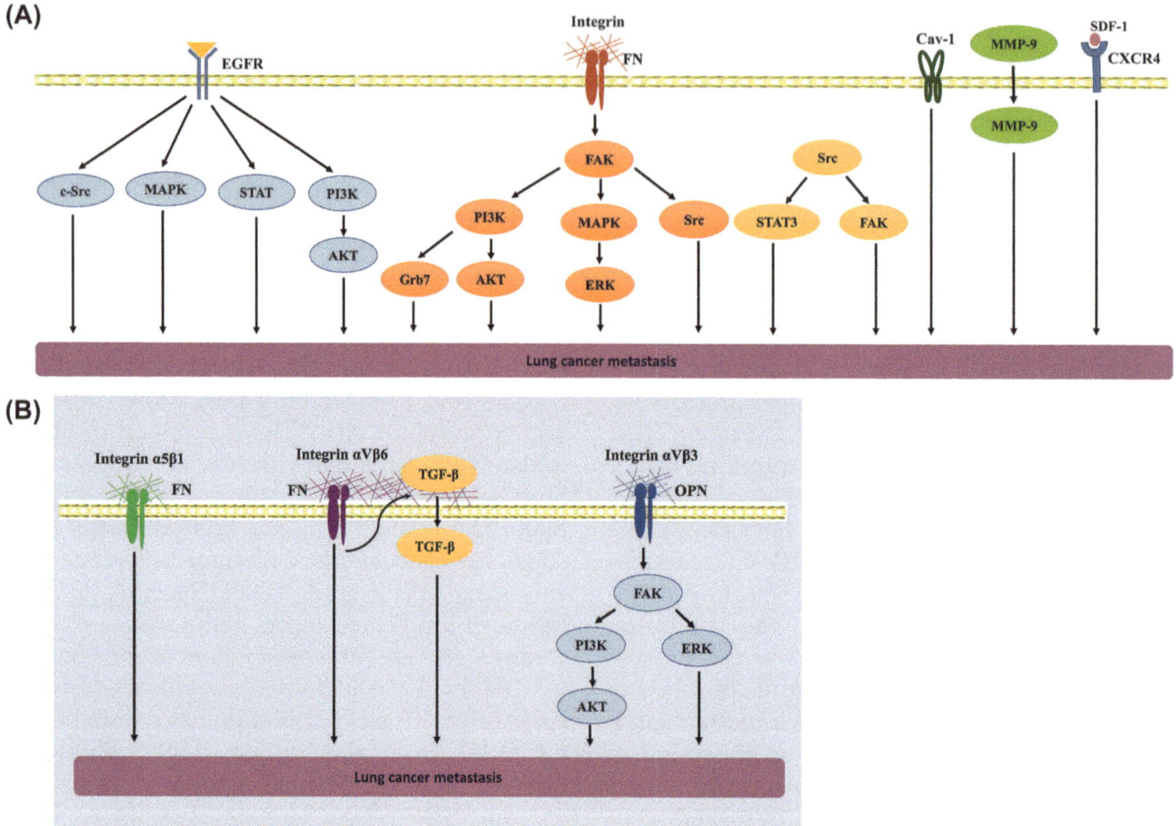


FIGURE 4.2 **Overview of the major pathways of metastasis.** (A) Cellular pathways that facilitate lung cancer cell dissemination. (B) Positive regulatory roles of certain integrins on lung cancer metastasis. *AKT*, protein kinase B; *EGFR*, epidermal growth factor receptor; *ERK*, extracellular signal-regulated kinase; *FAK*, focal adhesion kinase; *FN*, fibronectin; *MAPK*, mitogen-activated protein kinase; *MMP*, matrix metalloproteinases; *OPN*, osteopontin; *PI3K*, phosphatidylinositol 3′-kinase; *SDF*, stromal cell-derived factor-1; *Src*, proto-oncogene tyrosine-protein kinase Src; *STAT*, signal transducer and activator of transcription; *TGF*, transforming growth factor.

mesenchymal-like property also possess an ability to overcome anoikis.

Lung cancer cells undergo EMT and have been shown to express the mesenchymal transcription factors, such as, snail, slug, and twist. Together with the induction of vimentin and FN, alpha-smooth muscle actin, and switching of E-cadherin to N-cadherin, has been shown to render the change in cell morphology toward spindle shape and detachment of the cells. In terms of invasion, N-cadherin was shown to exhibit lesser cell to cell adhesion, and meanwhile encourages invasive and metastasis behaviors (Hazan et al., 2000). In addition, EMT increases the activity of MMPs, such as, MMP-2, MMP-3, and MMP-9 (Thiery and Sleeman, 2006).

Although EMT has been shown to positively correlate with progression and poor prognosis of several types of cancer including lung cancer, some evidence indicates otherwise. The induction of EMT genes was shown to be associated with poorly differentiated tumors relative to low-grade tumors in breast cancer (Teschendorff et al., 2007). Additionally, the E-cadherin to N-cadherin switching in prostate cancer is strongly linked to the cancer progression

(Gravdal et al., 2007). Certain studies have shown the positive correlation between EMT and poor prognosis in lung cancer. In NSCLC, the low level of vimentin was shown to be an independent predictor of better survival in primary NSCLC (Al-Saad et al., 2008). In terms of brain metastasis, the increase of long noncoding RNA MALAT1 was shown to promote brain metastasis via EMT induction (Shen et al., 2015). In addition, a novel EMT inducer PREP1 was demonstrated in NSCLC cells to trigger EMT and such an EMT inducer was shown to accumulate in brain metastases of NSCLC (Risolino et al., 2014). However, other studies showed that neither reduction of E-cadherin nor increase of N-cadherin is associated with poor prognosis in patients with NSCLC (Prudkin et al., 2009).

Focusing on molecular basis, the intrinsic characteristics of cancer cells like (KRAS) mutation or HER2 overexpression was shown to control the process of EMT (Edme et al., 2002; Jenndahl et al., 2005). Besides, EMT can be triggered by extracellular signal like cytokines including interleukin-6 and TGF-β1, nitric oxide, cancer-associated fibroblasts, and immune cells (Yadav et al., 2011; Ko, 2015; Yongsanguanchai et al., 2015). The induction of EMT involves the activation of several developmental signaling pathways such as Wnt (Mazieres et al., 2005), Notch (Yuan et al., 2014; Liu et al., 2014), JAK/STAT3 (Liu et al., 2014), and Hedgehog (Yue et al., 2014; Maitah et al., 2011). These signals thus increase the intracellular transcription factors of Snail1 and Snail2, Slug, Twist, ZEB1 and 2, and TCF3/E47/E12 (Eccles and Welch, 2007; Zeisberg et al., 2007; Moustakas and Heldin, 2007; Ishikawa et al., 2013; Casas et al., 2011). In particular, the Wnt3 was shown to upregulate Notch3 and potentiated metastatic potential in lung cancer (Li et al., 2015). Speaking of Notch, it has been shown to play a role along with Myc in the carcinogenesis of lung adenocarcinoma in transgenic mice model (Allen et al., 2011). Also, Notch has been shown to regulate EMT in many ways including downregulation of epithelial markers and upregulation of mesenchymal markers (Timmerman et al., 2004; Chang et al., 2014), and through Jagged-1 dependent mechanism (Nyhan et al., 2010). Notch ligand and receptor in NSCLC are linked with poor outcome (Donnem et al., 2010) and the overexpression of Notch1 and Notch3 is 30% and 40% in NSCLC tumors, respectively (Westhoff et al., 2009; Dang et al., 2000).

Recent studies have revealed the important roles of microRNA (miRNA) in regulating cancer metastasis. miRNAs are small noncoding RNAs that negatively affect gene expression by blocking translation of mRNA (He and Hannon, 2004). In lung adenocarcinoma, the miR-200 family and miR-205 have been shown to inhibit EMT by suppressing ZEB1, ZEB2, and SIP1 (Park et al., 2008; Gregory et al., 2008; Ahmad et al., 2013). Also, the MiR-126 was shown to suppress NSCLC by inhibiting tumor growth (Yang et al., 2012), adhesion, migration, and invasion (Crawford et al., 2008), and its low expression is associated with unfavorable prognosis in NSCLC patients (Yang et al., 2012).

Many of EMT-related pathways have been shown to control stemness in lung cancer (Huber et al., 2005; Sung et al., 2015). In NSCLC cells, the EMT phenotypes in NSCLC cells are associated with the increase in stem cell-like phenotypes after receiving nitric oxide (Yongsanguanchai et al., 2015). Likewise, the NSCLC cells that resist ionizing radiation treatment exhibit CSC and EMT phenotypes (Gomez-Casal et al., 2013).

## 5. CANCER STEM CELLS

CSCs are rare cancer subpopulations within tumor consequence of carcinogenesis. This group of cancer cells contains normal stem cell-like property, which was first described in 1997 by Bonnet and Dick (1997). Both CSCs and normal stem cells have self-renewal property and can produce differentiated progeny called pluripotency. However, unlike the normal stem cell, CSCs have an uncontrolled capacity to regenerate the phenotypic diversity of the original tumor and can indefinitely cause proliferation of tumor cells in the end. The small populations resist apoptosis and eventually possess metastasis ability to move to other tissues, and these are found in many types of cancer including lung. As a result of the stemness properties, CSCs are commonly known as tumor-initiating cells, or tumor-propagating cells. CSC theory has become the most interesting topic in both the research and clinical assessments, in particular in lung cancer.

According to their tumorigenic potential, CSCs are linked with the metastasis as well as cancer relapse. In lung cancer, the presence and survival of CSCs are believed to be associated with disease relapse. In lung cancer, potential CSC markers include the cell surface expression of CD133, CD166, and CD44, activity and expression of aldehyde dehydrogenase and ability to exclude the membrane permeable dyes correlated with CSC-like phenotypes in vitro and in vivo. In NSCLC, the side population, the cells with the ability to pump out membrane permeable dye were identified and shown to form anchorage-independent spheroids, possess great proliferative potential, exhibit enhanced tumorigenicity, and have drug-resistant characteristic (Ho et al., 2007; Shi et al., 2012). The putative lung CSCs isolated from lung adenocarcinoma cells (A549 and H2170) based on CD166, CD44, and EpCAM expressions exhibited multipotent properties of stem cells and self-renewal. Such cells also expressed high level of stem cell-related transcription factors Sox2 and Oct3/4 and showed the potential to generate new tumors in nude mice (Zakaria et al., 2015). In addition, analysis of several patient-derived lung cancer cells showed that the subpopulation with high level of ALHD and CD44 exhibits self-renewal, high invasiveness, tumorigenicity, and chemotherapeutic resistance (Liu et al., 2013). Regarding metastasis, interesting studies using CD15 and CD133 as markers have showed the presence of stem-like population in lung cancer metastasis to brain (Nolte et al., 2013).

The recent topic of interest is the origin of CSC. One speculation receiving most attention is origin of CSC from the malignant transformation of normal stem cells (Dean et al., 2005). However, its self-renewal, differentiation, as well as aggressive behaviors were shown to be inducible by several factors. The CSC population in cancers was shown to be induced by a number of environment factors, and the shift of self-renewal and differentiation homeostasis may result in the alteration of CSC population and metastasis potential. Like the normal stem cells, CSCs receive the signals from extracellular sources and response in terms of self-renewal to increase their stem cell population, or differentiate to functioning cells. Until now, a number of factors including TGF-β, fibroblast growth factor, platelet-derived growth factor, insulin-like growth factor, and epidermal growth factor have been shown to regulate stemness in cancer (Yang and Weinberg, 2008; Castano et al., 2012). In lung cancer cells, hypoxic condition (Murakami et al., 2014) and nitric oxide (Yongsanguanchai et al., 2015) were shown to enhance stem cell-like phenotypes. Interestingly, the study by Yongsanguanchai et al. suggested the reversible effect of nitric oxide-mediated stemness. By withdrawing nitric oxide after long-term treatment, the CSC-like phenotypes were found to reduce to the normal level comparable with the nontreated cells (Yongsanguanchai et al., 2015). This evidence highlights the complexity of the CSC biology that the CSC phenotypes can be regulated by many factors, and the cell response to such stimuli may be adjustable.

In terms of molecular views, as mentioned earlier, several pathways found in EMT cancer cells such as Wnt (Zhang et al., 2015a; Jiang et al., 2015), Notch (Liu et al., 2014; Hassan et al., 2013), and Hedgehog (Zhang et al., 2012) have been implicated in lung cancer stemness. The regulation of lung CSCs through Wnt signaling has been intensively demonstrated in lung cancer cells models (Zhang et al., 2015a,b; Teng et al., 2010). The impact of Wnt signaling in the regulation of cell proliferation has been addressed in many cell line models. In particular, Wnt was shown to regulate lung CSCs by controlling the expression of stem cell factors like Oct4 and nanog (Zhang et al., 2015a). Furthermore, the blocking of Wnt by pyrviniumpamoate results in a significant decrease in proliferation of lung CSCs, suggesting the role of Wnt in self-renewal control in lung CSCs (Zhang et al., 2015c). Regarding Notch pathway, limited information has been provided in lung cancer model; however, its activity has been correlated with growth of CSCs (Liu et al., 2014), stem cell-like properties, and poor survival in lung cancer (Yongsanguanchai et al., 2015). In SCLC, the Hedgehog pathway has been highlighted to play a crucial role in enhancing tumor growth, tumorigenic potential, and progression in vivo (Park et al., 2011). Besides, Akt has been shown to implicate lung CSC signaling (Hou et al., 2015; Singh et al., 2012). Study in multiple NSCLC cell lines (H1650, H1975, A549), and primary human tumor explants grown in nude mice indicated that Akt signal is critical for Sox-2 regulating stemness in lung CSCs (Yue et al., 2014). Also, miRNA-31 was shown to inhibit lung adenocarcinoma stem-like cells through suppression of Akt pathway (Liu et al., 2014). In NSCLC, the plant-derived compound gigantol was shown to reduce stemness through Akt suppression (Bhummaphan and Chanvorachote, 2015). Although several questions remain unanswered, including the key pathways in controlling EMT and CSC and the effect of environmental factors, the defining regulatory pathways underlying lung cancer metastasis are the critical basis for the development of novel strategies in overcoming lung cancer.

# 6. CONCLUDING REMARKS

As a well-recognized cause of lung cancer-related high mortality, metastasis has garnered increasing attention in both research and clinical fields. The molecular features as well as pathways that cancer cells use during metastasis are highly complex and inseparable with the regulatory effects of their environmental factors. Even though cancer cell dissemination is not a normal physiological process, the exciting biological pathway is utilized by the cells to reach their goal. Most of the knowledge in cancer cell biology may be useful for understanding the process of lung cancer metastasis; however, the unique characters, predominant pathways, and factors depending on cell types, phenotypes, and tumor environment are necessary to understand before a novel efficient therapy could be designed. So far, research has highlighted the importance of anoikis resistance, migration and invasion, EMT, and cancer stemness, together with the emphasis on the molecular mechanisms. Understanding all the key signals and potentiating factors in this cancer could be beneficial for the development of strategies to block such specific pathways and improve the clinical outcome of lung cancer patients.

## Acknowledgments

The authors wish to thank Prof. Krich Rajprasit and Ms. Chuanpit Ninsontia.

## References

Adachi, M., Taki, T., Higashiyama, M., Kohno, N., Inufusa, H., Miyake, M., 2000. Significance of integrin alpha5 gene expression as a prognostic factor in node-negative non-small cell lung cancer. Clin. Cancer Res. 6 (1), 96–101.

Ahmad, A., Maitah, M.Y., Ginnebaugh, K.R., et al., 2013. Inhibition of Hedgehog signaling sensitizes NSCLC cells to standard therapies through modulation of EMT-regulating miRNAs. J. Hematol. Oncol. 6 (1), 77.

Allen, T.D., Rodriguez, E.M., Jones, K.D., Bishop, J.M., 2011. Activated Notch1 induces lung adenomas in mice and cooperates with Myc in the generation of lung adenocarcinoma. Cancer Res. 71 (18), 6010–6018.

Al-Saad, S., Al-Shibli, K., Donnem, T., Persson, M., Bremnes, R.M., Busund, L.T., 2008. The prognostic impact of NF-kappaB p105, vimentin, E-cadherin and Par6 expression in epithelial and stromal compartment in non-small-cell lung cancer. Br. J. Cancer 99 (9), 1476–1483.

Bhummaphan, N., Chanvorachote, P., 2015. Gigantol suppresses cancer stem cell-like phenotypes in lung cancer cells. Evid. Based Complement. Alternat. Med. 2015, 836564.

Bleul, C.C., Farzan, M., Choe, H., et al., 1996. The lymphocyte chemoattractant SDF-1 is a ligand for LESTR/fusin and blocks HIV-1 entry. Nature 382 (6594), 829–833.

Bonnet, D., Dick, J.E., 1997. Human acute myeloid leukemia is organized as a hierarchy that originates from a primitive hematopoietic cell. Nat. Med. 3 (7), 730–737.

Boudreau, N.J., Jones, P.L., 1999. Extracellular matrix and integrin signalling: the shape of things to come. Biochem. J. 339 (Pt. 3), 481–488.

Casas, E., Kim, J., Bendesky, A., Ohno-Machado, L., Wolfe, C.J., Yang, J., 2011. Snail2 is an essential mediator of Twist1-induced epithelial mesenchymal transition and metastasis. Cancer Res. 71 (1), 245–254.

Castano, Z., Fillmore, C.M., Kim, C.F., McAllister, S.S., 2012. The bed and the bugs: interactions between the tumor microenvironment and cancer stem cells. Semin. Cancer Biol. 22 (5–6), 462–470.

Caswell, P.T., Chan, M., Lindsay, A.J., McCaffrey, M.W., Boettiger, D., Norman, J.C., 2008. Rab-coupling protein coordinates recycling of alpha5beta1 integrin and EGFR1 to promote cell migration in 3D microenvironments. J. Cell Biol. 183 (1), 143–155.

Chang, A.C., Garside, V.C., Fournier, M., et al., 2014. A Notch-dependent transcriptional hierarchy promotes mesenchymal transdifferentiation in the cardiac cushion. Dev. Dyn. 243 (7), 894–905.

Chanvorachote, P., Pongrakhananon, V., Halim, H., 2015. Caveolin-1 regulates metastatic behaviors of anoikis resistant lung cancer cells. Mol. Cell. Biochem. 399 (1–2), 291–302.

Chunhacha, P., Chanvorachote, P., 2012. Roles of caveolin-1 on anoikis resistance in non small cell lung cancer. Int. J. Physiol. Pathophysiol. Pharmacol. 4 (3), 149–155.

Chunhacha, P., Pongrakhananon, V., Rojanasakul, Y., Chanvorachote, P., 2012. Caveolin-1 regulates Mcl-1 stability and anoikis in lung carcinoma cells. Am. J. Physiol. Cell Physiol. 302 (9), C1284–C1292.

Cowden Dahl, K.D., Symowicz, J., Ning, Y., et al., 2008. Matrix metalloproteinase 9 is a mediator of epidermal growth factor-dependent e-cadherin loss in ovarian carcinoma cells. Cancer Res. 68 (12), 4606–4613.

Cox, G., Jones, J.L., O'Byrne, K.J., 2000. Matrix metalloproteinase 9 and the epidermal growth factor signal pathway in operable non-small cell lung cancer. Clin. Cancer Res. 6 (6), 2349–2355.

Crawford, M., Brawner, E., Batte, K., et al., 2008. MicroRNA-126 inhibits invasion in non-small cell lung carcinoma cell lines. Biochem. Biophys. Res. Commun. 373 (4), 607–612.

Dang, T.P., Gazdar, A.F., Virmani, A.K., et al., 2000. Chromosome 19 translocation, overexpression of Notch3, and human lung cancer. J. Natl. Cancer Inst. 92 (16), 1355–1357.

Dean, M., Fojo, T., Bates, S., 2005. Tumour stem cells and drug resistance. Nat. Rev. Cancer 5 (4), 275–284.

Donnem, T., Andersen, S., Al-Shibli, K., Al-Saad, S., Busund, L.T., Bremnes, R.M., 2010. Prognostic impact of Notch ligands and receptors in nonsmall cell lung cancer: coexpression of Notch-1 and vascular endothelial growth factor-A predicts poor survival. Cancer 116 (24), 5676–5685.

Eccles, S.A., Welch, D.R., 2007. Metastasis: recent discoveries and novel treatment strategies. Lancet 369 (9574), 1742–1757.

Edme, N., Downward, J., Thiery, J.P., Boyer, B., 2002. Ras induces NBT-II epithelial cell scattering through the coordinate activities of Rac and MAPK pathways. J. Cell Sci. 115 (Pt. 12), 2591–2601.

Elayadi, A.N., Samli, K.N., Prudkin, L., et al., 2007. A peptide selected by biopanning identifies the integrin alphavbeta6 as a prognostic biomarker for nonsmall cell lung cancer. Cancer Res. 67 (12), 5889–5895.

Fong, Y.C., Liu, S.C., Huang, C.Y., et al., 2009. Osteopontin increases lung cancer cells migration via activation of the alphavbeta3 integrin/FAK/Akt and NF-kappaB-dependent pathway. Lung Cancer 64 (3), 263–270.

Frisch, S.M., Francis, H., 1994. Disruption of epithelial cell-matrix interactions induces apoptosis. J. Cell Biol. 124 (4), 619–626.

Fukuoka, M., Wu, Y.L., Thongprasert, S., et al., 2011. Biomarker analyses and final overall survival results from a phase III, randomized, open-label, first-line study of gefitinib versus carboplatin/paclitaxel in clinically selected patients with advanced non-small-cell lung cancer in Asia (IPASS). J. Clin. Oncol. 29 (21), 2866–2874.

Goldstraw, P., Ball, D., Jett, J.R., et al., 2011. Non-small-cell lung cancer. Lancet 378 (9804), 1727–1740.

Gomez-Casal, R., Bhattacharya, C., Ganesh, N., et al., 2013. Non-small cell lung cancer cells survived ionizing radiation treatment display cancer stem cell and epithelial-mesenchymal transition phenotypes. Mol. Cancer 12 (1), 94.

Grandis, J.R., Sok, J.C., 2004. Signaling through the epidermal growth factor receptor during the development of malignancy. Pharmacol. Ther. 102 (1), 37–46.

Graus-Porta, D., Beerli, R.R., Daly, J.M., Hynes, N.E., 1997. ErbB-2, the preferred heterodimerization partner of all ErbB receptors, is a mediator of lateral signaling. EMBO J. 16 (7), 1647–1655.

Gravdal, K., Halvorsen, O.J., Haukaas, S.A., Akslen, L.A., 2007. A switch from E-cadherin to N-cadherin expression indicates epithelial to mesenchymal transition and is of strong and independent importance for the progress of prostate cancer. Clin. Cancer Res. 13 (23), 7003–7011.

Gregory, P.A., Bert, A.G., Paterson, E.L., et al., 2008. The miR-200 family and miR-205 regulate epithelial to mesenchymal transition by targeting ZEB1 and SIP1. Nat. Cell Biol. 10 (5), 593–601.

Han, J.Y., Kim, H.S., Lee, S.H., Park, W.S., Lee, J.Y., Yoo, N.J., 2003. Immunohistochemical expression of integrins and extracellular matrix proteins in non-small cell lung cancer: correlation with lymph node metastasis. Lung Cancer 41 (1), 65–70.

Hassan, K.A., Wang, L., Korkaya, H., et al., 2013. Notch pathway activity identifies cells with cancer stem cell-like properties and correlates with worse survival in lung adenocarcinoma. Clin. Cancer Res. 19 (8), 1972–1980.

Hazan, R.B., Phillips, G.R., Qiao, R.F., Norton, L., Aaronson, S.A., 2000. Exogenous expression of N-cadherin in breast cancer cells induces cell migration, invasion, and metastasis. J. Cell Biol. 148 (4), 779–790.

He, L., Hannon, G.J., 2004. MicroRNAs: small RNAs with a big role in gene regulation. Nat. Rev. Genet. 5 (7), 522–531.

Heidemann, F., Schildt, A., Schmid, K., et al., 2014. Selectins mediate small cell lung cancer systemic metastasis. PloS One 9 (4), e92327.

Hirsch, F., Hansen, H., 2001. New techniques for early diagnosis of lung cancer. Ugeskr. Laeg. 163 (33), 4321–4323.

Hirsch, F.R., Franklin, W.A., Veve, R., Varella-Garcia, M., Bunn Jr., P.A., 2002. HER2/neu expression in malignant lung tumors. Semin. Oncol. 29 (1 Suppl. 4), 51–58.

Hirsch, F.R., Varella-Garcia, M., Cappuzzo, F., 2009. Predictive value of EGFR and HER2 overexpression in advanced non-small-cell lung cancer. Oncogene 28 (Suppl. 1), S32–S37.

Ho, C.C., Huang, P.H., Huang, H.Y., Chen, Y.H., Yang, P.C., Hsu, S.M., 2002. Up-regulated caveolin-1 accentuates the metastasis capability of lung adenocarcinoma by inducing filopodia formation. Am. J. Pathol. 161 (5), 1647–1656.

Ho, M.M., Ng, A.V., Lam, S., Hung, J.Y., 2007. Side population in human lung cancer cell lines and tumors is enriched with stem-like cancer cells. Cancer Res. 67 (10), 4827–4833.

Ho, C.C., Kuo, S.H., Huang, P.H., Huang, H.Y., Yang, C.H., Yang, P.C., 2008. Caveolin-1 expression is significantly associated with drug resistance and poor prognosis in advanced non-small cell lung cancer patients treated with gemcitabine-based chemotherapy. Lung Cancer 59 (1), 105–110.

Hoikkala, S., Paakko, P., Soini, Y., Makitaro, R., Kinnula, V., Turpeenniemi-Hujanen, T., 2006. Tissue MMP-2 and MMP-9 [corrected] are better prognostic factors than serum MMP-2/TIMP-2–complex or TIMP-1 [corrected] in stage [corrected] I-III lung carcinoma. Cancer Lett. 236 (1), 125–132.

Hou, C., Sun, B., Jiang, Y., et al., 2015. MicroRNA-31 inhibits lung adenocarcinoma stem-like cells via down-regulation of MET-PI3K-Akt signaling pathway. Anticancer Agents Med. Chem. 16 (4), 501–518.

Hrabec, E., Strek, M., Nowak, D., Hrabec, Z., 2001. Elevated level of circulating matrix metalloproteinase-9 in patients with lung cancer. Respir. Med. 95 (1), 1–4.

Huber, M.A., Kraut, N., Beug, H., 2005. Molecular requirements for epithelial-mesenchymal transition during tumor progression. Curr. Opin. Cell Biol. 17 (5), 548–558.

Ishikawa, T., Shimizu, T., Ueki, A., et al., 2013. Twist2 functions as a tumor suppressor in murine osteosarcoma cells. Cancer Sci. 104 (7), 880–888.

Jemal, A., Bray, F., Center, M.M., Ferlay, J., Ward, E., Forman, D., 2011. Global cancer statistics. CA Cancer J. Clin. 61 (2), 69–90.

Jenndahl, L.E., Isakson, P., Baeckstrom, D., 2005. c-erbB2-induced epithelial-mesenchymal transition in mammary epithelial cells is suppressed by cell-cell contact and initiated prior to E-cadherin downregulation. Int. J. Oncol. 27 (2), 39–448.

Jiang, H.L., Jiang, L.M., Han, W.D., 2015. Wnt/beta-catenin signaling pathway in lung cancer stem cells is a potential target for the development of novel anticancer drugs. J. BUON 20 (4), 1094–1100.

Johnson, D.H., Schiller, J.H., Bunn Jr., P.A., 2014. Recent clinical advances in lung cancer management. J. Clin. Oncol 32 (10), 973–982.

Jung, M.J., Rho, J.K., Kim, Y.M., et al., 2013. Upregulation of CXCR4 is functionally crucial for maintenance of stemness in drug-resistant non-small cell lung cancer cells. Oncogene 32 (2), 209–221.

Karni, R., Jove, R., Levitzki, A., 1999. Inhibition of pp60c-Src reduces Bcl-XL expression and reverses the transformed phenotype of cells overexpressing EGF and HER-2 receptors. Oncogene 18 (33), 4654–4662.

Ko, H., 2015. Geraniin inhibits TGF-beta1-induced epithelial-mesenchymal transition and suppresses A549 lung cancer migration, invasion and anoikis resistance. Bioorg. Med. Chem. Lett. 25 (17), 3529–3534.

Laird, A.D., Li, G., Moss, K.G., et al., 2003. Src family kinase activity is required for signal transducer and activator of transcription 3 and focal adhesion kinase phosphorylation and vascular endothelial growth factor signaling in vivo and for anchorage-dependent and -independent growth of human tumor cells. Mol. Cancer Ther. 2 (5), 461–469.

Leung, E.L., Tam, I.Y., Tin, V.P., et al., 2009. SRC promotes survival and invasion of lung cancers with epidermal growth factor receptor abnormalities and is a potential candidate for molecular-targeted therapy. Mol. Cancer Res. 7 (6), 923–932.

Li, C., Song, G., Zhang, S., Wang, E., Cui, Z., 2015. Wnt3a increases the metastatic potential of non-small cell lung cancer cells in vitro in part via its upregulation of Notch3. Oncol. Rep. 33 (3), 1207–1214.

Liu, G., Meng, X., Jin, Y., et al., 2008. Inhibitory role of focal adhesion kinase on anoikis in the lung cancer cell A549. Cell Biol. Int. 32 (6), 663–670.

Liu, J., Xiao, Z., Wong, S.K., et al., 2013. Lung cancer tumorigenicity and drug resistance are maintained through ALDH(hi)CD44(hi) tumor initiating cells. Oncotarget 4 (10), 1698–1711.

Liu, J., Mao, Z., Huang, J., Xie, S., Liu, T., Mao, Z., 2014. Blocking the NOTCH pathway can inhibit the growth of CD133-positive A549 cells and sensitize to chemotherapy. Biochem. Biophys. Res. Commun. 444 (4), 670–675.

Liu, L., Chen, X., Wang, Y., et al., 2014. Notch3 is important for TGF-beta-induced epithelial-mesenchymal transition in non-small cell lung cancer bone metastasis by regulating ZEB-1. Cancer Gene Ther. 21 (9), 364–372.

Liu, R.Y., Zeng, Y., Lei, Z., et al., 2014. JAK/STAT3 signaling is required for TGF-beta-induced epithelial-mesenchymal transition in lung cancer cells. Int. J. Oncol. 44 (5), 1643–1651.

Maitah, M.Y., Ali, S., Ahmad, A., Gadgeel, S., Sarkar, F.H., 2011. Up-regulation of sonic hedgehog contributes to TGF-beta1-induced epithelial to mesenchymal transition in NSCLC cells. PloS One 6 (1), e16068.

Malin, D., Strekalova, E., Petrovic, V., et al., 2015. ERK-regulated alphaB-crystallin induction by matrix detachment inhibits anoikis and promotes lung metastasis in vivo. Oncogene 34 (45), 5626–5634.

Masaki, T., Igarashi, K., Tokuda, M., et al., 2003. pp60c-src activation in lung adenocarcinoma. Eur. J. Cancer 39 (10), 1447–1455.

Mazieres, J., He, B., You, L., Xu, Z., Jablons, D.M., 2005. Wnt signaling in lung cancer. Cancer Lett. 222 (1), 1–10.

Mazurenko, N.N., Kogan, E.A., Zborovskaia, I.B., Sukhova, N.M., Kiselev, F.L., 1991. The detection of the c-src protein gene product in human lung tumors. Vopr. Onkol. 37 (6), 683–690.

Mazurenko, N.N., Kogan, E.A., Zborovskaya, I.B., Kisseljov, F.L., 1992. Expression of pp60c-src in human small cell and non-small cell lung carcinomas. Eur. J. Cancer 28 (2–3), 372–377.

Mehlen, P., Puisieux, A., 2006. Metastasis: a question of life or death. Nat. Rev. Cancer 6 (6), 449–458.

Meng, X.N., Jin, Y., Yu, Y., et al., 2009. Characterisation of fibronectin-mediated FAK signalling pathways in lung cancer cell migration and invasion. Br. J. Cancer 101 (2), 327–334.

Merrick, D.T., Kittelson, J., Winterhalder, R., et al., 2006. Analysis of c-ErbB1/epidermal growth factor receptor and c-ErbB2/HER-2 expression in bronchial dysplasia: evaluation of potential targets for chemoprevention of lung cancer. Clin. Cancer Res. 12 (7 Pt 1), 2281–2288.

Moon, K.C., Lee, G.K., Yoo, S.H., et al., 2005. Expression of caveolin-1 in pleomorphic carcinoma of the lung is correlated with a poor prognosis. Anticancer Res. 25 (6C), 4631–4637.

Morello, V., Cabodi, S., Sigismund, S., et al., 2011. Beta1 integrin controls EGFR signaling and tumorigenic properties of lung cancer cells. Oncogene 30 (39), 4087–4096.

Moustakas, A., Heldin, C.H., 2007. Signaling networks guiding epithelial-mesenchymal transitions during embryogenesis and cancer progression. Cancer Sci. 98 (10), 1512–1520.

Muller, A., Homey, B., Soto, H., et al., 2001. Involvement of chemokine receptors in breast cancer metastasis. Nature 410 (6824), 50–56.

Murakami, A., Takahashi, F., Nurwidya, F., et al., 2014. Hypoxia increases gefitinib-resistant lung cancer stem cells through the activation of insulin-like growth factor 1 receptor. PloS One 9 (1), e86459.

Nolte, S.M., Venugopal, C., McFarlane, N., et al., 2013. A cancer stem cell model for studying brain metastases from primary lung cancer. J. Natl. Cancer Inst. 105 (8), 551–562.

Nyhan, K.C., Faherty, N., Murray, G., et al., 2010. Jagged/Notch signalling is required for a subset of TGFbeta1 responses in human kidney epithelial cells. Biochim. Biophys. Acta 1803 (12), 1386–1395.

Olayioye, M.A., Neve, R.M., Lane, H.A., Hynes, N.E., 2000. The ErbB signaling network: receptor heterodimerization in development and cancer. EMBO J. 19 (13), 3159–3167.

Pancotti, F., Roncuzzi, L., Maggiolini, M., Gasperi-Campani, A., 2012. Caveolin-1 silencing arrests the proliferation of metastatic lung cancer cells through the inhibition of STAT3 signaling. Cell. Signal. 24 (7), 1390–1397.

Park, S.M., Gaur, A.B., Lengyel, E., Peter, M.E., 2008. The miR-200 family determines the epithelial phenotype of cancer cells by targeting the E-cadherin repressors ZEB1 and ZEB2. Genes Dev. 22 (7), 894–907.

Park, K.S., Martelotto, L.G., Peifer, M., et al., 2011. A crucial requirement for Hedgehog signaling in small cell lung cancer. Nat. Med. 17 (11), 1504–1508.

Phillips, R.J., Burdick, M.D., Lutz, M., Belperio, J.A., Keane, M.P., Strieter, R.M., 2003. The stromal derived factor-1/CXCL12-CXC chemokine receptor 4 biological axis in non-small cell lung cancer metastases. Am. J. Respir. Crit. Care Med. 167 (12), 1676–1686.

Prudkin, L., Liu, D.D., Ozburn, N.C., et al., 2009. Epithelial-to-mesenchymal transition in the development and progression of adenocarcinoma and squamous cell carcinoma of the lung. Mod. Pathol. 22 (5), 668–678.

Rao, J.S., Gondi, C., Chetty, C., Chittivelu, S., Joseph, P.A., Lakka, S.S., 2005. Inhibition of invasion, angiogenesis, tumor growth, and metastasis by adenovirus-mediated transfer of antisense uPAR and MMP-9 in non-small cell lung cancer cells. Mol. Cancer Ther. 4 (9), 1399–1408.

Ravid, D., Maor, S., Werner, H., Liscovitch, M., 2005. Caveolin-1 inhibits cell detachment-induced p53 activation and anoikis by upregulation of insulin-like growth factor-I receptors and signaling. Oncogene 24 (8), 1338–1347.

Reiske, H.R., Kao, S.C., Cary, L.A., Guan, J.L., Lai, J.F., Chen, H.C., 1999. Requirement of phosphatidylinositol 3-kinase in focal adhesion kinase-promoted cell migration. J. Biol. Chem. 274 (18), 12361–12366.

Risolino, M., Mandia, N., Iavarone, F., et al., 2014. Transcription factor PREP1 induces EMT and metastasis by controlling the TGF-beta-SMAD3 pathway in non-small cell lung adenocarcinoma. Proc. Natl. Acad. Sci. U.S.A. 111 (36), E3775–E3784.

Sakuma, Y., Yamazaki, Y., Nakamura, Y., et al., 2012. WZ4002, a third-generation EGFR inhibitor, can overcome anoikis resistance in EGFR-mutant lung adenocarcinomas more efficiently than Src inhibitors. Lab. Invest. 92 (3), 371–383.

Shaw, A.T., Kim, D.W., Nakagawa, K., et al., 2013. Crizotinib versus chemotherapy in advanced ALK-positive lung cancer. N. Engl. J. Med. 368 (25), 2385–2394.

Shen, L., Chen, L., Wang, Y., Jiang, X., Xia, H., Zhuang, Z., 2015. Long noncoding RNA MALAT1 promotes brain metastasis by inducing epithelial-mesenchymal transition in lung cancer. J. Neurooncol. 121 (1), 101–108.

Sheppard, D., 2004. Roles of alphav integrins in vascular biology and pulmonary pathology. Curr. Opin. Cell Biol. 16 (5), 552–557.

Shi, Y., Fu, X., Hua, Y., Han, Y., Lu, Y., Wang, J., 2012. The side population in human lung cancer cell line NCI-H460 is enriched in stem-like cancer cells. PloS One 7 (3), e33358.

Siegel, R.L., Miller, K.D., Jemal, A., 2016. Cancer statistics, 2016. CA Cancer J. Clin. 66 (1), 7–30.

Singh, S., Trevino, J., Bora-Singhal, N., et al., 2012. EGFR/Src/Akt signaling modulates Sox2 expression and self-renewal of stem-like side-population cells in non-small cell lung cancer. Mol. Cancer 11, 73.

Sloan, K.A., Marquez, H.A., Li, J., et al., 2009. Increased PEA3/E1AF and decreased Net/Elk-3, both ETS proteins, characterize human NSCLC progression and regulate caveolin-1 transcription in Calu-1 and NCI-H23 NSCLC cell lines. Carcinogenesis 30 (8), 1433–1442.

Song, L., Turkson, J., Karras, J.G., Jove, R., Haura, E.B., 2003. Activation of Stat3 by receptor tyrosine kinases and cytokines regulates survival in human non-small cell carcinoma cells. Oncogene 22 (27), 4150–4165.

Strieter, R.M., 2001. Chemokines: not just leukocyte chemoattractants in the promotion of cancer. Nat. Immunol. 2 (4), 285–286.

Su, L., Zhang, J., Xu, H., et al., 2005. Differential expression of CXCR4 is associated with the metastatic potential of human non-small cell lung cancer cells. Clin. Cancer Res. 11 (23), 8273–8280.

Sung, W.J., Park, K.S., Kwak, S.G., Hyun, D.S., Jang, J.S., Park, K.K., 2015. Epithelial-mesenchymal transition in patients of pulmonary adenocarcinoma: correlation with cancer stem cell markers and prognosis. Int. J. Clin. Exp. Pathol. 8 (8), 8997–9009.

Teng, Y., Wang, X., Wang, Y., Ma, D., 2010. Wnt/beta-catenin signaling regulates cancer stem cells in lung cancer A549 cells. Biochem. Biophys. Res. Commun. 392 (3), 373–379.

Teschendorff, A.E., Journee, M., Absil, P.A., Sepulchre, R., Caldas, C., 2007. Elucidating the altered transcriptional programs in breast cancer using independent component analysis. PLoS Comput. Biol. 3 (8), e161.

Thiery, J.P., Sleeman, J.P., 2006. Complex networks orchestrate epithelial-mesenchymal transitions. Nat. Rev. Mol. Cell Biol. 7 (2), 131–142.

Timmerman, L.A., Grego-Bessa, J., Raya, A., et al., 2004. Notch promotes epithelial-mesenchymal transition during cardiac development and oncogenic transformation. Genes Dev. 18 (1), 99–115.

Tuomi, S., Mai, A., Nevo, J., et al., 2009. PKCepsilon regulation of an alpha5 integrin-ZO-1 complex controls lamellae formation in migrating cancer cells. Sci. Signal. 2 (77), ra32.

Tzahar, E., Waterman, H., Chen, X., et al., 1996. A hierarchical network of interreceptor interactions determines signal transduction by Neu differentiation factor/neuregulin and epidermal growth factor. Mol. Cell. Biol. 16 (10), 5276–5287.

Wald, O., Izhar, U., Amir, G., et al., 2006. CD4+CXCR4highCD69+ T cells accumulate in lung adenocarcinoma. J. Immunol. 177 (10), 6983–6990.

Wei, L., Yang, Y., Zhang, X., Yu, Q., 2004. Altered regulation of Src upon cell detachment protects human lung adenocarcinoma cells from anoikis. Oncogene 23 (56), 9052–9061.

Westhoff, B., Colaluca, I.N., D'Ario, G., et al., 2009. Alterations of the Notch pathway in lung cancer. Proc. Natl. Acad. Sci. U.S.A. 106 (52), 22293–22298.

Wipff, P.J., Rifkin, D.B., Meister, J.J., Hinz, B., 2007. Myofibroblast contraction activates latent TGF-beta1 from the extracellular matrix. J. Cell Biol. 179 (6), 1311–1323.

Yadav, A., Kumar, B., Datta, J., Teknos, T.N., Kumar, P., 2011. IL-6 promotes head and neck tumor metastasis by inducing epithelial-mesenchymal transition via the JAK-STAT3-SNAIL signaling pathway. Mol. Cancer Res. 9 (12), 1658–1667.

Yang, J., Weinberg, R.A., 2008. Epithelial-mesenchymal transition: at the crossroads of development and tumor metastasis. Dev. Cell 14 (6), 818–829.

Yang, J., Lan, H., Huang, X., Liu, B., Tong, Y., 2012. MicroRNA-126 inhibits tumor cell growth and its expression level correlates with poor survival in non-small cell lung cancer patients. PloS One 7 (8), e42978.

Yasuda, M., Tanaka, Y., Tamura, M., et al., 2001. Stimulation of beta1 integrin down-regulates ICAM-1 expression and ICAM-1-dependent adhesion of lung cancer cells through focal adhesion kinase. Cancer Research 61 (5), 2022–2030.

Ylisirnio, S., Hoyhtya, M., Turpeenniemi-Hujanen, T., 2000. Serum matrix metalloproteinases -2, -9 and tissue inhibitors of metalloproteinases -1, -2 in lung cancer–TIMP-1 as a prognostic marker. Anticancer Res. 20 (2b), 1311–1316.

Yongsanguanchai, N., Pongrakhananon, V., Mutirangura, A., Rojanasakul, Y., Chanvorachote, P., 2015. Nitric oxide induces cancer stem cell-like phenotypes in human lung cancer cells. Am. J. Physiol. Cell Physiol. 308 (2), C89–C100.

Yoo, S.H., Park, Y.S., Kim, H.R., et al., 2003. Expression of caveolin-1 is associated with poor prognosis of patients with squamous cell carcinoma of the lung. Lung Cancer 42 (2), 195–202.

Youlden, D.R., Cramb, S.M., Baade, P.D., 2008. The International Epidemiology of Lung Cancer: geographical distribution and secular trends. J. Thorac. Oncol. 3 (8), 819–831.

Yuan, X., Wu, H., Han, N., et al., 2014. Notch signaling and EMT in non-small cell lung cancer: biological significance and therapeutic application. J. Hematol. Oncol. 7 (1), 87.

Yue, D., Li, H., Che, J., et al., 2014. Hedgehog/Gli promotes epithelial-mesenchymal transition in lung squamous cell carcinomas. J. Exp. Clin. Cancer Res. 33, 34.

Zakaria, N., Yusoff, N.M., Zakaria, Z., et al., 2015. Human non-small cell lung cancer expresses putative cancer stem cell markers and exhibits the transcriptomic profile of multipotent cells. BMC Cancer 15, 84.

Zeisberg, E.M., Potenta, S., Xie, L., Zeisberg, M., Kalluri, R., 2007. Discovery of endothelial to mesenchymal transition as a source for carcinoma-associated fibroblasts. Cancer Res. 67 (21), 10123–10128.

Zhang, S., Wang, Y., Mao, J.H., et al., 2012. Inhibition of CK2alpha down-regulates Hedgehog/Gli signaling leading to a reduction of a stem-like side population in human lung cancer cells. PloS One 7 (6), e38996.

Zhang, J., Cao, J., Li, J., et al., 2014. A phase I study of AST1306, a novel irreversible EGFR and HER2 kinase inhibitor, in patients with advanced solid tumors. J. Hematol. Oncol. 7, 22.

Zhang, Y., Zhang, X., Huang, J., Dong, Q., 2015a. Wnt signaling regulation of stem-like properties in human lung adenocarcinoma cell lines. Med. Oncol. 32 (5), 157.

Zhang, X., Lou, Y., Wang, H., et al., 2015b. Wnt signaling regulates the stemness of lung cancer stem cells and its inhibitors exert anticancer effect on lung cancer SPC-A1 cells. Med. Oncol. 32 (4), 95.

Zhang, X., Lou, Y., Zheng, X., et al., 2015c. Wnt blockers inhibit the proliferation of lung cancer stem cells. Drug Des. Dev. Ther. 9, 2399–2407.

Zheng, Y., Ritzenthaler, J.D., Roman, J., Han, S., 2007. Nicotine stimulates human lung cancer cell growth by inducing fibronectin expression. Am. J. Respir. Cell Mol. Biol. 37 (6), 681–690.

CHAPTER

# 5

# Cervical Cancer Metastasis

*S.W. Aziz*[1], *M.H. Aziz*[2]

[1]IPC Healthcare, Inc., Johnson City, TN, United States; [2]East Tennessee State University, Johnson City, TN, United States

## Abbreviations

**AEG-1** Astrocyte elevated gene-1
**CHI3L1** Chitinase 3 like 1
**CIN** Cervical intraepithelial neoplasia
**COX-2** Cyclooxygenase-2
**CxC** Cervical cancer
**CXCR** CXC chemokine receptor
**EGFR** Epidermal growth factor receptor
**EMT** epithelial to mesenchymal transition
**FAK** Focal adhesion kinase
**FIGO** Federation of Gynecology and Obstetrics
**Gal-3** Galectin-3 protein
**HOX** Homeobox
**HOTAIR** HOX Antisense Intergenic RNA
**HPV** Human papillomavirus
**LAPTM4B** Lysosome-associated protein transmembrane 4 beta
**LMVD** Lymphatic microvessel density
**LNM** Lymph node metastasis
**LTBP-2** Latent transforming growth factor-beta 1 binding protein-2
**LVD** Lymphatic vessel density
**miRNAs** MicroRNAs
**MMP-9** Matrix metalloproteinase-9
**MVD** Microvessel density
**PAK-4** p21-activated kinase-4
**PDGFR** Platelet-derived growth factor receptors
**Sema4D** Semaphorin 4D
**SOCE** Store operated Ca (2+) entry
**STIM1** Stromal-interaction molecule 1
**TACC3** Transforming acidic coiled-coil protein family
**VEGF** Vascular endothelial growth factors
**WIF1** Wnt inhibitory factor 1

## 1. INTRODUCTION

Cervical cancer (CxC) is the third most commonly diagnosed and cause of death among gynecologic cancers in the United States (American Cancer Society, 2015; Siegel et al., 2015). In 2012, CxC accounted for an estimated 528,000 new cancer cases worldwide and for 266,000 deaths globally (Cervical cancer, 2015). Although the incidence of CxC is high, in the past three decades, there has been a more than 50% decline in the incidence of invasive CxC in the United States due to early detection and treatment of precancerous lesion with Papanicolaou (Pap) test screening. CxC in the early stages is frequently asymptomatic, underscoring the importance of screening. Global incidence and mortality rates depend upon the presence of human papillomavirus (HPV) vaccination and of screening programs for cervical precancer and cancer, which are most likely to be available in developed countries. Due to the availability of Pap screening programs and HPV vaccination, about 75% decline in the incidence and mortality of CxC has been seen over the past five decades in developed countries (Quinn et al., 1999; Willoughby et al., 2006). However, in countries with no or minimal access to


http://dx.doi.org/10.1016/B978-0-12-804003-4.00005-0

CxC screening and prevention programs, it remains the second most common type of cancer (17.8/100,000 women) and cause of cancer deaths (9.8/100,000) in women. In developing countries, CxC is the second most common type of cancer in women (15.7/100,000 women) and the third most common cause of cancer mortality (8.3/100,000) (WHO/ICO Information Center of HPV and Cervical Cancer (HPV Information Center), 2010). On the other hand, in developed countries, in 2012, CxC was the eleventh most common type of cancer in women (9.9/100,000 women) and the ninth most common cause of cancer mortality (3.3/100,000) (Torre et al., 2012).

It is estimated that if HPV vaccine rates of 70% worldwide are achieved, it would be expected to see a decrease of 344,520 new cases of CxC annually and 178,182 CxC-related deaths can be avoided (Van Kriekinge et al., 2014). However even after the implementation of vaccination programs, significant decreases in cervical dysplasia or cancer is not likely to be seen for many years due to the latency period of 10–15 years between HPV exposure and CxC development. As high grade dysplasia is a necessary precursor for squamous, adenocarcinoma, and adenosquamous carcinomas of the cervix, a decrease in high grade cervical dysplasia should translate into decreased incidence of CxC in the next decade (Wei, 2009).

CxC estimates are higher for certain racial and ethnic groups: white (incidence: 7.1/100,000 and mortality: 2.0/100,000), non-Hispanic black (10.2/100,000 and 4.2/100,000), Hispanic/Latino (10.5/100,000 and 2.8/100,000), Asian/Pacific Islander (6.4/100,000 and 1.8/100,000), and American Indian/Alaska Native (9.7/100,000 and 3.4/100,000) (Siegel et al., 2011).

National data from 2000 to 2004 showed the lifetime risk of developing CxC for United States women, was 0.76% (Ries et al., 2007). The average age of diagnosis was 48 years. Incidence of CxC in girls under age 20 was 0.1/100,000, rising to 1.5/100,000 in women age 20–24 years, and then ranging from 11.0 to 15.8/100,000 for women age 30–85 years. Incidence declined in women above 85 years of age with only 5.7% of cases (Ries et al., 2007).

Given high morbidity and mortality of CxC, its better understanding is of vital importance for its prevention and cure. In this chapter, the risk factors (Fig. 5.1), modes of metastasis and major signaling pathways (Fig. 5.2) involved in the progression and metastasis of CXC will be highlighted.

## 2. RISK FACTORS

**Human papillomavirus**: HPV is the key factor leading to the development of cervical neoplasia and can be detected in about 99.7% of CxC (Moscicki, 2008). Genital tract HPV infection is extremely common. It has been estimated that 75–80% of sexually active adults will acquire genital tract HPV before the age of 50 (Centers for Disease Control and Prevention et al., 2006). Apart from CxC, genital HPV infection causes other conditions, including anogenital warts, and cancer of the vulva, vagina, anus, and penis (Chin-Hong et al., 2004; El-Mofty, 2012). More than 40 genital mucosal HPV subtypes are identified. Out of these, approximately 15 are known to be oncogenic and subtypes HPV 16 and HPV 18 are found in over 70% of all CxC (Chin-Hong et al., 2004; El-Mofty, 2012).

Most HPV infections are transient, and the virus alone is not sufficient to cause cervical neoplasia. Other risk factors for development of CxC are associated with an increased risk of acquiring HPV infection or result in compromised immune response to the infection causing persistence of infection and progression to cancer. These risk factors are highlighted in the following sections:

**Early onset of sexual activity**: The risk is approximately twofold for women younger than 18 years and 1.5-fold for 18–20 years compared to women with age at first intercourse of 21 years or older (International Collaboration of Epidemiological Studies of Cervical Cancer, 2007).

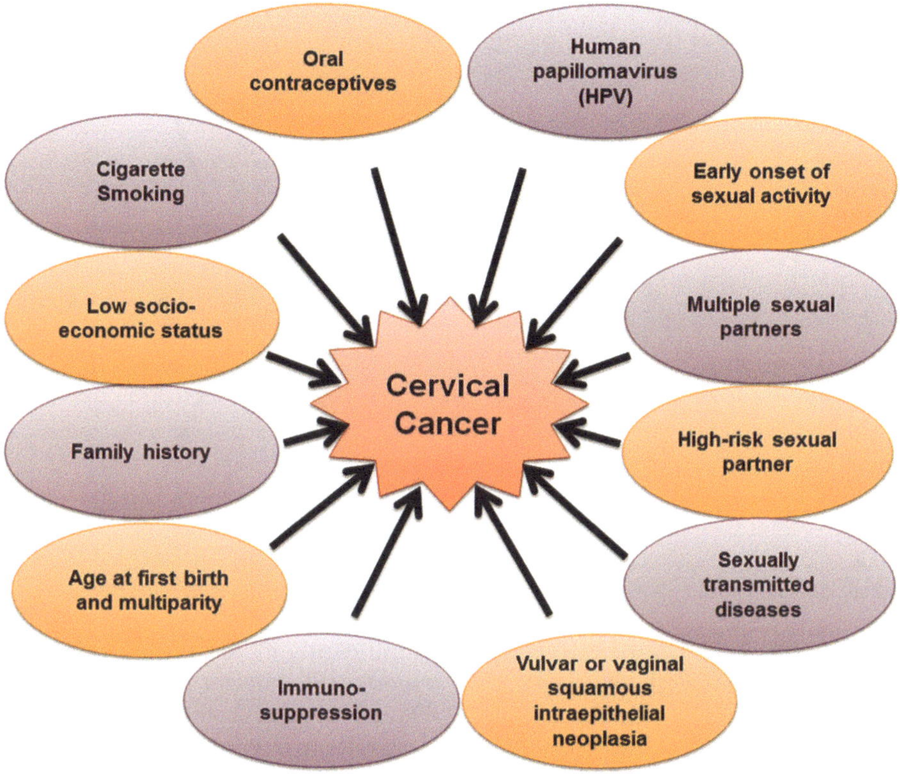


FIGURE 5.1 Major risk factors involved in cervical cancer development.

**Multiple sexual partners**: Women with multiple sexual partners are at higher risk of developing CxC. The risk is approximately threefold with six or more sexual partners and twofold with two partners compared to women with only one partner (International Collaboration of Epidemiological Studies of Cervical Cancer, 2007; Ogunwale et al., 2015; Liu et al., 2015; Soohoo et al., 2013; Ojiyi et al., 2013).

**High-risk sexual partner**: Women with high risk sexual partner (e.g., a partner with multiple sexual partners or with known HPV infection) are at higher risk of developing CxC (Soohoo et al., 2013; Tay and Oon, 2014; Sichanh et al., 2014; Sierra-Torres et al., 2003; Schiffman et al., 2011).

**Sexually transmitted diseases**: Women with history of sexually transmitted infections e.g., Chlamydia trachomatis or genital herpes have higher risk of developing CxC (Scholes et al., 1996).

**Vulvar or vaginal squamous intraepithelial neoplasia or cancer**: HPV infection is the etiology of most of these cases and therefore women with history of these conditions are at higher risk of developing CxC (van Beurden et al., 1995; Preti et al., 2005).

**Immunosuppression**: Studies have shown a higher prevalence of HPV infection in HIV-seropositive women than in seronegative women, and the severity of immunosuppression as measured by CD4+ T-cell counts is directly proportional to HPV prevalence. This translates to about five times increased prevalence of CxC in HIV-infected women (Leitao Jr et al., 2008; van der Burg et al., 2007; Schäfer et al., 1991; Maiman et al., 1997).

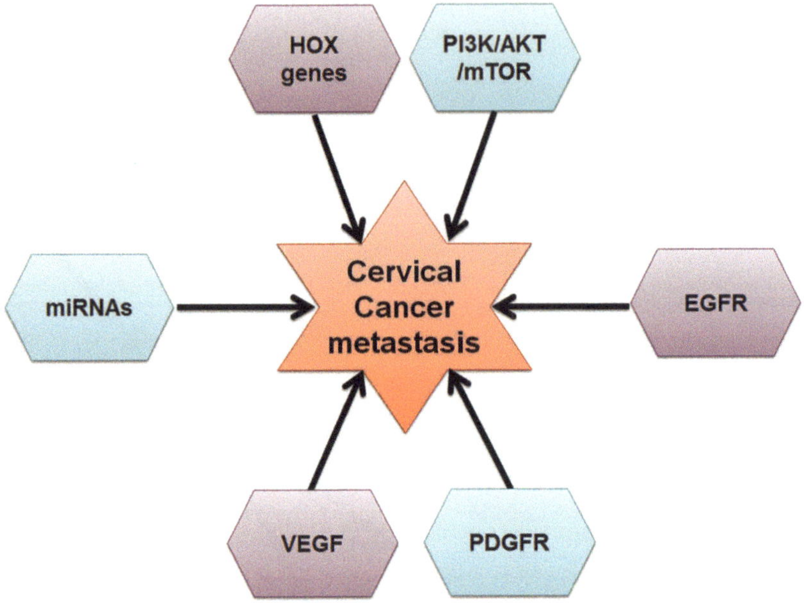


FIGURE 5.2 Major molecular factors involved in cervical cancer metastasis.

**Age at first birth and multiparity**: Age younger than 20 years at first birth and multiparity with three or more full term births are also associated with higher risk of CxC, likely due to increased exposure to HPV infection through sexual behavior (Brinton et al., 1987).

**Family history**: Risk of cervical cancer is two to three times higher in women with family history of CxC in first-degree relative (mother or sister) (Couto and Hemminki, 2006).

**Low socioeconomic status**: Higher incidence of CxC is noted in women from low socioeconomic status. In the United States, from 1988 to 1992, CxC incidence was higher in women who lived in communities with higher poverty levels. CxC mortality was also 71% higher in women in high compared to low poverty counties (Garner, 2003).

**Cigarette smoking**: Cigarette smoking is another risk factor for the development of CxC. The chemicals in cigarette smoke interact with the cells of the cervix, causing precancerous changes that may over time progress to cancer. In contrast to squamous cell cancer of the cervix, cigarette smoking is not associated with a significantly increased risk of cervical adenocarcinoma (Fletcher et al., 2014; Roura et al., 2014; Dmoch-Gajzlerska et al., 2014).

**Oral contraceptives**: A collaborative analysis of data from 24 epidemiological studies found that among current users of oral contraceptives, the risk of invasive CxC increased with increasing duration of use (5 or more years). The risk declined after use was discontinued, and by 10 or more years, had returned to that of women who have never used oral contraceptives (Franceschi, 2005; International Collaboration of Epidemiological Studies of Cervical Cancer et al., 2007).

## 3. PATHOGENESIS

The average time from initial HPV infection to development of high grade cervical intraepithelial neoplasia and, finally, invasive cancer is 15 years, although more rapid courses have been reported (Giorgi Rossi, 2015; Dong et al., 2015). The major steps in the development of CxC are:

- Infection of metaplastic epithelium at the cervical transformation zone (the junction between the squamous epithelium of the ectocervix and the glandular epithelium of the endocervical canal) with oncogenic HPV strain.
- Persistence of the HPV infection.
- Progression of a clone of epithelial cells to precancer.
- Progression of precancer lesion to carcinoma and invasion through the basement membrane.

## 4. HISTOPATHOLOGY

The diagnosis of CxC is made based upon histologic evaluation of cervical biopsy. The histopathologic types of CxC include:

- Squamous cell carcinoma (SCC) – 69%.
- Adenocarcinoma (including adenosquamous cancer) – 25%.
- Other histologies (small cell carcinoma, rhabdomyosarcoma, primary cervical lymphoma, cervical sarcoma) – 6%.

The two most common histologic types of CxC SCC and adenocarcinoma are caused primarily by HPV. In a study of over 30,000 CxC cases, the distribution of HPV subtypes in SCC was HPV 16 (59%), HPV 18 (13%), HPV 58 (5%), HPV 33 (5%), HPV 45 (4%) and in adenocarcinoma, HPV 16 (36%), HPV 18 (37%), HPV 45 (5%), HPV 31 (2%), HPV 33 (2%) of cases (Ries et al., 2007; Society of Gynecologic Oncologists Education Resource Panel Writing group et al., 2006).

There has been a significant increase in the incidence of invasive cervical adenocarcinoma and its variants over the past few decades, particularly in younger women (Smith et al., 2000; Look et al., 1996). Increased prevalence of specific HPV-16 and 18 variants that are associated more with adenocarcinoma than with SCC and increased exposure to endogenous (e.g., obesity) and exogenous (e.g., hormonal contraception, postmenopausal estrogen therapy) estrogens, have been proposed as causative factors for this rising trend. Adenosquamous tumors exhibit both glandular and squamous differentiation and are associated with a poorer outcome than squamous cell cancers or adenocarcinomas (Grisaru et al., 2001; Lea et al., 2003; Albores-Saavedra et al., 1997). Neuroendocrine or small cell carcinomas can also originate in the cervix in women, but are infrequent. Rhabdomyosarcoma of the cervix typically occurs in adolescents and young women of the cervix and is also rare as is primary cervical lymphoma and cervical sarcoma (Behtash et al., 2003; Villella et al., 2005).

## 5. ROUTES OF DISSEMINATION

There are three main routes for the metastasis of CxC, (1) direct local invasion, (2) lymphatic dissemination, and (3) hematogenous dissemination.

**Direct local invasion**: The malignant cervical cells usually spread by direct local invasion to the surrounding tissue. Direct extension of cancer may involve the uterine corpus, vagina, parametria, peritoneal cavity, bladder, or rectum. It mainly extends by lateral spread following a path of least resistance to involve the parametria. Another common pattern of spread is distal, to involve the vaginal fornices and later with larger-volume tumors, the midportion and distal third of the vagina. It may spread posteriorly to involve the rectum or the uterosacral ligaments. Rectal spread is usually associated with posterior vaginal involvement. Anterior spread to the bladder is usually with large-volume tumors with parametrial extension. The ureter being within 2 cm of the uterine cervix, is vulnerable to involvement by CxC. About 5% of patients with CxC have bilateral ureteral obstruction. The ureteral obstruction usually occurs in the distal third of the ureter. Ovarian involvement by direct extension of CxC is rare, with ovarian metastases occurring in approximately 0.5% of SCCs and 1.7% of adenocarcinomas (Gallup, 2008; Pilleron et al., 1974).

**Lymphatic dissemination**: Lymphatic spread of CxC involves the transport of the cancer cells through lymphatic system and is one of the major routes for the dissemination of cancer cells. Lymphatic spaces are present within 1–2mm of the basement membrane. Involvement of these spaces leads to lymphatic embolization to regional lymph nodes. This is more likely to occur with large-volume tumors and with involvement of more lymphatic spaces. Lymphovascular space invasion has been described as an independent factor predicting tumor aggressiveness (Balsat et al., 2014; Sevin et al., 1996). Sentinel lymph node mapping technique has shown that any of the pelvic lymph node groups, and even paraaortic lymph nodes, may contain the first draining lymph node and may be the first site of nodal metastasis in CxC (Levenback et al., 2002; Bader et al., 2007). The distribution of sites of nodal metastasis seen are: external iliac (43%), obturator (26%), parametrial (21%), common iliac (7%), presacral (1%), and paraaortic (1%). The risk of pelvic lymph node metastasis increases with increasing depth of invasion, whereas risk of paraaortic nodal involvement increases with increase in the extent of local disease (Levenback et al., 2002; Bader et al., 2007).

Recently, lymphatic vessel density (LVD) has been proposed as a new promising predictive variable of lymph node metastasis (Ji, 2006). Studies on cervical tissues showed that high levels of LVD as well as marked expression of vascular endothelial growth factor C (VEGF-C) and VEGF-D, the two main lymphangiogenic promoters, are already detected in preinvasive neoplasia (cervical intraepithelial neoplasia or CIN3) (Van Trappen et al., 2003; Jach et al., 2010; Cimpean et al., 2011). There occurs a gradual increase of LVD in the peritumoral stroma during cancer progression and in invasive CxC (Utrera-Barillas et al., 2010; Yang et al., 2009; Gombos et al., 2005). Gombos et al. (2005) were the first to link high peritumoral LVD with growth factor expression, lymphovascular space invasion, lymph node metastases, and overall survival (Gombos et al., 2005). Subsequent immunohistochemical studies on cervical neoplasms confirmed a positive association between LVD and lymph node metastases (Yang et al., 2009; Gao et al., 2006; Zhang et al., 2009; Yu et al., 2009; Liu et al., 2011; Cai et al., 2013). These studies therefore suggest an early lymphangiogenesis switch with an increased risk of lymphatic dissemination during CxC progression (Van der Auwera et al., 2006). Risk of lymphatic metastasis can therefore be assessed by LVD quantification. The global stromal evaluation also sheds light on topographical features of the lymphatic vascular network which may provide prognostic information. Topographical modifications with a shift of lymphatic vessels toward the neoplastic edges have been detected in CxC (Balsat et al., 2011; Burghardt and Ostor, 1983; Herfs et al., 2012).

**Hematogenous dissemination**: Hematogenous distant metastases are found in 5% of CxC patients (Gallup, 2008). The most common sites for hematogenous spread are the lungs, liver, and bone. The bowel, adrenal glands, spleen, and brain are less frequent sites of spread. Patients with large lesions and more advanced disease may have hematogenous dissemination by direct blood vessel invasion, by way of lacerated capillaries and veins, through the thoracic duct or through smaller lymphatic and venous channels (Gallup, 2008). Hematogenous metastasis is more commonly seen with poorly differentiated tumors and those with aggressive cell types like neuroendocrine cervical tumors and glassy cell tumors. The metastasis of CxC to the brain is usually associated with systemic spread to lungs, liver, and bones. However a few cases of isolated brain metastasis from CxC have been reported, including from neuroendocrine cell carcinoma (Pyeon et al., 2015; Dimitrakopoulos et al., 2011; DeEulis et al., 1987).

# 6. MAJOR MOLECULAR FACTORS

Genetic alterations in various cell signaling processes affect the decision of cell death or survival and are responsible for CxC progression and metastasis. An understanding of the alterations in these pathways leading to the development of CxC and its metastasis will further help in the development of potential therapeutic strategies targeting these pathways. Major molecular pathways involved in CxC metastasis are discussed below:

**HOX genes**: The homeobox (HOX) genes have been shown to play important role in the development of CxC (López-Romero et al., 2015; Gonzalez-Herrera et al., 2015; López et al., 2006; Kim et al., 2015). The HOX genes are a family of transcriptional factors that bind to specific DNA sequences and lead to cellular proliferation and differentiation. Several studies have reported that aberrant expression of these genes in CxC cell lines has been associated with tumorigenesis and metastasis (López-Romero et al., 2015; Kim et al., 2015; Li et al., 2015a,b; Alami et al., 1999). Studies on CxC cell lines, primary tumors and premalignant lesions have suggested the involvement of several isoforms of HOX genes, such as HOX-A1, C-5, C-6, C-8, C-10, D-3, D-9, D-13, in the development of CxC. Studies also suggested that de-novo expression of genes HOX B-2, B-4, B-13, C11-13 transformed cervical epithelial cells into malignant cells (López-Romero et al., 2015; Gonzalez-Herrera et al., 2015; López et al., 2006; Alami et al., 1999; Barba-de la Rosa et al., 2012; Andikyan and Taylor, 2009; Zhai et al., 2007; Hung et al., 2003).

Recent study suggested that overexpression of HOX Antisense Intergenic RNA (HOTAIR) in human CxC significantly correlates with CxC aggressiveness and progression by inhibiting the p21 expression (Jing et al., 2015; Aziz et al., 2013). Furthermore, Kim et al., reported that HOTAIR also regulates the expression of VEGF, matrix metalloproteinase 9 (MMP-9) and epithelial to mesenchymal transition (EMT)-related genes, which are important for cell motility and metastasis, during the progression of CxC (Kim et al., 2015).

Thus HOX genes may represent one of the important molecular factors for CxC progression and may serve as a promising therapeutic target.

Phosphatidylinositol-4,5-bisphosphate 3-kinase/protein kinase B (PKB), also known as AKT/mammalian target of rapamycin (**PI3K/AKT/mTOR) signaling pathway**: Several studies have reported that PI3K/AKT/mTOR signaling pathway is a potential diagnostic indicator and therapeutic target for various human cancers, including CxC (Aziz et al., 2006; Lu et al., 2015; Tsai et al., 2015; Bai et al., 2015; Hou et al., 2014; Eskander and Tewari, 2014; Roy et al., 2014; Wu et al., 2013; Kang et al., 2012; Zhang et al., 2008; Du et al., 2015). It regulates multiple molecular and cellular functions and alteration in the expression of its target genes leads to tumor initiation, development and progression (Tsai et al., 2015; Bai et al., 2015; Hou et al., 2014; Eskander and Tewari, 2014; Roy et al., 2014; Wu et al., 2013; Kang et al., 2012; Zhang et al., 2008; Du et al., 2015). Du et al. (2015) suggested that activation of PI3K/AKT signaling pathway occurs through upregulation of miR-221 expression via decreased phosphatase and tensin homologue expression levels in CxC cells (Shu et al., 2015). Study also suggested that PI3K/AKT pathway upregulates p21-activated kinase-4 in CxC tissues and correlates with Federation of Gynecology and Obstetrics (FIGO) stage, lymph node or distant metastasis, and poor histological grade (Ren et al., 2015).

It has been reported that PI3K/AKT signaling pathway is also regulated by latent transforming growth factor-beta 1 binding protein-2 (LTBP-2), a member of the fibrillin/LTBP superfamily of extracellular matrix proteins. Expression of LTBP-2, has been shown to correlate with clinical stage, cervical tumor size, depth of cervical stromal invasion, and lymph node metastasis, suggesting the involvement of this pathway in the development of CxC (Pim et al., 2005).

Studies suggest that HPV infection leads to activation and alteration of the PI3K/AKT pathway through inhibition of p53 and pRb genes, thus stimulating cell survival and proliferation. HPV infection accompanied by E6/E7 expression also activates PI3K/AKT pathway altering multiple cellular and molecular events to drive cervical carcinogenesis (Menges et al., 2006; Tinker et al., 2013). Furthermore, E7 protein from HPV-16 can modulate p27 localization through the PI3K/AKT pathway in CxC metastasis (Keysar et al., 2013; Ma et al., 2000; Bertelsen et al., 2006; Lui et al., 2013; Wu et al., 2013).

All these studies suggest that activation of PI3K/AKT/mTOR signaling pathway plays important role in proliferation, malignant transformation, invasion, and metastasis of CxC and selective targeting of PI3K/AKT/mTOR signaling pathway might be beneficial in preventing metastatic and recurrent CxC.

**Epidermal growth factor receptor (EGFR)**: EGFR is an important oncogenic receptor tyrosine kinase, which is overexpressed in several solid tumors (Shostak et al., 2014; Hu et al., 2015; Khalil et al., 2003). Alteration in EGFR expression has been shown to play an essential role in apoptosis inhibition, angiogenesis and metastasis (Shostak et al., 2014; Hu et al., 2015; Khalil et al., 2003). Hu et al. (2015) recently found a close association between the expression of EGFR and the autophagy-related markers Beclin1 and microtubule-associated protein 1 light chain 3 alpha (LC3) in cervical SCC and high-grade CIN (Hu et al., 2015). The study revealed that Beclin1-or LC3-negativity with EGFR-positivity was associated with higher FIGO stage ($P = .011$ and $P = .013$, respectively) and pelvic lymph node metastasis ($P = .036$ and $P = .092$, respectively), thus suggesting that autophagy inhibition combined with upregulation of EGFR promotes cervical SCC progression (Hu et al., 2015).

Ha et al. (2013), have shown that activated EGF/EGFR induces CxC invasion and metastasis through third member of transforming acidic coiled-coil protein family (TACC3), which is found to be deregulated in a variety of human cancers including CxC, induces EMT by activation of PI3K/AKT and extracellular signal-regulated protein kinases (ERKs) signal transduction pathways. The study also found that Snail1, a key transcription factor in EGF-mediated EMT, correlates with the expression of TACC3, in CxC progression (Ha et al., 2013; Carneiro et al., 2015).

Shostak et al. (2014) reported EGF-mediated EMT is promoted by cell migration-inducing and hyaluronan-binding protein (CEMIP), formerly known as KIAA1199, an oncogenic protein induced by HPV infection which is aberrantly expressed in cervical (pre)neoplastic lesions. Furthermore, EGF-dependent Src, mitogen-activated protein/ERK kinase 1, and ERK1/2 phosphorylation is impaired by KIAA1199 deficiency. Thus KIAA1199 promotes CxC survival and invasiveness through EGFR signaling (Shostak et al., 2014). Expression of EGFR also correlates with CXC chemokine receptor 7 (CXCR7) and CXCR4, which are associated with tumor growth and metastases in CxC (Schrevel et al., 2012).

The above evidences suggest the important role of EGFR in CxC development and metastasis and may be potential target for therapeutic intervention.

**Platelet-derived growth factor receptors (PDGFR)**: PDGFR, the cell surface tyrosine kinase receptors for members of the PDGF family, play important role in cell proliferation, development, and progression of CxC. Cheng et al., recently identified key genes (PDGFRA, caveolin 1, and gap junction protein, alpha 1), which are involved in CxC invasion and metastasis (Cheng et al., 2015). PDGFRA expression is observed in the stromal component, fibroblasts, endothelial cells, as well as in the cytoplasm of adenosquamous carcinoma, an aggressive subtype of CxC. Okino et al. (2005) further reported that PDGFR interacts with growth factor receptor-bound protein

10 gene, the expression of which is increased in CxC cell, thus providing additional evidence for its role in CxC development. These genes may be used for early diagnosis and prognosis assessment of CxC (Cheng et al., 2015; Candelaria et al., 2009; Longatto-Filho et al., 2009; Jain et al., 2008; Taja-Chayeb et al., 2006).

**Vascular endothelial growth factor**: VEGF is an important protein that helps to create blood and lymph vessels (Biedka et al., 2012; Botting et al., 2010; Zhang et al., 2010). There are seven isoforms in the VEGF family involved in various functions of vasculogenic, angiogenic, or lymphangiogenic pathways and have been shown to play important role in CxC progression (Chen et al., 2014a,b; Mathur et al., 2005; Landt et al., 2011; Kim et al., 2011).

Chen et al. (2014a,b) showed VEGF overexpression positively correlated with Fibulin-4 expression, a member of the fibulin family of extracellular glycoproteins, and is significantly associated with advanced stage, poor differentiation, lymph node metastasis, and poor prognosis in patients with CxC (Chen et al., 2014). Yu et al. (2015) showed that there is upregulation of VEGF expression by astrocyte elevated gene 1 (AEG-1) through activation of nuclear factor kappa B (NF-κB) signal pathway in CxC, which promotes tumor growth and angiogenesis (Yu et al., 2015). Long et al. (2013) also showed significantly decreased expression of angiogenesis related genes including VEGF by knockdown of AEG-1 expression by using specific small hairpin ribonucleic acid, further suggesting that overexpression of VEGF plays a significant role in the aggressiveness of CxC (Long et al., 2013).

VEGF expression has also been shown to positively correlate with CHI3L1, a glycoprotein, which is associated with cancer metastasis, poor prognosis and short survival time in CxC patients. Study also showed that CHI3L1 positively correlates with microvessel density promoting angiogenesis and contributing to the development and progression of CxC (Ngernyuang et al., 2014).

Liu et al. (2014a,b,c,d) showed interaction of VEGF receptors with galactin-3 protein (Gal-3), a glycan binding protein in CxC cell line. Study showed that VEGF C enhances CxC cell invasiveness via upregulation of Gal-3 through NF-κB, which is the key intermediate. Moreover, silencing of Gal-3 expression with specific small interfering ribonucleic acid (siRNA) largely impaired VEGF-C-enhanced cell invasion. In conclusion, these findings further confirm the role of VEGF in CxC metastasis and its potential usefulness as biologic marker of CxC (Liu et al., 2014b).

Liu et al. (2014a,b,c,d) further showed the association of VEGF-C/D with semaphorin 4D (Sema4D), a potent inducer of angiogenesis, and its overexpression is associated with tumor progression and poor clinical outcome in CxC patients. Study suggested that increased Sema4D expression positively correlated with VEGF-C/D, presence of lymphatic invasion, occurrence of lymph node metastasis, and FIGO stage. Furthermore the study also showed that Sema4D induced activation of GTPase Ras homolog gene family, member A (RhoA), mitogen-activated protein kinase (MAPK), and AKT suggesting that Sema4D mediated VEGF-C/D overexpression might contribute to enhance invasiveness and tumor progression of CxC (Liu et al., 2014a,c,d).

Chen et al. (2013a,b) showed in in vitro study that VEGF-C mediated enhanced focal adhesion kinase (FAK) phosphorylation and increased formation of focal adhesion complexes further promoting migration and invasion of CxC (Chen et al., 2013a). In addition, VEGF-C resulted in c-Src phosphorylation by increasing interaction of its receptor Flt-4 with nonreceptor tyrosine kinase c-Src. This study thus suggested that VEGF-C promotes CxC metastasis by activation of FAK protein through Flt-4/c-Src pathway (Zhang et al., 2010; Chen et al., 2013).

Ramachandran et al. (2012) showed that VEGF is one of the Wnt/β-catenin transcriptional target genes, which leads to multiple critical events such as cell proliferation, angiogenesis, and invasion in CxC. Furthermore downregulation of these genes by Wnt inhibitory factor 1 (WIF1), induces apoptosis and inhibits CxC growth, invasion and angiogenesis in vivo. Taken together, these data suggest that WNT1 and VEGF upregulation contribute to CxC progression and metastasis (Ramachandran et al., 2012).

Chen et al. (2011) showed that the production of VEGF in CxC cells is regulated by stromal-interaction molecule 1 (STIM1), which leads to CxC growth, angiogenesis and migration. STIM1 is an endoplasmic reticulum Ca(2+) sensor that triggers store operated Ca (2+) entry (SOCE) activation. Furthermore interference with STIM1 expression or blockade of SOCE activity inhibits tumor angiogenesis and growth in animal models, confirming the crucial role of STIM1-mediated Ca(2+) influx in CxC metastasis (Chen et al., 2011).

Studies have shown correlation of VEGF-C with cyclooxygenase 2 (COX-2) expression, lymphangiogenesis, and lymph node metastasis in human CxC (Liu et al., 2011; Shi et al., 2007, 2008). Zeng et al. (2014) showed close correlation between VEGF-C, COX-2, and heparanase, which may play important role in carcinogenesis, tumor growth, and metastasis via multiple mechanisms including inactivation of host antitumor immunity and promotion of tumor cell migration, invasiveness, and tumor-associated angiogenesis and lymphangiogenesis (Zeng et al., 2014). Yoysungnoen-Chintana et al. (2014) also supported that expression of VEGF and COX-2 were upregulated in CaSki, a cell line derived from a human CxC, and significantly attenuated tumor growth and angiogenesis in CxC line-implanted mice, which were mediated by the downregulation of VEGF and COX-2 after high dose treatment of curcumin (Yoysungnoen-Chintana et al., 2014). Liu et al. (2011) also showed there was strong association between the levels of COX-2 expression and the levels of VEGF-C expression and secretion. Furthermore, VEGF-C mRNA expression was inhibited by siRNA-mediated knockdown of COX-2 expression. Collectively, these data provide further evidence for association between COX-2 and VEGF-C expressions in CxC (Liu et al., 2011).

Lysosome-associated protein transmembrane 4 beta (LAPTM4B) is a novel oncogene with important functions in aggressive human carcinomas, including CxC. Studies suggest that LAPTM4B regulates expression of several important target genes such as CDK12, hypoxia-inducible factor 1-alpha (HIF-1alpha), MMP-2, MMP-9 and VEGF, which play significant role in the CxC growth, angiogenesis and metastasis. Meng et al. (2015a,b) showed that HeLa cells with down regulated LAPTM4B exhibited significantly reduced VEGF, CDK12, HIF-1α, MMP-2, and MMP-9 expression thus inhibiting cell growth and angiogenesis (Meng et al., 2015a,b). Recently, Meng et al. (2015a,b) also showed that coexpression of VEGF and lysosome-associated protein transmembrane 4β-35 (LAPTM4B-35) was significantly associated with FIGO stage ($P=.014$), histologic grade ($P=.033$), lymph node metastasis ($P=.045$), and recurrence ($P=.010$) in CxC patients. Also coexpression of LAPTM4B-35 and VEGF was associated with poor overall survival and disease-free survival ($P=.015$ and $P=.016$, respectively) (Meng et al., 2015a,b).

Taken together, all these studies suggest that VEGF promotes CxC progression by sustaining hyperactivity of multiple oncogenic signaling pathways, including NF-kB, CHI3L1, Gal-3, Sema4D, MAPK, Wnt/beta-catenin, STIM1, COX-2, and LAPTM4B. These studies also provide a basis for exploring VEGF as a diagnostic and therapeutic target in CxC.

**MicroRNAs (miRNAs)**: miRNA-mediated regulation of posttranscriptional gene expression has been recently identified in the development and metastasis of many cancer types including CxC (Banno et al., 2014; Wan et al., 2014). miRNAs are small RNA molecules of 20–23 nucleotides that do not encode a protein. They are endogenous noncoding short RNAs that inhibit gene expression by binding to target mRNAs at their 3′-untranslated region (Griffiths-Jones et al., 2006). miRNAs involved in carcinogenesis are classified into oncogenic and tumor suppressor miRNAs. They are increased or decreased in patients with CxC and are implicated in posttranscriptional regulation of gene expression and in the pathogenesis of CxC (Zeng and Li, 2014; Chu et al., 2014; Reshmi et al., 2011; Granados López and López, 2014; Servín-González et al., 2015).

Overexpression of several miRNAs are reported in CxC such as miR-9, miR-10a, miR-15a, miR-15b, miR-16, miR17-5p, miR-19a/b, miR-20a, miR-20b, miR-21, miR-31, miR-92a, miR-93, miR-106a, miR-133b, miR-146a, miR-146b-5p, miR-155, miR-183, miR-185, miR-193b, miR-196a, miR-199a, miR-200a, miR-205, miR-224, miR-590 and inhibition of these miRNA activity in vivo may be a therapeutic strategy for treatment of CxC (Chu et al., 2014; Granados López and López, 2014; Servín-González et al., 2015).

Downregulation of various microRNAs such as miR-1, miR-10b, miR-23b, miR-26a, miR-27b, miR-99a, miR-99b, miR-100, miR-125b, miR-126, miR-143, miR-145, miR-195, miR-196b, miR-199a-5p, miR-203, miR-214, miR-218, miR-372, miR-375, miR-424, and miR-497 is associated with advanced stage, progression, metastasis, and poor prognosis in human CxC (Granados López and López, 2014; Servín-González et al., 2015).

In vivo and in vitro studies by Song et al. (2015) showed that EGFR signaling is downregulated by miR-133a, which is tumor-suppressor, causing CxC growth inhibition. Aberrant expression of miR-133a correlated with lymph node metastasis, histological grade and FIGO stage of CxC suggesting that restoration of miR-133a inhibited EGFR expression can be a potential therapeutic target for CxC (Song et al., 2015).

Studies have also indicated that miR-125a is a novel antioncogene with low expression in several cancers (Cowden Dahl et al., 2009; Li et al., 2009). miR-125a inhibits CxC cell proliferation and invasion, and suppresses cell cycle. miR-125a overexpression reduces cervical tumor growth and metastasis in vivo by targeting signal transducer and activator of transcription 3 (STAT3) and inhibiting its expression and downstream genes. STAT3 upregulation has pro-tumorigenic functions that enhance tumor cell growth, invasion and metastasis in a variety of cancers including CxC (Li et al., 2009; Calin and Croce, 2006; Chen et al., 2013). Upon activation of the Janus kinase 2-STAT3 pathway, STAT3 promotes cyclin B1 and cdc2 expression, and inhibits p21 and p53 expression (Farha et al., 2014; Feng et al., 2014). Thus a high expression level of miR-125a correlates with less aggressive behavior of the tumor. miR-125a also downregulates expression of the oncogene c-myc through inhibition of STAT3, which might be another molecular mechanism through which miR-125a suppresses CxC cell growth. miR-125a overexpression modulates the invasion and metastasis of CxC cells by decreasing the expression of N-cadherin, MMP-2, MMP-9 and increased expression of E-cadherin by directly targeting STAT3 that acts as a transcription factor for these genes (Fan et al., 2015, 2016; Qin et al., 2015), suggesting that miR-125a regulates CxC metastasis by inhibiting STAT3-mediated EMT and MMPs expression. Activation of miR-125a or inhibition of STAT3 may therefore be useful strategies for CxC treatment. HPV infection also might regulate miR-125a expression by increasing DNA methylation of its promoter or interacting with transcription factors that regulate miR-125a expression. HPV18 E6/E7 and HPV16 E6/E7 have been shown to inhibit miR-125a expression and increase p53 expression, which is a transcription factor, bound to the promoter of miR-125a, and stimulates its transcription. Thus the reason of low expression of miR-125a in CxC is likely due to the suppression of p53 caused by HPV infection (Fan et al., 2015, 2016; Qin et al., 2015).

All these studies demonstrate that aberrant expression of miRNAs is associated with progression, metastasis, and poor prognosis of CxC and may provide novel therapeutic target for CxC.

# 7. CONCLUSION

In summary, various signaling pathways and molecular factors interact with each other and form a complex network for regulation of cellular functions. Disruption of this network leads to the development, progression, and metastasis of cervical cancer. Manipulating these molecular factors and signaling pathways including HOX gene, PI3K/AKT/mTOR, EGFR, PDGFR, VEGF pathways and miRNAs by targeting their expression levels and activity has potential for effective therapeutic strategies for CxC and its metastasis.

## References

Alami, Y., Castronovo, V., Belotti, D., Flagiello, D., Clausse, N., 1999. HOXC5 and HOXC8 expression are selectively turned on in human cervical cancer cells compared to normal keratinocytes. Biochem. Biophys. Res. Commun. 257, 738–745.

Albores-Saavedra, J., Gersell, D., Gilks, C.B., Henson, D.E., Lindberg, G., Santiago, H., et al., 1997. Terminology of endocrine tumors of the uterine cervix: results of a workshop sponsored by the College of American Pathologists and the National Cancer Institute. Arch. Pathol. Lab. Med. 121, 34–39.

American Cancer Society, 2015. Cancer Facts & Figures. American Cancer Society, Atlanta.

Andikyan, V., Taylor, H.S., 2009. WT1 represses HOX gene expression in the regulation of gynaecologic tumour histologic type. J. Cell. Mol. Med. 13, 4522–4531.

Aziz, M.H., Nihal, M., Fu, V.X., Jarrard, D.F., Ahmad, N., 2006. Resveratrol-caused apoptosis of human prostate carcinoma LNCaP cells is mediated via modulation of phosphatidylinositol 3′-kinase/Akt pathway and Bcl-2 family proteins. Mol. Cancer Ther. 5, 1335–1341.

Aziz, S.W., Aziz, M.H., 2013. Major Signaling Pathways Involved in Breast Cancer. Breast Cancer Metastasis and Drug Resistance. Springer Science Business Media, New York, pp. 47–63. http://dx.doi.org/10.1007/978-1-4614-5647-6_4.

Bader, A.A., Winter, R., Haas, J., Tamussino, K.F., 2007. Where to look for the sentinel lymph node in cervical cancer. Am. J. Obstet. Gynecol. 197, 678.e1–678.e7.

Bai, X., Ma, Y., Zhang, G., 2015. Butein suppresses cervical cancer growth through the PI3K/AKT/mTOR pathway. Oncol. Rep. 33, 3085–3092.

Balsat, C., Blacher, S., Signolle, N., et al., 2011. Whole slide quantification of stromal lymphatic vessel distribution and peritumoral lymphatic vessel density in early invasive cervical cancer: a method description. ISRN Obstet. Gynecol. 2011, 354861.

Balsat, C., Signolle, N., Goffin, F., Delbecque, K., Plancoulaine, B., Sauthier, P., et al., 2014. Improved computer-assisted analysis of the global lymphatic network in human cervical tissues. Mod. Pathol. 27, 887–898.

Banno, K., Iida, M., Yanokura, M., Kisu, I., Iwata, T., Tominaga, E., et al., 2014. MicroRNA in cervical cancer: oncomiRs and tumor suppressor miRs in diagnosis and treatment. Sci. World J. 2014, 178075.

Barba-de la Rosa, A.P., Briones-Cerecero, E., Lugo-Melchor, O., De León-Rodríguez, A., Santos, L., et al., 2012. Hox B4 as potential marker of non-differentiated cells in human cervical cancer cells. J. Cancer Res. Clin. Oncol. 138, 293–300.

Behtash, N., Mousavi, A., Tehranian, A., Khanafshar, N., Hanjani, P., 2003. Embryonal rhabdomyosarcoma of the uterine cervix: case report and review of the literature. Gynecol. Oncol. 91, 452–455.

Bertelsen, B.I., Steine, S.J., Sandvei, R., Molven, A., Laerum, O.D., 2006. Molecular analysis of the PI3K-AKT pathway in uterine cervical neoplasia: frequent PIK3CA amplification and AKT phosphorylation. Int. J. Cancer 118, 1877–1883.

Biedka, M., Makarewicz, R., Marszałek, A., Sir, J., Kardymowicz, H., Goralewska, A., 2012. Labeling of microvessel density, lymphatic vessel density and potential role of proangiogenic and lymphangiogenic factors as a predictive/prognostic factors after radiotherapy in patients with cervical cancer. Eur. J. Gynaecol. Oncol. 33, 399–405.

Botting, S.K., Fouad, H., Elwell, K., Rampy, B.A., Salama, S.A., Freeman, D.H., Diaz-Arrastia, C.R., 2010. Prognostic significance of peritumoral lymphatic vessel density and vascular endothelial growth factor receptor 3 in invasive squamous cell cervical cancer. Transl. Oncol. 3, 170–175.

Brinton, L.A., Hammen, R.F., Huggins, G.R., Lehman, H.F., Levin, R.S., Mallin, K., et al., 1987. Sexual and reproductive risk factors for invasive squamous cell cancer. J. Natl. Cancer Inst. 79, 23–30.

Burghardt, E., Ostor, A.G., 1983. Site and origin of squamous cervical cancer: a histomorphologic study. Obstet. Gynecol. 62, 117–127.

Cai, L., Yang, S., Ding, H., Cai, J., Wang, Z., 2013. Tumor-associated lymphatic endothelial cell promotes invasion of cervical cancer cells. APMIS 121, 1162–1168.

Calin, G.A., Croce, C.M., 2006. MicroRNA signatures in human cancers. Nat. Rev. Cancer 6, 857–866.

Candelaria, M., Arias-Bonfill, D., Chávez-Blanco, A., Chanona, J., Cantú, D., Pérez, C., Dueñas-González, A., 2009. Lack in efficacy for imatinib mesylate as second-line treatment of recurrent or metastatic cervical cancer expressing platelet-derived growth factor receptor alpha. Int. J. Gynecol. Cancer 19, 1632–1637.

Carneiro, B.A., Elvin, J.A., Kamath, S.D., Ali, S.M., Paintal, A.S., Restrepo, A., Berry, E., Giles, F.J., Johnson, M.L., 2015. FGFR3-TACC3: a novel gene fusion in cervical cancer. Gynecol. Oncol. Rep. 13, 53–56.

Centers for Disease Control and Prevention, Workowski, K.A., Berman, S.M., 2006. Sexually transmitted diseases treatment guidelines. MMWR Recomm. Rep. 55, 1–94.

Cervical Cancer, 2015. Estimated Incidence, Mortality and Prevalence Worldwide in 2012. http://globocan.iarc.fr/Pages/fact_sheets_cancer.aspx.

Chen, Y.F., Chiu, W.T., Chen, Y.T., Lin, P.Y., Huang, H.J., Chou, C.Y., et al., 2011. Calcium store sensor stromal-interaction molecule 1-dependent signaling plays an important role in cervical cancer growth, migration, and angiogenesis. Proc. Natl. Acad. Sci. U.S.A 108, 15225–15230.

Chen, H., Suo, K., Cheng, Y., Zheng, B., Xu, L., 2013a. Vascular endothelial growth factor C enhances cervical cancer migration and invasion via activation of focal adhesion kinase. Gynecol. Endocrinol. 29, 20–24.

Chen, Y., Wang, J., Wang, X., Liu, X., Li, H., Lv, Q., Zhu, J., Wei, B., Tang, Y., 2013b. STAT3, a poor survival predicator, is associated with lymph node metastasis from breast cancer. J. Breast Cancer 16, 40–49.

Chen, B., Zhang, C., Dong, P., Guo, Y., Mu, N., 2014. Molecular regulation of cervical cancer growth and invasion by VEGFa. Tumour Biol. 35, 11587–11593.

Chen, J., Zhang, J., Liu, X., Fang, R., Zhao, Y., Ma, D., 2014. Overexpression of fibulin-4 is associated with tumor progression and poor prognosis in patients with cervical carcinoma. Oncol. Rep. 31, 2601–2610.

Cheng, Y., Ma, D., Zhang, Y., Li, Z., Geng, L., 2015. Cervical squamous cancer mRNA profiles reveal the key genes of metastasis and invasion. Eur. J. Gynaecol. Oncol. 36, 309–317.

Chin-Hong, P.V., Vittinghoff, E., Cranston, R.D., Buchbinder, S., Cohen, D., Colfax, G., et al., 2004. Age-Specific prevalence of anal human papillomavirus infection in HIV-negative sexually active men who have sex with men: the EXPLORE study. J. Infect. Dis. 190, 2070–2076.

Chu, Y., Ouyang, Y., Wang, F., Zheng, A., Bai, L., Han, L., Chen, Y., Wang, H., 2014. MicroRNA-590 promotes cervical cancer cell growth and invasion by targeting CHL1. J. Cell. Biochem. 115, 847–853.

Cimpean, A.M., Mazuru, V., Cernii, A., et al., 2011. Detection of early lymphangiogenesis by lymphatic microvascular density and endothelial proliferation status in preneoplastic and neoplastic lesions of the uterine cervix. Pathol. Int. 61, 395–400.

Couto, E., Hemminki, K., 2006. Heritable and environmental components in cervical tumors. Int. J. Cancer 119, 2699–2701.

Cowden Dahl, K.D., Dahl, R., Kruichak, J.N., Hudson, L.G., 2009. The epidermal growth factor receptor responsive miR-125a represses mesenchymal morphology in ovarian cancer cells. Neoplasia 11, 1208–1215.

DeEulis, T.G., Markoe, A.M., Sedlacek, T.V., Brady, L.W., 1987. The influence of hematogenous and serosal metastases on advanced (FIGO IIA-IVB) squamous cell carcinoma of the cervix. Am. J. Clin. Oncol. 10, 10–20.

Dimitrakopoulos, I., Ntomouchtsis, A., Iordanidis, F., 2011. Infratemporal fossa metastasis from carcinoma of the uterine cervix. Oral Maxillofac. Surg. 15, 121–125.

Dmoch-Gajzlerska, E., Kozakiewicz, B., Chądzyńska, M., 2014. Women's knowledge regarding the effects of cigarette smoking and human papillomavirus infection on the development of cervical cancer. Clin. Oncol. (R. Coll. Radiol.) 26, 358.

Dong, Y., Kuang, Q., Dai, X., Li, R., Wu, Y., Leng, W., Li, Y., Li, M., 2015. Improving the understanding of pathogenesis of human papillomavirus 16 via mapping protein-protein interaction network. Biomed. Res. Int. 2015, 890381.

Du, J., Wang, L., Li, C., Yang, H., Li, Y., Hu, H., Li, H., Zhang, Z., 2015. MicroRNA-221 targets PTEN to reduce the sensitivity of cervical cancer cells to gefitinib through the PI3K/Akt signaling pathway. Tumour. Biol. PubMed PMID:26482612.

El-Mofty, S.K., 2012. HPV-related squamous cell carcinoma variants in the head and neck. Head Neck Pathol. 6, S55–S62.

Eskander, R.N., Tewari, K.S., 2014. Exploiting the therapeutic potential of the PI3K-AKT-mTOR pathway in enriched populations of gynecologic malignancies. Expert. Rev. Clin. Pharmacol. 7, 847–858.

Fan, Z., Cui, H., Xu, X., Lin, Z., Zhang, X., Kang, L., et al., 2015. MiR-125a suppresses tumor growth, invasion and metastasis in cervical cancer by targeting STAT3. Oncotarget 6, 25266–25280.

Fan, Z., Cui, H., Yu, H., Ji, Q., Kang, L., Han, B., et al., 2016. MiR-125a promotes paclitaxel sensitivity in cervical cancer through altering STAT3 expression. Oncogenesis 5, e197.

Farha, A.K., Dhanya, S.R., Mangalam, S.N., Geetha, B.S., Latha, P.G., Remani, P., 2014. Deoxyelephantopin impairs growth of cervical carcinoma SiHa cells and induces apoptosis by targeting multiple molecular signaling pathways. Cell. Biol. Toxicol. 30, 331–343.

Feng, M., Wang, Y., Chen, K., Bian, Z., Jinfang, W., Gao, Q., 2014. IL-17A promotes the migration and invasiveness of cervical cancer cells by coordinately activating MMPs expression via the p38/NF-kappaB signal pathway. PLoS One 9, e108502.

Fletcher, F.E., Vidrine, D.J., Tami-Maury, I., Danysh, H.E., King, R.M., Buchberg, M., et al., 2014. Cervical cancer screening adherence among HIV-positive female smokers from a comprehensive HIV clinic. AIDS Behav. 18, 544–554.

Franceschi, S., 2005. The IARC commitment to cancer prevention: the example of papillomavirus and cervical cancer. Recent Results Cancer Res. 166, 277–297.

Gallup, D.G., 2008. The spread and staging of cervical cancer. Glob. Libr. Women's Med.. ISSN: 1756-2228. http://dx.doi.org/10.3843/GLOWM.10231.

Gao, P., Zhou, G.Y., Yin, G., et al., 2006. Lymphatic vessel density as a prognostic indicator for patients with stage I cervical carcinoma. Hum. Pathol. 37, 719–725.

Garner, E.I., 2003. Cervical cancer: disparities in screening, treatment, and survival. Cancer Epidemiol. Biomark. Prev. 12, 242s–247s.
Giorgi Rossi, P., 2015. Recommendation without experts? Epistemological implications in the development of screening guidelines. Prev. Med. 83, 22–25.
Gombos, Z., Xu, X., Chu, C.S., et al., 2005. Peritumoral lymphatic vessel density and vascular endothelial growth factor C expression in early-stage squamous cell carcinoma of the uterine cervix. Clin. Cancer Res. 11, 8364–8371.
Gonzalez-Herrera, A., Salgado-Bernabe, M., Velazquez-Velazquez, C., Salcedo-Vargas, M., Andrade-Manzano, A., et al., 2015. Increased expression of HOXB2 and HOXB13 proteins is associated with HPV infection and cervical cancer progression. Asian Pac. J. Cancer Prev. 16, 1349–1353.
Granados López, A.J., López, J.A., 2014. Multistep model of cervical cancer: participation of miRNAs and coding genes. Int. J. Mol. Sci. 15, 15700–15733.
Griffiths-Jones, S., Grocock, R.J., van Dongen, S., Bateman, A., Enright, A.J., 2006. miRBase: microRNA sequences, targets and gene nomenclature. Nucleic Acids Res. 34, D140–D144.
Grisaru, D., Covens, A., Chapman, B., Shaw, P., Colgan, T., Murphy, J., et al., 2001. Does histology influence prognosis in patients with early-stage cervical carcinoma? Cancer 92, 2999–3004.
Ha, G.H., Kim, J.L., Breuer, E.K., 2013. TACC3 is essential for EGF-mediated EMT in cervical cancer. PLoS One 8, e70353.
Herfs, M., Yamamoto, Y., Laury, A., et al., 2012. A discrete population of squamocolumnar junction cells implicated in the pathogenesis of cervical cancer. Proc. Natl. Acad. Sci. U.S.A. 109, 10516–10521.
Hou, M.M., Liu, X., Wheler, J., Naing, A., Hong, D., Coleman, R.L., et al., 2014. Targeted PI3K/AKT/mTOR therapy for metastatic carcinomas of the cervix: a phase I clinical experience. Oncotarget 5, 11168–11179.
Hu, Y.F., Lei, X., Zhang, H.Y., Ma, J.W., Yang, W.W., Chen, M.L., Cui, J., Zhao, H., 2015. Expressions and clinical significance of autophagy-related markers Beclin1, LC3, and EGFR in human cervical squamous cell carcinoma. Onco Targets Ther. 8, 2243–2249.
Hung, Y.C., Ueda, M., Terai, Y., Kumagai, K., Ueki, K., Kanda, K., et al., 2003. Homeobox gene expression and mutation in cervical carcinoma cells. Cancer Sci. 94, 437–441.
International Collaboration of Epidemiological Studies of Cervical Cancer, 2007. Comparison of risk factors for invasive squamous cell carcinoma and adenocarcinoma of the cervix: collaborative reanalysis of individual data on 8,097 women with squamous cell carcinoma and 1,374 women with adenocarcinoma from 12 epidemiological studies. Int. J. Cancer 120, 885–891.
International Collaboration of Epidemiological Studies of Cervical Cancer, Appleby, P., Beral, V., et al., 2007. Cervical cancer and hormonal contraceptives: collaborative reanalysis of individual data for 16,573 women with cervical cancer and 35,509 women without cervical cancer from 24 epidemiological studies. Lancet 370, 1609–1621.
Jach, R., Dulinska-Litewka, J., Laidler, P., et al., 2010. Expression of VEGF, VEGF-C and VEGFR-2 in in situ and invasive SCC of cervix. Front. Biosci. (Elite Ed.) 2, 411–423.
Jain, R.K., Lahdenranta, J., Fukumura, D., 2008. Targeting PDGF signaling in carcinoma-associated fibroblasts controls cervical cancer in mouse model. PLoS Med. 5, e24.
Ji, R.C., 2006. Lymphatic endothelial cells, tumor lymphangiogenesis and metastasis: new insights into intratumoral and peritumoral lymphatics. Cancer Metastasis Rev. 25, 677–694.
Jing, L., Yuan, W., Ruofan, D., Jinjin, Y., Haifeng, Q., 2015. HOTAIR enhanced aggressive biological behaviors and induced radio-resistance via inhibiting p21 in cervical cancer. Tumour Biol. 36, 3611–3619.
Kang, S., Dong, S.M., Kim, B.R., Park, M.S., Trink, B., Byun, H.J., Rho, S.B., 2012. Thioridazine induces apoptosis by targeting the PI3K/Akt/mTOR pathway in cervical andendometrial cancer cells. Apoptosis 17, 989–997.
Keysar, S.B., Astling, D.P., Anderson, R.T., Vogler, B.W., Bowles, D.W., Morton, J.J., et al., 2013. A patient tumor transplant model of squamous cell cancer identifies PI3K inhibitors as candidate therapeutics in defined molecular bins. Mol. Oncol. 7, 776–790.
Khalil, M.Y., Grandis, J.R., Shin, D.M., 2003. Targeting epidermal growth factor receptor: novel therapeutics in the management of cancer. Expert. Rev. Anticancer. Ther. 3, 367–380.
Kim, N.S., Kang, Y.J., Jo, J.O., Kim, H.Y., Oh, Y.R., et al., 2011. Elevated expression of thymosin β4, vascular endothelial growth factor (VEGF), and hypoxia inducible factor (HIF)-1α in early-stage cervical cancers. Pathol. Oncol. Res. 17, 493–502.
Kim, H.J., Lee, D.W., Yim, G.W., Nam, E.J., Kim, S., Kim, S.W., Kim, Y.T., 2015. Long non-coding RNA HOTAIR is associated with human cervical cancer progression. Int. J. Oncol. 46, 521–530.
Landt, S., Wehling, M., Heidecke, H., Jeschke, S., Korlach, S., Stöblen, F., et al., 2011. Prognostic significance of angiogenic factors in uterine cervical cancer. Anticancer. Res. 31, 2589–2595.
Lea, J.S., Coleman, R.L., Garner, E.O., Duska, L.R., Miller, D.S., Schorge, J.O., 2003. Adenosquamous histology predicts poor outcome in low-risk stage IB1 cervical adenocarcinoma. Gynecol. Oncol. 91, 558–562.
Leitao Jr., M.M., White, P., Cracchiolo, B., 2008. Cervical cancer in patients infected with the human immunodeficiency virus. Cancer 112, 2683–2689.

Levenback, C., Coleman, R.L., Burke, T.W., Lin, W.M., Erdman, W., Deavers, M., Delpassand, E.S., 2002. Lymphatic mapping and sentinel node identification in patients with cervix cancer undergoing radical hysterectomy and pelvic lymphadenectomy. J. Clin. Oncol. 20, 688–693.
Li, W., Duan, R., Kooy, F., Sherman, S.L., Zhou, W., Jin, P., 2009. Germline mutation of microRNA-125a is associated with breast cancer. J. Med. Genet. 46, 358–360.
Li, J., Wang, Y., Yu, J., Dong, R., Qiu, H., 2015a. A high level of circulating HOTAIR is associated with progression and poor prognosis of cervical cancer. Tumour Biol. 36, 1661–1665.
Li, J., Wen, W., Zhao, S., Wang, J., Chen, J., Wang, Y., Zhang, Q., 2015b. Prognostic role of HOTAIR in four estrogen-dependent malignant tumors: a meta-analysis. Onco Targets Ther. 8, 1471–1482.
Liu, H., Xiao, J., Yang, Y., et al., 2011. COX-2 expression is correlated with VEGF-C, lymphangiogenesis and lymph node metastasis in human cervical cancer. Microvasc. Res. 82, 131–140.
Liu, D., Li, L., Zhang, X.X., Wan, D.Y., Xi, B.X., Hu, Z., et al., 2014a. SIX1 promotes tumor lymphangiogenesis by coordinating TGFβ signals that increase expression of VEGF-C. Cancer Res. 74, 5597–5607.
Liu, H., Yang, Y., Xiao, J., Yang, S., Liu, Y., Kang, W., Li, X., Zhang, F., 2014b. Semaphorin 4D expression is associated with a poor clinical outcome in cervical cancer patients. Microvasc. Res. 93, 1–8.
Liu, J., Cheng, Y., He, M., Yao, S., 2014c. Vascular endothelial growth factor C enhances cervical cancer cell invasiveness via upregulation of galectin-3 protein. Gynecol. Endocrinol. 30, 461–465.
Liu, Y.Q., Li, H.F., Han, J.J., Tang, Q.L., Sun, Q., Huang, Z.Q., Li, H.G., 2014d. CD44v3 and VEGF-C expression and its relationship with lymph node metastasis in squamous cell carcinomas of the uterine cervix. Asian Pac. J. Cancer Prev. 15, 5049–5053.
Liu, Z.C., Liu, W.D., Liu, Y.H., Ye, X.H., Chen, S.D., 2015. Multiple sexual partners as a potential independent risk factor for cervical cancer: a meta-analysis of epidemiological studies. Asian Pac. J. Cancer Prev. 16, 3893–3900.
Long, M., Dong, K., Gao, P., Wang, X., et al., 2013. Overexpression of astrocyte-elevated gene-1 is associated with cervical carcinoma progression and angiogenesis. Oncol. Rep. 30, 1414–1422.
Longatto-Filho, A., Pinheiro, C., Martinho, O., Moreira, M.A., Ribeiro, L.F., Queiroz, G.S., et al., 2009. Molecular characterization of EGFR, PDGFRA and VEGFR2 in cervical adenosquamous carcinoma. BMC Cancer 9, 212.
Look, K.Y., Brunetto, V.L., Clarke-Pearson, D.L., Averette, H.E., Major, F.J., Alvarez, R.D., et al., 1996. An analysis of cell type in patients with surgically staged stage IB carcinoma of the cervix: a Gynecologic Oncology Group study. Gynecol. Oncol. 63, 304–311.
López, R., Garrido, E., Piña, P., Hidalgo, A., Lazos, M., Ochoa, R., Salcedo, M., 2006. HOXB homeobox gene expression in cervical carcinoma. Int. J. Gynecol. Cancer 16, 329–335.
López-Romero, R., Marrero-Rodríguez, D., Romero-Morelos, P., Villegas, V., Valdivia, A., Arreola, H., et al., 2015. The role of developmental HOX genes in cervical cancer. Rev. Med. Inst. Mex Seguro Soc. 53, S188–S193.
Lu, D., Qian, J., Li, W., Feng, Q., Pan, S., Zhang, S., 2015. β-hydroxyisovaleryl-shikonin induces human cervical cancer cell apoptosis via PI3K/AKT/mTOR signaling. Oncol. Lett. 10, 3434–3442.
Lui, V.W., Hedberg, M.L., Li, H., Vangara, B.S., Pendleton, K., Zeng, Y., et al., 2013. Frequent mutation of the PI3K pathway in head and neck cancer defines predictive biomarkers. Cancer Discov. 3, 761–769.
Ma, Y.Y., Wei, S.J., Lin, Y.C., Lung, J.C., Chang, T.C., Whang-Peng, J., et al., 2000. PIK3CA as an oncogene in cervical cancer. Oncogene 19, 2739–2744.
Maiman, M., Fruchter, R.G., Clark, M., Arrastia, C.D., Matthews, R., Gates, E.J., 1997. Cervical cancer as an AIDS-defining illness. Obstet. Gynecol. 89, 76–80.
Mathur, S.P., Mathur, R.S., Gray, E.A., Lane, D., Underwood, P.G., Kohler, M., Creasman, W.T., 2005. Serum vascular endothelial growth factor C (VEGF-C) as a specific biomarker for advanced cervical cancer: relationship to insulin-like growth factor II (IGF-II), IGF binding protein 3 (IGF-BP3) and VEGF-A. Gynecol. Oncol. 98, 467–483.
Meng, F., Chen, X., Song, H., Lou, G., 2015a. LAPTM4B down regulation inhibits the proliferation, invasion and angiogenesis of HeLa cells in vitro. Cell. Physiol. Biochem. 37, 890–900.
Meng, F., Tan, S., Liu, T., Song, H., Lou, G., 2015b. Predictive significance of combined LAPTM4B and VEGF expression in patients with cervical cancer. Tumour Biol. PubMed PMID:26526574.
Menges, C.W., Baglia, L.A., Lapoint, R., McCance, D.J., 2006. Human papillomavirus type 16 E7 up-regulates AKT activity through the retinoblastoma protein. Cancer Res. 66, 5555–5559.
Moscicki, A.B., 2008. Hpv vaccines: today and in the future. J. Adolesc. Health 43, S26–S40.
Ngernyuang, N., Francescone, R.A., Jearanaikoon, P., Daduang, J., Supoken, A., Yan, W., et al., 2014. Chitinase 3 like 1 is associated with tumor angiogenesis in cervical cancer. Int. J. Biochem. Cell. Biol. 51, 45–52.
Ogunwale, A.N., Coleman, M.A., Sangi-Haghpeykar, H., Valverde, I., Montealegre, J., Jibaja-Weiss, M., Anderson, M.L., 2015. Assessment of factors impacting cervical cancer screening among low-income women living with HIV-AIDS. AIDS Care 1–4.
Ojiyi, E., Dike, I., Okeudo, C., Ejikem, C., Nzewuihe, A., Agbata, A., 2013. Local risk factors in genital human papilloma virus infection in cervical smears. Ann. Med. Health Sci. Res. 3, 529–535.

Okino, K., Konoshi, H., Doi, D., Takeshita, T., 2005. Up-regulation of growth factor receptor-bound protein 10 in cervical squamous cell carcinoma. Oncol. Rep. 13(6), 1069–1074.

Pilleron, J.P., Durand, J.C., Hamelin, J.P., 1974. Prognostic value of node metastasis in cancer of the uterine cervix. Am. J. Obstet. Gynecol. 119, 458–462.

Pim, D., Massimi, P., Dilworth, S.M., Banks, L., 2005. Activation of the protein kinase B pathway by the HPV-16 E7 oncoprotein occurs through a mechanism involving interaction with PP2A. Oncogene 24, 7830–7838.

Preti, M., Van Seters, M., Sideri, M., Van Beurden, M., 2005. Squamous vulvar intraepithelial neoplasia. Clin. Obstet. Gynecol. 48, 845–861.

Pyeon, S.Y., Park, J.Y., Ulak, R., Seol, H.J., Lee, J.M., 2015. Isolated brain metastasis from uterine cervical cancer: a case report and review of literature. Eur. J. Gynaecol. Oncol. 36, 602–604.

Qin, X., Wan, Y., Wang, S., Xue, M., 2015. MicroRNA-125a-5p modulates human cervical carcinoma proliferation and migration by targeting ABL2. Drug Des. Devel. Ther. 10, 71–79.

Quinn, M., Babb, P., Jones, J., Allen, E., 1999. Effect of screening on incidence of and mortality from cancer of cervix in England: evaluation based on routinely collected statistics. BMJ 318, 904–908.

Ramachandran, I., Thavathiru, E., Ramalingam, S., Natarajan, G., Mills, W.K., Benbrook, D.M., et al., 2012. Wnt inhibitory factor 1 induces apoptosis and inhibits cervical cancer growth, invasion and angiogenesis in vivo. Oncogene 31, 2725–2737.

Ren, Y., Lu, H., Zhao, D., Ou, Y., Yu, K., Gu, J., et al., 2015. LTPB2 acts as a prognostic factor and promotes progression of cervical adenocarcinoma. Am. J. Transl. Res. 7, 1095–1105.

Reshmi, G., Surya, R., Jissa, V.T., Babu, P.S., Preethi, N.R., Santhi, W.S., et al., 2011. C-T variant in a miRNA target site of BCL2 is associated with increased risk of human papilloma virus related cervical cancer–an in silico approach. Genomics 98, 189–193.

Ries, L.A.G., Melbert, D., Krapcho, M., et al., 2007. SEER Cancer Statistics Review, 1975–2004. National Cancer Institute, Bethesda, MD.

Roura, E., Castellsagué, X., Pawlita, M., Travier, N., Waterboer, T., Margall, N., Bosch, F.X., et al., 2014. Smoking as a major risk factor for cervical cancer and pre-cancer: results from the EPIC cohort. Int. J. Cancer 135, 453–466.

Roy, B., Pattanaik, A.K., Das, J., Bhutia, S.K., Behera, B., Singh, P., Maiti, T.K., 2014. Role of PI3K/Akt/mTOR and MEK/ERK pathway in Concanavalin A induced autophagy in HeLa cells. Chem. Biol. Interact. 210, 96–102.

Schäfer, A., Friedmann, W., Mielke, M., Schwartländer, B., Koch, M.A., 1991. The increased frequency of cervical dysplasia-neoplasia in women infected with the human immunodeficiency virus is related to the degree of immunosuppression. Am. J. Obstet. Gynecol. 164, 593–599.

Schiffman, M., Wentzensen, N., Wacholder, S., Kinney, W., Gage, J.C., Castle, P.E., 2011. Human papillomavirus testing in the prevention of cervical cancer. J. Natl. Cancer Inst. 103, 368–383.

Scholes, D., Stergachis, A., Heidrich, F.E., Andrilla, H., Holmes, K.K., Stamm, W.E., 1996. Prevention of pelvic inflammatory disease by screening for cervical chlamydial infection. N. Engl. J. Med. 334, 1362–1366.

Schrevel, M., Karim, R., ter Haar, N.T., van der Burg, S.H., Trimbos, J.B., Fleuren, G.J., et al., 2012. CXCR7 expression is associated with disease-free and disease-specific survival in cervical cancer patients. Br. J. Cancer 106, 1520–1525.

Sevin, B.U., Lu, Y., Bloch, D.A., et al., 1996. Surgically defined prognostic parameters in patients with early cervical carcinoma. A multivariate survival tree analysis. Cancer 78, 1438–1446.

Servín-González, L.S., Granados-López, A.J., López, J.A., 2015. Families of microRNAs expressed in clusters regulate cell signaling in cervical cancer. Int. J. Mol. Sci. 16, 12773–12790.

Shi, X., Chen, G., Xing, H., Weng, D., Bai, X., Ma, D., 2007. VEGF-C, VEGFR-3, and COX-2 enhances growth and metastasis of human cervical carcinoma cell lines in vitro. Oncol. Rep. 18, 241–247.

Shi, X., Xi, L., Weng, D., Chen, G., Song, X., Wu, P., Wang, B., Wei, J., Wang, S., Zhou, J., Ma, D., 2008. Clinicopathological significance of VEGF-C, VEGFR-3 and Cyclooxygenase-2 in early-stage cervical cancer. Int. J. Biomed. Sci. 4, 58–63.

Shostak, K., Zhang, X., Hubert, P., Göktuna, S.I., Jiang, Z., Klevernic, I., et al., 2014. NF-κB-induced KIAA1199 promotes survival through EGFR signaling. Nat. Commun. 5, 5232.

Shu, X.R., Wu, J., Sun, H., Chi, L.Q., Wang, J.H., 2015. PAK4 confers the malignance of cervical cancers and contributes to the cisplatin-resistance in cervical cancer cells via PI3K/AKT pathway. Diagn. Pathol. 10, 177.

Sichanh, C., Quet, F., Chanthavilay, P., Diendere, J., Latthaphasavang, V., Longuet, C., Buisson, Y., 2014. Knowledge, awareness and attitudes about cervical cancer among women attending or not an HIV treatment center in Lao PDR. BMC Cancer 14, 161.

Siegel, R., Ward, E., Brawley, O., Jemal, A., 2011. Cancer statistics, 2011: the impact of eliminating socioeconomic and racial disparities on premature cancer deaths. CA Cancer J. Clin. 61, 212–236.

Siegel, R.L., Miller, K.D., Jemal, A., 2015. Cancer statistics. CA Cancer J. Clin. 2015 (65), 5–29.

Sierra-Torres, C.H., Tyring, S.K., Au, W.W., 2003. Risk contribution of sexual behavior and cigarette smoking to cervical neoplasia. Int. J. Gynecol. Cancer 13, 617–625.

Smith, H.O., Tiffany, M.F., Qualls, C.R., Key, C.R., 2000. The rising incidence of adenocarcinoma relative to squamous cell carcinoma of the uterine cervix in the United States–a 24-year population-based study. Gynecol. Oncol. 78, 97–105.

Society of Gynecologic Oncologists Education Resource Panel Writing group, Collins, Y., Einstein, M.H., Gostout, B.S., Herzog, T.J., Massad, L.S., Rader, J.S., Wright, J., 2006. Cervical cancer prevention in the era of prophylactic vaccines: a preview for gynecologic oncologists. Gynecol. Oncol. 102, 552–562.

Song, X., Shi, B., Huang, K., Zhang, W., 2015. miR-133a inhibits cervical cancer growth by targeting EGFR. Oncol. Rep. 34, 1573–1580.

Soohoo, M., Blas, M., Byraiah, G., Carcamo, C., Brown, B., 2013. Cervical HPV infection in female sex workers: a global perspective. Open AIDS J. 7, 58–66.

Taja-Chayeb, L., Chavez-Blanco, A., Martínez-Tlahuel, J., González-Fierro, A., Candelaria, M., et al., 2006. Expression of platelet derived growth factor family members and the potential role of imatinib mesylate for cervical cancer. Cancer Cell. Int. 6, 22.

Tay, S.K., Oon, L.L., 2014. Prevalence of cervical human papillomavirus infection in healthy women is related to sexual behaviours and educational level: a cross-sectional study. Int. J. STD AIDS 25, 1013–1021.

Tinker, A.V., Ellard, S., Welch, S., Moens, F., Allo, G., Tsao, M.S., et al., 2013. Phase II study of temsirolimus (CCI-779) in women with recurrent, unresectable, locally advanced or metastatic carcinoma of the cervix. A trial of the NCIC Clinical Trials Group (NCIC CTG IND 199). Gynecol. Oncol. 130, 269–274.

Torre, L.A., Bray, F., Siegel, R.L., Ferlay, J., Lortet-Tieulent, J., Jemal, A., 2012. Global cancer statistics. CA Cancer J. Clin. 2015 (65), 87–108.

Tsai, J.P., Lee, C.H., Ying, T.H., Lin, C.L., Lin, C.L., Hsueh, J.T., Hsieh, Y.H., 2015. Licochalcone A induces autophagy through PI3K/Akt/mTOR inactivation and autophagy suppression enhances Licochalcone A-induced apoptosis of human cervical cancer cells. Oncotarget 6, 28851–28866.

Utrera-Barillas, D., Castro-Manrreza, M., Castellanos, E., et al., 2010. The role of macrophages and mast cells in lymphangiogenesis and angiogenesis in cervical carcinogenesis. Exp. Mol. Pathol. 89, 190–196.

van Beurden, M., ten Kate, F.J., Smits, H.L., Berkhout, R.J., de Craen, A.J., van der Vange, N., et al., 1995. Multifocal vulvar intraepithelial neoplasia grade III and multicentric lower genital tract neoplasia is associated with transcriptionally active human papillomavirus. Cancer 75, 2879–2884.

Van der Auwera, I., Cao, Y., Tille, J.C., et al., 2006. First international consensus on the methodology of lymphangiogenesis quantification in solid human tumours. Br. J. Cancer 95, 1611–1625.

van der Burg, S.H., Piersma, S.J., de Jong, A., van der Hulst, J.M., Kwappenberg, K.M., et al., 2007. Association of cervical cancer with the presence of CD4+ regulatory T cells specific for human papillomavirus antigens. Proc. Natl. Acad. Sci. U.S.A. 104, 12087–12092.

Van Kriekinge, G., Castellsagué, X., Cibula, D., Demarteau, N., 2014. Estimation of the potential overall impact of human papillomavirus vaccination on cervical cancer cases and deaths. Vaccine 32, 733–739.

Van Trappen, P.O., Steele, D., Lowe, D.G., et al., 2003. Expression of vascular endothelial growth factor (VEGF)-C and VEGF-D, and their receptor VEGFR-3, during different stages of cervical carcinogenesis. J. Pathol. 201, 544–554.

Villella, J.A., Bogner, P.N., Jani-Sait, S.N., Block, A.M., Lele, S., 2005. Rhabdomyosarcoma of the cervix in sisters with review of the literature. Gynecol. Oncol. 99, 742–748.

Wan, H.Y., Li, Q.Q., Zhang, Y., Tian, W., et al., 2014. MiR-124 represses vasculogenic mimicry and cell motility by targeting amotL1 in cervical cancer cells. Cancer Lett. 355, 148–158.

Wei, J., 2009. Glob. Libr. Women's Med. http://dx.doi.org/10.3843/GLOWM.10230.

WHO/ICO Information Center of HPV and Cervical Cancer (HPV Information Center), 2011. Human Papillomavirus and Related Cancers in the World. Summary Report 2010 http://www.who.int/hpvcentre/en.

Willoughby, B.J., Faulkner, K., Stamp, E.C., Whitaker, C.J., 2006. A descriptive study of the decline in cervical screening coverage rates in the North East and Yorkshire and the Humber regions of the UK from 1995 to 2005. J. Public Health (Oxf.) 28, 355–360.

Wu, J., Chen, C., Zhao, K.N., 2013. Phosphatidylinositol 3-kinase signaling as a therapeutic target for cervical cancer. Curr. Cancer Drug Targets 13, 143–156.

Yang, S., Cheng, H., Cai, J., et al., 2009. PlGF expression in pre-invasive and invasive lesions of uterine cervix is associated with angiogenesis and lymphangiogenesis. APMIS 117, 831–838.

Yoysungnoen-Chintana, P., Bhattarakosol, P., Patumraj, S., 2014. Antitumor and antiangiogenic activities of curcumin in cervical cancer xenografts in nude mice. Biomed. Res. Int. 2014, 817972.

Yu, H., Zhang, S., Zhang, R., et al., 2009. The role of VEGF-C/D and Flt-4 in the lymphatic metastasis of early-stage invasive cervical carcinoma. J. Exp. Clin. Cancer Res. 28, 98.

Yu, J.Q., Zhou, Q., Zhu, H., Zheng, F.Y., Chen, Z.W., 2015. Overexpression of astrocyte elevated gene-1 (AEG-1) in cervical cancer and its correlation with angiogenesis. Asian Pac. J. Cancer Prev. 16, 2277–2281.

Zeng, T., Li, G., 2014. MicroRNA-10a enhances the metastatic potential of cervical cancer cells by targeting phosphatase and tensin homologue. Mol. Med. Rep. 10, 1377–1382.

Zeng, C., Chen, L., Yang, Z., Sun, S., 2014. The close correlation between heparanase and COX-2 expression in lymphangiogenesis of cervical cancer. Med. Oncol. 31, 314.

Zhai, Y., Kuick, R., Nan, B., Ota, I., Weiss, S.J., Trimble, C.L., et al., 2007. Gene expression analysis of preinvasive and invasive cervical squamous cell carcinomas identifies HOXC10 as a key mediator of invasion. Cancer Res. 67, 10163–10172.

Zhang, X.Y., Zhang, H.Y., Zhang, P.N., Lu, X., Sun, H., 2008. Elevated phosphatidylinositol 3-kinase activation and its clinicopathological significance in cervical cancer. Eur. J. Obstet. Gynecol. Reprod. Biol. 139, 237–244.

Zhang, S.Q., Yu, H., Zhang, L.L., 2009. Clinical implications of increased lymph vessel density in the lymphatic metastasis of early-stage invasive cervical carcinoma: a clinical immunohistochemical method study. BMC Cancer 9, 64.

Zhang, S., Yu, H., Zhang, L., 2010. Role of vascular endothelial growth factor receptor-3/Flt-4 in early-stage cervical cancer. Oncol. Lett. 1, 453–456.

CHAPTER

# 6

# Colorectal Cancer Metastasis

M.A. *Aziz*

**King Saud Bin Abdulaziz University for Health Sciences, Riyadh, Kingdom of Saudi Arabia**

## 1. INTRODUCTION

Colorectal cancer (CRC) is the development of malignant cells in the large intestine (colon or rectum). It is mostly the disease of elderly but young adults and children are not excluded (Goldberg and Furman, 2012; Durno et al., 2005). CRC presents a challenging scenario for health care systems with increase in aging population in certain parts of the world. CRC is the third most commonly diagnosed malignancy and the fourth leading cause of cancer-related deaths in the world. Its burden is expected to increase by 60% to more than 2.2 million new cases and 1.1 million cancer deaths by 2030 (Arnold et al., 2016). The success stories of some countries are being narrated by declining graphs in incidence as well as mortality owing to multiple factors. Most notable of these are good prevention strategies executed through screening and surveillance programs at population level. Early detection of CRC remains the most effective measure. Surgical removal of polyps from colon at an early stage has very good prognosis. Once the cancer cells from the large intestine start spreading to other organs like liver and lungs, it is classified as grade III/IV and it becomes metastatic CRC (mCRC). Therapeutic regimens are most effective against early stage cancers. The survival rate of cancers detected at stage I and II is far better than stage III and IV by using surgical interventions, chemotherapy (CT), adjuvant therapy, and neoadjuvant therapy. Although the cancer community has progressed significantly in dealing with early stage CRC, there have been limited advancements for treating mCRC. The prognosis of mCRC is bleak with poor survival rates and increased morbidity and mortality. Twenty-five percent patients presenting with conventionally staged node negative disease (American Joint Committee on Cancer Stage I and II), and over 50% of patients with Stage III disease will develop local recurrence and/or metastases (Bilchik et al., 2006; Manfredi et al., 2006a).

The molecular level changes in CRC are subtle and have been linked to different stages. These are well-established causal gene mutations, copy number changes, as well as significantly altered gene expression profile that characterize the progression of CRC. Molecular evidence has proved to be of significant predictive value in genetically predisposed CRC. Mechanistic pathways involved in

*Introduction to Cancer Metastasis*
http://dx.doi.org/10.1016/B978-0-12-804003-4.00006-2

CRC have been studied in extensive detail that revealed targets for therapeutic agents. The efficacy of these agents for targeted therapy in general population has been debated. The use of biomarkers in predicting the patient cohort that will respond to targeted therapies is now in clinical use.

The ability of tumor to metastasize is dependent on inherent properties of the cancer cell. The role of tumor microenvironment and angiogenesis is very crucial (Balkwill et al., 2012; Leibovici et al., 2011). More than 90% of the colorectal carcinomas are adenocarcinomas originating from epithelial cells of colon mucosa (Fleming et al., 2012). Other CRCs include gastrointestinal carcinoid tumors, gastrointestinal stromal tumors, primary colorectal lymphoma, leiomyosarcoma, melanoma, and squamous cell carcinoma. Pathological examination and staging of tumor decides the therapeutic strategies. The formation of metastases is a multistep process, in which malignant cells disseminate from the primary tumor to colonize distant organs. This is a highly inefficient and complex process, which involves early steps of tumor cell invasion of the microenvironment, entering the bloodstream, survival during migration, and extravasation into distant organs. Subsequent steps including proliferation, induction of angiogenesis, and evading apoptotic death are crucial for colonization of the secondary site. Tumor cells need to be proficient in all these processes to produce metastatic outgrowth. Therefore, each step in metastasis requires specific genetic and epigenetic changes (de Krijger et al., 2011). In mCRC liver, peritoneum and lung are the most affected organs.

This chapter focuses on molecular level changes associated with CRC metastases, organs that are mostly affected and therapeutic options. Challenge to increase the survival rates for these patients is enormous but solutions in the form of personalized medicine and immunotherapy are on the horizon.

## 2. MOLECULAR ASPECTS OF COLORECTAL CANCER METASTASIS

CRC can be defined in terms of molecular changes that mark different stages of its initiation and progression. The transformation of normal cell into adenoma followed by carcinoma and metastasis is linked to molecular changes at gene level. Bert Vogelstein and Fearon's (1990) model of sequential progress of CRC suggests a threshold of gene mutations for each stage to progress. Other models have also been proposed (Halazonetis et al., 2008; Markowitz and Bertagnolli, 2009) which provide a wealth of information about molecular changes causing a normal cell to become tumorous. Based on Vogelstein's model, there are four main molecular pathways that cause CRC—chromosomal instability, CpG island methylator phenotype, microsatellite instability (MSI), and serrated pathways (Worthley and Leggett, 2010). Most of the CRCs are sporadic cancers but a small percentage has genetic aspect. Hereditary nonpolyposis colorectal carcinoma, familial adenomatous polyposis, and Gardner's syndrome are caused by genetically regulated mechanism. Mechanistic studies using molecular biology tools have provided extensive details about the molecular aspects of the CRC initiation and progression. Involvement of molecular pathways and genetic alterations has been studied and linked to therapeutic outcome (Sadanandam et al., 2013). Global analyses of DNA and RNA level changes that characterize the molecular perturbations in CRC have been investigated in detail for different populations (Aziz et al., 2014a,b; Eldai et al., 2013). In these well-characterized pathways and models of CRC development, there is no conclusive evidence on which molecular changes are responsible for tipping the balance toward metastasis. A holistic view of the molecules associated with mCRC gives us a highly interconnected network, which

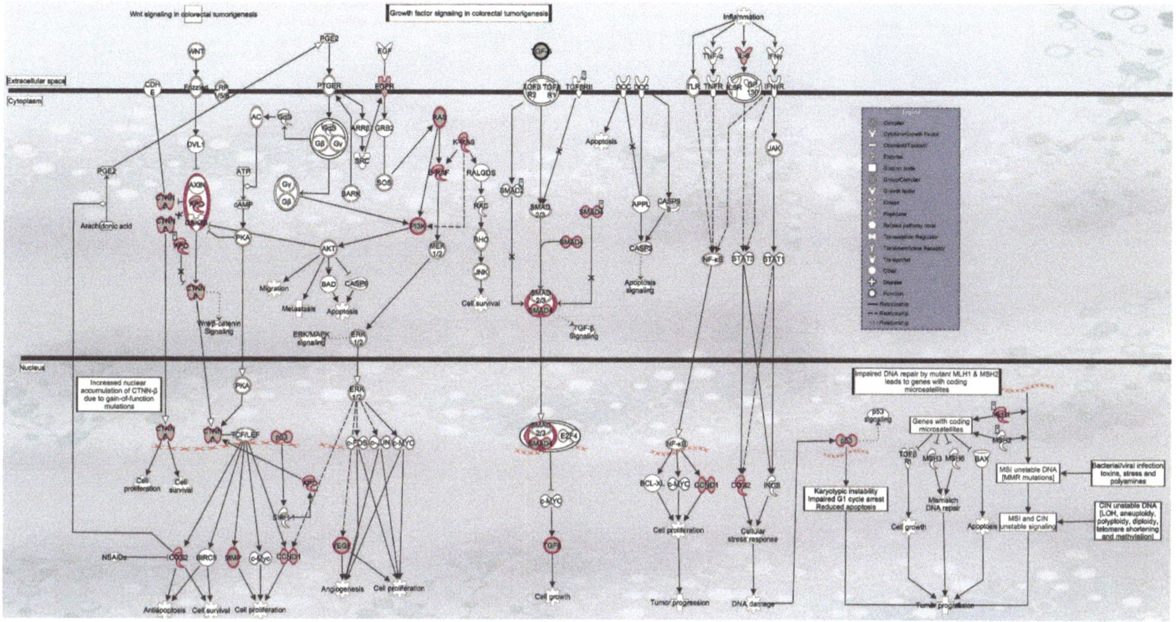


FIGURE 6.1 **Metastatic colorectal cancer (mCRC) pathways.** Pathways involved in CRC are interconnected. Molecules colored red are known biomarkers. The complexity of the network of molecules makes mCRC a challenging problem for cancer research. *This figure was prepared using Ingenuity Pathway analysis software licensed from Qiagen.*

needs to be dealt as such to develop treatment strategies that are effective and nonrecurrent, and with least side effects (Fig. 6.2).

Metastasis involves genetic changes in tumor cells that make them capable to detach and migrate. Unexpectedly, there are no new genetic mutations that are exclusive for CRC metastases (Jones et al., 2008; Kim et al., 2015) suggesting other critical factors that may help the migration of tumor cells. This migration is facilitated by change in the microenvironment, which includes blood and oxygen supplies. Increased expression of hypoxia inducing factor-1 and vascular endothelial growth factor (VEGF) have been reported in CRC (Weis and Cheresh, 2011; Ellis et al., 2000) and have been used for developing therapeutic agents. In particular, VEGF189 mRNA isoform expression pattern suggests a poor prognosis for patients with CRC metastasizing to the liver (Tokunaga et al., 1998). Gene expression level changes have been linked to metastasis and can be used as prognostic markers (Jorissen et al., 2009). A systematic approach is to integrate data from various aspects associated with a gene, for example, copy number changes, mutations, methylation, and expression, to suggest their involvement in metastasis (Lee et al., 2013). Increased clinical value of genes reported through such integrated analyses is yet to be validated.

The involvement of signaling pathways in mCRC has been studied with an aim to find novel therapeutic targets and biomarkers. The involvement of Wnt-β catenin, p53, Notch, transforming growth factor beta (TGF-β), inflammation, and DNA repair pathways has been well studied in mCRC (Sonoshita et al., 2011; Liu et al., 2015) (Fig. 6.1). As we continue to develop tools to better study the molecular changes in metastasized stages, there would be opportunities to discover other molecules and pathways that play a causal

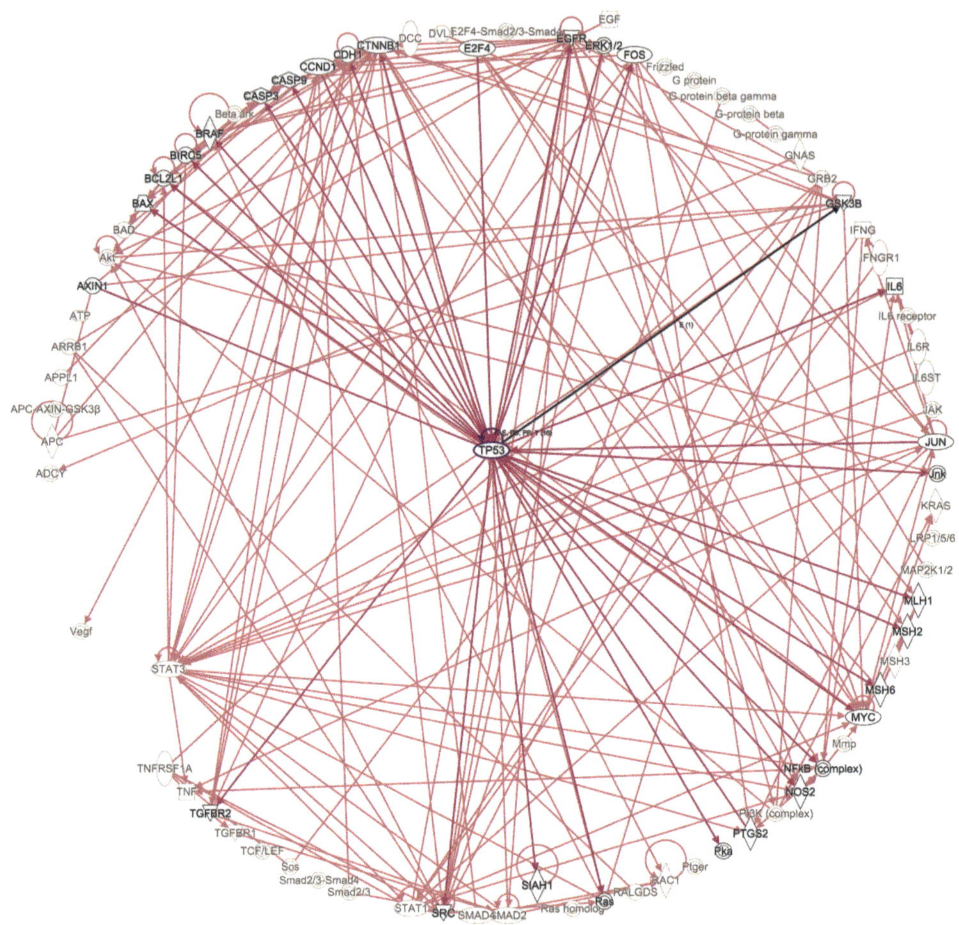


FIGURE 6.2 **Network of molecules associated with metastatic colorectal cancer.** A network of molecules with p53 as the central node was generated. This network shows the direct connection of p53 with several molecules (indicated by *solid arrows*). Other molecules in the background are also highly interconnected. Understanding mCRC as a network with a systems biology approach is therefore needed. *This figure was prepared using Ingenuity Pathway analysis software licensed from Qiagen.*

role (Gulhati et al., 2011; Stein et al., 2009; Weidle et al., 2015). Characterization of these novel molecules and pathways would allow their assessment as potential biomarkers and/or therapeutic targets that are of translational value.

## 2.1 Biomarkers for Colorectal Cancer Metastasis

For CRC metastasis, there are about 43 known biomarkers that can be used in diagnosis, assessing disease progression, efficacy, safety, prognosis, and response to therapy (Table 6.1). **Biomarkers** are valuable tools to develop well-informed management of cancer patients (Newton et al., 2012; Gonzalez-Pons and Cruz-Correa, 2015; Henry and Hayes, 2012). A statistically significant measurable indicator of a patient could serve as a biomarker. This indicator could be a clinical characteristic [e.g., carcinoembryonic antigen (CEA) levels], gene mutation (e.g., point substitution in codon 12 and 13 of KRAS gene), DNA copy number changes (e.g., loss of heterozygosity of 18q chromosome), or altered gene expression (e.g.,

TABLE 6.1 Biomarkers for CRC Metastasis

| Symbol | Entrez Gene Name | Location | Type(s) | Application(s) | Drug(s) |
|---|---|---|---|---|---|
| APC | Adenomatous polyposis coli | Nucleus | Enzyme | Diagnosis, disease progression, efficacy, prognosis, safety | |
| BRAF | B-Raf proto-oncogene, serine/threonine kinase | Cytoplasm | Enzyme | Diagnosis, prognosis, response to therapy | Vemurafenib, dabrafenib, regorafenib, bortezomib/sorafenib, dexamethasone/lenalidomide/sorafenib, bevacizumab/sorafenib, LGX-818, dabrafenib/trametinib, CEP-32496, cytarabine/idarubicin/sorafenib, 5-azacytidine/sorafenib, decitabine/sorafenib, MLN2480, BRAF inhibitor, ARQ 736, cobimetinib/vemurafenib, everolimus/sorafenib, sorafenib |
| CCND1 | Cyclin D1 | Nucleus | Transcription regulator | Diagnosis, efficacy, prognosis, response to therapy | Arsenic trioxide/tretinoin, arsenic trioxide/daunorubicin/tretinoin, arsenic trioxide/gemtuzumab ozogamicin/tretinoin, arsenic trioxide/idarubicin/tretinoin, arsenic trioxide |
| CD44 | CD44 molecule (Indian blood group) | Plasma membrane | Enzyme | Diagnosis, disease progression, prognosis, | |
| CDKN2A | Cyclin-dependent kinase inhibitor 2A | Nucleus | Transcription regulator | Diagnosis, disease progression, efficacy, prognosis, response to therapy, safety | |
| CTNNB1 | Catenin beta 1 | Nucleus | Transcription regulator | Diagnosis, disease progression, efficacy, prognosis | PRI-724 |
| CXCL8 | Chemokine (C-X-C motif) ligand 8 | Extracellular space | Cytokine | Diagnosis, efficacy, prognosis | |
| CXCL12 | Chemokine (C-X-C motif) ligand 12 | Extracellular space | Cytokine | Diagnosis, efficacy, prognosis | |
| DPYD | Dihydropyrimidine dehydrogenase | Cytoplasm | Enzyme | Disease progression, efficacy, prognosis, safety | Eniluracil, gimeracil/oxonic acid/tegafur, gimeracil, dihydropyrimidine dehydrogenase inhibitor |

*Continued*

TABLE 6.1 Biomarkers for CRC Metastasis—cont'd

| Symbol | Entrez Gene Name | Location | Type(s) | Application(s) | Drug(s) |
|---|---|---|---|---|---|
| EGFR | Epidermal growth factor receptor | Plasma membrane | Kinase | Diagnosis, disease progression, efficacy, prognosis, response to therapy, safety | Cetuximab, AEE 788, panitumumab, BMS-599626, varlitinib, tesevatinib, bevacizumab/erlotinib, afatinib, nimotuzumab, necitumumab, icotinib, sapitinib, cetuximab/irinotecan, lapatinib/pazopanib, irinotecan/panitumumab, erlotinib/vismodegib, erlotinib/gemcitabine, lapatinib/letrozole, capecitabine/lapatinib, bevacizumab/panitumumab, bevacizumab/cetuximab, EGFR tyrosine kinase inhibitor, capecitabine/erlotinib, lapatinib/paclitaxel, rociletinib, AP26113, MEHD7945A, cabozantinib/erlotinib, osimertinib, poziotinib, HM61713, sym004, IMGN289, EGFR antisense DNA, EGF816, ASP8273, selatinib, 18F-PEG6-IPQA, lapatinib/trastuzumab, ABT-414, anti-EGFR monoclonal antibody GT-MAB 5.2-GEX, AZD3759, avitinib, PF-06747775, LY3164530, RG 14620, TAK-285, PF-06459988, tyrphostin A30, RG 13022, afatinib/cetuximab, canertinib, gefitinib, neratinib, PD 153035, lapatinib, vandetanib, erlotinib |
| ERBB2 | Erb-b2 receptor tyrosine kinase 2 | Plasma membrane | Kinase | Diagnosis, disease progression, efficacy, prognosis, safety | Trastuzumab, BMS-599626, varlitinib, tesevatinib, CP-724,714, afatinib, pertuzumab, ertumaxomab, sapitinib, trastuzumab emtansine, lapatinib/pazopanib, lapatinib/letrozole, paclitaxel/trastuzumab, capecitabine/lapatinib, cyclophosphamide/docetaxel/epirubicin/5-fluorouracil/trastuzumab, docetaxel/trastuzumab, paclitaxel/pertuzumab/trastuzumab, trastuzumab/vinorelbine, capecitabine/trastuzumab, lapatinib/paclitaxel, pertuzumab/trastuzumab, MGAH22, osimertinib, 89Zr-trastuzumab, ARRY-380, poziotinib, selatinib, MM-111, lapatinib/trastuzumab, MM-302, cyclophosphamide/epirubicin/5-fluorouracil/paclitaxel/trastuzumab, cyclophosphamide/docetaxel/trastuzumab, SYD985, TAK-285, afatinib/cetuximab, neratinib, lapatinib, erlotinib |
| ERCC1 | Excision repair cross-complementation group 1 | Nucleus | Enzyme | Diagnosis, disease progression, efficacy, prognosis, response to therapy | |

| | | | | | |
|---|---|---|---|---|---|
| FLT1 | Fms related tyrosine kinase 1 | Plasma membrane | Kinase | Diagnosis, disease progression, efficacy, prognosis, response to therapy, safety | Sunitinib, pazopanib, axitinib, CEP 7055, lenvatinib, motesanib, cabozantinib, nintedanib, regorafenib, bortezomib/sorafenib, lapatinib/pazopanib, dexamethasone/lenalidomide/sorafenib, bevacizumab/sorafenib, MGCD-265, cabozantinib/erlotinib, famitinib, lucitanib, cytarabine/idarubicin/sorafenib, 5-azacytidine/sorafenib, decitabine/sorafenib, fruquintinib, MGCD516, sulfatinib, everolimus/sorafenib, sorafenib, vandetanib |
| FLT4 | Fms related tyrosine kinase 4 | Plasma membrane | Transmembrane receptor | Diagnosis, disease progression, efficacy, prognosis | Sunitinib, pazopanib, axitinib, CEP 7055, lenvatinib, tivozanib, motesanib, telatinib, cabozantinib, nintedanib, regorafenib, bortezomib/sorafenib, lapatinib/pazopanib, dexamethasone/lenalidomide/sorafenib, bevacizumab/sorafenib, MGCD-265, cabozantinib/erlotinib, famitinib, lucitanib, cytarabine/idarubicin/sorafenib, 5-azacytidine/sorafenib, decitabine/sorafenib, fruquintinib, MGCD516, anlotinib, sulfatinib, everolimus/sorafenib, sorafenib, vandetanib |
| GDF15 | Growth differentiation factor 15 | Extracellular space | Growth factor | Diagnosis | |
| GSTP1 | Glutathione S-transferase pi 1 | Cytoplasm | Enzyme | Diagnosis, efficacy, prognosis, safety, unspecified application | |
| IGF1 | Insulin like growth factor 1 | Extracellular space | Growth factor | Diagnosis, efficacy, prognosis, safety | BI 836845 |
| IGF1R | Insulin like growth factor 1 receptor | Plasma membrane | Transmembrane receptor | Diagnosis, efficacy, prognosis | Picropodophyllin, linsitinib, cixutumumab, ganitumab, AVE1642, BIIB022, IGF1, dalotuzumab, ceritinib, MM-141, IGF-methotrexate conjugate, PL225B, KW-2450 |
| IL6 | Interleukin 6 | Extracellular space | Cytokine | Diagnosis, disease progression, efficacy, prognosis, response to therapy, safety | Tocilizumab, siltuximab |
| KDR | Kinase insert domain receptor | Plasma membrane | Kinase | Disease progression, efficacy, prognosis, response to therapy, safety | AEE 788, sunitinib, cediranib, pazopanib, axitinib, tesevatinib, CEP 7055, brivanib alaninate, CHIR-265, lenvatinib, tivozanib, motesanib, OSI-930, telatinib, ramucirumab, cabozantinib, nintedanib, regorafenib, bortezomib/sorafenib, lapatinib/pazopanib, dexamethasone/lenalidomide/sorafenib, bevacizumab/sorafenib, MGCD-265, cabozantinib/erlotinib, famitinib, lucitanib, paclitaxel/ramucirumab, docetaxel/ramucirumab, cytarabine/idarubicin/sorafenib, 5-azacytidine/sorafenib, decitabine/sorafenib, fruquintinib, altiratinib, MGCD516, anlotinib, pexmetinib, sulfatinib, everolimus/sorafenib, vatalanib, sorafenib, vandetanib, pegaptanib |

*Continued*

TABLE 6.1 Biomarkers for CRC Metastasis—cont'd

| Symbol | Entrez Gene Name | Location | Type(s) | Application(s) | Drug(s) |
|---|---|---|---|---|---|
| KIT | KIT proto-oncogene receptor tyrosine kinase | Plasma membrane | Transmembrane receptor | Diagnosis, efficacy, prognosis, safety | Nilotinib, dasatinib, sunitinib, pazopanib, lenvatinib, tivozanib, motesanib, OSI-930, telatinib, tandutinib, cabozantinib, quizartinib, regorafenib, flumatinib, ponatinib, bortezomib/sorafenib, lapatinib/pazopanib, dexamethasone/lenalidomide/sorafenib, bevacizumab/sorafenib, imatinib/sirolimus, PLX3397, cabozantinib/erlotinib, famitinib, imatinib/nilotinib/pegintron, cytarabine/idarubicin/sorafenib, 5-azacytidine/sorafenib, decitabine/sorafenib, LOP628, MGCD516, AKN-028, chiauranib, everolimus/sorafenib, imatinib, sorafenib |
| KRAS | Kirsten rat sarcoma viral oncogene homolog | Cytoplasm | Enzyme | Diagnosis, efficacy, prognosis, response to therapy | |
| MET | MET proto-oncogene, receptor tyrosine kinase | Plasma membrane | Kinase | Diagnosis, disease progression, efficacy, prognosis, response to therapy | Crizotinib, tivantinib, cabozantinib, BMS 777607, capmatinib, onartuzumab, MGCD-265, cabozantinib/erlotinib, MSC2156119J, AMG 337, LY2875358, ARGX-111, SAR125844, volitinib, LY3164530, altiratinib, MGCD516, merestinib, EMD 1204831, MK8033 |
| mir-8 | MicroRNA 200c | Cytoplasm | MicroRNA | Diagnosis | |
| mir-17 | MicroRNA 17 | Cytoplasm | MicroRNA | Prognosis | |
| MLH1 | MutL homolog 1 | Nucleus | Enzyme | Diagnosis, efficacy, prognosis | |
| MMP2 | Matrix metallopeptidase 2 | Extracellular space | Peptidase | Diagnosis, disease progression, efficacy, prognosis, response to therapy | MMP2 MMP9 inhibitor, marimastat |
| MTHFR | Methylenetetrahydrofolate reductase (NAD(P)H) | Cytoplasm | Enzyme | Diagnosis, efficacy, response to therapy, safety | |
| PDGFRA | Platelet derived growth factor receptor alpha | Plasma membrane | Kinase | Efficacy, prognosis, safety | Sunitinib, pazopanib, axitinib, lenvatinib, telatinib, nintedanib, regorafenib, lapatinib/pazopanib, imatinib/sirolimus, crenolanib, imatinib, becaplermin |
| PDGFRB | Platelet derived growth factor receptor beta | Plasma membrane | Kinase | Prognosis, response to therapy | Dasatinib, sunitinib, pazopanib, axitinib, tivozanib, telatinib, tandutinib, nintedanib, regorafenib, bortezomib/sorafenib, lapatinib/pazopanib, dexamethasone/lenalidomide/sorafenib, bevacizumab/sorafenib, imatinib/sirolimus, cytarabine/idarubicin/sorafenib, 5-azacytidine/sorafenib, decitabine/sorafenib, everolimus/sorafenib, imatinib, sorafenib, becaplermin |

| | | | | | |
|---|---|---|---|---|---|
| PGF | Placental growth factor | Extracellular space | Growth factor | Diagnosis, disease progression, efficacy, prognosis | Aflibercept, aflibercept/irinotecan |
| PIK3CA | Phosphatidylinositol-4,5-bisphosphate 3-kinase catalytic subunit alpha | Cytoplasm | Kinase | Efficacy, prognosis, response to therapy | SF 1126, PX-866, dactolisib, pictilisib, buparlisib, XL147, PKI-179, MLN1117, taselisib, AZD8835, PWT33597, BAY1082439, PQR309 |
| PTEN | Phosphatase and tensin homolog | Cytoplasm | Phosphatase | Diagnosis, disease progression, efficacy, prognosis, response to therapy | |
| PTGS2 | Prostaglandin-endoperoxide synthase 2 (prostaglandin G/H synthase and cyclooxygenase) | Cytoplasm | Enzyme | Diagnosis, disease progression, efficacy, prognosis | Acetaminophen/pentazocine, acetaminophen/clemastine/pseudoephedrine, aspirin/butalbital/caffeine, acetaminophen/caffeine/dihydrocodeine, aspirin/hydrocodone, aspirin/oxycodone, acetaminophen/aspirin/caffeine, aspirin/pravastatin, acetaminophen/dexbrompheniramine/pseudoephedrine, aspirin/meprobamate, aspirin/caffeine/propoxyphene, aspirin/butalbital/caffeine/codeine, aspirin/caffeine/dihydrocodeine, chlorpheniramine/ibuprofen/pseudoephedrine, licofelone, menatetrenone, icosapent, suprofen, lornoxicam, tiaprofenic acid, lumiracoxib, tenoxicam, naproxen/sumatriptan, parecoxib, ibuprofen/phenylephrine, acetaminophen/aspirin/codeine, esomeprazole/naproxen, aspirin/esomeprazole, aspirin/dipyridamole/telmisartan, famotidine/ibuprofen, aspirin/dabigatran etexilate, diclofenac/omeprazole, chlorpheniramine/ibuprofen/phenylephrine, dexamethasone/pomalidomide, sulindac/tamoxifen, sulindac/toremifene, raloxifene/sulindac, ketorolac/phenylephrine, aspirin/bivalirudin, diclofenac/hyaluronic acid, COX2 inhibitor, diclofenac/misoprostol, acetaminophen/butalbital/caffeine, hydrocodone/ibuprofen, acetaminophen/hydrocodone, acetaminophen/tramadol, acetaminophen/codeine, acetaminophen/oxycodone, acetaminophen/propoxyphene, niflumic acid, nitroaspirin, ketoprofen, diclofenac, etoricoxib, naproxen, meclofenamic acid, pomalidomide, meloxicam, celecoxib, dipyrone, nimesulide, acetaminophen, mefenamic acid, diflunisal, ibuprofen, GW406381X, phenylbutazone, indomethacin, sulfasalazine, piroxicam, valdecoxib, aspirin, carprofen, zomepirac, rofecoxib, aspirin/caffeine/orphenadrine, acetaminophen/butalbital, balsalazide, aspirin/dipyridamole, acetaminophen/butalbital/caffeine/codeine, racemic flurbiprofen, phenacetin, sulindac, nabumetone, etodolac, tolmetin, ketorolac, oxaprozin, mesalamine, salsalate, fenoprofen, salicylic acid, acetaminophen/caffeine/chlorpheniramine/hydrocodone/phenylephrine, bromfenac |

*Continued*

TABLE 6.1 Biomarkers for CRC Metastasis—cont'd

| Symbol | Entrez Gene Name | Location | Type(s) | Application(s) | Drug(s) |
|---|---|---|---|---|---|
| RB1 | Retinoblastoma 1 | Nucleus | Transcription regulator | Diagnosis, efficacy, prognosis | |
| SMAD4 (similar to the gene products of the Drosophila gene 'mothers against decapentaplegic') | SMAD family member 4 | Nucleus | Transcription regulator | Prognosis | |
| TGFB1 | Transforming growth factor beta 1 | Extracellular space | Growth factor | Diagnosis, efficacy, prognosis, safety | Dalantercept |
| TOP1 | Topoisomerase (DNA) I | Nucleus | Enzyme | Efficacy | Elsamitrucin, T 0128, polyglutamate camptothecin, elomotecan, tafluposide, lurtotecan, belotecan, cetuximab/irinotecan, capecitabine/irinotecan, 5-fluorouracil/irinotecan, irinotecan/panitumumab, cisplatin/topotecan, cyclophosphamide/topotecan, cyclophosphamide/topotecan/vincristine, irinotecan/temozolomide, irinotecan/vincristine, cyclophosphamide/irinotecan/temozolomide, irinotecan/temozolomide/vincristine, bevacizumab/irinotecan, bevacizumab/irinotecan/oxaliplatin, aflibercept/irinotecan, irinotecan/oxaliplatin, cyclophosphamide/temozolomide/topotecan, docetaxel/irinotecan, capecitabine/irinotecan/oxaliplatin, bevacizumab/capecitabine/irinotecan/oxaliplatin, ifosfamide/topotecan, 5-fluorouracil/irinotecan/oxaliplatin, gemcitabine/irinotecan, bevacizumab/paclitaxel/topotecan, paclitaxel/topotecan, cyclophosphamide/ifosfamide/irinotecan/temozolomide, cytarabine/topotecan, simmitecan, mureletecan, camptothecin-20-O-propionate, indotecan, Genz-644282, indimitecan, topoisomerase I inhibitor, TAS-103, beta-lapachone, irinotecan, SN-38, topotecan, camptothecin derivative, camptothecin, IDEC-132, stibogluconic acid, rubitecan, gimatecan, karenitecin |
| TP53 | Tumor protein p53 | Nucleus | Transcription regulator | Diagnosis, disease progression, efficacy, prognosis, response to therapy | CGM097, kevetrin |

| | | | | | |
|---|---|---|---|---|---|
| TYMS | Thymidylate synthetase | Nucleus | Enzyme | Diagnosis, disease progression, efficacy, prognosis, response to therapy | Flucytosine, 5-fluorouracil, tegafur, trimethoprim, cyclophosphamide/epirubicin/5-fluorouracil, pralatrexate, 5-fluorouracil/oxaliplatin, capecitabine/irinotecan, 5-fluorouracil/irinotecan, capecitabine/docetaxel, capecitabine/lapatinib, capecitabine/ixabepilone, bevacizumab/5-fluorouracil, bevacizumab/capecitabine/oxaliplatin, cyclophosphamide/docetaxel/epirubicin/5-fluorouracil/trastuzumab, capecitabine/oxaliplatin, bevacizumab/pemetrexed, capecitabine/trastuzumab, capecitabine/gemcitabine, docetaxel/5-fluorouracil/oxaliplatin, epirubicin/5-fluorouracil/oxaliplatin, capecitabine/docetaxel/gemcitabine, capecitabine/epirubicin/oxaliplatin, capecitabine/erlotinib, capecitabine/irinotecan/oxaliplatin, bevacizumab/capecitabine/irinotecan/oxaliplatin, bevacizumab/capecitabine, capecitabine/temozolomide, 5-fluorouracil/irinotecan/oxaliplatin, leucovorin/methotrexate, 5-fluorouracil/gemcitabine, 5-fluorouracil/imiquimod, gimeracil/oxonic acid/tegafur, gemcitabine/pemetrexed, cyclophosphamide/epirubicin/5-fluorouracil/vinorelbine, cyclophosphamide/epirubicin/5-fluorouracil/paclitaxel/trastuzumab, cyclophosphamide/epirubicin/5-fluorouracil/paclitaxel, cyclophosphamide/docetaxel/epirubicin/5-fluorouracil, capecitabine/5-fluorouracil, capecitabine/5-fluorouracil/oxaliplatin, thymidylate synthase inhibitor, DFP-11207, raltitrexed, plevitrexed, nolatrexed, leucovorin, capecitabine, trifluridine, floxuridine, pemetrexed |
| VEGFA | Vascular endothelial growth factor A | Extracellular space | Growth factor | Diagnosis, efficacy, prognosis, safety | Dalteparin, bevacizumab, ranibizumab, aflibercept, bevacizumab/erlotinib, bevacizumab/sorafenib, bevacizumab/5-fluorouracil, bevacizumab/temozolomide, bevacizumab/irinotecan, bevacizumab/carmustine/lomustine, bevacizumab/paclitaxel, bevacizumab/irinotecan/oxaliplatin, aflibercept/irinotecan, bevacizumab/capecitabine/oxaliplatin, bevacizumab/panitumumab, bevacizumab/cetuximab, bevacizumab/pemetrexed, bevacizumab/gemcitabine, bevacizumab/capecitabine/irinotecan/oxaliplatin, bevacizumab/capecitabine, bevacizumab/paclitaxel/topotecan, bevacizumab/oxaliplatin, bevacizumab/docetaxel, bevacizumab/doxorubicin/paclitaxel, pegaptanib |
| VEGFB | Vascular endothelial growth factor B | Extracellular space | Growth factor | Efficacy, prognosis | Aflibercept, aflibercept/irinotecan |
| VEGFC | Vascular endothelial growth factor C | Extracellular space | Growth factor | Diagnosis, disease progression, efficacy, prognosis | |

decreased MLH1 gene expression). These biomarkers can be used for diagnostic or therapeutic purposes. Often, these biomarkers could also be the therapeutic target owing to their significant association with the disease. We will discuss here "molecular" biomarkers that are associated with CRC metastasis. Molecular biomarkers could be diagnostic, prognostic, or predictive depending on their established role. Same molecule may qualify for more than one role, for example, thymidylate synthase could be described as predictive biomarker for 5-fluorouracil (5-FU) based therapy and prognostic marker for surgical resections. Increased levels of thymidylate synthase show better response rates for 5-FU therapy (Karlberg et al., 2010).

**Diagnostic biomarker** is defined as a marker that can be used to diagnose a particular disease or disease state. Epithelial cell adhesion molecule (CD36, epitheilal cell adhesion molecule (EpCAM)) could serve as prognostic and diagnostic marker as well as target for immunotherapy (Patriarca et al., 2012). CEA is a well-established biomarker that is routinely tested in suspected CRC patients for diagnosis (Tabuchi et al., 1991). CEA levels are also indicative of recurrence after resection of liver metastases (Takamoto et al., 2016). Netrin-1 is a diagnostic biomarker in many cancers and has been reported to increase cell survival, affect microenvironment, and angiogenesis in CRC (Ko et al., 2014). The expression levels of several microRNAs (miRNAs) have been linked to particular stages in the progression of CRC leading to metastasis. These miRNAs could serve as biomarkers for diagnosing, staging, and prognosis of CRC (Hur et al., 2015a,b; Hur, 2015). miR-141 could be used alongside CEA as a diagnostic biomarker for CRC with distant metastases (Cheng et al., 2011).

**Predictive biomarker** is defined as a biomarker that can be used to predict the efficacy of a particular therapeutic regimen. KRAS, neuroblastoma RAS viral oncogene homolog (NRAS), BRAF, and PIK3CA gene mutations in specific codons are well-established biomarkers in epidermal growth factor receptor (EGFR) directed mCRC combination therapy using cetuximab and panitumumab (Douillard et al., 2013; Peeters et al., 2013a,b; De Roock et al., 2010a,b). Mutations in KRAS gene were initially found to confer resistance to these therapies. Nonresponding patients with wild-type KRAS gene were found to harbor other mutations in genes like NRAS, BRAF, and PIK3CA. These molecules are part of a network. Other molecules from the same network like PTEN have also been used as prognostic marker and propensity to develop liver metastases (Sawai et al., 2008). KRAS mutation is reported in 35–45% of mCRC (Andreyev et al., 1998, 2001; Lievre et al., 2006; Richman et al., 2009). KRAS has recently been reported as independent prognostic marker for recurrence-free survival in mCRC patients receiving liver resection (Coulson, 2015). Loss of SMAD4 (similar to the gene products of the Drosophila gene 'mothers against decapentaplegic') gene (located on chromosome 18q) is a predictive biomarker for 5-FU therapy failure (Alhopuro et al., 2005).

**Prognostic biomarker** is defined as a marker that is used to monitor and predict the progression of a disease. Recent reports suggest that coexpression of chemokine $CX_3CL1$ along with its specific receptor $CX_3CR1$ at low level could mean poor prognosis with high potential of metastatic progression (Erreni et al., 2016). High levels of miRNAs (miR-141, -21, -200c) detected in plasma are indicative of poor prognosis in mCRC. miR-21 is a promising biomarker for early detection and prognosis of CRC (Toiyama et al., 2013). miR-200c is an independent biomarker for lymph node metastases (Toiyama et al., 2014). S100A4 is a prognostic biomarker and inducer for CRC metastasis (Dahlmann et al., 2014). Microsatellite status is a prognostic marker. MSI tumors have better prognosis than microsatellite stable (MSS) tumors with lesser chances of distant metastases (Malesci et al., 2007; Boland and Goel, 2010). MSS tumors do not respond well to 5-FU based treatments and hence their microsatellite status can be used as predictive biomarker.

# 3. MAIN SITES OF COLORECTAL CANCER METASTASIS

The inherent nature of the tumor cells along with their adhesion to the neighboring cells and the pattern of blood circulation decide the spread of a tumor. We still do not know the absolute mechanism by which the tumor cells favor a particular organ to settle and develop metastases. Famous "seed and soil" theory (Qian and Teh, 2012; Dai et al., 2006) and blood circulation pattern between primary and metastatic site, provide some answers but the molecular details are elusive. Presence of somatic mutations in KRAS, BRAF, PIK3CA, and NRAS oncogenes has been correlated with the site of metastasis (Lipsyc and Yaeger, 2015; Hayashi et al., 1995). Lung and liver metastases have been found to have lower incidence in BRAF gene mutant tumors (Tran et al., 2011). Approximately 50% of patients diagnosed with CRC will die because of the complications of distant metastases. Nearly 20% of mCRC patients, and a significant proportion of previously treated patients have disease recurrence typically involving the liver or lungs (Koukourakis and Sotiropoulou-Lontou, 2011; Aranda et al., 2011). The occurrence rate of colorectal metastases at the different sites is as follows: liver 60–71%, lung 25–40%, bone 5–10%, ovary 3–5%, adrenal gland 1%, and central nervous system 1% (Aranda et al., 2011). About one-third of CRC patients have liver metastasis either at the time of diagnosis (synchronous in 1/3 of the cases) or during the disease course (metachronous in 2/3 of the cases) (Wang and Li, 2012; Cai and Cai, 2012). We will discuss here three main sites of CRC metastases which differ in incidence from different reports: liver, peritoneum, and lung (Marz and Piso, 2015; Thomassen et al., 2013; Ibrahim et al., 2016; Villeneuve and Sundaresan, 2009).

## 3.1 Liver

The most common site of CRC metastasis is liver. In 30–40% patients, it is also the only site of metastasis (Weiss et al., 1986; Hugh et al., 1997a,b). The prognosis of patients with mCRC of the liver has improved significantly over the last few years. The management has become multidisciplinary using modern CT regimens and more developed surgical techniques. There is a huge difference in 5-year survival with and without surgical intervention. There is a tenfold increase in percentage of patients (from 6% to 60%) with 5-year survival in surgically resected cases versus untreated (Manfredi et al., 2006b; Choti et al., 2002; Brouquet et al., 2010, 2011). However, not all patients are suitable for surgical resection. There are at least three categories of colorectal liver metastasis (CLM) patients (Ismaili, 2011):

1. The hepatic lesion(s) are clearly resectable at the time of presentation
2. The hepatic lesion(s) could become resectable after neoadjuvant therapy
3. Unresectable hepatic lesion(s).

Surgical resection of liver metastases of CRC greatly improves the clinical outcome of patients with advanced disease. This remains the gold standard for the treatment of resectable CLM (Ismaili, 2011). Systemic postoperative, preoperative, conversion, targeted combination, and locoregional chemotherapies are some of the options that have been used with varied levels of success (Wang and Li, 2012). Different combinations of 5-FU, leucovorin (LV), irinotecan, and oxaliplatin had been used in many clinical trials with no clear verdict favoring one regimen over the other as first line therapy in down staging of CLM in terms of progression free survival (PFS) and overall survival (OS) (Wang and Li, 2012). "Conversion CT" is the term used to describe the administration of CT in patients with liver metastasis to convert them from unresectable to resectable (Wang and Li, 2012). The modality of perioperative CT over surgery alone has resulted in more patients with initially unresectable metastasis receiving a complete resection and enjoying a prolonged survival after surgery. Radiofrequency ablation has also helped in converting unresectable cases to resectable (Berber et al., 2008; Gillams and Lees, 2008; Oshowo et al., 2003; Solbiati et al., 2001). For patients with resectable metastases treated with primary surgery, adjuvant CT is a reasonable option. Emerging data have suggested that preoperative CT, as well as postoperative CT, could be advantageous compared to surgery alone in terms of disease-free survival for patients with resectable CLM (Wang and Li, 2012). For patients with resectable metastases, perioperative CT based on (oxaliplatin with fluorouracil (5FU) and folinic acid chemotherapy) FOLFOX4 is the reference treatment according to the results of the (European Organisation for Research and Treatment of Cancer) EORTC40983 randomized trial. This trial exhibits the involvement of immune system in providing better PFS (Tanis et al., 2015; Nordlinger et al., 2008, 2013). The use of selective internal radiation therapy with yttrium-90 is being discussed with optimism to be included in treatment regimen for CLM (Lau et al., 2016). If metastases are unresectable and unlikely to be resectable, palliative CT with or without targeted agents is the mainstay treatment (Ismaili, 2011).

## 3.2 Peritoneum

Another most frequent metastatic site for cancer of colorectal origin is peritoneum (Segelman et al., 2012; Jayne et al., 2002). CRC peritoneal metastasis (CPM) is diagnosed in 10–15% of CRC (Sluiter et al., 2016). It is estimated that more than 40% of CRC patients will develop peritoneal metastasis (PM) (Chua et al., 2013). Most of the CRC deaths were found to have PM (Weiss et al., 1986; Welch and Donaldson, 1979; Gilbert, 1983). Often it is considered as a terminal stage cancer and systemic palliative therapies are administered. PM is established in a sequential manner through the following five steps (Lambert, 2015):

- Release of cancer cells from primary tumor into abdominal cavity
- Transport in abdominal cavity
- Adhesion to the peritoneum
- Invasion into submesothelial layers
- Access to systemic circulation

Molecular biology of PM is not well understood. The initial event of detachment of cells from primary site may involve decreased cell junction proteins like cadherins. Initial tumor–mesothelial interaction appears to involve several adhesion molecules, including CD44, the selectins, and various leukocyte associated antigens. The exact molecules involved are probably determined by the nature of the metastatic tumor cell. Invasion of the mesothelial monolayer appears to occur by tumor-induced mesothelial apoptosis, at least in part via the Fas/FasL system, although invasion between intercellular spaces may also play a role (Qian and Teh, 2012). Presence of free floating cancer cells in the peritoneal cavity is known to increase the risk of peritoneal carcinomatosis (Nakanishi et al., 2000; Takebayashi et al., 2014; Sugarbaker, 1995). Patients undergoing abdominal surgery also have increased risk of PM. Based on current data Sialyl Lewis a ($_{s}Le^{a}$) and mucin 16 are the most promising prognostic biomarkers for CPM (Sluiter et al., 2016).

The therapeutic options for surgeons and oncologists treating CPM are not well supported by the evidence generated through randomized clinical trials. Use of combining aggressive cytoreductive surgery, hyperthermic intraperitoneal CT (HIPEC) has been advised by some groups with no clear supporting evidence (Lambert, 2015). With several ongoing clinical trials and multimodality approaches available, PM is no longer a terminal illness. There is a hope to get better treatment strategies with the advent of genomic medicine. It will be interesting to develop therapeutic targets and biomarkers that are specific for peritoneal metastases in CRC.

### 3.3 Lung

Lung is considered to be the second most common site of CRC metastasis after liver (Weiss et al., 1986; Galandiuk et al., 1992; Kobayashi et al., 2007; Mitry et al., 2010; Pihl et al., 1987; Vatandoust et al., 2015) but isolated metastases are less common (Mitry et al., 2010; Tan et al., 2009). Surgical resection of PM is done with OS rates of approximately 50% and it has improved over the years (Lumachi et al., 2016; Goya et al., 1989). Similar to liver metastases, lung metastases can be resectable and nonresectable. The suitability criteria for the resection of pulmonary metastases include control of the primary tumor, possible complete resection, and adequate pulmonary reserve to tolerate the planned resection (Warwick and Page, 2007). Image-guided radiofrequency ablation (RFA) has been used to reduce morbidity and increase probability of success for subsequent surgical intervention, if necessary. In this technique, radiofrequency energy is applied via a needle electrode inserted into the target tissue. As the cells are agitated by the applied energy, they release heat, causing denaturation and cell death (Baba et al., 2014). Its use is more successful with small metastasis, low CEA levels, and/or no extrapulmonary metastasis. The notable advantages of RFA are that it is simple and minimally invasive; preserves pulmonary function; can be repeated; and is applicable regardless of previous treatments (Hiraki et al., 2014).

## 4. THERAPEUTIC OPTIONS FOR METASTATIC COLORECTAL CANCER

Due to advancements in therapeutic options that began with 5-FU monotherapy, the survival for metastatic patients has increased from 12 months to about 2 years (Cunningham et al., 2010). An assessment for therapeutic options for CRC patients in stage III or IV is based on the general health of the patient, organ function, concomitant nonmalignant diseases, site and extent of metastasis, involvement of multiple sites, and recurrence. To identify the optimal treatment strategy for patients with mCRC, the staging should include at least clinical examination, blood counts, liver, and renal function tests, CEA and CT scan of the abdomen and chest (or alternatively magnetic resonance imaging) (Van Cutsem et al., 2014). These patients can be administered salvage surgery, maintenance therapy, local therapy, and CT or combination therapy. The therapeutic option depends on the profile of the patient and a customized approach is desirable over one-size-fits-all (Weinberg et al., 2016). First line therapy for mCRC has progressed with advancements in surgical procedures as well as CT. Precision medicine and immunotherapy are promising approaches that may spell the future of treatment strategies.

### 4.1 Surgical Intervention

In colorectal metastases, multiple strategies are employed to allow possible and sustainable surgical resection. In case of resectable metastatic organ sites, surgery remains the first choice. In unresectable cases where the metastasized organ has been involved with multiple lesions and at sites where surgical procedures are not possible, the use of ablative and systemic therapies is explored. Mostly the use of radiotherapy and CT in an adjuvant setting can achieve this objective. The aim of these therapies is to halt the progression of cancer and make it resectable. Surgery of primary CRC tumor with synchronized mCRC is controversial, although data suggest that resection might be a positive prognostic factor for survival. Randomized controlled trials are needed to substantiate these findings (Clancy et al., 2014; Ichikawa et al., 2013; Verhoef et al., 2011; Stillwell et al., 2011).

Lung or liver metastasectomy has proved to be effective in improving OS. Successful surgery would involve a resection good enough to remove "all" metastatic nodes but leaving behind adequate parenchyma to carry out the function. Patient selection for surgical resection in metastatic sites is crucial. Complete removal of metastases with least effect on function is a challenge that can be met with careful analysis of risk factors and prognostic markers (Gonzalez and Gervaz, 2015).

In many cases it is desirable to reduce the metastasized tumor by using systemic CT, percutaneous ablation (Ridge and Solomon, 2015), and HIPEC. Percutaneous treatment using RFA has been successfully used in liver metastases but more evidence from randomized controlled trials is needed to include it in the treatment regimen of lung metastases (Lyons et al., 2015; Van Raemdonck, 2015; Kim et al., 2014).

## 4.2 Chemotherapy

Surgical interventions are promising but the morbidity associated with these procedures makes it less attractive in some patients. Systemic treatment in the form of CT is often used along with surgery to increase survival rates and lower morbidity. Radiation is also used to assist both the strategies. It is difficult to conclude the advantage of one approach over others. A combination of these approaches is often preferred over stand-alone therapy. CT has been the mainstay in battle against cancer of all stages in CRC. CT can be administered as primary, adjuvant, or neoadjuvant therapy. In the recent past, combination of CT with biological agents (bevacizumab, cetuximab, panitumumab) has been successfully employed in treating mCRC. Bevacizumab is a monoclonal antibody (mAb) that blocks angiogenesis by binding to VEGF (Sonpavde, 2004). Cetuximab is a chimeric (mouse-human) IgG1 mAb against EGFR. EGFR is overexpressed in CRC, which makes it a selective target for metastatic cells (Spano et al., 2005). Panitumumab is also an anti-EGFR mAb but unlike cetuximab it is completely human and does not contain murine portions of IgG (Bouché et al., 2010). New therapeutic agents are being reported which are effective in combination with existing regimens. The discovery of predictive and prognostic biomarkers for CT made it more customizable than surgical resection.

With current 5-year survival of 10% for patients with mCRC, it is hard to believe that less than 15 years ago, single agent 5-FU, an intravenous fluoropyrimidine, with LV was the only treatment that demonstrated benefit for the treatment of mCRC since 1950s (Davies and Goldberg, 2011). Fourteen most widely used chemotherapeutic regimens involve different combinations of 5-FU, oxaliplatin, irinotecan, VEGF, and EGFR inhibitors (Ismaili, 2011). First line of therapy still has 5-FU in intravenous or oral form as the backbone. FOLFOX (5-FU + LV + oxaliplatin) and FOLFIRI (5-FU + LV + irinotecan) are well established to provide better response and survival rates in mCRC patients (Douillard et al., 2013; de Gramont et al., 2000). With the advent of combination regimens and biological therapy, mCRC has become a quasichronic condition that requires long-term management. Median survival has improved to 20–24 months with combination cytotoxic regimens containing bevacizumab or an anti-EGFR agent (Cartwright, 2012).

# 5. CHALLENGES AND PROBABLE SOLUTIONS

There has been significant progress in identifying molecular drug targets using molecular biology techniques and appropriate laboratory models. Several therapeutic regimens have been tried with drugs developed out of these fragmented molecular evidences. Novel therapeutic and predictive molecular markers were discovered by several laboratories across the world. These targets were subsequently tested in clinical trials worldwide with poorly understood variables associated with the treatment outcome. Empirical testing of different combinations of these drugs by clinical trials has unsuccessfully addressed the underlying problems of treating mCRC. The inherent complexity of genomic alterations in late-stage cancers, coupled with numerous heterotypic interactions that occur between tumor and stromal cells, represents fundamental challenges in our quest to understand metastatic disease. The incorporation of genomic and systems level approaches, as well as technological breakthroughs in imaging and animal modeling, has galvanized the effort to overcome gaps in our understanding of metastasis (Sethi and Kang, 2011). Personalized medicine is providing new hopes and immunotherapy is adding to the arsenal against mCRC. Both these approaches are promising but are not yet mature at the clinical level.

## 5.1 Personalized Medicine

Personalized medicine is a practicing modality where prevention, diagnosis, or treatment is designed based on characteristics of an individual. At present, this characterization can be done at the gene level and combined with other factors to decide the treatment regimen. The biggest single contributor to the success of personalized medicine is the finding that KRAS mutant patients are refractory to anti-EGFR therapy. With the phenomenal increase in our capability to interrogate the genome, the excitement has built up to practice personalized medicine in the clinic. However, several technical, ethical, and safety concerns remain. This is an ongoing area of research in mCRC that has potential to change the disease statistics in a significant manner.

Cancer arises from changes in the DNA sequence in the genomes of cancer cells. These accelerating somatic mutations dysregulate signaling pathways, including EGFR, Wnt/Notch, Hedgehog, and others, with a central role in cell growth, proliferation, survival, angiogenesis, and metastasis. All of these genetic alterations can now be discovered using next-generation DNA sequencing technology. This high-throughput technology can achieve two major goals: first, to complete the catalogue of driver mutations, including point mutations, rearrangements and copy number changes, by full and targeted sequencing; and second, to explore the functional role of cancer genes and their interactions by genome-wide RNA, serial analysis of gene expression, miRNAs, protein–DNA interactions, and comprehensive analyses of transcriptomes and interactomes (Roukos, 2010a,b). One important question under debate is to define in mCRC patients, the most appropriate tissue on which DNA analysis should be performed in the context of targeted therapies. Indeed, most studies published so far are based on the analysis of genetic material extracted from primary tumors, whereas targeted therapies are mainly directed at metastatic disease treatment (Di Fiore et al., 2010). The data generated through various -omics using microarray and next generation sequencing technologies give this opportunity to understand cancer genome in its entirety and study it as a complete system. The current limitations of the therapies for mCRC are clearly evident from the patient outcome data. Recent wealth of data has highlighted the gaps, which could be responsible for this inefficacy. Inability to generalize a treatment for mass population is now well accepted among clinicians and scientists. Different populations with varying genetic landscape need a paradigm shift in the approach toward developing novel therapeutic strategies.

## 5.2 Immunotherapy

Using our own immune system against the tumor cells has been found to be effective in clinic in different types of cancer. Most success stories have come from hematological malignancies and melanoma. It was approved by United States Food and Drug Administration for bladder cancer in 1990. In CRC initial evidence was in the form of infiltrating CD8+ cells in CRC (Naito et al., 1998). Immunotherapy has affected all types of cancer in more than one way. From preventive vaccines to adoptive cell immunotherapies, checkpoint inhibitors, mAbs, cytokines and oncolytic virus therapy, the options with immunotherapy are growing. In mCRC, there is no major success story but clinical trials are ongoing and success story from other cancers could be replicated (Zhao et al., 2016). Use of immunoscore by pathologists in classifying and staging cancers could help toward applying personalized medicine with the aid of immunotherapy (Galon et al., 2014).

# 6. CONCLUSIONS

The current scenario for mCRC is clinically disappointing but the progress made in the last few decades has given us hope. Any success in treating mCRC will have an amplified effect on associated health care costs, quality of life, and survival statistics. There is a pertinent need to seek biomarkers that can help in practicing personalized medicine. Treating liver, lung, and PM will cover a major population of affected patients. Technological advancements, new additions in therapeutic agents along with paradigm shift in the clinical practices could mark a new era in the life of mCRC patients. They may live longer to witness it.

## References


Alhopuro, P., Alazzouzi, H., Sammalkorpi, H., Davalos, V., Salovaara, R., Hemminki, A., et al., 2005. SMAD4 levels and response to 5-fluorouracil in colorectal cancer. Clin. Cancer Res. 11 (17), 6311–6316.

Andreyev, H.J., Norman, A.R., Cunningham, D., Oates, J.R., Clarke, P.A., 1998. Kirsten ras mutations in patients with colorectal cancer: the multicenter "RASCAL" study. J. Natl. Cancer Inst. 90 (9), 675–684.

Andreyev, H.J., Norman, A.R., Cunningham, D., Oates, J., Dix, B.R., Iacopetta, B.J., et al., 2001. Kirsten ras mutations in patients with colorectal cancer: the "RASCAL II" study. Br. J. Cancer 85 (5), 692–696.

Aranda, E., Abad, A., Carrato, A., Cervantes, A., Garcia-Foncillas, J., Garcia Alfonso, P., et al., 2011. Treatment recommendations for metastatic colorectal cancer. Clin. Transl. Oncol. 13 (3), 162–178.

Arnold, M., Sierra, M.S., Laversanne, M., Soerjomataram, I., Jemal, A., Bray, F., 2016. Global patterns and trends in colorectal cancer incidence and mortality. Gut.

Aziz, M.A., Periyasamy, S., Al Yousef, Z., AlAbdulkarim, I., Al Otaibi, M., Alfahed, A., et al., 2014a. Integrated exon level expression analysis of driver genes explain their role in colorectal cancer. PLoS One 9 (10), e110134.

Aziz, M.A., Periyasamy, S., Yousef, Z., Deeb, A., AlOtaibi, M., 2014b. Colorectal cancer driver genes identified by patient specific comparison of cytogenetic microarray. Genom. Data 2, 29–31.

Baba, Y., Watanabe, M., Yoshida, N., Kawanaka, K., Yamashita, Y., Baba, H., 2014. Radiofrequency ablation for pulmonary metastases from gastrointestinal cancers. Ann. Thorac. Cardiovasc. Surg. 20 (2), 99–105.

Balkwill, F.R., Capasso, M., Hagemann, T., 2012. The tumor microenvironment at a glance. J. Cell Sci. 125 (Pt 23), 5591–5596.

Berber, E., Tsinberg, M., Tellioglu, G., Simpfendorfer, C.H., Siperstein, A.E., 2008. Resection versus laparoscopic radiofrequency thermal ablation of solitary colorectal liver metastasis. J. Gastrointest. Surg. 12 (11), 1967–1972.

Bilchik, A.J., DiNome, M., Saha, S., Turner, R.R., Wiese, D., McCarter, M., et al., 2006. Prospective multicenter trial of staging adequacy in colon cancer: preliminary results. Arch. Surg. 141 (6), 527–533 discussion 533–534.

Boland, C.R., Goel, A., 2010. Microsatellite instability in colorectal cancer. Gastroenterology 138 (6), 2073–2087 e3.

Bouché, O., Beretta, G.D., Alfonso, P.G., Geissler, M., 2010. The role of anti-epidermal growth factor receptor monoclonal antibody monotherapy in the treatment of metastatic colorectal cancer. Cancer Treat. Rev. 36, S1–S10.

Brouquet, A., Mortenson, M.M., Vauthey, J.N., Rodriguez-Bigas, M.A., Overman, M.J., Chang, G.J., et al., 2010. Surgical strategies for synchronous colorectal liver metastases in 156 consecutive patients: classic, combined or reverse strategy? J. Am. Coll. Surg. 210 (6), 934–941.

Brouquet, A., Abdalla, E.K., Kopetz, S., Garrett, C.R., Overman, M.J., Eng, C., et al., 2011. High survival rate after two-stage resection of advanced colorectal liver metastases: response-based selection and complete resection define outcome. J. Clin. Oncol. 29 (8), 1083–1090.

Cai, G.X., Cai, S.J., 2012. Multi-modality treatment of colorectal liver metastases. World J. Gastroenterol. 18 (1), 16–24.

Cartwright, T.H., 2012. Treatment decisions after diagnosis of metastatic colorectal cancer. Clin. Colorectal Cancer 11 (3), 155–166.

Cheng, H., Zhang, L., Cogdell, D.E., Zheng, H., Schetter, A.J., Nykter, M., et al., 2011. Circulating plasma MiR-141 is a novel biomarker for metastatic colon cancer and predicts poor prognosis. PLoS One 6 (3), e17745.

Choti, M.A., Sitzmann, J.V., Tiburi, M.F., Sumetchotimetha, W., Rangsin, R., Schulick, R.D., et al., 2002. Trends in long-term survival following liver resection for hepatic colorectal metastases. Ann. Surg. 235 (6), 759–766.

Chua, T.C., Esquivel, J., Pelz, J.O., Morris, D.L., 2013. Summary of current therapeutic options for peritoneal metastases from colorectal cancer. J. Surg. Oncol. 107 (6), 566–573.

Clancy, C., Burke, J.P., Barry, M., Kalady, M.F., Calvin Coffey, J., 2014. A meta-analysis to determine the effect of primary tumor resection for stage IV colorectal cancer with unresectable metastases on patient survival. Ann. Surg. Oncol. 21 (12), 3900–3908.

Coulson, R., 2015. Molecular profiling in resectable colorectal liver metastases: the role of KRAS mutation status in assessing prognosis in the preoperative setting. J. Adv. Pract. Oncol. 6 (5), 470–474.

Cunningham, D., Atkin, W., Lenz, H.J., Lynch, H.T., Minsky, B., Nordlinger, B., et al., 2010. Colorectal cancer. Lancet 375 (9719), 1030–1047.

Dahlmann, M., Okhrimenko, A., Marcinkowski, P., Osterland, M., Herrmann, P., Smith, J., et al., 2014. RAGE mediates S100A4-induced cell motility via MAPK/ERK and hypoxia signaling and is a prognostic biomarker for human colorectal cancer metastasis. Oncotarget 5 (10), 3220–3233.

Dai, C.Y., Haqq, C.M., Puzas, J.E., 2006. Molecular correlates of site-specific metastasis. Semin. Radiat. Oncol. 16 (2), 102–110.

Davies, J.M., Goldberg, R.M., 2011. Treatment of metastatic colorectal cancer. Semin. Oncol. 38 (4), 552–560.

De Roock, W., Jonker, D.J., Di Nicolantonio, F., Sartore-Bianchi, A., Tu, D., Siena, S., et al., 2010a. Association of KRAS p.G13D mutation with outcome in patients with chemotherapy-refractory metastatic colorectal cancer treated with cetuximab. JAMA 304 (16), 1812–1820.

De Roock, W., Claes, B., Bernasconi, D., De Schutter, J., Biesmans, B., Fountzilas, G., et al., 2010b. Effects of KRAS, BRAF, NRAS, and PIK3CA mutations on the efficacy of cetuximab plus chemotherapy in chemotherapy-refractory metastatic colorectal cancer: a retrospective consortium analysis. Lancet Oncol. 11 (8), 753–762.

Di Fiore, F., Sesboue, R., Michel, P., Sabourin, J.C., Frebourg, T., 2010. Molecular determinants of anti-EGFR sensitivity and resistance in metastatic colorectal cancer. Br. J. Cancer 103 (12), 1765–1772.

Douillard, J.Y., Oliner, K.S., Siena, S., Tabernero, J., Burkes, R., Barugel, M., et al., 2013. Panitumumab-FOLFOX4 treatment and RAS mutations in colorectal cancer. N. Engl. J. Med. 369 (11), 1023–1034.

Durno, C., Aronson, M., Bapat, B., Cohen, Z., Gallinger, S., 2005. Family history and molecular features of children, adolescents, and young adults with colorectal carcinoma. Gut 54 (8), 1146–1150.

Eldai, H., Periyasamy, S., Al Qarni, S., Al Rodayyan, M., Muhammed Mustafa, S., Deeb, A., et al., 2013. Novel genes associated with colorectal cancer are revealed by high resolution cytogenetic analysis in a patient specific manner. PLoS One 8 (10), e76251.

Ellis, L.M., Takahashi, Y., Liu, W., Shaheen, R.M., 2000. Vascular endothelial growth factor in human colon cancer: biology and therapeutic implications. Oncologist 5 (Suppl. 1), 11–15.

Erreni, M., Siddiqui, I., Marelli, G., Grizzi, F., Bianchi, P., Morone, D., et al., 2016. The fractalkine-receptor axis improves human colorectal cancer prognosis by limiting tumor metastatic dissemination. J. Immunol. 196 (2), 902–914.

Fearon, E.R., Vogelstein, B., 1990. A genetic model for colorectal tumorigenesis. Cell 61 (5), 759–767.

Fleming, M., Ravula, S., Tatishchev, S.F., Wang, H.L., 2012. Colorectal carcinoma: pathologic aspects. J. Gastrointest. Oncol. 3 (3), 153–173.

Galandiuk, S., Wieand, H.S., Moertel, C.G., Cha, S.S., Fitzgibbons Jr., R.J., Pemberton, J.H., et al., 1992. Patterns of recurrence after curative resection of carcinoma of the colon and rectum. Surg. Gynecol. Obstet. 174 (1), 27–32.

Galon, J., Mlecnik, B., Bindea, G., Angell, H.K., Berger, A., Lagorce, C., et al., 2014. Towards the introduction of the "Immunoscore" in the classification of malignant tumours. J. Pathol. 232 (2), 199–209.

Gilbert, J.M., 1983. Distribution of metastases at necropsy in colorectal cancer. Clin. Exp. Metastasis 1 (2), 97–101.

Gillams, A.R., Lees, W.R., 2008. Five-year survival following radiofrequency ablation of small, solitary, hepatic colorectal metastases. J. Vasc. Interv. Radiol. 19 (5), 712–717.

Goldberg, J., Furman, W.L., 2012. Management of colorectal carcinoma in children and young adults. J. Pediatr. Hematol. Oncol. 34 (Suppl. 2), S76–S79.

Gonzalez, M., Gervaz, P., 2015. Risk factors for survival after lung metastasectomy in colorectal cancer patients: systematic review and meta-analysis. Future Oncol. 11 (2 Suppl.), 31–33.

Gonzalez-Pons, M., Cruz-Correa, M., 2015. Colorectal cancer biomarkers: where are we now? Biomed. Res. Int. 2015, 149014.

Goya, T., Miyazawa, N., Kondo, H., Tsuchiya, R., Naruke, T., Suemasu, K., 1989. Surgical resection of pulmonary metastases from colorectal cancer. 10-year follow-up. Cancer 64 (7), 1418–1421.

de Gramont, A., Figer, A., Seymour, M., Homerin, M., Hmissi, A., Cassidy, J., et al., 2000. Leucovorin and fluorouracil with or without oxaliplatin as first-line treatment in advanced colorectal cancer. J. Clin. Oncol. 18 (16), 2938–2947.

Gulhati, P., Bowen, K.A., Liu, J., Stevens, P.D., Rychahou, P.G., Chen, M., et al., 2011. mTORC1 and mTORC2 regulate EMT, motility, and metastasis of colorectal cancer via RhoA and Rac1 signaling pathways. Cancer Res. 71 (9), 3246–3256.

Halazonetis, T.D., Gorgoulis, V.G., Bartek, J., 2008. An oncogene-induced DNA damage model for cancer development. Science 319 (5868), 1352–1355.

Hayashi, N., Ito, I., Nakamura, Y., Yanagisawa, A., Kato, Y., Nakamori, S., et al., 1995. Genetic diagnosis of lymph-node metastasis in colorectal cancer. Lancet 345 (8960), 1257–1259.

Henry, N.L., Hayes, D.F., 2012. Cancer biomarkers. Mol. Oncol. 6 (2), 140–146.

Hiraki, T., Gobara, H., Iguchi, T., Fujiwara, H., Matsui, Y., Kanazawa, S., 2014. Radiofrequency ablation as treatment for pulmonary metastasis of colorectal cancer. World J. Gastroenterol. 20 (4), 988–996.

Hugh, T.J., Kinsella, A.R., Poston, G.J., 1997a. Management strategies for colorectal liver metastases–part I. Surg. Oncol. 6 (1), 19–30.

Hugh, T.J., Kinsella, A.R., Poston, G.J., 1997b. Management strategies for colorectal liver metastases–part II. Surg. Oncol. 6 (1), 31–48.

Hur, K., Toiyama, Y., Okugawa, Y., Ide, S., Imaoka, H., Boland, C.R., et al., 2015a. Circulating microRNA-203 predicts prognosis and metastasis in human colorectal cancer. Gut.

Hur, K., Toiyama, Y., Schetter, A.J., Okugawa, Y., Harris, C.C., Boland, C.R., et al., 2015b. Identification of a metastasis-specific microRNA signature in human colorectal cancer. J. Natl. Cancer Inst. 107 (3).

Hur, K., 2015. MicroRNAs: promising biomarkers for diagnosis and therapeutic targets in human colorectal cancer metastasis. BMB Rep. 48 (4), 217–222.

Ibrahim, T., Tselikas, L., Yazbeck, C., Kattan, J., 2016. Systemic versus local therapies for colorectal cancer pulmonary metastasis: what to choose and when? J. Gastrointest. Cancer 47 (3), 223–231.

Ichikawa, Y., Goto, A., Kobayashi, N., Ota, M., Tokuhisa, M., Ishibez, A., et al., 2013. Does resection of primary lesions show survival benefit for stage IV colorectal cancer patients with unresectable metastases? Hepatogastroenterology 60 (128), 1945–1949.

Ismaili, N., 2011. Treatment of colorectal liver metastases. World J. Surg. Oncol. 9, 154.

Jayne, D.G., Fook, S., Loi, C., Seow-Choen, F., 2002. Peritoneal carcinomatosis from colorectal cancer. Br. J. Surg. 89 (12), 1545–1550.

Jones, S., Chen, W.D., Parmigiani, G., Diehl, F., Beerenwinkel, N., Antal, T., et al., 2008. Comparative lesion sequencing provides insights into tumor evolution. Proc. Natl. Acad. Sci. U. S. A. 105 (11), 4283–4288.

Jorissen, R.N., Gibbs, P., Christie, M., Prakash, S., Lipton, L., Desai, J., et al., 2009. Metastasis-associated gene expression changes predict poor outcomes in patients with Dukes stage B and C colorectal cancer. Clin. Cancer Res. 15 (24), 7642–7651.

Karlberg, M., Ohrling, K., Edler, D., Hallstrom, M., Ullen, H., Ragnhammar, P., 2010. Prognostic and predictive value of thymidylate synthase expression in primary colorectal cancer. Anticancer Res. 30 (2), 645–651.

Kim, H.K., Cho, J.H., Lee, H.Y., Lee, J., Kim, J., 2014. Pulmonary metastasectomy for colorectal cancer: how many nodules, how many times? World J. Gastroenterol. 20 (20), 6133–6145.

Kim, R., Schell, M.J., Teer, J.K., Greenawalt, D.M., Yang, M., Yeatman, T.J., 2015. Co-evolution of somatic variation in primary and metastatic colorectal cancer may expand biopsy indications in the molecular era. PLoS One 10 (5), e0126670.

Ko, S.Y., Blatch, G.L., Dass, C.R., 2014. Netrin-1 as a potential target for metastatic cancer: focus on colorectal cancer. Cancer Metastasis Rev. 33 (1), 101–113.

Kobayashi, H., Mochizuki, H., Sugihara, K., Morita, T., Kotake, K., Teramoto, T., et al., 2007. Characteristics of recurrence and surveillance tools after curative resection for colorectal cancer: a multicenter study. Surgery 141 (1), 67–75.

Koukourakis, G.V., Sotiropoulou-Lontou, A., 2011. Targeted therapy with bevacizumab (Avastin) for metastatic colorectal cancer. Clin. Transl. Oncol. 13 (10), 710–714.

de Krijger, I., Mekenkamp, L.J., Punt, C.J., Nagtegaal, I.D., 2011. MicroRNAs in colorectal cancer metastasis. J. Pathol. 224 (4), 438–447.

Lambert, L.A., 2015. Looking up: recent advances in understanding and treating peritoneal carcinomatosis. CA Cancer J. Clin. 65 (4), 284–298.

Lau, W.Y., Teoh, Y.L., Win, K.M., Lee, R.C., de Villa, V.H., Kim, Y.H., et al., 2016. Current role of selective internal radiation with yttrium-90 in liver tumors. Future Oncol. 12 (9), 1193–1204.

Lee, H., Flaherty, P., Ji, H.P., 2013. Systematic genomic identification of colorectal cancer genes delineating advanced from early clinical stage and metastasis. BMC Med. Genomics 6, 54.

Leibovici, J., Itzhaki, O., Huszar, M., Sinai, J., 2011. The tumor microenvironment: part 1. Immunotherapy 3 (11), 1367–1384.

Lievre, A., Bachet, J.B., Le Corre, D., Boige, V., Landi, B., Emile, J.F., et al., 2006. KRAS mutation status is predictive of response to cetuximab therapy in colorectal cancer. Cancer Res. 66 (8), 3992–3995.

Lipsyc, M., Yaeger, R., 2015. Impact of somatic mutations on patterns of metastasis in colorectal cancer. J. Gastrointest. Oncol. 6 (6), 645–649.

Liu, X., Ji, Q., Fan, Z., Li, Q., 2015. Cellular signaling pathways implicated in metastasis of colorectal cancer and the associated targeted agents. Future Oncol. 11 (21), 2911–2922.

Lumachi, F., Chiara, G.B., Tozzoli, R., Del Contea, A., Basso, S.M., 2016. Factors affecting survival in patients with lung metastases from colorectal cancer. A short meta-analysis. Anticancer Res. 36 (1), 13–19.

Lyons, N.J., Pathak, S., Daniels, I.R., Spiers, A., Smart, N.J., 2015. Percutaneous management of pulmonary metastases arising from colorectal cancer; a systematic review. Eur. J. Surg. Oncol. 41 (11), 1447–1455.

Malesci, A., Laghi, L., Bianchi, P., Delconte, G., Randolph, A., Torri, V., et al., 2007. Reduced likelihood of metastases in patients with microsatellite-unstable colorectal cancer. Clin. Cancer Res. 13 (13), 3831–3839.

Manfredi, S., Bouvier, A.M., Lepage, C., Hatem, C., Dancourt, V., Faivre, J., 2006a. Incidence and patterns of recurrence after resection for cure of colonic cancer in a well defined population. Br. J. Surg. 93 (9), 1115–1122.

Manfredi, S., Lepage, C., Hatem, C., Coatmeur, O., Faivre, J., Bouvier, A.M., 2006b. Epidemiology and management of liver metastases from colorectal cancer. Ann. Surg. 244 (2), 254–259.

Markowitz, S.D., Bertagnolli, M.M., 2009. Molecular origins of cancer: molecular basis of colorectal cancer. N. Engl. J. Med. 361 (25), 2449–2460.

Marz, L., Piso, P., 2015. Treatment of peritoneal metastases from colorectal cancer. Gastroenterol. Rep. (Oxf.) 3 (4), 298–302.

Mitry, E., Guiu, B., Cosconea, S., Jooste, V., Faivre, J., Bouvier, A.M., 2010. Epidemiology, management and prognosis of colorectal cancer with lung metastases: a 30-year population-based study. Gut 59 (10), 1383–1388.

Naito, Y., Saito, K., Shiiba, K., Ohuchi, A., Saigenji, K., Nagura, H., et al., 1998. CD8+ T cells infiltrated within cancer cell nests as a prognostic factor in human colorectal cancer. Cancer Res. 58 (16), 3491–3494.

Nakanishi, H., Kodera, Y., Yamamura, Y., Ito, S., Kato, T., Ezaki, T., et al., 2000. Rapid quantitative detection of carcinoembryonic antigen-expressing free tumor cells in the peritoneal cavity of gastric-cancer patients with real-time RT-PCR on the lightcycler. Int. J. Cancer 89 (5), 411–417.

Newton, K.F., Newman, W., Hill, J., 2012. Review of biomarkers in colorectal cancer. Colorectal Dis. 14 (1), 3–17.

Nordlinger, B., Sorbye, H., Glimelius, B., Poston, G.J., Schlag, P.M., Rougier, P., et al., 2008. Perioperative chemotherapy with FOLFOX4 and surgery versus surgery alone for resectable liver metastases from colorectal cancer (EORTC Intergroup trial 40983): a randomised controlled trial. Lancet 371 (9617), 1007–1016.

Nordlinger, B., Sorbye, H., Glimelius, B., Poston, G.J., Schlag, P.M., Rougier, P., et al., 2013. Perioperative FOLFOX4 chemotherapy and surgery versus surgery alone for resectable liver metastases from colorectal cancer (EORTC 40983): long-term results of a randomised, controlled, phase 3 trial. Lancet Oncol. 14 (12), 1208–1215.

Oshowo, A., Gillams, A., Harrison, E., Lees, W.R., Taylor, I., 2003. Comparison of resection and radiofrequency ablation for treatment of solitary colorectal liver metastases. Br. J. Surg. 90 (10), 1240–1243.

Patriarca, C., Macchi, R.M., Marschner, A.K., Mellstedt, H., 2012. Epithelial cell adhesion molecule expression (CD326) in cancer: a short review. Cancer Treat Rev. 38 (1), 68–75.

Peeters, M., Oliner, K.S., Parker, A., Siena, S., Van Cutsem, E., Huang, J., et al., 2013a. Massively parallel tumor multigene sequencing to evaluate response to panitumumab in a randomized phase III study of metastatic colorectal cancer. Clin. Cancer Res. 19 (7), 1902–1912.

Peeters, M., Douillard, J.Y., Van Cutsem, E., Siena, S., Zhang, K., Williams, R., et al., 2013b. Mutant KRAS codon 12 and 13 alleles in patients with metastatic colorectal cancer: assessment as prognostic and predictive biomarkers of response to panitumumab. J. Clin. Oncol. 31 (6), 759–765.

Pihl, E., Hughes, E.S., McDermott, F.T., Johnson, W.R., Katrivessis, H., 1987. Lung recurrence after curative surgery for colorectal cancer. Dis. Colon Rectum 30 (6), 417–419.

Qian, C.-N., Teh, B.T., 2012. "Seed and soil" theory of metastasis. In: Schwab, M. (Ed.), Encyclopedia of Cancer. Springer, Berlin, Heidelberg, pp. 3354–3355.

Richman, S.D., Seymour, M.T., Chambers, P., Elliott, F., Daly, C.L., Meade, A.M., et al., 2009. KRAS and BRAF mutations in advanced colorectal cancer are associated with poor prognosis but do not preclude benefit from oxaliplatin or irinotecan: results from the MRC FOCUS trial. J. Clin. Oncol. 27 (35), 5931–5937.

Ridge, C.A., Solomon, S.B., 2015. Percutaneous ablation of colorectal lung metastases. J. Gastrointest. Oncol. 6 (6), 685–692.

Roukos, D.H., 2010a. Systems medicine: a real approach for future personalized oncology? Pharmacogenomics 11 (3), 283–287.

Roukos, D.H., 2010b. Novel clinico-genome network modeling for revolutionizing genotype-phenotype-based personalized cancer care. Expert Rev. Mol. Diagn. 10 (1), 33–48.

Sadanandam, A., Lyssiotis, C.A., Homicsko, K., Collisson, E.A., Gibb, W.J., Wullschleger, S., et al., 2013. A colorectal cancer classification system that associates cellular phenotype and responses to therapy. Nat. Med. 19 (5), 619–625.

Sawai, H., Yasuda, A., Ochi, N., Ma, J., Matsuo, Y., Wakasugi, T., et al., 2008. Loss of PTEN expression is associated with colorectal cancer liver metastasis and poor patient survival. BMC Gastroenterol. 8, 56.

Segelman, J., Granath, F., Holm, T., Machado, M., Mahteme, H., Martling, A., 2012. Incidence, prevalence and risk factors for peritoneal carcinomatosis from colorectal cancer. Br. J. Surg. 99 (5), 699–705.

Sethi, N., Kang, Y., 2011. Unravelling the complexity of metastasis - molecular understanding and targeted therapies. Nat. Rev. Cancer 11 (10), 735–748.

Sluiter, N., de Cuba, E., Kwakman, R., Kazemier, G., Meijer, G., Te Velde, E.A., 2016. Adhesion molecules in peritoneal dissemination: function, prognostic relevance and therapeutic options. Clin. Exp. Metastasis 33 (5), 401–416.

Solbiati, L., Livraghi, T., Goldberg, S.N., Ierace, T., Meloni, F., Dellanoce, M., et al., 2001. Percutaneous radio-frequency ablation of hepatic metastases from colorectal cancer: long-term results in 117 patients. Radiology 221 (1), 159–166.

Sonoshita, M., Aoki, M., Fuwa, H., Aoki, K., Hosogi, H., Sakai, Y., et al., 2011. Suppression of colon cancer metastasis by Aes through inhibition of Notch signaling. Cancer Cell 19 (1), 125–137.

Sonpavde, G., 2004. Bevacizumab in colorectal cancer. N. Engl. J. Med. 351 (16), 1690–1691 author reply 1690–1691.

Spano, J.P., Lagorce, C., Atlan, D., Milano, G., Domont, J., Benamouzig, R., et al., 2005. Impact of EGFR expression on colorectal cancer patient prognosis and survival. Ann. Oncol. 16 (1), 102–108.

Stein, U., Walther, W., Arlt, F., Schwabe, H., Smith, J., Fichtner, I., et al., 2009. MACC1, a newly identified key regulator of HGF-MET signaling, predicts colon cancer metastasis. Nat. Med. 15 (1), 59–67.

Stillwell, A.P., Ho, Y.H., Veitch, C., 2011. Systematic review of prognostic factors related to overall survival in patients with stage IV colorectal cancer and unresectable metastases. World J. Surg. 35 (3), 684–692.

Sugarbaker, P.H., 1995. Patient selection and treatment of peritoneal carcinomatosis from colorectal and appendiceal cancer. World J. Surg. 19 (2), 235–240.

Tabuchi, Y., Deguchi, H., Imanishi, K., Saitoh, Y., 1991. Colorectal cancer patients with high risk of hematogenous metastasis: correlation with CEA levels in peripheral and draining venous blood during the period of operation. J. Surg. Oncol. 47 (2), 87–91.

Takamoto, T., Sugawara, Y., Hashimoto, T., Shimada, K., Inoue, K., Maruyama, Y., et al., 2016. Dynamic assessment of carcinoembryonic antigen in the first month after liver resection for colorectal liver metastases as a rapid-recurrence predictor. J. Surg. Oncol. 113 (4), 463–468.

Takebayashi, K., Murata, S., Yamamoto, H., Ishida, M., Yamaguchi, T., Kojima, M., et al., 2014. Surgery-induced peritoneal cancer cells in patients who have undergone curative gastrectomy for gastric cancer. Ann. Surg. Oncol. 21 (6), 1991–1997.

Tan, K.K., Lopes Gde Jr., L., Sim, R., 2009. How uncommon are isolated lung metastases in colorectal cancer? A review from database of 754 patients over 4 years. J. Gastrointest. Surg. 13 (4), 642–648.

Tanis, E., Julie, C., Emile, J.F., Mauer, M., Nordlinger, B., Aust, D., et al., 2015. Prognostic impact of immune response in resectable colorectal liver metastases treated by surgery alone or surgery with perioperative FOLFOX in the randomised EORTC study 40983. Eur. J. Cancer 51 (17), 2708–2717.

Thomassen, I., van Gestel, Y.R., Lemmens, V.E., de Hingh, I.H., 2013. Incidence, prognosis, and treatment options for patients with synchronous peritoneal carcinomatosis and liver metastases from colorectal origin. Dis. Colon Rectum 56 (12), 1373–1380.

Toiyama, Y., Takahashi, M., Hur, K., Nagasaka, T., Tanaka, K., Inoue, Y., et al., 2013. Serum miR-21 as a diagnostic and prognostic biomarker in colorectal cancer. J. Natl. Cancer Inst. 105 (12), 849–859.

Toiyama, Y., Hur, K., Tanaka, K., Inoue, Y., Kusunoki, M., Boland, C.R., et al., 2014. Serum miR-200c is a novel prognostic and metastasis-predictive biomarker in patients with colorectal cancer. Ann. Surg. 259 (4), 735–743.

Tokunaga, T., Oshika, Y., Abe, Y., Ozeki, Y., Sadahiro, S., Kijima, H., et al., 1998. Vascular endothelial growth factor (VEGF) mRNA isoform expression pattern is correlated with liver metastasis and poor prognosis in colon cancer. Br. J. Cancer 77 (6), 998–1002.

Tran, B., Kopetz, S., Tie, J., Gibbs, P., Jiang, Z.Q., Lieu, C.H., et al., 2011. Impact of BRAF mutation and microsatellite instability on the pattern of metastatic spread and prognosis in metastatic colorectal cancer. Cancer 117 (20), 4623–4632.

Van Cutsem, E., Cervantes, A., Nordlinger, B., Arnold, D., 2014. Metastatic colorectal cancer: ESMO Clinical Practice Guidelines for diagnosis, treatment and follow-up. Ann. Oncol. 25 (Suppl. 3), iii1–iii9.

Van Raemdonck, D., 2015. Pulmonary metastasectomy: common practice but is it also best practice? Future Oncol. 11 (2 Suppl.), 11–14.

Vatandoust, S., Price, T.J., Karapetis, C.S., 2015. Colorectal cancer: metastases to a single organ. World J. Gastroenterol. 21 (41), 11767–11776.

Verhoef, C., de Wilt, J.H., Burger, J.W., Verheul, H.M., Koopman, M., 2011. Surgery of the primary in stage IV colorectal cancer with unresectable metastases. Eur. J. Cancer 47 (Suppl. 3), S61–S66.

Villeneuve, P.J., Sundaresan, R.S., 2009. Surgical management of colorectal lung metastasis. Clin. Colon Rectal Surg. 22 (4), 233–241.

Wang, C.C., Li, J., 2012. An update on chemotherapy of colorectal liver metastases. World J. Gastroenterol. 18 (1), 25–33.

Warwick, R., Page, R., 2007. Resection of pulmonary metastases from colorectal carcinoma. Eur. J. Surg. Oncol. 33 (Suppl. 2), S59–S63.

Weidle, U.H., Birzele, F., Kruger, A., 2015. Molecular targets and pathways involved in liver metastasis of colorectal cancer. Clin. Exp. Metastasis 32 (6), 623–635.

Weinberg, B.A., Marshall, J.L., Hartley, M., Salem, M.E., 2016. A paradigm shift from one-size-fits-all to tailor-made therapy for metastatic colorectal cancer. Clin. Adv. Hematol. Oncol. 14 (2), 116–128.

Weis, S.M., Cheresh, D.A., 2011. Tumor angiogenesis: molecular pathways and therapeutic targets. Nat. Med. 17 (11), 1359–1370.

Weiss, L., Grundmann, E., Torhorst, J., Hartveit, F., Moberg, I., Eder, M., et al., 1986. Haematogenous metastatic patterns in colonic carcinoma: an analysis of 1541 necropsies. J. Pathol. 150 (3), 195–203.

Welch, J.P., Donaldson, G.A., 1979. The clinical correlation of an autopsy study of recurrent colorectal cancer. Ann. Surg. 189 (4), 496–502.

Worthley, D.L., Leggett, B.A., 2010. Colorectal cancer: molecular features and clinical opportunities. Clin. Biochem. Rev. 31 (2), 31–38.

Zhao, H., Wang, Y., Yu, J., Wei, F., Cao, S., Zhang, X., et al., 2016. Autologous cytokine-induced killer cells improves overall survival of metastatic colorectal cancer patients: results from a phase II clinical trial. Clin. Colorectal Cancer.

CHAPTER

# 7

# Metastatic Pancreatic Cancer: Current State and Future Directions

*P. Martinelli*[1], *E. Lonardo*[2,3]

[1]Medical University Vienna, Austria; [2]Institute for Research in Biomedicine, Barcelona, Spain; [3]Institute of Genetics and Biophysics 'A. Buzzati-Traverso', Naples, Italy

## 1. INTRODUCTION

### 1.1 Pancreatic Cancer Epidemiology, Demographics, and Etiology

Pancreatic ductal adenocarcinoma (PDAC) is the most common type of pancreatic cancer, accounting for around 85–90% of all malignant pancreatic tumors (Maitra and Hruban, 2008). Although it is not particularly frequent (less than 3% of all cancer cases), PDAC ranks fourth among the causes of cancer-related death in the Western countries, because it is almost invariably lethal, with a median 5-year overall survival of around 5% (Li et al., 2004; Siegel et al., 2015). Such a dismal prognosis stems from a number of features, including the lack of reliable methods for early diagnosis, the resistance to therapy, and the extremely high metastatic potential.

The causes of PDAC remain unknown, and although several environmental factors have been linked to PDAC development, evidence for a causative role exists only for tobacco (Raimondi et al., 2007). Chronic pancreatitis and diabetes are also rather well-established risk factors. Only 5–10% of cases have a family history of the disease and in some families with increased risk of PDAC, a genetic cause is suspected but still not identified (Hidalgo, 2010). In most cases, PDAC does not cause any specific symptoms: as a result, most patients are diagnosed with already advanced metastatic disease and are not eligible for surgical resection of the tumor, which is to date the only curative option, when combined with adjuvant chemotherapy (Neoptolemos et al., 2004, 2010; Oettle et al., 2007). Only less than 20% of patients present with localized disease and can undergo surgery, which extends the survival up to 24% (Oettle et al., 2007), whereas the vast majority of patients present already with locally advanced or metastatic disease and receive largely palliative chemotherapeutic treatments, with an extension of the overall survival by only few month (Conroy et al., 2011; Von Hoff et al., 2013). Furthermore, around 80% of patients undergoing surgery eventually relapse with metastatic lethal disease. As a result, around 90% of PDAC patients have metastases at the time of death, and most patients succumb to metastatic disease

*Introduction to Cancer Metastasis*
http://dx.doi.org/10.1016/B978-0-12-804003-4.00007-4

rather than to the primary tumor (Yachida and Iacobuzio-Donahue, 2009). The most common site of distant metastasis is the liver, followed by the peritoneum, lung and pleura, bones, and adrenal glands; however, metastases from pancreatic cancer have been reported in virtually every organ site (Yachida and Iacobuzio-Donahue, 2009). Although the liver is the most common metastatic site, up to one-third of patients do not have hepatic metastases at autopsy, instead they have metastases mostly in the lungs and in the abdominal lymph nodes, suggesting that multiple routes of spreading are present in PDAC. The extent of metastatic disease is highly variable, ranging from less than 10 to more than 1000 lesions (Iacobuzio-Donahue et al., 2009).

## 1.2 Genetics and Biology of Pancreatic Ductal Adenocarcinoma

Histologically, the typical PDAC shows duct-like structures embedded within abundant desmoplastic stroma. Due to its histological appearance and to the fact that it expresses some markers of pancreatic ductal cells (i.e., KRT7, KRT8, KRT18, KRT19), PDAC is classically considered to originate from normal ductal cells, where multiple genetic alterations accumulate and contribute to the progressive cell transformation. However, evidence from mouse models support that normal acinar cells can also give rise to ductal tumors through a metaplastic process called acinar-ductal metaplasia (ADM) (Guerra et al., 2007). ADM is often observed in patients with pancreatic cancer (Brune et al., 2006).

Molecular pathology studies and extensive genomic analyses have established a multistep progression model for PDAC, which is still widely accepted in the field (Fig. 7.1) (Maitra and Hruban, 2008; Hezel et al., 2006), whereby the accumulation of genetic alterations causes the appearance of preneoplastic lesions of increasing grade, which then progresses to full-blown invasive carcinoma. The most common preneoplastic lesions are pancreatic intraepithelial neoplasias (PanINs), which go from minimal dysplasia (PanIN-1A and 1B) to more severe dysplasia with nuclear atypia (PanIN-2), to in situ carcinoma (PanIN-3), and finally to PDAC (Maitra and Hruban, 2008). Other less common preneoplastic lesions that were described in PDAC patients are intraductal papillary mucinous neoplasms, and mucinous cystic neoplasm (Maitra and Hruban, 2008; Hezel et al., 2006).

Genetic profiling studies have revealed that the most common alterations present in PDACs are activating mutations of the *KRAS* oncogene (>95% of cases) and inactivation of the tumor suppressors *CDKN2A* (>95%), *TP53* (50–75%), and *SMAD4* (55%) (Maitra and Hruban, 2008; Hezel et al., 2006). Activating *KRAS* mutations are often detected in PanINs of the lowest grade (up to 90%) and also in tissues without PDAC (Luttges et al., 1999; Kanda et al., 2012) and therefore they are considered an initiating event. Consistently, pancreas-specific expression of mutant *KRAS* in engineered mouse models is sufficient to initiate murine PDAC (mPDAC), although additional genetic alterations or induction of pancreatitis are required for tumor progression (Guerra et al., 2007; Bardeesy et al., 2006; De La et al., 2008; Guerra et al., 2003; Hingorani et al., 2005). Inactivation of *CDKN2A* was also observed in PanINs of the lowest grade, even in the absence of *KRAS* mutations, supporting the hypothesis that this is also an early event during PDAC initiation (Kanda et al., 2012). On the contrary, inactivating mutations of the *TP53* and *SMAD4* tumor suppressors were detected only in higher grade PanIN-2 and PanIN-3, suggesting that they are late event, which contribute to the progression of PDAC rather than to its initiation.

It has to be noted that the available evidence supporting the linearity of PDAC progression is not extremely strong, and alternative routes could be leading to the development of PDAC from distinct low-grade lesions or even directly from

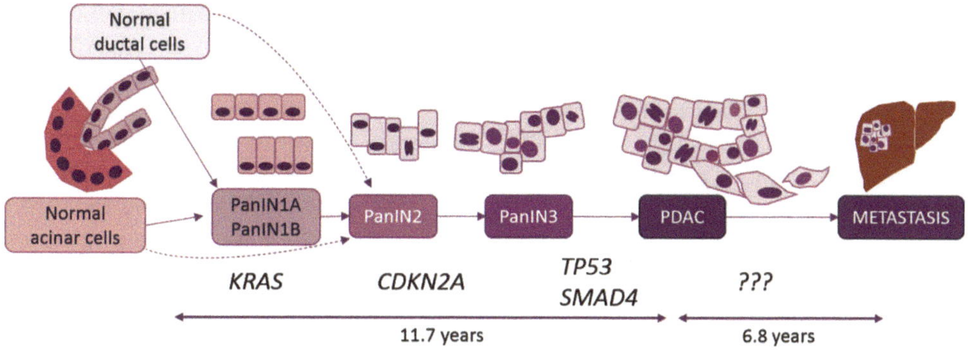


FIGURE 7.1 **Genetic progression model of pancreatic adenocarcinoma.** Pancreatic ductal adenocarcinoma can originate from normal ductal or acinar cells. The progression from histologically normal epithelium to low-grade pancreatic intraepithelial neoplasia (PanIN), to high-grade PanIN, to invasive carcinoma (left to right) is associated with the accumulation of specific genetic alterations. The hallmark genetic alterations are depicted here. Activating mutation of the KRAS oncogene is an early event and considered to be the initiating alteration. Loss of the CDKN2A tumor suppressor is an intermediate event, whereas the loss of TP53 and SMAD4 are late events.

high-grade PanINs (Real et al., 2008) (Fig. 7.1). Single-cell sequencing technologies might shed light on this issue in the future.

More recently, whole exome and whole genome sequencing efforts have revealed a much more complex pattern of genomic alterations, affecting multiple core signaling pathways (Jones et al., 2008; Waddell et al., 2015). Although PDAC genetics is now rather well known, the translation of such knowledge into a clinically relevant advance is still hard, because the few genes that are frequently altered have proven difficult to target therapeutically, whereas the remaining alterations occur in <10% of cases, and therefore are not ideal targets for therapy.

# 2. MANAGEMENT OF PANCREATIC DUCTAL ADENOCARCINOMA

## 2.1 Screening

Considering that an earlier diagnosis of PDAC could improve the outcome (Yachida et al., 2010), the International Cancer of the Pancreas Screening Society recently took an important step toward the creation of a rational approach to screen high-risk individuals for early, curable, pancreatic neoplasia (Canto et al., 2013) with a study on pancreatic cancer surveillance in high-risk individuals (either because of germline mutations in PDAC susceptibility genes or because of strong family history), in which endoscopic ultrasonography (EUS) and/or magnetic resonance (MRI) were used to screen 139 asymptomatic patients. Among them, nine (6%) were found to have at least one clinically relevant pancreatic lesion (defined as a solid lesion or cystic lesions) at baseline screening (Tanaka et al., 2012). Of note, the two imaging procedures were shown to be complementary: EUS was superior at detecting solid lesions and MRI was better at detecting cystic lesions. Remarkably, 135 out of 139 patients underwent a second observation 1 year later, during which 12 clinically relevant lesions were observed in eight patients. Overall, these discoveries highlighted the potential of EUS and MRI for early diagnosis and the necessity to follow-up high-risk patients after their baseline studies. The best approach to screening high-risk individuals will most likely combine screening technologies (e.g., imaging, circulating tumor DNA and novel molecular analyses of blood, pancreatic juice, and pancreatic cyst fluid; Bettegowda et al., 2014) and clinical data such as a detailed family history. There is currently no screening program for PDAC and we look forward to the development of a body of clinical work that will guide clinicians in the management of individuals at high risk of pancreatic cancer.

## 2.2 Diagnosis

### *2.2.1 Presenting Symptoms*

Around 65% of PDAC lesions are spotted in the head of the pancreas (HOP), 15% in the body and tail and the remainder involves the whole gland (Artinyan et al., 2008). Tumor localization in the HOP can cause biliary obstruction, thus leading to early diagnosis. Nonetheless, the symptoms (e.g., back pain, obstructive jaundice, cachexia, and weight loss; Stapley et al., 2012) tend to manifest once the tumor has invaded surrounding tissues or metastasized to other organs. Tumor development is accompanied by a plethora of systemic changes affecting many aspects of physiology, particularly global metabolism. Diabetes or hyperglycemia can also occur due to damage to the pancreatic parenchyma. Furthermore, obstruction of the pancreatic duct by the tumor may lead to acute pancreatitis (Stapley et al., 2012). A better understanding of the biological processes occurring in the early stages of PDAC could facilitate preventive and accurate diagnosis.

### *2.2.2 Laboratory Tests and Radiological Imaging*

Where there is a suspicion for pancreatic cancer, further investigations must be undertaken such as blood tests including tumor markers. In the clinical setting of pancreatic cancer, the most commonly used and validated tumor marker is carbohydrate antigen 19-9 (CA 19-9) (Kannagi, 2007). Some limitations include a considerable number of false positive (Wu et al., 2013), the lack of sensitivity for early pancreatic cancers (Steinberg, 1990), and the lower expression in poorly differentiated cancers (Steinberg, 1990). The serological marker ADAM-9 was also found to have diagnostic and prognostic value for PDAC patients (Grutzmann et al., 2004). Recently, Mayer et al. have identified a prediagnostic signature of branched-chain amino acids (BCAAs) that could improve risk prediction up to 5 years before the disease appears (Mayers et al., 2014). They screened plasma samples from four large-scale prospective public health cohorts, including hundreds of subjects who subsequently developed pancreatic cancer. High blood levels of isoleucine, leucine, and valine were associated with increased risk of developing PDAC. The structure of this study is remarkable for its use of prospective samples collected from cohort studies years before a cancer diagnosis was made. Moreover, the team confirmed this correlation in vivo using a mouse model for PDAC, where they also investigated the mechanisms connecting changes in BCAAs with disease appearance.

Another diagnostic approach is the radiological imaging of the pancreas, which also aids staging of PDAC and may include computed tomography (CT, with a sensitivity of 91%), EUS (with a diagnostic sensitivity of 98%; De Angelis et al., 2013), MRI (with a diagnostic sensitivity of 84%; Bipat et al., 2005), and/or positron emission tomography. As part of the initial investigation of abdominal symptoms, an ultrasound scan of the pancreas may be performed but this imaging modality is unable to definitively diagnose and stage patients. So far, a triphasic CT protocol (i.e., arterial, late arterial, and venous phases) seems to be the gold standard system.

## 2.3 Staging

Patients with pancreatic cancer are staged according to the American Joint Committee on Cancer tumor-lymph node-metastasis (TNM) staging system, which provides a detailed summary of the tumor size and the extent of spreading within and outside the pancreas (Varadhachary et al., 2006). Tumor grading, which accounts for the histological appearance of the tumor, is also usually included in the diagnosis, since it is important in determining the prognosis.

For treatment purpose, however, clinicians use a simpler staging system, based on whether the tumor can be surgically removed or not. According to this system, tumors can be "resectable"—when the tumor is confined to the pancreas or has only partially invaded the peritoneal cavity; "borderline resectable"—when the tumor has reached some nearby blood vessel but might still be removed surgically; and "unresectable"—when the tumor has grown around major blood vessels in the peritoneal cavity (locally advanced) or has spread to distal organs (metastatic).

## 2.4 Treatment and Clinical Trials

The critical choice for pancreatic cancer treatment is its surgical removal. Nevertheless, in advanced metastatic stages, the treatment (chemo- or radiotherapy) aims to minimize the patient's disease-related symptoms and to increase the survival by optimal control of metastases. Currently, a limited number of drugs are accessible for the treatment of metastatic PDAC. Gemcitabine has been the reference chemotherapeutic regimen for patients with metastatic pancreatic cancer since 1997 (Burris et al., 1997). Recently, different combinations such as FOLFIRINOX (folinic acid, 5-fluorouracil [5-FU], irinotecan, and oxaliplatin) and Abraxane (albumin-bound paclitaxel also named nab-paclitaxel) plus gemcitabine have been attempted (Conroy et al., 2011; Von Hoff et al., 2013) and are currently undergoing testing in the adjuvant setting in the Partenariat de Recherche en Oncologie Digestive (PRODIGE) (Avan et al., 2013) and metastatic pancreatic cancer trial (MPACT) (Luo et al., 2013) studies, respectively. Some other combinations, such as gemcitabine with gamma-secretase inhibitor (MK-0752) or connective tissue growth factor inhibitor (FG-3019), have also shown encouraging results (Yu et al., 2014). Another promising drug combination is erlotinib (epidermal growth factor inhibitor) plus gemcitabine that was evaluated in the AIO PK0104 trial (Ormanns et al., 2014).

Chemotherapy combined with radiotherapy is used in patients with unresectable disease; nevertheless, the advantage of chemo/radiotherapy in the adjuvant setting, examined in the -ESPAC-1) trial (Neoptolemos et al., 2004), was found to decrease the median survival of patients by 2 months compared with the chemotherapy only group (Neoptolemos et al., 2004). These findings where confirmed in the 2000–01 SFRO study (Chauffert et al., 2008).

Given the poor clinical outcome of patients with PDAC, novel treatment strategies are urgently needed.

## 2.5 Novel Treatment Opportunities in Pancreatic Ductal Adenocarcinoma

### 2.5.1 *Next-Generation Sequencing*

This is performed in patients with advanced-stage cancer and is an approach that is rapidly becoming common in clinical practice (Iyer et al., 2012). The roadmap to attaining advances in this area will include initiatives (e.g., the National Cancer Institute (NCI) Exceptional Responders Initiative) intended to improve the comprehension of the unique responses to treatment at the molecular level (Takebe et al., 2015). These initiatives will provide the rationale to run trials based on biomarkers or histology.

### 2.5.2 *Drugs Targeting Tumor Metabolism*

A factor restraining the efficacy of therapies is the impaired drug delivery due to the desmoplastic reaction that occurs in PDAC (Jacobetz et al., 2013). Moreover, the flow of oxygen and nutrients to cancer cells is limited by the desmoplastic reaction. In this microenvironment, cancer cells reprogram metabolic pathways and activate autophagy (Le et al., 2012). Currently, hydroxychloroquine (an autophagy inhibitor) in combination with nab-paclitaxel and gemcitabine is undergoing evaluation both in the neoadjuvant setting (NCT01978184) as well as in advanced-stage disease (NCT01506973) (Garrido-Laguna and Hidalgo, 2015).

### 2.5.3 *Drugs Targeting the Stromal Compartment*

A growing body of evidence indicates a strong communication between epithelial cancer cells and the microenvironment, resulting in cancer cell proliferation, survival, and resistance to therapy (Garrido-Laguna et al., 2011). Sadly, a recent clinical trial based on the use of Sonic hedgehog (SHH) pathway inhibitor failed. Indeed, treatment with SHH inhibitors increased tumor progression through a mechanism that seems to involve the release of the tumor-restraining stimuli of the stroma (Madden, 2012).

Recently, a trial based on the use of PEGPH20 in combination with nab-paclitaxel and gemcitabine was reopened (HALO-109-202) (Medicine, UNLo, 2014). Finally, the S1313 trial is evaluating PEGPH20 in combination with FOLFIRINOX, also in untreated patients with metastatic PDAC (NCT01959139) (Medicine, UNLo, 2014).

### 2.5.4 *Immunotherapy and Cancer Vaccines*

PDAC is characterized by an immunosuppressive microenvironment (Wormann et al., 2014). Disrupting this immunosuppressive system might provide new opportunities in the treatment of this deadly disease.

A method used to enhance the immune response is through activation of CD40 in tumor-associated macrophages. A recent phase I study (Beatty et al., 2013) provided preliminary evidence of activity of gemcitabine combined with CP-870,893 (a CD40 agonist) in chemotherapy-naïve patients with advanced-stage PDAC.

In 2008, the allogenic cancer vaccine GVAX pancreas was developed, which is currently in phase II of clinical trial (Laheru et al., 2008) and is giving encouraging results.

## 3. MOLECULAR MECHANISMS DRIVING METASTASIS

Although the first events of pancreatic carcinogenesis have been extensively studied and rather well elucidated, the molecular mechanisms responsible for tumor spread and formation of metastases are relatively unexplored. One reason for this is that the management of advanced-stage pancreatic cancer usually does not involve surgical resection of either the primary tumor or the metastasis, therefore only a limited amount of samples are available for medical research beyond the initial biopsy. Therefore, the current knowledge about pathology and molecular features of advanced-stage pancreatic cancer comes mostly from a handful of studies where PDAC samples were collected at the time of autopsy. More recently, rapid autopsy programs were established, which allow for the simultaneous collection and comparison of matched primary tumor metastasis samples (Iacobuzio-Donahue et al., 2009; Campbell et al., 2010; Embuscado et al., 2005).

These studies revealed the existence of a subgroup of patients (12%) that did not present any metastasis at autopsy and an additional group (18%) that only had limited metastatic burden (less than 10 lesions), whereas the rest had massive metastatic spread, with multiple gross deposits, often more than 100 and even more than 1000 in some patients. Interestingly, the histopathological features of the tumors at diagnosis did not significantly differ between oligo- or nonmetastatic and multimetastatic patients. These observations indicate that pancreatic cancer could include two distinct entities, sharing similar morphologic features at diagnosis, but with different metastatic efficiency. Furthermore, only a very small percentage of tumor cells will form distal metastases, which indicates that the development of metastatic disease requires the acquisition of one or more genetic or epigenetic events that may promote survival of disseminated cells within the circulation and in the target organ site.

Following the classical "seed and soil" hypothesis that was proposed more than a century ago, whereby metastatic cancer cells (the seed) only form metastases when they reach a favorable environment (the soil), such (epi) genetic events would generate a cell with metastatic potential (the seed), whereas additional events will likely take place in the target organ (the soil) for the metastatic cell to survive and grow. A complex crosstalk between the two compartments (tumor cells/environment) allows for the formation of metastases.

## 3.1 The Seed: Tumor Cell-Specific Events Promoting Metastasis

### *3.1.1 Genetic Events*

As mentioned before, four genetic alterations, namely activating mutations in the *KRAS* gene and inactivation of the *CDKN2A*, *TP53*, and *SMAD4* tumor suppressor genes, are extremely common in PDAC and are considered its hallmarks. These four genes were systematically analyzed in matched pairs of samples collected from primary tumors and metastases of the same patients (Embuscado et al., 2005). In addition to this, their role in PDAC initiation and progression was extensively investigated with different mouse models of mutant *KRas*-driven PDAC (Mazur and Siveke, 2012).

*KRAS* activating mutations do not seem to be a prerequisite for metastasis formation, as they are present universally in both primary and metastatic cancers from the same patient (Embuscado et al., 2005). However, the wild-type *KRAS* allele is lost more frequently in PDAC cell lines generated from metastatic tissue compared to cells generated from primary lesions (Qiu et al., 2011), suggesting that the enforcement of KRAS-dependent signaling pathways is important for metastasis. Likewise, no differences were observed in the inactivation rate of *CDKN2A* in metastatic versus primary lesions from the same patient (Embuscado et al., 2005), but its deletion accelerates metastasis formation in a mouse model of PDAC (Mazur and Siveke, 2012). *SMAD4* (also known as *DPC4*) was found to be inactivated significantly more frequently in metastatic cancers compared with early stage PDACs, and loss of SMAD4 protein expression correlates with a shorter survival in patients with resectable PDAC (Embuscado et al., 2005; Biankin et al., 2002). Furthermore, *SMAD4* loss promotes the formation of metastases in mouse models of PDAC (Mazur and Siveke, 2012). Inactivating mutations in *TP53* also correlated with higher metastatic spread, although not to the extent of *SMAD4*.

SMAD4 is a central mediator of both the tumor growth factor beta (TGF-β) and bone morphogenetic protein (BMP) signaling pathways (Massague, 2008), and in pancreatic cancer cell lines, *SMAD4* alterations universally impair part of the TGF-β-dependent response, including the transcriptional regulation of the p21 and p15 cell cycle inhibitors (Peng et al., 2002). Furthermore, the TGF-β-dependent pathway is known to have a broad impact on cell differentiation and invasive potential (see below).

Iacobuzio-Donahue et al. deeply investigated also the reciprocal correlation between *TP53* and *SMAD4* inactivating mutations and observed that *SMAD4* loss of function frequently occurred in the presence of *TP53* mutations, whereas the opposite was not observed, suggesting that *SMAD4* alterations occur later during tumor progression (Iacobuzio-Donahue et al., 2009). The type of *TP53* alteration that coexisted with *SMAD4* inactivation also showed some interesting differences: missense mutations were enriched in *SMAD4*-null tumors, whereas null mutations were found more frequently in *SMAD4*-positive tumors, suggesting that inactivation of *SMAD4* might be selected to remove the residual cell cycle inhibitory function of *TP53* missense mutant proteins. Indeed, the distinct combinations of *TP53* and *SMAD4* alterations were both highly correlated with massive metastatic burden (Iacobuzio-Donahue et al., 2009).

Beyond the four hallmark alterations mentioned above, next generation sequencing revealed that a complex pattern of mutations and genomic rearrangements is acquired during tumor spreading to distal sites, but the evolution of metastatic subclones from the founders clones seems to follow very heterogeneous paths, which sometimes can even converge (Yachida et al., 2010; Campbell et al., 2010). Recently, a phylogenetic tree following the evolution of each founder clone from the primary tumor to the distal metastatic site was created. These studies revealed that multiple clones within the primary tumor, carrying distinct combinations of genetic alterations, have similar capacity to form metastases; however, specific metastasis-driver mutations could not be identified. Additionally, metastatic clones undergo further molecular evolution, which, in some cases, seems to be organ specific, supporting a crosstalk between tumor cells and the microenvironment in the distal organ (consistent with the "seed and soil" theory). In summary, these studies showed that (1) there is considerable genetic heterogeneity among cells capable of initiating metastasis, even within the same primary tumor; (2) a metastasis-promoting genomic signature could be present in some tumor clones, but the specific genetic drivers are hard to identify; (3) metastatic clones further evolve after they have reached and colonized the distal site, and this evolution can depend on the targeted organ.

### 3.1.2 *Loss of Differentiation and Epithelial-to-Mesenchymal Transition*

Histological examination of primary PDAC in patients with and without associated metastatic disease showed that patients with distant metastases tended to have more poorly differentiated tumors (Kamisawa et al., 1995). Furthermore, advanced-stage pancreatic cancers more commonly showed high-grade (anaplastic) histology at autopsy, independently of the clinical stage and grade at initial diagnosis. Finally, loss of E-cadherin expression is almost universally associated with anaplastic features and with poor outcome in PDAC patients. Altogether, these observations indicate that loss of epithelial differentiation occurs during tumor progression and is important for the acquisition of metastatic potential (Iacobuzio-Donahue et al., 2009; Winter et al., 2008; Hong et al., 2011), but the underlying molecular mechanisms are still not fully clear.

Loss of epithelial traits such as cell–cell contacts and cell polarity, accompanied by the progressive acquisition of more mesenchymal features, including motility and invasiveness, constitutes an epithelial-to-mesenchymal transition (EMT). EMT is a key process during embryo development that can be abnormally reactivated during tumor progression, to endow carcinoma cells with metastatic potential (Brabletz, 2012). During EMT, epithelial cells reorganize their cytoskeleton, undergo a change in the signaling programs that define the cell shape, and reprogram gene expression; this enables them to enter the bloodstream (a process known as intravasation) and also confers them some stem cell properties (Mani et al., 2008). At this point, metastatic cells are typically referred to as circulating tumor cells/cancer stem cells (CTCs/CSCs) (Hoffman, 1999). To form metastases, the CTCs/CSCs must exit the bloodstream (extravasation) and reach favorable niches in distant organs (Tuveson et al., 2004). Interestingly, the metastatic colonization step requires a reverse mesenchymal-to-epithelial transition (MET), which facilitates the differentiation of CTCs/CSCs into non-CTCs/CSCs (Ocana et al., 2012; Tsai et al., 2012). The association of EMT and CTCs/CSCs in metastasis has open potential opportunities for new therapeutic intervention (Hingorani et al., 2003). Currently, over 500 registered clinical trials involve inhibition of EMT and CTCs to block tumor cell invasion in early stage carcinomas (Rhim et al., 2012). Nevertheless, in patients with spread micrometastases, eliminating mesenchymal cancer cells or preventing MET should be the goal. For instance, salinomycin was identified as a compound that induced selective elimination of mesenchymal-type breast cancer cells and reduced the proportion of breast CSCs (Gu et al., 2002).

EMT is mostly controlled by three families of transcription factors, SNAI, ZEB, and TWIST, which repress E-cadherin expression and the epithelial differentiation program (Thiery et al., 2009). Recently, a previously unrecognized anti-EMT function is emerging for a few transcription factors that are major regulators of pancreatic development and maintenance. Among them is GATA6, which is essential for the maintenance of the acinar compartment in the adult mouse pancreas (Martinelli et al., 2013) and was recently described to act as a tumor suppressor in a mouse model of PDAC (Martinelli et al., 2015) as well as in patients (Martinelli et al., 2016), thanks to its proepithelial and antimesenchymal function. In particular, loss of GATA6 in tumors was associated with altered differentiation and its downregulation enhanced the metastatic potential of PDAC cells (Martinelli et al., 2016). Similarly, FOXA1/2 (also important during pancreas development) represses EMT in PDAC cells through the direct induction of E-cadherin expression (Song et al., 2010). A tumor suppressive function was described also for Ptf1a, a master regulator of pancreatic development, whose loss is sufficient to induce ADM in the mouse pancreas (Krah et al., 2015). Although it appears that PTF1A is lost early during tumor onset, it is possible that its loss favors the appearance of a poorly differentiated tumor clone with higher metastatic potential.

### 3.1.3 *Cancer Stem Cells Concept*

It is becoming progressively evident that cancer heterogeneity is not only confined to distinct evolutionary subclones within each tumor (Yachida et al., 2010), but it is also driven by phenotypic and functional heterogeneity and plasticity within each subclone (Garcia-Silva et al., 2013). The cancer stem cells (CSCs) or tumor-initiating cells are responsible for intraclonal heterogeneity (Hermann et al., 2007, 2008, 2009; Li et al., 2007). Specifically, CSCs represent a subset of cancer cells for which many conclusive evidence have provided that they represent the root of the disease by giving rise to all differentiated progenies within each cancer subclone (Hermann et al., 2007; Kreso and Dick, 2014; Kawaguchi et al., 2002). More importantly, these cells are essential for metastatic behavior and also represent an important source for disease relapse following treatment, even when relatively effective drugs capable of inducing initial tumor regression (e.g., nab-paclitaxel) are used (Hermann et al., 2007, 2010; Gallmeier et al., 2011; Mueller et al., 2009). It is important to note that CSCs do not necessarily represent bona fide stem cells and neither do they arise from tissue stem cells in many instances, but rather they have acquired certain features of stem cells. Most of these are functionally defined; for example, CSCs possess indefinite self-renewal capacity, are resistant to conventional chemotherapy, and show increased invasiveness, which promote metastatic activity. CSCs are also believed to be resistant to chemotherapy due to high expression or activation of ATP-binding cassette transporter proteins (Obata et al., 2001), aldehyde dehydrogenase, prosurvival signaling components (Kawaguchi et al., 2002), and DNA repair molecules (Lonardo et al., 2015).

## 3.2 The Soil: Tumor Microenvironment

PDAC is characterized by the presence of a dense stroma, which is very heterogeneous and comprises cellular and noncellular components, such as fibroblasts, myofibroblasts, pancreatic stellate cells, immune cells, blood vessels, and extracellular matrix (ECM). Additionally, fibronectin, fibrillar collagen, hyaluronic acid, as well as soluble growth factors (e.g., TGF-β, fibroblast growth factor-FGF, and vascular endothelial growth factor-VEGF) are secreted and stored in a highly dynamic location, which is termed the tumor microenvironment (Cioffi et al., 2015). The function and composition of desmoplastic (or fibrotic) stroma is highly dynamic during cancer progression, and changes in relation to the evolving genetic landscape of cancer cells. The fibrotic stroma in PDAC forms an environment that promotes cancer progression by enhancing pancreatic tumor growth as well as regional and distant metastasis (Waghray et al., 2013). Furthermore, the stroma has been shown to contribute to resistance to chemotherapy and radiation therapy (Hwang et al., 2008) and to constitute a barrier to the delivery of therapeutic agents (Calon et al., 2015). Based on the observations that the administration of a Hedgehog (HH) signaling inhibitor enhanced drug delivery and prognosis in a mouse model, a phase 2 clinical trial was initiated, which, however, had to be stopped early due to the poor survival in the patients who received this inhibitor (Miranda-Lorenzo et al., 2014). The difference between the preclinical study and clinical trial can be explained by the fact that desmoplasia can also have a protective role. Even though a significant body of literature supports the advantages in targeting the stroma, very recently two different groups have provided a new opposite point of view. They demonstrated, using genetically engineered mouse models (GEMMs) in which alpha smooth muscle actin (a-SMA) + cancer-associated fibroblasts (CAFs) were depleted or by selective ablation of Sonic-HH signaling, that not only the stroma exerts a protective role but also its targeting can lead to a more biologically aggressive form of PDAC (Lonardo et al., 2013; Guardiola et al., 2012). Analogously, patients with PDAC and elevated stromal activation have a better prognosis than patients with less stromal activity (Lonardo et al., 2013). In light of these findings, it is highly demanded to reassess various therapeutic approaches and to perform additional analyses to discern the utility to target specific stromal cell types and/or their secreted factors.

In order to study the role of the tumor stroma in pancreatic cancer biology, researchers have utilized stellate cells from rat and human pancreata (PSC) (Hwang et al., 2008; Ikenaga et al., 2010; Erkan et al., 2012). During desmoplasia, quiescent, lipid droplet-storing PSCs become activated via TGF-β and platelet-derived growth factor-mediated transcriptional programs, and elicit ECM synthesis and deposition (Lonardo et al., 2012). Whether all the fibroblasts within the stroma are derived from stellate cells is yet to be defined. A strong relation between the activation of ECM-producing-PSCs and enhanced cancer growth has been proposed in several models (De Conto et al., 2012). However, it is reasonable to expect that other components might contribute to the fibroblast population and cancer progression, including perivascular fibroblasts that become activated upon tissue injury and bone-marrow derived cells. Moreover, the presence of other stromal components such as tumor associated macrophages (TAMs) in the tumor microenvironment has been shown to be a poor prognostic marker (Pollard, 2004). TAMs promote tumor cell invasiveness by breaking down ECM in PDAC (Benson et al., 2012). Although some studies have focused on the characterization of different cell types within the pancreatic tumor microenvironment to identify potential prognostic and therapeutic markers, others have focused on the proteins secreted by these different cell types.

## 3.3 Tumor-Stroma Crosstalk

### 3.3.1 *The TGF-β Pathway*

The TGF-β signaling pathway plays a context-dependent dual role in pancreatic cancer, being either a tumor suppressor or tumor promoter. Three TGF-β isoforms, TGF-β1, 2, and 3 have been identified. Of these, TGF-β1 is the most abundant in humans. TGF-β signaling activation starts with the binding of TGF-βs to three high affinity receptors (TGF-βRI, II, and III). Consequently, TGF-βRI and TGF-βRII dimerize and induce the phosphorylation of the intracellular effectors SMAD2/SMAD3, which in turn bind to the common transducer SMAD4, translocate to the nucleus, and activate the transcription of target genes. SMAD2/3 can be negatively modulated by SMAD7 and SMURFs (Drabsch and ten Dijke, 2012). Multivariate analysis showed that low-level expression of SMAD7 and SMAD4 correlates with lymph node metastasis and poor prognosis in patients with pancreatic cancer (Wang et al., 2009; Jiang et al., 2012). Mutation in at least one of the TGF-β pathway genes occurs in 100% of the pancreatic tumors. A recent exome sequencing analysis showed that *TGF-βR2* is one of the most commonly mutated genes in PDAC (Capocefalo et al., 2010). Moreover, loss of *SMAD4* occurs in 50% of patients with pancreatic cancer and has been used to define pancreatic origin in cases of metastases of unknown primary (Dorado et al., 2011). Jonson et al. showed that the structural rearrangement of *SMAD4* occurs in 42% of pancreatic tumor cells. Moreover, the study suggested TGF-βR1-SMAD4 as a major target for inactivation in PDAC, whereas the expression of TGF-βR2, TGF-βR3, and of the receptors involved in TGF-β activation is preserved (Dorado et al., 2011). Deregulated TGF-β signaling has crucial roles in both the development of tumors and metastasis, and its influence extends beyond the cancer cells themselves, to the tumor microenvironment and the whole organism. The role of the TGF-β pathway as a tumor promoter or suppressor within cancer cells is still a matter of debate, due to its differential effects at the early and late stages of carcinogenesis. In contrast, it is now rather well established that the TGF-β pathway contributes to the generation of a favorable microenvironment for tumor growth and metastasis. Therefore, targeting the TGF-β pathway for cancer therapy may be considered primarily as a microenvironment-targeted strategy.

#### 3.3.2 *Tumor-Derived Exosomes and the Metastatic Niche*

As mentioned above, metastatic cells need a favorable environment to home and grow in distant sites. Accumulating evidence supports the existence of metastasis-supportive environments, called "metastatic niches," which are formed in target organs before the colonization and growth of metastatic cells (Peinado et al., 2011). Tumor cells can orchestrate the formation of distant metastatic niches through the generation of exosomes, small membrane-bound vesicles containing multiple nucleic acids and peptides that are produced and shed by many different cells, including both normal and cancer cells. Exosomes can fuse with the membrane of target cells, thereby transferring their peculiar content of nucleic acids and proteins. This is a recently described mode of cell-to-cell communication, which is not exclusive of cancer cells, but is particularly active in a subset of them (Peinado et al., 2011; ELA et al., 2013). A recent report described that PDAC cells produce exosomes that are specifically taken up by liver Kupffer cells, which subsequently activate a signaling cascade resulting in a fibrotic metastatic niche where tumor cells can efficiently home and grow (Costa-Silva et al., 2015). Authors described the protein macrophage migration inhibitory factor as a key mediator of this process, thus opening new opportunities for antimetastatic therapies.

## 4. IN VIVO EXPERIMENTAL MODELS

State-of-the-art models that best recapitulate aspects of human disease are patient-derived xenografts (PDX), GEMMs, and tumor organoids (Fig. 7.2).

### 4.1 Patient-Derived Xenografts

Xenotransplantation of human cancer cells or tumor tissue (freshly obtained from surgery) into immunodeficient mice has been used for several years as a main tool to study cancer biology and for preclinical screening to develop novel cancer therapeutics. Two well-established PDX model

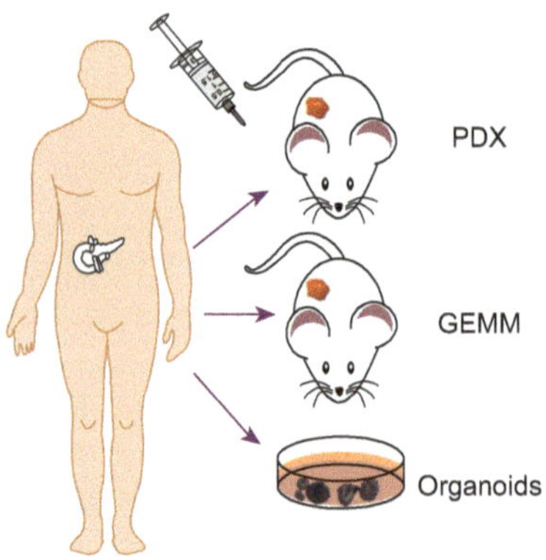


| Model | Cost | Clinical Relevance | Reproducibility | Advantages | Limitations |
|---|---|---|---|---|---|
| **PDX** | | | | | |
| subcutaneous | $ | ☺ | √√√ | Simple generation, collection and tumor measurements<br>Excellent for tumor biology studies and drug screenings | Localized tumor growth and no metastasis formation<br>Failure to resemble the genetic heterogeneity of the original tumors<br>Lack of histopathological features (i.e., microenvironment complexity)<br>Absence of immune response |
| orthotopic | $$ | ☺☺☺ | √ | More realistic biology of the tumor<br>Pontent tools for preclinical studies (i.e., diagnostic, therapeutic, and prevention)<br>Excellent for metastasis studies | Technically difficult<br>Absence of immune response |
| **GEMM** | $$$ | ☺☺ | √√√ | Recapitulates human disease<br>Specific genetic alterations | Lack of specificity in the promoters driving Cre expression<br>Lack of time specific gene ablation/activation<br>Time consuming |
| **Organoids** | $$ | ☺ | √√ | Simple generation and collection<br>Recapitulates in vitro human disease<br>Pontent tools for preclinical studies (i.e., diagnostic and therapeutic) | Absence of immune response |

FIGURE 7.2 **Mouse models of pancreatic adenocarcinoma: patient-derived xenograft (PDX), genetically engineered mouse models (GEMM), and patient-derived organoids.** PDXs are developed directly from implantation of surgical, biopsy, or tumor cell specimens into immunocompromised mice. GEMMs of human cancer are developed through the organ-specific expression of driver mutations in immunocompetent mice. New in vitro techniques to develop tumor cell organoid cultures are also under active investigation as preclinical models that might more accurately predict the efficacy of targeted therapeutic agents. Collectively, these different approaches provide a series of tools for studying preclinical modalities including diagnostic, therapeutic, and prevention studies and reveal information about the tumor biology of pancreatic ductal adenocarcinoma as reported in the table.

have been widely used for studying PDAC: the subcutaneous and the orthotopic.

Subcutaneous inoculation is the most commonly used and is a valuable tool for tumor biology studies and drug screenings, thanks to its easy generation, low cost, and simplicity of tumor measurements. Nevertheless, it fails to resemble the genetic heterogeneity of the original tumors and lack histopathological features, such as the microenvironment complexity, of human tumors. Another limitation is the localized tumor growth, which provides a natural barrier to metastatic spread. Orthotopic implantation directly into the pancreas is more complex and challenging to generate, but offers a more realistic biology of the tumor, including the capacity to give rise to metastasis at different rates (Hoffman, 1999). Remarkably, the xenografts generated from tumor tissue better preserve the genetic heterogeneity and recapitulate the histological features and stroma compartment of the original tumor (Rubio-Viqueira and Hidalgo, 2009).

A major limitation of both PDX models is the absence of immune response from the host, nevertheless PDXs still provide a potent tools for both preclinical (i.e., diagnostic, therapeutic, and prevention) and tumor biology studies of PDAC.

## 4.2 Genetically Engineered Mouse Models

The development of GEMMs for PDAC has improved from initial use of transgenic expression of oncogenes to more recent refined methods based on the Cre-Lox system to generate compound mutant mice with pancreas lineage-specific activation or inactivation of genes involved in the disease. One of the most remarkable model resembling the human PDAC is based on the expression of a mutant KRas oncogene in pancreatic progenitor cells, thanks to the combination of the conditional $KRas^{LSLG12D}$ or $KRas^{LSLG12V}$ knock-in alleles with a pancreas-specific Cre-expressing allele, such as *Pdx1-Cre* or *Ptf1a-Cre* (Tuveson et al., 2004; Hingorani et al., 2003). These GEMMs form PanIN with 100% penetrance and develop PDAC at an advanced age. As mutant *KRas*-based GEMMs develop tumors very late, additional GEMMs with concomitant loss or mutation of *Tp53* and loss of *p16Ink4a/p19Arf* were developed. *Tp53*-based GEMM recapitulates many aspects of the human disease, including histopathology, incidence of metastasis to liver and lung, comorbidities, activation of distinctive signaling pathways, and genome instability (Hingorani et al., 2005). One important limitation is due to the tissue-specific promoters driving Cre expression, which are often not fully specific. For instance, the Pdx1 and the Ptf1a promoters are active not only in the pancreatic progenitor compartment (Gu et al., 2002; Kawaguchi et al., 2002) but also in the developing foregut and epidermis (Pdx1) (Mazur et al., 2010), and in the cerebellum (Ptf1a) (Obata et al., 2001), which can potentially lead to confounding results.

Another important issue of the GEMM that has been used in the past is that all Cre-dependent genetic changes take place in one single event, usually at the time of tumor initiation. This does not fully recapitulate the multistep progression that is observed in patients, and might also give rise to inconsistencies, especially when considering the late events in tumor progression. The new dual recombinase model that was developed recently, where two independent recombinases are activated in sequence, will help in the study of metastatic disease.

## 4.3 Organoid Models

Cells obtained from patients' tumors are usually grown in monolayer (two dimensional), for subsequent studies. These culture conditions, however, do recapitulate the whole tumor biology. Three-dimensional (3D) culture strategies have been developed to mainly study metastatic cells such as CSCs (Lonardo et al., 2015). Recently, Boj et al. developed 3D organoids from normal and malignant murine pancreatic tissues and used this model system to investigate PDAC pathogenesis (Boj et al., 2015). They showed that following orthotopic transplantation, pancreatic organoids derived from either normal pancreata or murine PDAC perfectly recapitulate the normal ductal architecture and the relevant features of disease progression in vitro, respectively. Similar results were obtained with organoids generated from human tissue, both normal and tumor. Finally, the authors demonstrated the utility of organoids to identify molecular pathways that correlate with disease progression and that represent therapeutic and diagnostic opportunities. Importantly, authors showed that organoids can be generated even from as little tissue as what is obtained from a fine needle biopsy. This will hopefully help researchers to obtain samples from patients who do not undergo surgery.

# 5. OPEN QUESTION: ONSET AND DYNAMICS OF METASTASES

An important question that has not been fully answered yet is whether the extremely poor prognosis of PDAC and the high prevalence of metastasis at the time of diagnosis are a consequence of early metastatic dissemination or the result of late diagnosis. This issue has a strong clinical relevance, as late formation of metastases would leave a wider window for more effective treatment of pancreatic cancer. Rhim et al. used genetically engineered mice to explore this question, where they performed a genetic lineage tracing of the pancreatic cells during tumor initiation and progression (Rhim et al., 2012). Within this model, they observed that cells of pancreatic origin were already present in the circulation before tumors were detectable. These metastatic cells displayed features of mesenchymal cells, including loss of E-cadherin and expression of Zeb1 and N-cadherin, indicating that EMT and cell spreading are early events, which can even precede the formation of a primary tumor. However, sequencing studies using patients' samples collected at autopsy have revealed that the majority of genetic events are shared between primary tumor and metastasis, supporting a model whereby metastatic dissemination is a late event (Yachida et al., 2010; Campbell et al., 2010). A rigorous mathematical model was developed from these data, which proposed a temporal line of events, with an average interval of 11.7 years between the initiating event and the development of a parental clone with full blown tumorigenic potential, followed by a 6.8-year gap until a metastatic subclone is formed, with patients dying an average of 2.7 years later (Yachida et al., 2010). To understand this apparent contradiction between mouse- and human-derived data, it is important to consider that the mouse model used by Rhim et al. was characterized by the loss of *Tp53* at the time of tumor initiation, whereas this is considered a late event during PDAC progression in patients, as inactivating mutations in *TP53* were not observed in low-grade PanINs (Kanda et al., 2012), therefore the KPC mouse model does not fully recapitulate the stepwise progression of PDAC. In particular, p53 was shown to inhibit EMT in normal pancreatic acinar cells (Pinho et al., 2011), which might explain the early induction of EMT in *Tp53*-null pancreatic cells. Furthermore, the detection of circulating pancreatic cancer cells at an early stage does not directly prove that these cells can give rise to metastases, and it might thereby underestimate the time that is required for the formation of a metastatic lesion. The newly established dual recombination-based mouse model for PDAC, where two genetic events can be uncoupled in time thanks to the use of two independent recombinases (Schonhuber et al., 2014) will help researchers develop better models to study the metastatic progression of PDAC.

As mentioned above, the issue of when does metastasis develop has clinical relevance. A recent study developed a computational model for studying the kinetics of metastatic disease (Haeno et al., 2012), which supported this idea. In this report, Haeno et al. investigated PDAC progression with a mathematical framework based on comprehensive data of 228 patients, which included data on the primary tumor size and metastatic burden at diagnosis, at autopsy, and at least one intermediate time. This stochastic mathematical model revealed that metastatic subclones are most likely generated during the exponential growth of the tumor, and that all patients are expected to harbor metastases or at least metastasis-enabled cells at the time of diagnosis. Based on this prediction, the early administration of neoadjuvant chemotherapeutic agents that efficiently reduce the growth rate of these metastatic cells is predicted to be more beneficial than upfront resection of the primary tumor.

## 6. CONCLUSIONS

- Up to 90% of PDAC patients have metastasis at autopsy, ranging from less than 10 to more than 1000 lesions. Most patients succumb to metastatic disease, rather than to the primary tumor.
- Liver is the most common site of metastasis in PDAC patients.
- The most frequent genetic alterations in PDAC are activating mutations in the KRAS oncogene and loss of the tumor suppressors CDKN2A, TP53, and SMAD4. Plenty of additional alterations were identified with lower frequency. Among the major genetic alterations, only loss of SMAD4 showed a significant association with metastatic disease.
- The detailed molecular mechanisms underlying metastatic conversion of PDAC cells are largely unknown, as a result no targeted therapy is available, and patients with metastatic PDAC are given mostly palliative treatments.
- Loss of differentiation, EMT, and the acquisition of CSC properties increase the metastatic potential of pancreatic cancer cells.

- A complex and dynamic crosstalk is present between cancer cells and the various component of the tumor stroma. TGF-β signaling is among the most important pathways mediating this crosstalk in the short distance, whereas exosomes are involved in the long distance communication between the primary tumor and the distal metastatic sites.
- The available in vivo models (PDXs, GEMMs, and organoids) do not fully recapitulate the whole metastatic process taking place in PDAC patients, but are nevertheless valuable tools for research and preclinical studies. Newly developed models have the potential to be more informative.
- The onset and dynamics of metastatic disease in PDAC patients are still a matter of debate, but recent next-generation sequencing-based studies coupled to mathematical modeling suggests that every patient carries cells with metastatic potential at diagnosis, supporting the clinical value of neoadjuvant therapy.
- Despite the recent advances in our knowledge about metastatic PDAC, more investigation is still required to fully understand how a PDAC cell converts into a metastatic cell to be able to therapeutically tackle the essential events.

## References

Artinyan, A., Soriano, P.A., Prendergast, C., Low, T., Ellenhorn, J.D., Kim, J., 2008. The anatomic location of pancreatic cancer is a prognostic factor for survival. HPB 10 (5), 371–376.

Avan, A., Quint, K., Nicolini, F., et al., 2013. Enhancement of the antiproliferative activity of gemcitabine by modulation of c-Met pathway in pancreatic cancer. Curr. Pharm. Des. 19 (5), 940–950.

Bardeesy, N., Aguirre, A.J., Chu, G.C., et al., 2006. Both p16(Ink4a) and the p19(Arf)-p53 pathway constrain progression of pancreatic adenocarcinoma in the mouse. Proc. Natl. Acad. Sci. U.S.A. 103 (15), 5947–5952.

Beatty, G.L., Torigian, D.A., Chiorean, E.G., et al., 2013. A phase I study of an agonist CD40 monoclonal antibody (CP-870,893) in combination with gemcitabine in patients with advanced pancreatic ductal adenocarcinoma. Clin. Cancer Res. 19 (22), 6286–6295.

Benson, D.D., Meng, X., Fullerton, D.A., et al., 2012. Activation state of stromal inflammatory cells in murine metastatic pancreatic adenocarcinoma. Am. J. Physiol. Regul. Integr. Comp. Physiol. 302 (9), R1067–R1075.

Bettegowda, C., Sausen, M., Leary, R.J., et al., 2014. Detection of circulating tumor DNA in early- and late-stage human malignancies. Sci. Transl. Med. 6 (224), 224ra24.

Biankin, A.V., Morey, A.L., Lee, C.S., et al., 2002. DPC4/Smad4 expression and outcome in pancreatic ductal adenocarcinoma. J. Clin. Oncol. 20 (23), 4531–4542.

Bipat, S., Phoa, S.S., van Delden, O.M., et al., 2005. Ultrasonography, computed tomography and magnetic resonance imaging for diagnosis and determining resectability of pancreatic adenocarcinoma: a meta-analysis. J. Comput. Assist. Tomogr. 29 (4), 438–445.

Boj, S.F., Hwang, C.I., Baker, L.A., et al., 2015. Organoid models of human and mouse ductal pancreatic cancer. Cell 160 (1–2), 324–338.

Brabletz, T., 2012. To differentiate or not–routes towards metastasis. Nat. Rev. Cancer 12 (6), 425–436.

Brune, K., Abe, T., Canto, M., et al., 2006. Multifocal neoplastic precursor lesions associated with lobular atrophy of the pancreas in patients having a strong family history of pancreatic cancer. Am. J. Surg. Pathol. 30 (9), 1067–1076.

Burris 3rd, H.A., Moore, M.J., Andersen, J., et al., 1997. Improvements in survival and clinical benefit with gemcitabine as first-line therapy for patients with advanced pancreas cancer: a randomized trial. J. Clin. Oncol. 15 (6), 2403–2413.

Calon, A., Lonardo, E., Berenguer-Llergo, A., et al., 2015. Stromal gene expression defines poor-prognosis subtypes in colorectal cancer. Nat. Genet. 47 (4), 320–329.

Campbell, P.J., Yachida, S., Mudie, L.J., et al., 2010. The patterns and dynamics of genomic instability in metastatic pancreatic cancer. Nature 467 (7319), 1109–1113.

Canto, M.I., Harinck, F., Hruban, R.H., et al., 2013. International cancer of the pancreas screening (CAPS) consortium summit on the management of patients with increased risk for familial pancreatic cancer. Gut 62 (3), 339–347.

Capocefalo, A., Franceschi, V., Mertens, P.P., et al., 2010. Expression and secretion of bluetongue virus serotype 8 (BTV-8)VP2 outer capsid protein by mammalian cells. J. Virol. Methods 169 (2), 420–424.

Chauffert, B., Mornex, F., Bonnetain, F., et al., 2008. Phase III trial comparing intensive induction chemoradiotherapy (60 Gy, infusional 5-FU and intermittent cisplatin) followed by maintenance gemcitabine with gemcitabine alone for locally advanced unresectable pancreatic cancer. Definitive results of the 2000-01 FFCD/SFRO study. Ann. Oncol. 19 (9), 1592–1599.

Cioffi, M., Trabulo, S.M., Sanchez-Ripoll, Y., et al., 2015. The miR-17-92 cluster counteracts quiescence and chemoresistance in a distinct subpopulation of pancreatic cancer stem cells. Gut 64 (12).

Conroy, T., Desseigne, F., Ychou, M., et al., 2011. FOLFIRINOX versus gemcitabine for metastatic pancreatic cancer. N. Engl. J. Med. 364 (19), 1817–1825.

Costa-Silva, B., Aiello, N.M., Ocean, A.J., et al., 2015. Pancreatic cancer exosomes initiate pre-metastatic niche formation in the liver. Nat. Cell. Biol. 17 (6), 816–826.

De Angelis, C., Brizzi, R.F., Pellicano, R., 2013. Endoscopic ultrasonography for pancreatic cancer: current and future perspectives. J. Gastrointest. Oncol. 4 (2), 220–230.

De Conto, F., Di Lonardo, E., Arcangeletti, M.C., Chezzi, C., Medici, M.C., Calderaro, A., 2012. Highly dynamic microtubules improve the effectiveness of early stages of human influenza A/NWS/33 virus infection in LLC-MK2 cells. PLos One 7 (7), e41207.

De La, O.J., Emerson, L.L., Goodman, J.L., et al., 2008. Notch and Kras reprogram pancreatic acinar cells to ductal intraepithelial neoplasia. Proc. Natl. Acad. Sci. U.S.A. 105 (48), 18907–18912.

Dorado, J., Lonardo, E., Miranda-Lorenzo, I., Heeschen, C., 2011. Pancreatic cancer stem cells: new insights and perspectives. J. Gastroenterol. 46 (8), 966–973.

Drabsch, Y., ten Dijke, P., 2012. TGF-beta signalling and its role in cancer progression and metastasis. Cancer Metastasis Rev. 31 (3–4), 553–568.

EL Andaloussi, S., Mager, I., Breakefield, X.O., Wood, M.J., 2013. Extracellular vesicles: biology and emerging therapeutic opportunities. Nat. Rev. Drug. Discov. 12 (5), 347–357.

Embuscado, E.E., Laheru, D., Ricci, F., et al., 2005. Immortalizing the complexity of cancer metastasis: genetic features of lethal metastatic pancreatic cancer obtained from rapid autopsy. Cancer Biol. Ther. 4 (5), 548–554.

Erkan, M., Adler, G., Apte, M.V., et al., 2012. StellaTUM: current consensus and discussion on pancreatic stellate cell research. Gut 61 (2), 172–178.

Gallmeier, E., Hermann, P.C., Mueller, M.T., et al., 2011. Inhibition of ataxia telangiectasia- and Rad3-related function abrogates the in vitro and in vivo tumorigenicity of human colon cancer cells through depletion of the CD133(+) tumor-initiating cell fraction. Stem Cells 29 (3), 418–429.

Garcia-Silva, S., Frias-Aldeguer, J., Heeschen, C., 2013. Stem cells & pancreatic cancer. Pancreatology 13 (2), 110–113.

Garrido-Laguna, I., Hidalgo, M., 2015. Pancreatic cancer: from state-of-the-art treatments to promising novel therapies. Nat. Rev. Clin. Oncol. 12 (6), 319–334.

Garrido-Laguna, I., Uson, M., Rajeshkumar, N.V., et al., 2011. Tumor engraftment in nude mice and enrichment in stroma-related gene pathways predict poor survival and resistance to gemcitabine in patients with pancreatic cancer. Clin. Cancer Res. 17 (17), 5793–5800.

Grutzmann, R., Luttges, J., Sipos, B., et al., 2004. ADAM9 expression in pancreatic cancer is associated with tumour type and is a prognostic factor in ductal adenocarcinoma. Br. J. Cancer 90 (5), 1053–1058.

Gu, G., Dubauskaite, J., Melton, D.A., 2002. Direct evidence for the pancreatic lineage: NGN3+ cells are islet progenitors and are distinct from duct progenitors. Development 129 (10), 2447–2457.

Guardiola, O., Lafuste, P., Brunelli, S., et al., 2012. Cripto regulates skeletal muscle regeneration and modulates satellite cell determination by antagonizing myostatin. Proc. Natl. Acad. Sci. U.S.A. 109 (47), E3231–E3240.

Guerra, C., Mijimolle, N., Dhawahir, A., et al., 2003. Tumor induction by an endogenous K-ras oncogene is highly dependent on cellular context. Cancer Cell 4 (2), 111–120.

Guerra, C., Schuhmacher, A.J., Canamero, M., et al., 2007. Chronic pancreatitis is essential for induction of pancreatic ductal adenocarcinoma by K-Ras oncogenes in adult mice. Cancer Cell 11 (3), 291–302.

Haeno, H., Gonen, M., Davis, M.B., Herman, J.M., Iacobuzio-Donahue, C.A., Michor, F., 2012. Computational modeling of pancreatic cancer reveals kinetics of metastasis suggesting optimum treatment strategies. Cell 148 (1–2), 362–375.

Hermann, P.C., Huber, S.L., Herrler, T., et al., 2007. Distinct populations of cancer stem cells determine tumor growth and metastatic activity in human pancreatic cancer. Cell Stem Cell 1 (3), 313–323.

Hermann, P.C., Huber, S.L., Heeschen, C., 2008. Metastatic cancer stem cells: a new target for anti-cancer therapy? Cell Cycle 7 (2), 188–193.

Hermann, P.C., Mueller, M.T., Heeschen, C., 2009. Pancreatic cancer stem cells–insights and perspectives. Expert Opin. Biol. Ther. 9 (10), 1271–1278.

Hermann, P.C., Bhaskar, S., Cioffi, M., Heeschen, C., 2010. Cancer stem cells in solid tumors. Semin. Cancer Biol. 20 (2), 77–84.

Hezel, A.F., Kimmelman, A.C., Stanger, B.Z., Bardeesy, N., Depinho, R.A., 2006. Genetics and biology of pancreatic ductal adenocarcinoma. Genes. Dev. 20 (10), 1218–1249.

Hidalgo, M., 2010. Pancreatic cancer. N. Engl. J. Med. 362 (17), 1605–1617.

Hingorani, S.R., Petricoin, E.F., Maitra, A., et al., 2003. Preinvasive and invasive ductal pancreatic cancer and its early detection in the mouse. Cancer Cell 4 (6), 437–450.

Hingorani, S.R., Wang, L., Multani, A.S., et al., 2005. Trp53R172H and KrasG12D cooperate to promote chromosomal instability and widely metastatic pancreatic ductal adenocarcinoma in mice. Cancer Cell 7 (5), 469–483.

Hoffman, R.M., 1999. Orthotopic metastatic mouse models for anticancer drug discovery and evaluation: a bridge to the clinic. Invest. New Drugs 17 (4), 343–359.

Hong, S.M., Li, A., Olino, K., et al., 2011. Loss of E-cadherin expression and outcome among patients with resectable pancreatic adenocarcinomas. Mod. Pathol. 24 (9), 1237–1247.

Hwang, R.F., Moore, T., Arumugam, T., et al., 2008. Cancer-associated stromal fibroblasts promote pancreatic tumor progression. Cancer Res. 68 (3), 918–926.

Iacobuzio-Donahue, C.A., Fu, B.J., Yachida, S., et al., 2009. DPC4 gene status of the primary carcinoma correlates with patterns of failure in patients with pancreatic cancer. J. Clin. Oncol. 27 (11), 1806–1813.

Ikenaga, N., Ohuchida, K., Mizumoto, K., et al., 2010. CD10+ pancreatic stellate cells enhance the progression of pancreatic cancer. Gastroenterology 139 (3), 1041–1051 51 e1– e8.

Iyer, G., Hanrahan, A.J., Milowsky, M.I., et al., 2012. Genome sequencing identifies a basis for everolimus sensitivity. Science 338 (6104), 221.

Jacobetz, M.A., Chan, D.S., Neesse, A., et al., 2013. Hyaluronan impairs vascular function and drug delivery in a mouse model of pancreatic cancer. Gut 62 (1), 112–120.

Jiang, H., He, C., Geng, S., et al., 2012. RhoT1 and Smad4 are correlated with lymph node metastasis and overall survival in pancreatic cancer. Plos One 7 (7), e42234.

Jones, S., Zhang, X., Parsons, D.W., et al., 2008. Core signaling pathways in human pancreatic cancers revealed by global genomic analyses. Science 321 (5897), 1801–1806.

Kamisawa, T., Isawa, T., Koike, M., Tsuruta, K., Okamoto, A., 1995. Hematogenous metastases of pancreatic ductal carcinoma. Pancreas 11 (4), 345–349.

Kanda, M., Matthaei, H., Wu, J., et al., 2012. Presence of somatic mutations in most early-stage pancreatic intraepithelial neoplasia. Gastroenterology 142 (4) 730–733 e9.

Kannagi, R., 2007. Carbohydrate antigen sialyl Lewis a–its pathophysiological significance and induction mechanism in cancer progression. Chang. Gung. Med. J. 30 (3), 189–209.

Kawaguchi, Y., Cooper, B., Gannon, M., Ray, M., MacDonald, R.J., Wright, C.V., 2002. The role of the transcriptional regulator Ptf1a in converting intestinal to pancreatic progenitors. Nat. Genet. 32 (1), 128–134.

Krah, N.M., De La, O.J., Swift, G.H., et al., 2015. The acinar differentiation determinant PTF1A inhibits initiation of pancreatic ductal adenocarcinoma. eLife 4.

Kreso, A., Dick, J.E., 2014. Evolution of the cancer stem cell model. Cell Stem Cell 14 (3), 275–291.

Laheru, D., Lutz, E., Burke, J., et al., 2008. Allogeneic granulocyte macrophage colony-stimulating factor-secreting tumor immunotherapy alone or in sequence with cyclophosphamide for metastatic pancreatic cancer: a pilot study of safety, feasibility, and immune activation. Clin. Cancer Res. 14 (5), 1455–1463.

Le, A., Rajeshkumar, N.V., Maitra, A., Dang, C.V., 2012. Conceptual framework for cutting the pancreatic cancer fuel supply. Clin. Cancer Res. 18 (16), 4285–4290.

Li, D., Xie, K., Wolff, R., Abbruzzese, J.L., 2004. Pancreatic cancer. Lancet 363 (9414), 1049–1057.

Li, C., Heidt, D.G., Dalerba, P., et al., 2007. Identification of pancreatic cancer stem cells. Cancer Res. 67 (3), 1030–1037.

Lonardo, E., Frias-Aldeguer, J., Hermann, P.C., Heeschen, C., 2012. Pancreatic stellate cells form a niche for cancer stem cells and promote their self-renewal and invasiveness. Cell Cycle 11 (7), 1282–1290.

Lonardo, E., Cioffi, M., Sancho, P., et al., 2013. Metformin targets the metabolic Achilles heel of human pancreatic cancer stem cells. PloS One 8 (10), e76518.

Lonardo, E., Cioffi, M., Sancho, P., Crusz, S., Heeschen, C., 2015. Studying pancreatic cancer stem cell characteristics for developing new treatment strategies. J. Vis. Exp. (100), e52801.

Luo, G., Long, J., Cui, X., et al., 2013. Highly lymphatic metastatic pancreatic cancer cells possess stem cell-like properties. Int. J. Oncol. 42 (3), 979–984.

Luttges, J., Reinecke-Luthge, A., Mollmann, B., et al., 1999. Duct changes and K-ras mutations in the disease-free pancreas: analysis of type, age relation and spatial distribution. Virchows Arch. 435 (5), 461–468.

Madden, J.I., 2012. Infinity Reports Update From Phase 2 Study of Saridegib Plus Gemcitabine in Patients With Metastatic Pancreatic Cancer. Infinity Pharmaceuticals, Inc. (online).

Maitra, A., Hruban, R.H., 2008. Pancreatic cancer. Annu. Rev. Pathol. 3, 157–188.

Mani, S.A., Guo, W., Liao, M.J., et al., 2008. The epithelial-mesenchymal transition generates cells with properties of stem cells. Cell 133 (4), 704–715.

Martinelli, P., Canamero, M., del Pozo, N., Madriles, F., Zapata, A., Real, F.X., 2013. Gata6 is required for complete acinar differentiation and maintenance of the exocrine pancreas in adult mice. Gut 62 (10), 1481–1488.

Martinelli, P., Madriles, F., Canamero, M., et al., 2015. The acinar regulator Gata6 suppresses KrasG12V-driven pancreatic tumorigenesis in mice. Gut 65 (3).

Martinelli, P., Carrillo-de Santa Pau, E., Cox, T., Sainz, B. Jr, Dusetti, N., Greenhalf, W., Rinaldi, L., Costello, E., Ghaneh, P., Malats, N., Büchler, M., Pajic, M., Biankin, A.V., Iovanna, J., Neoptolemos, J., Real, F.X., 2016. GATA6 regulates EMT and tumour dissemination, and is a marker of response to adjuvant chemotherapy in pancreatic cancer. Gut. http://dx.doi.org/10.1136/gutjnl-2015-311256. pii: gutjnl-2015-311256, [Epub ahead of print].

Massague, J., 2008. TGFbeta in cancer. Cell 134 (2), 215–230.

Mayers, J.R., Wu, C., Clish, C.B., et al., 2014. Elevation of circulating branched-chain amino acids is an early event in human pancreatic adenocarcinoma development. Nat. Med. 20 (10), 1193–1198.

Mazur, P.K., Siveke, J.T., 2012. Genetically engineered mouse models of pancreatic cancer: unravelling tumour biology and progressing translational oncology. Gut 61 (10), 1488–1500.

Mazur, P.K., Gruner, B.M., Nakhai, H., et al., 2010. Identification of epidermal Pdx1 expression discloses different roles of Notch1 and Notch2 in murine Kras(G12D)-induced skin carcinogenesis in vivo. Plos One 5 (10), e13578.

Medicine, UNLo, 2014. ClinicalTrials.gov (online). https://clinicaltrials.gov/ct2/show/NCT01959139.

Miranda-Lorenzo, I., Dorado, J., Lonardo, E., et al., 2014. Intracellular autofluorescence: a biomarker for epithelial cancer stem cells. Nat. Methods 11 (11), 1161–1169.

Mueller, M.T., Hermann, P.C., Witthauer, J., et al., 2009. Combined targeted treatment to eliminate tumorigenic cancer stem cells in human pancreatic cancer. Gastroenterology 137 (3), 1102–1113.

Neoptolemos, J.P., Stocken, D.D., Friess, H., et al., 2004. A randomized trial of chemoradiotherapy and chemotherapy after resection of pancreatic cancer. N. Engl. J. Med. 350 (12), 1200–1210.

Neoptolemos, J.P., Stocken, D.D., Bassi, C., et al., 2010. Adjuvant chemotherapy with fluorouracil plus folinic acid vs gemcitabine following pancreatic cancer resection a randomized controlled trial. JAMA 304 (10), 1073–1081.

Obata, J., Yano, M., Mimura, H., et al., 2001. p48 subunit of mouse PTF1 binds to RBP-Jkappa/CBF-1, the intracellular mediator of Notch signalling, and is expressed in the neural tube of early stage embryos. Genes Cells 6 (4), 345–360.

Ocana, O.H., Corcoles, R., Fabra, A., et al., 2012. Metastatic colonization requires the repression of the epithelial-mesenchymal transition inducer Prrx1. Cancer Cell 22 (6), 709–724.

Oettle, H., Post, S., Neuhaus, P., et al., 2007. Adjuvant chemotherapy with gemcitabine vs observation in patients undergoing curative-intent resection of pancreatic cancer – a randomized controlled trial. JAMA 297 (3), 267–277.

Ormanns, S., Siveke, J.T., Heinemann, V., et al., 2014. pERK, pAKT and p53 as tissue biomarkers in erlotinib-treated patients with advanced pancreatic cancer: a translational subgroup analysis from AIO-PK0104. BMC Cancer 14, 624.

Peinado, H., Lavotshkin, S., Lyden, D., 2011. The secreted factors responsible for pre-metastatic niche formation: old sayings and new thoughts. Semin. Cancer Biol. 21 (2), 139–146.

Peng, B., Fleming, J.B., Breslin, T., et al., 2002. Suppression of tumorigenesis and induction of p15(ink4b) by Smad4/DPC4 in human pancreatic cancer cells. Clin Cancer Res. 8 (11), 3628–3638.

Pinho, A.V., Rooman, I., Real, F.X., 2011. p53-dependent regulation of growth, epithelial-mesenchymal transition and stemness in normal pancreatic epithelial cells. Cell Cycle 10 (8), 1312–1321.

Pollard, J.W., 2004. Tumour-educated macrophages promote tumour progression and metastasis. Nat. Rev. Cancer 4 (1), 71–78.

Qiu, W., Sahin, F., Iacobuzio-Donahue, C.A., et al., 2011. Disruption of p16 and activation of Kras in pancreas increase ductal adenocarcinoma formation and metastasis in vivo. Oncotarget 2 (11), 862–873.

Raimondi, S., Maisonneuve, P., Lohr, J.M., Lowenfels, A.B., 2007. Early onset pancreatic cancer: evidence of a major role for smoking and genetic factors. Cancer Epidemiol. Biomarkers Prev. 16 (9), 1894–1897.

Real, F.X., Cibrian-Uhalte, E., Martinelli, P., 2008. Pancreatic cancer development and progression: remodeling the model. Gastroenterology 135 (3), 724–728.

Rhim, A.D., Mirek, E.T., Aiello, N.M., et al., 2012. EMT and dissemination precede pancreatic tumor formation. Cell 148 (1–2), 349–361.

Rubio-Viqueira, B., Hidalgo, M., 2009. Direct in vivo xenograft tumor model for predicting chemotherapeutic drug response in cancer patients. Clin. Pharmacol. Ther. 85 (2), 217–221.

Schonhuber, N., Seidler, B., Schuck, K., et al., 2014. A next-generation dual-recombinase system for time- and host-specific targeting of pancreatic cancer. Nat. Med. 20 (11), 1340–1347.

Siegel, R.L., Miller, K.D., Jemal, A., 2015. Cancer statistics, 2015. CA Cancer J. Clin. 65 (1), 5–29.

Song, Y., Washington, M.K., Crawford, H.C., 2010. Loss of FOXA1/2 is essential for the epithelial-to-mesenchymal transition in pancreatic cancer. Cancer Res. 70 (5), 2115–2125.

Stapley, S., Peters, T.J., Neal, R.D., Rose, P.W., Walter, F.M., Hamilton, W., 2012. The risk of pancreatic cancer in symptomatic patients in primary care: a large case-control study using electronic records. Br. J. Cancer 106 (12), 1940–1944.

Steinberg, W., 1990. The clinical utility of the CA 19-9 tumor-associated antigen. Am. J. Gastroenterol. 85 (4), 350–355.

Takebe, N., McShane, L., Conley, B., 2015. Biomarkers: exceptional responders-discovering predictive biomarkers. Nat. Rev. Clin. Oncol. 12 (3), 132–134.

Tanaka, M., Fernandez-del Castillo, C., Adsay, V., et al., 2012. International consensus guidelines 2012 for the management of IPMN and MCN of the pancreas. Pancreatology 12 (3), 183–197.

Thiery, J.P., Acloque, H., Huang, R.Y., Nieto, M.A., 2009. Epithelial-mesenchymal transitions in development and disease. Cell 139 (5), 871–890.

Tsai, J.H., Donaher, J.L., Murphy, D.A., Chau, S., Yang, J., 2012. Spatiotemporal regulation of epithelial-mesenchymal transition is essential for squamous cell carcinoma metastasis. Cancer Cell 22 (6), 725–736.

Tuveson, D.A., Shaw, A.T., Willis, N.A., et al., 2004. Endogenous oncogenic K-ras(G12D) stimulates proliferation and widespread neoplastic and developmental defects. Cancer Cell 5 (4), 375–387.

Varadhachary, G.R., Tamm, E.P., Abbruzzese, J.L., et al., 2006. Borderline resectable pancreatic cancer: definitions, management, and role of preoperative therapy. Ann. Surg. Oncol. 13 (8), 1035–1046.

Von Hoff, D.D., Ervin, T., Arena, F.P., et al., 2013. Increased survival in pancreatic cancer with nab-paclitaxel plus gemcitabine. N. Engl. J. Med. 369 (18), 1691–1703.

Waddell, N., Pajic, M., Patch, A.M., et al., 2015. Whole genomes redefine the mutational landscape of pancreatic cancer. Nature 518 (7540), 495–501.

Waghray, M., Yalamanchili, M., di Magliano, M.P., Simeone, D.M., 2013. Deciphering the role of stroma in pancreatic cancer. Curr. Opin. Gastroenterol. 29 (5), 537–543.

Wang, P., Fan, J., Chen, Z., et al., 2009. Low-level expression of Smad7 correlates with lymph node metastasis and poor prognosis in patients with pancreatic cancer. Ann. Surg. Oncol. 16 (4), 826–835.

Winter, J.M., Ting, A.H., Vilardell, F., et al., 2008. Absence of E-cadherin expression distinguishes noncohesive from cohesive pancreatic cancer. Clin. Cancer Res. 14 (2), 412–418.

Wormann, S.M., Diakopoulos, K.N., Lesina, M., Algul, H., 2014. The immune network in pancreatic cancer development and progression. Oncogene 33 (23), 2956–2967.

Wu, E., Zhou, S., Bhat, K., Ma, Q., 2013. CA 19–9 and pancreatic cancer. H&O 11 (1), 53–55.

Yachida, S., Iacobuzio-Donahue, C.A., 2009. The pathology and genetics of metastatic pancreatic cancer. Arch. Pathol. Lab. Med. 133 (3), 413–422.

Yachida, S., Jones, S., Bozic, I., et al., 2010. Distant metastasis occurs late during the genetic evolution of pancreatic cancer. Nature 467 (7319), 1114–1117.

Yu, E., Saif, M.W., Huber, K., 2014. Phase 1 trials in pancreatic cancer. JOP 15 (4), 326–328.

CHAPTER

# 8

# Gastric Cancer Metastasis

*K. Debiec, J. Wydmanski*

**Maria Sklodowska-Curie Memorial Cancer Center and Institute of Oncology, Gliwice, Poland**

## 1. EPIDEMIOLOGY AND SCREENING PROGRAMS

Gastric cancer (GC) is the fifth most common cancer in the world after lung, breast, colorectal, and prostate cancer. In 2012, 952,000 new cases were estimated, accounting for 6.8% of total cancer cases worldwide, and a half of them were reported in east Asia (mainly in China). GC is the third most frequent cause of neoplasm-related deaths in both sexes (723,000 deaths, 8.8% of the total). The highest mortality rates were reported in east Asia (24/100,000 for men and 9.8/100,000 for women), whereas the lowest rates were reported in North America (2.8 and 1.5, respectively). In developing countries, the incidence of mortality exceeds the values present in developed countries (362,300 for men, 186,100 for women vs. 106,700 for men and 68,000 for women, respectively). High mortality rates in Central and Eastern Europe, as well as Central and South America reflect this situation (Ferlay et al., 2013).

Despite the gradually decreasing incidence of GC, its diagnosis has a 75% mortality rate, one of the highest worldwide (Fock, 2014). In most countries, the 5-year survival rate is between 10% and 30%. In the United States, it ranges from 20% to 30%. Across the European countries, 5-year survival rates show substantial differences (Matsuda and Saika, 2013). Iceland, Germany, Switzerland, and southern Europe report among the highest survival rates, at around 30%. By contrast, the lowest survival rates are reported in Denmark, Ireland, the United Kingdom, Poland, and eastern Europe, ranging from 17% to 19%. For comparison, the highest 5-year survival rate in the world is found in Japan, ranging from 50% to 70% for both sexes (Coleman et al., 2003; Anderson et al., 2015). This result can be attributed to, among other things, the implementation of a national screening program, which allows for a significant increase in the frequency of diagnosing early gastric cancer (EGC), achieving a diagnostic rate of approximately 60% (Asaka and Mabe, 2014). When EGC is diagnosed, 5-year survival rates reach as high as 95% (Pasechnikov et al., 2014). In Western countries, the diagnostic rate for EGC is much lower, yet the exact percentage is not really known. However, some studies show it is between 10% and 37% (Koessler et al., 2014; Bollschweiler et al., 2014). Due to the lower incidence of stomach cancer in Western countries, large-scale screening programs for this form of cancer are not being carried out there, which may result in delayed detection

*Introduction to Cancer Metastasis*
http://dx.doi.org/10.1016/B978-0-12-804003-4.00008-6

of the disease when compared to Eastern countries. Therefore, large-scale implementation of early detection programs seems to be the most effective way to improve the odds of diagnosing EGC when compared to advanced or metastatic gastric cancer (AGC), as well as to improve the survival rates of people diagnosed with this disease. Accordingly, the following chapter also includes issues related to GC screening.

Japan and South Korea are examples of high-risk GC areas where screening programs have been successfully implemented. In Japan, this program was introduced in 1960. Until recently, the preferred method of screening has been photofluorography after ingestion of a barium meal (Pasechnikov et al., 2014). However, over the last decade, endoscopic screening has been conducted in clinical settings in 18.3% of municipalities in Japan, and it seems to be three to five times more effective than radiography. Since 2000, both radiographic and endoscopic screening have been readily available in South Korea. Although patients can freely select a screening method, endoscopy is the preferred method of choice (Hamashima, 2014). Findings from a case–control study published in 2013 suggest that endoscopic screening decreased GC mortality rates by 30% when compared to no screening within a 36-month period prior to GC diagnosis (Hamashima et al., 2013). Despite this promising data, a qualified endoscopist is necessary to successfully perform an endoscopic procedure. Moreover, this screening method is costly and can have adverse effects. Therefore, mass endoscopic screening is difficult to conduct (Compare et al., 2014). However, a study by Dan et al. (2006) revealed that endoscopic screening in moderate and high-risk groups is cost effective.

Serologic testing is another method that can be used to target a high-risk group for GC. This includes screening the concentration of serum pepsinogen (PG), which is inversely related to atrophic gastritis and *Helicobacter pylori* antibody. Diagnostic cut-off values for PG vary. Therefore, regional approval of tests and additional pilot studies in screening settings are required (Pasechnikov et al., 2014; Hamashima, 2014). In 2004, "screen-and-treat" programs involving chemoprevention of *H. pylori* in asymptomatic patients were introduced in areas of high incidence for GC, for example, the Matsu Islands and Taiwan. As a result of these programs, the incidence of GC has declined by 25% (Hamashima, 2014). Recent studies have confirmed the cost effectiveness of this strategy in preventing GC, even in areas with low incidence of the disease (Lansdorp-Vogelaar and Sharp, 2013). In 2013, the Japanese national health program employed *H. pylori* eradication therapy for chronic gastritis. Patients who received this treatment were classified as a "pseudo-low-risk group" for both atrophy and *H. pylori* infection (Hamashima, 2014).

Other GC screening methods are still subject to assessment and include the GastroPanel test (a combined set of PGI, PGII, Gastrin-17, and *H. pylori* IgG antibody detection), risk stratification by means of staging systems before and after *H. pylori* eradication, that is, operative link on gastritis and operative link on intestinal metaplasia, molecular biology approaches, as well as gene panel-testing to detect cancer-specific antibodies (Pasechnikov et al., 2014). Also, cancer screening in asymptomatic Korean and Japanese subjects is being conducted by utilizing 18F-fluorodeoxyglucose positron emission tomography /computed tomography (18F-FDG PET/CT). According to a preliminary report, this method seems to be a useful cancer screening tool for the detection of high-risk groups and appropriate for combination with other screening methods (Lee et al., 2009; Minamimoto et al., 2013). However, the effects of cancer screening should not be expected within a short period of time following its implementation. A long-term follow-up is necessary to evaluate any decrease in incidence or mortality rate. It is recommended that targeted screening strategies for stomach cancer should be explored.

# 2. METASTATIC GC—CLINICAL ASPECTS

## 2.1 Symptoms

GC symptoms are nonspecific, and this makes it difficult to establish an early diagnosis. Dyspepsia and abdominal pain are often mistaken for the symptoms of peptic ulcer disease or indigestion. Nevertheless, it is important to note that the presence of concurrent dyspeptic and alarm symptoms may imply an advanced, and often inoperable, disease. Alarm symptoms in GC, such as dysphagia, weight loss, and a palpable abdominal mass, are independently related to survival and an accumulation of these symptoms is correlated to a higher risk of death (Maconi et al., 2008).

Furthermore, previously identified GC symptoms in a patient, including ascites, jaundice, or gastrointestinal tract obstruction, may indicate that the disease has already spread. A large metastatic ovarian tumor (Krukenberg tumor or KT) (Yada-Hashimoto et al., 2008) or a large peritoneal implant in the pelvis (Blumer's shelf) can cause peritoneal fluid accumulation or colorectal obstruction. These pelvic masses are palpable on direct rectal examination (Layke and Lopez, 2004). The superficial lymph nodes should always be examined, including the supraclavicular lymph nodes (Virchow's node; Kanagawa et al., 2013) as well as any palpable nodules bulging into the umbilicus (Sister Mary Joseph nodule), since, in the presence of metastases, they are readily observable during clinical examination and histopathological biopsy (Fratellone and Holowecki, 2009).

Unfortunately, patients with locally advanced or metastatic GC who fail to have a radical gastrectomy performed represent a significant group of patients. A study by Lello comprised 356 subjects. The total resection rate was 43% (154/356). R0 resection was performed in 97 (27%) patients, R1 resection in 16 (4%) patients, palliative R2 resection in 41 (12%) patients, and other palliative procedures in 59 (17%) patients. Palliative care alone was provided to 143 (40%) subjects (Lello et al., 2007). Metastatic cancer diagnosis is immensely important since it helps to avoid aggressive and extensive surgery, as well as it allows palliative systemic therapy to be employed.

## 2.2 GC—Metastatic Patterns

Metastatic GC (M1) is characterized by distant metastasis to specific organs other than the regional lymph nodes (Santiago et al., 2011). The most common sites of distant metastases from GC are peritoneal metastases, ovarian metastases, hematogenous metastases that spread to the liver, lungs, and bone, as well as metastases to the distant lymph nodes, such as para-aortic, mesenteric, mediastinal, or neck lymph nodes (Japanese Gastric Cancer Association, 2011).

In a study by Wu et al., 611 patients were examined to assess the incidence and patterns of recurrence after intended curative surgery for GC. In this study, the following five patterns of recurrence were reported: lymph nodes, the remnant stomach, local, peritoneal, and hematogenous spread. The results showed that 40.1% of patients experienced disease recurrence. Approximately 80% of relapses were cumulative within the first 2 years. Fifty percent of patients were reported to have a single recurrence. However, in the group of patients with single and multiple relapses, distant metastases reached as much as 86.9%. Out of this 86.9%, 44.9% involved local recurrence and 49.8% involved both local and distant recurrence. With regard to distant spread, 53% of patients had peritoneal dissemination, 43.3% had hematogenous metastases, and 28.6% had distant lymphatic recurrence (Wu et al., 2003).

Other studies have also shown that peritoneal spread is the most frequent pattern of failure after curative resection (Landry et al., 1990; Boku et al., 1990; Bando et al., 1999). According to the seventh edition of TNM classification for

GC, positive peritoneal cytology is categorized as a stage IV disease (M1) and is associated with poor outcomes (Santiago et al., 2011; Broll et al., 2001). About 10–20% of patients are diagnosed with macro/microscopic peritoneal seeding detected intraoperatively after they have failed to be diagnosed preoperatively (Broll et al., 2001; Sugarbaker and Yonemura, 2000; Mezhir et al., 2011). At present, laparoscopy with/without peritoneal washings for malignant cells is recommended in all stages of IB–III GC before surgery according to the latest guidelines developed by the European Society for Medical Oncology (ESMO), the European Society of Surgical Oncology, and the European Society of Radiotherapy and Oncology (Waddell et al., 2014).

Lauren's criteria for histological classification, which differentiate between intestinal-type and diffuse-type gastric carcinomas, determine patterns of spread for GC. This view was confirmed in an Italian multicenter longitudinal study that included 412 surgically treated patients. The intestinal-type group of patients was found to have the following recurrence rates: locoregional spread (20%), hematogenous spread (19%), and peritoneal spread (9%). The recurrence rates for the diffuse-type group of patients were locoregional spread (27%), hematogenous spread (16%), and peritoneal spread (34%). The difference between these two groups was statistically significant for peritoneal recurrence ($P < .0001$). Substantially high rates with regard to peritoneal dissemination were revealed in the diffuse-type group of subjects with serosal infiltration, which amounted to approximately 50%, whereas in the intestinal-type group with T3 and T4 tumors, it amounted to only 16%. Involvement of second level lymph nodes or tumors larger than 4 cm in diffuse-type cancer was also associated with a high percentage of peritoneal relapse. These data suggest that superextended lymphadenectomy may play a more important role in diffuse-type cancer (Marrelli et al., 2002).

Hepatic metastases occur in approximately 5–14% of patients with GC at the time of diagnosis. Hepatic lesions are very rarely isolated and they most frequently occur as a part of general spread, involving the peritoneum, lungs, and lymph nodes (Ueda et al., 2009; Kerkar et al., 2010). However, the incidence of hepatic metastases may also depend on the histology type. Marelli reported that after curative gastrectomy of the intestinal-type GC, metastases primarily metastasize to the liver via a hematogenous route, whereas in the case of the diffuse-type cancer, metastases to the liver represent only half of the cases of distant hematogeonus dissemination (Marrelli et al., 2002).

Sixteen percent of patients with GC develop pulmonary metastases (Kemp et al., 2010). This spread usually manifests as lymphangitic carcinomatosis or numerous lesions, and, as in any other cases of distant dissemination, it is associated with a poor prognosis (Shiono et al., 2013).

Metastatic tumors account for 21% of malignant ovarian tumors. KT refer to gastrointestinal cancer that metastasizes to the ovaries. GC is the most frequent primary source of KT and constitutes about 29% of all metastatic ovarian tumors (Yada-Hashimoto et al., 2008). In Western countries, KT is rather rare and accounts for 3–4% of all metastatic ovarian tumor cases (Latour et al., 1956). However, in high-risk areas for GC, such as Japan, the incidence rate is significantly greater (Yakushiji et al., 1987; Hale, 1968). All in all, ovarian metastases from GC are not infrequent and it is important to keep in mind that KT, especially in high-risk areas, is a significant cause of treatment failure in female patients (Duarte and Llanos, 1981).

Nevertheless, treatment efficacy has improved in recent years. This is connected with the introduction of neoadjuvant chemotherapy, chemoradiotherapy (CRT), and postoperative CRT. As a result, survival in patients with later-stage metastatic cancer is greater. Not only is this improvement seen in common metastasis sites, but it is also true for rare sites, including the central nervous system (Kraszkiewicz and Wydmanski, 2014). Some unusual cases have been described in literature and they include, for example, GC metastasis to the gingiva (Shimoyama et al., 2004), uterine cervix (Imachi et al., 1993), testicle (Li et al., 2015), ureters (George et al., 2015), and skin (Ahmad et al., 2015).

## 2.3 Importance of Diagnostics

Detailed diagnostics allow correct staging of GC and create the possibility for individualization of treatment. It is a key factor in distinguishing patients with locally advanced, operational GC from metastatic patients. Also, in groups of patients with inoperable or unresectable GC, distant metastasis should be excluded before implementing radical radiotherapy or chemoradiotherapy (Wydmanski et al., 2014). Routine diagnostic methods do not allow detection of metastases in about 30% of patients (Smyth et al., 2012).

This important problem requires more attention to improve diagnostic methods and explore prognostic and predictive factors, which allow individualization of treatment. This chapter looks at trends in research with regard to the above issues, paying special attention to the role of PET/CT in GC diagnostics and its prognostic value. Aspects related to markers and treatment of metastatic GC are also discussed.

# 3. DIAGNOSIS OF METASTATIC GC WITH A PARTICULAR FOCUS ON 18-F-FDG PET/CT

## 3.1 Current Role of 18-F-FDG PET/CT Imaging in GC Diagnostics

According to the guidelines of the National Comprehensive Cancer Network (NCCN), each patient with GC should be tested by means of the following diagnostic tools: history taking, physical examination, complete blood count, biochemical profile, endoscopy of the upper gastrointestinal tract with biopsy and histopathological examination of the primary lesion, CT of the chest, abdomen, and pelvis with intravenous and oral contrast media. Other diagnostic tools that are recommended by the NCCN include endoscopic ultrasound (EUS), PET/CT, as well as laparoscopic surgery of the abdomen to detect metastatic disease. PET/CT is particularly recommended for patients with locally advanced tumors (T3–4), and for those who failed to have dissemination detected by standard methods. Also, PET/CT is an efficient diagnostic tool for restaging of patients with primarily unresectable GC after neoadjuvant treatment. Following primary detection of metastatic disease by PET/CT, histopathological verification is necessary to exclude false-positive results. Having confirmed the diagnosis of metastatic disease, evaluation of human epidermal growth factor receptor-2 (HER2) protein expression by immunohistochemistry should be carried out to qualify patients for chemotherapy combined with immunotherapy (Trastuzumab) (NCCN Clinical Practice Guidelines in Oncology (NCCN Guidelines), 2015). Recommendations of the ESMO also suggest that PET/CT can improve the staging of GC due to more accurate detection of metastases (Waddell et al., 2014).

Numerous studies suggest that PET/CT is a useful diagnostic method to detect various types of cancer, for instance, lung, melanoma, head and neck, colorectal, or gynecological cancer (Moon et al., 2013). In recent years, several studies on the use of this method for GC have been published. However, there is a controversy about the use of 18F-FDG-PET/CT for GC. This technique has low sensitivity to detect early GC and the histological subtype of signet-ring cell carcinoma (Park et al., 2014). For EGC, method sensitivity ranges from 26% to 63%, whereas for AGC, it is much higher and ranges from 84.9% to 98% (Dassen et al., 2009; Namikawa et al., 2014).

## 3.2 Usefulness of PET/CT in the Preoperative Staging of GC Compared With Other Diagnostic Methods

Among AGC patients who have undergone preoperative staging with the use of standard methods, disease dissemination goes undetected in one-third of cases (Smyth et al., 2012). Patients with T3–4 gastric tumors, who have received cancer treatment with curative intent, have been reported to experience recurrence of metastases to the peritoneum, retroperitoneal lymph nodes (LNs), or by the hematogenous route. This recurrence usually develops between 2 and 3 years postoperatively (Wu et al., 2003; Yoo et al., 2000; Smyth and Shah, 2011). This situation confirms that a proper diagnostic process is not employed when qualifying patients for radical treatment. Therefore, PET/CT gives hope to improve the quality of diagnosis and detection of patients with both dissemination and poor prognosis (Wu and Zhu, 2014) Figs. 8.1 and 8.2 show 18-FDG-PET/CT scans of patients with gastric cancer. In these cases, the method allows for either detection or confirmation of metastatic disease which could not be done using the standard methods.

The results of the meta-analysis by Cui et al. on the role of PET/CT in the preoperative staging of GC confirmed the usefulness of this method. The authors determined the performance measures, such as sensitivity, specificity,

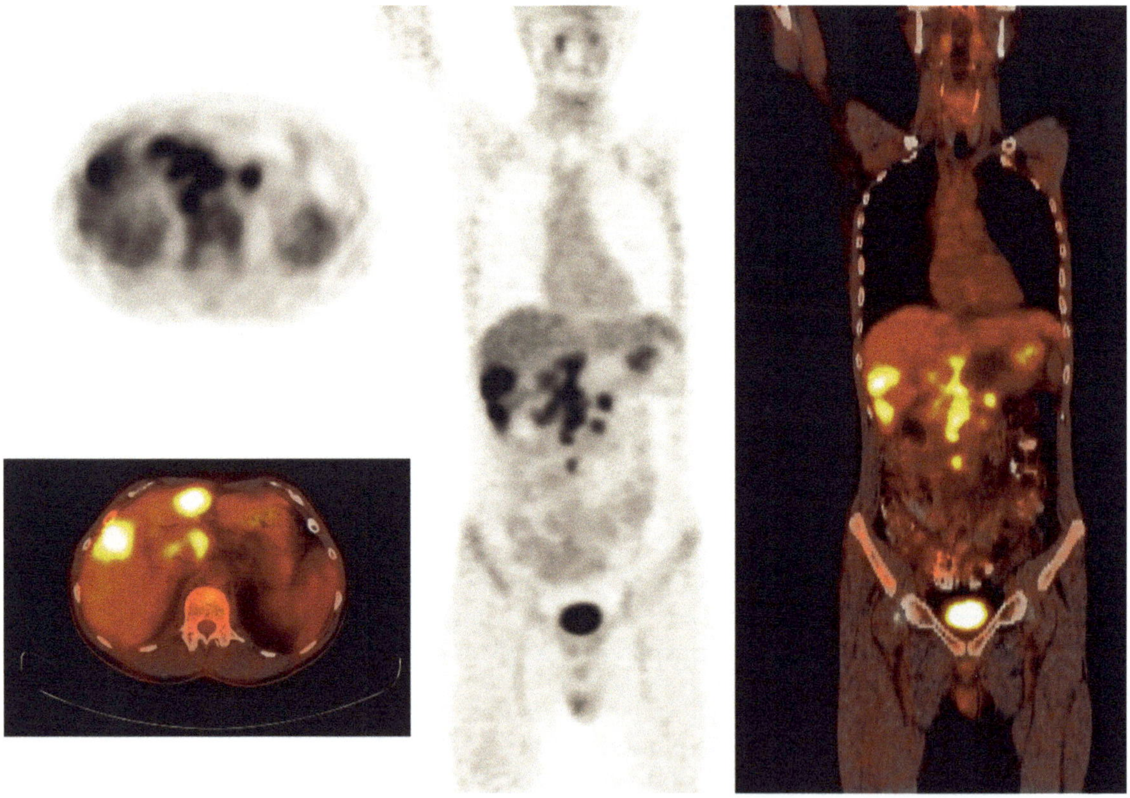

FIGURE 8.1 A 40-year-old man with previously untreated histologically confirmed GC. 18-FDG-PET/CT was performed to determine the stage of the disease. PET/CT examination revealed multiple metastases of the retroperitoneal lymph nodes and both lobes of the liver. *Images from Maria Skłodowska-Curie Memorial Cancer Centre and Institute of Oncology, with permission.*

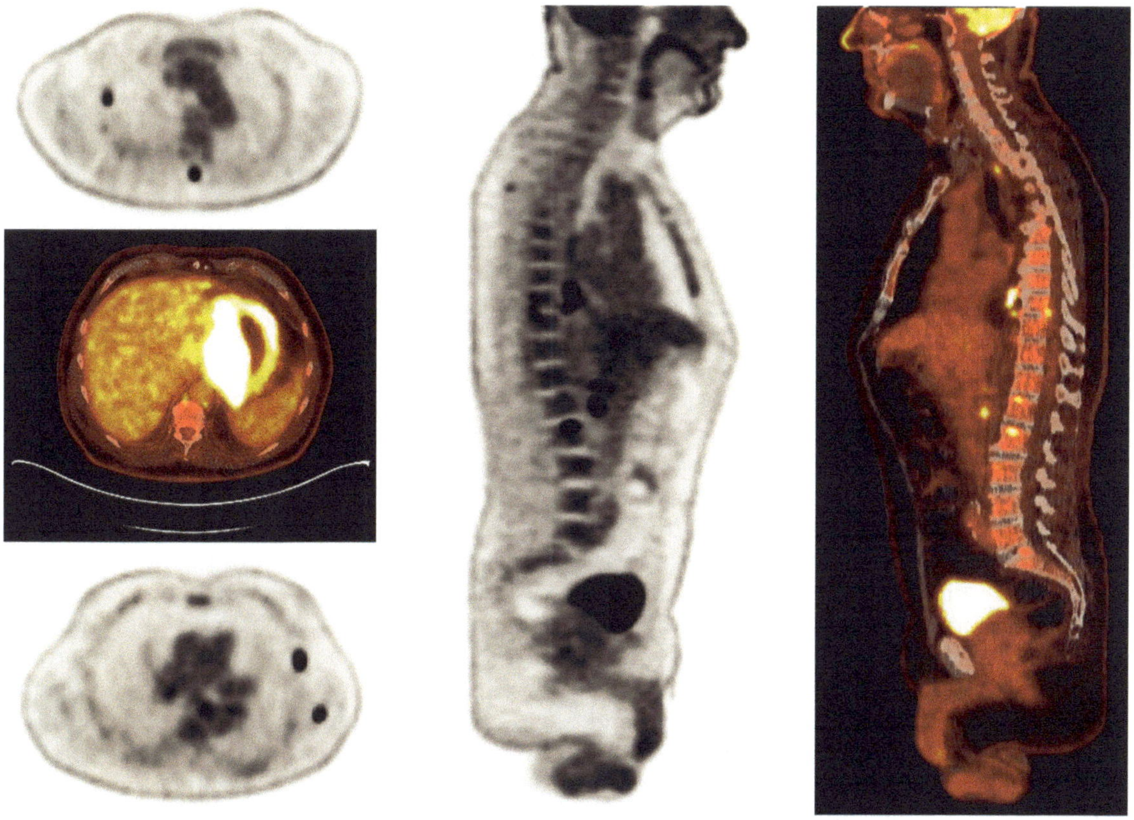

FIGURE 8.2 A 75-year-old man with previously untreated histologically confirmed GC. Based on the previous conventional tests, the patient was diagnosed with locally advanced cancer. 18-FDG-PET/CT had been performed as an additional examination before planned radical chemoradiation. Findings: strong uptake within the entire gastric wall with predominance in the lesser curvature; multiple metastases to the thoracic and lumbar vertebrae, ribs, and a number of the nonregional lymph nodes (paratracheal, paraesophageal, retrocrural, retroperitoneal, and mesenteric). *Images from Maria Skłodowska-Curie Memorial Cancer Centre and Institute of Oncology, with permission.*

positive predictive value (PPV), negative predictive value (NPV), and area under curve (AUC). The results for the primary lesion were 0.921 (sensitivity), 0.895 (specificity), 5.12 (PPV), 0.13 (NPV), 0.9520 (AUC); for the regional LNs: 0.6 (sensitivity), 0.986 (specificity), 11.68 (PPV), 0.36 (NPV), 0.9343 (AUC); and for distant metastasis: 0.703 (sensitivity), 0.976 (specificity), 16.7 (PPV), 0.26 (NPV), 0.9454 (AUC) (Cui et al., 2013).

However, the meta-analysis that compared the sensitivity between abdominal ultrasonography, EUS, multidetector CT, magnetic resonance imaging (MRI), and PET/CT revealed that none of these methods are suitable to either confirm or exclude metastatic LNs when used alone. Both the sensitivity and specificity of FDG-PET/CT in this study were lower than in the previously cited analysis, and ranged from 33.3% to 64.6% (median 34.3%) and from 85.7% to 97% (median 93.2%), respectively. The authors of this study suggest that high-resolution PET/CT combined with functional MRI (fMRI) can be a promising method for LNs detection (Kwee and Kwee, 2009).

Relatively low sensitivity of PET/CT compared with CECT (contrast enhancement CT) with regard to the regional LNs metastasis was found, and it amounted to 56.6% in a study by Choi and to 64.5% in a study by Namikawa. This finding is often considered a significant restriction of this method. The problem with LNs detection is linked to, among others, low resolution of most scanners used in research. However, more sophisticated equipments to detect changes with accuracy of approximately 5mm are now available (Namikawa et al., 2014; Choi et al., 2014). Another problematic aspect involves FDG uptake in the primary tumor that can merge with the surrounding pathological lymph nodes, as well as the presence of the tumor histological subtype with low FDG uptake (Park et al., 2012). Nevertheless, the specificity higher than in CECT provides substantial information that helps distinguish enlarged metastatic LNs from benign lymphadenopathy (Namikawa et al., 2014).

Several reports have been published and looked at the correlation between the primary tumor maximum standardized uptake value ($SUV_{max}$) parameter and the presence of metastatic lymph nodes, as well as at its predictive role. In the previously cited study by Choi, it was revealed that the $SUV_{max}$ of the primary tumor exceeded the value of 3.75, and the sensitivity and specificity of PET/CT for nodal metastases detection increased by 73.5% and 74.5%, respectively. However, the $SUV_{max}$ of the gastric tumor greater than 4.35 correlated with positive FDG uptake in LNs. In addition, the sensitivity and specificity for predicting tumor unresectability were 58.8% and 91.6%, respectively (Choi et al., 2014). In a study by Oh et al. (2011) it was found that the $SUV_{max}$ of the primary lesion higher than 3.4 was an independent indicator of spread to the regional LNs. Moreover, in a study by Coupe, positive LNs in PET/CT were said to be an independent predictor of shorter overall survival (OS), which was not observed for CT (Coupe et al., 2014).

In their study, Li et al. proposed a preoperative staging that improves the sensitivity of PET/CT in the assessment of the regional LNs, and also allows detection of distant metastases. The authors combined EUS with PET/CT, which proved to be much more effective than utilizing these methods separately. Since PET/CT reveals distant metastasis, a change in a therapeutic approach from radical to palliative was made with regard to 31 out of 124 examined patients (27.4%). EUS, in turn, had an impact on qualification of 37 patients (29.8%) for neoadjuvant chemotherapy (Li et al., 2012).

## 3.3 Usefulness of PET/CT in the Assessment of Distant Metastasis (M1) and Its Prognostic Role

There have been very few studies to date that have evaluated the utility of PET in patients with distant dissemination. One of them was a study by Yoshioka and it was based on PET alone instead of fused PET/CT imaging. This method occurred to be effective in staging of metastatic and recurrent GC. High sensitivity, specificity, and accuracy of this method were revealed and they amounted to, respectively: 91%, 100%, 92% (the primary lesion), 85%, 74%, 79% (liver metastases), 73%, 78%, 75% (metastases to the regional and nonregional LNs), and 67%, 88%, 86% (metastases to the lungs). Significantly lower values were recorded for spread to the peritoneum (50%, 63%, 57%, respectively), pleura (4%, 100%, 83%, respectively), and bones (30%, 82%, 57%, respectively). These results were said to be due to, among others, a small size of the lesion and poor FDG uptake in these locations. The peritoneal and pleural nodules were visualized only when their size reached 2cm. The author suggested that this could be linked with low FDG excretion from the blood vessels into the pleural or peritoneal fluid. With respect to bone metastases, the osteolytic lesions were visible on both X-ray and PET, whereas small metastases were visible neither on X-ray nor on PET. Perhaps, these were osteoblastic metastases, and because of lower cellularity and poor blood supply the lesions failed to absorb an adequate amount of tracer (Yoshioka et al., 2003).

Slightly different conclusions were reached by Chung et al. who studied 35 patients with metastatic GC. The bone lesions, as well as metastases to the liver, lungs, and adrenal glands, were classified as solid organ metastases. For these locations, PET/CT showed very high sensitivity (95.2%), specificity (100%), and accuracy (97.1%). However, for nonsolid metastases relating to dissemination to both the distant LNs and the peritoneum, the cost effectiveness of PET/CT was lower. Nevertheless, peritoneal dissemination was reported to have higher sensitivity (66.7%), specificity (100%), and accuracy (92%) than in other published studies. Such effective detection of metastatic lesions might have resulted from the fact that patients with signet-ring cell carcinoma had been disqualified from the study, and this made it possible to exclude some false-negative results. Studies comparing the sensitivity and specificity of 18-F-FDG-PET/CT for detection of gastric cancer metastasis are presented in Table 8.1. Not only did the study reveal that PET/CT was an effective method to detect metastases, but it also found that the primary tumor $SUV_{max}$ of >8 in the group of metastatic patients was an independent prognostic factor, and it was associated with shorter OS. No association between the parameters measured by PET/CT and chemotherapeutic response was revealed (Chung et al., 2010).

Several other studies have also found that PET/CT may have prognostic significance. A study by Park et al. involved 82 previously untreated patients with metastatic GC. It revealed that the primary tumor $SUV_{max}$ of >6, which had been measured prior to palliative chemotherapy, was an independent prognostic factor associated with shorter progression-free survival (PFS). Moreover, patients with lower initial $SUV_{max}$ were observed to have longer PFS and a more satisfactory chemotherapeutic response (Park et al., 2012). Di Fabio studied a group of disseminated patients who had received systemic therapy. As a result, an early metabolic response assessed by 18F-FDG PET, according to the Response Evaluation Criteria in Solid Tumor criteria, was found to be of prognostic value. Responders, namely patients with CR (complete response), PR (partial response), SD (stable disease), who underwent Cetuximab plus FOLFIRI-based chemotherapy, were observed to have longer TTP (time to progression) than nonresponders (with progressive disease) TTP for responders versus nonresponders was 11 versus 5 months respectively ($P=.0016$). (Di Fabio et al., 2007). Also, in other publications, the $SUV_{max}$ of the primary gastric tumor provided information about clinical significance, metastatic potential, and unresectability of this tumor (Namikawa et al., 2014).

In addition to the $SUV_{max}$, volume-based PET parameters for the primary lesion, such as metabolic tumor volume (MTV—tumor volume above a certain SUV threshold) and total lesion glycolysis ($TLG = SUV_{mean} \times MTV$) can also differentiate between patients with both locally advanced and metastatic GC. Moreover, they can be both prognostic and predictive factors for OS and TTM (time to metastasis) (Grabinska et al., 2014). Larger-scale studies on significance of these prognostic factors have been carried out with regard to other cancer sites, such as the lungs, esophagus, and pancreas (Van de Wiele et al., 2013; Zaizen et al., 2012; Lee et al., 2014; Li et al., 2014; Larson et al., 1999). However, as far as stomach cancer is concerned, only single publications are available (Grabinska et al., 2014; Kim et al., 2014).

In a study by Kim et al., a group of GC patients had undergone PET/CT before they received treatment. The images showed that both the $SUV_{max}$ and the $MTV_{2.5}$ parameters were important prognostic factors for PFS ($P=.034$, hazard ratio 3.59, 95% confidence interval (CI) 1.10–11.71) (Kim et al., 2014). However, the Polish study has revealed that MTV, TLG, and $SUV_{mean}$ are parameters that differentiate between groups of patients with both locally advanced and metastatic GC. In addition, such values as $TLG_{30}$ and $MTV_{2.5}$ have been confirmed to be prognostic factors for OS and TTM (Grabinska et al., 2014). A summary of studies related to the prognostic value of 18-FDG-PET/CT in gastric cancer is included in Table 8.2.

TABLE 8.1 Studies Comparing the Sensitivity and Specificity of 18-F-FDG-PET/CT for Detection of GC Metastasis (M1)

| Author | Number of Patients | Region of Metastasis | Sensitivity (%) | Specificity (%) |
|---|---|---|---|---|
| Yoshioka et al. (2003) | 42[a] | • Liver | 78–85% | 82–74% |
| | | • Lung | 67% | 88% |
| | | • Peritoneum | 9–50% | 63–79% |
| | | • Ascites | 2–24% | 76–93% |
| | | • Pleural | 4% | 100% |
| | | • Bone | 30% | 82% |
| Chung et al. (2010) | 35 | • Solid organ metastasis[b] | 95.2% | 100% |
| | | • Peritoneum | 66.7% | 100% |
| Smyth et al. (2012) | 113[c] 76[d] | • Distant metastasis | 35% | 99% |
| | | • Distant metastasis | 50% | 98% |
| Zhou et al. (2013) | 39 | • Liver | 100% | |
| | | • Bone | 100% | |
| | | • Peritoneum | 94.7% | |
| | | • Ovary | 90.9% | |
| | | • Lymph node M1 | 93.5% | |
| Cui et al. (2013) | [e] | • Distant metastasis | 79% | 97% |
| Lee et al. (2014) | 46 | • Distant metastasis | 100% | 93% |
| Altini et al. (2015) | 45 | • Distant metastases | 60% | 88.57% |

[a] *Only PET method, without fusion with CT.*
[b] *Lung, liver, bone or adrenal gland.*
[c] *The entire patient population.*
[d] *Patients with FDG-avid primary tumor.*
[e] *Meta-analyses of 15 studies were included, no informationin abstract about the number of patients, full text in Chinese.*

## 3.4 Limitations of 18-FDG-PET/CT in GC Diagnosis

As mentioned above, it is crucial to conduct research on the use of PET/CT scanning in preoperative staging of GC and follow-up of patients after gastrectomy. On the other hand, not all authors share the same optimism about the applicability of PET/CT. According to some studies that have investigated the usefulness of both PET and PET/CT in cancer staging, these methods have failed to surpass standard CT scanning (Park et al., 2014; Ha et al., 2011; Kim et al., 2011). However, high specificity of this method for the detection of regional lymph nodes and distant metastases can be of complementary value when the CT images are not conclusive (Altini et al., 2015; Kim et al., 2011). One of the problematic issues is relatively poor sensitivity of PET/CT for the detection of metastases to the peritoneum, being one of the most common metastatic sites. This is an influential factor responsible for disqualifying patients from radical treatment. In approximately one-third of patients with locally advanced GC, peritoneal metastasis goes undetected preoperatively

TABLE 8.2 Predictive and Prognostic Value of 18-F-FDG PET/CT in GC

| Author (year) | *N* | Stage of Disease | Parameter | Univariate Analysis | Parameter | Multivariate Analysis |
|---|---|---|---|---|---|---|
| Di Fabio et al. (2007) | 22 | IV | [a]Early metabolic response in PET/CT after chemotherapy CR, PR, SD[b] | | [a]Longer time to disease progression (response evaluated by CT was not predicted for the clinical outcome) | |
| Chung et al. (2010) | 35 | IV | Stomach $SUV_{max}$[c] <8 | • Longer OS[d] | Stomach $SUV_{max}$ <8 | • Longer OS |
| Oh et al. (2011) | 147 | I–III | Stomach $SUV_{max}$ >3.2 | • LN[e] metastasis | Stomach $SUV_{max}$ >3.2 | • LN metastasis |
| Pak et al. (2011) | 41 | I–IV (only SRC[f]) | Stomach $SUV_{max}$ >3.8 | • More postoperative recurrence<br>• Shorter relapse-free survival<br>• Lower 30-months CSS[g] rates | | |
| Park et al. (2012) | 82 | IV | Total $SUV_{max}$[h] <11.5<br><br>Stomach $SUV_{max}$ <6 | • Longer PFS[i]<br>• Better chemotherapeutic response<br>• Longer OS, PFS<br>• Better chemotherapeutic response | Stomach $SUV_{max}$ <6 | • Longer OS, PFS |
| Lee et al. (2012) | 271 | I–III | FDG-PET-positive Tu[j] and stomach $SUV_{max}$ >8.2 | • More postoperative recurrences | FDG-PET-positive Tu and stomach $SUV_{max}$ >8.2 | • Inferior 24-month RFS[k] (especially in tubular adenocarcinoma and poorly differentiated AC) |
| Kim et al. (2014) | 97<br>50 | I–IV[l] | Stomach $SUV_{max}$ >5.74 | • Inferior PFS<br>• More postoperative recurrences | Stomach $SUV_{max}$ >5.74<br>$MTV_{2.5}$[m] >16.42 $cm^3$ | • Inferior PFS<br>• Inferior PFS |
| Choi et al. (2014) | 156 | I–IV | | | Stomach $SUV_{max}$ >4.35 | • Noncurative surgery<br>• LN metastasis |
| Coupe et al. (2014) | 97 | I–IV | | | FDG-PET-positive Tu<br>FDG-PET-positive LN | • Inferior OS<br>• Inferior OS |

*Continued*

Table 8.2 Predictive and Prognostic Value of 18-F-FDG PET/CT in GC—cont'd

| Author (year) | *N* | Stage of Disease | Parameter | Univariate Analysis | Parameter | Multivariate Analysis |
|---|---|---|---|---|---|---|
| Grabinska et al. (2014) | 40 | I–IV | Tumor $TLG_{30}$ >319.4 $cm^3$ | • Inferior OS | Higher tumor $TLG_{30}$[n] | • Inferior OS |
| | 23 | I–III | Tumor $TLG_{30}$ >149.9 $cm^3$ | • Inferior OS | Higher tumor $TLG_{30}$ | • Inferior OS and TTM[o] |
| | | | | | | • Inferior OS |
| | | | | | Higher tumor $MTV_{2.5}$ | |

*AC*, adenocarcinoma; *PFS*, progression-free survival; *RFS*, recurrence-free survival; $SUV_{max}$, maximum standardized uptake value.

[a] *No information about univariate or multivariate analysis.*

[b] *Complete response, partial response, stable disease.*

[c] *$SUV_{max}$ in the stomach before treatment.*

[d] *Overal survival.*

[e] *Lymph nodes.*

[f] *Signet ring cell carcinoma.*

[g] *Cancer-specific survival.*

[h] *The highest $SUV_{max}$ among all target lesions (primary tumor or regions of metastasis).*

[i] *Progression-free survival.*

[j] *Tumor.*

[k] *RFS.*

[l] *Subgroup analysis, MTV was measured in 50 patients with clearly distinguishable primary lesion.*

[m] *metabolic tumor volume.*

[n] *Total lesion glycolysis.*

[o] *Time-to-metastasis.*

(Smyth et al., 2012). The sensitivity of PET/CT for this metastatic site ranges from 28% to 66.7%, and is lower when compared to standard CT scanning (76–80%) (Smyth and Shah, 2011; Yoshioka et al., 2003; Wang and Chen, 2011). Low sensitivity of PET/CT reported in the literature may be due to the fact that it is difficult for the scanner to detect small peritoneal nodules since they are below its spatial resolution. Also, there has been a greater tendency for dissemination of poorly differentiated carcinomas with very low FDG uptake. Some studies have been based on the use of PET alone instead of fused PET/CT, and also on the use of the equipment with technical parameters that are not up to the present standard, and this additionally reduces the sensitivity of this method (Smyth et al., 2012). It is important to bear in mind that false-positive results may be obtained in the case of inflammatory conditions, such as gastritis and anastomositis, as well as in the case of increased FDG uptake in the ovaries during the periovulatory phase, which may suggest the presence of KT (Lerman et al., 2004).

## 3.5 Usefulness of Laparoscopy Alone and in Combination With PET/CT in the Assessment of GC Metastases

Laparoscopy is a method that allows direct evaluation of the peritoneal cavity, aspiration of the peritoneal fluid for cytology, and detection of microdissemination. According to the NCCN guidelines, this method is recommended (category 2B) prior to radical treatment for patients who have undergone neoadjuvant therapy, or for those with locally advanced GC (≥T3, N+) (NCCN Clinical Practice Guidelines in Oncology (NCCN Guidelines), 2015). Currently, it is being used only in selected centers, mainly in developed countries. Unfortunately, it seems that this diagnostic method will not be widely implemented.

Since both PET/CT and CECT have been reported to be insufficiently effective in detecting the spread to the peritoneum, laparoscopy is claimed to be an important tool to exclude peritoneal metastasis preoperatively. Also, it is recommended that this method should be combined with PET/CT (Mezhir et al., 2011; Smyth et al., 2012; Standby et al., 2015).

A large prospective study was conducted, involving 1241 patients. Thirty-six percent of them suffered from gastroesophageal junction cancer, and the rest of the subjects had locally advanced GC. Prior standard testing had not confirmed metastasis in these patients. The subjects underwent laparoscopic peritoneal lavage. Following laparoscopy, 291 (23%) patients were diagnosed with peritoneal metastasis, including 198 (68%) subjects with lesions (M1) visible at laparoscopy, and 93 (32%) subjects with positive cytology (M1 Cyt+). It was shown that the median disease-specific survival (DSS) for the entire group was 1 year. For the M1 group, DSS was 0.8 years, and for the M1Cyt + group, it was 1.3 years. Survival was significantly lower in the group with visible changes when compared to patients at the same disease stage of the disease without macroscopic metastasis. Also, prolonged survival was reported in M1Cyt + patients who, after receiving palliative chemotherapy, obtained negative cytology at the follow-up laparoscopy ($n=27$) versus patients without cytological regression ($n=21$). For these subgroups, DSS was 2.5 years versus 1.4 years, respectively ($P=.0003$). Although DSS prolongation was achieved after successful chemotherapy, the prognosis for these patients is still unfavorable and complete recovery is rare. Laparoscopy has proved to be a useful diagnostic and prognostic tool. The author recommends that this method should be a part of routine preoperative staging of patients with gastric tumor (≥T3 and N+) (Mezhir et al., 2011).

In the study by Smyth et al., the authors investigated a group of 113 patients with locoregionally advanced GC (≥T3, N+) diagnosed by CT and EUS to improve the preoperative staging of such patients. All subjects had an additional 18-FDG PET/CT and laparoscopy performed. Following the diagnostic process, 31 (27%) patients were detected with OMD (occult metastatic disease) that disqualified them from radical surgery. PET/CT showed distant metastasis in 11 (10%) patients, and laparoscopy confirmed peritoneal dissemination in 21 (19%) patients, five of whom had positive peritoneal fluid cytology. Only one patient had metastasis detected by both methods at the same time. PET/CT revealed metastases to bones ($n=2$), the liver ($n=4$), and the nonregional lymph nodes ($n=6$), which failed to be detected by CT, EUS, or laparoscopy. A false-positive result was revealed by PET/CT for only one lesion, namely a peritoneal lump that turned out to be an exophytic infiltration from the primary tumor. Most OMD cases of the diffuse histological subtype were not visible on PET/CT, however, they were detected at laparoscopy. All in all, 27% of the study population avoided unnecessary, aggravating, and cost-intensive radical gastrectomy as a result of laparoscopy and PET/CT, and 10% of them by means of only PET/CT. It was estimated that combining PET/CT with CT, EUS and laparoscopy saved approximately $10,000 per patient. Performing PET/CT prior to laparoscopy reduced costs an additional $2000 per patient by limiting the number of laparoscopic procedures performed in M1 subjects. This was because biopsy had been previously used to verify PET changes in these patients. The need to obtain histopathological confirmation among subjects with clinical suspicion of GC metastases revealed by PET/CT was also emphasized. Based on the results of this study, laparoscopy and PET/CT should be included in standard diagnostic procedures used for AGC patients (Smyth et al., 2012).

## 3.6 Follow-Up of Patients After Radical Treatment of GC

The most common sites of dissemination after radical treatment include the peritoneum, retroperitoneal lymph nodes, and also hematogenous metastasis, therefore both PET/CT and laparoscopy seem to be useful in monitoring patients after laparotomy (Smyth et al., 2012; Lee et al., 2014). In a pilot study by Shim et al., a second-look laparoscopy proved a valuable method to detect any intra-abdominal spread that was not clearly shown on CT or PET/CT. Eleven out of 12 subjects, with inconclusive CT and/or PET/CT results had metastasis confirmed at laparoscopy. Potential technical problems that could have occurred due to postoperative adhesions and possible perforation were not identified. A control ultrasound examination was performed during trocar insertion by a highly experienced laparoscopic surgeon (Shim et al., 2011). In the meta-analyses on the utility of PET/CT in detecting GC recurrence, the sensitivity ranged between 78% and 86% and the specificity between 82% and 88% (Smyth and Shah, 2011; Wu and Zhu, 2014; Zhou et al., 2013).

However, for subjects with GC recurrence suspected in CT results and with positive tumor markers, the sensitivity of PET/CT was significantly higher and could reach between 94% and 100% with a specificity between 70% and 100%(Lee et al., 2014; Patriti et al., 2007; Jadvar et al., 2003). It should be noted that inflammation in the area of postoperative anastomosis may be the reason for inconclusive CT results and false-positive PET/CT results. Therefore, endoscopy accompanied by histopathological examination should always be performed to verify the results (Lee et al., 2014). Moreover, a study by Lee et al. (2012) that involved a group of 271 GC patients found that preoperative FDG uptake in the primary site was an independent predictor of relapse, and an $SUV_{max}$ over 8.2 was associated with a significantly shorter survival rate and a smaller percentage of patients with a 24-month recurrence-free survival rate than those with $SUV_{max}$ below the cutoff value (58% vs. 79%, $P=.001$). However, there is still no consensus on the role of PET/CT in the postoperative monitoring of GC patients. Several reports indicate that combining PET with CT does not increase the chance for recurrence detection (Shim et al., 2011).

It appears that both laparoscopy and PET/CT can be used as complementary diagnostic methods in the evaluation of disease recurrence in a group of patients with either inconclusive CT results or positive tumor markers.

## 3.7 18F-FLT PET/CT in Metastatic GC Diagnosis

Based on the limitations associated with 18F-FDG uptake and metabolism in GC, several attempts have been made to use 18F-FLT (3-deoxy-3-18F-fluorothymidine) as a proliferation imaging tracer. This marker seems to be more sensitive than 18F-FDG when assessing the primary lesion (Herrmann et al., 2007). However, the greatest disadvantage is very low sensitivity for the detection of metastasis, particularly in the liver and bones. This is associated with high physiological uptake of 18F-FLT in the liver and bone marrow, which makes it difficult to interpret imaging findings. In a study by Zhou et al., the authors evaluated the sensitivity of both 18F-FLT PET/CT and 18F-FDG-PET/CT in the detection of hepatic and bone metastases. The results were 30% versus 100% for hepatic metastasis, and 20% versus 100% for bone metastasis, respectively ($P < .05$). When detecting spread to the lymph nodes, peritoneum, and ovaries, the difference between these two methods was not significant (Zhou et al., 2013).

## 3.8 Usefulness of Magnetic Resonance Imaging in the Evaluation of GC Metastasis

In addition to CT, EUS, PET/CT, and laparoscopy, several studies on the usefulness of magnetic resonance imaging (MRI) in the diagnosis of GC have also been published. In most cases, static MRI has not been found to be more effective than CT scanning when staging the disease in question (Anzidei et al., 2009; Kim et al., 2000; Kinkel et al., 2002). The focus has now shifted toward whole-body diffusion-weighted magnetic resonance (WB-MR DWI) versus PET/CT. As the name suggests, this method allows whole-body scanning by measuring the diffusion of water molecules in various tissues within a short time. Changes depend on diffusion and they differ between healthy and cancerous tissues. This contributes to the contrast between tissue types visible on structural images. Some studies on small groups of patients have been published and they suggest that diagnostic functional MRI may be an alternative for PET/CT when staging GC. It has similar sensitivity for primary tumor detection, and also similar limitations with regard to tumor size and infiltration depth. However, further studies are necessary to determine the role of the WB-MR DWI with regard to this tumor location (Cafagna et al., 2012; Stecco et al., 2009; Manenti et al., 2012; Tomizawa et al., 2015). Although an increasing number of studies on fused PET/MRI are being published, no significant reports have been presented so far with regard to GC. Fused PET/MR appears to be a promising tool, but it has not been widely used as yet. Nevertheless, hybrid PET/MR systems are still being improved (Zaidi and Del Guerra, 2011).

## 3.9 Perspective Directions for Improvement and Development of PET Imaging

Methods to improve PET/CT imaging are constantly being explored. One of them is to increase the resolution of this apparatus. Another simple way is to improve its accuracy. This may be done by filling the stomach with fluid (e.g., water or milk) before the examination. When the stomach wall is stretched, the density of physiological uptake in this organ is reduced, and negative oral contrast agents make it easy to visualize the tumor and the regional lymph nodes (Wu and Zhu, 2014; Kamimura et al., 2009). Another alternative mentioned in several publications is an additional assessment of late FDG uptake, that is, after approximately

2–3h (dual-time imaging). Physiological areas and benign lesions are characterized by the FDG uptake that either decreases or remains at the same level and pathological foci are characterized by increased FDG. Observing the metabolic rate of the tracer can differentiate between these benign and malignant lesions (Zhuang et al., 2001). In a study by Jeong et al., PET/CT combined with multidetector CECT (hybrid PET/CT Gastrography) with regard to GC patients was evaluated. As a result, the sensitivity of this method was greatly improved when compared to the standard PET/CT (Jeong et al., 2013). The most important direction of development of PET seems to be searching for target-specific tracers. [18F]-FDG is the most widely used PET radiotracer in oncology, but it is not a target-specific tracer. Increased metabolic rates associated with cancer cannot be distinguished from other etiologies, such as inflammation. Nowadays, bioinformatic methods provide more information than before about biological data-mining tools, targets linked to cancer, as well as locations and distributions of the molecular targets (intercellular, cell membrane, extracellular matrix, etc.). Numerous attempts are being made to use PET imaging for biological processes important to the pathogenesis of cancer. A few examples of such processes are angiogenesis, apoptosis, hypoxia, and cell proliferation, among others. Moreover, this modality is also used to identify targets like molecular biomarkers (e.g., protein kinases and epidermal growth factor receptors). It will be critical to better understand the processes that occur during cancer progression for new targets to be discovered (Chen and Chen, 2011).

It should be noted that the more specific the tracer, the higher the need for anatomical information. Hybrid PET/MRI technology is still being improved, and it seems to be a powerful tool for in vivo molecular imaging. This fusion can give excellent anatomical information, particularly with regard to soft tissue. It also allows examination of MR spectroscopy (MRS) extensive data (Zaidi and Del Guerra, 2011). All this can be a large source of information for bioinformatic tools as well. Since improved PET methods provide an increasing amount of information, the systems will need to be maximally automated in the near future so that they are easy and clear to operate. Most likely, PET imaging will lead the way toward personalized cancer medicine.

## 4. PROGNOSTIC AND PREDICTIVE FACTORS

In addition to the PET/CT parameters mentioned above which seem to be of prognostic value, a number of studies relating to several other prognostic and predictive factors are available in the literature. Lymph node metastasis and depth of tumor invasion belong to the most important and well proven prognostic factors for GC patients who have undergone curative gastrectomy. They are associated with poor prognosis, an increased risk of recurrence and shorter OS (Maruyama, 2009). Presence of vascular invasion—lymphatic vessel invasion and/or blood vessel invasion—is also related to these factors (Li et al., 2015). In addition, it has been reported that tumor location has an impact on further prognosis. Saito et al. found that adenocarcinoma of the gastric cardia is an independent prognostic factor. With regard to the site of recurrence, both lymph node and hematogenous recurrences were observed more frequently in the cardia than in the remaining parts of the stomach (Saito et al., 2006).

In other studies, the following independent prognostic and predictive factors were found to influence distant metastases free survival: body weight loss, age, performance status of patients, tumor location, cancer (AJCC) staging, ratio of involved to resected lymph nodes, degree of tumor differentiation, and operation radicality (Wydmanski, 2008).

Several studies have investigated prognostic indices (PIs) in GC. They are based on inflammation markers, such as the neutrophil to lymphocyte ratio (NLR), and on the nutritional status determined by serum albumin concentration. These PIs have been reported as useful noninvasive tools that may play a role in GC surveillance programs as a means of delivering more personalized cancer care (Hu et al., 2015; Chen et al., 2015; Sun et al., 2015). However, more studies with larger sample sizes are needed to explore the PIs in GC and evaluate the benefits of intervention to improve both immunological and nutritional status of GC patients.

## 5. BIOMARKERS

GC biomarkers, such as carcinoembryonic antigen (CEA), cancer antigens (CA 19-9, CA 72, and CA 125) are widely utilized. However, since these biomarkers show poor sensitivity and specificity, they should not be used in the primary diagnosis of stomach cancer (Jin et al., 2015; Sun and Zhang, 2014). Nevertheless, at the monitoring stage of GC patients, their growth may suggest progression of the disease. In this case, it is recommended that these biomarkers be taken into account (Ychou et al., 2000). The combination of several markers, for example, CEA with CA 19-9, increases the accuracy of this method (Jin et al., 2015; Sun and Zhang, 2014). However, the prognostic role of these markers has not yet been clearly demonstrated. Several reports have been published on the positive correlation between high levels of these biomarkers before curative surgery and tumor stage, LNs and distant metastasis, as well as on the prognostic role with regard to both an increased risk of recurrence and poor survival (Sun and Zhang, 2014; Ychou et al., 2000; Kim et al., 2014). Furthermore, these markers can be considered to predict treatment response. In a study by Sun et al., it was found that high pretreatment levels of CEA (>50 ng/mL) may predict clinical disease progression after neoadjuvant chemotherapy, and decreases (>70%) in CA 72-4 may predict pathological response after this treatment (Sun and Zhang, 2014).

Further sensitive and specific biomarkers, as well as prognostic factors, are still needed. They would help detect unfavorable prognostic groups and allow individualization of treatment. Detection and prediction of cancer spread are of great importance when qualifying GC patients for a particular treatment procedure. They also allow application of the most effective method. Numerous genetic and epigenetic alterations are involved in the mechanism of GC pathogenesis and dissemination. Therefore, intensive research into new molecular biomarkers has recently been conducted by utilizing molecular diagnostics, genomics, and proteomics (Jin et al., 2015; Abramowicz et al., 2015; Tuan et al., 2012; Nobili et al., 2011).

MicroRNAs (miRNAs) are small, noncoding RNA molecules that regulate gene expression posttranscriptionally and are reported to be involved in the multistep process of tumor metastasis like epithelial–mesenchymal transitions, anoikis, angiogenesis, invasion, and migration (Shi et al., 2014). In a study by Zhang et al. (2015) high levels of hemolysis-free plasma miRNA-214 assessed in the group of GC patients before planned gastrectomy correlated significantly with higher incidence of distant metastasis. In turn, Chen showed that microRNA-217 expression can inhibit cell proliferation, migration, and invasion in vitro, as well as tumor progression and metastasis in vivo. Therefore, miR-217 levels were lower in gastric tumor tissue when compared to adjacent normal tissue, and were associated with aggressive tumor phenotypes and poor OS in GC (Chen et al., 2015). In like manner, miR-144 inhibits GC metastasis. MicroRNA-144 suppresses GC progression by direct down-regulation of MET proto-oncogene expression, and this subsequently prevents activation of the pro-oncogenic Akt-pathway. Liu et al. (2015) revealed that during GC dissemination, expression of miR-144 decreased and miR-144 levels were significantly lower in GC metastatic tissues.

Also, long noncoding RNAs (lncRNAs) have an important role in different gene regulatory pathways by epigenetic modification and by both transcriptional and posttranscriptional regulation. Researchers have demonstrated that lncRNAs participate in neovascular formation, cell-to-cell tight junction, adhesion regulation, and extracellular matrix degranulation. Expression of some lncRNAs plays a crucial role in GC dissemination. It was found that lncRNAs can be both promoters of GC metastasis, such as lncRNAs H19 and HOTAIR, and inhibitors of GC metastasis, such as lncRNA maternally expressed gene 3 and FENDRR (Yang et al., 2015; Song et al., 2015; Zhu et al., 2015).

Several studies indicate that the risk of dissemination is significantly associated with the alterations of genes involved in the metastatic molecular pathways. This may be exemplified by growth factor genes, such as HER2 (Zhang et al., 2009; Yonemura et al., 1991), vascular endothelial growth factor (VEGF) (Zhao et al., 2014; Kosaka et al., 2007), or c-MET (Peng et al., 2014). Other examples include apoptosis regulators, such as BAX (Anagnostopoulos et al., 2007) and Fas ligand (Yang et al., 2015), signal transducer and activator of transcription 3 (Yang et al., 2015; Chen et al., 2015), cell adhesion molecules, such as E cadherin (Li et al., 2014), and also other factors associated with the cell cycle and cytokines (Jin et al., 2015).

Metabolomisc of GC metastasis can be detected by chromatography and mass spectrometry (Chen et al., 2010). Proteomic studies could lead to the molecular characterization of cellular events associated with GC progression and metastasis (Ryu et al., 2003). The study by Chen showed that metabolite profiles of GC tumors differentiate metastatic from nonmetastatic cases. Moreover, the study revealed that metabolism of both serine and proline was highlighted during the progression of metastasis (Chen et al., 2010). In a study by Abramowicz et al. (2015) the molecular differences between locally advanced and metastatic GC, as well as between cancer patients and healthy individuals, had an influence on the level of serum proteome.

Such information should be considered pivotal in cancer prognosis, diagnosis, prevention, and therapy. This could be a good starting point to discover new therapeutic targets and identify new biomarkers, as well as prognostic factors.

## 6. TREATMENT METHODS OF METASTATIC GC

Following the recommendations of the NCCN, systemic chemotherapy is considered the standard of care for patients with metastatic GC (NCCN Clinical Practice Guidelines in Oncology (NCCN Guidelines), 2015). Wagner et al. performed a meta-analysis for the Cochrane collaboration. The authors compared chemotherapy (CHT) combined with best supportive care (BSC) versus best supportive care alone. As a result, an improvement in the median survival was observed, that is, from 4.3 months for BSC to approximately 11 months for CHT. There was also an improvement in TTP (7 months for CHT vs. 2.5 months for BSC). This comparison showed that chemotherapy combined with best supportive care brought a greater survival benefit than BSC alone (Wagner et al., 2010). The NCCN also recommends combination-agent chemotherapy. Two-drug cytotoxic regimens are preferred for patients with advanced GC and with poor performance status (PS), and three-drug cytotoxic regimens for medically fit patients with good PS. Trastuzumab can be combined with first-line chemotherapy for patients with HER2-neu overexpressing adenocarcinoma (NCCN Clinical Practice Guidelines in Oncology (NCCN Guidelines), 2015). The ToGa study demonstrated that implementation of immunotherapy is associated with a positive survival benefit. For patients who received trastuzumab plus chemotherapy, the median overall survival was 13.8 months (95% CI 12-16). However, patients who received chemotherapy alone were reported with an OS median of 11.1 months (Bang et al., 2010). When selecting second-line regimens, it is important to consider PS and prior therapy. The anti-VEGFR2 antibody Ramucirumab improves survival in second line treatment both as a monotherapy and in combination with paclitaxel. Also, it represents a novel and recommended treatment method (NCCN Clinical Practice Guidelines in Oncology (NCCN Guidelines), 2015; Hacker and Lordick, 2015). New cytotoxic regimens and directions of targeted therapy are still being explored (Kanagavel et al., 2015).

Best supportive care is an integral part of each type of treatment. In some cases, for example, patients with poor PS, it may be the only method available. Radiotherapy can be used for symptomatic treatment of bleeding and obstruction, and also for pain control. In these cases, endoscopic palliative procedures should also be taken into account. Palliative surgery (gastrojejunostomy or gastrectomy) for obstruction can be considered in selected patients (NCCN Clinical Practice Guidelines in Oncology (NCCN Guidelines), 2015).

The therapeutic benefit of surgery for metastatic GC patients is controversial. There are reports on promising treatment results with gastrectomy and/or metastasectomy in this group of patients (Kerkar et al., 2009), for example, resection of the liver (Gadde et al., 2015; Sun et al., 2013; Oki et al., 2015), lung (Kemp et al., 2010), ovary (Lu et al., 2012; Cheong et al., 2004), or peritoneal stripping in combination with systemic and/or intraperitoneal chemotherapy (Rudloff et al., 2014). It seems that this treatment strategy may substantially affect the prolongation of survival in patients with a good general condition, improve their quality of life, and also their response to chemotherapy (Kanagavel et al., 2015; Kim et al., 2011; Li et al., 2015; Yang et al., 2015; Shin et al., 2015). However, the studies are not without criticism. This includes lack of randomization, a retrospective form, a small group of patients studied, and a high probability of selection bias. In groups of patients eligible for metastasectomy and/or gastrectomy in combination with chemotherapy, they were frequently in a better general condition and with more limited disease. On the other hand, patients treated with chemotherapy alone had worse PS and were more likely to have unresectable, multiple metastatic lesions. Oftentimes, they were not suitable for aggressive surgical procedures (Gadde et al., 2015; Oki et al., 2015). Well-designed prospective trials are needed to confirm good outcomes of metastasectomy and palliative gastrectomy in GC and to determine a properly selected group of patients suitable for this treatment.

## 7. CONCLUSION

Patients with locally advanced or metastatic GC represent a significant group of patients, mainly in the Western world. Metastatic cancer diagnosis is immensely important since it helps to avoid aggressive and extensive surgery, as well as it allows systemic therapy to be administered.

Several studies suggest that using 18F-FDG-PET/CT can be effective in staging of GC. In addition, parameters obtained from this diagnostic method can provide extensive information about clinical significance, metastatic potential, or unresectability of gastric tumor. Also, they can play a role as both prognostic and predictive factors for OS and TTM. Aside from PET as a useful technique in searching for target-specific tracers, other bioinformatic methods also provide extensive information about biological data-mining tools, targets linked to cancer, as well as locations and distributions of the molecular targets. It should be noted that the more specific the tracer, the higher the need for anatomical information. Thus, PET/MR appears to be a promising tool.

MRS and WB-MR DWI are also effective tools to obtain comprehensive data that can be used for bioinformatic tools. However, they have not been commonly used as yet and no significant reports have been presented so far with regard to GC. Therefore, further studies are necessary to evaluate their role with respect to this tumor location.

In addition to systemic chemotherapy and best supportive care, other methods, such as immunotherapy, palliative surgery, or radiotherapy may be also considered in metastatic GC treatment. The anti-VEGFR2 antibody Ramucirumab improves survival in second line treatment, both as a monotherapy and in combination with paclitaxel, being a novel and recommended treatment method. Moreover, new cytotoxic regimens and directions of targeted therapy are still being explored.

## Acknowledgments

All pictures come from Maria Skłodowska-Curie Memorial Institute of Oncology, Gliwice Branch, and refer to patients who participated in the research conducted at the Institute on the usefulness of PET/CT in GC. All patients provided written informed consent to participate in the research.

## References

Abramowicz, A., Wojakowska, A., Gdowicz-Klosok, A., et al., 2015. Identification of serum proteome signatures of locally advanced and metastatic gastric cancer: a pilot study. J. Transl. Med. 13, 304.

Ahmad, B., Pierson, N., Adnan, M.M., 2015. Distant skin metastases as primary presentation of gastric cancer. J. Community Support. Oncol. 13 (4), 156–158.

Altini, C., Niccoli Asabella, A., Di Palo, A., et al., 2015. 18F-FDG PET/CT role in staging of gastric carcinomas: comparison with conventional contrast enhancement computed tomography. Medicine (Baltimore) 94 (20), e864.

Anagnostopoulos, G.K., Stefanou, D., Arkoumani, E., et al., 2007. Expression of Bax protein in gastric carcinomas. A clinicopathological and immunohistochemical study. Acta Gastroenterol. Belg. 70 (3), 285–289.

Anderson, L.A., Tavilla, A., Brenner, H., EUROCARE-5 Working Group, et al., 2015. Survival for oesophageal, stomach and small intestine cancers in Europe 1999–2007: results from EUROCARE-5. Eur. J. Cancer (15), S0959–S8049. http://dx.doi.org/10.1016/j.ejca.2015.07.026 00706-6.

Anzidei, M., Napoli, A., Zaccagna, F., et al., 2009. Diagnostic performance of 64-MDCT and 1.5-T MRI with high-resolution sequences in the T staging of gastric cancer: a comparative analysis with histopathology. Radiol. Med. 114 (7), 1065–1079.

Asaka, M., Mabe, K., 2014. Strategies for eliminating death from gastric cancer in Japan. Proc. Jpn. Acad. Ser. B Phys. Biol. Sci. 90 (7), 251–258.

Bando, E., Yonemura, Y., Takeshita, Y., et al., 1999. Intraoperative lavage for cytological examination in 1,297 patients with gastric carcinoma. Am. J. Surg. 178 (3), 256–262.

Bang, Y.J., Van Cutsem, E., Feyereislova, A., et al., 2010. Trastuzumab in combination with chemotherapy versus chemotherapy alone for treatment of HER2-positive advanced gastric or gastro-oesophageal junction cancer (ToGA): a phase 3, open-label, randomised controlled trial. Lancet 376 (9742), 687–697.

Boku, T., Nakane, Y., Minoura, T., et al., 1990. Prognostic significance of serosal invasion and free intraperitoneal cancer cells in gastric cancer. Br. J. Surg. 77 (4), 436–439.

Bollschweiler, E., Berlth, F., Baltin, C., Mönig, S., Hölscher, A.H., 2014. Treatment of early gastric cancer in the Western world. World J. Gastroenterol. 20 (19), 5672–5678.

Broll, R., Weschta, M., Windhoevel, U., et al., 2001. Prognostic significance of free gastrointestinal tumor cells in peritoneal lavage detected by immunocytochemistry and polymerase chain reaction. Langenbecks Arch. Surg. 386 (4), 285–292.

Cafagna, D., Rubini, G., Iuele, F., et al., 2012. Whole-body MR-DWIBS vs. [18F]-FDG-PET/CT in the study of malignant tumors: a retrospective study. Radiol. Med. 117 (2), 293–311.

Chen, K., Chen, X., 2011. Positron emission tomography imaging of cancer biology: current status and future prospects. Semin. Oncol. 38 (1), 70–86.

Chen, J.L., Tang, H.Q., Hu, J.D., Fan, J., Hong, J., Gu, J.Z., 2010. Metabolomics of gastric cancer metastasis detected by gas chromatography and mass spectrometry. World J. Gastroenterol. 16 (46), 5874–5880.

Chen, J., Hong, D., Zhai, Y., Shen, P., 2015. Meta-analysis of associations between neutrophil-to-lymphocyte ratio and prognosis of gastric cancer. World J. Surg. Oncol. 13, 122.

Chen, D.L., Zhang, D.S., Lu, Y.X., et al., 2015. microRNA-217 inhibits tumor progression and metastasis by downregulating EZH2 and predicts favorable prognosis in gastric cancer. Oncotarget 6 (13), 10868–10879.

Chen, J., Liu, X., Jiao, H., et al., 2015. Prognostic and clinical significance of STAT3 and MMP9 in patients with gastric cancer: a meta-analysis of a Chinese cohort. Int. J. Clin. Exp. Med. 8 (1), 546–557.

Cheong, J.H., Hyung, W.J., Chen, J., Kim, J., Choi, S.H., Noh, S.H., 2004. Survival benefit of metastasectomy for Krukenberg tumors from gastric cancer. Gynecol. Oncol. 94 (2), 477–482.

Choi, J.Y., Shim, K.N., Kim, S.E., Jung, H.K., Jung, S.A., Yoo, K., 2014. The clinical value of 18F-fluorodeoxyglucose uptake on positron emission tomography/computed tomography for predicting regional lymph node metastasis and non-curative surgery in primary gastric carcinoma. Korean J. Gastroenterol. 64 (6), 340–347.

Chung, H.W., Lee, E.J., Cho, Y.H., et al., 2010. High FDG uptake in PET/CT predicts worse prognosis in patients with metastatic gastric adenocarcinoma. J. Cancer Res. Clin. Oncol. 136 (12), 1929–1935.

Coleman, M.P., Gatta, G., Verdecchia, A., et al., 2003. EUROCARE Working Group. EUROCARE-3 summary: cancer survival in Europe at the end of the 20th century. Ann. Oncol. (Suppl. 5), v128–49.

Compare, D., Rocco, A., Nardone, G., 2014. Screening for and surveillance of gastric cancer. World J. Gastroenterol. 20 (38), 13681–13691.

Coupe, N.A., Karikios, D., Chong, S., et al., 2014. Metabolic information on staging FDG-PET-CT as a prognostic tool in the evaluation of 97 patients with gastric cancer. Ann. Nucl. Med. 28 (2), 128–135.

Cui, J.X., Li, T., Xi, H.Q., Wei, B., Chen, L., 2013. Evaluation of (18)F-FDG PET/CT in preoperative staging of gastric cancer: a meta-analysis. Zhonghua Wei Chang Wai Ke Za Zhi 16 (5), 418–424.

Dan, Y.Y., So, J.B., Yeoh, K.G., 2006. Endoscopic screening for gastric cancer. Clin. Gastroenterol. Hepatol. 4 (6), 709–716.

Dassen, A.E., Lips, D.J., Hoekstra, C.J., Pruijt, J.F., Bosscha, K., 2009. FDG-PET has no definite role in preoperative imaging in gastric cancer. Eur. J. Surg. Oncol. 35 (5), 449–455.

Di Fabio, F., Pinto, C., Rojas Llimpe, F.L., et al., 2007. The predictive value of 18F-FDG-PET early evaluation in patients with metastatic gastric adenocarcinoma treated with chemotherapy plus cetuximab. Gastric Cancer 10 (4), 221–227.

Duarte, I., Llanos, O., 1981. Patterns of metastases in intestinal and diffuse types of carcinoma of the stomach. Hum. Pathol. 12 (3), 237–242.

Ferlay, J., Soerjomataram, I., Ervik, M., et al., 2013. GLOBOCAN 2012 v1.0, Cancer Incidence and Mortality Worldwide: IARC Cancer Base No. 11 [Internet]. International Agency for Research on Cancer, Lyon, France. Available from: URL http://globocan.iarc.fr.

Fock, K.M., 2014. Review article: the epidemiology and prevention of gastric cancer. Aliment. Pharmacol. Ther. 40 (3), 250–260.

Fratellone, P.M., Holowecki, M.A., 2009. Forgotten node: a case report. World J. Gastroenterol. 15 (39), 4974–4975.

Gadde, R., Tamariz, L., Hanna, M., Avisar, E., Livingstone, A., Franceschi, D., Yakoub, D., 2015. Metastatic gastric cancer (MGC) patients: can we improve survival by metastasectomy? A systematic review and meta-analysis. J. Surg. Oncol. 112 (1), 38–45.

George, K., Al Hooti, Q.M., Al Busaidy, S.S., Joseph, M., Kamona, A., 2015. Synchronous bilateral ureteric metastasis from gastric cancer. Urol. Ann. 7 (3), 408–409.

Grabinska, K., Pelak, M., Wydmanski, J., Tukiendorf, A., d'Amico, A., 2014. Prognostic value and clinical correlations of 18-fluorodeoxyglucose metabolism quantifiers in gastric cancer. Nucl. Med. Mol. Imaging 48 (1), 33–40.

Ha, T.K., Choi, Y.Y., Song, S.Y., Kwon, S.J., 2011. F18-fluorodeoxyglucose-positron emission tomography and computed tomography is not accurate in preoperative staging of gastric cancer. J. Korean Surg. Soc. 81 (2), 104–110.

Hacker, U., Lordick, F., 2015. Current standards in the treatment of gastric cancer. Dtsch. Med. Wochenschr. 140 (16), 1202–1205.

Hale, R.W., 1968. Krukenberg tumor of the ovaries. A review of 81 records. Obstet. Gynecol. 32 (2), 221–225.

Hamashima, C., Ogoshi, K., Okamoto, M., Shabana, M., Kishimoto, T., Fukao, A., 2013. A community-based, case-control study evaluating mortality reduction from gastric cancer by endoscopic screening in Japan. PLoS One 8 (11), e79088.

Hamashima, C., 2014. Current issues and future perspectives of gastric cancer screening. World J. Gastroenterol. 20 (38), 13767–13774.

Herrmann, K., Ott, K., Buck, A.K., et al., 2007. Imaging gastric cancer with PET and the radiotracers 18F-FLT and 18F-FDG: a comparative analysis. J. Nucl. Med. 48 (12), 1945–1950.

Hu, Z.D., Huang, Y.L., Qin, B.D., et al., 2015. Prognostic value of neutrophil to lymphocyte ratio for gastric cancer. Ann. Transl. Med. 3 (4), 50.

Imachi, M., Tsukamoto, N., Amagase, H., Shigematsu, T., Amada, S., Nakano, H., 1993. Metastatic adenocarcinoma to the uterine cervix from gastric cancer. A clinicopathologic analysis of 16 cases. Cancer 71 (11), 3472–3477.

Jadvar, H., Tatlidil, R., Garcia, A.A., Conti, P.S., March 2003. Evaluation of recurrent gastric malignancy with [F-18]-FDG positron emission tomography. Clin. Radiol. 58 (3), 215–221.

Japanese Gastric Cancer Association, 2011. Japanese classification of gastric carcinoma: 3rd English edition. Gastric Cancer 14 (2), 101–112. http://dx.doi.org/10.1007/s10120-011-0041-5.

Jeong, J., Cho, I., Kong, E., et al., 2013. Evaluation of hybrid PET/CT gastrography in gastric cancer. Nuklearmedizin 52 (3), 107–112.

Jin, Z., Jiang, W., Wang, L., 2015. Biomarkers for gastric cancer: progression in early diagnosis and prognosis (review). Oncol. Lett. 9 (4), 1502–1508.

Kamimura, K., Nagamachi, S., Wakamatsu, H., et al., 2009. Role of gastric distention with additional water in differentiating locally advanced gastric carcinomas from physiological uptake in the stomach on 18F-fluoro-2-deoxy-D-glucose PET. Nucl. Med. Commun. 30 (6), 431–439.

Kanagavel, D., Fedyanin, M., Tryakin, A., Tjulandin, S., 2015. Second-line treatment of metastatic gastric cancer: current options and future directions. World J. Gastroenterol. 21 (41), 11621–11635.

Kanagawa, T., Ookawa, T., Nogami, H., Kobatake, T., 2013. A case of gastric cancer with virchow and para-aortic node metastasis treated successfully by neo-adjuvant chemotherapy with S-1/CDDP. Gan To Kagaku Ryoho 40 (9), 1217–1220.

Kemp, C.D., Kitano, M., Kerkar, S., et al., 2010. Pulmonary resection for metastatic gastric cancer. J. Thorac. Oncol. 5 (11), 1796–1805.

Kerkar, S.P., Kemp, C.D., Duffy, A., et al., 2009. The GYMSSA trial: a prospective randomized trial comparing gastrectomy, metastasectomy plus systemic therapy versus systemic therapy alone. Trials 10, 121.

Kerkar, S.P., Kemp, C.D., Avital, I., 2010. Liver resections in metastatic gastric cancer. HPB (Oxford) 12 (9), 589–596.

Kim, A.Y., Han, J.K., Seong, C.K., Kim, T.K., Choi, B.I., 2000. MRI in staging advanced gastric cancer: is it useful compared with spiral CT? J. Comput. Assist. Tomogr. 24 (3), 389–394.

Kim, K.H., Lee, K.W., Baek, S.K., et al., 2011. Survival benefit of gastrectomy ± metastasectomy in patients with metastatic gastric cancer receiving chemotherapy. Gastric Cancer 14 (2), 130–138.

Kim, E.Y., Lee, W.J., Choi, D., et al., 2011. The value of PET/CT for preoperative staging of advanced gastric cancer: comparison with contrast-enhanced CT. Eur. J. Radiol. 79 (2), 183–188.

Kim, J.H., Jun, K.H., Jung, H., Park, I.S., Chin, H.M., 2014. Prognostic value of preoperative serum levels of five tumor markers (carcinoembryonic antigen, CA19-9, alpha-fetoprotein, CA72-4, and CA125) in gastric cancer. Hepatogastroenterology 61 (131), 863–869.

Kim, J., Lim, S.T., Na, C.J., et al., 2014. Pretreatment F-18 FDG PET/CT parameters to evaluate progression-free survival in gastric cancer. Nucl. Med. Mol. Imaging 48 (1), 33–40.

Kinkel, K., Lu, Y., Both, M., Warren, R.S., Thoeni, R.F., 2002. Detection of hepatic metastases from cancers of the gastrointestinal tract by using noninvasive imaging methods (US, CT, MR imaging, PET): a meta-analysis. Radiology 224 (3), 748–756.

Koessler, T., Roth, A., Cacheux, W., 2014. Early gastric cancer: epidemiology, diagnostic and management. Rev. Med. Suisse 10 (431), 1118–1122.

Kosaka, Y., Mimori, K., Fukagawa, T., et al., 2007. Identification of the high-risk group for metastasis of gastric cancer cases by vascular endothelial growth factor receptor-1 overexpression in peripheral blood. Br. J. Cancer 96 (11), 1723–1728.

Kraszkiewicz, M., Wydmanski, J., 2014. Brain metastases from stomach cancer – the role of different treatment modalities and efficacy of palliative radiotherapy. Rep. Pract. Oncol. Radiother. 20 (1), 32–37.

Kwee, R.M., Kwee, T.C., 2009. Imaging in assessing lymph node status in gastric cancer. Gastric Cancer 12 (1), 6–22.

Landry, J., Tepper, J.E., Wood, W.C., Moulton, E.O., Koerner, F., Sullinger, J., 1990. Patterns of failure following curative resection of gastric carcinoma. Int. J. Radiat. Oncol. Biol. Phys. 19 (6), 1357–1362.

Lansdorp-Vogelaar, I., Sharp, L., 2013. Cost-effectiveness of screening and treating *Helicobacter pylori* for gastric cancer prevention. Best Pract. Res. Clin. Gastroenterol. 27 (6), 933–947.

Larson, S.M., Erdi, Y., Akhurst, T., et al., 1999. Tumor treatment response based on visual and quantitative changes in global tumor glycolysis using PET-FDG imaging. The visual response score and the change in total lesion glycolysis. Clin. Positron Imaging 2 (3), 159–171.

Latour, J.P., Soloway, I., Young, M.H., 1956. Krukenberg tumors of the ovary. Obstet. Gynecol. 8 (5), 636–638.

Layke, J.C., Lopez, P.P., 2004. Gastric cancer: diagnosis and treatment options. Am. Fam. Physician 69 (5), 1133–1140.

Lee, J.W., Kang, K.W., Paeng, J.C., et al., 2009. Cancer screening using 18F-FDG PET/CT in Korean asymptomatic volunteers: a preliminary report. Ann. Nucl. Med. 23 (7), 685–691.

Lee, J.W., Lee, S.M., Lee, M.S., Shin, H.C., 2012. Role of 18F-FDG PET/CT in the prediction of gastric cancer recurrence after curative surgical resection. Eur. J. Nucl. Med. Mol. Imaging 39 (9), 1425–1434.

Lee, D.Y., Lee, C.H., Seo, M.J., Lee, S.H., Ryu, J.S., Lee, J.J., 2014. Performance of (18)F-FDG PET/CT as a postoperative surveillance imaging modality for asymptomatic advanced gastric cancer patients. Ann. Nucl. Med. 28 (8), 789–795.

Lee, J.W., Kang, C.M., Choi, H.J., Lee, W.J., Song, S.Y., Lee, J.H., Lee, J.D., 2014. Prognostic value of metabolic tumor volume and total lesion glycolysis on preoperative 18F-FDG PET/CT in patients with pancreatic cancer. J. Nucl. Med. 55, 898–904.

Lello, E., Furnes, B., Edna, T.H., 2007. Short and long-term survival from gastric cancer. A population-based study from a county hospital during 25 years. Acta Oncol. 46 (3), 308–315.

Lerman, H., Metser, U., Grisaru, D., Fishman, A., Lievshitz, G., Even-Sapir, E., 2004. Normal and abnormal 18F-FDG endometrial and ovarian uptake in pre- and postmenopausal patients: assessment by PET/CT. J. Nucl. Med. 45 (2), 266–271.

Li, B., Zheng, P., Zhu, Q., Lin, J., 2012. Accurate preoperative staging of gastric cancer with combined endoscopic ultrasonography and PET-CT. Tohoku J. Exp. Med. 228 (1), 9–16.

Li, T., Chen, J., Liu, Q.L., Huo, Z.H., Wang, Z.W., 2014. Meta-analysis: E-cadherin immunoexpression as a potential prognosis biomarker related to gastric cancer metastasis in Asian patients. Eur. Rev. Med. Pharmacol. Sci. 18 (18), 2693–2703.

Li, Y.M., Lin, Q., Zhao, L., et al., 2014. Pre-treatment metabolic tumor volume and total lesion glycolysis are useful prognostic factors for esophageal squamous cell cancer patients. Asian Pac. J. Cancer Prev. 15, 1369–1373.

Li, P., Ling, Y.H., Zhu, C.M., et al., April 1, 2015. Vascular invasion as an independent predictor of poor prognosis in nonmetastatic gastric cancer after curative resection. Int. J. Clin. Exp. Pathol. 8 (4), 3910–3918.

Li, Z., Fan, B., Shan, F., et al., 2015. Gastrectomy in comprehensive treatment of advanced gastric cancer with synchronous liver metastasis: a prospectively comparative study. World J. Surg. Oncol. 13, 212.

Li, B., Cai, H., Kang, Z.C., 2015. Testicular metastasis from gastric carcinoma: a case report. World J. Gastroenterol. 21 (21), 6764–6768.

Liu, J., Xue, H., Zhang, J., et al., 2015. MicroRNA-144 inhibits the metastasis of gastric cancer by targeting MET expression. J. Exp. Clin. Cancer Res. 34, 35.

Lu, L.C., Shao, Y.Y., Hsu, C.H., et al., 2012. Metastasectomy of Krukenberg tumors may be associated with survival benefits in patients with metastaticgastric cancer. Anticancer Res. 32 (8), 3397–3401.

Maconi, G., Manes, G., Porro, G.B., 2008. Role of symptoms in diagnosis and outcome of gastric cancer. World J. Gastroenterol. 14 (8), 1149–1155.

Manenti, G., Cicciò, C., Squillaci, E., et al., 2012. Role of combined DWIBS/3D-CE-T1w whole-body MRI in tumor staging: comparison with PET-CT. Eur. J. Radiol. 81 (8), 1917–1925.

Marrelli, D., Roviello, F., de Manzoni, G., et al., 2002. Italian Research Group for gastric cancer. Different patterns of recurrence in gastric cancer depending on Lauren's histological type: longitudinal study. World J. Surg. 26 (9), 1160–1165.

Maruyama, K., 2009. The most important prognostic factors for gastric cancer patients: a study using univariate and multivariate analyses. Scand. J. Gastroenterol. 22 (s133), 63–68.

Matsuda, T., Saika, K., 2013. The 5-year relative survival rate of stomach cancer in the USA, Europe and Japan. Jpn. J. Clin. Oncol. 43 (11), 1157–1158.

Mezhir, J.J., Shah, M.A., Jacks, L.M., Brennan, M.F., Coit, D.G., Strong, V.E., 2011. Positive peritoneal cytology in patients with gastric cancer: natural history and outcome of 291 patients. Indian J. Surg. Oncol. 2 (1), 16–23.

Minamimoto, R., Senda, M., Jinnouchi, S., et al., 2013. The current status of an FDG-PET cancer screening program in Japan, based on a 4-year (2006–2009) nationwide survey. Ann. Nucl. Med. 27 (1), 46–57.

Moon, S.H., Hyun, S.H., Choi, J.Y., 2013. Prognostic significance of volume-based PET parameters in cancer patients. Korean J. Radiol. 14 (1), 1–12.

Namikawa, T., Okabayshi, T., Nogami, M., Ogawa, Y., Kobayashi, M., Hanazaki, K., 2014. Assessment of (18) F-fluorodeoxyglucose positron emission tomography combined with computed tomography in the preoperative management of patients with gastric cancer. Int. J. Clin. Oncol. 19 (4), 649–655.

Nccn clinical practice guidelines in oncology (NCCN guidelines). Gastric Cancer2015. Available from: URL. http://www.nccn.org/professionals/physician_gls/pdf/gastric.pdf.

Nobili, S., Bruno, L., Landini, I., et al., 2011. Genomic and genetic alterations influence the progression of gastric cancer. World J. Gastroenterol. 17 (3), 290–299.

Oh, H.H., Lee, S.E., Choi, I.S., et al., 2011. The peak-standardized uptake value (P-SUV) by preoperative positron emission tomography-computed tomography (PET-CT) is a useful indicator of lymph node metastasis in gastric cancer. J. Surg. Oncol. 104 (5), 530–533.

Oki, E., Tokunaga, S., Emi, Y., et al., August 11, 2015. Surgical treatment of liver metastasis of gastric cancer: a retrospective multicenter cohort study (KSCC1302). Gastric Cancer Epub ahead of print.

Pak, K.H., Yun, M., Cheong, J.H., Hyung, W.J., Choi, S.H., Noh, S.H., 2011. Clinical implication of FDG-PET in advanced gastric cancer with signet ring cell histology. J. Surg. Oncol. 104 (6), 566–570.

Park, J.C., Lee, J.H., Cheoi, K., et al., 2012. Predictive value of pretreatment metabolic activity measured by fluorodeoxyglucose positron emission tomography in patients with metastatic advanced gastric cancer: the maximal SUV of the stomach is a prognostic factor. Eur. J. Nucl. Med. Mol. Imaging 39 (7), 1107–1116.

Park, K., Jang, G., Baek, S., Song, H., 2014. Usefulness of combined PET/CT to assess regional lymph node involvement in gastric cancer. Tumori 100 (2), 201–206.

Pasechnikov, V., Chukov, S., Fedorov, E., Kikuste, I., Leja, M., 2014. Gastric cancer: prevention, screening and early diagnosis. World J. Gastroenterol. 20 (38), 13842–13862.

Patriti, A., Graziosi, L., Baffa, N., 2007. Postoperative follow-up of gastric adenocarcinoma with neoplastic markers and 18-FDG-PET/CT. Ann. Ital. Chir. 78 (6), 481–485.

Peng, Z., Zhu, Y., Wang, Q., et al., 2014. Prognostic significance of MET amplification and expression in gastric cancer: a systematic review with meta-analysis. PLoS One 9 (1), e84502.

Rudloff, U., Langan, R.C., Mullinax, J.E., et al., 2014. Impact of maximal cytoreductive surgery plus regional heated intraperitoneal chemotherapy (HIPEC) on outcome of patients with peritoneal carcinomatosis of gastric origin: results of the GYMSSA trial. J. Surg. Oncol. 110 (3), 275–284.

Ryu, J.W., Kim, H.J., Lee, Y.S., et al., 2003. The proteomics approach to find biomarkers in gastric cancer. J. Korean Med. Sci. 18 (4), 505–509.

Saito, H., Fukumoto, Y., Osaki, T., et al., 2006. Distinct recurrence pattern and outcome of adenocarcinoma of the gastric cardia in comparison with carcinoma of other regions of the stomach. World J. Surg. 30 (10), 1864–1869.

Santiago, J.M., Sasako, M., Osorio, J., 2011. TNM-7th edition 2009 (UICC/AJCC) and Japanese Classification 2010 in Gastric Cancer. Towards simplicity and standardisation in the management of gastric cancer. Cir. Esp. 89 (5), 275–281.

Shi, Z., Wei, Q., She, J., 2014. MicroRNAs in gastric cancer metastasis. Crit. Rev. Eukaryot. Gene Expr. 24 (1), 39–53.

Shim, J.H., Yoo, H.M., Lee, H.H., et al., 2011. Use of laparoscopy as an alternative to computed tomography (CT) and positron emission tomography (PET) scans for the detection of recurrence in patients with gastric cancer: a pilot study. Surg. Endosc. 25 (10), 3338–3344.

Shimoyama, S., Seto, Y., Aoki, F., et al., 2004. Gastric cancer with metastasis to the gingiva. J. Gastroenterol. Hepatol. 19 (7), 831–835.

Shin, H.B., Lee, S.H., Son, Y.G., Ryu, S.W., Sohn, S.S., 2015. Chemoresponse after non-curative gastrectomy for M1 gastric cancer. World J. Surg. Oncol. 13, 13.

Shiono, S., Sato, T., Horio, H., Metastatic Lung Tumor Study Group of Japan, 2013. Outcomes and prognostic factors of survival after pulmonary resection for metastatic gastric cancer. Eur. J. Cardiothorac. Surg. 43 (1), e13–6.

Smyth, E.C., Shah, M.A., 2011. Role of $^{1}$F 2-fluoro-2-deoxyglucose positron emission tomography in upper gastrointestinal malignancies. World J. Gastroenterol. 17 (46), 5059–5074.

Smyth, E., Schöder, H., Strong, V.E., et al., 2012. A prospective evaluation of the utility of 2-deoxy-2-[(18) F] fluoro-D-glucose positron emission tomography and computed tomography in staging locally advanced gastric cancer. Cancer 118 (22), 5481–5488.

Song, B., Guan, Z., Liu, F., Sun, D., Wang, K., Qu, H., 2015. Long non-coding RNA HOTAIR promotes HLA-G expression via inhibiting miR-152 in gastric cancer cells. Biochem. Biophys. Res. Commun. 464 (3), 807–813.

Standby, R.B., Svendsen, L.B., Fallentin, E., Egeland, C., Achiam, M.P., 2015. The multidisciplinary team conference's decision on M-staging in patients with gastric- and gastroesophageal cancer is not accurate without staging laparoscopy. Scand. J. Surg. http://dx.doi.org/10.1177/1457496915598760. Epub ahead of print.

Stecco, A., Romano, G., Negru, M., et al., 2009. Whole-body diffusion-weighted magnetic resonance imaging in the staging of oncological patients: comparison with positron emission tomography computed tomography (PET-CT) in a pilot study. Radiol. Med. 114 (1), 1–17.

Sugarbaker, P.H., Yonemura, Y., 2000. Clinical pathway for the management of resectable gastric cancer with peritoneal seeding: best palliation with a ray of hope for cure. Oncology 58 (2), 96–107.

Sun, Z., Zhang, N., 2014. Clinical evaluation of CEA, CA19-9, CA72-4 and CA125 in gastric cancer patients with neoadjuvant chemotherapy. World J. Surg. Oncol. 12, 397.

Sun, J., Song, Y., Wang, Z., Chen, X., Gao, P., Xu, Y., Zhou, B., Xu, H., 2013. Clinical significance of palliative gastrectomy on the survival of patients with incurable advanced gastric cancer: a systematic review and meta-analysis. BMC Cancer 13, 577.

Sun, K.Y., Xu, J.B., Chen, S.L., et al., 2015. Novel immunological and nutritional-based prognostic index for gastric cancer. World J. Gastroenterol. 21 (19), 5961–5971.

Tomizawa, M., Shinozaki, F., Uchida, Y., et al., 2015. Diffusion-weighted whole-body imaging with background body signal suppression/T2 image fusion and positron emission tomography/computed tomography of upper gastrointestinal cancers. Abdom. Imaging 40 (8), 3012–3019.

Tuan, T.F., Chung, C.T., Tsou, H.H., et al., 2012. Putative tumor metastasis-associated genes in human gastric cancer. Int. J. Oncol. 41 (3), 1068–1084.

Ueda, K., Iwahashi, M., Nakamori, M., et al., 2009. Analysis of the prognostic factors and evaluation of surgical treatment for synchronous liver metastases from gastric cancer. Langenbecks Arch. Surg. 394 (4), 647–653.

Van de Wiele, C., Kruse, V., Smeets, P., Sathekge, M., Maes, A., 2013. Predictive and prognostic value of metabolic tumour volume and total lesion glycolysis in solid tumours. Eur. J. Nucl. Med. Mol. Imaging 40, 290–301.

Waddell, T., Verheij, M., Allum, W., Cunningham, D., Cervantes, A., Arnold, D., European Society for Medical Oncology (ESMO), European Society of Surgical Oncology (ESSO), European Society of Radiotherapy and Oncology (ESTRO), 2014. Gastric cancer: ESMO-ESSO-ESTRO clinical practice guidelines for diagnosis, treatment and follow-up. Radiother. Oncol. 110 (1), 189–194.

Wagner, A.D., Unverzagt, S., Grothe, W., et al., 2010. Chemotherapy for advanced gastric cancer. Cochrane Database Syst. Rev. (3). http://dx.doi.org/10.1002/14651858.CD004064.pub3. CD004064.

Wang, Z., Chen, J.Q., 2011. Imaging in assessing hepatic and peritoneal metastases of gastric cancer: a systematic review. BMC Gastroenterol. 11, 19.

Wu, C.X., Zhu, Z.H., 2014. Diagnosis and evaluation of gastric cancer by positron emission tomography. World J. Gastroenterol. 20 (16), 4574–4585.

Wu, C.W., Lo, S.S., Shen, K.H., et al., 2003. Incidence and factors associated with recurrence patterns after intended curative surgery for gastric cancer. World J. Surg. 27 (2), 153–158.

Wydmanski, J., Grabinska, K., Polanowski, P., 2014. Radiotherapy and chemoradiotherapy as a novel option for the treatment of locally advanced inoperable gastric adenocarcinoma: a phase II study. Mol. Clin. Oncol. 2 (6), 1150–1154.

Wydmanski, J., 2008. Study of effectiveness and tolerance of pre and postoperative chemoradiotherapy for patients with stomach cancer. Nowotwory J. Oncol. 58, 5–47.

Yada-Hashimoto, N., Yamamoto, T., Kamiura, S., et al., 2008. Metastatic ovarian tumors: a review of 64 cases. Gynecol. Oncol. 89 (2), 314–317.

Yakushiji, M., Tazaki, T., Nishimura, H., Kato, T., 1987. Krukenberg tumors of the ovary: a clinicopathologic analysis of 112 cases. Nihon Sanka Fujinka Gakkai Zasshi 39 (3), 479–485.

Yang, Z.G., Gao, L., Guo, X.B., Shi, Y.L., 2015. Roles of long non-coding RNAs in gastric cancer metastasis. World J. Gastroenterol. 21 (17), 5220–5230.

Yang, Y., Zhao, Q., Cai, Z., et al., 2015. Fas signaling promotes gastric cancer metastasis through STAT3-dependent upregulation of fascin. PLoS One 10 (5), e0125132.

Yang, K., Liu, K., Zhang, W.H., et al., 2015. The value of palliative gastrectomy for gastric cancer patients with intraoperatively proven peritoneal seeding. Medicine (Baltimore) 94 (27), e1051.

Ychou, M., Duffour, J., Kramar, A., Gourgou, S., Grenier, J., 2000. Clinical significance and prognostic value of CA72-4 compared with CEA and CA19-9 in patients with gastric cancer. Dis. Markers 16 (3–4), 105–110.

Yonemura, Y., Ninomiya, I., Yamaguchi, A., et al., 1991. Evaluation of immunoreactivity for erbB-2 protein as a marker of poor short term prognosis in gastric cancer. Cancer Res. 51 (3), 1034–1038.

Yoo, C.H., Noh, S.H., Shin, D.W., Choi, S.H., Min, J.S., 2000. Recurrence following curative resection for gastric carcinoma. Br. J. Surg. 87 (2), 236–242.

Yoshioka, T., Yamaguchi, K., Kubota, K., et al., 2003. Evaluation of 18F-FDG PET in patients with advanced, metastatic, or recurrent gastric cancer. J. Nucl. Med. 44 (5), 690–699.

Zaidi, H., Del Guerra, A., 2011. An outlook on future design of hybrid PET/MRI systems. Med. Phys. 38 (10), 5667–5689.

Zaizen, Y., Azuma, K., Kurata, S., et al., 2012. Prognostic significance of total lesion glycolysis in patients with advanced non-small cell lung cancer receiving chemotherapy. Eur. J. Radiol. 81, 4179–4184.

Zhang, X.L., Yang, Y.S., Xu, D.P., et al., October 2009. Comparative study on overexpression of HER2/neu and HER3 in gastric cancer. World J. Surg. 33 (10), 2112–2118.

Zhang, K.C., Xi, H.Q., Cui, J.X., et al., 2015. Hemolysis-free plasma miR-214 as novel biomarker of gastric cancer and is correlated with distant metastasis. Am. J. Cancer Res. 5 (2), 821–829.

Zhao, D.Q., Chen, J., Wu, Y.F., Tian, D., Zhou, R.X., 2014. Correlation between vascular endothelial growth factor and somatostatin receptor with progression and prognosis in gastric cancer. Hepatogastroenterology 61 (132), 1154–1158.

Zhou, M., Wang, C., Hu, S., et al., 2013. 18F-FLT PET/CT imaging is not competent for the pretreatment evaluation of metastatic gastric cancer: a comparison with 18F-FDG PET/CT imaging. Nucl. Med. Commun. 34 (7), 694–700.

Zhu, L., Liu, J., Ma, S., Zhang, S., 2015. Long noncoding RNA MALAT-1 can predict metastasis and a poor prognosis: a meta-analysis. Pathol. Oncol. Res. 21 (4), 1259–1264.

Zhuang, H., Pourdehnad, M., Lambright, E.S., et al., 2001. Dual time point 18F-FDG PET imaging for differentiating malignant from inflammatory processes. J. Nucl. Med. 42 (9), 1412–1417.

CHAPTER

# 9

# Hepatocellular Carcinoma: Metastatic Disease

*S. Chandradas, C.B. Skillin, C.T. Frenette*

**Scripps Clinic/Green Hospital, La Jolla, CA, United States**

## 1. INTRODUCTION

Hepatocellular carcinoma (HCC) is the most common primary liver malignancy. The incidence and prevalence of HCC have been steadily increasing since 1975, and is projected to continue to rise over the next one to two decades (Altekruse et al., 2009). Unfortunately, HCC carries one of the highest mortality rates of solid organ cancers, with a 5-year survival of less than 15% (El-Serag, 2004; Mittal and El-Serag, 2013). One of the main reasons for the poor prognosis is that liver cancer is often diagnosed at an advanced stage. Over two-thirds of patients with HCC present with advanced stage or metastatic disease, and data indicate underuse of curative or palliative therapy for many patients with HCC (Wong and Corley, 2009; El-Serag et al., 2006). It is critical to understand the background of HCC to appropriately evaluate and treat metastatic disease, but even in patients with metastatic disease, treatment has been shown to improve survival over best supportive care (Altekruse et al., 2009).

## 2. EPIDEMIOLOGY

### 2.1 Incidence

HCC is responsible for 90% of all documented cases of liver cancer in the United States. The incidence of HCC has tripled from 1975 to 2005, likely due to the high rate of hepatitis C virus (HCV) infection from 1960 to 1990, as well as the significant increase in fatty liver disease in the United States (Altekruse et al., 2009). HCV infection escalated due to an increase in blood transfusions, and HCV was not screened for until 1992. Hepatitis C can also be contracted through intravenous (IV) drug use, and occasionally through casual contact with blood. Since 2005, HCC incidence has continued to rise, and in 2010 6 out of every 100,000 people in the United States were diagnosed with HCC (El-Serag and Kanwal, 2014). This number is projected to continue to rise over the next two decades. Despite currently having a more significant control over HCV infection rates, the infections that occurred before 1990 will be the main contributing factor to this rise in HCC prevalence. This is due to the

*Introduction to Cancer Metastasis*
http://dx.doi.org/10.1016/B978-0-12-804003-4.00009-8

20- to 30-year lag time between initial infection with HCV and the development of HCC. HCC is typically diagnosed between 50 and 60 years of age and because most cases of HCC are due to chronic liver disease over a long period of time. It is very rare to see HCC in younger, healthier patients (Tsukuma et al., 1993).

More exhaustive screening and an increased understanding of HCC have also contributed to the observed increase in incidence. For example, patients with a history of IV drug use and pre-1990 transfusions are now being screened for hepatitis C at a higher frequency. This has led to an increase in documented cases of hepatitis C–related cirrhosis. Cirrhotic patients are subsequently screened for HCC development every 6 months via imaging, and alfa-fetoprotein (AFP) serum levels. Thanks to this screening, HCC is continually diagnosed earlier and earlier in its progression, leading to an increase in 1-year survival rates. From 1992 to 2004, the 1-year survival rate doubled, and the 4- to 5-year survival rate also improved significantly (Altekruse et al., 2009). The consistent improvement of HCC diagnosis contributes to the rise in survival with lead time bias, but there has also been an increase in treatment options available for patients at all stage of disease. Despite these improvements, the 1-year survival rate remains below 50%, and the mortality rate of HCC has grown faster than any other type of cancer. Today, HCC causes between 250,000 and 1 million deaths per year worldwide (Muñoz and Bosch, 1987).

HCC is more common in men than in women. Generally, the incidence in men is up to three times higher. For men, HCC is the fifth most common cancer diagnosis and it is the second leading cause of cancer-related death worldwide (Ferlay et al., 2010). For women HCC is the fifth most diagnosed cancer and the sixth leading cause of cancer-related death worldwide.

The incidence of HCC is not distributed evenly throughout the globe. It is more common in developing countries due to unhealthy life styles, inferior medical facilities/techniques, and restricted access to care. Furthermore, the disparity in HCC development between men and women is significantly greater in high incidence regions. HCC incidence is the highest in East Asia with an incidence of 35.5 per 100,000 (Parkin et al., 2002). The vast majority of HCC cases in Eastern Asia are due to hepatitis B infection, with exception of Japan where hepatitis C is a more frequent underlying risk factor. Although still very high, these rates have been decreasing due to the increased use of the hepatitis B vaccine. Sub-Saharan Africa has the second highest incidence of HCC in the world, at 24.2 per 100,000 (Parkin et al., 2002).

Within a country there are also racial/ethnic variations in the development of HCC. Within the United States, the incidence is highest among Asian/Pacific Islanders, followed by Hispanics, African Americans, and American Indians/Alaskan Natives (Altekruse et al., 2009). It is worth noting that many of the Asian Americans who developed HCC were born outside of the United States, and likely contracted hepatitis B virus (HBV) vertically during birth or childhood.

## 2.2 Underlying Disease/Risk Factors

As mentioned above, two of the largest risk factors for HCC are hepatitis B and hepatitis C. These viruses damage the liver and often lead to cirrhosis if left untreated. Hepatitis B may also lead to the development of HCC without the presence of cirrhosis (Beasley, 1988). Beyond the development of cirrhosis, there are other factors within the context of HBV that can increase the risk of HCC tumorigenesis. These include an elevated viral HBV DNA load, HBV e antigen presence, and the presence of the hepatitis B surface antigen (Tsai et al., 1996; Chen et al., 2006).

HBV is a heterogeneous virus consisting of eight different genotypes. These genotypes are typically found exclusively in separate regions of the world. Genotype A is generally found in North America, Africa, and northern Europe.

Genotypes B and C are commonly found in Asia, and genotype D is found in southern Europe and the Middle East (Chemin and Zoulim, 2009). Genotypes B and C are most common among those chronically infected, and genotype A is the most common for acute infections. Studies measuring the effect of these different genotypes on the development of HCC have yielded varying results, but it is generally accepted that having genotype C-hepatitis B is associated with an increased risk of cirrhosis and cancer development. Coinfection with hepatitis C will also increase the risk of cancer development in HBV patients compared to those who are infected with a single virus.

Like HBV, hepatitis C increases the risk of HCC. In the United States, 36% of the reported HCC cases are due to the underlying chronic HCV infection (Yang et al., 2011). HCC due to HCV almost always occurs in patients who also have advanced fibrosis or cirrhosis. Similar to HBV, a higher viral load increases the risk of malignancy. Unlike HBV, the successful eradication of the HCV decreases the risk of developing HCC and can even lead to a regression in cirrhosis/fibrosis (although the risk of HCC development never completely normalizes) (George et al., 2009). Interestingly, among patients with HCC secondary to HCV, there is no statistical difference in incidence between men and women.

Cirrhosis is the most alarming and consistent risk factor for the development of liver cancer. Each of the above factors leads to the development of cirrhosis before the initiation of HCC tumorigenesis the majority of the time. Because of growing obesity and diabetes problems in the United States, an increasing percentage of these patients have cirrhosis stemming from nonalcoholic fatty liver disease (NAFLD) and nonalcoholic steatohepatitis (NASH). In fact, NAFLD is now the most common cause of liver disease in the Western world (Younossi et al., 2011). Furthermore, around 13% of the reported HCC cases in the United States are secondary to underlying NAFLD or NASH (Yang et al., 2011). The mechanism of HCC development via NASH is not yet entirely clear, however, NASH is most likely an indirect cause of HCC through the development of cirrhosis.

Alcoholism-related HCC is also on the rise. Alcohol-related liver disease accounts for 29% of HCC in the United States (Yang et al., 2011). There is no precise amount, or duration, of alcohol intake that causes a direct increase in the development of HCC; it is clear that risk increases as a function of the amount of alcohol consumed. The increased risk is independent of the patient's age at consumption, and the duration of consumption. Also, lower levels of intake over a longer period of time are likely less damaging than similar levels of intake over a short period of time (Donato et al., 2002). Like the other risk factors for HCC, the development of cirrhosis is an important indicator. Studies have shown that individuals who drink more than 40 g/day of alcohol are at risk for developing cirrhosis and thus more likely to develop HCC.

## 2.3 Development of Metastatic Hepatocellular Carcinoma

Approximately 15% of HCC patients will have extrahepatic spread (EHS) upon initial HCC diagnosis, or experience a metastatic event at some point in the course of the disease (Uka et al., 2007). EHS of HCC is closely tied to intrahepatic tumor stage. Tumor node metastasis (TNM) staging is often used to classify the severity of HCC, and the criteria used in this system can be seen in Table 9.1. A study examining 403 HCC patients in the United States found that 76% of cases involving metastases occurred in patients with a TNM stage IVA intrahepatic tumor. Eleven percent of the patients that experienced metastases had a TNM stage III tumor (Katyal et al., 2000).

The most common location of metastasis in this same study was the lungs. Fifty-five percent of metastatic patients have EHS in the lungs, usually in the lower lobes. The lungs are also

TABLE 9.1 TNM Staging for Hepatocellular Carcinoma

| | | | |
|---|---|---|---|
| ***PRIMARY TUMOR (T)*** | | | |
| TX | Primary tumor cannot be assessed | | |
| T0 | No evidence of primary tumor | | |
| T1 | Solitary tumor without vascular invasion | | |
| T2 | Solitary tumor with vascular invasion or multiple tumors none more than 5cm | | |
| T3a | Multiple tumors more than 5cm | | |
| T3b | Single tumor or multiple tumors of any size involving a major branch of the portal vein or hepatic vein | | |
| T4 | Tumor(s) with direct invasion of adjacent organs other than the gallbladder or with perforation of visceral peritoneum | | |
| ***REGIONAL LYMPH NODES (N)*** | | | |
| NX | Regional lymph nodes cannot be assessed | | |
| N0 | No regional lymph node metastasis | | |
| N1 | Regional lymph node metastasis | | |
| ***DISTANT METASTASIS (M)*** | | | |
| M0 | No distant metastasis | | |
| M1 | Distant metastasis | | |
| ***FIBROSIS SCORE (F)***[a] | | | |
| F0 | Fibrosis score 0–4 (none to moderate fibrosis) | | |
| F1 | Fibrosis score 5–6 (severe fibrosis or cirrhosis) | | |
| ***ANATOMIC STAGE/PROGNOSTIC GROUPS*** | | | |
| Stage I | T1 | N0 | M0 |
| Stage II | T2 | N0 | M0 |
| Stage IIIA | T3a | N0 | M0 |
| Stage IIIB | T3b | N0 | M0 |
| Stage IIIC | T4 | N0 | M0 |
| Stage IVA | Any T | N1 | M0 |
| Stage IVB | Any T | Any N | M1 |

[a] *The fibrosis score as defined by Ishak is recommended because of its prognostic value in overall survival. This scoring system uses a 0–6 scale.*

the most common location for initial presentation of EHS. For all patients with metastases to the lungs, it was the first presentation of extrahepatic HCC 39% of the time. This is the primary reason why patients diagnosed with HCC receive routine chest imaging. Forty-one percent of metastatic patients experience spread to the regional lymph nodes. The third most common location of metastasis in HCC patients is bone at 28%. In patients with spread to bone, greater than 60% experience spread in either the lumbosacral or thoracic spine. Bone is the initial presentation of EHS 9% of the time and thus bone scans are also routinely performed in HCC patients. The fourth most common location of metastatic disease is the adrenal glands at 12%; 75% of the time this EHS affected the adrenals unilaterally (Katyal et al., 2000). There are other, less common, sites of metastases including the head and neck, however, these are infrequent and are almost never the initial presentation of EHS. Consequently, tests are not routinely performed on these areas.

# 3. DIAGNOSIS

## 3.1 Imaging

Imaging plays a significant role in the classification and management planning of metastatic HCC. When deciding on the appropriate plan for staging and managing metastatic HCC, considerations must be made for which type of imaging modality will be used—ultrasound (US), computed tomography (CT), magnetic resonance imaging (MRI), nuclear medicine studies, or angiography—as well as the intervals and specific targets to image.

The basic imaging modality for diagnosing HCC in the liver is either a CT abdomen with contrast (dynamic and multiphase) or an MRI abdomen with contrast (multiphase). Aside from these techniques, newer advances in CT and MRI such as dual-energy CT, perfusion CT, diffusion-weighted imaging (DWI) MRI, and hepatocyte-specific MR contrast agents (such as gadolinium-EOB-DTPA) have increased sensitivity of diagnosis (Mintorovitch et al., 2000). Newer technologies such as MR elastography (MRE) and acoustic radiation force impulse imaging allow for novel approaches to the evaluation of liver nodules in general.

Prior to diagnosis of HCC, the current recommendation for screening modality is periodic US in patients at risk for developing HCC. Patients with cirrhosis should undergo screening at 6-month intervals with US. If a nodule is found that is >1 cm then further evaluation should be performed with CT or MRI. At this time, the American Association for the Study of Liver Disease (AASLD) does not recommend monitoring AFP in conjunction with US at 6-month intervals for screening purposes, although many experts still advocate the use of this biomarker for screening and risk assessment. At this time US and contrast-enhanced US are not recommended for diagnosis of HCC, but only for screening purposes, with all concerning findings referred for more definitive imaging with modalities such as MRI or CT (Hennedige et al., 2012).

CT has previously been the predominant imaging modality used in the diagnosis of HCC. Classically HCC will demonstrate arterial phase enhancement followed by washout in the portal and delayed phase. There is often a pseudocapsule noted around the nodule during the delayed or portal phase. The appearance of HCC on noncontrast imaging is highly variable, from isodense to hypodense to hyperdense. If portal vein thrombosis is present, the typical features of HCC may not be present due to presence of arterioportal shunting (Ichikawa et al., 2010). Aside from basic triple-phase CT imaging of the liver, more recent advances in CT have led to the introduction of dynamic multidetector CT and quadruple phase CT, incorporating early and late arterial phase, which have led to increased sensitivity in detecting HCC.

MRI allows for a diagnostic tool that does not involve radiation exposure. Gadolinium-based scans can be used to evaluate HCC and diagnosis is mainly based on the enhancement patterns of the postgadolinium-enhanced scans. Recently, there has been the introduction of hepatocyte-specific contrast agents, such as gadolinium-EOB-DTPA, which show early uptake in the arterial phase, washout patterns in the equilibrium phase, and lower intensity in the hepatobiliary phase. Another contrast agent, superparamagnetic iron oxide is selectively taken up by Kupffer cells in the liver, which are not usually present in HCCs. This contrast leads to the reduction in signal intensity in normal liver tissue in T2-weighted images, thus making tumor nodules appear hyperintense in comparison (Ichikawa et al., 2010; Xu et al., 2009). Compared to multidetector CT and unenhanced MRI, gadolinium-EOB-DTPA in particular has been shown to be equal or even better in identifying small HCC nodules, or HCC in cirrhotic livers. Contrast-enhanced MRI imaging can be combined with DWI to increase sensitivity for detection of HCC. DWI looks at water composition at the cellular level. Tumors tend to be highly cellular, which limits water mobility. This translates to hyperintense HCCs on DWI. When combined with contrast-enhanced MRI, DWI allowed for 96% detection rate, compared to 68% with contrast-enhanced only. On the horizon are the use of both MRE and spectroscopy for the diagnosis of liver tumors, and currently research is ongoing (Ichikawa et al., 2010).

Staging of HCC is very important at the time of diagnosis as it will allow for better formulation of an appropriate treatment plan whether that be local therapy, systemic chemotherapy, resection, or possible liver transplantation. The basic workup for staging of HCC involves some kind of liver imaging as outlined above, CT chest, and a bone scan. Recently, the use of positron emission tomography (PET) scan as an adjunct to staging has been explored. PET has been shown to be quite useful in identifying extrahepatic lesions but it is not as sensitive for intrahepatic HCC. PET has been reported to miss up to 30–50% of HCC lesions in the liver, although contradictory studies have shown up to 88% accuracy for intrahepatic HCC lesions (Lee et al., 2012). Recent studies have also demonstrated a higher sensitivity to extrahepatic lesions compared to standard imaging techniques such as bone scans and CT chest (Sneag et al., 2011). Various tracers have been utilized such as $^{18}$F-methylcholine, $^{11}$C-acetate, and fluorodeoxyglucose ($^{18}$F-FDG) to identify HCC and extrahepatic lesions, with ongoing studies looking at combinations of tracers versus single tracer studies. Some studies have shown that PET/CT detects extrahepatic lesions missed by standard imaging in up to 50% of patients (Lee et al., 2012; Bieze et al., 2014). More recently, the role of PET in the identification of microvascular invasion has been suggested. The presence of both micro- and macrovascular invasion is clearly important, especially when considering liver transplantation. Studies have looked retrospectively at imaging done pretransplant to determine evidence of microvascular invasion that was later discovered on explanted organs. In these studies, positive PET lesions using $^{18}$F-FDG had a sensitivity of 64.3% and a specificity of 86.7% for detecting microvascular invasion. When combined with findings of peri-tumoral enhancement seen on contrast-enhanced MRI, sensitivity increased to 78.6% with a specificity of 80% (Ahn et al., 2015). These findings would significantly alter treatment management and allow for better stratification of patients, although further data are needed before this can be incorporated into selection of patients for liver transplantation. Finally, PET/CT offers a low radiation test alternative to standard imaging. Currently, there is not enough data to support using PET scan for routine clinical practice in the evaluation of patients with HCC, although these recommendations may change as more data become available.

Once diagnosed with HCC, case-by-case treatment will determine further surveillance intervals. Initially, regardless of treatment modality utilized, a 3-month interval surveillance scan is recommended. This interval should be maintained for at least 2 years, after which, the interval could be extended at the treating physician's discretion depending on response to therapy. The modality of imaging recommended by the AASLD for follow-up of patients with HCC is contrast-enhanced cross-sectional imaging, either CT or MRI (Bruix and Sherman, 2011).

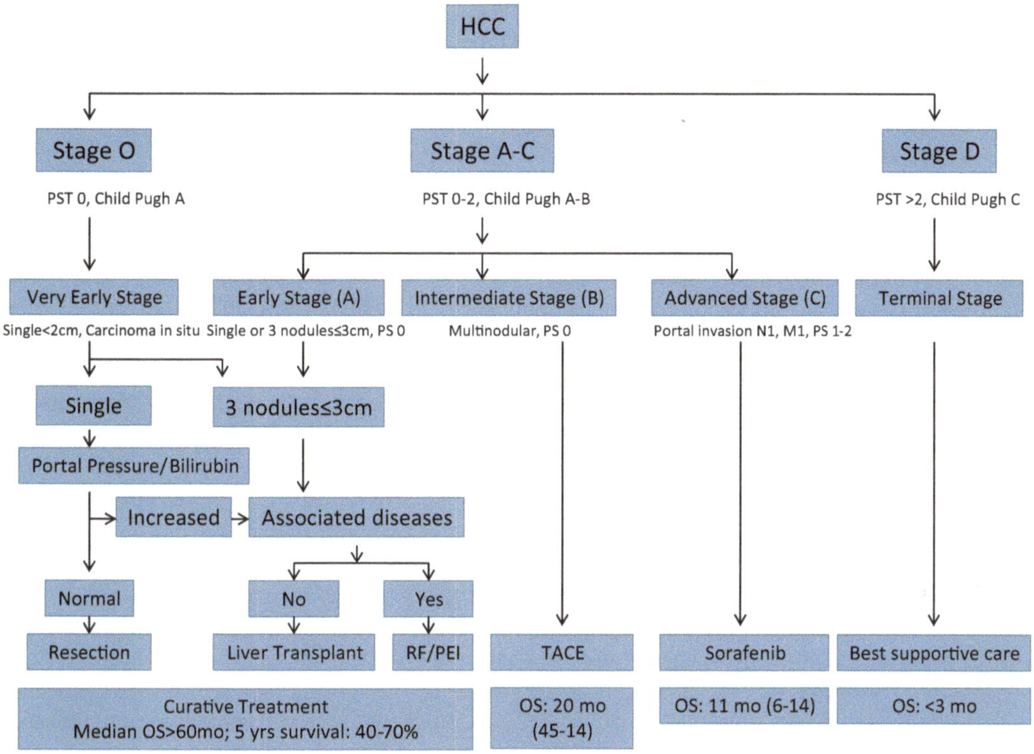


FIGURE 9.1 The Barcelona clinic liver cancer staging system of hepatocellular carcinoma (HCC).

## 3.2 Biopsy

Intrahepatic HCC is a diagnosis made by imaging, and 90% of HCCs do not require a biopsy to make the diagnosis (Bruix and Sherman, 2011). Interpretation of biopsies of liver masses and distinction between high-grade dysplastic nodules and HCC is challenging. Expert pathologic diagnosis is reinforced by staining for glypican 3, heat shock protein 70, and glutamine synthetase. Positivity for two of these three stains confirms HCC (Bruix and Sherman, 2011). Sensitivity and specificity of biopsy for HCC are 91.5% and 100%, respectively, but there is a 0.4% rate of bleeding (5 of the 11 patients who bled died) and 0.2% rate of implantation metastases (Wang et al., 2008). A negative finding after biopsy does not rule out HCC, since the false negative rate can reach 30% because of sampling error or absence of specific histological hallmarks for diagnosis of this cancer (Forner et al., 2012). If the biopsy is negative for HCC, patients should be followed with imaging at 3- to 6-month intervals until the nodule disappears, enlarges, or displays diagnostic characteristics of HCC.

Extrahepatic HCC is generally biopsied to prove the diagnosis before subjecting noncurable patients to therapies. The risk of bleeding is not insignificant, as HCC is a vascular tumor and many of these patients have underlying coagulopathy due to liver disease. With intrahepatic tumor, biopsies often require a core needle sample to maintain the architecture of the biopsy to distinguish malignancy, but with extrahepatic disease a fine needle aspiration is often sufficient to make the diagnosis.

## 4. PROGNOSIS AND STAGING

Given that more than 90% of patients with HCC have underlying cirrhosis, staging for HCC is more complicated than other cancers. Underlying liver function in patients with cirrhosis is closely tied to prognosis. Because of this concomitant disease process contributing to the difficulties of staging HCC, there is currently no universally accepted staging system that applies well for all HCC patients. For example, the TNM staging system does not take underlying liver function into account. Ideally, an HCC staging system takes tumor characteristics, liver function, and overall health into account. The system should also function as a tool used to guide the treatment for patients with different HCC classifications. The Barcelona Clinic Liver Cancer (BCLC) staging system is generally considered the best system with regards to the requirements above (Fig. 9.1) (Bruix and Sherman, 2011). The AASLD and the European Association for the Study of the Liver-European Organization for Research and Treatment of Cancer (EASL-EORTC) guidelines support the use of the BCLC staging and classification system for staging HCC. This classification system incorporates tumor status, synthetic function of the liver, performance status, and treatment-dependent variables (Xu et al., 2009).

In order for curative treatment to be considered in HCC patients, the tumor must be caught early (BCLC stage 0, A, or B). Liver resection, transplantation, and/or tumor ablation with radiofrequency, microwaves, or ethanol, and transarterial chemoembolization (TACE) are treatments that can be used in combination to treat HCC in early stage patients. Patients who are classified as stage C according to BCLC have fewer treatment options, and much worse survival rates. Within this group, those who have metastatic disease along with poor underlying liver function have an even worse prognosis. A study conducted in South Korea examined the comprehensiveness of the BCLC staging system and the prognosis of stage C patients. They found the median survival rate for patients with stage C disease to be 6.4 months. Survival rates were 51% at 6 months, 33% at 1 year, and 13% at 3 years (Sinn et al., 2015).

Prognosis also differed based on the extent of portal vein invasion (PVI) and the type of EHS. Sinn et al. found that patients who had more extensive PVI had shortened survival (Sinn et al., 2015). Patients with PVI in branches of the portal vein had an average survival time of 9 months, whereas those who had invasion in the main trunk of the portal vein survived for 4.6 months on average. With respect to the type of EHS, those with distant metastasis had shorter survival time than those experiencing only nodal metastasis. Patients with nodal metastasis survived for an average of 7 months, whereas those with distant metastasis survived for an average of only 4.2 months (Sinn et al., 2015).

Sinn et al. examined several factors in their study to critique the BCLC system. They found that underlying liver function (determined by Child-Pugh score), tumor size, extent of PVI, type of EHS, and AFP levels were all independent factors associated with survival. Even within stage C patients, these different factors led to significantly different survival times. The authors use this evidence to suggest that there needs to be substages specifically within stage C patients in the BCLC system to give a more accurate prognosis for all types of patients.

For advanced HCC, a new model—the Advanced Liver Cancer Prognostic System has been proposed, which incorporates ascites, abdominal pain, weight loss, Child-Pugh score, alkaline phosphatase, total bilirubin, urea, and tumor characteristics—portal vein thrombosis, tumor size, lung metastases, and AFP (Yau et al., 2008). Based on these factors this model provides a survival rate percentage over the next 3 months. With this information, clinicians are better able to determine the most appropriate patients with advanced HCC for treatment. In addition to risk-stratifying patients, we must also develop a system to assess response to therapy, so that risks and side effects of therapy may be considered. One emerging guideline for determining response to therapy is the modified Response Evaluation Criteria in Solid Tumors, which provides criteria for responses—stable disease, progressive disease, partial response, and complete response (Eisenhauer et al., 2009). Survival with extrahepatic metastases in HCC has been noted around 20–25% at 1 year, 6% at 3 years, and 0% at 5 years (Wang et al., 2015). Median survival assuming no therapy is around 4.6 months once extrahepatic metastases are present (Wang et al., 2015).

## 5. TREATMENT

Until the more recent years the approach to treatment for advanced HCC was quite limited and centered mainly on palliation. With the advent of new therapies such as sorafenib the environment is actively evolving. We will focus our examination on options for advanced HCC only rather than discuss treatment options for early stage HCC.

Surgical options in the setting of advanced HCC are more focused on the extrahepatic lesions and whether intervention in them will ultimately lead to some kind of benefit. More than 50% of extrahepatic lesions tend to be pulmonary. There have been no randomized controlled trials looking at the approach to surgical versus nonsurgical therapy for extrahepatic metastases. For pulmonary metastases, surgical approach would involve either a wedge resection or lobectomy depending on the number, size, and distribution of nodules. Obviously consideration must be made to the overall health of the patient, as there would be a reasonable amount of concern for decompensation in the postoperative period. Patients with more than one pulmonary nodule, high AFP, a history of prior recurrence, and a shorter disease-free interval before appearance of lesions is associated with worse outcomes with surgical resection of pulmonary metastases. Over the course of the last several years with the advent of new surgical techniques such as the last several years with the advent of new surgical techniques such as video-assisted thoracoscopic surgery and minimally invasive approaches, survival at 3 years is reported to range between 24% and 78% with median survival between 21 and 29 months (Mizuguchi et al., 2015; Xiang et al., 2015).

Techniques that were initially limited to intrahepatic disease have now been investigated for the application to extrahepatic manifestations of HCC. One option in particular is the utilization of radiofrequency ablation (RFA) for metastases. Using a percutaneous approach some research has shown a benefit in RFA of pulmonary metastases. Recent studies have shown improved survival to 87% at 1 year and 57% at 3 years with median survival being 37.7 months after RFA of metastatic disease (Hiraki et al., 2011). Factors that predicted better survival post-RFA treatment included absence of viable intrahepatic recurrence, Child Class A cirrhosis or absence of liver cirrhosis, absence of HPV infection, and an AFP no greater than 10 mg/mL at the time of RFA. RFA utilizes a combination of nonthermal and thermal energy to heat up the targeted tissue. Once the tissue temperature surpasses 60°C, the cells begin to die, which lead to necrosis of the tumor. There are limitations with this treatment option, in particular, the size of the nodule. Lesions larger than 4 cm tend to not respond as well to this therapy (Hiraki et al., 2011).

Along the same lines of exploring minimally invasive procedures to treat extrahepatic disease, some investigators have discussed the role of $^{125}$I brachytherapy in the treatment of metastatic lesions similar to the role of Y90 in hepatic lesions. These radioactive seeds could be introduced to specific lesions to provide local radiation therapy. $^{125}$I brachytherapy has been studied in non-small-cell lung cancer, and Y90 has been using in primary HCC lesions in recent years (Zhang et al., 2014). Studies are currently ongoing in the application of these modalities to treat extrahepatic HCC.

External beam radiation therapy (EBRT) has been used for several years in the management of metastatic malignancies, but over recent years its role in HCC has been explored more extensively, particularly for a use outside of palliation alone. EBRT delivers energy in a particularly targeted area that result in ionization of atoms in DNA resulting in double-strand breaks. These breaks result in cell death. Even though EBRT is targeted to specific areas there will be some overlap with normal tissue. Normal cells have a stronger ability to handle the ill effects of radiation therapy and repair the DNA breaks and mutations, as opposed to malignant cells that do not have these facilities. Thus, there is a threshold of radiation exposure where even healthy tissue cannot recover. The balance between under and overexposure must be calculated prior to treatment delivery. Over the last several years various studies have looked into the usefulness of EBRT in management of more advanced HCC. A slew of retrospective studies demonstrated improved response in patients with portal vein/inferior vena cava (PV/IVC) tumor thrombus, lymph node, brain, lung, bone, and adrenal metastases with complete response noted around 26% and survival at 1 year up to 50.1% and on average up to 10.7 months posttherapy. In particular with bone metastases, the benefit is seen more with pain relief, which has been known and studied in other malignancies with metastatic spread to bone. Median survival with bone metastases of HCC is only 7.4 months (Jiang et al., 2013; Havashi, 2014; Jiang, 2012).

Systemic chemotherapy has been used on a wide range of malignancies, but has had less use in HCC for certain reasons. For one, HCC is considered a chemotherapy refractory tumor, which is likely related to the high expression of drug resistance genes. Other factors that have prevented its adoption are that systemic chemotherapy does not work well with baseline hepatic dysfunction and studies that have been done were performed on very select populations. There is no one specific regimen that has been shown to be most beneficial for HCC. Proposed regimens that have been studied in limited randomized control trials with modest benefit include doxorubicin, 5-fluorouracil, or gemcitabine as monotherapy regimens. A few combination regimens incorporating two or three chemotherapeutic agents have had similar modest results. Combination therapies are usually cisplatin, gemcitabine, or oxaliplatin based. All patients who received some kind of chemotherapeutic agent had a survival of <12 months and when compared to supportive care alone, at best showed an improvement by less than 3 months. On the forefront now is the incorporation of interferon alfa (IFNa) therapy with chemotherapy. IFNa functions as an immunomodulatory cytokine that in in vitro models demonstrated antitumor effect against a wide range of malignancies one of which was HCC. Its ability to provide benefit is usually tied to the expression of the type I interferon receptor in the tumor as has been suggested in some reports. There have been mixed results in studies comparing IFNa to single agent chemotherapy regimens or in combination (Nagano et al., 2007). Ongoing research may reveal some benefit but the issues with poor response and an inability of patients to tolerate the medication remain barriers. One major concern with any of these immunosuppressive systemic treatments is the concern for reactivation of hepatitis B, which make up a significant portion of this patient population. It is necessary to make sure that patients are on appropriate therapy prior to starting any systemic chemotherapy or immunotherapy.

The major player in the management and treatment of advanced HCC is molecularly targeted therapy, specifically with sorafenib. Earlier studies have demonstrated the role of epidermal growth factor, vascular endothelial growth factor (VEGF), and tyrosine kinases (TK) in the proliferation and malignant transformation of HCC. HCC is a very vascular tumor that these factors play some role, thus further research in this area has been looking at inhibitors of these specific molecules as potential targets for treatment of HCC. Sorafenib has been shown to be the most successful in this arena. Sorafenib is an oral TK inhibitor that targets receptor activation factor (Raf) kinase and the VEGF receptor intracellular kinase pathways. VEGF is involved with vascular proliferation of HCC and Raf kinase, which is part of the mitogen-activating protein kinase pathway and plays a role in the regulation of cell growth, transformation, and apoptosis. In this way, sorafenib affects tumor blood supply and helps limit spread. Sorafenib was the first systemic medication to demonstrate survival benefit in patients with advanced HCC and was approved by the FDA for unresectable HCC in 2007 (Wu et al., 2003; Cabibbo et al., 2015).

One of the concerns with the use of sorafenib is whether it can be applied to all patient groups with advanced HCC. The major phase III study that resulted in sorafenib approval was the SHARP trial, which was a European-based study that showed a survival benefit of 10.7 months versus 7.9 months in the placebo group (Llovet et al., 2008). The majority of patients in this study had hepatitis C-associated HCC. When a similar study was performed in an Asian population where the majority of patients had hepatitis B associated HCC (73% vs. 18% in the SHARP trial), a benefit was still seen with sorafenib but to a lesser degree—6.5 versus 4.2 months in placebo (Cheng et al., 2009). This study highlighted the importance of considering the etiology of the underlying HCC as a factor contributing to response to sorafenib. Ongoing studies are specifically looking at this issue and future research will shed more light on this issue (Cabibbo et al., 2015).

Aside from efficacy, the issues of safety and side effects are a concern with any therapy. The majority of patients with HCC have underlying hepatic dysfunction related to cirrhosis. Retrospective studies that have looked at subgroup analysis of cirrhosis response to therapy have shown overall worse survival benefit in Child Class B and C patients compared to Child Class A. It is hard to decipher whether that survival benefit is related to progression of cirrhosis or less of an antitumor effect, but findings such as these must be a consideration when deciding on any therapy for patients with advanced HCC. Side effects specific to sorafenib and other antiangiogenic TK inhibitors include hypertension, rash, arterial thromboembolism, hand–foot skin reaction, cardiotoxicity, bleeding, and thyroid dysfunction. There have been some reports of sorafenib resulting in hepatotoxicity with more of a hepatocellular pattern, so that concern must be taken into consideration with this therapy (Wu et al., 2003; Cabibbo et al., 2015).

More recently, there has been a discussion regarding the role of sorafenib in combination therapy. Given the effect on vascular inhibition in HCC, the role in combination with TACE is actively being investigated. In a recent retrospective that looked at patients being treated with sorafenib with advanced HCC, half of the patients received additional therapy despite metastatic disease with TACE and Y90. The group of patients that received combination therapy had an overall survival rate of 25 months compared to 11 months for sorafenib alone (Schmidt et al., 2014). Based on these findings and other studies, future endeavors are looking to better define the role of local hepatic therapy in combination with systemic therapy in patients who previously based on the BCLC classification system would not have been candidates.

Since the discovery of sorafenib, the role of other VEGF inhibitor in advanced HCC has been explored. Studies have looked at bevacizumab, sunitinib, and axitinib and have provided mixed results. Bevacizumab when compared to placebo alone or in combination with systemic chemotherapy has shown a survival benefit, but no randomized trials have looked at combination therapy versus bevacizumab alone (Frenette, 2012). Thus, at this point no conclusions can really be made regarding benefit over current regimens. In the registration trials, there was also concern raised with bevacizumab potentially causing an increased risk of variceal bleeding (Frenette, 2012). Sunitinib was recently compared directly to sorafenib in a phase III study and was shown to be not as beneficial and with increased toxicity (Cheng et al., 2013). Axitinib, which is a second generation TK inhibitor, has been studied as an option for second-line treatment for patients who have failed prior TK inhibitor therapy (Kang et al., 2015). In this patient population some benefit was seen and further studies are ongoing.

New therapeutics for the management of advanced HCC are currently being explored. Given the benefit seen with VEGF inhibitors, study has now shifted toward the role of endothelial growth factor receptors (EGFR) inhibitors, given the role of endothelial growth factor (EGF) in HCC proliferation. Ongoing studies are comparing these EGFR inhibitors to sorafenib, placebo, and use in combination with other therapies such as bevacizumab. New molecular targeted medications are looking at inhibitors of hepatocyte growth factor, mechanistic target of rapamycin inhibitors, and immunotherapy.

# 6. CONCLUSIONS

The treatment of advanced HCC represents a complex problem that often times require a multidisciplinary team. Without the aid of various experts an appropriate treatment plan is difficult to compose. Multiple studies have shown the benefit of multidisciplinary teams over individual providers in the management of these complex patients (Naugler et al., 2015). At this point, there is a limited array of options for these patients, but with ongoing research the options are quickly changing. Future therapeutics will likely involve a multipronged approach.

## References

Altekruse, S.F., McGlynn, K.A., Reichman, M.E., 2009. Hepatocellular carcinoma incidence, mortality, and survival trends in the United States from 1975 to 2005. J. Clin. Oncol. 27 (9), 1485–1491.

Ahn, S.Y., et al., 2015. Prediction of microvascular invasion of hepatocellular carcinoma using gadoxetic acid-enhanced MR and $^{18}$F-FDG PET/CT. Abdom. Imaging 40, 843–851.

Beasley, R., 1988. Hepatitis B virus: the major etiology of hepatocellular carcinoma. Cancer 61 (10), 1942–1956.

Bieze, M., et al., 2014. Diagnostic accuracy of $^{18}$F-methylcholine PET/CT for intra- and extrahepatic hepatocellular carcinoma. Hepatology 49 (3), 996–1006.

Bruix, J., Sherman, M., March 2011. Management of hepatocellular carcinoma: an update. AASLD practice guidelines. Hepatology 53 (3), 1020–1022.

Chen, C.J., et al., 2006. Risk of hepatocellular carcinoma across a biological gradient of serum hepatitis B virus DNA level. JAMA 295 (1), 65–73.

Chemin, I., Zoulim, F., 2009. Hepatitis B induced hepatocellular carcinoma. Cancer Lett. 286 (1), 52–59.

Cabibbo, G., et al., 2015. Sorafenib for hepatocellular carcinoma: from randomized controlled trials to clinical practice. Dig. Dis. 33, 668–674.

Cheng, A.L., Kang, Y.K., Chen, Z., et al., 2009. Efficacy and safety of sorafenib in patients in the Asia-Pacific region with advanced hepatocellular carcinoma: a phase III randomized, double-blind, placebo-controlled trial. Lancet Oncol. 10 (1), 25–34.

Cheng, A.L., Kang, Y.K., Lin, D.Y., et al., 2013. Sunitinib versus sorafenib in advanced hepatocellular cancer: results of a randomized phase III trial. J. Clin. Oncol. 31 (32), 4067–4075.

Donato, F., et al., 2002. Alcohol and hepatocellular carcinoma: the effect of lifetime intake and hepatitis virus infections in men and women. Am. J. Epidemiol. 155 (4), 323–331.

Eisenhauer, E.A1., Therasse, P., Bogaerts, J., Schwartz, L.H., Sargent, D., Ford, R., Dancey, J., Arbuck, S., Gwyther, S., Mooney, M., Rubinstein, L., Shankar, L., Dodd, L., Kaplan, R., Lacombe, D., Verweij, J., 2009. New response evaluation criteria in solid tumours: revised RECIST guideline (version 1.1). Eur J Cancer. 45 (2), 228–247. http://dx.doi.org/10.1016/j.ejca.2008.10.026.

El-Serag, H.B., 2004. Hepatocellular carcinoma: recent trends in the United States. Gastroenterology 127 (5 Suppl. 1), S27–S34.

El-Serag, H.B., Seigel, A.B., Davila, J.A., et al., 2006. Treatment and outcomes of treating hepatocellular carcinoma among Medicare recipients in the United States: a population-based study. J. Hepatol. 44, 158–166.

El-Serag, H.B., Kanwal, F., 2014. Epidemiology of hepatocellular carcinoma in the United States: where are we? Where do we go? Hepatology 60 (5), 1767–1775.

Ferlay, J., Shin, H.R., Bray, F., et al., 2010. Estimates of worldwide burden of cancer in 2008: GLOBOCAN 2008. Int. J. Cancer 127 (12), 11–30.

Forner, A., Llovet, J.M., Bruix, J., 2012. Hepatocellular carcinoma. Lancet 379 (9822), 1245–1255.

Frenette, C.T., 2012. Current status of bevacizumab for advanced hepatocellular carcinoma. Chin. Clin. Oncol. 1, 13.

George, S.L., et al., 2009. Clinical, virologic, histologic, and biochemical outcomes after successful HCV therapy: a 5-year follow-up of 150 patients. Hepatology 49 (3), 729–738.

Hennedige, T., et al., 2012. Imaging of hepatocellular carcinoma: diagnosis, staging and treatment monitoring. Cancer Imaging 12 (3), 530–547.

Hiraki, T., et al., 2011. Percutaneous radiofrequency ablation for pulmonary metastases from hepatocellular carcinoma: results of a multicenter study in Japan. J. Vasc. Interv. Radiol. 22, 741–748.

Havashi, S., et al., December 27, 2014. Palliative external-beam radiotherapy for bone metastases from hepatocellular carcinoma. World J. Hepatol. 6 (12), 923–929.

Ichikawa, T., et al., 2010. Detection and characterization of focal liver lesions: a Japanese phase III, multicenter comparison between gadoxetic acid disodium enhanced magnetic resonance imaging and contrast-enhanced computed tomography predominantly in patients with hepatocellular carcinomas and chronic liver disease. Investig. Radiol. 45, 133–141.

Jiang, W., et al., 2013. Is it time to adopt external beam radiotherapy in the NCCN guidelines as a therapeutic strategy for intermediate/advanced hepatocellular carcinoma? Oncology 84 (Suppl. 1), 69–74.

Jiang, W., et al., March 2012. Palliative radiation therapy for pulmonary metastases from hepatocellular carcinoma. Clin. Exp. Metastasis 29 (3), 197–205.

Katyal, S., et al., 2000. Extrahepatic metastases of hepatocellular carcinoma. Radiology 216 (3), 698–703.

Kang, Y.-K., et al., September 18, 2015. Randomized phase II study of axitinib versus placebo plus best supportive care in second-line treatment of advanced hepatocellular carcinoma. Ann. Oncol. (Epub ahead of print).

Lee, J.E., et al., June 21, 2012. Diagnostic value for extrahepatic metastases for hepatocellular carcinoma in positron emission tomography/computed tomography scan. World J. Gastroenterol. 18 (23), 2979–2987.

Llovet, J.M., Ricci, S., Mazzaferro, V., et al., 2008. Sorafenib in advanced hepatocellular carcinoma. NEJM 359 (4), 378–390.

Mittal, S., El-Serag, H.B., 2013. Epidemiology of hepatocellular carcinoma: consider the population. J. Clin. Gastroenterol. 47 (1 Suppl.), S2–S6.

Muñoz, N., Bosch, X., 1987. Epidemiology of hepatocellular carcinoma. In: Neoplasms of the Liver. Springer, Japan, pp. 3–19.

Mintorovitch, J., et al., May 31, 2000. Eovist injection and resovist injection: two new liver-specific contrast agents for MRI. Oncology 14 (Suppl. 3), 37–40.

Mizuguchi, S., et al., August 2015. Clinical significance of multiple pulmonary metastasectomy for hepatocellular carcinoma. World J. Surg. 26 (Epub ahead of print).

Nagano, H., et al., 2007. Interferon-alfa and 5-fluorouracil combination therapy after palliative hepatic resection in patients with advanced hepatocellular carcinoma, portal venous tumor thrombus in the major trunk and multiple nodes. Cancer 110 (11), 2493.

Naugler, W.E., Alsina, A.E., Frenette, C.T., et al., 2015. Building the multidisciplinary team for management of patients with hepatocellular carcinoma. Clin. Gastroenterol. Hepatol. 13 (5), 827–835.

Parkin, D.M., et al., 2002. IARC Scientific Publication 155. In: Cancer Incidence in Five Continents, vol. VIII. IARC, Lyon. ISBN: 92-832-2155-9.

Sneag, D.B., et al., February 2011. Extrahepatic spread of hepatocellular carcinoma: spectrum of imaging findings. Gastrointest. Imaging 197, 658–664.

Sinn, S.H., et al., 2015. Different survival of Barcelona Clinic Liver Cancer stage C hepatocellular carcinoma patients by the extent of portal vein invasion and the type of extrahepatic spread. PloS One 10 (4).

Schmidt, L., et al., 2014. Addition of local hepatic therapy to sorafenib in patients with advanced hepatocellular carcinoma (stage BCLC C). Digestion 90, 219–228.

Tsukuma, H., et al., 1993. Risk factors for hepatocellular carcinoma among patients with chronic liver disease. NEJM 328 (25), 1797–1801.

Tsai, J.F., et al., 1996. Additive effect modification of hepatitis B surface antigen and e antigen on the development of hepatocellular carcinoma. Br. J. Cancer 73 (12), 1498.

Uka, K., et al., 2007. Clinical features and prognosis of patients with extrahepatic metastases from hepatocellular carcinoma. World J. Gastroenterol. 13 (3), 414–420.

Wong, R.J., Corley, D.A., 2009. Survival differences by race/ethnicity and treatment for localized hepatocellular carcinoma within the United States. Dig. Dis. Sci. 54, 2031–2039.

Wang, P., Meng, Z.Q., Chen, Z., et al., 2008. Diagnostic value and complications of fine needle aspiration for primary liver cancer and its influence on treatment outcome – a study based on 3011 patients in China. Eur. J. Surg. Oncol. 34 (5), 541–546.

Wang, C.-H., et al., 2015. Current trends and recent advances in diagnosis, therapy, and prevention of hepatocellular carcinoma. Asian Pac. J. Cancer Prev. 16 (9), 3595–3604.

Wu, B.W., et al., 2003. Study on the mechanism of epidermal growth factor-induced proliferation of hepatoma cells. World J. Gastroenterol. 9 (2), 271.

Xu, P.J., et al., 2009. The value of breath-hold diffusion-weighted imaging in small hepatocellular carcinoma lesions detection. Zhonghua Yi Xue Za Zhi 89, 592–596.

Xiang, Z.W., et al., September 18, 2015. Treatment of pulmonary metastases after liver transplantation. World J. Hepatol. 7 (20), 2309–2314.

Yang, J.D., et al., 2011. Factors that affect risk for hepatocellular carcinoma and effects of surveillance. Clin. Gastroenterol. Hepatol. 9 (7), 617–623.

Younossi, Z.M., et al., 2011. Changes in the prevalence of the most common causes of chronic liver diseases in the United States from 1988 to 2008. Clin. Gastroenterol. Hepatol. 9 (6), 524–530.

Yau, T., Yao, T.J., Chan, P., et al., 2008. A new prognostic score system in patients with advanced hepatocellular carcinoma not amenable to locoregional therapy. Cancer 113 (10), 2742–2751.

Zhang, T., et al., 2014. CT-guided implantation of radioactive $^{125}$I seed in advanced non-small cell lung cancer after failure of first line chemotherapy. J. Cancer Res. Clin. Oncol. 140, 1383–1390.

CHAPTER

# 10

# Metastatic Bladder Cancer

*J.L. Chin, K.M. Siddiqui, K.-C. Tran*

**Western University, London, ON, Canada**

## 1. INTRODUCTION

Metastatic urothelial cancer is the 13th most common cause of cancer-related deaths and accounts for approximately 145,000 deaths per year worldwide (Ferlay et al., 2010). The majority of bladder tumors originate from the urothelium lining the bladder. Urothelial cancer is strongly associated with exposure to environmental toxins. A number of these carcinogens are excreted in the urine. Men and women are more likely to develop bladder cancer and have 1 in 26 chance of developing bladder cancer compared to women who have 1 in 86 chance of having bladder cancer in their lifetime (Ploeg et al., 2009). Family members of patients diagnosed with bladder cancer also are at a higher risk. Possible explanation includes both exposure to a common environmental carcinogen or a genetic anomaly. Cigarette smokers are at higher risk and are twice more likely to develop bladder cancer than nonsmokers. A number of chemical compounds such as aniline dyes, aromatic amines (benzidine, beta-naphthylamine), arsenic, byproducts of hydrocarbon, and chemotherapeutic agents such as cyclophosphamide have been attributed to bladder cancer development. Industrial workers of rubber, leather, textiles, paint manufacturing, and printing industry are especially vulnerable (Clapp et al., 2008). Urothelial carcinoma is a disease more of the elderly and the average age at diagnosis is 73 years.

## 2. EPIDEMIOLOGY

Bladder cancer is the ninth most common malignancy in the world. It is estimated that 401,000 new cases will be diagnosed worldwide and 174,000 patients will die secondary to bladder cancer in 2013 (Fitzmaurice et al., 2015). The predominant histological type of bladder cancer in the Western world is urothelial or transitional cell carcinoma (TCC). In northern Africa, mainly near the Nile Delta, where schistosomiasis is prevalent, there is a higher incidence of squamous cell carcinoma (SCC); the incidence of SCC is rapidly decreasing due to improvements in eradication of schistosomiasis (Parkin, 2008). Rare histologic variants are discussed under "Section 7"—see below.

The majority of cases are diagnosed in industrialized countries but the incidence of urothelial carcinoma is rising in Asian countries with rapid industrial development. The highest incidence of urothelial cancer is reported in Belgium with an age-standardized rate (ASR) of 17.5 per 100,000. In the United States, the ASR of bladder cancer

*Introduction to Cancer Metastasis*
http://dx.doi.org/10.1016/B978-0-12-804003-4.00010-4

is 11.6 per 100,000 (Ferlay et al., 2014). Table 10.1 illustrates the incidence in the top 20 countries.

The majority of patients diagnosed with bladder cancer have disease confined to the urinary bladder. According to the SEER database, 85% are diagnosed at the local stage, with 51% having the disease confined to the mucosa or invading the lamina propria but not yet invading the detrusor muscle layer. These cancers are categorized as nonmuscle invasive bladder cancer (NMIBC) and are also sometimes referred to (although a somewhat outdated term) as "superficial" cancer. The presence of metastasis to regional lymph nodes at diagnosis is detected in 7% and distant metastasis at presentation is reported in only 4%.

For metastatic bladder cancer, the worldwide incidence is unknown. It is estimated that 10–15% of patients are already metastatic at diagnosis (Rosenberg et al., 2005). Metastatic disease usually develops during the course of treatment in patients diagnosed with muscle invasive bladder cancer (MIBC). One-third of patients with urothelial cancer present with muscle-invasive disease confined to the bladder and, depending on the pathological stage, at least one-third further progress to metastatic disease (Sternberg and Vogelzang, 2003). Patients with nonmuscle invasive disease at presentation may also progress to metastatic stage, especially those who have high-risk features such as: size of tumor >3cm, number of tumors >6, T1 stage, presence of concurrent carcinoma in situ (CIS), prior recurrence, high-grade tumor,

TABLE 10.1 Countries With the Highest Age-Standardized Rate (ASR) per 100,000 of Bladder Cancer

| Rank | Country | ASR per 100,000 |
|---|---|---|
| 1 | Belgium | 17.5 |
| 2 | Lebanon | 16.6 |
| 3 | Malta | 15.8 |
| 4 | Turkey | 15.2 |
| 5 | Denmark | 14.4 |
| 6 | Hungary | 14.1 |
| 7 | Spain | 13.9 |
| 8 | Norway | 13.5 |
| 9 | Germany | 13.4 |
| 10 | Egypt | 13.1 |
| 11 | Israel | 12.6 |
| 12 | Armenia | 12.3 |
| 13 | Switzerland | 12.2 |
| 14 | Portugal | 12.0 |
| 15 | Italy | 11.8 |
| 16 | Cyprus | 11.6 |
| **17** | **United States of America** | **11.6** |
| 18 | Canada | 11.5 |
| 18 | Czech Republic | 11.5 |
| 18 | Iceland | 11.5 |
| 18 | Macedonia | 11.5 |

TABLE 10.2 The Risk of Progression of Bladder Cancer and Mortality According to World Health Organization/International Society of Urologic Pathologists classification system—2004

| Tumor | Frequency (%) | Progression (%) | Deaths (%) |
|---|---|---|---|
| Noninvasive | | | |
| Papilloma | 10 | 0–1 | 0 |
| Papillary urothelial neoplasm of low malignant potential | 20 | 3 | 0–1 |
| Papillary cancer low-grade (TaG1) | 20 | 5–10 | 1–5 |
| Papillary cancer high-grade (TaG3) | 30 | 15–40 | 10–25 |
| Invasive | | | |
| Papillary cancer (T1G3) | 20 | 30–50 | 33 |
| Carcinoma in situ (primary) | 10 | >50 | |

variant histology (see below), lymphovascular invasion, and presence of hydronephrosis. A systematic review of 19 trials showed that progression from NMIBC to MIBC occurs in 21%, and 65% of these patients die of metastatic bladder cancer (van den Bosch and Alfred Witjes, 2011). Table 10.2 illustrates the risk of progression and death according to the grade of cancer (Wein et al., 2010). The global age-adjusted death rates of bladder cancer per 100,000 populations, presumably with metastatic disease, are listed in Table 10.3 (Ferlay et al., 2014).

# 3. STAGING

The clinical stage of bladder cancer is based on histologic findings of the transurethral resection of bladder tumor (TURBT), physical exam (including bimanual exam under anesthesia), and radiologic findings on imaging studies (see below for details). Pathological staging is based on the histologic findings of surgically resected bladder specimen as illustrated in Fig. 10.1 along with pathologic evaluation of the pelvic lymph nodes.

TABLE 10.3 Global Age Adjusted Death Rates of Bladder Cancer per 100,000

| Rank | Country | Rate | Rank | Country | Rate | Rank | Country | Rate |
|---|---|---|---|---|---|---|---|---|
| 1 | Armenia | 7.92 | 59 | Iceland | 2.96 | 117 | Somalia | 1.41 |
| 2 | Egypt | 7.51 | 60 | New Zealand | 2.93 | 118 | South Sudan | 1.40 |
| 3 | Iraq | 6.90 | 61 | Ukraine | 2.86 | 119 | Jamaica | 1.40 |
| 4 | Turkey | 6.86 | 62 | Algeria | 2.83 | 120 | Venezuela | 1.39 |
| 5 | Lebanon | 6.09 | 63 | Rwanda | 2.82 | 121 | Niger | 1.38 |
| 6 | Spain | 5.89 | 64 | Moldova | 2.79 | 122 | Mauritania | 1.34 |
| 7 | Syria | 5.70 | 65 | Argentina | 2.73 | 123 | Brunei | 1.33 |
| 8 | Malawi | 5.28 | 66 | North Korea | 2.62 | 124 | Bolivia | 1.32 |
| 9 | Denmark | 5.27 | 67 | Oman | 2.62 | 125 | Yemen | 1.32 |
| 10 | Uruguay | 5.26 | 68 | Ireland | 2.60 | 126 | Benin | 1.30 |
| 11 | Hungary | 5.23 | 69 | Comoros | 2.60 | 127 | Malaysia | 1.29 |
| 12 | Slovenia | 5.16 | 70 | Madagascar | 2.47 | 128 | Chad | 1.28 |
| 13 | Poland | 5.12 | 71 | Finland | 2.46 | 129 | Uzbekistan | 1.27 |
| 14 | Latvia | 5.03 | 72 | Equ. Guinea | 2.43 | 130 | Kyrgyzstan | 1.27 |
| 15 | Luxembourg | 4.90 | 73 | Burkina Faso | 2.40 | 131 | Paraguay | 1.27 |
| 16 | Mali | 4.85 | 74 | Belarus | 2.37 | 132 | Belize | 1.25 |
| 17 | Macedonia | 4.78 | 75 | Brazil | 2.35 | 133 | Panama | 1.24 |
| 18 | Serbia | 4.68 | 76 | Pakistan | 2.34 | 134 | Colombia | 1.21 |
| 19 | Croatia | 4.67 | 77 | Arab Emirates | 2.33 | 135 | Mexico | 1.16 |
| 20 | France | 4.67 | 78 | Chile | 2.29 | 136 | Bahamas | 1.14 |
| 21 | Kuwait | 4.59 | 79 | Senegal | 2.26 | 137 | Kenya | 1.11 |

*Continued*

TABLE 10.3 Global Age Adjusted Death Rates of Bladder Cancer per 100,000—cont'd

| Rank | Country | Rate | Rank | Country | Rate | Rank | Country | Rate |
|---|---|---|---|---|---|---|---|---|
| 22 | Netherlands | 4.57 | 80 | Zambia | 2.23 | 138 | Bangladesh | 1.08 |
| 23 | Malta | 4.54 | 81 | Tanzania | 2.21 | 139 | India | 1.07 |
| 24 | Montenegro | 4.54 | 82 | Timor-Leste | 2.20 | 140 | Mozambique | 1.04 |
| 25 | United Kingdom | 4.49 | 83 | Indonesia | 2.18 | 141 | Ecuador | 1.04 |
| 26 | Belgium | 4.38 | 84 | Qatar | 2.18 | 142 | New Guinea | 1.04 |
| 27 | Portugal | 4.37 | 85 | Barbados | 2.15 | 143 | Sri Lanka | 1.02 |
| 28 | Iran | 4.27 | 86 | Costa Rica | 2.06 | 144 | Bhutan | 1.01 |
| 29 | Greece | 4.23 | 87 | South Korea | 2.01 | 145 | Singapore | 1.01 |
| 30 | Israel | 4.15 | 88 | Japan | 2.00 | 146 | Cote d'Ivoire | 1.00 |
| 31 | Albania | 4.14 | 89 | Djibouti | 1.89 | 147 | Cambodia | 0.98 |
| 32 | Romania | 4.13 | 90 | Ethiopia | 1.89 | 148 | Cape Verde | 0.96 |
| 33 | Czech Republic | 4.08 | 91 | Georgia | 1.82 | 149 | Central Africa | 0.96 |
| 34 | Tunisia | 4.07 | 92 | Sudan | 1.81 | 150 | Lesotho | 0.93 |
| 35 | Italy | 4.07 | 93 | Burundi | 1.80 | 151 | Nigeria | 0.87 |
| 36 | Morocco | 3.98 | 94 | Gabon | 1.79 | 152 | Liberia | 0.82 |
| 37 | Zimbabwe | 3.97 | 95 | Turkmenistan | 1.79 | 153 | Guyana | 0.80 |
| 38 | Cuba | 3.90 | 96 | China | 1.78 | 154 | Haiti | 0.80 |
| 39 | Bulgaria | 3.87 | 97 | Angola | 1.73 | 155 | El Salvador | 0.80 |
| 40 | Canada | 3.85 | 98 | Eritrea | 1.71 | 156 | Solomon Isl. | 0.73 |
| 41 | Libya | 3.82 | 99 | South Africa | 1.70 | 157 | Sierra Leone | 0.72 |
| 42 | Jordan | 3.76 | 100 | Guinea-Bissau | 1.69 | 158 | Guinea | 0.71 |
| 43 | Lithuania | 3.74 | 101 | Tajikistan | 1.67 | 159 | Nicaragua | 0.66 |
| 44 | Norway | 3.71 | 102 | DR Congo | 1.65 | 160 | Suriname | 0.65 |
| 45 | Slovakia | 3.70 | 103 | Saudi Arabia | 1.64 | 161 | Botswana | 0.64 |
| 46 | Switzerland | 3.65 | 104 | Thailand | 1.61 | 162 | Uganda | 0.63 |
| 47 | Cyprus | 3.62 | 105 | Swaziland | 1.60 | 163 | Philippines | 0.63 |
| **48** | **United States** | **3.55** | 106 | Trinidad/Tob. | 1.57 | 164 | Honduras | 0.61 |
| 49 | Sweden | 3.53 | 107 | Namibia | 1.55 | 165 | Vietnam | 0.60 |
| 50 | Austria | 3.46 | 108 | Ghana | 1.54 | 166 | Mongolia | 0.59 |
| 51 | Russia | 3.35 | 109 | Azerbaijan | 1.51 | 167 | Gambia | 0.56 |
| 52 | Estonia | 3.32 | 110 | Peru | 1.51 | 168 | Dominican Rep. | 0.55 |
| 53 | Bahrain | 3.28 | 111 | Myanmar | 1.50 | 169 | Cameroon | 0.54 |
| 54 | Bosnia/Herzeg. | 3.26 | 112 | Laos | 1.50 | 170 | Maldives | 0.48 |
| 55 | Germany | 3.12 | 113 | Fiji | 1.50 | 171 | Guatemala | 0.45 |
| 56 | Kazakhstan | 3.06 | 114 | Mauritius | 1.49 | 172 | Congo | 0.32 |
| 57 | Australia | 3.04 | 115 | Togo | 1.48 | | | |
| 58 | Afghanistan | 3.00 | 116 | Nepal | 1.44 | | | |

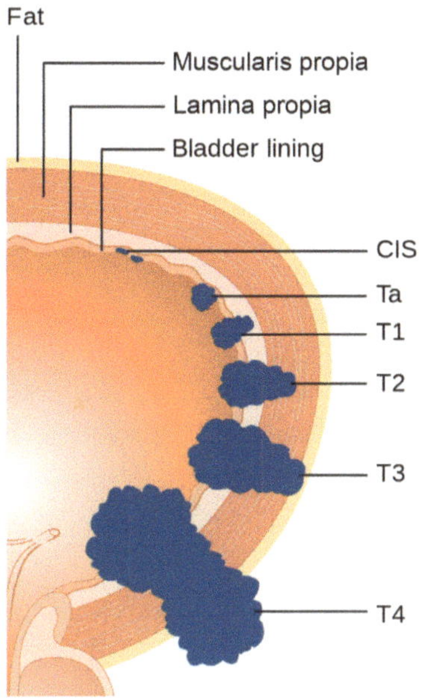


FIGURE 10.1 Diagram showing the pathologic T stages of bladder cancer. *Image from Cancer Research UK, via Wikimedia Commons. Available at: https://commons.wikimedia.org/wiki/File%3ADiagram_showing_the_T_stages_of_bladder_cancer_CRUK_372.svg.*

Tumor-node-metastasis (TNM) classification, the staging system of the American Joint Committee on Cancer, is most commonly used (Greene et al., 2002). Table 10.4 lists the details of the TNM classification system of bladder cancer.

TABLE 10.4 AJCC Staging of Bladder Cancer

| | |
|---|---|
| Tx | Primary tumor cannot be assessed |
| Ta | Noninvasive papillary carcinoma Tis: Carcinoma in situ |
| T1 | Tumor invades lamina propria |
| T2 | Tumor invades muscularis propria |
| T2a | Invades superficial muscularis propria (inner half) |
| T2b | Invades deep muscularis propria (outer half) |
| T3 | Tumor invades perivesical tissue/fat |
| T3a | Invades perivesical tissue fat microscopically |
| T3b | Invades perivesical tissue fat macroscopically |
| T4 | Tumor invades prostate, uterus, vagina, pelvic wall, or abdominal wall |
| T4a | Invades adjacent organs (uterus, ovaries, prostate stoma) |
| T4b | Invades pelvic wall and/or abdominal wall |
| Nx | Regional lymph nodes cannot be assessed due to lack of information |
| N0 | There is no regional lymph node spread |
| N1 | The cancer has spread to a single lymph node in the true pelvis |
| N2 | The cancer has spread to 2 or more lymph nodes in the true pelvis |
| N3 | The cancer has spread to lymph nodes along the common iliac artery |
| M0 | There are no signs of distant spread |
| M1 | The cancer has spread to distant parts of the body. The most common sites are distant lymph nodes, bones, lungs, and liver |

## 3.1 Carcinoma In Situ

This is a unique subtype of nonmuscle invasive TCC. CIS appears as a flat erythematous or "velvety" lesions of the mucosa. Fig. 10.2 illustrates the typical cystoscopic appearance of different types of bladder cancer. The majority of CIS cases occur in association with high-grade nodular tumors and only 3–5% occur as isolated CIS. CIS can also present as an occult lesion that is not easily visualized on standard white light cystoscopy. Histologically, CIS lesions are flat with severe cytological atypia and nuclear aplasia. Typically, urine cytology examination can detect malignant cells from CIS. CIS is a high-grade lesion and if it is present as an isolated lesion, it is often difficult to readily identify, survey, and treat (Koch et al., 1996). This potentially aggressive tumor has a high (83%) propensity for progression to invasive cancer and eventually metastatic disease (Hudson and Herr, 1995).

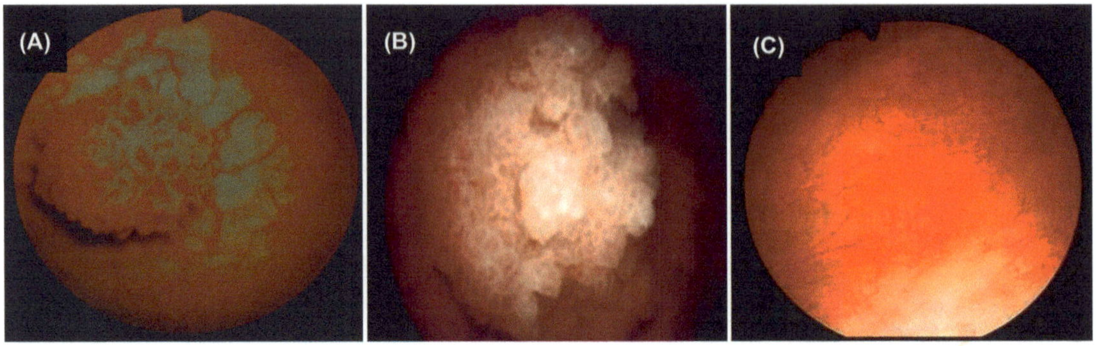


FIGURE 10.2 Classic cystoscopic appearance of (A) low-grade nonmuscle invasive tumor, (B) high-grade muscle invasive tumor, and (C) carcinoma in situ.

## 4. GRADING

Tumor grade is the most well-recognized prognostic indicators of bladder cancer and is one of the most important determinants of disease progression to advanced and eventually metastatic disease (Epstein et al., 1998). In 2004, the International Society of Urologic Pathologists published a revision of the previously used World Health Organization (WHO) classification system abolishing the "WHO Grade 2" or "intermediate grade" and reclassified them as "low-grade" (Eble et al., 2004). Table 10.5 describes this classification system.

TABLE 10.5 Classification of Urothelial Neoplasia According to World Health Organization/International Society of Urologic Pathologists—2004

| |
|---|
| Hyperplasia (flat and papillary) |
| Reactive atypia |
| Atypia of unknown significance |
| Urothelial dysplasia |
| Urothelial carcinoma in situ |
| Urothelial papilloma |
| Papillary urothelial neoplasm of low malignant potential |
| Nonmuscle invasive low-grade papillary urothelial carcinoma |
| Nonmuscle invasive high-grade papillary urothelial carcinoma |
| Muscle invasive papillary urothelial carcinoma |

## 5. STAGING TOOLS

The American Urologic Association and National Comprehensive Cancer Network (NCCN) guidelines recommend a contrast-enhanced computed tomography (CT) of the abdomen and pelvis with excretory phase known as CT urography or magnetic resonance (MR) urography for the staging of MIBC (Rais-Bahrami et al., 2015). Unfortunately, despite such recommendations from the guidelines, CT scan has significant limitations for providing information on local staging of bladder cancer. It cannot reliably discern muscle layers and relies on indirect findings of perivesical fat stranding or presence of an adjacent mass or visible extension beyond the bladder wall to predict the depth of invasion. The overall accuracy of CT scan for local staging is only 50% (Hoosein and Rajesh, 2014). Conventional MR imaging (MRI) performs better than CT scan for local staging due to its enhanced ability to delineate soft tissue. Dynamic contrast-enhanced MRI (DCE-MRI)

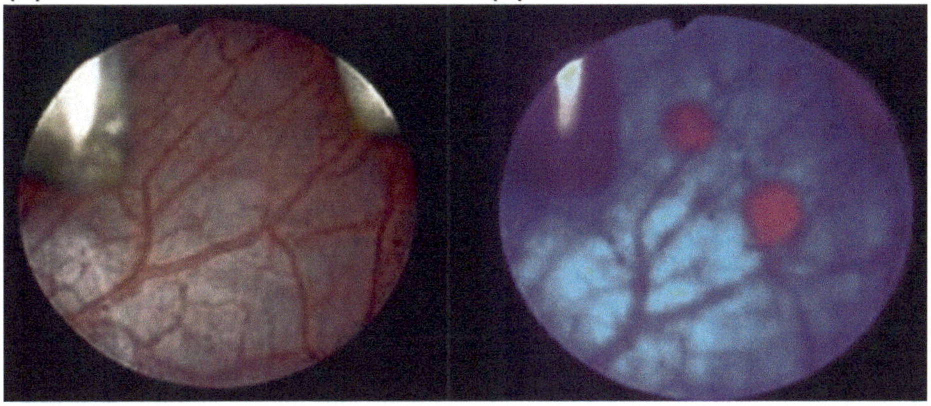


FIGURE 10.3 (A) Illustrates a cystoscopic view of bladder tumor under standard white light and (B) shows the same view in narrow band imaging.

with gadolinium-based contrast helps to visualize fat and thereby diagnose perivesical extension of the disease, however, it is still difficult to visualize tumor versus inflamed bladder wall (Hoosein and Rajesh, 2014).

The assessment lymph nodes on CT scan is based on identification of anatomically enlarged lymph node, which may or may not be cancerous. The performance of CT scan in diagnosing metastatic lymph nodes is better than local staging but still 25% of patients with lymph node metastases found on final pathology are not visualized on preoperative CT scan (Gray et al., 2014). The sensitivity of CT scan to detect lymph node positivity is good for lymph node more than 10 mm. For smaller lymph nodes, MRI performs significantly better (Saokar et al., 2010). DCE-MRI uses contrast to enhance the metastatic lymph node and is reported to have high accuracy.

TURBT, which is the corner stone of local staging for bladder cancer also under-stages the tumor due to technical-related or sampling errors. A second restaging TURBT has been shown to upstage the disease in up to 40% (Dobruch et al., 2014). The NCCN guidelines recommend restaging TURBT within 2- to 6-week of primary resection for high-risk NMIBC (Clark et al., 2013). Adjunctive aids used during transurethral resection, that is, narrow band imaging and fluorescence cystoscopy, have helped in better identification of local disease allowing better sampling and more complete resection, thus providing more accurate staging and prognostication information. Narrow band imaging uses optical enhancement by using different wavelengths of light to enhance the contrast between mucosal surfaces and microvascular structures (Herr and Donat, 2011). Fluorescence cystoscopy uses of hexaminolevulinate fluorescence to detect tumors and permits better visualization of residual tumor for complete eradication of the local disease (Witjes et al., 2014). Fig. 10.3 illustrates a lesion that is not easily identifiable on standard white light cystoscopy but becomes easily apparent with optical enhancement.

Functional imaging based on positron emission tomography (PET) has shown promise for bladder cancer staging. Fluorine-18 2-fluoro-2-deoxy-D-glucose (FDG) radiotracer is more commonly used to stage other cancers, although its excretion via urine has been a major limitation in applying it to bladder tumors imaging. For instance, Jeong et al. (2015) compared $^{18}$FDG PET-CT with conventional CT and reported no advantage with similar, positive, and negative predictive values. Radiotracers such as

$^{11}$C-choline, with minimal urinary excretion, have shown improved bladder cancer detection rate. Maurer et al. have reported an accuracy rate of 81% for nodal staging (Maurer et al., 2012). Picchio et al. have reported lower diagnostic accuracy (62%) of lymph node involvement by $^{11}$C-choline PET (Picchio et al., 2006). Currently, PET is not considered as standard imaging modality for bladder cancer staging.

Multiparametric MRI (mpMRI) has proven beneficial in the diagnosis and staging of other urological cancer but its incorporation in the diagnostic armamentarium of bladder cancer requires further standardization (Giannarini et al., 2012). mpMRI incorporates multiple anatomic imaging parameters to measures tissue density and composition. Studies have shown that diffusion weighted MRI (DW-MRI) can discriminate between metastatic versus nonmetastatic lymph nodes although further work is required to refine the apparent diffusion coefficient (ADC) cut-off value. When using an ADC cut-off value of $1.30 \times 10^3 mm^2/s$, Eiber et al. reported a sensitivity of 86% and specificity of 85% to diagnose metastatic disease (Eiber et al., 2010).

A combination of DW-MRI with ultrasmall superparamagnetic particles of iron oxide (USPIO) is also a promising technique. In a pilot study matching histologic findings with preoperative scans reported 90% accuracy in detecting lymph node metastases (Thoeny et al., 2009).

These advances in imaging hold promise to improve the preoperative staging and survival but need validation from large prospective trials. Currently, PET is not recommended for routine staging of MIBC and USPIO is not commercially used.

## 6. EMERGING BIOMARKERS

Biomarkers are molecules produced by the tumor that can be both measured and detected in tissue samples or fluids. The importance of biomarkers lies in the fact that they may be able to help with patient risk stratification, prognostication regarding likelihood of progression and developing metastatic disease, predicting response to therapy, counseling, and thereby treatment decision making (Bolenz and Lotan, 2008).

Currently, there is no clinically successful biomarker that has been identified for bladder cancer. The difficulty in finding a biomarker lies secondary to the lack of predictive accuracy. A wide range of biomarkers have been identified, and herein some of these will be discussed.

To understand what role these biomarkers play in the disease process of urothelial carcinoma, it is important to view low-grade and high-grade urothelial carcinoma as having different genetic mutations, despite the fact that alterations in chromosome 9 are common between both. Most urothelial carcinomas are low-grade, and these arise mainly from HRAS and fibroblast growth factor receptor (FGFR3) mutations. These low-grade tumors are typically noninvasive and though they tend to recur, they are less likely to progress. Conversely, high-grade urothelial carcinomas exhibit more p53 and pRb alterations.

p53 is the most extensively studied biomarker in bladder cancer. There has been much controversy regarding the prognostic outcomes of the p53 protein expression, and as such there is no conclusive data. It has also been found that the absence of caspase-3, an increase in survivin, and increase in Bcl-2 may indicate that there is an increased rate of recurrence of bladder cancer and thereby also a decreased risk of bladder cancer survival. With respect to low-grade disease, altered FGFR3 has been found to be associated with increased recurrence, but without any apparent effect upon disease progression or survival (Poyet et al., 2015).

The presence of high Ki-67 expression in tissue specimens may possibly predict a favorable survival in those undergoing bladder-sparing trimodal therapy (TMT) (aggressive "deep" transurethral local tumor resection, chemotherapy and external beam radiotherapy—see below for more detailed discussion). This outcome was found to be especially applicable in those with cT3 disease (Tanabe et al., 2015).

Combinations of biomarkers are being evaluated as well. The higher the total number of pRB, p21, p27, p53 alterations in bladder cancer specimens, the higher the patient risk stratification and thus risk of progression to metastatic disease and death (Shariat et al., 2007). Ranzi et al. also reviewed 38 biomarker publications and found that with a combination of biomarkers (e.g., p53, Ki-67, and CK20) one may be able to provide a more accurate prediction of outcome compared with using any single marker (Ranzi et al., 2015). This may pave the way toward improving risk stratification and clinical management of patients with bladder cancer.

# 7. PROGNOSTIC FACTORS

## 7.1 Histologic Variants of Urothelial Bladder Cancer and Nonurothelial Bladder Cancer

Ten percent of diagnosed MIBC display variant histology (i.e., atypical component) (Chalasani et al., 2009). Of these variants, some have been shown to portend a poorer prognosis than their nonvariant urothelial carcinoma counterpart (Xylinas et al., 2013).

*Micropapillary variant bladder* cancer is known to be associated with a poor prognosis (Xylinas et al., 2013; Kamat et al., 2007; Wang and Wang, 2015). Its incidence is estimated to be between 0.7% and 2.2% of all urothelial malignancies, and it generally presents at an advanced stage. At diagnosis, over half of the patients will have muscle invasive, nodal, or metastatic disease. The estimated 5-year survival is 51% for these patients, with only 24% surviving to 10 years from time of diagnosis (Xylinas et al., 2013; Wang and Wang, 2015). The clinical progression rate from non-MIBC to muscle invasion is high at 70%. There may also be a high rate of occult metastatic disease upon initial diagnosis. At this point in time, neoadjuvant chemotherapy is not recommended for these patients as it may delay time to receiving definitive treatment. Overall, the benefit of chemotherapy has yet to be proven in the setting of micropapillary urothelial carcinoma.

*Nested variant urothelial* cancer has also been shown to be aggressive (Beltran et al., 2014). Patients tend to present with advanced disease, and prognosis is poor regardless of initial stage of presentation (Linder et al., 2013). Nested variant affects more men than women, and most patients progress to metastases or die within 2 years of diagnosis (Beltran et al., 2014). In examined specimens, there was a high rate of adverse pathology and nodal involvement (Linder et al., 2013).

*Neuroendocrine nonurothelial variant* can also be termed small cell carcinoma of the bladder and it accounts for less than 1% of all diagnosed bladder cancers (Bhatt et al., 2014). It is associated with a poorer prognosis, and 5-year survival is estimated to be around 8% (Rogers et al., 2006). There is also a higher likelihood for extravesical extension or nodal disease upon diagnosis, and 50% are classified as having Stage IV disease (Rogers et al., 2006). Given the advanced stage, the initial treatment of neuroendocrine bladder cancers is for patients to undergo chemotherapy prior to considering consolidative surgery (Siefker-Radtke et al., 2004). The chemotherapy typically employed is platinum-based chemotherapy, which is based upon small-cell lung cancer studies, given the rarity of the entity and thus difficulty in obtaining prospective data (Bhatt et al., 2014; Siefker-Radtke et al., 2004).

*Squamous variant* is typically found in regions endemic with schistosomiasis, such as the Nile Delta in Egypt, or in patients with chronically inflamed bladders (Rogers et al., 2006). Rogers and Zargar-Shoshtari found that the mortality and survival curves of urothelial carcinoma versus squamous carcinoma of the bladder were relatively similar. This was corroborated in a later study examining the effects of chemotherapy on pure urothelial cancer versus mixed squamous and urothelial histology, whereby it was found that the squamous component did not confer any additional resistance to therapy (Scosyrev et al., 2011). In contrast, other studies have found that there is decreased survival in mixed squamous pathology groups, and yet further lower rates of survival in pure squamous histology (Gluck et al., 2014). Regardless, consensus remains that if patients are fit for chemotherapy, standard perioperative regimens should continue to be employed (see Sections 11.2 and 11.3) (Zargar-Shoshtari et al., 2016).

## 8. CLINICAL PRESENTATION OF METASTATIC BLADDER DISEASE

More than half the patients who present with MIBC eventually die of metastatic disease. Lymphatic spread (to lymph nodes) is more common than hematogenous spread (to distal organ sites). The most common metastatic sites are lymph nodes (69%), bone (47%), lung (37%), liver (26%), and peritoneum (16%) (Shinagare et al., 2011). The more common clinical presentations of metastatic disease include manifestations of obstructive uropathy and ureteral obstruction from lymph node enlargement (hydronephrosis, renal insufficiency, flank pain); lymphatic obstruction (lower limb edema, genital edema) bony involvement (pain, and less commonly, pathologic fractures and hypercalcemia); lung involvement (hemoptysis, dyspnea with pleural effusion, cough), liver enzyme elevation and dysfunction; and bowel obstruction.

## 9. MANAGEMENT OF CLINICALLY LOCALIZED BLADDER CANCER, POSITIVE PELVIC NODES AND METASTATIC DISEASE

The preoperative evaluation of a patient with MIBCs includes CT scan or MRI of the abdomen and pelvis. Clinically localized MIBC, in patients with acceptable anesthetic and surgical risks, is usually managed with radical cystectomy and urinary diversion (see Section 9.3). However, occult metastasis is not infrequent and may be missed at the time of initial diagnosis of MIBC, due to limitations of imaging modalities.

Meticulous bilateral pelvic lymphadenectomy is an essential component of the surgical management of MIBC and is a critical part of the accurate pathologic staging. Several pathoanatomic studies have been performed to map the lymphatic drainage of the bladder (Leissner et al., 2004; Vazina et al., 2004). The internal iliac, external iliac, and obturator groups of lymph nodes are the primary drainage regions. If these sites are negative, there is a very low risk of skip metastasis to secondary nodal sites. Three lymphadenectomy levels have been proposed: level 1 is comprised of all lymphatic tissue below the common iliac bifurcation, level 2 includes lymphatic tissue above the common iliac bifurcation and below the aortic bifurcation, and level 3 is defined as the lymphatic tissue above the aortic bifurcation up to the inferior mesenteric artery. Isolated positive modes at the level 3, that is, skip metastasis to level 3, have rarely been encountered.

A "standard" pelvic lymphadenectomy includes external and internal iliac nodes including the nodes in the fossa of Marcille distal to the common iliac bifurcation and complete dissection of the obturator fossa. An extended lymph node dissection, in addition to the above, also includes dissection of the common iliac and presacral nodes.

As noted previously, limitations of imaging modality in identifying occult metastatic disease, especially in patients found to have positive lymph node involvement at surgery or locally advanced (spread beyond the bladder), have prompted consideration for adjuvant chemotherapy in addition to radical cystectomy, in those deemed to have high-risk for subsequent development of clinically evident metastatic disease.

Locally advanced bladder cancer may be treated with neoadjuvant chemotherapy, followed by radical cystectomy (see below for discussion on other management options). Occasionally, some cancers are rendered from "surgically unresctable" to "resectable" by neoadjuvant chemotherapy (see below).

Management of patients with gross pelvic lymphadenopathy is, likewise, multimodal and relies on systemic chemotherapy in addition to surgery or radiation for the locoregional disease control. Neoadjuvant chemotherapy is preferable as tolerability of chemotherapy and patient compliance is expected to be better before, rather than after, surgery. Patients might respond to neoadjuvant therapy and reveal a favorable pathological status, determined mainly by achieving pT0, a negative lymph node status, and negative surgical margins (Advanced Bladder Cancer (ABC) Meta-analysis Collaboration, 2005). However, if a patient is not a suitable candidate, or neoadjuvant chemotherapy is anticipated to delay surgery, adjuvant chemotherapy may also be an option. A delay in surgery exceeding 90 days from the time of diagnosis to definitive treatment has been associated with poor outcome (Chang et al., 2003).

If metastasis is suspected outside the true pelvis, a biopsy may be recommended to confirm the diagnosis. Radical cystectomy as "monotherapy" in the presence of proven metastatic disease is not recommended since it is not curative and chemotherapy is the mainstay of therapy for proven metastatic disease.

## 9.1 Radical Cystectomy in Conjunction With Perioperative Chemotherapy

In 2001, Herr et al. looked at 207 patients with unresectable or regionally metastatic urothelial carcinoma. Of these patients, 39% received neoadjuvant cisplatin-based chemotherapy (see Sections 11.2 and 11.3) and then subsequently underwent radical cystectomy and urinary diversion. Thirty percent of postchemotherapy surgical specimens did not demonstrate any viable cancer, thereby pathologically confirming a complete response to chemotherapy. For those with viable cancer, 61% of the population had their cancer completely resected, and at the time of publication 41% of patients were alive. Thus, this was the first study to demonstrate a role for perioperative chemotherapy. It was further demonstrated that optimal candidates for this approach were those who had disease restricted locoregionally to the bladder and pelvis or local lymph nodes, and who exhibited a significant response to chemotherapy (Herr et al., 2001).

A multicenter study from United States and Europe was undertaken on 304 patients, with clinically positive nodal disease. The effects of neoadjuvant chemotherapy prior to radical cystectomy were examined. Depending upon initial nodal status (cN1-N3), the pN0 rate was around 48%. The complete and partial pathological response rates for the entire cohort were 14.5% and 27%. Survival outcomes were improved in male patients' post-cisplatin chemotherapy with negative surgical margins and a complete nodal response. Survival was also improved for those who had more than 15 nodes resected, and complete response was an important prognostic indicator (Zargar-Shoshtari et al., 2016).

## 9.2 Postchemotherapy Lymphadenectomy

In a study where postcystectomy patients with relapsing locally advanced or metastatic disease received high intensity MVAC, half went on to have a full retroperitoneal and pelvic lymph node dissection, and the remainder either went on to have radiation or were observed. These patients were followed for 88 months and it was found that median progression-free survival was improved in the lymphadenectomy group (18 vs. 11 months). The median overall survival was 37 months in the surgery group versus 19 months in the radiation or observation group (Necchi et al., 2015).

## 9.3 Trimodal Bladder-Sparing Therapy

TMT has been described as using surgery, radiosensitizing chemotherapy, and radiation to attempt to preserve the bladder and maintain quality of life in appropriately selected patients, who, otherwise, would also be candidates for radical cystectomy.

Surgery consists of an aggressive, but safe, transurethral resection of the bladder to debulk the majority of the tumor. Radiosensitizing chemotherapy is discussed elsewhere in this chapter. Radiation itself is applied locally to the bladder as well as pelvic lymph nodes, though retroperitoneal lymph nodes are not traditionally targeted. Patients are then examined and monitored by cystoscopy serially, and if there is any sign of failure (i.e., tumor recurrence), patients would undergo prompt salvage radical cystectomy. Initial studies were conducted on patients with T2–T4a disease, pathologic nodal status was unknown and patients did not have any known metastases (Kaufman et al., 2000).

It is difficult to compare TMT directly with the gold standard of radical cystectomy as there is a lack of randomized clinical trials. Studies have shown that following through the TMT regimen, including prompt salvage cystectomy, the 5-year overall survival rates can approach 48–60%, as compared to 62–68% for radical cystectomy (Ploussar et al., 2014).

In the metastatic or inoperable setting, it is also possible to employ TMT, however, it is worth noting that overall survival was poor. At 4 years, studies quote survival to be between 30% and 42%, and as such, for inoperable disease, it should be mentioned that results are modest (Ploussar et al., 2014; Hussain et al., 2001).

# 10. CLINICAL PROGNOSTICATORS

Horn et al. studied 88 cystectomy patients with node-positive disease to determine postsurgical and pathologic prognostic factors (Horn et al., 2015). Not surprisingly, patients with a lower pathological stage had a better prognosis. Those who had an ileal conduit had a better prognosis than those who underwent orthotopic diversion, possibly due to a negative margin rates and better local control.

Extra-nodal extension of lymph node disease was a poorer prognosticator, whereas a lower density of diseased lymph nodes, either fewer than two positive nodes or less than 4% of total lymph node specimen, conferred a better prognosis (Bruins et al., 2009; Horn et al., 2015). Neoadjuvant chemotherapy has been shown to be a favorable prognosticator.

## 10.1 Prechemotherapy Prognosticators

It has been found that the presence of liver metastases, a poor performance status (ECOG >2), and higher leukocyte count (>8000 μL), are all poor prognostic indicators in patients who have not received chemotherapy. Furthermore, these three factors can be combined to risk-stratify patients. Those without any of the above listed factors are classified as "favorable" risk, with an average 22.7-month survival. "Intermediate" risk patients (one factor) have a 9.3-month survival, and "poor" risk patients with two or more factors have a 4.9-month survival. It has additionally been found that those with liver metastases would die on average 9.3 months from their original diagnosis (Taguchi et al., 2013).

## 10.2 Postchemotherapy Prognosticators

Risk score models and nomograms have identified several factors postplatinum-based chemotherapy that affect prognosis in patients with advanced local or metastatic urothelial cancer. In 1999, Bajorin et al. found that the presence of visceral metastases to the lung, liver, or bone, and a Karnofsky performance score of less than 80% was each associated with a poorer prognosis. The median survival for those with 0, 1, or 2 of these factors was 33, 13.4, and 9.3 months, respectively. This model has been further validated (Bajorin et al., 1999).

Another model has found that postchemotherapy patients exhibited a poorer prognosis if they had an ECOG of 1 or more, a hemoglobin less than 10 g/dL, or liver metastases. Four risk groups were constructed based upon 0, 1, 2, or 3 prognostic factors, with a median overall survival of 14.2, 7.3, 3.8, and 1.7 months, respectively. A shorter time from when patients received first-line chemotherapy and when they were enrolled for second-line chemotherapy also portended a poorer prognosis (Sonpavde et al., 2013).

A nomogram was produced by Apolo et al. whereby visceral metastases, albumin level, performance status, and hemoglobin were considered (Apolo et al., 2013). This nomogram was initially based upon the MSKCC risk factors, and using number univariate and multivariate tests further variables were elucidated (Bajorin et al., 1999). The nomogram has been externally validated, and some consider it to be more accurate in terms of predicting 1-, 2-, and 5-year survival (Apolo et al., 2013).

# 11. MANAGEMENT OF METASTATIC BLADDER CANCER

## 11.1 Standard Chemotherapy Regimes

Platinum-based chemotherapy has long been considered the standard of care for locally advanced and metastatic bladder cancer (Calabrò and Sternberg, 2012; Fletcher et al., 2011). Original treatment consisted of cisplatin monotherapy, however, it was found that the addition of other agents improved outcomes (Loehrer et al., 1992).

## 11.2 Methotrexate, Vinblastine, Doxorubicin, Cisplatin

Traditional MVAC was initially the standard first-line chemotherapy regime in medically fit patients (Calabrò and Sternberg, 2012; Yagoda, 1983). It was found that patients experienced significant morbidity and toxicity from therapy. Side effects included myleosuppression, neutropenic fever and sepsis, nausea, vomiting, mucositis, and a 3–4% rate of death (Sternberg et al., 2001).

An EORTC study was then conducted whereby patients were randomized in a 1:1 ratio to standard MVAC versus high-dose intensity MVAC and granulocyte colony stimulating factor in patients with advanced urothelial carcinoma (Sternberg et al., 2001). At 7 years the dose-dense MVAC cohort had higher complete response rates (21% vs. 9%), borderline improved progression-free survival (9.5 vs. 8.1 months), improved overall survival (24.6% vs. 13.2%), and less treatment toxicity (leukopenia, febrile neutropenia, mucositis). The standard MVAC arm had more bladder cancer specific deaths at 76% compared to 64.9% in the high-dose MVAC group. In summary, high-dose intensity MVAC was found to be superior to traditional MVAC regimens as it allowed for increased delivery of cisplatin and doxorubicin in less time, leading to less toxicity, and delays in dosing (Calabrò and Sternberg, 2012; Fletcher et al., 2011; Sternberg et al., 2001). High-dose MVAC is now considered a treatment of choice, with benefits in progression-free survival, complete and overall response rates (Calabrò and Sternberg, 2012; Fletcher et al., 2011; Sternberg et al., 2001).

## 11.3 Gemcitabine and Cisplatin

A phase III randomized control trial of neoadjuvant gemcitabine and cisplatin versus MVAC was conducted in 2005 (von der Maase et al., 2005). Gemcitabine and cisplatin were shown to have similar efficacy and progression-free survival. Median survival in the gemcitabine and cisplatin (GC) group was 14 versus 15.2 months in the MVAC group. At 2 years, 25% of GC patients were alive compared to 31% of

MVAC patients. By 5 years, 13% of GC patients and 15.3% of MVAC patients were alive (von der Maase et al., 2005; Shelley et al., 2011). The safety profile of the GC regimen was found to be greater, with less episodes of febrile neutropenia, sepsis, and mucositis (von der Maase et al., 2005; Shelley et al., 2011). Due to this, the combination of GC is now also commonly employed as first-line therapy for bladder cancer. Further studies have shown that in those who cannot tolerate cisplatin due to age or comorbidities, the combination of gemcitabine and carboplatin may be a reasonable alternative for those with unresectable or locally advanced or metastatic disease.

## 11.4 Neoadjuvant Chemotherapy

An initial clinical concern was whether or not neoadjuvant chemotherapy would be of benefit to patients. Grossman randomized patients with locally advanced bladder cancer to three cycles of neoadjuvant MVAC as opposed to upfront radical cystectomy (Grossman et al., 2003). It was found that chemotherapy led to a reduced risk of death of 33% and an increased median survival of 77 months (as compared to 46 months for upfront surgery). This increase in survival was theorized to be due to the treatment of occult micro-metastases prior to cystectomy. Thirty-eight percent of the MVAC group was downstaged to pT0 at the time of cystectomy, as opposed to only 15% of the cystectomy group (Grossman et al., 2003). This work has been further confirmed by other studies, whereby neoadjuvant chemotherapy has continued to improve overall survival and decreased lymph node positivity rate (Mertens et al., 2014).

It has also been found that patients with certain variant histology, namely mixed glandular or squamous differentiation, also derive benefit from neoadjuvant chemotherapy (Scosyrev et al., 2011).

## 11.5 Radiosensitizing Chemotherapy

Platinum-based regimes have been the mainstay of radiosensitization for bladder cancer treatment, however, since the advent of TMT (bladder-preserving surgery, radiation, and chemotherapy) to treat bladder cancer, other radiosensitizers have been studied. An initial trial of mitomycin and 5-fluorouracil (5-FU) demonstrated acceptable toxicities (Hussain et al., 2001). A phase 3 trial has shown superior disease-free survival in those receiving mitomycin/5-FU chemoradiotherapy as opposed to radiotherapy alone (67% vs. 54%). More adverse effects were seen secondary to radiosensitization; however, these side effects did not appear to persist long term in the studied group (James et al., 2012).

## 11.6 Alternative Chemotherapeutic Agents

The substitution of cisplatin by carboplatin in elderly patients or those with renal insufficiency has been deemed as safe, although it has been proven to be less effective (Yoneyama et al., 2015). However, in patients who fail gemcitabine and carboplatin there has been no benefit to adding docetaxel to treatment (Yoneyama et al., 2015).

Docetaxel itself though is currently being investigated in poor prognosis patients for second-line efficacy in those failing platinum-based chemotherapy. Preliminary results have shown safe toxicity profiles and good tolerability, and docetaxel has been shown to exhibit modest antitumor activity (Albany and Sonpavde, 2015).

Vinflunine is another promising agent for second-line chemotherapy. In Europe it has been approved based upon a phase 3 trial that demonstrated an increased median overall survival of 6.9 versus 4.3 months when compared with best supportive care (Sonpavde et al., 2013).

# 12. NEW AGENTS

## 12.1 Tyrosine Kinase Receptor Inhibitors

FGF are implicated in wound healing, cellular proliferation, and angiogenesis. Within bladder cancer itself there has been a high frequency of genetic alterations in FGF or an overexpression of FGF without alteration. As such, phase 1 testing has begun on oral FGF receptor inhibitors. Preliminary results currently demonstrate tolerable toxicity profiles (Fletcher et al., 2011; Mazzola et al., 2014).

The phosphatidylinositol 3-kinase/v-akt murine thymoma viral oncogene homolog 1/mammalian target of rapamycin (PI3K/AKT/mTOR) pathway has been implicated in cancer survival, motility, and metabolism. The genetic defects seen in this pathway are typically due to activating mutations or deletions of pathway regulators. PI3K blockade is currently being investigated as second-line treatment, as initial results have shown these drugs to exhibit antiproliferative tendencies on bladder cancer cells. AKT inhibitors have not been found to be effective thus far in clinical trials, but given that it is activated by Human epidermal growth factor receptor 2 (HER2), there may be a future role for AKT inhibitors in HER2 positive bladder cancers.

mTOR inhibitors are being investigated as they have been commonly employed in the treatment of tuberous sclerosis; the hope is that they may be able to confer a good response in those with TSC1 bladder cancer mutations. Current studies involve the comparison of GEM/CIS versus everolimus or temsirolimus (Fletcher et al., 2011; Carneiro et al., 2015).

Epidermal growth factor 2 (HER2) has been found to correlate with higher frequencies of nodal metastases. Typically there is either an overexpression of the gene or a gene mutation. The presence of HER2 RNA is also implicated with poorer recurrence free survival rates postcystectomy. Currently, there are no HER2 inhibitors that are in use, but they are undergoing investigation (Fletcher et al., 2011; Carneiro et al., 2015; Peyromaure et al., 2005).

Vascular endothelial growth factor inhibitors have been studied, however, to date clinical results have been disappointing for advanced bladder cancer (Mazzola and Chin, 2015).

## 12.2 Cell Cycle Regulators

Aurora kinases are enzymes that help the cell cycle through the regulation of mitotic spindles. Mutations in this enzyme lead to aneuploidy or genomic instability. The increased expression of aurora kinases has been found in breast cancer, colon cancer, and neural malignancies. For bladder cancer in particular, its expression has been correlated with higher grade and stage. These regulators are currently undergoing clinical trials (Fletcher et al., 2011; Carneiro et al., 2015; Fenner, 2013).

Polo-like kinase 1 plays a similar role as aurora kinases as they also assist with the cell cycle and mitosis. Overexpression of this kinase occurs in 41% of bladder cancer. Initial drug trials have found that the drug is well tolerated, and initial therapy effects are promising (Fletcher et al., 2011; Carneiro et al., 2015; Brassesco et al., 2013).

Heat shock proteins (HSP) regulate both the formation and breakdown of proteins. The overexpression of HSPs may confer resistance to chemotherapy and radiation. Thus, preclinical studies have shown that by targeting HSP it is tentatively possible to increase responsiveness to certain chemotherapies, all the while decreasing growth in bladder cancer cell lines (Fletcher et al., 2011; Carneiro et al., 2015).

# 13. IMMUNOTHERAPY

## 13.1 PD-L1 Inhibitors

PD-L1 is expressed on the surface of tumor cells and it is able to bind to PD-1 on the surface of activated T cells and B cells. The binding of PD-L1 to PD-1 leads to an immunosuppressive effect and allows the tumor to evade immune destruction (Swaika et al., 2015). PD-L1 is expressed by 12–50% of all bladder tumor cells (Rouanne et al., 2015). In a phase 1 study there were no serious adverse events, and approximately half of PD-L1 tumor-positive patients responded. This is a promising new agent that is currently undergoing extensive clinical trials.

## 13.2 Anticytotoxic T Lymphocyte Associated Antigen (Anti-CTLA-4)

CTLA-4 was the first immune-checkpoint receptor to be clinically targeted (2015). The blockage of CTLA-4 enhances T-lymphocyte action, and this has led to ongoing research with respect to its application in bladder cancer (Carneiro et al., 2015).

# 14. ROLE OF SURGERY IN METASTATIC BLADDER CANCER

## 14.1 Extended Lymph Node Dissection

A proper lymph node dissection with at least nine lymph nodes on pathological specimen was initially found to allow for proper patient staging and stratification. Patients also exhibited improved local control and survival when more lymph nodes were removed (Herr et al., 2002). A retrospective study looking at almost 300 patients found that the more extended the dissection with at least 14 lymph nodes were removed, the better the patient outcomes (Brunocilla et al., 2013). Looking at limited dissections versus extended node dissections, a lower number of positive node rates was found in the limited dissection group (13% vs. 26%), although 5-year recurrence free survival was significantly better for the extended node dissection group (Dhar et al., 2008). Thus, it would appear as though a more extended resection translates into a better outcome regardless of extent of disease.

## 14.2 Palliative Treatment Options in Metastatic Bladder Cancer

### *14.2.1 Palliative Chemotherapy*

Chemotherapy has been the mainstay of treatment, even in palliative bladder cancer cases (Calabrò and Sternberg, 2012). Please see the above discussion.

### *14.2.2 Palliative Radiation*

Palliative radiation has been employed for the use of symptom relief, and typically for the control of hematuria or treatment of bone metastases (Fletcher et al., 2011; Abt et al., 2013). As compared to curative intent, palliative treatments are employed over a shorter period of time, and side effects are limited (Chan et al., 1979). The most common side effect has been radiation cystitis, which leads to bladder irritability and smaller bladder capacity, resulting in urinary frequency. There is generally a good short-term tumor response with some local control. Reports vary, but radiation is adequate to control hematuria between 50% and 92% of the time (Fletcher et al., 2011; Abt et al., 2013).

### *14.2.3 Palliative Surgery*

Even in the presence of metastatic bladder cancer, urinary diversion has been used occasionally, in cases of intractable gross hematuria. It has been thought that through the diversion of urine, "internal tamponade" may occur. As well, urokinase in the urine is also thought to enhance bleeding and thus by limiting its contact with bladder mucosa, clotting may occur, thus alleviating the hematuria problem. In addition, other treatment options such as intravesical instillation of formalin can be conducted with a reduced risk to the patient.

### 14.3 Supportive Measures in Patients with Advanced and/or Metastatic Bladder Cancer

Bisphosphonates have been used to treat the hypercalcemia of malignancy as well as bony pain secondary to metastases from urothelial cancer. They have also been found to prolong time to skeletal related events, decrease the incidence of skeletal related events, and improve survival (Zaghloul et al., 2010).

For patients with metastatic cancer, including bladder cancer, reduced appetite and hypoalimentation are often concerns for caregivers of patients under palliative care. It has been found that weight loss leads the patient to rate their quality of life and overall disease outlook lower. As appetite wanes, social contentment likewise decreases. Cachexia affects most palliative cancer patients and is responsible for 22% of cancer-related deaths. The pathogenesis of cancer-induced cachexia is multifactorial, involving the disease process itself, psychological and social factors (Poole and Froggatt, 2002). Due to this, no single treatment for cancer-induced cachexia is typically adequate.

Pharmacologically, progesterone derivatives such as megestrol acetate have been found to stimulate appetite, though most weight gain is due to fat. Studies have been conflicting as to whether progesterone is able to improve quality of life or overall survival (Laviano et al., 2005). Cannaboids have been used in the past as well to stimulate appetite, however, results vary and efficacy is questioned (Strasser et al., 2006). Corticosteroids have been used in treating other cancer-related symptoms, with an added benefit that it decreases malignancy-related weight loss and promotes an overall sense of well-being. Steroids have not been shown to improve overall survival or assist with weight gain (Argilés et al., 2010). Certain other pharmacological agents have been investigated, but thus far without clear effect.

Total parenteral nutrition is not usually adequate to replace daily requirements, and this likely indicates an underlying metabolic disturbance in advanced disease. As such, therapy targeting cytokines has been researched, and there is some evidence that thalidomide may decrease weight loss rates and preserve lean muscle mass.

Other medications have been used off-label to assist with cachexia. Beta-blockers are thought to slow down metabolic demand and can partially reverse cancer-related weight loss. Fish oil is an anti-inflammatory and may also assist with weight maintenance (Ryan et al., 2009).

Last, constipation can be a great source of discomfort for palliative cancer patients. The etiology of this may be due to decreased fluid intake, decreased mobility, increased opioid consumption, or structurally due to growth of the tumor itself. A study was conducted which found that the most important factors that led to constipation of palliative care patients was insufficient fluid intake, inadequate conditions of privacy, caregiver dependence, and a low performance status. Keeping these factors in mind, it may be possible to avoid or abate issues with constipation (Dzierżanowski and Ciałkowska-Rysz, 2015). Certain pharmaceutical agents such as laxatives and stool softeners are also commonly employed.

## 15. CONCLUSIONS

Bladder cancer is becoming more prevalent as more of the world becomes industrialized. Approximately 1 in 10 patients will present with metastatic disease, and in those who present with nonmuscle invasive disease, at least a third will progress to metastatic cancer. Proper staging is of utmost importance in this disease, and in conjunction with a thorough history and physical examination, either CT or MRI urography is recommended. In patients with high-risk NMIBC, it has been widely recommended that a second bladder resection at 2–6 weeks after the initial resection should be performed to confirm pathological staging. In those with locally advanced disease, various treatment modalities may be offered to the patient. Options include trimodal bladder-sparing therapy, radical surgery with or without neoadjuvant chemotherapy, and palliation. In-depth counseling of these patients in a multidisciplinary setting is essential, as patient comorbidities, potential outcomes of respective options, acceptance of risk, and a thorough discussion are important in determining the patient's plan of care.

Certain histologic variants, node-positive disease, degree of surgical resection, extra-nodal extension of cancer, and neoadjuvant chemotherapy have all been described as prognosticators in patients with advanced bladder cancer. Methotrexate/vinblastine/doxorubicin/cisplatin and gemcitabine/cisplatin have been considered standard chemotherapy regimens for bladder cancer, although other chemotherapeutic options are discussed. In terms of radiosensitization for TMT, platinum-based chemotherapies are the gold standard. New therapies are underway for bladder cancer, and these studies are ongoing.

Last, the palliation of bladder cancer may consist of chemotherapy, radiation, or surgery. The goals are to optimize patient comfort and to minimize symptoms. Paraneoplastic symptoms may also be temporized through the use of medication, and cachexia associated with advanced disease may be offset through pharmacological means as well.

## References

Abt, D., Bywater, M., Engeler, D.S., Schmid, H.P., 2013. Therapeutic options for intractable hematuria in advanced bladder cancer. Int. J. Urol. 20, 651–660. http://dx.doi.org/10.1111/iju.12113.

Advanced Bladder Cancer (ABC) Meta-analysis Collaboration, 2005. Neoadjuvant chemotherapy in invasive bladder cancer: update of a systematic review and meta-analysis of individual patient data advanced bladder cancer (ABC) meta-analysis collaboration. Eur. Urol. 48, 202–205.

Albany, C., Sonpavde, G., 2015. Docetaxel for the treatment of bladder cancer. Expert. Opin. Investig. Drugs 24, 1657–1664. http://dx.doi.org/10.1517/13543784.2015.1109626.

Apolo, A.B., Ostrovnaya, I., Halabi, S., et al., 2013. Prognostic model for predicting survival of patients with metastatic urothelial cancer treated with cisplatin-based chemotherapy. J. Natl. Cancer Inst. 105, 499–503. http://dx.doi.org/10.1093/jnci/djt015.

Argilés, J.M., Olivan, M., Busquets, S., López-Soriano, F.J., 2010. Optimal management of cancer anorexia-cachexia syndrome. Cancer Manag. Res. 2, 27–38.

Bajorin, D.F., Dodd, P.M., Mazumdar, M., et al., 1999. Long-term survival in metastatic transitional-cell carcinoma and prognostic factors predicting outcome of therapy. J. Clin. Oncol. 17, 3173–3181.

Beltran, A.L., Cheng, L., Montironi, R., et al., 2014. Clinicopathological characteristics and outcome of nested carcinoma of the urinary bladder. Virchows Arch. 465, 199–205. http://dx.doi.org/10.1007/s00428-014-1601-y.

Bhatt, V.R., Loberiza, F.R., Tandra, P., Krishnamurthy, J., Shrestha, R., Wang, J., 2014. Risk factors, therapy and survival outcomes of small cell and large cell neuroendocrine carcinoma of urinary bladder. Rare Tumors 6, 5043. http://dx.doi.org/10.4081/rt.2014.5043.

Bolenz, C., Lotan, Y., 2008. Molecular biomarkers for urothelial carcinoma of the bladder: challenges in clinical use. Nat. Clin. Pract. Urol. 5, 676–685.

van den Bosch, S., Alfred Witjes, J., 2011. Long-term cancer-specific survival in patients with high-risk, non-muscle-invasive bladder cancer and tumour progression: a systematic review. Eur. Urol. 60, 493–500. http://dx.doi.org/10.1016/j.eururo.2011.05.045.

Brassesco, M.S., Pezuk, J.A., Morales, A.G., et al., 2013. In vitro targeting of polo-like kinase 1 in bladder carcinoma: comparative effects of four potent inhibitors. Cancer Biol. Ther. 14, 648–657. http://dx.doi.org/10.4161/cbt.25087.

Bruins, H.M., Huang, G.J., Cai, J., Skinner, D.G., Stein, J.P., Penson, D.F., 2009. Clinical outcomes and recurrence predictors of lymph node positive urothelial cancer after cystectomy. J. Urol. 182, 2182–2187. http://dx.doi.org/10.1016/j.juro.2009.07.017.

Brunocilla, E., Pernetti, R., Schiavina, R., et al., 2013. The number of nodes removed as well as the template of the dissection is independently correlated to cancer-specific survival after radical cystectomy for muscle-invasive bladder cancer. Int. Urol. Nephrol. 45, 711–719. http://dx.doi.org/10.1007/s11255-013-0461-8.

Calabrò, F., Sternberg, C.N., 2012. Metastatic bladder cancer: anything new? Curr. Opin. Support. Palliat. Care 6, 304–309. http://dx.doi.org/10.1097/SPC.0b013e3283552d19.

Carneiro, B.A., Meeks, J.J., Kuzel, T.M., Scaranti, M., Abdulkadir, S.A., Giles, F.J., 2015. Emerging therapeutic targets in bladder cancer. Cancer Treat. Rev. 41, 70–78. http://dx.doi.org/10.1016/j.ctrv.2014.11.003.

Chalasani, V., Chin, J.L., Izawa, J.I., 2009. Histologic variants of urothelial bladder cancer and nonurothelial histology in bladder cancer. Can. Urol. Assoc. J. 3, 193–198.

Chan, R.C., Bracken, R.B., Johnso, D.E., 1979. Single dose whole pelvis megavoltage irradiation for palliative control of hematuria or ureteral obstruction. J. Urol. 122, 750–751.

Chang, S.S., Hassan, J.M., Cookson, M.S., et al., 2003. Delaying radical cystectomy for muscle invasive bladder cancer results in worse pathological stage. J. Urol. 170, 1085–1087.

Clapp, R.W., Jacobs, M.M., Loechler, L.L., 2008. Environmental and occupational causes of cancer new evidence, 2005–2007. Rev. Environ. Health 23, 1–37.

Clark, P.E., Agarwal, N., Biagioli, M.C., Eisenberger, M.A., Greenberg, R.E., Herr, H.W., et al., 2013. Bladder Cancer National Comprehensive Cancer Network (NCCN). J. Natl. Compr. Cancer Netw. 11, 446–475.

Dhar, N.B., Klein, E.A., Reuther, A.M., Thalmann, G.N., Madersbacher, S., Studer, U.E., 2008. Outcome after radical cystectomy with limited or extended pelvic lymph node dissection. J. Urol. 179, 873–878. http://dx.doi.org/10.1016/j.juro.2007.10.076.

Dobruch, J., Borówka, A., Herr, H.W., 2014. Clinical value of transurethral second resection of bladder tumor: systematic review. Urology 84, 881–885.

Dzierżanowski, T., Ciałkowska-Rysz, A., 2015. Behavioral risk factors of constipation in palliative care patients. Support Care Cancer 23, 1787–1793. http://dx.doi.org/10.1007/s00520-014-2495-6.

Eble, J.N., Sauter, G., Epstein, J.I., Sesterhenn, I.A., 2004. World Health Organization Classification of Tumours: Pathology and Genetics of Tumours of the Urinary and Male Genital Organs. IARC Press, Lyon.

Eiber, M., Beer, A.J., Holzapfel, K., et al., 2010. Preliminary results for characterization of pelvic lymph nodes in patients with prostate cancer by diffusion-weighted MR-imaging. Investig. Radiol. 45, 15–23.

Epstein, J.I., Amin, M.B., Reuter, V.R., Mostofi, F.K., 1998. The World Health Organization/International Society of Urological Pathology consensus classification of urothelial (transitional cell) neoplasms of the urinary bladder. Bladder Consensus Conference Committee. Am. J. Surg. Pathol. 22, 1435.

Fenner, A., 2013. Bladder cancer: aurora kinase inhibitors light up the therapeutic horizon in bladder cancer. Nat. Rev. Urol. 10, 184. http://dx.doi.org/10.1038/nrurol.2013.35.

Ferlay, J., Shin, H.R., Bray, F., Forman, D., Mathers, C., Parkin, D.M., 2010. Estimates of worldwide burden of cancer in 2008: GLOBOCAN 2008. Int. J. Cancer 127, 2893–2917. http://dx.doi.org/10.1002/ijc.25516.

Ferlay, J., Soerjomataram, I., Ervik, M., et al., 2014. GLOBOCAN 2012 v1.1, Cancer Incidence and Mortality Worldwide: IARC CancerBase No. 11 [Internet]. International Agency for Research on Cancer, Lyon, France. Available from: http://globocan.iarc.fr.

Fitzmaurice, C., Dicker, D., Pain, A., Hamavid, H., Moradi-Lakeh, M., MacIntyre, M.F., Allen, C., Global Burden of Disease Cancer Collaboration, et al., 2015. The Global Burden of Cancer 2013. JAMA Oncol. 1, 505–527. http://dx.doi.org/10.1001/jamaoncol.2015.0735.

Fletcher, A., Choudhury, A., Alam, N., 2011. Metastatic bladder cancer: a review of current management. ISRN Urol. 2011, 545241. http://dx.doi.org/10.5402/2011/545241.

Giannarini, G., Petralia, G., Thoeny, H.C., 2012. Potential and limitations of diffusion-weighted magnetic resonance imaging in kidney, prostate, and bladder cancer including pelvic lymph node staging: a critical analysis of the literature. Eur. Urol. 61, 326–340.

Gluck, G., Hortopan, M., Stănculeanu, D., Chiriţă, M., Stoica, R., Sinescu, I., 2014. Comparative study of conventional urothelial carcinoma, squamous differentiation carcinoma and pure squamous carcinoma in patients with invasive bladder tumors. J. Med. Life 7, 211–214.

Gray, P.J., Lin, C.C., Jemal, A., Shipley, A.U., Fedewa, F.A., et al., 2014. Clinical-pathologic stage discrepancy in bladder cancer patients treated with radical cystectomy: results from the national cancer data base. Int. J. Radiat. Oncol. Biol. Phys. 88, 1048–1056.

Greene, F.L., Page, D.L., Fleming, I.D., Fritz, A., Balch, C.M., Haller, D.G., et al., 2002. AJCC Cancer Staging Manual, sixth ed. Springer-Verlag, New York.

Grossman, H.B., Natale, R.B., Tangen, C.M., et al., 2003. Neoadjuvant chemotherapy plus cystectomy compared with cystectomy alone for locally advanced bladder cancer. N. Engl. J. Med. 349, 859–866. http://dx.doi.org/10.1056/NEJMoa022148.

Herr, H.W., Donat, S.M., 2011. Reduced bladder tumour recurrence rate associated with narrow-band imaging surveillance cystoscopy. BJU Int. 107, 396–398.

Herr, H.W., Donat, S.M., Bajorin, D.F., 2001. Post-chemotherapy surgery in patients with unresectable or regionally metastatic bladder cancer. J. Urol. 165, 811–814.

Herr, H.W., Bochner, B.H., Dalbagni, G., Donat, S.M., Reuter, V.E., Bajorin, D.F., 2002. Impact of the number of lymph nodes retrieved on outcome in patients with muscle invasive bladder cancer. J. Urol. 167, 1295–1298.

Hoosein, M.M., Rajesh, A., 2014. MR imaging of the urinary bladder. Magn. Reson. Imaging Clin. N. Am. 22, 129–134. http://dx.doi.org/10.1016/j.mric.2014.01.001.

Horn, T., Schmid, S.C., Seitz, A.K., Grab, J., Wolf, P., Haller, B., Gschwend, J.E., 2015. Clinical prognosticators of survival in patients with urothelial carcinoma of the bladder and lymph node metastases after cystectomy with curative intent. World J. Urol. 33, 813–819. http://dx.doi.org/10.1007/s00345-014-1355-9.

Hudson, M.A., Herr, H.W., 1995. Carcinoma in situ of the bladder. J. Urol. 153, 564.

Hussain, S.A., Moffitt, D.D., Glaholm, J.G., Peake, D., Wallace, D.M., James, N.D., 2001. A phase I-II study of synchronous chemoradiotherapy for poor prognosis locally advanced bladder cancer. Ann. Oncol. 12, 929–935.

James, N.D., Hussain, S.A., Hall, E., et al., 2012. Radiotherapy with or without chemotherapy in muscle-invasive bladder cancer. N. Engl. J. Med. 366, 1477–1488.

Jeong, I.G., Hong, S., You, D., Hong, J.H., Ahn, H., Kim, C.S., 2015. FDG PET-CT for lymph node staging of bladder cancer: a prospective study of patients with extended pelvic lymphadenectomy. Ann. Surg. Oncol. 22, 3150–3156.

Kamat, A.M., Dinney, C.P., Gee, J.R., et al., 2007. Micropapillary bladder cancer: a review of the University of Texas M.D. Anderson cancer center experience with 100 consecutive patients. Cancer 110, 62–67. http://dx.doi.org/10.1002/cncr.22756.

Kaufman, D.S., Winter, K.A., Shipley, W.U., et al., 2000. The initial results in muscle-invading bladder cancer of RTOG 95-06: phase I/II trial of transurethral surgery plus radiation therapy with concurrent cisplatin and 5-fluorouracil followed by selective bladder preservation or cystectomy depending on the initial response. Oncologist 5, 471–476.

Koch, M.O., Smith Jr., J.A., 1996. Natural history and surgical management of superficial bladder cancer (stages Ta/T1/CIS). In: Vogelzang, N.J., Debruyne, F.M., Shipley, W.U., Scardino, P. (Eds.), Comprehensive Textbook of Genitourinary Oncology. Lippincott Williams and Wilkins, Baltimore, pp. 405–415 (Chapter 26).

Laviano, A., Meguid, M.M., Inui, A., Muscaritoli, M., Rossi-Fanelli, F., 2005. Therapy insight: cancer anorexia–cachexia syndrome when all you can eat is yourself. Nat. Clin. Pract. Oncol. 2, 158–165.

Leissner, J., Ghoneim, M.A., Abol-Enein, H., et al., 2004. Extended radical lymphadenectomy in patients with urothelial bladder cancer: results of a prospective multicenter study. J. Urol. 171, 139–144.

Linder, B.J., Frank, I., Cheville, J.C., et al., 2013. Outcomes following radical cystectomy for nested variant of urothelial carcinoma: a matched cohort analysis. J. Urol. 189, 1670–1675. http://dx.doi.org/10.1016/j.juro.2012.11.006.

Loehrer, P.J., Einhorn, L.H., Elson, P.J., et al., 1992. A randomized comparison of cisplatin alone or in combination with methotrexate, vinblastine, and doxorubicin in patients with metastatic urothelial carcinoma: a cooperative group study. J. Clin. Oncol. 10, 1066–1073.

von der Maase, H., Sengelov, L., Roberts, J.T., et al., 2005. Long-term survival results of a randomized trial comparing gemcitabine plus cisplatin, with methotrexate, vinblastine, doxorubicin, plus cisplatin in patients with bladder cancer. J. Clin. Oncol. 23, 4602–4608. http://dx.doi.org/10.1200/JCO.2005.07.757.

Maurer, T., Souvatzoglou, M., Kübler, H., et al., 2012. Diagnostic efficacy of [11C]choline positron emission tomography/computed tomography compared with conventional computed tomography in lymph node staging of patients with bladder cancer prior to radical cystectomy. Eur. Urol. 61, 1031–1038. http://dx.doi.org/10.1016/j.eururo.2011.12.009.

Mazzola, C.R., Chin, J., 2015. Targeting the VEGF pathway in metastatic bladder cancer. Expert Opin. Investig. Drugs 24, 913–927. http://dx.doi.org/10.1517/13543784.2015.104158.

Mazzola, C.R., Siddiqui, K.M., Billia, M., Chin, J., 2014. Dovitinib: rationale, preclinical and early clinical data in urothelial carcinoma of the bladder. Expert Opin. Investig. Drugs 23, 1553–1562. http://dx.doi.org/10.1517/13543784.2014.966900.

Mertens, L.S., Meijer, R.P., Meinhardt, W., et al., 2014. Occult lymph node metastases in patients with carcinoma invading bladder muscle: incidence after neoadjuvant chemotherapy and cystectomy vs after cystectomy alone. BJU Int. 114, 67–74. http://dx.doi.org/10.1111/bju.12447.

Necchi, A., Giannatempo, P., Lo Vullo, S., Farè, E., Raggi, D., Nicolai, N., et al., 2015. Postchemotherapy lymphadenectomy in patients with metastatic urothelial carcinoma: long-term efficacy and implications for trial design. Clin. Genitourin. Cancer 13, 80–86. http://dx.doi.org/10.1016/j.clgc.2014.06.003.

Parkin, D.M., 2008. The global burden of urinary bladder cancer. Scand. J. Urol. Nephrol. Suppl. 218, 12–20.

Peyromaure, M., Scotté, F., Amsellem-Ouazana, D., Vieillefon, A., Oudard, S., Beuzeboc, P., 2005. Trastuzumab (herceptin) in metastatic transitional cell carcinoma of the urinary tract: report on six patients. Eur. Urol. 48, 771–775. http://dx.doi.org/10.1016/j.eururo.2005.07.01.

Picchio, M., Treiber, U., Beer, A.J., et al., 2006. Value of 11C-choline PET and contrast-enhanced CT for staging of bladder cancer: correlation with histopathologic findings. J. Nucl. Med. 47, 938–944.

Ploeg, M., Aben, K.K., Kiemeney, L.A., 2009. The present and future burden of urinary bladder cancer in the world. World J. Urol. 27, 289–293. http://dx.doi.org/10.1007/s00345-009-0383-3.

Ploussar, G., Daneshmand, S., Efstathiou, J.A., et al., 2014. Critical analysis of bladder sparing with trimodal therapy in muscle-invasive bladder cancer: a systematic review. Eur. Urol. 66, 120–137. http://dx.doi.org/10.1016/j.eururo.2014.02.038.

Poole, K., Froggatt, K., 2002. Loss of weight and loss of appetite in advanced cancer: a problem for the patient, the carer, or the health professional? Palliat. Med. 16, 499–506. http://dx.doi.org/10.1191/0269216302pm593oa.

Poyet, C., Hermanns, T., Zhong, Q., et al., 2015. Positive fibroblast growth factor receptor 3 immunoreactivity is associated with low-grade non-invasive urothelial bladder cancer. Oncol. Lett. 10, 2753–2760. http://dx.doi.org/10.3892/ol.2015.3691.

Rais-Bahrami, S., Pietryga, J.A., Nix, J.W., 2015. Contemporary role of advanced imaging for bladder cancer staging. Urol. Oncol. 28, S1078–S1439.

Ranzi, A.D., da Silva, J.N., Graziottin, T.M., Annels, N., Bica, C.G., November 16, 2015. Immunohistochemistry biomarkers in nonmuscle invasive bladder cancer. Appl. Immunohistochem. Mol. Morphol.

Rogers, C.G., Palapattu, G.S., Shariat, S.F., et al., 2006. Clinical outcomes following radical cystectomy for primary nontransitional cell carcinoma of the bladder compared to transitional cell carcinoma of the bladder. J. Urol. 75, 2048–2053. http://dx.doi.org/10.1016/S0022-5347(06)00317-X.

Rosenberg, J.E., Carroll, P.R., Small, E.J., 2005. Update on chemotherapy for advanced bladder cancer. J. Urol. 174, 14–20.

Rouanne, M., Loriot, Y., Lebret, T., Soria, J.C., November 9, 2015. Novel therapeutic targets in advanced urothelial carcinoma. Crit. Rev. Oncol. Hematol. http://dx.doi.org/10.1016/j.critrevonc.2015.10.021 pii:S1040–8428(15)30070-6.

Ryan, A.M., Reynolds, J.V., Healy, L., et al., 2009. Enteral nutrition enriched with eicosapentaenoic acid (EPA) preserves lean body mass following esophageal cancer surgery: results of a double-blinded randomized controlled trial. Ann. Surg. 249, 355–363. http://dx.doi.org/10.1097/SLA.0b013e31819a4789.

Saokar, A., Islam, T., Jantsch, M., Saksena, M.A., Hahn, P.F., Harisinghani, M.G., 2010. Detection of lymph nodes in pelvic malignancies with Computed Tomography and Magnetic Resonance Imaging. Clin. Imaging 34, 361–366. http://dx.doi.org/10.1016/j.clinimag.2009.07.004.

Scosyrev, E., Ely, B.W., Messing, E.M., et al., 2011. Do mixed histological features affect survival benefit from neoadjuvant platinum-based combination chemotherapy in patients with locally advanced bladder cancer? A secondary analysis of southwest oncology group-directed intergroup study (S8710). BJU Int. 108, 693–699. http://dx.doi.org/10.1111/j.1464-410X.2010.09900.x.

Shariat, S.F., Ashfaq, R., Sagalowsky, A.I., Lotan, Y., 2007. Predictive value of cell cycle biomarkers in nonmuscle invasive bladder transitional cell carcinoma. J. Urol. 177, 481–487. http://dx.doi.org/10.1016/j.juro.2006.09.038.

Shelley, M., Cleves, A., Wilt, T.J., Mason, M., 2011. Gemcitabine for unresectable, locally advanced or metastatic bladder cancer. Cochrane Database Syst. Rev. 4, CD008976. http://dx.doi.org/10.1002/14651858.CD008976.pub2.

Shinagare, A.B., Ramaiya, N.H., Jagannathan, J.P., et al., 2011. Metastatic pattern of bladder cancer: correlation with the characteristics of the primary tumor. Am. J. Roentgenol. 196, 117–122. http://dx.doi.org/10.2214/AJR.10.5036.

Siefker-Radtke, A.O., Dinney, C.P., Abrahams, N.A., et al., 2004. Evidence supporting preoperative chemotherapy for small cell carcinoma of the bladder: a retrospective review of the M.D. Anderson cancer experience. J. Urol. 172, 481–484.

Sonpavde, G., Pond, G.R., Fougera, R., et al., 2013. Time from prior chemotherapy enhances prognostic risk grouping in the second-line setting of advanced urothelial carcinoma: a retrospective analysis of pooled, prospective phase 2 trials. Eur. Urol. 63, 717–723. http://dx.doi.org/10.1016/j.eururo.2012.11.042.

Sternberg, C.N., Vogelzang, N.J., 2003. Gemcitabine, paclitaxel, pemetrexed and other newer agents inurothelial and kidney cancers. Crit. Rev. Oncol. Hematol. 46, S105–S115.

Sternberg, C.N., de Mulder, P.H., Schornagel, J.H., European Organization for Research and Treatment of Cancer Genitourinary Tract Cancer Cooperative Group, et al., 2001. Randomized phase III trial of high-dose-intensity methotrexate, vinblastine, doxorubicin, and cisplatin (MVAC) chemotherapy and recombinant human granulocyte colony-stimulating factor versus classic MVAC in advanced urothelial tract tumors: European organization for research and treatment of cancer protocol no. 30924. J. Clin. Oncol. 2001 (19), 2638–2646.

Strasser, F., Luftner, D., Possinger, K., et al., 2006. Comparison of orally administered cannabis extract and delta-9-tetrahydrocannabinol in treating patients with cancer-related anorexia-cachexia syndrome: a multicenter, phase III, randomized, double-blind, placebo-controlled clinical trial from the cannabis-in-cachexia-study-group. J. Clin. Oncol. 24, 3394–3400. http://dx.doi.org/10.1200/JCO.2005.05.184.

Swaika, A., Hammond, W.A., Joseph, R.W., 2015. Current state of anti-pd-l1 and anti-pd-1 agents in cancer therapy. Mol. Immunol. 67, 4–17. http://dx.doi.org/10.1016/j.molimm.2015.02.009.

Taguchi, S., Nakagawa, T., Hattori, M., Niimi, A., et al., 2013. Prognostic factors for metastatic urothelial carcinoma undergoing cisplatin-based salvage chemotherapy. Jpn. J. Clin. Oncol. 43, 923–928. http://dx.doi.org/10.1093/jjco/hyt096.

Tanabe, K., Yoshida, S., Koga, F., Inoue, M., Kobayashi, S., Ishioka, J., et al., 2015. High ki-67 expression predicts favorable survival in muscle-invasive bladder cancer patients treated with chemoradiation-based bladder-sparing protocol. Clin. Genitourin. Cancer 13, 243–251. http://dx.doi.org/10.1016/j.clgc.2015.03.002.

Thoeny, H.C., Triantafyllou, M., Birkhaeuser, F.D., et al., 2009. Combined ultrasmall superparamagnetic particles of iron oxide–enhanced and diffusion-weighted magnetic resonance imaging reliably detect pelvic lymph node metastases in normal-sized nodes of bladder and prostate cancer patients. Eur. Urol. 55, 761–769.

Vazina, A., Dugi, D., Shariat, S.F., Evans, J., Link, R., Lerner, S.P., 2004. Stage specific lymph node metastasis mapping in radical cystectomy specimens. J. Urol. 171, 1830–1834.

Wang, J., Wang, F.W., 2015. The natural history, treatment pattern, and outcomes of patients with micropapillary bladder carcinoma. Am. J. Clin. Oncol. 38, 472–478. http://dx.doi.org/10.1097/COC.0b013e3182a5329.

Wein, A.J., Kavoussi, L.R., Novick, A.C., Partin, A.W., Peters, C.A. (Eds.), 2010. Campbell-Walsh Urology, tenth ed. Saunders, Philadelphia.

Witjes, J.A., Babjuk, M., Gontero, P., et al., 2014. Clinical and cost effectiveness of hexaminolevulinate-guided blue-light cystoscopy: evidence review and updated expert recommendations. Eur. Urol. 66, 863–871. http://dx.doi.org/10.1016/j.eururo.2014.06.037. (Epub July 4, 2014. Review).

Xylinas, E., Rink, M., Robinson, B.D., et al., 2013. Impact of histological variants on oncological outcomes of patients with urothelial carcinoma of the bladder treated with radical cystectomy. Eur. J. Cancer 49, 1889–1897. http://dx.doi.org/10.1016/j.ejca.2013.02.001.

Yagoda, A., 1983. Phase-II trials in patients with urothelial tract tumors. Memorial sloan-kettering cancer center. Cancer Chemother. Pharmacol. 11, 9–12.

Yoneyama, T., Tobisawa, Y., Yoneyam, T., et al., 2015. Carboplatin-based combination chemotherapy for elderly patients with advanced bladder cancer. Int. J. Clin. Oncol. 20, 369–374. http://dx.doi.org/10.1007/s10147-014-0706-2.

Zaghloul, M.S., Boutrus, R., El-Hossieny, H., Kader, Y.A., El-Attar, I., Nazmy, M., 2010. A prospective, randomized, placebo-controlled trial of zoledronic acid in bony metastatic bladder cancer. Int. J. Clin. Oncol. 15, 382–389. http://dx.doi.org/10.1007/s10147-010-0074-5.

Zargar-Shoshtari, K., Zargar, H., Lotan, Y., et al., 2016. A multi-institutional analysis of outcomes of patients with clinically node positive urothelial bladder cancer treated with induction chemotherapy and radical cystectomy. J. Urol. 195, 53–59. http://dx.doi.org/10.1016/j.juro.2015.07.0.

CHAPTER

# 11

# Metastatic Cutaneous Squamous Cell Carcinomas

*I. Garcia-Diez, A. Toll*

**Hospital del Mar. Parc de Salut Mar, Barcelona, Spain**

## 1. INTRODUCTION

Cutaneous squamous cell carcinoma (cSCC) is the second most common skin cancer and its incidence is increasing all over the world. A recent study estimated that between 186,157 and 419,843 white people were diagnosed of cSCC in the United States in 2012 (Karia et al., 2013). This malignant tumor mainly affects Caucasians and elderly, although its incidence among youth is increasing. Males have more risk of developing the disease than females (Karia et al., 2013; Lomas et al., 2012; Christenson et al., 2005).

Cutaneous SCC originates from the malignant proliferation of keratinocytes located on the epidermis or its appendages. In most cases, it arises from precursor lesions called actinic keratosis and Bowen's disease, although it can also appear the novo or in chronically injured skin. The most important risk factor for the development of cSCC is the chronic exposure to ultraviolet radiation (UVR), thus the majority of cSCC appear on sun-exposed areas such as head, neck, and the back of hands and arms (de Vries et al., 2012; Revenga Arranz et al., 2004). UVR-related mutations inactivating TP53 tumor suppressor gene are very common, both in actinic keratosis and in cSCC (Ratushny et al., 2012). As these mutations also appear in microscopic clusters of keratinocytes of apparently normal sun-exposed skin, they are believed to occur in an early stage of malignant transformation (Hussein, 2005; Jonason et al., 1996). Other risk factors are immunosuppression, genetical predisposition, and human papilloma virus (HPV) infection. However, HPVs role in cSCC development is controversial, especially in immunocompetent patients (Nindl et al., 2007; Arron et al., 2011).

The treatment of choice for cSCCs is surgery and, in some cases, radiotherapy. In most cSCCs, the prognosis is very good, with a 5-year survival rate after surgical excision of 90% (Martorell-Calatayud et al., 2013) and a disease-specific mortality of approximately 1.5% (Karia et al., 2013). Although the metastasis rate of cSCC is relatively low, ranging from 2% to 5% (Martorell-Calatayud et al., 2013), these, which are mainly to the lymph nodes, are associated with a bad prognosis and a median survival time of less than 2 years (Stratigos et al., 2015). More than 60% of metastases from cSCC arising on the head and

*Introduction to Cancer Metastasis*
http://dx.doi.org/10.1016/B978-0-12-804003-4.00011-6

neck spread to parotid lymph nodes. A subtype of cSCC known as high-risk cSCC is associated with a more aggressive clinical behavior, higher frequency of recurrences, metastases, and mortality (Navarrete-Dechent et al., 2015). Recognizing this high-risk cSCC variant is relevant because patients with this subtype of tumor will need quicker and more aggressive treatments as well as closer follow-ups than those with the more common low-risk cSCC.

# 2. DEFINING THE HIGH-RISK VARIANT

There is a series of clinical, histological, and molecular variables that have been associated with high-risk cSCC (Table 11.1). However, there is still a need of large prospective studies to expand our knowledge about the prognostic factors associated with this variant of cSCC.

## 2.1 Clinical

### *2.1.1 Immunosuppression*

Immunosuppression is the patient-dependent factor with the strongest impact on prognosis of cSCC. Patients with human immunodeficiency virus (HIV), hematological malignancy, and solid organ transplantation or other medical conditions, such as inflammatory bowel disease or chronic autoimmune diseases, requiring long-term immunosuppressive therapy, have an increased risk of developing cSCC and these are more aggressive (Tomaszewski et al., 2014; Brewer et al., 2015; NCCN, 2015).

TABLE 11.1 Clinical, Histological Criteria and Molecular Criteria Defining the High-Risk Cutaneous Squamous Cell Carcinoma (cSCC)

| Clinical Features | Histologic Features | Molecular Markers |
|---|---|---|
| Immunosuppression:<br>• Solid organ transplantation (heart and lung > kidney > liver)<br>• Human immunodeficiency virus<br>• Hematological malignancies (chronic lymphocytic leukemia) | Tumor thickness > 4 mm | • EGFR<br>• CKS1B<br>• PTPRD<br>• EMT<br>• IKKalpha |
| Personal history of:<br>• Dystrophic epidermolysis bullosa<br>• Xeroderma pigmentosum<br>• Dyskeratosis congenita<br>• Epidermodysplasia verruciformis<br>• Oculocutaneous albinism | Certain SCC variants:<br>• Desmoplastic<br>• Acantholytic<br>• Spindle cell<br>• Adenosquamous<br>• Pseudoangiosarcomatous | |
| Tumor diameter > 20 mm | Poorly differentiated tumor | |
| Tumor recurrence | Positive surgical margins | |
| Tumor site:<br>• Ear, lip, nose, eyelid, scalp and anogenital<br>• Non-sun-exposed areas | Perineural invasion of nerves with a diameter ≥0.1 mm | |
| Tumor arising on chronic injured skin (scar, burn, ulcers, radiation dermatitis) | Extension beyond dermis into or through subcutaneous fate | |
| | Lymphovascular invasion | |

*EGFR*, Epidermal growth factor receptor; *EMT*, epithelial to mesenchymal transition; *PTPRD*, protein tyrosine phosphatase receptor type D.

#### 2.1.1.1 HEMATOLOGIC MALIGNANCIES

Regarding hematologic malignancies, patients with chronic lymphatic leukemia have the highest risk of developing cSCCs, with an incidence of 2224.9 per 100,000 person-year in a recent study conducted in 282 patients with chronic lymphocytic leukemia (Brewer et al., 2015). Furthermore, their cSCC shows an aggressive behavior, with increased metastatic and mortality rates (Tomaszewski et al., 2014; Mehrany et al., 2005a). They also have a high risk of local recurrences after both standard and micrographic surgery, with a reported cumulative recurrence incidence after 1 and 2 years after standard surgery of 9.5% and 15%, respectively (Tomaszewski et al., 2014; Brewer et al., 2015; Mehrany et al., 2005b). Recently, an increased incidence and recurrence rates in cSCC have also been reported in patients with non-Hodgkin lymphoma in a series of 435 patients, although these rates were lower in comparison with patients with chronic lymphocytic leukemia (Brewer et al., 2015).

#### 2.1.1.2 SOLID ORGAN TRANSPLANT

The incidence of cSCC is increased by 65 times in organ transplant recipients (OTRs) compared to normal population, and the risk of metastasis is increased to 8% (Fig. 11.1). Their tumors are associated with increased morbidity and mortality rates and some patients develop a very high number of cSCC. The risk of developing a cSCC increases along with the time posttransplant, with incidence rates in the United States and Western Europe that increase to 10–27% after 10 years and to 40–70% after 20 years. Risk factors for developing a cSCC in OTRs include Fitzpatrick skin type I to III, cumulative sun exposure, greater ages at transplantation, and the intensity and duration of the immunosuppressive drug regimen (Zwald and Brown, 2011; Chockalingam et al., 2015; Vajdic and van Leeuwen, 2009). Individuals with heart and lung transplants have the higher risk of cSCC development, followed by individuals with kidney transplant, and then, individuals with liver transplant (Zwald and Brown, 2011; Krynitz et al., 2013). The molecular phenotype of cSCC from OTRs is different from that of cSCC appearing in immunocompetent patients. It also seems that the infection by HPV plays a role in their pathogenesis, although the mechanisms involved are still being elucidated (Zwald and Brown, 2011; Chockalingam et al., 2015). Certain immunosuppressive medications (azathioprine, tacrolimus, and cyclosporine) are associated with an increased risk of developing nonmelanoma skin cancer in OTRs. The use of voriconazole has also been associated with more aggressive cSCC in OTRs (Zwald and Brown, 2011; Goyal, 2015; Cowen et al., 2010). In recent years, a new family of immunosuppressants called the mammalian target of rapamycin inhibitors (e.g., sirolimus and everolimus), has been associated with a reduced risk of developing skin cancer (Zwald and Brown, 2011; Chockalingam et al., 2015).

#### 2.1.1.3 HEMATOPOIETIC STEM CELL TRANSPLANT

The incidence of cutaneous malignancies in hematopoietic stem cell transplant (HSCT) recipients has not been so well studied. In a

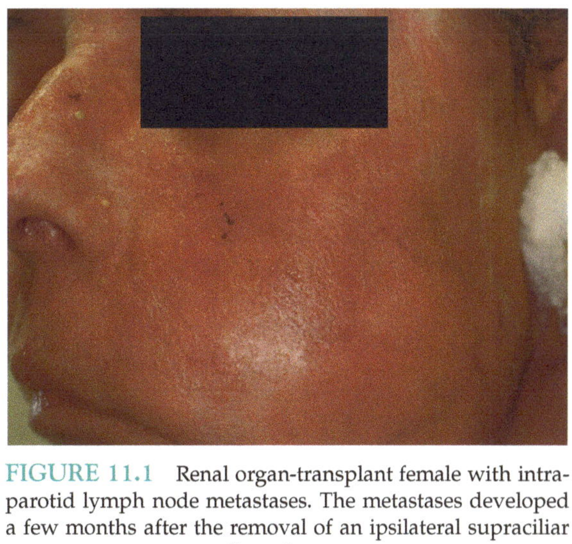

FIGURE 11.1 Renal organ-transplant female with intraparotid lymph node metastases. The metastases developed a few months after the removal of an ipsilateral supraciliar cutaneous squamous cell carcinoma.

recent publication based on a Danish cohort of 3302 patients who underwent HSCT and 4789 patients who underwent renal transplant, an increased risk of cSCC was found in allogenic HSCT recipients, but not in autologous recipients. HSCT recipients had an increased risk of developing cSCC of 18.3 (hazard ratio) in comparison with general population, although they had a lower risk than renal transplant patients (hazard ratio of 34.3 in comparison with general population) (Omland et al., 2015).

#### 2.1.1.4 HUMAN IMMUNODEFICIENCY VIRUS INFECTION

Recently, different studies have reported an increased incidence of nonmelanoma skin cancer in HIV individuals in comparison with HIV-negative individuals. In a population-based cohort study of 2008 the adjusted incidence rate ratio of cSCC in HIV-positive patients was 2.6 (Zhao et al., 2016; Silverberg et al., 2013). In a recent meta-analysis, individuals who received highly active antiretroviral therapy (HAART) had a lower risk of developing nonmelanoma skin cancer than those who had not received HAART (Zhao et al., 2016). Furthermore, HIV infection appears to be related to a more aggressive cSCC, at younger ages and with a higher rate of recurrences (Nguyen et al., 2002; Hausauer et al., 2013). In a small series of 10 HIV-positive patients, with ages ranging from 31 to 54 and with all stages of HIV disease, Nguyen et al. (2002) found that the metastasis, local recurrence, and survival did not correlate with the level of immunosuppression of the patients, and thus, suggested that cSCC in HIV-infected patients should be treated aggressively despite the patient's degree of immunosuppression and HIV prognosis.

### 2.1.2 *Genetic Disorders*

There are many genetic disorders associated to an increased risk of cSCCs, which predispose to cSCCs appearing at younger ages and with an increased aggressivity. Oculocutaneous albinism, xeroderma pigmentosum, epidermodysplasia verruciformis, dyskeratosis congenita, and epidermolysis bullosa (EB) can be included in this category (Martorell-Calatayud et al., 2013; Schaffer and Orlow, 2011; Fine et al., 2009).

Among EB, dystrophic EB has the highest rate of cSCC, especially the Hallopeau-Siemens subtype (Fig. 11.2). The risk of developing a cSCC increases with age and nearly all patients with this variant of dystrophic EB will have at least one cSCC by mid-adulthood (cumulative risk of 90.1% at 55 years old) according to a recent study performed by the National Registry of EB of the United States during years 1986–2002 in 3280 EB patients (Fine et al., 2009). cSCCs of EB patients tend to have a more aggressive behavior with considerable morbidity and mortality. In patients with recessive dystrophic EB, cSCC is the first cause of death (Mellerio et al., 2016). In the same study of the National registry of EB, SCC deaths occurred only in the recessive dystrophic EB variant, with a cumulative risk of death in patients with the Hallopeau variant of 38.7%, 70.0%, and 78.7% by ages 35, 45, and 55, respectively (Fine et al., 2009). Although not as prevalent as in recessive dystrophic epidermolysis bullosa (RDEB), risk of cSCC is also increased in junctional EB. cSCC in RDEB usually appears in areas of chronic cutaneous damage due to

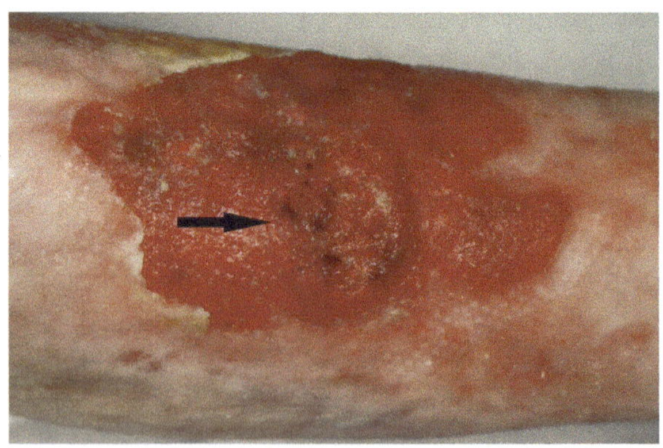

FIGURE 11.2 Squamous cell carcinoma (*arrow*) appearing over a chronic ulcer on the leg of a patient with recessive dystrophic epidermolysis bullosa.

blisters, wounds, or scars. Therefore, the diagnosis of cSCC in these patients may represent a considerable challenge. Taking into account all the above, a close monitoring and an aggressive diagnostic strategy, with the performance of cutaneous biopsies of any suspect lesion, is strongly recommended (Fine et al., 2009; Mellerio et al., 2016; Yuen and Jonkman, 2011).

### 2.1.3 *Cutaneous Squamous Cell Carcinomas Arising at the Site of a Preexisting Lesion*

cSCCs arising in sites of long lasting chronic inflammation such as scars, old burns, venous ulcers, sinus tracks of hidradenitis suppurativa, and radiation dermatitis, have an increased risk of metastasis. This is believed to occur due to a reduction in E-cadherin levels in the tumoral cells that favors their invasion of the dermis (Martorell-Calatayud et al., 2013; Stratigos et al., 2015; SIGN, 2014).

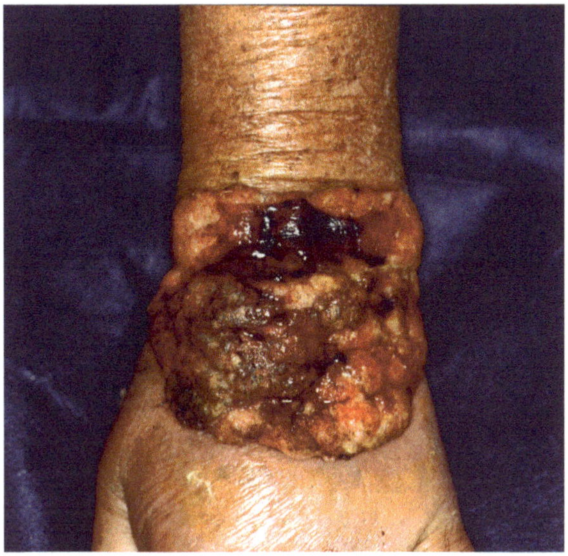

FIGURE 11.3 Large cutaneous squamous cell carcinoma on the wrist.

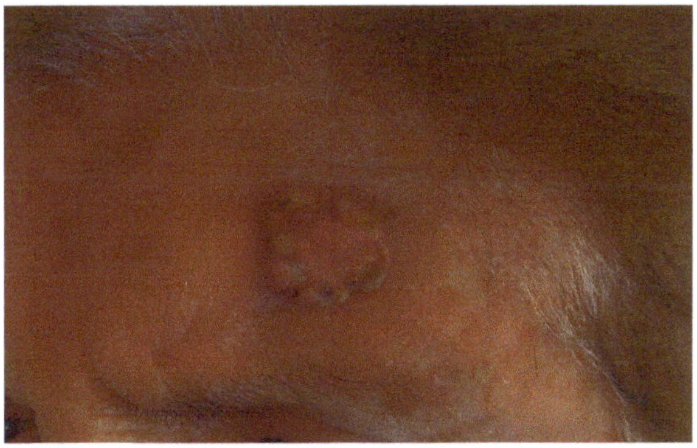

FIGURE 11.4 Large cutaneous squamous cell carcinoma on the frontal area.

### 2.1.4 *Clinical Features of the Tumors*

#### 2.1.4.1 LESION SIZE

Tumors with a diameter above 2cm are considered to have an increased risk of nodal metastasis, recurrences, and death (Figs. 11.3 and 11.4; Brantsch et al., 2008; Ch'ng et al., 2013; Roozeboom et al., 2013; Schmults et al., 2013). In a recent multicenter prospective analysis of perineural invasion in 753 cSCC, the presence of perineural invasion was significantly associated with tumor size, with 70% of tumors with perineural invasion (of nerves with a diameter greater than 0.1mm) showing a diameter greater than 2cm (Campoli et al., 2014).

#### 2.1.4.2 LESION SITE

Recurrence and metastatic risk vary according to the location of the cSCC. However, it is difficult to ascertain the exact risk that the anatomical site confers as far as most published studies include a low number of cases or do not correlate the location risk with the tumor diameters or depth of invasion (Bonerandi et al., 2011).

Tumors involving the lip and ear are considered to have the highest risk of metastasis with a 14% and 9%, respectively, according to a study (Figs. 11.5 and 11.6; Rowe et al., 1992). Anogenital SCCs also have a high metastatic risk (Schmults et al., 2013; Jennings and Schmults, 2010).

Other locations considered of high risk are the scalp, periorificial areas such as nose and eyelids,

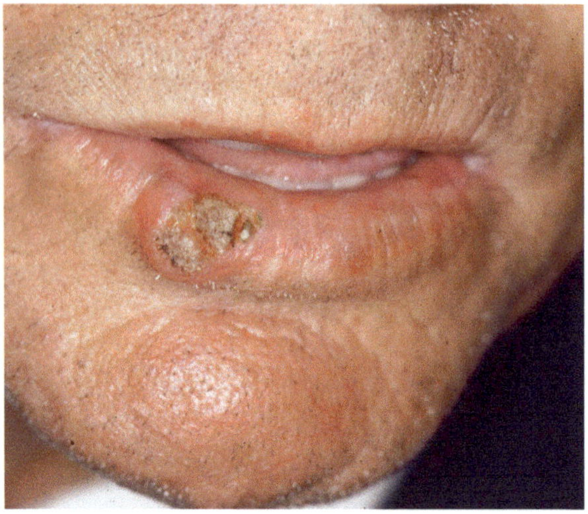

FIGURE 11.5 Cutaneous squamous cell carcinoma in the lip, a high-risk area.

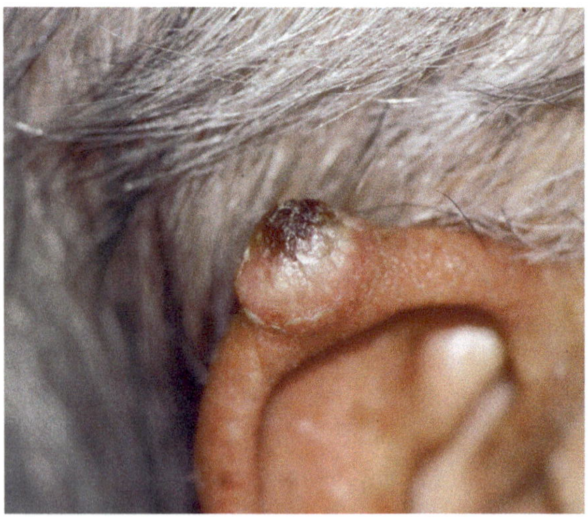

FIGURE 11.6 Cutaneous squamous cell carcinoma appearing on the ear, a high-risk area.

and non-sun-exposed areas like soles of feet, sacrum, and perineum (Bonerandi et al., 2011).

#### 2.1.4.3 RECURRENCE

Recurrent tumors have been associated with an elevated risk of nodal and distance metastasis. Metastatic rates following relapse vary from 25% to 45% depending on the region (Bonerandi et al., 2011; Jennings and Schmults, 2010; Krediet et al., 2015). In addition, recurrent cSCC have been associated with higher rates of perineural, lymphovascular, and subcutaneous tissue invasion. Approximately 75% of recurrences appear during the first 2 years after the initial diagnosis, and this percentage increases to 95% at 5 years (Stratigos et al., 2015). Recurrence has been associated with tumor size and with positive surgical margins (Martorell-Calatayud et al., 2013; Clayman et al., 2005). Desmoplastic cSCC variant and tumoral thickness have also been identified as independent risk factors for local recurrence (Brantsch et al., 2008). In rare cases the recurrence is associated with local lymphotropic micrometastasis (Bonerandi et al., 2011).

### 2.1.5 *Human Papillomavirus*

Beta-HPV is the most common HPV type found in cSCC. Although the etiologic role in the development of cSCC in patients with epidermodysplasia verruciformis and in immunosuppressed patients seems to be clear, its etiological paper in immunocompetent patients is controversial (Nindl et al., 2007). Beta-HPV infection is more commonly detected in cSCC and its precursor lesions in immunocompromised patients than in general immunocompetent population. Because of its presence in precursor lesions of cSCC, it has been suggested that it may play a role in early steps of carcinogenesis. Similarly, beta-HPV-positive cSCC do not seem to have a worse prognosis (Nguyen et al., 2002; McLaughlin-Drubin, 2015; Harwood et al., 2000; Toll et al., 2014).

In contrast, high-risk alpha-HPVs, mainly HPV 16 and HPV 18, are associated to cSCC arising in genital regions, acral sites, and head and neck (Riddel et al., 2011; Barzon et al., 2014; van de Nieuwenhof et al., 2009). The exact mechanisms behind this association remain unknown, but it is believed that these could be related with the capacity of the virus for altering cell cycle regulatory proteins such as p53 and retinoblastoma (pRb) tumor suppressors (Arron et al., 2011; McLaughlin-Drubin, 2015). Several studies suggest that alpha HPV confer a better prognosis in SCC arising in penile and head and neck areas (Ferrandiz-Pulido et al., 2013; Nevens and Nuyts, 2015).

## 2.2 Histologic Features of High RISK

### 2.2.1 *Tumor Thickness*

Tumor thickness is one of the most relevant independent predictor of metastasis in cSCC, with greater thickness associated with higher risk (Brantsch et al., 2008; Alam and Ratner, 2001; Carter et al., 2013). A prospective study by Brantsch et al. (2008) found that no tumors less than 2 mm in thickness metastasized, while metastases were observed in 4% of 2.1–6.0 mm thick tumors and in 16% of tumors with thickness >6.0 mm. Many other retrospective studies support these findings and show that tumor depth is of critical importance in detecting high-risk tumors (Schmults et al., 2013; Jensen et al., 2010; Kyrgidis et al., 2010). While the American Joint Committee on Cancer (AJCC) uses >2 mm tumor depth as one indicator of high-risk cSCC, the Brigham and Women's hospital (BWH) uses the invasion into subcutaneous fat as a risk factor (Jambusaria-Pahlajani et al., 2013). Tumor depth and anatomical level (invasion of subcutaneous fat) should be reported by pathologists.

### 2.2.2 *Perineural Invasion*

Perineural invasion (PNI) is found in 2.5–5.0% of primary cSCC, when tumor cells invade the perineurium surrounding a peripheral nerve (Fig. 11.7; Leibovitch et al., 2005). PNI is a prognostic marker for both local recurrence (Campoli et al., 2014; Kyrgidis et al., 2010; Jambusaria-Pahlajani et al., 2013), metastatic risk, and disease-specific survival (Clayman et al., 2005; Cherpelis et al., 2002; Moore et al., 2005; Garcia-Serra et al., 2003; Brougham et al., 2012; Goepfert et al., 1984) and it is included in the BWH classification system and in the AJCC revised TNM staging system for cSCC as a high-risk feature. Campoli et al. (2014) showed incidental PNI to be associated with palpable lymphadenopathy in a prospective study including 753 cSCC.

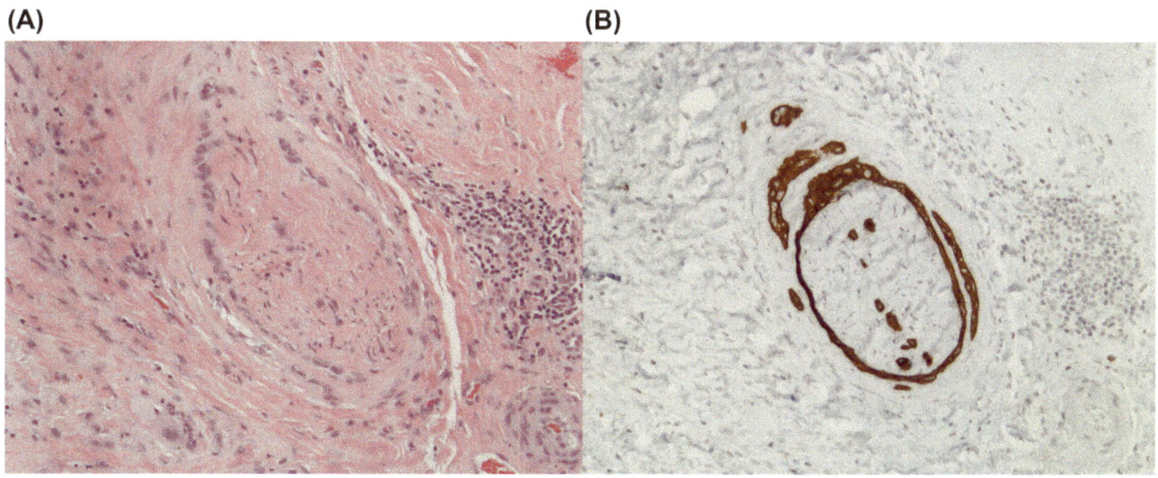


FIGURE 11.7 (A) Perineural invasion of a large nerve by squamous cell carcinoma. (B) S-100 immunostaining showing perineural invasion in the same patient.

Although there is some consensus that PNI of large caliber and/or named nerves is associated with risk of recurrence and metastases, the diameter of the nerves that allows classifying PNI in low or high risk is still controversial. In a small retrospective cohort study, Ross et al. (2009) found that invasion of nerves less than 0.1 mm in diameter was associated with significantly lower risks of recurrence, metastasis, and disease-specific and overall survival. Similarly, a nerve caliber >0.1 mm has been associated with an increased risk of metastasis on univariate analysis in a retrospective cohort study by Carter et al. (2013). However, this association was not found on the multivariate analysis and may be due in part to multiple other risk factors associated with large-caliber nerve invasion. In a prospective work by Campoli et al. (2014), the observation of microscopic invasion of large diameter nerves (>0.1) was also associated with other high-risk variables such as tumor diameter of 2 cm or greater, but not directly with metastatic risk. Taking into account a retrospective study including 1980 cSCC by Karia et al. (2014), only large-caliber nerve invasion has been included as a risk factor in the BWH staging system.

### 2.2.3 *Lymphovascular Invasion*

Lymphovascular invasion has been implicated as a high-risk factor in cSCC (Moore et al., 2005) and is associated with PNI and desmoplasia (Brantsch et al., 2008; Carter et al., 2013), but its impact on survival is controversial (Clayman et al., 2005; Brougham et al., 2012; Ebrahimi et al., 2013) and has not been considered for the revised AJCC staging system.

cSCC usually spread to regional lymphatics before reaching other organ systems and the presence of micro- or macrometastasis is prognostically significant. The demonstration of tumor-associated lymphangiogenesis has been demonstrated to be a prognostic indicator in cancer (Frech et al., 2009; Dadras et al., 2003). On the other hand, intravascular/lymphatic invasion also correlates with metastatic risk. However, distinguishing invasion of blood vessels from invasion of lymphatic vessels (LVs) had classically been extremely difficult.

New markers for LVs, such as lymphatic endothelial hyaluron receptor-1 (LYVE-1), podoplanin (D2-40), vascular endothelial growth factor-C (VEGF-C), and prospero homeobox protein 1 (PROX1) are now available and may aid in detecting intralymphatic tumor invasion or lymphatic density. Peritumoral lymphangiogenesis, which can be measured with LV density, is significantly associated with lymph node metastasis and poor prognosis in several cancers (Gao et al., 2008; Nakamura et al., 2005; Roma et al., 2006; Pastushenko et al., 2015). Moreover, increasing lymphangiogenesis in regional lymph nodes may also increase the ability of cancer cells to spread, as shown in melanomas (Morton et al., 1992; Liersch et al., 2012).

More recently, several lymphatic endothelial markers, such as LYVE-1, have been used to enhance the detection of LVs in SCC (Moussai et al., 2011). Moussai et al. (2011) found increased lymphatic density and reorganized lymphatic endothelial vessels in the dermis immediately adjacent to SCC nests. LYVE-1 has successfully been used to detect an increase in lymphatic density as a significant prognostic marker of metastatic risk in head and neck squamous cell carcinoma (Frech et al., 2009).

Several studies in animal tumor models have provided direct experimental evidence that increased levels of VEGF-C or VEGF-D, which mainly act through activation of VEGFR-3, promote tumor lymphangiogenesis and tumor spread through LVs to regional lymph nodes (Alitalo and Carmeliet, 2002). Moussai et al. (2011) found that VEGF-C may be produced by peritumoral macrophages in primary cSCC. In a transgenic mouse model of chemically induced skin carcinogenesis, Hirakawa et al.

(2007) showed that VEGF-C favors the expansion of lymphatic networks in sentinel lymph nodes. Moreover, tumors that expressed VEGF-C were more prone to metastasize to additional organs. This suggests that lymphangiogenic factors may be transported to lymph nodes from the primary tumor and induce intranodal lymphangiogenesis, to facilitate the spread of tumor cells (Zhao et al., 2012). Moreover, increased LV density and VEGF-C expression have been found to be predictors of lymph node metastasis in supraglottic and oral SCC (Baek et al., 2009; Sugiura et al., 2009).

Prox1 is a homeodomain transcription factor that induces the reprogramming of cell fate from venous endothelial cells to lymphatic progenitors (Oliver et al., 1993). Prox1 induces lymphangiogenesis by activating VEGFR-3 (Flister et al., 2010). However, the positivity of Prox1 in epithelial tumors precludes its utility as a marker of lymphatic density (Sasahira et al., 2014). Similarly to what is observed with D2-40, Prox1 expression correlates with prognosis in some tumors such as oral squamous cell carcinoma (Sasahira et al., 2014).

Podoplanin is a mucin-type glycoprotein expressed in lymphatic but not blood vascular endothelial cells. Mab D2-40 is a monoclonal antibody against human podoplanin and may help in identifying tumor emboli within LVs in cSCC (Toll et al., 2012). However, nonlymphatic tissues such as peritoneal mesothelial cells, osteocytes, glandular myoepithelial cells, and ependymal cells may also express podoplanin (Schacht et al., 2005). Moreover, neoplastic cSCC cells also show D2-40 expression frequently (Schacht et al., 2005), thus hampering the visualization of lymphatics and precluding this technique as an adequate tool to predict lymphatic metastasis in cSCC. More importantly, the intensity of D2-40 expression in cSCC cells, especially in the infiltrative areas of the tumor, correlates with metastatic risk (Toll et al., 2012; Kreppel et al., 2013). In a series of 101 cSCC, including 51 cases that had developed lymphatic metastases, we found moderate to intense immunohistochemical expression of D2-40 in 49% of metastatic primary cSCC, significantly increased when compared with 16% of nonmetastatic cSCC (Fig. 11.8; Toll et al., 2012). Similar findings were further confirmed by Kreppel et al. (2013) in cutaneous cSCC and animal models have also shown that podoplanin overexpression correlates with malignant progression (Durchdewald et al., 2008). These results not only suggest that podoplanin may be a good marker of metastatic risk in cSCC but also that it plays an important role in lymphatic spread, tumor cell invasion, and tumor progression. A hypothetical transendothelial function or transdifferentiation of keratinocytes, a phenomenon that has been observed in melanoma (Hendrix et al., 2002), might underlay the overexpression of podoplanin in metastatic cSCC. However, it is well-known that podoplanin is involved in the remodeling of the cytoskeleton of tumor cells, thus promoting cell invasion (Thiery et al., 2009; Friedl and Alexander, 2011; Fernandez-Munoz et al., 2011). Podoplanin interacts with actin through ezrin and moesin. Podoplanin also increases the activity of Rho

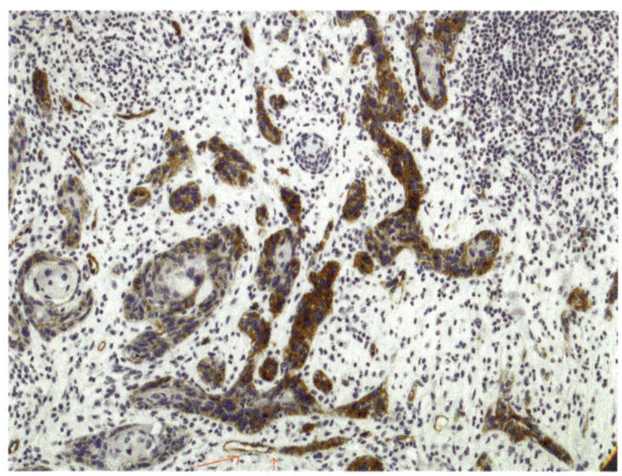

FIGURE 11.8 D2-40-expressing cutaneous squamous cell carcinoma. *Red arrows* in the bottom show positive lymphatic vessels.

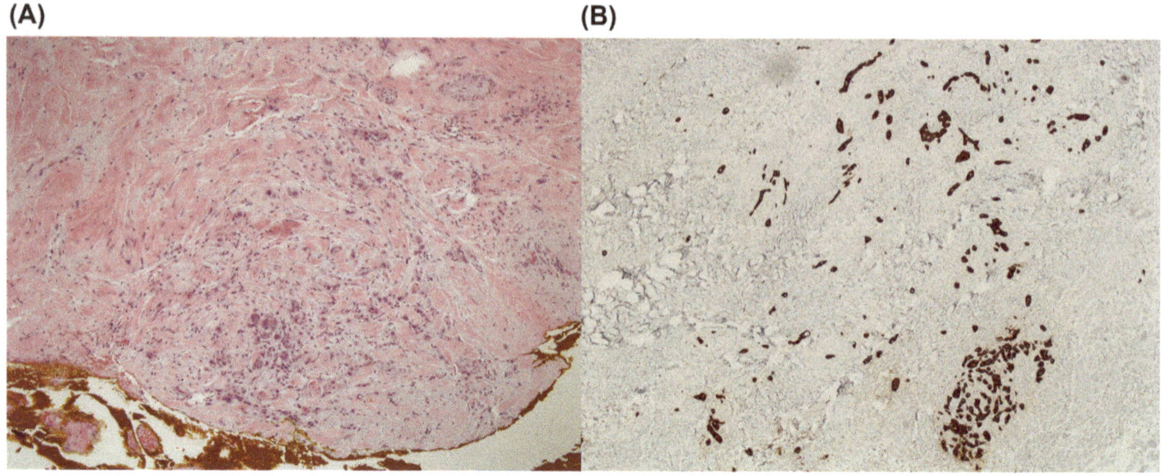


FIGURE 11.9 (A) Poorly differentiated cutaneous squamous cell carcinoma. (B) Immunohistochemical overexpression of the pan-cytokeratin panel AE1-AE3.

GTPases such as RhoA and is able to mediate tumor cell invasion in presence or absence of epithelial–mesenchymal transition (EMT).

### *2.2.4 Histologic Differentiation*

Poorly differentiated cSCC (Broders classes 3 and 4) correlates with metastatic risk (Fig. 11.9; Toll et al., 2012; Kyrgidis et al., 2010; Jambusaria-Pahlajani et al., 2013; Breuninger et al., 1990; Petter and Haustein, 2000). Rowe et al. (1992) showed that the metastatic rate triples with worsening histologic grade, reporting 32.8% and 9.2% for poorly differentiated and well-differentiated cSCC, respectively. A three-item system (well differentiated, moderately differentiated, and poorly differentiated) to report tumor differentiation on the most poorly differentiated area in the tumor in primary cSCC is encouraged by many pathology associations such as the Royal College of Pathologists.

### *2.2.5 Histologic Subtype*

The presence of desmoplasia (Fig. 11.10) has been associated to metastatic risk by several authors (Breuninger et al., 1997, 2012) and the prospective study by Brantsch et al. (2008) found it to be significantly associated

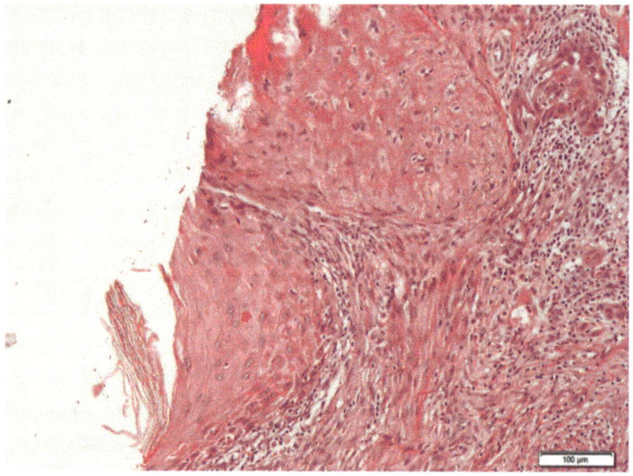

FIGURE 11.10 Cutaneous squamous cell carcinoma showing areas of desmoplasia.

to recurrence in the multivariate analysis. To classify a cSCC as being of the desmoplastic subtype, at least one-third of the tumor should show this phenotype. The World Health Organisation Classification and the Royal College of Pathologists Dataset consider the adenosquamous, pseudoangiosarcomatous, acantholytic, and spindle cell SCC arising after radiotherapy as high-risk subtypes (Fig. 11.11; LeBoit et al., 2008; Slater, 2012).

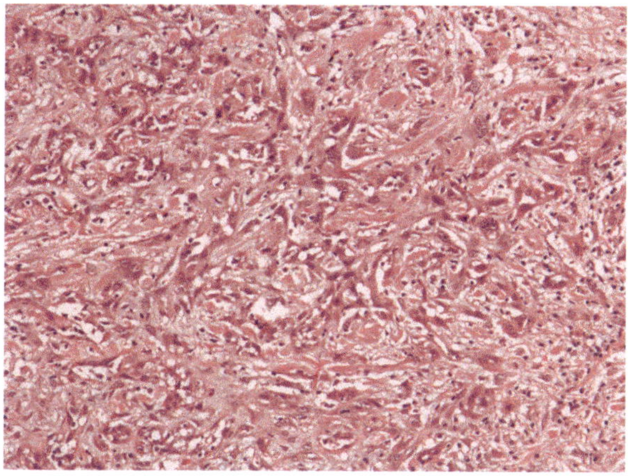

FIGURE 11.11 Spindle cell cutaneous squamous cell carcinoma.

## 2.3 Molecular High-Risk Markers

Although there are several candidate molecular markers of metastatic risk, such as the above-mentioned D2-40, these have to be validated in larger series and up to now their study may better be considered as an aid to better understand the mechanisms that favor the metastatic process in cSCC. Potential molecular surrogates of metastatic risk are EGFR, CDC28 protein kinase regulatory subunit 1B (CKS1B), EMT markers, and IKappaB kinase alpha (IKKalpha).

### 2.3.1 *Epidermal Growth Factor Receptor*

Epidermal growth factor receptor (EGFR) overexpression has been associated with metastatic potential in cSCC (Ch'ng et al., 2008), which might have relevant implications in the response to EGFR-targeted therapies. However, EGFR overexpression is also frequently observed in nonmetastatic cSCC and its utility as a prognostic marker is controversial (Sweeny et al., 2012).

### 2.3.2 *Cytogenetic Alterations: CKS1B and Protein Tyrosine Phosphatase Receptor Type D*

Cytogenetic alterations and chromosomal aberrations are more prevalent in the late stages of carcinogenesis. Our group found that amplification in CKS1B, a member of the highly conserved cyclin kinase subunit 1, determined by fluorescence in situ hybridization, is associated with an increased metastatic risk in cSCC (Salgado et al., 2010). CKS1B facilitates the proteolysis of p27, a protein that is involved in DNA double-strand break repair. Frequent deletions and somatic mutations (37%, respectively) in the protein tyrosine phosphatase receptor type D (PTPRD) gene have also been observed in metastatic cSCC (Lambert et al., 2012). PTPRD belongs to a family of 21 transmembrane protein tyrosine phosphatase receptors implicated in ligand-controlled protein dephosphorylation, cell–cell adhesion, and cell–matrix adhesion. PTPRD dephosphorylates the pStat3 protein and its inactivation may contribute to tumorigenesis.

### 2.3.3 *Epithelial–Mesenchymal Transition*

Although cells can reach the lymphatic circulation collectively, the optimal hematogenous dissemination is achieved when they migrate individually, through an EMT. EMT is an essential process required for organizing the cell layers during embryogenesis and also relevant in wound healing (Yan et al., 2010) and cancer progression (Friedl and Alexander, 2011). EMT is characterized by a functional loss of E-cadherin and the acquisition by epithelial tumor cells of traits of motile mesenchymal cells with an increased capacity of invasion, intravasation, and vascular dissemination. Cells that have undergone EMT can reverse this process in what is known as mesenchymal epithelial transition (Toll et al., 2013); it has been proposed that this conversion may be common once the tumor cells have seeded in the metastatic organ (Biddle and Mackenzie, 2012). Expression of EMT markers has been associated with the induction of metastasis in head and neck (Yang and Wu, 2008; Mandal et al., 2008) colorectal (Xiong et al., 2012), esophageal (Usami et al., 2008), and breast cancer tumors (Gjerdrum et al., 2010; Martin et al., 2005).

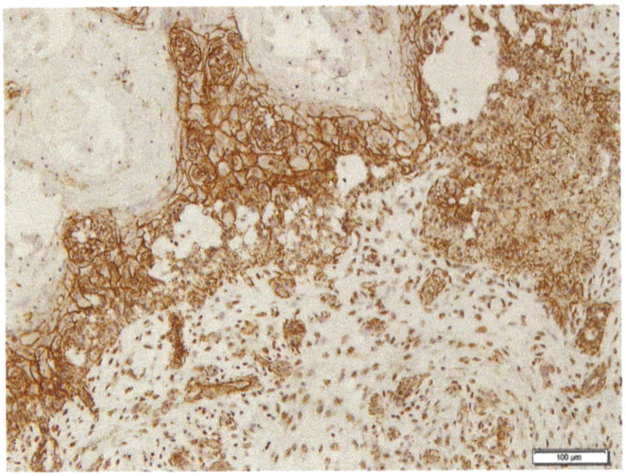

FIGURE 11.12 Membrane E-cadherin is being lost in deeper areas of the tumor.

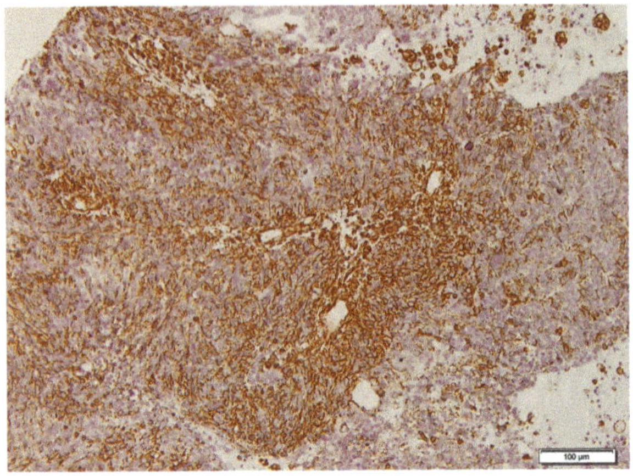

FIGURE 11.13 Cutaneous squamous cell carcinoma showing strong immunohistochemical expression of vimentin.

E-cadherin downregulation is the typical trait of an EMT (Fig. 11.12; Barrette et al., 2014). E-cadherin, a 120-kDa calcium-binding transmembrane glycoprotein, is the main component of the adhesion molecules that mediate anchoring intercellular junctions. The cytoplasmic domain of E-cadherin interacts with beta-catenin that participates in E-cadherin association to the actin cytoskeleton and the formation of adherens junctions (Takayama et al., 1996). Moreover, E-cadherin retains beta-catenin in the cell membrane, preventing its translocation to the nucleus and its activity as a transcriptional activator. The proteolytic cleavage of E-cadherin leads to loss of cell–cell adhesion and increases the cytosolic and nuclear pools of beta-catenin. During EMT, E-cadherin expression is downregulated by the action of several EMT-associated transcription repressors, such as members of the Snail, Twist, and Zeb families, that also promote the gain of mesenchymal markers, such as vimentin.

Loss of membrane E-cadherin and its translocation from the membrane to the nucleus can be observed in metastatic SCC (Toll et al., 2013). This translocation may be mediated by Twist. However, the expression of vimentin is a better indicator of EMT and metastatic risk (Fig. 11.13; Arumugam et al., 2012; Martins et al., 2009; Walsh et al., 2011). The complete loss of cell–cell junctions, or "complete" EMT, is not necessary to induce aggressive forms of cSCC and vimentin expression can be observed in tumors showing strong keratinocyte cohesion (Toll et al., 2013).

### 2.3.4 *IKKalpha*

Activation of nuclear factor kappa B (NF-κB) and the IKK complex is commonly found associated with inflammation-related cancers and is responsible for the promotion of tumor cell survival and cancer progression in different models (Bradford and Baldwin, 2014). The IKK complex is composed of IKKα, IKKβ, and NEMO, being the IKKβ subunit essential for NF-κB activation and for the regulation of the immune system and organism fitness (Baldwin, 2012; Delhase et al., 2000; Karin and Delhase, 2000). IKKα is not required for these functions, but it is necessary to signal through the alternative NF-κB pathway (that signals through the p52/RelB heterodimer), and to phosphorylate specific nuclear substrates such as histone H3, silencing mediator for retinoid and thyroid hormone receptors, and N-CoR thus facilitating NF-κB-dependent and NF-κB-independent transcription (Anest et al., 2003,

2004; Yamamoto et al., 2003; Margalef et al., 2012; Fernandez-Majada et al., 2007a,b). We found that the immunohistochemical overexpression of nuclear active IKK is a good indicator of metastatic risk both as an independent parameter and together with the degree of differentiation and recurrence (Toll et al., 2015). Moreover, the fact that nuclear IKK activation is maintained in the lymph node metastases opens the possibility of using IKK inhibitors to treat patients with metastatic cSCCs. Nuclear active IKKα regulates the expression and function of the metastasis-related gene *Maspin* (Luo et al., 2007). We found that Maspin protein, previously detected in approximately 28% of cSCC, is consistently reduced in metastatic cSCC negatively correlating with nuclear IKK activation (Toll et al., 2015).

## 3. STAGING SYSTEMS

Revised tumor (T) staging systems for cSCC were published by the AJCC and the International Union Against Cancer (UICC) in 2010 based on expert consensus (Edge et al., 2010; Breuninger et al., 2012; Sobin and Wittekind, 2009) (Table 11.2). A recent alternative staging system by the group of BWH has been proposed that would better predict outcome (Jambusaria-Pahlajani et al., 2013; Karia et al., 2014). This group found that most poor outcomes occurred in AJCC T2 tumors and that subdividing them in low-risk T2a and high-risk T2b categories offered improved homogeneity and monotonicity over AJCC staging (Table 11.2). The factors that better predicted

TABLE 11.2 Summary of the AJCC, UICC, and BWH Tumor (T) Staging Systems

| Tumor Staging System | Definition |
|---|---|
| *AJCC* | |
| T1 | Tumor ≤2 cm in greatest dimension with fewer than two high-risk factors[a] |
| T2 | Tumor >2 cm in greatest dimension or with two or more high-risk factors[a] |
| T3 | Tumor with invasion of orbit, maxilla, mandible, or temporal bones |
| T4 | Tumor with invasion of other bones or direct perineural invasion of skull base |
| *UICC* | |
| T1 | Tumor ≤2 cm in greatest dimension |
| T2 | Tumor >2 cm in greatest dimension |
| T3 | Tumor with invasion of deep structures (e.g., muscle, cartilage, bone —excluding axial skeleton, orbit) |
| T4 | Tumor with invasion of axial skeleton or direct perineural invasion of skull base |
| *BWH* | |
| T1 | 0 high-risk factors[b] |
| T2a | 1 high-risk factor |
| T2b | Two to three high-risk factors |
| T3 | ≥4 high-risk factors |

*AJCC*, American Joint Committee on Cancer; *BWH*, Brigham and Women´s Hospital; *UICC*, International Union Against Cancer.
[a] *AJCC high-risk factors: >2 mm thickness, Clark level ≥IV, perineural invasion, primary site nonhair bearing lip, or poorly differentiated histology.*
[b] *BWH high-risk factors: tumor diameter ≥2 cm, poorly differentiated histology, perineural invasion ≥0.1 mm, or tumor invasion beyond fat (excluding invasion which upgrades tumor to stage T3).*

high-risk were tumor diameter ≥2 cm, poorly differentiated histology, perineural invasion ≥0.1 mm, or tumor invasion beyond fat. In a work that reviewed the literature, the BWH staging system also better predicted sentinel lymph node (SLNB) positivity (Schmitt et al., 2014). T2b category of the BWH staging system appears to more precisely delineate lesions that may warrant consideration for SLNB, with a 29.4% of positivities when compared with only 7.1% of positive SLNB in T2a tumors (Schmitt et al., 2014).

## 4. TREATING HIGH-RISK CUTANEOUS SQUAMOUS CELL CARCINOMA

Patients with cSCC considered as high risk should be referred to a multidisciplinary committee, including dermatologists, oncologists, radiotherapists, and surgeons, to coordinate the best strategy of treatment.

Surgical excision, with direct or delayed repair, is the first line treatment for the high-risk variant of cSCC. It allows the confirmation of the complete excision of the lesion (free histological peripheral and deep margins) and provides histological information of the type of tumor (Stratigos et al., 2015; NCCN, 2015; SIGN, 2014; Jennings and Schmults, 2010).

For high-risk tumors a clinical peripheral margin of at least 6 mm is considered adequate, in cases in which this is surgically possible and clinically appropriate (Martorell-Calatayud et al., 2013; SIGN, 2014; Jennings and Schmults, 2010; Brodland and Zitelli, 1992).

The excision should be aimed to preserve functional and cosmetic results. In cases of recurrent tumors or tumors in a location with important aesthetic or functional risk, Mohs micrographic surgery is considered the treatment of choice. In this surgical technique frozen horizontal sections of the peripheral and deep margins are obtained, allowing to check 100% of tumor margins. This technique ensures complete clearance on the margins and on the other hand it enables sparing nontumoral tissue (Stratigos et al., 2015; NCCN, 2015; SIGN, 2014; Pugliano-Mauro and Goldman, 2010).

No evidence has been identified to assess an optimum deep margin, although findings suggest that a thinner margin could be associated with an increased number of recurrences (Jenkins et al., 2014; Khan et al., 2013). The deep excision should include the hypodermis, without removing the aponeurosis, perichondrium and periosteum when these are not invaded by the tumor (Stratigos et al., 2015).

Primary radiotherapy should not be considered as a first-line treatment in patients with high-risk cSCC because it carries poorer results than surgical excision with a high rate of early and aggressive recurrences. It can also induce iatrogenic carcinogenesis in the irradiated area. Therefore, it should be reserved for individual patients in whom a complete surgical excision would be extremely difficult to perform, especially in older patients with associated comorbidities, and for cases in which the surgery would entail unacceptable aesthetic or functional results (Stratigos et al., 2015; SIGN, 2014).

Adjuvant radiotherapy should be considered in patients with involved or very close surgical margins, in whom a new surgical excision could lead to an increased risk of complications, and in cases with extensive perineural or large nerve involvement (Stratigos et al., 2015; NCCN, 2015; SIGN, 2014).

With respect to adjuvant systemic treatment, there is insufficient evidence to support its use in high-risk cSCC (Stratigos et al., 2015; SIGN, 2014; Jennings and Schmults, 2010; DeConti, 2012; Apisarnthanarax et al., 2011; Tanvetyanon et al., 2015). The use of systemic retinoids alone or in combination with interferon alpha, showed no benefits in randomized trials (Jennings and Schmults, 2010; Brewster et al., 2007). The information regarding the use of chemotherapy in high-risk cutaneous SCC comes mainly from small case series. As there are no solid data to support the role of chemotherapy in primary high-risk cSCC, this is recommended to be used only in clinical trials (SIGN, 2014; DeConti, 2012).

## 5. SENTINEL NODE BIOPSY

Currently, the use of sentinel node biopsy in cSCC is controversial (Stratigos et al., 2015; Navarrete-Dechent et al., 2015; NCCN, 2015; SIGN, 2014; Krediet et al., 2015; Schmitt et al., 2014; Takahashi et al., 2014; Renzi et al., 2007; Gore et al., 2015). Although there is a lack of randomized controlled clinical trials evaluating sentinel node biopsy in high-risk cSCC, there is some evidence that supports considering sentinel node biopsy in this type of cSCC, especially in tumors classified as T2 in AJCC-7 or T2b in the alternative TNM staging system (Navarrete-Dechent et al., 2015; Schmitt et al., 2014). In these cases it could help identifying occult nodal metastasis and complete the patients staging. However, in a recent retrospective study, 6 out of 17 patients that underwent sentinel node biopsy developed metastasis (three nodal and three extranodal) despite a negative result (Krediet et al., 2015). Therefore, they concluded that sentinel node biopsy had a low sensitivity for predicting metastasis risk. The majority of patients who developed metastasis had cSCC located on the head area. Performing sentinel node biopsy of cSCC located in the head and neck (75–90% of cases) implies certain specific difficulties associated with this anatomic location, which might compromise the utility of the technique (Navarrete-Dechent et al., 2015; Krediet et al., 2015). Sentinel node biopsy has in its favor that is a secure technique with a low rate of complications (Navarrete-Dechent et al., 2015; Gore et al., 2015). In conclusion, more studies are needed to prove the benefits of sentinel node biopsy in the prognosis and treatment outcomes in high-risk cSCC. Meanwhile, this technique should be conducted in high-risk cSCC patients in the context of a clinical trial.

## 6. TREATMENT OF LOCALLY ADVANCED OR METASTATIC CUTANEOUS SQUAMOUS CELL CARCINOMA

In cases with lymph node metastasis, regional lymph node resection is recommended as the first choice treatment. If parotid nodes are involved, a superficial parotidectomy concomitantly with the nodal dissection should be performed. After lymphadenectomy, adjuvant radiotherapy should be considered in all patients with regional disease of the head and neck and the trunk and extremities. In these two last areas it is particularly advised when multiple positive nodes or extracapsular tumor extension is observed. In patients in whom surgical lymph node removal is not possible, local radiotherapy, sometimes with concurrent chemotherapy, is considered as the second-line treatment (Stratigos et al., 2015; NCCN, 2015). Radiotherapy is used as a palliative treatment for pain and hemorrhage control as well as for limiting the extension of the tumor to potentially critical areas such as the ocular orbit (Stratigos et al., 2015).

Satellite or in-transit metastases should be removed when possible, when this is not an option, radiotherapy alone or in combination with chemotherapy could be used (Stratigos et al., 2015).

Electrochemotherapy, a technique in which an electrical field, is generated inside of the tumor for increasing tumoral cells permeability to drugs, might be an option in locally advanced cSCC with cutaneous and subcutaneous lesions. Commonly used drugs in electrochemotherapy are bleomycin and cisplatin. It has the advantage of having a low rate of side effects and toxicity.

Its results in the treatment of advanced cSCC are variable among series, with local responses in a range of 20–70% (Stratigos et al., 2015; Testori et al., 2010; Kreuter et al., 2015).

Treatment of cSCC with distance metastasis or unresectable locally advanced disease (stage IV cSCC) is not well standardized, there are few reports in the literature and it is usually treated following the same treatment guidelines used for head and neck SCC. Although palliative chemotherapy is indicated in these patients, there is not an established chemotherapy regimen and the same regimens used for treating head and neck SCC are usually recommended. Most information available about the use of chemotherapy in this patients comes from adjuvant phase III trials, phase II trials, and a number of case and series reports (Cranmer et al., 2010). Chemotherapy agents that had been reported in cSCC include 5-fluorouracil, bleomycin, methotrexate, adriamycin, taxans, gemcitabine, capecitabine, ifosfamide, and platin derivatives such as cisplatin and carboplatin, alone or in multiple combinations (Behshad et al., 2011). Chemotherapy has been used alone or in association with radiotherapy or surgery (Stratigos et al., 2015; DeConti, 2012; Cranmer et al., 2010). As most patients with cSCC are elderly, chemotherapy's toxic effects are an important limitation in these patients. Thus, in general, monotherapy is preferred to polychemotherapy because of its safer toxic effects profile, and polychemotherapy should be reserved for selected cases requiring a more aggressive treatment. In most of the cases reported in the literature chemotherapy is not curative, being its effect temporary and followed by rapid recurrences (Stratigos et al., 2015; NCCN, 2015; DeConti, 2012). In addition to intravenous chemotherapy, intra-arterial chemotherapy has also been studied. This modality of chemotherapy has fewer side effects because it delivers the drug directly to the tumor site. It has been used principally in cSCC of the lip via superficial temporary artery (myticin/plazomicin) and via facial artery (methotrexate), and in cSCC of the lower extremities, via hyperthermic isolated limb perfusion (TNF-alpha, INFgamma, and melphalan; Stratigos et al., 2015; Behshad et al., 2011). In a 2010 review of the available evidence for the treatment of metastatic and unresectable cSCC, they concluded that there was insufficient evidence for the recommendation of a specific chemotherapy regimen, although they recommended using fluoropyrimidine (e.g., 5-fluorouracil and capecitabine) and platinum-based regimens due to their more secure and available profile (Cranmer et al., 2010).

Other systemic treatments that have been investigated include 13-*cis*-retinoic acid in combination with interferon 2-alpha and EGFR inhibitors (gefitinib, cetuximab, erlotinib, panitumumab, and dasatinib; Cranmer et al., 2010). The combination of 13-*cis*-retinic acid plus interferon 2-alpha with or without cisplatin, was performed in 2 phase II studies, without good results in distant metastatic disease but some activity in locally advanced disease (Lippman et al., 1992; Shin et al., 2002). EGFR inhibitors have been approved for the treatment of head and neck SCC. There are studies of phase II and III and case series/case reports evaluating the use of erlotinib, gefitinib, panitumumab, and cetuximab for cSCC, mostly alone but also in combination with chemotherapy or radiotherapy (Stratigos et al., 2015; Cranmer et al., 2010; Foote et al., 2014; Falivene et al., 2014; Reigneau et al., 2015; Mecca et al., 2012). Given that there is still not enough evidence for their use, they should be considered when chemotherapy fails, as a second line treatment (Stratigos et al., 2015; DeConti, 2012).

## 7. CONCLUDING REMARKS

The incidence of cSCC is increasing. There are several histological and clinical markers of high-risk cSCC and recently some authors have re-evaluated them to propose a new staging system that better predicts metastases. However, there is still a lack of reliable molecular markers that aid in better predicting those lesions at risk of metastasizing. Similarly, future studies that improve our knowledge of the pathways that drive the metastatic process in cSCC may aid in the development of specific targeted treatments.

## References

Arron, S.T., Ruby, J.G., Dybbro, E., Ganem, D., Derisi, J.L., 2011. Transcriptome sequencing demonstrates that human papillomavirus is not active in cutaneous squamous cell carcinoma. J. Invest. Dermatol. 131 (8), 1745–1753.

Alam, M., Ratner, D., 2001. Cutaneous squamous-cell carcinoma. N. Engl. J. Med. 344 (13), 975–983.

Alitalo, K., Carmeliet, P., 2002. Molecular mechanisms of lymphangiogenesis in health and disease. Cancer Cell 1 (3), 219–227.

Arumugam, A., Walsh, S.B., Xu, J., Afaq, F., Elmets, C.A., Athar, M., 2012. Combined inhibition of p38 and Akt signaling pathways abrogates cyclosporine A-mediated pathogenesis of aggressive skin SCCs. Biochem. Biophys. Res. Commun. 425 (2), 177–181.

Anest, V., Cogswell, P.C., Baldwin Jr., A.S., 2004. IkappaB kinase alpha and p65/RelA contribute to optimal epidermal growth factor-induced c-fos gene expression independent of IkappaBalpha degradation. J. Biol. Chem. 279 (30), 31183–31189.

Anest, V., Hanson, J.L., Cogswell, P.C., Steinbrecher, K.A., Strahl, B.D., Baldwin, A.S., 2003. A nucleosomal function for IkappaB kinase-alpha in NF-kappaB-dependent gene expression. Nature 423 (6940), 659–663.

Apisarnthanarax, S., Dhruva, N., Ardeshirpour, F., Tepper, J.E., Shores, C.G., Rosenman, J.G., et al., 2011. Concomitant radiotherapy and chemotherapy for high-risk nonmelanoma skin carcinomas of the head and neck. Int. J. Surg. Oncol. 2011, 464829.

Brewer, J.D., Shanafelt, T.D., Khezri, F., Sosa Seda, I.M., Zubair, A.S., Baum, C.L., et al., 2015. Increased incidence and recurrence rates of nonmelanoma skin cancer in patients with non-Hodgkin lymphoma: a Rochester epidemiology project population-based study in Minnesota. J. Am. Acad. Dermatol. 72 (2), 302–309.

Brantsch, K.D., Meisner, C., Schonfisch, B., Trilling, B., Wehner-Caroli, J., Rocken, M., et al., 2008. Analysis of risk factors determining prognosis of cutaneous squamous-cell carcinoma: a prospective study. Lancet Oncol. 9 (8), 713–720.

Bonerandi, J.J., Beauvillain, C., Caquant, L., Chassagne, J.F., Chaussade, V., Clavere, P., et al., 2011. Guidelines for the diagnosis and treatment of cutaneous squamous cell carcinoma and precursor lesions. J. Eur. Acad. Dermatol. Venereol. 25 (Suppl. 5), 1–51.

Barzon, L., Cappellesso, R., Peta, E., Militello, V., Sinigaglia, A., Fassan, M., et al., 2014. Profiling of expression of human papillomavirus-related cancer miRNAs in penile squamous cell carcinomas. Am. J. Pathol. 184 (12), 3376–3383.

Brougham, N.D., Dennett, E.R., Cameron, R., Tan, S.T., 2012. The incidence of metastasis from cutaneous squamous cell carcinoma and the impact of its risk factors. J. Surg. Oncol. 106 (7), 811–815.

Baek, S.K., Jung, K.Y., Lee, S.H., Woo, J.S., Kwon, S.Y., Chung, E.J., et al., 2009. Prognostic significance of vascular endothelial growth factor-C expression and lymphatic vessel density in supraglottic squamous cell carcinoma. Laryngoscope 119 (7), 1325–1330.

Breuninger, H., Black, B., Rassner, G., 1990. Microstaging of squamous cell carcinomas. Am. J. Clin. Pathol. 94 (5), 624–627.

Breuninger, H., Schaumburg-Lever, G., Holzschuh, J., Horny, H.P., 1997. Desmoplastic squamous cell carcinoma of skin and vermilion surface: a highly malignant subtype of skin cancer. Cancer 79 (5), 915–919.

Breuninger, H., Brantsch, K., Eigentler, T., Hafner, H.M., 2012. Comparison and evaluation of the current staging of cutaneous carcinomas. J. Dtsch Dermatol. Ges. 10 (8), 579–586.

Biddle, A., Mackenzie, I.C., 2012. Cancer stem cells and EMT in carcinoma. Cancer Metastasis Rev.

Barrette, K., Van Kelst, S., Wouters, J., Marasigan, V., Fieuws, S., Agostinis, P., et al., 2014. Epithelial-mesenchymal transition during invasion of cutaneous squamous cell carcinoma is paralleled by AKT activation. Br. J. Dermatol. 171 (5), 1014–1021.

Bradford, J.W., Baldwin, A.S., 2014. IKK/Nuclear factor-kappaB and oncogenesis: roles in tumor-initiating cells and in the tumor microenvironment. Adv. Cancer Res. 121, 125–145.

Baldwin, A.S., 2012. Regulation of cell death and autophagy by IKK and NF-kappaB: critical mechanisms in immune function and cancer. Immunol. Rev. 246 (1), 327–345.

Brodland, D.G., Zitelli, J.A., 1992. Surgical margins for excision of primary cutaneous squamous cell carcinoma. J. Am. Acad. Dermatol. 27 (2 Pt 1), 241–248.

Brewster, A.M., Lee, J.J., Clayman, G.L., Clifford, J.L., Reyes, M.J., Zhou, X., et al., 2007. Randomized trial of adjuvant 13-cis-retinoic acid and interferon alfa for patients with aggressive skin squamous cell carcinoma. J. Clin. Oncol. 25 (15), 1974–1978.

Behshad, R., Garcia-Zuazaga, J., Bordeaux, J.S., 2011. Systemic treatment of locally advanced nonmetastatic cutaneous squamous cell carcinoma: a review of the literature. Br. J. Dermatol. 165 (6), 1169–1177.

Christenson, L.J., Borrowman, T.A., Vachon, C.M., Tollefson, M.M., Otley, C.C., Weaver, A.L., et al., 2005. Incidence of basal cell and squamous cell carcinomas in a population younger than 40 years. JAMA 294 (6), 681–690.

Chockalingam, R., Downing, C., Tyring, S.K., 2015. Cutaneous squamous cell carcinomas in organ transplant recipients. J. Clin. Med. 4 (6), 1229–1239.

Cowen, E.W., Nguyen, J.C., Miller, D.D., McShane, D., Arron, S.T., Prose, N.S., et al., 2010. Chronic phototoxicity and aggressive squamous cell carcinoma of the skin in children and adults during treatment with voriconazole. J. Am. Acad. Dermatol. 62 (1), 31–37.

Ch'ng, S., Clark, J.R., Brunner, M., Palme, C.E., Morgan, G.J., Veness, M.J., 2013. Relevance of the primary lesion in the prognosis of metastatic cutaneous squamous cell carcinoma. Head Neck 35 (2), 190–194.

Campoli, M., Brodland, D.G., Zitelli, J., 2014. A prospective evaluation of the clinical, histologic, and therapeutic variables associated with incidental perineural invasion in cutaneous squamous cell carcinoma. J. Am. Acad. Dermatol. 70 (4), 630–636.

Clayman, G.L., Lee, J.J., Holsinger, F.C., Zhou, X., Duvic, M., El-Naggar, A.K., et al., 2005. Mortality risk from squamous cell skin cancer. J. Clin. Oncol. 23 (4), 759–765.

Carter, J.B., Johnson, M.M., Chua, T.L., Karia, P.S., Schmults, C.D., 2013. Outcomes of primary cutaneous squamous cell carcinoma with perineural invasion: an 11-year cohort study. JAMA Dermatol. 149 (1), 35–41.

Cherpelis, B.S., Marcusen, C., Lang, P.G., 2002. Prognostic factors for metastasis in squamous cell carcinoma of the skin. Dermatol. Surg. 28 (3), 268–273.

Ch'ng, S., Low, I., Ng, D., Brasch, H., Sullivan, M., Davis, P., et al., 2008. Epidermal growth factor receptor: a novel biomarker for aggressive head and neck cutaneous squamous cell carcinoma. Hum. Pathol. 39 (3), 344–349.

Cranmer, L.D., Engelhardt, C., Morgan, S.S., 2010. Treatment of unresectable and metastatic cutaneous squamous cell carcinoma. The Oncologist 15 (12), 1320–1328.

Dadras, S.S., Paul, T., Bertoncini, J., Brown, L.F., Muzikansky, A., Jackson, D.G., et al., 2003. Tumor lymphangiogenesis: a novel prognostic indicator for cutaneous melanoma metastasis and survival. Am. J. Pathol. 162 (6), 1951–1960.

Durchdewald, M., Guinea-Viniegra, J., Haag, D., Riehl, A., Lichter, P., Hahn, M., et al., 2008. Podoplanin is a novel fos target gene in skin carcinogenesis. Cancer Res. 68 (17), 6877–6883.

Delhase, M., Li, N., Karin, M., 2000. Kinase regulation in inflammatory response. Nature 406 (6794), 367–368.

DeConti, R.C., 2012. Chemotherapy of squamous cell carcinoma of the skin. Semin. Oncol. 39 (2), 145–149.

Edge, S., Byrd, D.R., Compton, C.C., Fritz, A.G., Greene, F.L., Trotti, A., et al., (Eds.), 2010 AJCC Cancer Staging Manual, seventh ed. Springer, New York.

Ebrahimi, A., Clark, J.R., Ahmadi, N., Palme, C.E., Morgan, G.J., Veness, M.J., 2013. Prognostic significance of disease-free interval in head and neck cutaneous squamous cell carcinoma with nodal metastases. Head Neck 35 (8), 1138–1143.

Fine, J.D., Johnson, L.B., Weiner, M., Li, K.P., Suchindran, C., 2009. Epidermolysis bullosa and the risk of life-threatening cancers: the national EB Registry experience, 1986–2006. J. Am. Acad. Dermatol. 60 (2), 203–211.

Ferrandiz-Pulido, C., Masferrer, E., de Torres, I., Lloveras, B., Hernandez-Losa, J., Mojal, S., et al., 2013. Identification and genotyping of human papillomavirus in a Spanish cohort of penile squamous cell carcinomas: correlation with pathologic subtypes, p16(INK4a) expression, and prognosis. J. Am. Acad. Dermatol. 68 (1), 73–82.

Frech, S., Hormann, K., Riedel, F., Gotte, K., 2009. Lymphatic vessel density in correlation to lymph node metastasis in head and neck squamous cell carcinoma. Anticancer Res. 29 (5), 1675–1679.

Flister, M.J., Wilber, A., Hall, K.L., Iwata, C., Miyazono, K., Nisato, R.E., et al., 2010. Inflammation induces lymphangiogenesis through up-regulation of VEGFR-3 mediated by NF-kappaB and Prox1. Blood 115 (2), 418–429.

Friedl, P., Alexander, S., 2011. Cancer invasion and the microenvironment: plasticity and reciprocity. Cell 147 (5), 992–1009.

Fernandez-Munoz, B., Yurrita, M.M., Martin-Villar, E., Carrasco-Ramirez, P., Megias, D., Renart, J., et al., 2011. The transmembrane domain of podoplanin is required for its association with lipid rafts and the induction of epithelial-mesenchymal transition. Int. J. Biochem. Cell Biol. 43 (6), 886–896.

Fernandez-Majada, V., Aguilera, C., Villanueva, A., Vilardell, F., Robert-Moreno, A., Aytes, A., et al., 2007a. Nuclear IKK activity leads to dysregulated notch-dependent gene expression in colorectal cancer. Proc. Natl. Acad. Sci. U.S.A. 104 (1), 276–281.

Fernandez-Majada, V., Pujadas, J., Vilardell, F., Capella, G., Mayo, M.W., Bigas, A., et al., 2007b. Aberrant cytoplasmic localization of N-CoR in colorectal tumors. Cell Cycle 6 (14), 1748–1752.

Foote, M.C., McGrath, M., Guminski, A., Hughes, B.G., Meakin, J., Thomson, D., et al., 2014. Phase II study of single-agent panitumumab in patients with incurable cutaneous squamous cell carcinoma. Ann. Oncol. 25 (10), 2047–2052.

Falivene, S., Giugliano, F.M., Grimaldi, A.M., Di Franco, R., Toledo, D., Muto, M., et al., 2014. Tomotherapy concomitant with cetuximab, followed by cetuximab as single-agent therapy for unresectable squamous cell carcinoma of the skin: a case report. BMC Dermatol. 14, 15.

Goyal, R.K., 2015. Voriconazole-associated phototoxic dermatoses and skin cancer. Expert Rev. Anti-Infective Ther. 13 (12), 1537–1546.

Garcia-Serra, A., Hinerman, R.W., Mendenhall, W.M., Amdur, R.J., Morris, C.G., Williams, L.S., et al., 2003. Carcinoma of the skin with perineural invasion. Head Neck 25 (12), 1027–1033.

Goepfert, H., Dichtel, W.J., Medina, J.E., Lindberg, R.D., Luna, M.D., 1984. Perineural invasion in squamous cell skin carcinoma of the head and neck. Am. J. Surg. 148 (4), 542–547.

Gao, P., Zhou, G.Y., Zhang, Q.H., Xiang, L., Zhang, S.L., Li, C., et al., 2008. Clinicopathological significance of peritumoral lymphatic vessel density in gastric carcinoma. Cancer Lett. 263 (2), 223–230.

Gjerdrum, C., Tiron, C., Hoiby, T., Stefansson, I., Haugen, H., Sandal, T., et al., 2010. Axl is an essential epithelial-to-mesenchymal transition-induced regulator of breast cancer metastasis and patient survival. Proc. Natl. Acad. Sci. U.S.A. 107 (3), 1124–1129.

Gore, S.M., Shaw, D., Martin, R.C., Kelder, W., Roth, K., Uren, R., et al., 2015. Prospective study of sentinel node biopsy for high-risk cutaneous squamous cell carcinoma of the head and neck. Head Neck 38 (Suppl. 1), E884–E889.

Hussein, M.R., 2005. Ultraviolet radiation and skin cancer: molecular mechanisms. J. Cutan. Pathol. 32 (3), 191–205.

Hausauer, A.K., Maurer, T., Leslie, K.S., Parvataneni, R., Stuart, S.E., Chren, M.M., 2013. Recurrence after treatment of cutaneous basal cell and squamous cell carcinomas in patients infected with human immunodeficiency virus. JAMA Dermatol. 149 (2), 239–241.

Harwood, C.A., Surentheran, T., McGregor, J.M., Spink, P.J., Leigh, I.M., Breuer, J., et al., 2000. Human papillomavirus infection and non-melanoma skin cancer in immunosuppressed and immunocompetent individuals. J. Med. Virol. 61 (3), 289–297.

Hirakawa, S., Brown, L.F., Kodama, S., Paavonen, K., Alitalo, K., Detmar, M., 2007. VEGF-C-induced lymphangiogenesis in sentinel lymph nodes promotes tumor metastasis to distant sites. Blood 109 (3), 1010–1017.

Hendrix, M.J., Seftor, R.E., Seftor, E.A., Gruman, L.M., Lee, L.M., Nickoloff, B.J., et al., 2002. Transendothelial function of human metastatic melanoma cells: role of the microenvironment in cell-fate determination. Cancer Res. 62 (3), 665–668.

Jonason, A.S., Kunala, S., Price, G.J., Restifo, R.J., Spinelli, H.M., Persing, J.A., et al., 1996. Frequent clones of p53-mutated keratinocytes in normal human skin. Proc. Natl. Acad. Sci. U.S.A. 93 (24), 14025–14029.

Jennings, L., Schmults, C.D., 2010. Management of high-risk cutaneous squamous cell carcinoma. J. Clin. Aesthet Dermatol. 3 (4), 39–48.

Jensen, V., Prasad, A.R., Smith, A., Raju, M., Wendel, C.S., Schmelz, M., et al., 2010. Prognostic criteria for squamous cell cancer of the skin. J. Surg. Res. 159 (1), 509–516.

Jambusaria-Pahlajani, A., Kanetsky, P.A., Karia, P.S., Hwang, W.T., Gelfand, J.M., Whalen, F.M., et al., 2013. Evaluation of AJCC tumor staging for cutaneous squamous cell carcinoma and a proposed alternative tumor staging system. JAMA Dermatol. 149 (4), 402–410.

Jenkins, G., Smith, A.B., Kanatas, A.N., Houghton, D.R., Telfer, M.R., 2014. Anatomical restrictions in the surgical excision of scalp squamous cell carcinomas: does this affect local recurrence and regional nodal metastases? Int. J. Oral Maxillofac. Surg. 43 (2), 142–146.

Karia, P.S., Han, J., Schmults, C.D., 2013. Cutaneous squamous cell carcinoma: estimated incidence of disease, nodal metastasis, and deaths from disease in the United States, 2012. J. Am. Acad. Dermatol. 68 (6), 957–966.

Krynitz, B., Edgren, G., Lindelof, B., Baecklund, E., Brattstrom, C., Wilczek, H., et al., 2013. Risk of skin cancer and other malignancies in kidney, liver, heart and lung transplant recipients 1970 to 2008—a Swedish population-based study. Int. J. Cancer 132 (6), 1429–1438.

Krediet, J.T., Beyer, M., Lenz, K., Ulrich, C., Lange-Asschenfeldt, B., Stockfleth, E., et al., 2015. Sentinel lymph node biopsy and risk factors for predicting metastasis in cutaneous squamous cell carcinoma. Br. J. Dermatol. 172 (4), 1029–1036.

Kyrgidis, A., Tzellos, T.G., Kechagias, N., Patrikidou, A., Xirou, P., Kitikidou, K., et al., 2010. Cutaneous squamous cell carcinoma (SCC) of the head and neck: risk factors of overall and recurrence-free survival. Eur. J. Cancer 46 (9), 1563–1572.

Karia, P.S., Jambusaria-Pahlajani, A., Harrington, D.P., Murphy, G.F., Qureshi, A.A., Schmults, C.D., 2014. Evaluation of American Joint Committee on Cancer, International Union Against Cancer, and Brigham and Women's Hospital tumor staging for cutaneous squamous cell carcinoma. J. Clin. Oncol. 32 (4), 327–334.

Kreppel, M., Krakowezki, A., Kreppel, B., Drebber, U., Wedemeyer, I., Mauch, C., et al., 2013. Podoplanin expression in cutaneous head and neck squamous cell carcinoma—prognostic value and clinicopathologic implications. J. Surg. Oncol. 107 (4), 376–383.

Karin, M., Delhase, M., 2000. The I kappa B kinase (IKK) and NF-kappa B: key elements of proinflammatory signalling. Semin. Immunol. 12 (1), 85–98.

Khan, A.A., Potter, M., Cubitt, J.J., Khoda, B.J., Smith, J., Wright, E.H., et al., 2013. Guidelines for the excision of cutaneous squamous cell cancers in the United Kingdom: the best cut is the deepest. J. Plast. Reconstr. Aesthet Surg. 66 (4), 467–471.

Kreuter, A., van Eijk, T., Lehmann, P., Fischer, M., Horn, T., Assaf, C., et al., 2015. Electrochemotherapy in advanced skin tumors and cutaneous metastases—a retrospective multicenter analysis. J. Dtsch. Dermatol. Ges. 13 (4), 308–315.

LeBoit, P.E., Burg, G., Weedon, D., et al., 2008. Pathology and Genetics Skin Tumours. IARC Press (World Health Organization Classification of Tumours), Lyon.

Lomas, A., Leonardi-Bee, J., Bath-Hextall, F., 2012. A systematic review of worldwide incidence of nonmelanoma skin cancer. Br. J. Dermatol. 166 (5), 1069–1080.

Leibovitch, I., Huilgol, S.C., Selva, D., Hill, D., Richards, S., Paver, R., 2005. Cutaneous squamous cell carcinoma treated with Mohs micrographic surgery in Australia II. Perineural invasion. J. Am. Acad. Dermatol. 53 (2), 261–266.

Liersch, R., Hirakawa, S., Berdel, W.E., Mesters, R.M., Detmar, M., 2012. Induced lymphatic sinus hyperplasia in sentinel lymph nodes by VEGF-C as the earliest premetastatic indicator. Int. J. Oncol. 41 (6), 2073–2078.

Lambert, S.R., Harwood, C.A., Purdie, K.J., Gulati, A., Matin, R.N., Romanowska, M., et al., 2012. Metastatic cutaneous squamous cell carcinoma shows frequent deletion in the protein tyrosine phosphatase receptor Type D gene. Int. J. Cancer 131 (3), E216–E226.

Luo, J.L., Tan, W., Ricono, J.M., Korchynskyi, O., Zhang, M., Gonias, S.L., et al., 2007. Nuclear cytokine-activated IKKalpha controls prostate cancer metastasis by repressing Maspin. Nature 446 (7136), 690–694.

Lippman, S.M., Parkinson, D.R., Itri, L.M., Weber, R.S., Schantz, S.P., Ota, D.M., et al., 1992. 13-cis-retinoic acid and interferon alpha-2a: effective combination therapy for advanced squamous cell carcinoma of the skin. J. Natl. Cancer Inst. 84 (4), 235–241.

Martorell-Calatayud, A., Sanmartin Jimenez, O., Cruz Mojarrieta, J., Guillen Barona, C., 2013. Cutaneous squamous cell carcinoma: defining the high-risk variant. Actas Dermosifiliogr 104 (5), 367–379.

Mehrany, K., Weenig, R.H., Lee, K.K., Pittelkow, M.R., Otley, C.C., 2005a. Increased metastasis and mortality from cutaneous squamous cell carcinoma in patients with chronic lymphocytic leukemia. J. Am. Acad. Dermatol. 53 (6), 1067–1071.

Mehrany, K., Weenig, R.H., Pittelkow, M.R., Roenigk, R.K., Otley, C.C., 2005b. High recurrence rates of squamous cell carcinoma after Mohs' surgery in patients with chronic lymphocytic leukemia. Dermatol. Surg. 31 (1), 38–42 (discussion).

Mellerio, J.E., Robertson, S.J., Bernardis, C., Diem, A., Fine, J.D., George, R., et al., 2016. Management of cutaneous squamous cell carcinoma in patients with epidermolysis bullosa: best clinical practice guidelines. Br. J. Dermatol. 174 (1), 56–57.

McLaughlin-Drubin, M.E., 2015. Human papillomaviruses and non-melanoma skin cancer. Semin. Oncol. 42 (2), 284–290.

Moore, B.A., Weber, R.S., Prieto, V., El-Naggar, A., Holsinger, F.C., Zhou, X., et al., 2005. Lymph node metastases from cutaneous squamous cell carcinoma of the head and neck. Laryngoscope 115 (9), 1561–1567.

Morton, D.L., Wen, D.R., Wong, J.H., Economou, J.S., Cagle, L.A., Storm, F.K., et al., 1992. Technical details of intraoperative lymphatic mapping for early stage melanoma. Arch. Surg. 127 (4), 392–399.

Moussai, D., Mitsui, H., Pettersen, J.S., Pierson, K.C., Shah, K.R., Suarez-Farinas, M., et al., 2011. The human cutaneous squamous cell carcinoma microenvironment is characterized by increased lymphatic density and enhanced expression of macrophage-derived VEGF-C. J. Invest. Dermatol. 131 (1), 229–236.

Mandal, M., Myers, J.N., Lippman, S.M., Johnson, F.M., Williams, M.D., Rayala, S., et al., 2008. Epithelial to mesenchymal transition in head and neck squamous carcinoma: association of Src activation with E-cadherin down-regulation, vimentin expression, and aggressive tumor features. Cancer 112 (9), 2088–2100.

Martin, T.A., Goyal, A., Watkins, G., Jiang, W.G., 2005. Expression of the transcription factors snail, slug, and twist and their clinical significance in human breast cancer. Ann. Surg. Oncol. 12 (6), 488–496.

Martins, V.L., Vyas, J.J., Chen, M., Purdie, K., Mein, C.A., South, A.P., et al., 2009. Increased invasive behaviour in cutaneous squamous cell carcinoma with loss of basement-membrane type VII collagen. J. Cell Sci. 122 (Pt 11), 1788–1799.

Margalef, P., Fernandez-Majada, V., Villanueva, A., Garcia-Carbonell, R., Iglesias, M., Lopez, L., et al., 2012. A truncated form of IKKalpha is responsible for specific nuclear IKK activity in colorectal cancer. Cell Rep. 2 (4), 840–854.

Mecca, C., Ponzetti, A., Caliendo, V., Ciuffreda, L., Lista, P., 2012. Complete response of metastatic cutaneous squamous cell carcinoma to cetuximab plus paclitaxel. Eur. J. Dermatol. 22 (6), 758–761.

National Comprehensive Cancer Network Clinical Paractice Guidelines in Oncology (NCCN Guidelines). 2015. Squamous Cell Skin Cancer. Version 1, 10/24/2014. NCCN.org.

Nindl, I., Gottschling, M., Stockfleth, E., 2007. Human papillomaviruses and non-melanoma skin cancer: basic virology and clinical manifestations. Dis. Markers 23 (4), 247–259.

Navarrete-Dechent, C., Veness, M.J., Droppelmann, N., Uribe, P., 2015. High-risk cutaneous squamous cell carcinoma and the emerging role of sentinel lymph node biopsy: a literature review. J. Am. Acad. Dermatol. 73 (1), 127–137.

Nguyen, P., Vin-Christian, K., Ming, M.E., Berger, T., 2002. Aggressive squamous cell carcinomas in persons infected with the human immunodeficiency virus. Arch. Dermatol. 138 (6), 758–763.

van de Nieuwenhof, H.P., van Kempen, L.C., de Hullu, J.A., Bekkers, R.L., Bulten, J., Melchers, W.J., et al., 2009. The etiologic role of HPV in vulvar squamous cell carcinoma fine tuned. Cancer Epidemiol. Biomarkers Prev. 18 (7), 2061–2067.

Nevens, D., Nuyts, S., 2015. HPV-positive head and neck tumours, a distinct clinical entity. B-ENT 11 (2), 81–87.

Nakamura, Y., Yasuoka, H., Tsujimoto, M., Imabun, S., Nakahara, M., Nakao, K., et al., 2005. Lymph vessel density correlates with nodal status, VEGF-C expression, and prognosis in breast cancer. Breast Cancer Res. Treat. 91 (2), 125–132.

Omland, S.H., Gniadecki, R., Haedersdal, M., Helweg-Larsen, J., Omland, L.H., 2015. Skin cancer risk in hematopoietic stem-cell transplant recipients compared with background population and renal transplant recipients: a population-based cohort study. JAMA Dermatol. 1–7.

Oliver, G., Sosa-Pineda, B., Geisendorf, S., Spana, E.P., Doe, C.Q., Gruss, P., 1993. Prox 1, a prospero-related homeobox gene expressed during mouse development. Mech. Dev. 44 (1), 3–16.

Pastushenko, I., Vermeulen, P.B., Vicente-Arregui, S., Van den Eynden, G.G., Alvarez-Alegret, R., Querol, I., et al., 2015. Peritumoral D2-40 Chalkley score independently predicts metastases and survival in patients with cutaneous malignant melanoma. J. Cutan. Pathol. 42 (10), 699–711.

Petter, G., Haustein, U.F., 2000. Histologic subtyping and malignancy assessment of cutaneous squamous cell carcinoma. Dermatol. Surg. 26 (6), 521–530.

Pugliano-Mauro, M., Goldman, G., 2010. Mohs surgery is effective for high-risk cutaneous squamous cell carcinoma. Dermatol. Surg. 36 (10), 1544–1553.

Revenga Arranz, F., Paricio Rubio, J.F., Mar Vazquez Salvado, M., del Villar Sordo, V., 2004. Descriptive epidemiology of basal cell carcinoma and cutaneous squamous cell carcinoma in Soria (north-eastern Spain) 1998–2000: a hospital-based survey. J. Eur. Acad. Dermatol. Venereol. 18 (2), 137–141.

Ratushny, V., Gober, M.D., Hick, R., Ridky, T.W., Seykora, J.T., 2012. From keratinocyte to cancer: the pathogenesis and modeling of cutaneous squamous cell carcinoma. J. Clin. Invest. 122 (2), 464–472.

Roozeboom, M.H., Lohman, B.G., Westers-Attema, A., Nelemans, P.J., Botterweck, A.A., van Marion, A.M., et al., 2013. Clinical and histological prognostic factors for local recurrence and metastasis of cutaneous squamous cell carcinoma: analysis of a defined population. Acta Dermato-Venereologica 93 (4), 417–421.

Rowe, D.E., Carroll, R.J., Day Jr., C.L., 1992. Prognostic factors for local recurrence, metastasis, and survival rates in squamous cell carcinoma of the skin, ear, and lip. Implications for treatment modality selection. J. Am. Acad. Dermatol. 26 (6), 976–990.

Riddel, C., Rashid, R., Thomas, V., 2011. Ungual and periungual human papillomavirus-associated squamous cell carcinoma: a review. J. Am. Acad. Dermatol. 64 (6), 1147–1153.

Ross, A.S., Whalen, F.M., Elenitsas, R., Xu, X., Troxel, A.B., Schmults, C.D., 2009. Diameter of involved nerves predicts outcomes in cutaneous squamous cell carcinoma with perineural invasion: an investigator-blinded retrospective cohort study. Dermatol. Surg. 35 (12), 1859–1866.

Roma, A.A., Magi-Galluzzi, C., Kral, M.A., Jin, T.T., Klein, E.A., Zhou, M., 2006. Peritumoral lymphatic invasion is associated with regional lymph node metastases in prostate adenocarcinoma. Mod. Pathol. 19 (3), 392–398.

Renzi, C., Caggiati, A., Mannooranparampil, T.J., Passarelli, F., Tartaglione, G., Pennasilico, G.M., et al., 2007. Sentinel lymph node biopsy for high risk cutaneous squamous cell carcinoma: case series and review of the literature. Eur. J. Surg. Oncol. 33 (3), 364–369.

Reigneau, M., Robert, C., Routier, E., Mamelle, G., Moya-Plana, A., Tomasic, G., et al., 2015. Efficacy of neoadjuvant cetuximab alone or with platinum salt for the treatment of unresectable advanced nonmetastatic cutaneous squamous cell carcinomas. Br. J. Dermatol. 173 (2), 527–534.

Sobin, L.G.M., Wittekind, C. (Eds.), 2009. UICC International Union Against Cancer TNM Classification of Malignant Tumors, seventh ed. Wiley-Blackwell, West Sussex, United Kingdom.

Stratigos, A., Garbe, C., Lebbe, C., Malvehy, J., del Marmol, V., Pehamberger, H., et al., 2015. Diagnosis and treatment of invasive squamous cell carcinoma of the skin: European consensus-based interdisciplinary guideline. Eur. J. Cancer 51 (14), 1989–2007.

Silverberg, M.J., Leyden, W., Warton, E.M., Quesenberry Jr., C.P., Engels, E.A., Asgari, M.M., 2013. HIV infection status, immunodeficiency, and the incidence of non-melanoma skin cancer. J. Natl. Cancer Inst. 105 (5), 350–360.

Schaffer, J.V., Orlow, S.J., 2011. Radiation therapy for high-risk squamous cell carcinomas in patients with xeroderma pigmentosum: report of two cases and review of the literature. Dermatology 223 (2), 97–103.

Schmults, C.D., Karia, P.S., Carter, J.B., Han, J., Qureshi, A.A., 2013. Factors predictive of recurrence and death from cutaneous squamous cell carcinoma: a 10-year, single-institution cohort study. JAMA Dermatol. 149 (5), 541–547.

Scottish Intercollegiate Guidelines Network (SIGN). 2014. Management of primary cutaneous squamous cell carcinoma. Edinburgh: SIGN. (SIGN publication no. 140). Available from: http://www.sign.ac.uk.

Sugiura, T., Inoue, Y., Matsuki, R., Ishii, K., Takahashi, M., Abe, M., et al., 2009. VEGF-C and VEGF-D expression is correlated with lymphatic vessel density and lymph node metastasis in oral squamous cell carcinoma: implications for use as a prognostic marker. Int. J. Oncol. 34 (3), 673–680.

Sasahira, T., Ueda, N., Yamamoto, K., Kurihara, M., Matsushima, S., Bhawal, U.K., et al., 2014. Prox1 and FOXC2 act as regulators of lymphangiogenesis and angiogenesis in oral squamous cell carcinoma. PLoS One 9 (3), e92534.

Schacht, V., Dadras, S.S., Johnson, L.A., Jackson, D.G., Hong, Y.K., Detmar, M., 2005. Up-regulation of the lymphatic marker podoplanin, a mucin-type transmembrane glycoprotein, in human squamous cell carcinomas and germ cell tumors. Am. J. Pathol. 166 (3), 913–921.

Slater, D.W.M., 2012. Dataset for the Histological Reporting of Primary Cutaneous Squamous Cell Carcinoma and Regional Lymph Nodes, second ed. Royal College of Pathologists.

Sweeny, L., Dean, N.R., Magnuson, J.S., Carroll, W.R., Helman, E.E., Hyde, S.O., et al., 2012. EGFR expression in advanced head and neck cutaneous squamous cell carcinoma. Head Neck 34 (5), 681–686.

Salgado, R., Toll, A., Alameda, F., Baro, T., Martin-Ezquerra, G., Sanmartin, O., et al., 2010. CKS1B amplification is a frequent event in cutaneous squamous cell carcinoma with aggressive clinical behaviour. Genes Chromosomes Cancer 49 (11), 1054–1061.

Schmitt, A.R., Brewer, J.D., Bordeaux, J.S., Baum, C.L., 2014. Staging for cutaneous squamous cell carcinoma as a predictor of sentinel lymph node biopsy results: meta-analysis of American Joint Committee on Cancer criteria and a proposed alternative system. JAMA Dermatol. 150 (1), 19–24.

Shin, D.M., Glisson, B.S., Khuri, F.R., Clifford, J.L., Clayman, G., Benner, S.E., et al., 2002. Phase II and biologic study of interferon alfa, retinoic acid, and cisplatin in advanced squamous skin cancer. J. Clin. Oncol. 20 (2), 364–370.

Tomaszewski, J.M., Gavriel, H., Link, E., Boodhun, S., Sizeland, A., Corry, J., 2014. Aggressive behavior of cutaneous squamous cell carcinoma in patients with chronic lymphocytic leukemia. Laryngoscope 124 (9), 2043–2048.

Toll, A., Lloveras, B., Masferrer, E., Ferrandiz-Pulido, C., Garcia-Patos, V., Gheit, T., et al., 2014. Human beta papillomavirus DNA study in primary cutaneous squamous cell carcinomas and their corresponding metastases. Arch. Dermatol. Res. 306 (1), 93–95.

Toll, A., Gimeno-Beltran, J., Ferrandiz-Pulido, C., Masferrer, E., Yebenes, M., Jucgla, A., et al., 2012. D2-40 immunohistochemical overexpression in cutaneous squamous cell carcinomas: a marker of metastatic risk. J. Am. Acad. Dermatol. 67 (6), 1310–1318.

Thiery, J.P., Acloque, H., Huang, R.Y., Nieto, M.A., 2009. Epithelial-mesenchymal transitions in development and disease. Cell 139 (5), 871–890.

Toll, A., Masferrer, E., Hernandez-Ruiz, M.E., Ferrandiz-Pulido, C., Yebenes, M., Jaka, A., et al., 2013. Epithelial to mesenchymal transition markers are associated with an increased metastatic risk in primary cutaneous squamous cell carcinomas but are attenuated in lymph node metastases. J. Dermatol. Sci. 72 (2), 93–102.

Takayama, T., Shiozaki, H., Shibamoto, S., Oka, H., Kimura, Y., Tamura, S., et al., 1996. Beta-catenin expression in human cancers. Am. J. Pathol. 148 (1), 39–46.

Toll, A., Margalef, P., Masferrer, E., Ferrandiz-Pulido, C., Gimeno, J., Pujol, R.M., et al., 2015. Active nuclear IKK correlates with metastatic risk in cutaneous squamous cell carcinoma. Arch. Dermatol. Res. 307, 721–729.

Tanvetyanon, T., Padhya, T., McCaffrey, J., Kish, J.A., Deconti, R.C., Trotti, A., et al., 2015. Postoperative concurrent chemotherapy and radiotherapy for high-risk cutaneous squamous cell carcinoma of the head and neck. Head Neck 37 (6), 840–845.

Takahashi, A., Imafuku, S., Nakayama, J., Nakaura, J., Ito, K., Shibayama, Y., 2014. Sentinel node biopsy for high-risk cutaneous squamous cell carcinoma. Eur. J. Surg. Oncol. 40 (10), 1256–1262.

Testori, A., Tosti, G., Martinoli, C., Spadola, G., Cataldo, F., Verrecchia, F., et al., 2010. Electrochemotherapy for cutaneous and subcutaneous tumor lesions: a novel therapeutic approach. Dermatol. Ther. 23 (6), 651–661.

Usami, Y., Satake, S., Nakayama, F., Matsumoto, M., Ohnuma, K., Komori, T., et al., 2008. Snail-associated epithelial-mesenchymal transition promotes oesophageal squamous cell carcinoma motility and progression. J. Pathol. 215 (3), 330–339.

de Vries, E., Trakatelli, M., Kalabalikis, D., Ferrandiz, L., Ruiz-de-Casas, A., Moreno-Ramirez, D., et al., 2012. Known and potential new risk factors for skin cancer in European populations: a multicentre case-control study. Br. J. Dermatol. 167 (Suppl. 2), 1–13.

Vajdic, C.M., van Leeuwen, M.T., 2009. Cancer incidence and risk factors after solid organ transplantation. Int. J. Cancer 125 (8), 1747–1754.

Walsh, L., Gillham, C., Dunne, M., Fraser, I., Hollywood, D., Armstrong, J., et al., 2011. Toxicity of cetuximab versus cisplatin concurrent with radiotherapy in locally advanced head and neck squamous cell cancer (LAHNSCC). Radiother. Oncol. 98 (1), 38–41.

Xiong, H., Hong, J., Du, W., Lin, Y.W., Ren, L.L., Wang, Y.C., et al., 2012. Roles of STAT3 and ZEB1 proteins in E-cadherin down-regulation and human colorectal cancer epithelial-mesenchymal transition. J. Biol. Chem. 287 (8), 5819–5832.

Yuen, W.Y., Jonkman, M.F., 2011. Risk of squamous cell carcinoma in junctional epidermolysis bullosa, non-Herlitz type: report of 7 cases and a review of the literature. J. Am. Acad. Dermatol. 65 (4), 780–789.

Yan, C., Grimm, W.A., Garner, W.L., Qin, L., Travis, T., Tan, N., et al., 2010. Epithelial to mesenchymal transition in human skin wound healing is induced by tumor necrosis factor-alpha through bone morphogenic protein-2. Am. J. Pathol. 176 (5), 2247–2258.

Yang, M.H., Wu, K.J., 2008. TWIST activation by hypoxia inducible factor-1 (HIF-1): implications in metastasis and development. Cell Cycle 7 (14), 2090–2096.

Yamamoto, Y., Verma, U.N., Prajapati, S., Kwak, Y.T., Gaynor, R.B., 2003. Histone H3 phosphorylation by IKK-alpha is critical for cytokine-induced gene expression. Nature 423 (6940), 655–659.

Zwald, F.O., Brown, M., 2011. Skin cancer in solid organ transplant recipients: advances in therapy and management: part I. Epidemiology of skin cancer in solid organ transplant recipients. J. Am. Acad. Dermatol. 65 (2), 253–261 quiz 62.

Zhao, H., Shu, G., Wang, S., 2016. The risk of non-melanoma skin cancer in HIV-infected patients: new data and meta-analysis. Int. J. STD AIDS 27 (7), 568–575.

Zhao, Y.C., Ni, X.J., Wang, M.H., Zha, X.M., Zhao, Y., Wang, S., 2012. Tumor-derived VEGF-C, but not VEGF-D, promotes sentinel lymph node lymphangiogenesis prior to metastasis in breast cancer patients. Med. Oncol. 29 (4), 2594–2600.

CHAPTER

# 12

# Osteosarcoma Metastasis—Prognostic Factors and Treatment Strategies

S. Iwata

Chiba Cancer Center, Chiba, Japan

## 1. INTRODUCTION

Osteosarcoma (OS) is the most common primary bone tumor with a predilection for males and has a dual peak incidence in the second to third decade and after the fifth decade (Meyers and Gorlick, 1997). Although OS predominantly arises in the long bones, it may also occur in other bones. Despite numerous advancements in diagnostic and treatment techniques, the survival rate among localized OS patients has remained at 60–70% (Meyers and Gorlick, 1997; Bielack et al., 2002; Yonemoto et al., 1998), although it has yet to further improve in the last 30 years. Distant metastasis is the leading cause of disease-related death among OS patients. From the perspective of metastasis, OS has unique characteristics compared to other cancers. In OS, the vast majority of distant metastases develop in the lung, whereas occurrence of lymph metastasis is uncommon, and repeated metastasectomy may result in long-term survival.

The aim of this article is to describe the current understanding of the characteristics, prognostic factors, and treatment strategies for metastasis from OS.

## 2. CLINICAL FEATURES OF METASTASIS FROM OSTEOSARCOMA

### 2.1 Incidence

The incidence of distant metastasis in OS patients at initial presentation is around 20% (Bacci et al., 2008; Kager et al., 2003; Meyers et al., 1993; Miller et al., 2013). The lung is the most common site, accounting for >80% of all initial metastases, and bone metastasis is found in >30% cases (Kager et al., 2003). Additionally, 25–30% of all patients who complete initial treatment may develop distant metastasis. In comparison, lymph node metastasis is extremely rare in OS patients with an incidence of only 0.8% (Bielack et al., 2002).

Axial tumor location and large tumor size are both associated with a high incidence of distant metastasis. Miller et al. attempted to identify factors of patients and tumor characteristics associated with an increased rate of distant metastasis at presentation using large database. They concluded that an age of 60 years or more, a tumor located in the axial skeleton and a larger tumor size were significant risk factors for metastasis

*Introduction to Cancer Metastasis*
http://dx.doi.org/10.1016/B978-0-12-804003-4.00012-8

at presentation (Miller et al., 2013). Especially, lesions of OS arising in the pelvis tend to enlarge, probably resulting in a high rate of pulmonary metastasis at diagnosis as well as poor outcome (Grimer et al., 1999; Kawai et al., 1998; Ozaki et al., 2002).

Several reports have described that elderly patients diagnosed with OS are more prone to initial metastasis with an incidence of 28–33% (Bacci et al., 1998; Iwata et al., 2014; Longhi et al., 2008; Manoso et al., 2005; Nishida et al., 2009). Several specific clinicopathological characteristics are frequently observed in the elderly patients with OS compared to pediatric/adolescent patients such as an unusual pathological subtype, axial location, and distant metastasis at presentation, suggesting a possibility for distinct biology of elderly OS from pediatric/adolescent OS (Iwata et al., 2014; Nishida et al., 2009; Carsi and Rock, 2002; Grimer et al., 2003).

Secondary OS, mainly occurring in bones affected by Paget disease or in irradiated bones, has an extremely higher incidence of metastasis than primary OS. Generally, 35–65% of secondary OS patients will develop distant metastasis during the disease course. This must be a reason why secondary OS patients endure catastrophic outcomes (Mangham et al., 2009; Mankin and Hornicek, 2005; Ruggieri et al., 2010).

## 2.2 Detection

To date, high-resolution computed tomography (HRCT) is the standard modality to detect pulmonary metastasis in OS patient (Meyer et al., 2008). HRCT performed using the spiral technique and collimation of ≤5 mm has a great advantage over conventional radiography for the detection of pulmonary metastasis (Vanel et al., 1984). Typically, a pulmonary metastatic nodule appears to be round or oval with sharp demarcation. Contrast-enhanced CT scan is not required in routine diagnosis.

Nonetheless, it remains difficult to radiologically distinguish between a benign and malignant nodule. Several criteria include a tumor with a diameter of >5 mm as "true pulmonary metastasis" of OS (Grampp et al., 2000; A Randomized Trial of the European and American Osteosarcoma; Robertson et al., 1988). On the other hand, in a study by McCarville et al., three radiologists attempted to identify predictors to distinguish a benign from a malignant nodule in pediatric malignant solid tumors by reviewing CT scans but failed to find a threshold value for either the size or number of nodules (McCarville et al., 2006).

Another dilemma lies in the discrepancy between the number of radiographic pulmonary nodules and resected pulmonary metastases (Fig. 12.1). Kayton et al. found that preoperative evaluation of the pulmonary nodules by CT underestimates the number of pathologically confirmed metastatic nodules resected by surgeons in 35% of all thoracotomies (Kayton et al., 2006). Meanwhile, Salah et al. described that 91% of all resected nodules from pulmonary metastatic OS were confirmed as metastatic, whereas the remaining were either inflammation (5%), fibrosis (3%), or benign lung tissue (1%) (Salah and Toubasi, 2015).

## 2.3 Prognostic Factors

Numerous reports have attempted to identify prognostic factors for OS patients with pulmonary metastasis (Table 12.1). Among these, the number of pulmonary nodules was repeatedly described as one of the most important prognostic factors (Bacci et al., 2001; Bielack et al., 2009; Hawkins and Arndt, 2003; Iwata et al., 2015; Kager et al., 2006). The Cooperative Osteosarcoma Study Group (COSS), which is one of the largest study groups of OS, reported that of a total of 379 OS patients with pulmonary metastasis, solitary and multiple lesions were discovered in 145 (38.3%) and 234 (61.7%) patients, respectively (Bielack et al., 2009). Many reports emphasized a difference in the impact on survival between solitary and

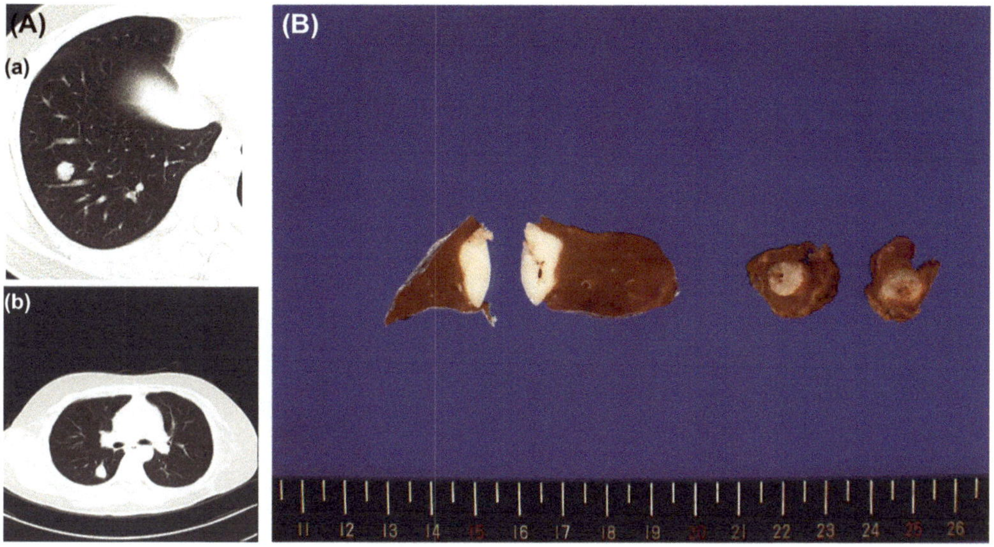


FIGURE 12.1 Radiological (A) and histological (B) appearance of pulmonary metastases from osteosarcoma.

multiple pulmonary metastases (Kager et al., 2003; Bielack et al., 2009; Hawkins and Arndt, 2003; Iwata et al., 2015; Kempf-Bielack et al., 2005; Briccoli et al., 2005), whereas other reports insisted that a smaller number of nodules (up to two; Bacci et al., 2001; Ferrari et al., 2003; Wu et al., 2009, or three, Iwata et al., 2015; Leary et al., 2013; Tsuchiya et al., 2002) had a more favorable prognosis, as compared with the presence of multiple nodules. From this point of view, curative treatment should be the goal of not only cases with solitary nodules but also those with limited pulmonary metastases.

Timing of occurrence of pulmonary metastasis is recognized as an important prognostic factor. Tsuchiya et al. described that pulmonary metastasis identified after treatment completion has a markedly better prognoses than those identified at the initial presentation of a primary tumor and during perioperative chemotherapy (Tsuchiya et al., 2002). This idea has been confirmed by a recent report describing that detection of pulmonary metastasis during perioperative chemotherapy was correlated with inferior overall survival (OS) (Salah and Toubasi, 2015). On the other hand, much effort has been made in attempts to save OS patients with pulmonary metastasis at initial presentation (M1 status) by administration of aggressive treatment. Bacci et al. reported a 5-year event-free survival rate of 27.4% among 91 M1 patients following complete surgical resection of all metastatic nodules and systemic chemotherapy (Bacci et al., 2008). Likewise, Yonemoto et al. reported a 5-year OS rate of 64.8% among nine M1 patients following complete resection of an average of 12.7 pulmonary nodules (Yonemoto et al., 1998).

The interval from previous complete remission is also a strong prognostic factor. Generally, a longer interval from previous resection of a metastatic lesion is associated with a favorable outcome, and the median interval tended to be shorter with subsequent relapse (Leary et al., 2013). Bacci et al. examined the outcomes of 235 patients with relapsed OS and found that the median interval from initial treatment to the first relapse was 25.4 months, whereas an interval from initial diagnosis to a first relapse of >24 months was correlated with poorer

TABLE 12.1 Reported Poor Prognostic Factors for Osteosarcoma Patients With Pulmonary Metastasis

| Factors | | Authors | N | *p*-Value |
|---|---|---|---|---|
| Number of nodules | >1 | Kager et al. (2003) | 198 | <.001 |
| | | Briccoli et al. (2005) | 94 | .046 |
| | >2 | Bacci et al. (2005) | 202 | <.0001 |
| | | Wu et al. (2009) | 91 | .002 |
| | >3 | Tsuchiya et al. (2002) | 280 | <.0001 |
| | | Iwata et al. (2015) | 151 | .011 |
| Timing of metastasis | Initial presentation/ during chemotherapy | Tsuchiya et al. (2002) | 280 | <.0001 |
| | | Salah and Toubasi (2015) | 25 | .0036 |
| Interval from previous treatment | <12 months | Iwata et al. (2015) | 151 | .012 |
| | | Harting et al. (2006) | 137 | <.001 |
| | <18 months | Leary et al. (2013) | 110 | <.0001 |
| | <24 months | Ferrari et al. (2003) | 114 | <.0001 |
| | | Hawkins and Arndt (2003) | 59 | .05 |
| Localization | Central | Letourneau et al. (2011) | 47 | .008 |
| Laterality | Bilateral | Kempf-Bielack et al. (2005) | 436 | <.0001 |
| Number of involved lobes | >1 | Wu et al. (2009) | 91 | .006 |
| Pleural disruption | Yes | Kempf-Bielack et al. (2005) | 436 | <.0001 |

event-free and OS after relapse (Bacci et al., 2005). Similar observations found that a longer interval to relapse was associated with better survival with a threshold of 12 (Iwata et al., 2015; Harting et al., 2006), 18 (Leary et al., 2013), and 24 months (Bacci et al., 2001; Hawkins and Arndt, 2003; Ferrari et al., 2003; Chou et al., 2005).

Metastatic distribution inside the lung (central or peripheral) was also identified as a predictor of survival of metastatic OS patients. Letourneau et al. clarified the impact of the location of the metastatic nodule in pediatric OS patients (Letourneau et al., 2011). According to their results, the median OS period after the first pulmonary metastasis was 1.06 years among patients with a peripheral location as compared with 0.38 years among those with a centrally located lesion ($p = .008$). The authors presumed that there might be an association between central metastasis and occurrence of extra-pulmonary metastasis.

Other reported prognostic factors in the literature are laterality (unilateral or bilateral; Bielack et al., 2009; Kempf-Bielack et al., 2005), number of involved lobes (Wu et al., 2009; Carter et al., 1991), and pleural disruption (Kempf-Bielack et al., 2005).

Considering extra-pulmonary metastasis, bone metastasis is associated with poor prognosis. Some reports suggested that relapse with bone metastasis tended to have a poorer prognosis than pulmonary metastasis alone (Kager et al., 2003; Ferrari et al., 2003; Leary et al., 2013). Kager et al. reviewed their data from the COSS and reported that only 8 (6.8%) of 117 patients with bone metastasis at diagnosis survived more than 3 years after diagnosis (Kager et al., 2003). Practically, the American Joint Committee of Cancer (AJCC) staging for bone tumor separated stage IVB (extra-pulmonary metastasis) from IVA (lung-only metastasis) based on apparent differences in prognosis.

Skip metastasis, defined as a smaller discontinuous lesion occurring synchronously in the same bone, is a rare event observed in only 1.4–6.5% of all OS patients (Kager et al., 2003, 2006; Sajadi et al., 2004) and has also been reported as an indicator of poor prognosis (Kager et al., 2006; Sajadi et al., 2004). Nevertheless, descriptions vary regarding the degree of impact of skip metastasis on poor prognosis. Sajadi et al. suggested that survival of patients with skip metastasis is significantly poorer than for distant metastasis (Sajadi et al., 2004), whereas Kager et al. reported a superior outcome of a 5-year survival rate of 62% among patients with skip metastasis, compared to these other dismal results (Kager et al., 2006). The most recent AJCC staging system defines OS patients with only skip metastasis and no other detectable metastatic disease at presentation as stage III, which was placed between stage II (localized disease) and stage IV (distant metastatic disease).

Lymph node metastasis is also a very rare event in OS with a reported incidence of 0.7–10% (Bielack et al., 2002; Thampi et al., 2013; Caceres et al., 1990). According to a large cohort study conducted in the United States, 74 (2.7%) patients with high-grade OS and regional lymph node involvement at diagnosis had a tendency for more aggressive behavior, such as an extraskeletal tumor, association with distant metastasis, and larger tumors >10cm. As expected, the outcome of those patients was dismal, showing a 5-year OS rate of only 10.9% (Thampi et al., 2013).

# 3. TREATMENT FOR METASTASIS FROM OSTEOSARCOMA

## 3.1 Surgical Treatment

Surgical resection is a mainstay of local treatment for metastasis from OS, especially for pulmonary metastasis. In general, OS is regarded as a tumor that can be cured with complete remission of all pulmonary metastases. From this point of view, achievement of complete resection (CR) of all pulmonary metastases strongly affects prognosis after relapse (Kager et al., 2003; Bielack et al., 2009; Hawkins and Arndt, 2003; Iwata et al., 2015; Ferrari et al., 2003; Leary et al., 2013; Chou et al., 2005; Salah et al., 2013), whereas treatment without surgical CR never obtains long survival. The reported 5-year OS rate after surgical CR of pulmonary metastases ranges from 32% to 39%, whereas that without surgical CR ranges from 0% to 2% (Bielack et al., 2009; Ferrari et al., 2003). Leary et al. reported a statistically significant difference among 41 (80%) of 51 patients with unilateral pulmonary metastases and 12 (39%) of 31 patients with bilateral pulmonary metastases who achieved CR ($P=.0002$) Mankin and Hornicek (2005). Other factors involved in the achievement of surgical CR were patient age, number of relapsed nodules, recurrence-free interval, and histologic response to preoperative chemotherapy (Bielack et al., 2009; Kempf-Bielack et al., 2005; Bacci et al., 2005). Recently, patients with only limited metastases were recognized as eligible for local treatment since they might achieve long-term survival or cure. Niibe et al. proposed the concept of "oligo-recurrence" (Niibe and Chang, 2012; Niibe and Hayakawa, 2010). In this state, cancer patients with one to five metastatic or recurrent lesions with a controlled primary lesion could be treated with local therapy, which might result in a more favorable prognosis. Iwata et al. attempted to determine the oligo-recurrence status of pulmonary metastatic OS patients through prognostic factor analysis. They proposed that surgical CR, an interval of more than 1year from previous thoracotomy, up to three resected nodules, and a total size of less than 30mm for pulmonary metastases were conditions that determined oligo-recurrence status in those patients and that the disease should be treated more aggressively (Iwata et al., 2015).

Concerning the type of thoracotomy, wedge resection is most frequently performed for pulmonary metastatic OS. Although this surgical procedure is dependent on the number or distribution of nodules, in 80–85% of the cases, the procedure involves wedge resection, whereas lobectomy and pneumonectomy are involved in 10–15% and 5% of the cases, respectively (Briccoli et al., 2005; Bacci et al., 2005; Letourneau et al., 2011; Salah et al., 2013; Blackmon et al., 2009). Letourneau et al. demonstrated that the prevalence of morbidity following wedge resection was similar to that for lobectomy/pneumonectomy (9% vs. 8%), with no statistical significance between median survival rates of the patients undergoing lobectomy and those undergoing wedge resection (0.61 years vs. 1.14 years; Letourneau et al., 2011).

Video-assisted thoracoscopic surgery (VATS) is increasingly used for the resection of pulmonary metastasis because it is less invasive compared to other procedures. As noted by Patrick et al., VATS enable better visualization of the pleura as well as of the pulmonary surface by magnification using an optic system that enables the surgeon to easily identify nodules on the pleural surface (Partrick et al., 2002). On the other hand, as described by a large number of studies in the literature, intraoperative palpation of a pulmonary metastatic nodule is a useful technique to detect smaller or deeper nodules undetectable by CT. Previous studies reported that 11–47% of patients had nonimaged but palpated lung metastasis following thoracotomy (Kayton et al., 2006; Parsons et al., 2007).

The benefit of contralateral exploratory thoracotomy for unilateral metastasis is debatable. Su et al. stated that the patients with unilateral nodules detected by CT within 2 years from diagnosis demonstrated a high incidence of subsequent contralateral metastasis, and therefore, bilateral exploratory thoracotomy should be considered (Su et al., 2004). In contrast, Karplus et al. argued that there is no statistical significant difference between the incidences of subsequent pulmonary relapse in the ipsilateral and contralateral lung (Karplus et al., 2009).

## 3.2 Radiotherapy

In general, OS is considered to be resistant to radiotherapy. Thus, limited data are available describing the efficacy of radiotherapy for the treatment of metastatic lesions in OS. No benefit of elective prophylactic lung radiation therapy has been confirmed in the past literature. Rab et al. reported the results of a randomized study of OS patients without pulmonary metastasis who were assigned into two groups, namely patients who received elective prophylactic lung radiation therapy and those who did not (Rab et al., 1976). The results showed no significant difference between groups. Similarly, the efficacy of adjuvant lung radiation therapy for the OS patients after thoracotomy remains unclear (Burgers et al., 1988).

Stereotactic body radiotherapy (SBRT) is a potential treatment option for pulmonary metastasis of OS because this technique enables delivery of higher dose radiation to metastatic lesions than conventional radiotherapy. Some recent reports described the utility of SBRT in either a definitive or palliative setting (Yu et al., 2014; Mehta et al., 2013; Brown et al., 2014). Yu et al. reported similar outcomes between SBRT and thoracotomy for OS (Yu et al., 2014). Favorable outcomes have also been described in other reports, although toxicity was rather severe, resulting in pneumonitis, myonecrosis, avascular necrosis, and neuropathy (Mehta et al., 2013; Brown et al., 2014).

## 3.3 Emerging Treatment Options

Besides the above-described conventional treatment methods for metastatic disease, other emerging treatment techniques have also been reported. Of these, radiofrequency ablation (RFA) has been documented as an effective option for the treatment for metastasis (Fig. 12.2).

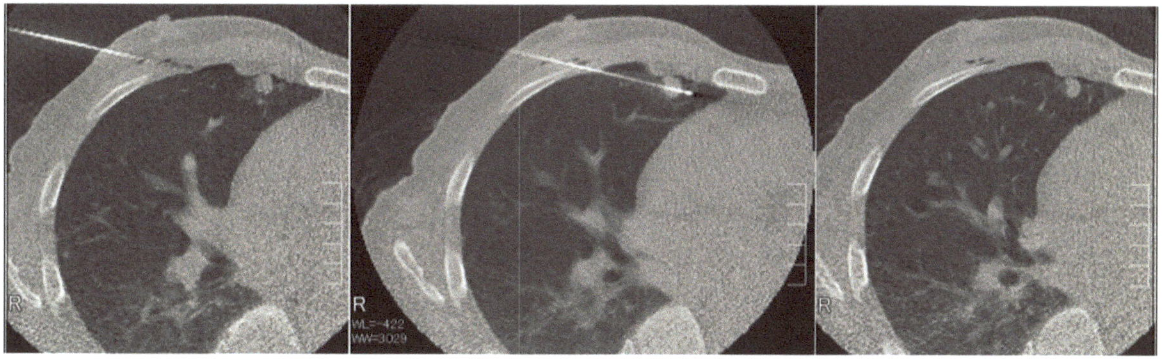


FIGURE 12.2 Radiofrequency ablation for the pulmonary metastasis from osteosarcoma. *Case provided by Dr. Tomoki Nakamura, Mie University Graduate School of Medicine, Mie, Japan.*

RFA has been performed using a percutaneously inserted electrode under image guidance, which deposits heat energy to cause focal coagulation necrosis. Because it is less invasive relative to thoracotomy, RFA is acceptable as a minimally invasive treatment for multiple metastases, for which it is impossible to achieve a complete resection. To date, there are relatively few reports regarding the therapeutic outcomes of RFA for the treatment of metastatic sarcoma (Palussiere et al., 2011; Yamakado et al., 2014; Yevich et al., 2015). Yamakado et al. reviewed 52 patients with unresectable pulmonary and extra-pulmonary metastasis from various sarcomas, including OS, and found that complete ablation of all metastatic lesions was achieved in 21 patients (40.4%), whereas 15 patients (28.8%) were disease free at the end of follow-up (Yamakado et al., 2014).

Targeted radiation therapy using samarium-153 ethylenediaminetetramethylene phosphonic acid ($^{153}$Sm-EDTMP) is reportedly a valid alternative to conventional surgery or radiation therapy for bone metastasis of OS. $^{153}$Sm-EDTMP is a radiopharmaceutical agent that selectively delivers radiation to osteoblastic skeletal lesions and shows rapid blood clearance of unbound $^{153}$Sm-EDTMP, although residual amounts are detectable on bone surfaces for months after treatment (Anderson et al., 2002; Goeckeler et al., 1987). Some reports have stated that high-dose $^{153}$Sm-EDTMP treatment is feasible to accompany hematopoietic stem cell support for advanced OS, although its efficacy is debatable (Anderson et al., 2002; Loeb et al., 2010; Berger et al., 2012).

## 4. CONCLUSION

Since pulmonary metastasis is a primary cause of mortality among patients with OS, clinicians have attempted to control these lesions for many years, and increasing evidence supports several appropriate treatment strategies for these patients. To achieve both better care and a potential cure, we should continue making every effort to pursue the best treatment options for OS patients.

## References

A Randomized Trial of the European and American Osteosarcoma Study Group to Optimize Treatment Strategies for Resectable Osteosarcoma Based on Histological Response to Pre-Operative Chemotherapy Study Protocol.

Anderson, P.M., Wiseman, G.A., Dispenzieri, A., et al., 2002. High-dose samarium-153 ethylene diamine tetramethylene phosphonate: low toxicity of skeletal irradiation in patients with osteosarcoma and bone metastases. J. Clin. Oncol. 20 (1), 189–196.

Bacci, G., Ferrari, S., Donati, D., et al., 1998. Neoadjuvant chemotherapy for osteosarcoma of the extremity in patients in the fourth and fifth decade of life. Oncol. Rep. 5 (5), 1259–1263.

Bacci, G., Ferrari, S., Longhi, A., et al., 2001. Pattern of relapse in patients with osteosarcoma of the extremities treated with neoadjuvant chemotherapy. Eur. J. Cancer 37 (1), 32–38.

Bacci, G., Briccoli, A., Longhi, A., et al., 2005. Treatment and outcome of recurrent osteosarcoma: experience at Rizzoli in 235 patients initially treated with neoadjuvant chemotherapy. Acta. Oncol. 44 (7), 748–755.

Bacci, G., Rocca, M., Salone, M., et al., 2008. High grade osteosarcoma of the extremities with lung metastases at presentation: treatment with neoadjuvant chemotherapy and simultaneous resection of primary and metastatic lesions. J. Surg. Oncol. 98 (6), 415–420.

Berger, M., Grignani, G., Giostra, A., et al., 2012. 153Samarium-EDTMP administration followed by hematopoietic stem cell support for bone metastases in osteosarcoma patients. Ann. Oncol. 23 (7), 1899–1905.

Bielack, S.S., Kempf-Bielack, B., Delling, G., et al., 2002. Prognostic factors in high-grade osteosarcoma of the extremities or trunk: an analysis of 1,702 patients treated on neoadjuvant Cooperative Osteosarcoma Study Group protocols. J. Clin. Oncol. 20 (3), 776–790.

Bielack, S.S., Kempf-Bielack, B., Branscheid, D., et al., 2009. Second and subsequent recurrences of osteosarcoma: presentation, treatment, and outcomes of 249 consecutive Cooperative Osteosarcoma Study Group patients. J. Clin. Oncol. 27 (4), 557–565.

Blackmon, S.H., Shah, N., Roth, J.A., et al., 2009. Resection of pulmonary and extrapulmonary sarcomatous metastases is associated with long-term survival. Ann. Thorac. Surg. 88 (3), 877–884 discussion 84–5.

Briccoli, A., Rocca, M., Salone, M., et al., 2005. Resection of recurrent pulmonary metastases in patients with osteosarcoma. Cancer 104 (8), 1721–1725.

Brown, L.C., Lester, R.A., Grams, M.P., et al., 2014. Stereotactic body radiotherapy for metastatic and recurrent Ewing sarcoma and osteosarcoma. Sarcoma 2014 418270.

Burgers, J.M., van Glabbeke, M., Busson, A., et al., 1988. Osteosarcoma of the limbs. Report of the EORTC-SIOP 03 trial 20781 investigating the value of adjuvant treatment with chemotherapy and/or prophylactic lung irradiation. Cancer 61 (5), 1024–1031.

Caceres, E., Zaharia, M., Calderon, R., 1990. Incidence of regional lymph node metastasis in operable osteogenic sarcoma. Semin. Surg. Oncol. 6 (4), 231–233.

Carsi, B., Rock, M.G., 2002. Primary osteosarcoma in adults older than 40 years. Clin. Orthop. Relat. Res. 397, 53–61.

Carter, S.R., Grimer, R.J., Sneath, R.S., Matthews, H.R., 1991. Results of thoracotomy in osteogenic sarcoma with pulmonary metastases. Thorax 46 (10), 727–731.

Chou, A.J., Merola, P.R., Wexler, L.H., et al., 2005. Treatment of osteosarcoma at first recurrence after contemporary therapy: the Memorial Sloan-Kettering Cancer Center experience. Cancer 104 (10), 2214–2221.

Ferrari, S., Briccoli, A., Mercuri, M., et al., 2003. Postrelapse survival in osteosarcoma of the extremities: prognostic factors for long-term survival. J. Clin. Oncol. 21 (4), 710–715.

Goeckeler, W.F., Edwards, B., Volkert, W.A., Holmes, R.A., Simon, J., Wilson, D., 1987. Skeletal localization of samarium-153 chelates: potential therapeutic bone agents. J. Nucl. Med. 28 (4), 495–504.

Grampp, S., Bankier, A.A., Zoubek, A., et al., 2000. Spiral CT of the lung in children with malignant extra-thoracic tumors: distribution of benign vs malignant pulmonary nodules. Eur. Radiol. 10 (8), 1318–1322.

Grimer, R.J., Carter, S.R., Tillman, R.M., Spooner, D., Mangham, D.C., Kabukcuoglu, Y., 1999. Osteosarcoma of the pelvis. J. Bone Jt. Surg. Br. 81 (5), 796–802.

Grimer, R.J., Cannon, S.R., Taminiau, A.M., et al., 2003. Osteosarcoma over the age of forty. Eur. J. Cancer 39 (2), 157–163.

Harting, M.T., Blakely, M.L., Jaffe, N., et al., 2006. Long-term survival after aggressive resection of pulmonary metastases among children and adolescents with osteosarcoma. J. Pediatr. Surg. 41 (1), 194–199.

Hawkins, D.S., Arndt, C.A., 2003. Pattern of disease recurrence and prognostic factors in patients with osteosarcoma treated with contemporary chemotherapy. Cancer 98 (11), 2447–2456.

Iwata, S., Ishii, T., Kawai, A., et al., 2014. Prognostic factors in elderly osteosarcoma patients: a multi-institutional retrospective study of 86 cases. Ann. Surg. Oncol. 21 (1), 263–268.

Iwata, S., Yonemoto, T., Iizasa, T., Niibe, Y., Kamoda, H., Ishii, T., 2015. Oligo-recurrence of osteosarcoma patients: treatment strategies for pulmonary metastases. Ann. Surg. Oncol. 22 (Suppl. 3).

Kager, L., Zoubek, A., Potschger, U., et al., 2003. Primary metastatic osteosarcoma: presentation and outcome of patients treated on neoadjuvant Cooperative Osteosarcoma Study Group protocols. J. Clin. Oncol. 21 (10), 2011–2018.

Kager, L., Zoubek, A., Kastner, U., et al., 2006. Skip metastases in osteosarcoma: experience of the Cooperative Osteosarcoma Study Group. J. Clin. Oncol. 24 (10), 1535–1541.

Karplus, G., McCarville, M.B., Smeltzer, M.P., et al., 2009. Should contralateral exploratory thoracotomy be advocated for children with osteosarcoma and early unilateral pulmonary metastases? J. Pediatr. Surg. 44 (4), 665–671.

Kawai, A., Huvos, A.G., Meyers, P.A., Healey, J.H., 1998. Osteosarcoma of the pelvis. Oncologic results of 40 patients. Clin. Orthop. Relat. Res. (348), 196–207.

Kayton, M.L., Huvos, A.G., Casher, J., et al., 2006. Computed tomographic scan of the chest underestimates the number of metastatic lesions in osteosarcoma. J. Pediatr. Surg. 41 (1), 200–206 discussion -6.

Kempf-Bielack, B., Bielack, S.S., Jurgens, H., et al., 2005. Osteosarcoma relapse after combined modality therapy: an analysis of unselected patients in the Cooperative Osteosarcoma Study Group (COSS). J. Clin. Oncol. 23 (3), 559–568.

Leary, S.E., Wozniak, A.W., Billups, C.A., et al., 2013. Survival of pediatric patients after relapsed osteosarcoma: the St. Jude Children's Research Hospital experience. Cancer 119 (14), 2645–2653.

Letourneau, P.A., Xiao, L., Harting, M.T., et al., 2011. Location of pulmonary metastasis in pediatric osteosarcoma is predictive of outcome. J. Pediatr. Surg. 46 (7), 1333–1337.

Loeb, D.M., Hobbs, R.F., Okoli, A., et al., 2010. Tandem dosing of samarium-153 ethylenediamine tetramethylene phosphoric acid with stem cell support for patients with high-risk osteosarcoma. Cancer 116 (23), 5470–5478.

Longhi, A., Errani, C., Gonzales-Arabio, D., Ferrari, C., Mercuri, M., 2008. Osteosarcoma in patients older than 65 years. J. Clin. Oncol. 26 (33), 5368–5373.

Mangham, D.C., Davie, M.W., Grimer, R.J., 2009. Sarcoma arising in Paget's disease of bone: declining incidence and increasing age at presentation. Bone 44 (3), 431–436.

Mankin, H.J., Hornicek, F.J., 2005. Paget's sarcoma: a historical and outcome review. Clin. Orthopaedics Relat. Res. 438, 97–102.

Manoso, M.W., Healey, J.H., Boland, P.J., et al., 2005. De novo osteogenic sarcoma in patients older than forty: benefit of multimodality therapy. Clin. Orthop. Relat. Res. 438, 110–115.

McCarville, M.B., Lederman, H.M., Santana, V.M., et al., 2006. Distinguishing benign from malignant pulmonary nodules with helical chest CT in children with malignant solid tumors. Radiology 239 (2), 514–520.

Mehta, N., Selch, M., Wang, P.C., et al., 2013. Safety and efficacy of stereotactic body radiation therapy in the treatment of pulmonary metastases from high grade sarcoma. Sarcoma 2013, 360214.

Meyer, J.S., Nadel, H.R., Marina, N., et al., 2008. Imaging guidelines for children with Ewing sarcoma and osteosarcoma: a report from the Children's Oncology Group Bone Tumor Committee. Pediatr. Blood Cancer 51 (2), 163–170.

Meyers, P.A., Gorlick, R., 1997. Osteosarcoma. Pediatr. Clin. North Am. 44 (4), 973–989.

Meyers, P.A., Heller, G., Healey, J.H., et al., 1993. Osteogenic sarcoma with clinically detectable metastasis at initial presentation. J. Clin. Oncol. 11 (3), 449–453.

Miller, B.J., Cram, P., Lynch, C.F., Buckwalter, J.A., 2013. Risk factors for metastatic disease at presentation with osteosarcoma: an analysis of the SEER database. J. Bone Jt. Surg. Am. 95 (13), e89.

Niibe, Y., Chang, J.Y., 2012. Novel insights of oligometastases and oligo-recurrence and review of the literature. Pulm. Med. 2012, 261096.

Niibe, Y., Hayakawa, K., 2010. Oligometastases and oligo-recurrence: the new era of cancer therapy. Jpn. J. Clin. Oncol. 40 (2), 107–111.

Nishida, Y., Isu, K., Ueda, T., et al., 2009. Osteosarcoma in the elderly over 60 years: a multicenter study by the Japanese Musculoskeletal Oncology Group. J. Surg. Oncol. 100 (1), 48–54.

Ozaki, T., Flege, S., Liljenqvist, U., et al., 2002. Osteosarcoma of the spine: experience of the Cooperative Osteosarcoma Study Group. Cancer 94 (4), 1069–1077.

Palussiere, J., Italiano, A., Descat, E., et al., 2011. Sarcoma lung metastases treated with percutaneous radiofrequency ablation: results from 29 patients. Ann. Surg. Oncol. 18 (13), 3771–3777.

Parsons, A.M., Ennis, E.K., Yankaskas, B.C., Parker Jr., L.A., Hyslop, W.B., Detterbeck, F.C., 2007. Helical computed tomography inaccuracy in the detection of pulmonary metastases: can it be improved? Ann. Thorac. Surg. 84 (6), 1830–1836.

Partrick, D.A., Bensard, D.D., Teitelbaum, D.H., Geiger, J.D., Strouse, P., Harned, R.K., 2002. Successful thoracoscopic lung biopsy in children utilizing preoperative CT-guided localization. J. Pediatr. Surg. 37 (7), 970–973 discussion -3.

Rab, G.T., Ivins, J.C., Childs Jr., D.S., Cupps, R.E., Pritchard, D.J., 1976. Elective whole lung irradiation in the treatment of osteogenic sarcoma. Cancer 38 (2), 939–942.

Robertson, P.L., Boldt, D.W., De Campo, J.F., 1988. Paediatric pulmonary nodules: a comparison of computed tomography, thoracotomy findings and histology. Clin. Radiol. 39 (6), 607–610.

Ruggieri, P., Calabro, T., Montalti, M., Mercuri, M., 2010. The role of surgery and adjuvants to survival in Pagetic osteosarcoma. Clin. Orthop. Relat. Res. 468 (11), 2962–2968.

Sajadi, K.R., Heck, R.K., Neel, M.D., et al., 2004. The incidence and prognosis of osteosarcoma skip metastases. Clin. Orthop. Relat. Res. 426, 92–96.

Salah, S., Toubasi, S., 2015. Factors predicting survival following complete surgical remission of pulmonary metastasis in osteosarcoma. Mol. Clin. Oncol. 3 (1), 157–162.

Salah, S., Fayoumi, S., Alibraheem, A., et al., 2013. The influence of pulmonary metastasectomy on survival in osteosarcoma and soft-tissue sarcomas: a retrospective analysis of survival outcomes, hospitalizations and requirements of home oxygen therapy. Interact. Cardiovasc. Thorac. Surg. 17 (2), 296–302.

Su, W.T., Chewning, J., Abramson, S., et al., 2004. Surgical management and outcome of osteosarcoma patients with unilateral pulmonary metastases. J. Pediatr. Surg. 39 (3), 418–423 discussion -23.

Thampi, S., Matthay, K.K., Goldsby, R., DuBois, S.G., 2013. Adverse impact of regional lymph node involvement in osteosarcoma. Eur. J. Cancer 49 (16), 3471–3476.

Tsuchiya, H., Kanazawa, Y., Abdel-Wanis, M.E., et al., 2002. Effect of timing of pulmonary metastases identification on prognosis of patients with osteosarcoma: the Japanese Musculoskeletal Oncology Group study. J. Clin. Oncol. 20 (16), 3470–3477.

Vanel, D., Henry-Amar, M., Lumbroso, J., et al., 1984. Pulmonary evaluation of patients with osteosarcoma: roles of standard radiography, tomography, CT, scintigraphy, and tomoscintigraphy. AJR Am. J. Roentgenol. 143 (3), 519–523.

Wu, P.K., Chen, W.M., Chen, C.F., Lee, O.K., Haung, C.K., Chen, T.H., 2009. Primary osteogenic sarcoma with pulmonary metastasis: clinical results and prognostic factors in 91 patients. Jpn. J. Clin. Oncol. 39 (8), 514–522.

Yamakado, K., Matsumine, A., Nakamura, T., et al., 2014. Radiofrequency ablation for the treatment of recurrent bone and soft-tissue sarcomas in non-surgical candidates. Int. J. Clin. Oncol. 19 (5), 955–962.

Yevich, S., Gaspar, N., Tselikas, L., et al., 2015. Percutaneous computed tomography-guided thermal ablation of pulmonary osteosarcoma metastases in children. Ann. Surg. Oncol. 23 (4).

Yonemoto, T., Tatezaki, S., Ishii, T., Satoh, T., Kimura, H., Iwai, N., 1998. Prognosis of osteosarcoma with pulmonary metastases at initial presentation is not dismal. Clin. Orthop. Relat. Res. (349), 194–199.

Yu, W., Tang, L., Lin, F., et al., 2014. Stereotactic radiosurgery, a potential alternative treatment for pulmonary metastases from osteosarcoma. Int. J. Oncol. 44 (4), 1091–1098.

PART II

# CANCER METASTASIS TO DISTANT ORGANS

CHAPTER

# 13

# Lymph Node Metastasis*

*S.D. Nathanson, K. Rosso, D. Chitale, M. Burke*

**Henry Ford Health System, Detroit, MI, United States**

## 1. INTRODUCTION

In vertebrates, malignant tumors of ectodermal or endodermal origin may spontaneously metastasize to the regional lymph nodes (RLNs), starting first in the sentinel lymph node (SLN) (Nathanson et al., 2015). There is an anatomic and mechanical logic to this pattern of spread (Reintgen et al., 1994): tumor cells enter the initial lymphatics (Mohammed et al., 2007; Hoshida et al., 2006), aided by increased pressure in the primary tumor (Nathanson, 2003; Butler et al., 1975; Boucher and Jain, 1992; Jain et al., 2014; Nathanson and Nelson, 1994) and increased flow rates through the collecting lymphatics, and the lymphatic trunks allow more tumor cells to flow per unit time into the afferent lymphatics (Hoshida et al., 2006; Nathanson et al., 1997; Avery et al., 1992) and into the subcapsular sinus of the SLN (Sainte-Marie et al., 1982; Kelly, 1975; Belisle and Sainte-Marie, 1990; Gretz et al., 1997). However, the anatomy is not the only factor influencing the journey that tumor cells follow to reach the SLN. An interacting array of complex biochemical, genetic, and other molecular events facilitates the process.

Our current understanding of the mechanisms of lymph node metastasis (LNMets) is based upon a combination of clinical and basic research, in vitro and in vivo, that includes detailed studies of lymphatic embryology (Tammela and Alitalo, 2010), anatomy (Gray, 1939), microanatomy, pathology (Sainte-Marie et al., 1982), physiology, molecular biology (Alitalo, 2011), and mechanobiology (Hoshida et al., 2006; Butler et al., 1975; Boucher and Jain, 1992; Jain et al., 2014; Nathanson and Nelson, 1994; Avery et al., 1992; Jonietz, 2012; Pathak and Kumar, 2011; Padera et al., 2004; Friedl and Alexander, 2011; Mitchell and King, 2013; Shieh, 2011; McDonald, 2008). Many of the biochemical and molecular studies in systemic metastasis (Sleeman et al., 2011; Fidler, 1973) probably apply in some way to LNMets as well.

Shear stresses and biomechanical signals from altered tumor vascular pathophysiology influence cellular function (Jonietz, 2012; Ingber, 1998; Resnick et al., 2003; Topper and Gimbrone Jr, 1999; Ng et al., 2005; Pedersen et al., 2010), exert changes in peritumoral stroma (Swartz et al., 2014; Kalluri and Zeisberg, 2006; Kessenbrock et al., 2012; Orimo et al., 2002;

*Funded in part by the Nathanson/Rands Chair in Breast Cancer Research. Artwork by Kelly Rosso, MD, and Dhananjay Chitale, MD.


http://dx.doi.org/10.1016/B978-0-12-804003-4.00013-X

Adams et al., 2002), alter entry sites into lymphatic capillaries (McDonald, 2008; Hirakawa et al., 2003; Baluk et al., 2007), induce changes in the tumor microenvironment (Kuhn et al., 2015), influence neoplastic progression (Swartz et al., 2014), induce angiogenesis and lymphangiogenesis (Orimo et al., 2002; Stacker et al., 2014; Alitalo and Carmeliet, 2002), produce an increase in matrix stiffness (Shieh, 2011), induce an increase in focal adhesion molecules (Shieh, 2011; Pang et al., 2015), enhance migration of tumor cells (Joyce and Pollard, 2009), and increase the likelihood of invasion.

The tumor microenvironment plays a vital role in the pathway to LNMets (Nathanson et al., 2015; Swartz et al., 2014; Stacker et al., 2014). Tumor cells preferentially invade new peri tumoral lymphatic capillaries (McDonald, 2008; He et al., 2005; Skobe et al., 2001) by both active and passive (Nathanson et al., 2015; Hoshida et al., 2006; Nathanson, 2003; Nathanson et al., 2009a; Padera et al., 2004; Karaman and Detmar, 2014; Baluk et al., 2003) processes. There is indirect evidence for occasional direct invasion into blood vessels (BVs) at the primary tumor site, but no direct observation in human tumors. Epidemiological, clinical, and pathology studies strongly suggest that tumors invade peritumoral lymphatics and gain access to the systemic circulation through the SLN (Reintgen et al., 1994; Baluk et al., 2003; Nathanson et al., 1986, 2009; Carr, 1983). In contrast, tumors arising from mesodermal structures tend to primarily invade BVs and rarely metastasize to LNs (Stelzner et al., 2013). Metastasis genes and molecular factors may play a role in this pathogenesis.

The natural history of LNMets is vital to the management of the common human tumors and, although there are unique aspects to the LNMets of the most common tumors seen by clinicians [lung, breast, prostate, colon, rectum, pancreas, melanoma (skin), bladder, uterus (cervix and endometrium), ovaries, thyroid, kidney, upper aero-digestive tract], there are many similarities among the patterns of spread in those primary sites, which we will focus upon in this chapter.

The clinical importance of LNMets is apparent by the significant impact this event has on staging and prognosis (Fisher, 1980). Staging is incomplete without a determination of the clinical and pathologic status of the RLNs. The clinical diagnosis is heavily dependent upon clinical and radiological observations but the gold standard, the ultimate documentation that provides the important information necessary for staging, is the pathologic evaluation of the biopsied node(s).

Clinical palpation of RLNs is, at best, a relatively crude method for estimating the likelihood of metastasis. The clinician can only determine whether the node(s) are enlarged and (usually) firm, with a significant likelihood of false negative and false positive conclusions. Radiologic evaluation of LNs is more sensitive and specific than clinical examination of RLNs but it is also subject to fairly high rates of false negative and false positive findings.

Prevention of LNMets has not been possible so far although there is some hope that molecular therapy, possibly aimed at unique tumor targets, might one day be uniquely available for treatment of these metastases. Current methods of treating RLNs harboring metastases depend primarily on surgery, radiation, chemotherapy, and targeted therapy.

The surgical management of LNMets has changed dramatically in breast cancer over the past few years (Giuliano et al., 2011; Rutgers et al., 2013; Veronesi et al., 2010; Gentilini and Veronesi, 2012), where the likelihood of complete axillary lymph node dissection (CALND), even when the SLN contains metastases, is markedly decreased. The same cannot be said for tumors arising from other anatomic sites. Although still vigorously debated, complete lymph node dissection (CLND) is strongly advocated for patients with SLN metastasis in melanoma (Morton et al., 2014) and penile cancer. The use of SLN biopsy is still not widely practiced for most of the other primary sites, where CLND is still the standard approach. The rationale for CLND is thought to be important for both staging and treatment in many (if not most) gynecologic, gastrointestinal, aero-digestive, genitourinary, and thyroid primary cancers.

## 2. HISTORY

LNMets in cancer patients were undoubtedly recognized before the nineteenth century when autopsy evaluation of patients dying of cancer revealed massive tumors, often indistinguishable from their enlarged and involved RLNs. Virchow first noted a distinct separation between the primary tumor and some draining LNs (Virchow, 1860) when he identified supraclavicular LNMets from gastric cancer and postulated that the LNs filtered tumor cells and harmful particles from gaining access to the systemic circulation.

Surgeons developed operations to excise primary cancers and their draining LNs, initially with gastric (Weil and Buchberger, 1999) and breast cancers (Halsted, 1907; Rosenberg, 2008). Halsted based his radical mastectomy on the belief that breast cancers spread via lymphatics to the axillary LNs. He also believed that the tumor needed to be removed together with the entire breast plus all the LNs in the axilla. Because involved LNs were found behind the pectoralis muscles it seemed likely that lymphatics from the breast directly traversed the muscles to enter the LNs. It was assumed that cutting across those lymphatics would release tumor cells into the wound and result in the local chest wall recurrence. Therefore, these muscles were removed completely with the breast and the LNs in one piece ("en bloc").

Halsted's procedure "cured" about one-third of his patients (Halsted, 1907), a major advance for the time. Radical operations soon became the aim of surgeries for all cancers of epithelial origin. By the 1930s radical "en bloc" resection of cancers of the rectum, colon, stomach, uterus, lung, and pancreas was the standard of practice globally (Rosenberg, 2008). Some surgeons advocated even more radical operations (Urban and Baker, 1952), believing that such physically and physiologically deforming operations were justified because they would likely cure more people of their disease.

The development of more modern techniques of epidemiological and clinical research in the second half of the twentieth century encouraged surgeons to question some of the more radical surgical approaches to cancer. In the United Kingdom, possibly based on the demonstration in Detroit that lymphatic vessels (LVs) did not gain access to axillary LNs by passing through but around the lateral edge of the pectoralis muscles (Gray, 1939), modified radical mastectomy (MRM) was developed (Patey and Dyson, 1948). Surgeons were initially not convinced that MRM would be as effective as radical mastectomy at controlling locoregional breast cancer and it took another 25 years before the MRM became the standard of care (Rosenberg, 2008). It was soon recognized that, in addition to the better cosmetic and physical results with more intact anatomy, the cancer outcomes were not compromised.

LN surgery for melanoma, breast, and penile cancer paved the way for even less radical operative intervention when the era of sentinel lymph node biopsy (SLNB) was introduced. The sentinel node concept, first in human penile cancer (Cabanas, 1977), then in mice with footpad melanomas (Nathanson et al., 1986), then in cats, followed shortly thereafter in human melanomas and breast cancer (Giuliano et al., 2011; Rutgers et al., 2013; Veronesi et al., 2010; Morton et al., 1992), demonstrated an obvious truth: there is no logical rationale for radical surgical techniques to remove LNs not harboring tumor. Within the decade of the 1990s, it was accepted that SLNB was an accurate way to stage patients with these malignant neoplasms. The prior dogma that CLND surgery was universally therapeutic, especially in breast cancer (Halsted, 1907), changed. In breast cancer, SLNB without CALND has proved to be very effective at controlling local, regional, and systemic disease (Giuliano et al., 2011; Rutgers et al., 2013; Veronesi et al., 2010), probably because of the added effectiveness of radiation plus systemic therapy (chemotherapy, targeted therapy, and hormonal manipulation in selected patients). The same is not yet true for other cancers where the use of SLNB is still being investigated but, with more spectacular research findings in molecular and immuno-therapies, it seems likely that highly effective systemic therapy will soon be available for those tumors as well, at which time CLND may not be necessary for the majority of patients.

The SLN era has prompted advances in not only surgery but also radiological techniques to diagnose metastasis, focused pathological examination of LNs, and efforts to understand the mechanisms whereby tumors metastasize to SLNs (Nathanson et al., 2015).

## 3. MECHANISMS—HOW TUMORS METASTASIZE TO LYMPH NODES

### 3.1 Molecular Aspects

The precise mechanisms by which tumors metastasize to LNs have yet to be fully identified. Lymphatics play a key role in pathologic inflammatory responses and the progression of cancer metastasis. Molecular receptors unique to lymphatic endothelial cells (LECs) include Prox1, podoplanin, VEGF-receptors (VEGF-R), endothelial hyaluron receptor-1 (LYVE-1), integrin α9β1, LYP-1, neuropilin-2, angiopoietin-1, and CCL21 and probably play an important role in guiding tumor cells to LVs and enhance LNMets (Nathanson et al., 2015). Clinical and basic science studies have suggested that the rich and dynamic microenvironment of the LN is influenced genetically by changes in the surrounding biochemical and mechanical milieu. Understanding the role of LECs in cancer metastasis has been advanced by knowledge of their many receptor-mediated activation pathways, downstream signaling, and crosstalk with cancer cells.

The VEGF family promotes and regulates angiogenesis during many important physiological as well as pathological processes. Five structurally related ligands (VEGF-A, -B, -C, -D, and placental growth factor) bind their unique coreceptors and induce different biological responses through both autocrine and paracrine signaling pathways (Ferrara et al., 2003). In animal models, VEGF-C secreted by the primary tumor has been shown to induce expansion and hyperplasia of the lymphatic network within the premetastatic SLN. These findings suggest that growth factors secreted from the primary tumor can remodel the SLN in preparation for metastasis, the so-called premetastatic niche (Nathanson et al., 2015; Sleeman et al., 2011; Qian et al., 2006). Lymphangiogenesis increased LNMets and the presence of nodal metastasis further increased lymphangiogenesis (Hoshida et al., 2006; Qian et al., 2006). Tumors expressing VEGF-C are more likely to metastasize systemically from the tumor-involved LN. VEGF-C, therefore, may play an important role in the production of lymphatic-blood vessel capillary connections that may allow cancer cells to travel from the LN to the systemic circulation (Hirakawa et al., 2007; Liersch et al., 2012).

Extracellular matrix growth factors and proteins may further play a role in the differentiation of LECs and lymphatic dissemination of cancer. The extracellular transforming growth factor-β induced protein (TGFβIp) induces LEC sprouting, downstream phosphorylation of signal transduction pathways, and subsequent increase in CCL21 expression on LECs (Maeng et al., 2015). TGFβIp also initiates the "epithelial to mesenchymal transition" of tumor cells that involves the inactivation of epithelial proteins and cellular adhesion molecules (such as E-cadherin) and activation of mesenchymal proteins (such as N-cadherin and vimentin) (Pang et al., 2015). Following this transcriptional modification of the tumor cell into a mesenchymal phenotype, CCR7/CCL21-mediated chemotaxis leads to the targeted recruitment and migration of cancer cells through the lymphatic system (Pang et al., 2015; Tutunea-Fatan et al., 2015). The CCR7-CCL21 ligand–receptor pair has been associated with increased metastatic tumor formation, increased lymphatic invasion, tumor staging, histological grade, and poorer prognosis in many solid tumor types (Tutunea-Fatan et al., 2015; Xia et al., 2015; Liu et al., 2015).

Chemokines, a superfamily of cytokine-like molecules that can induce cytoskeletal rearrangement and directional migration, and their receptors play a prominent role in the traffic of immune cells to lymph nodes from central and peripheral sites. They also play a vital role in the metastasis of tumor cells to SLNs (Wiley et al., 2001; Muller et al., 2001; Zlotnik, 2006; Takeuchi et al., 2004) by directing cancer cells to LVs at the primary tumor site and attracting metastatic cells to secondary sites after they invade BVs or LVs.

Signaling through the CXCR4 (receptor CXCL12) and CCR7 (receptors CCL21 and CCL19) enhances actin polymerization and pseudopodia formation and acts as potent chemoattractant signals in breast cancer cells (Muller et al., 2001), resulting in increased motility and invasion.

Metastasis of murine cancer cells to SLNs was increased when expression of chemokine receptor CCR7 was increased by retroviral vector coding (Wiley et al., 2001). CCL21, a chemokine constitutively produced by LECs, binds CCR7. Blockade of this chemoattractant pathway decreased metastasis. SLN metastases are also directly related to expression of chemokine receptors CXCR3 and CXCR4 in animal models of oral squamous carcinomas, and carcinomas of the stomach, breast, lung, and colon (Stacker et al., 2014).

CCR8, the receptor for CCL1, is strongly expressed in human malignant melanoma. Migration of tumor cells to LECs in vitro is inhibited by blocking CCR8 or CCL1. Recombinant CCL1 promotes motility of CCR8-expressing tumor cells. Tumor cells expressing VEGF-C and -D attract LECs that express the receptor VEGFR-3.

## 3.2. Mechanical (Anatomic) Aspects

### 3.2.1 *Angiogenesis*

Diffusible proangiogenic molecules, released into the tumor microenvironment by proliferating tumor and host cells, induce tip, stalk, and phalanx endothelial cells of neighboring BVs to proliferate and form a new tumor-associated vasculature by sprouting and vascular remodeling (Carmeliet and Jain, 2000). The new peri- and intratumoral BVs are dependent upon the ongoing presence of VEGF, promitotic through binding to VEGFRs, and/or other soluble cytokines for survival. Further primary tumor growth is angiogenesis dependent. The increased permeability induced by VEGF and the plasma that leaks from BVs into the interstitial tissues provides extracellular matrix components such as fibronectin, fibrinogen, vitronectin, and growth factors such as TGF-β, fibroblast growth factor (FGF), and platelet-derived growth factor receptors (PDGFR), which further upregulate VEGF (Shieh, 2011; Swartz et al., 2014; Stacker et al., 2014). Structural defects in the growing vasculature result in impaired endothelial barrier function and increased interstitial fluid pressure (IFP) (Jain et al., 2014).

New BVs provide oxygen and nutrients to growing tumor cells, which may outgrow their blood supply and become hypoxic. Hypoxia stimulates the overexpression of a number of genes, including VEGF, via the hypoxia-inducible transcription factors (Gruber and Simon, 2006). Since the original description of angiogenesis, a so-called "Hallmark" of cancer growth (Hanahan and Weinberg, 2011), the obvious deduction is that tumor cells have access to, and therefore may invade, the structurally defective new BVs and thereby gain access to the systemic circulation. Despite this attractive hypothesis, there has never been direct observation of invasion of intra- or peritumoral BVs. In contrast, green fluorescent dye-labeled syngeneic melanoma cells have been observed by time-sequenced video microscopy to invade into lymphatics in a mouse model (Hoshida et al., 2006). In addition, the natural history of clinical and animal tumors overwhelmingly confirms that most tumors metastasize to SLNs first and then to systemic sites (Fig. 13.1).

### 3.2.2 *Lymphangiogenesis*

Mechanical, enzymatic, and molecular integration results in peri tumoral lymphangiogenesis, a necessary anatomic event for metastasis to the SLNs in most animal models studied (Nathanson et al., 2009a). Tumor-associated lymphangiogenesis is a coordinated and complex series of molecular, cellular, and stromal events that incorporate proliferation and migration of LECs, tube formation, and sprouting driven by VEGF-C and VEGF-D through the VEGFR-3 receptor (Tammela and Alitalo, 2010; Alitalo, 2011) and other growth factors (Tammela and Alitalo, 2010; Podgrabinska et al., 2002). Other molecular mechanisms, such as DLL4, notch 1, prostaglandins, angiopoetins, FGF, HIF1α, erythropoietin, adrenomedullin, EGF, integrins, and neuropilin 2, identified in animal models (Nathanson et al., 2009a), may also play a role in the lymphatic growth (Xu et al., 2010). Increased peritumoral lymphatics coupled with elevated IFP results in increased fluid fluxes from the tumor into the surrounding tumor microenvironment with greater fluid shear stress and significant concentration gradients of soluble factors (Fleury et al., 2006).

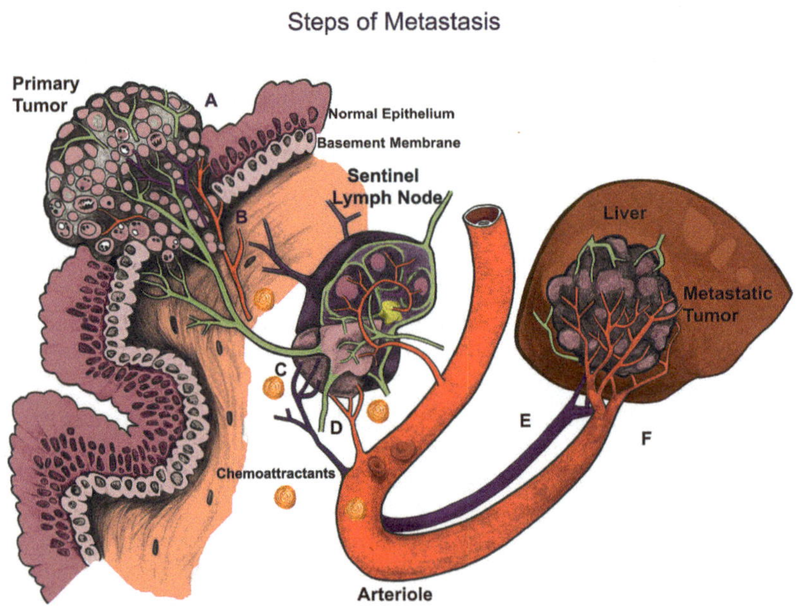


FIGURE 13.1 Steps in LN and systemic metastasis. Tumor cells invade lymphatics and (perhaps) blood vessels at the primary tumor site and travel to the SLN via lymphatic trunks. Tumor cells gain access to the systemic circulation through lymphovascular anastomoses, in the periphery (maybe in the SLN?) and/or in the neck. Once in the systemic blood vessels they may end up growing in a secondary site, like the liver.

### 3.2.3 *Lymphatic Pathways From Primary Tumor to Sentinel Lymph Node*

A crucial step in the development of distant metastases is invasion of the lumens of initial lymphatics at the primary tumor site. Invasion of tumor cells into the peritumoral microenvironment is the first stage of intravasation into functional vessels (Padera et al., 2002), either preexisting or new, crossing physical barriers such as basement membranes of the organ of origin, stromal elements, host cells, and pericytes.

Fluids, dissolved macromolecules, leukocytes, and tumor cells migrate through inter- and intraendothelial gaps and invade into new or existing functional peritumoral, VEGFR3-activated (He et al., 2005) LVs (Hoshida et al., 2006; Stacker et al., 2014; Achen et al., 2005) and/or BVs, guided by increased intratumoral pressure (Butler et al., 1975; Boucher and Jain, 1992; Jain et al., 2014; Nathanson and Nelson, 1994), mechanical flow, increased surface area of the lymphatics (Witte et al., 2006), chemokine gradients (Wiley et al., 2001; Muller et al., 2001; Zlotnik, 2006; Takeuchi et al., 2004; Das et al., 2013), tumor cell invadopodia and motility (Wang et al., 2005; Condeelis and Segall, 2003; Wyckoff et al., 2004; Yamaguchi et al., 2005; Weaver Am Page et al., 2013), and the presence of macrophages and platelets (Skobe et al., 2001).

Once in the LV lumen, tumor cells are carried from the initial lymphatics to the collecting lymphatic trunks. Several forces drive the lymphatic flow (Gashev, 2008). There is no single pump in the lymphatic system as there is in the cardiovascular system, where cardiac contractions are enough to move blood through the systemic circulation. Instead the lymphatic system possesses its own numerous short-distance intrinsic pumps, chains of valve-protected and connected lymphangions, whose smooth and skeletal muscle contractions create a positive pressure gradient, which drives a bolus of lymph only to fill the next or perhaps a few next lymphangions, propelling the fluid, and cells centripetally. This gradient generates a short-lasting local reversed flow, which shuts the valve at the entrance to the lymphangion. The force of contractions increases intraluminal pressure and propels lymph during "lymphangion systole." Pressures range from −5 to 0 cm $H_2O$ during diastole, creating a suction-like effect, in essence pulling fluid (and cells) from the upstream lymphangion. Lymph can potentially be propelled by the driving force of lymph formation (*vis a tergo*) and by extrinsic forces, such as cardiac and arterial pulsations and skeletal muscle contractions.

There is no constant positive pressure gradient along the lymphatic system but, at rest, the side pressure in human leg lymphatics averages about 13 mmHg, with a range of about 5–21 mmHg (Olszewski and Engeset, 1980). Lymphangions contract about 2.5 times per minute, with a range of about 0.5–4.2 times per minute, which increase when rising to a vertical position and with flexion of the foot. Because of the difficulties in measuring lymphatic pressures in collecting trunks draining common tumors, such information is not yet known. However, lymphatic flow rates in animal tumor models (Hoshida et al., 2006; Avery et al., 1992) are increased, possibly reflecting increased lymph formation, VEGF-C and -D induced lymphatic dilatation (Hoshida et al., 2006; Nathanson et al., 1997) and increased intralymphatic pressure.

Lymphangion contractions are driven by $Ca^{++}$ fluxes from extracellular and intracellular stores, which operate through myosin light chain kinase to allow the binding of actin with the myosin light chain cross bridges, generating force (Wang et al., 2009).

LV contractions are influenced by mechanical signals. These vessels, like BVs, respond to increased intraluminal pressure by changing diameters. Increased lymph production in the rat mesentery resulted in increased diastolic diameter, contraction frequency, and amplitude, probably because of intrinsic stretch responsive mechanisms (Benoit et al., 1989). Coordinated contractions occur under pacemaker control, dependent upon calcium fluxes (spontaneous transient depolarizations—STDs) and pathways that affect calcium levels in the smooth muscle cytosol (Munn, 2015). It is still not clear how calcium fluctuations progress to become sexually transmitted diseases in an individual pacemaker cell. Inflammatory mediators like histamine have been shown to affect lymphatic contractility. Many biochemical pathways are also likely to influence lymphatic contractility, including substances derived from endothelial cells, such as endothelin-1.

### 3.2.4 *Mechanobiology and Tumor Progression*

Mechanical forces in a primary cancer create increased firmness (Jonietz, 2012), a clinical finding with most tumors. Pressures inside a growing tumor are usually increased because of increased IFP (Butler et al., 1975; Boucher and Jain, 1992; Jain et al., 2014; Nathanson and Nelson, 1994), closely packed and proliferating host and malignant cells (Padera et al., 2004), increased collagen deposition (desmoplasia) (Ng et al., 2005; Pedersen et al., 2010), alterations in cytoskeletal structures (Shieh, 2011), and "stiffening" of the tumor microenvironment (Jonietz, 2012; Pathak and Kumar, 2011; Friedl and Alexander, 2011; Shieh, 2011). Stress forces also exert changes in the peritumoral stroma, entry sites into lymphatic capillaries (McDonald, 2008; Baluk et al., 2007), in lymphatic trunks (Hoshida et al., 2006; Avery et al., 1992), and in the SLN (Nathanson and Mahan, 2011; Nathanson et al., 2014).

Increased matrix stiffness results from collagen fiber realignment, which enhances the migration and invasion of tumor cells (durotaxis) (Shieh, 2011). Another invasion pathway is through P13K and focal adhesion molecules.

Cells are continuously exposed to physical forces, such as compression, tension, hydrostatic pressure, and shear stress (Jonietz, 2012). These physical forces may also affect tumor and host cell cytoskeleton and the surrounding extracellular matrix (Davies et al., 1997). Activation of shear stress responsive genes (Resnick et al., 2003; Topper and Gimbrone Jr, 1999) can trigger angiogenesis and, possibly, lymphangiogenesis (Ingber, 1998; Davies et al., 1997). Fibroblasts exposed to increased fluid flow undergo differentiation of myofibroblasts (Ng et al., 2005), which produce more force when they contract. Increased mechanical stress also induces the expression of invasion/metastasis genes caveolin, β1integrin, phosphatase and tensin homolog (PTEN), and Rac 1 (Nathanson et al., 2015), modulates cell–cell and cell-to-matrix adhesion, and increases tumor growth rates.

The cellular responses to shear stress include endothelial swelling and secretion of chemokines from the endothelial cell surface that may aid in the recruitment of monocytes/macrophages and the production of matrix metalloproteases that are crucial for basement membrane digestion during tumor cell invasion (Genis et al., 2006).

The biomechanics of the tumor microenvironment increases the intratumoral pressure (Boucher and Jain, 1992; Jain et al., 2014; Nathanson and Nelson, 1994). The tumor exudes more interstitial fluid, often seen intraoperatively and by magnetic resonance imaging (MRI) as peritumoral edema, and a resultant increase in lymphatic flow, which carries more cells to the SLN per unit time (Hoshida et al., 2006).

### 3.2.5 *The Role of the Lymph Node Microenvironment in Progression of Metastatic Cancer*

Cellular and molecular changes occur in the SLN microenvironment prior to the arrival of tumor cells in the tumor-draining LNs (Fig. 13.2). They participate in cancer progression by undergoing stromal remodeling, recruitment of immune cells, and alterations in the vasculature in response to lymph-transported growth factors, chemokines, and other signaling molecules secreted by tumor and host cells in the tumor microenvironment (Stacker

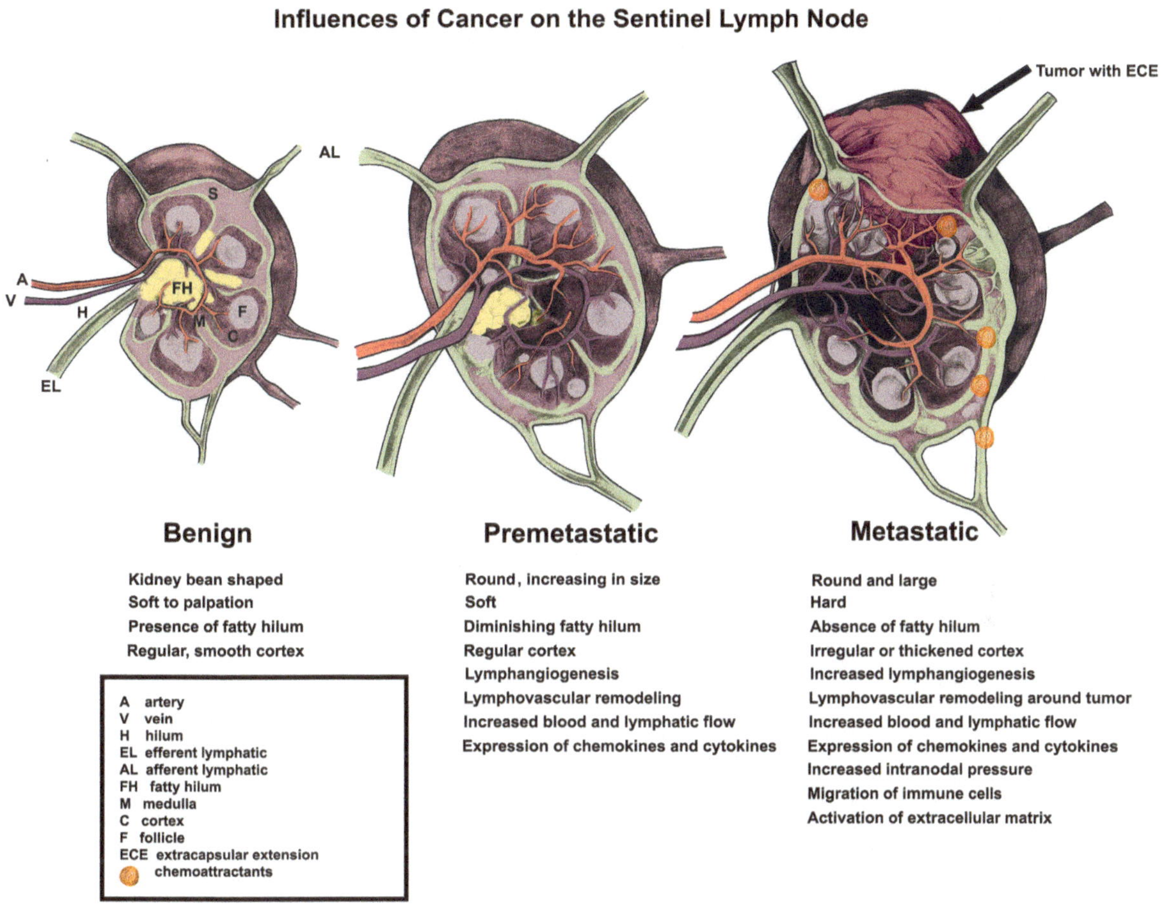


FIGURE 13.2 Primary cancers influence the morphology and function of SLNs both before and after tumor cells arrive and grow in the node.

et al., 2014). Some distinctive features of the premetastatic SLN include increased lymphangiogenesis and lymph flow (Harrell et al., 2007), remodeling of high endothelial venules (Qian et al., 2006), and alteration in the host immune cells (Lee et al., 2005). Immune and stromal cell populations expand and seem to express a largely immunosuppressive cytokine environment, such as IL-10, TGF-β, GM–CSF, and $PGE_2$ (Leong et al., 2002).

The best documented aspect of the premetastatic niche in the SLN is lymphangiogenesis driven by VEGF-A and/or -C and -D. VEGF-C–P13K mediated signaling is important for remodeling the LN prior to tumor cell arrival by stimulating LEC proliferation, leading to tumor-associated lymphangiogenesis (Pereira et al., 2015). This signaling leads to activation of integrin α4β1 on LN lymphatic endothelium, which attracts VCAM-1-expressing tumor cells, thereby promoting LNMets. Many growth factors may directly or indirectly lead to lymphangiogenesis via upregulation of VEGF-C.

High endothelial venule remodeling, associated with loss of bone morphogenetic protein-4 and peripheral node addressin, plays an important role in both lymphocyte homing and LNMets. Primary tumor cells secrete cytokines, which initiate these responses, suggesting that they may inhibit immune responses while also providing nutrients to metastatic tumor cells in the LN (Pereira et al., 2015).

The immune microenvironment in the SLN consists of stromal (fibroblasts, blood, and LECs), myeloid (macrophages and dendritic cells), and lymphoid (T and B) cell populations. SLNs are sites where foreign, pathogenic, or self-antigens accumulate. It is here where dendritic cells and effector T lymphocytes interact to initiate immune responses to tumor cells. Malignant cells have evolved ways to evade the immune surveillance mechanisms in the LN and thereby colonize and proliferate at the metastatic site (Ribas, 2015). Smaller numbers of effector T cells were found in premetastatic and metastatic LNs (Kohrt et al., 2005). B cells also play a role by presenting antigen to naïve $CD4^+$ and $CD8^+$ T cells in the SLNs (Li et al., 2009).

# 4. CLINICAL ASPECTS OF LYMPH NODE METASTASIS

## 4.1 Clinical Diagnosis

Solid tumors such as melanoma, breast, stomach, colon, lung, and those of the head and neck often metastasize to RLNs prior to or in concert with systemic dissemination. Clinical examination involves systematic preoperative or intraoperative palpation of the expected regional nodal basin or basins that drain the primary tumor. Suspicious nodes are those that are enlarged, firm (hard), palpable, and/or matted.

In cases of breast cancer, the potential sites of LNMets are the ipsilateral axilla and/or supraclavicular fossa. Patients with melanoma of the head and neck may have drainage to parotid, occipital, submandibular, and/or neck LNs. Melanoma of the trunk can metastasize to the neck, axillae, or groin. Melanoma of the lower limbs may metastasize to the ipsilateral inguinal nodes or (rarely) to the popliteal node. Melanoma in the arms may metastasize to the axillary nodes or (rarely) to the epitrochlear LN.

Lymphatic drainage tends to accompany the BVs that supply the organ where the primary cancer arises. This is true of intra-abdominal primary cancers in the stomach, pancreas, colon, prostate, bladder, kidney, uterus, cervix, ovaries, and small bowel; it is also true of primary tumors of endodermal origin in the chest, such as lung or esophagus; and primary cancers of the head and neck such as the tongue, floor of the mouth, pharynx, larynx, sinuses, and thyroid.

Another type of LNMet pattern that does not appear to follow the blood supply of the organ is the so-called Virchow's node (Virchow, 1860), metastasis traveling through the thoracic duct, arising from primary cancers in the thoracic cavity, abdomen, or pelvis and lodging in the left supraclavicular node.

LN status is one of the most important prognostic factors in the management of cancer. Positive RLNs are independently associated with recurrence, systemic metastasis, and overall decreased survival in many solid tumor types (Carter et al., 1989; Gershenwald et al., 1999; Ferrone et al., 2002). For instance, in breast cancer, axillary LN involvement has been shown to predict local and distant recurrence as well as disease-specific survival (Carter et al., 1989). Clinical variables that predict LNMets generally include larger size of the primary tumor, depth of invasion, presence of lymphovascular invasion, and higher grade (Table 13.1). In addition to prognosis, LN status also informs treatment decisions, as patients with LNMets usually require systemic chemotherapy and/or radiation.

## 4.2 Radiological Imaging

The contribution of radiological imaging to the pretreatment diagnosis of LNMets continues to improve with advances in existing technology

TABLE 13.1 Clinical and Pathologic Variables That Predict LNMet

| | Clinical and Pathologic Variables That Predict Lymph Node Metastasis |
|---|---|
| Melanoma | Younger age, higher mitotic rate, increasing Breslow thickness (>1 mm), higher Clark level (III, IV, V), presence of ulceration, absence of tumor infiltrating lymphocytes, tumor location (truncal, head/neck, subungual), lymphovascular invasion, desmoplasia, neural invasion, immunosuppression |
| Breast | Increasing tumor size, lymphovascular invasion, high nuclear grade, age (younger), race (African-American) |
| Lung | Large primary tumor size, central location of primary tumor, lymphangiogenesis, grade of differentiation, pleural involvement |
| Stomach | Large primary tumor size, lymphatic vessel involvement, and invasion into the submucosal layer |
| Colorectal | Increased depth of submucosal invasion, location in the lower third of the rectum, lymphovascular invasion, tumor budding, cell differentiation, and tumor cell dissociation |

and development of new imaging concepts (Kyzas et al., 2008). The objective is to identify ever smaller metastatic deposits in LNs. Much of radiological diagnosis of LNMets depends upon identifying changes in morphology and size of the LN. The challenge is to distinguish benign from malignant changes (Thoeny et al., 2014). Functional imaging techniques permit more specific identification of metastases in solid organs and in LNs. New lines of investigation in the laboratory offer promise for improving sensitivity and specificity (Luke et al., 2014).

### 4.2.1 Mammography

Level I axillary LNs are seen in about half of all mammograms. Normal axillary LNs are round or oval shaped and are variable in size. A radiolucent fatty hilum is frequently visible. A combination of features that suggest LNMets include round shape, irregular margins, increased node density, and loss of the fatty hilum. Size measurements alone are not useful in determining if a node is abnormally enlarged, unless the node size is of out of proportion to other axillary nodes. In the absence of primary breast cancer, enlargement greater than 1 mm per year is usually reactive to a benign process (Lee et al., 1997). However, the meticulous evaluation of the mammograms for findings of primary breast cancer is imperative. If an abnormally enlarged LN is identified on mammography, ultrasound (US) is the procedure or choice for further evaluation of the node.

Mammographically visible calcifications in axillary LNs may indicate metastases from breast, papillary ovarian, or thyroid cancer (Helvie et al., 1988; Singer et al., 2001; Chen et al., 1998). However, this is an uncommon presentation of metastasis and must be distinguished from benign intranodal calcifications.

### 4.2.2 Ultrasound

High-resolution US is the major imaging tool for evaluating LNs in the initial diagnostic setting for breast, rectal, anal, esophageal, gastric, head and neck, and pancreatic cancers and, occasionally, for melanoma. It is indicated for evaluation of clinically suspicious LNs in the neck, axilla, groins, and for evaluation for possible LNMets in patients with a newly diagnosed invasive breast cancer.

US evaluation for axillary LNMets in breast cancer is performed from the axillary tail superiorly until no additional LNs are identified. The SLN is the first node that will show changes related to breast metastasis. It does not have a fixed position or number, but is most often found at the level of the lowest axillary hair follicles (Nathanson et al., 2007) or in the distal axillary tail (Abe et al., 2007). By targeting the largest or

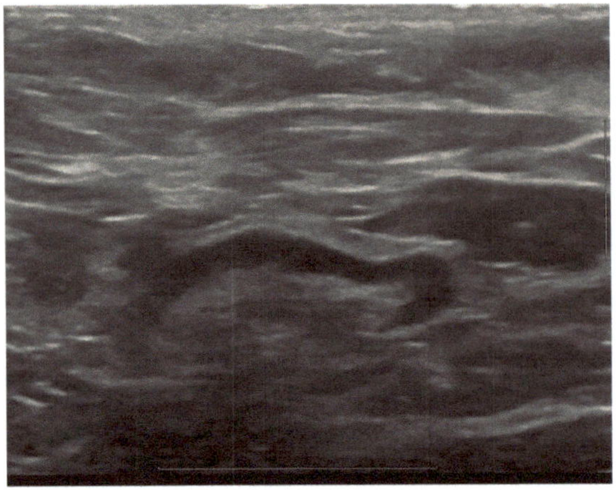

FIGURE 13.3 US of a normal axillary LN. Oval shape, thin uniform cortex, preserved fatty hilum.

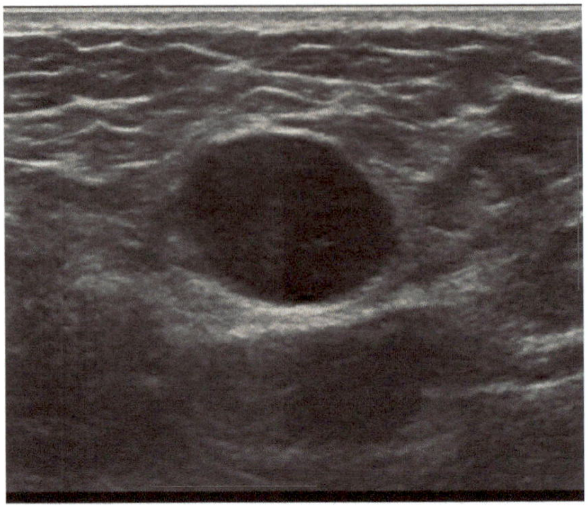

FIGURE 13.4 US of axillary lymph node with metastasis. Enlarged, round shape, loss of fatty hilum.

most abnormal node in these areas, a significant number of patients with axillary LNMets were identified by needle biopsy. In a number of patients the biopsied LN was the only node with metastasis, confirming it was the SLN.

The US features of normal LNs include a sharp circumscribed margin, an oval or reniform shape, a uniformly thin peripheral hypoechoic cortex, and a central hyperechoic fatty hilum (Fig. 13.3). Blood flow is usually via a single artery and vein traversing the hilum and branching into the cortex. The earliest detectable finding for LNMets is a slight uniform increase in cortical thickness or focal nodular thickening of the cortex. A threshold of 3mm for identifying an abnormal node in patients with suspicious breast findings or known primary malignancy is commonly used, although other measurements have been used. It is highly sensitive though of low specificity. Focal nodular thickening of 3mm or more in the presence of a <3mm thick cortex is also considered suspicious. Cortical nodules can bulge the contour of the node or indent the fatty hilum. With larger tumor deposits, eccentric cortical thickening, transformation to a round shape, further compression or obliteration of the fatty hilum, and node enlargement are seen (Fig. 13.4). Blood flow may be recruited through the cortex and be visible with color Doppler US. The margins of the LN may become irregular as the metastasis invades outside the capsule.

Although high-frequency endoscopic US (EUS) is the main technique for staging esophageal cancers, nodes further than 2cm from the esophagus are difficult to image. Fine needle aspiration of nodes can establish staging and avert thoracoscopy or laparoscopy. EUS assessment of N stage is from 38% to 73% accurate (Krasna, 2013).

### 4.2.3 *Magnetic Resonance Imaging*

LN size, using the short axis diameter, is the most common crude parameter for diagnosing LNMets with MR and CT. Small nodes may contain metastasis and benign nodes may enlarge due to reactive changes and hyperplasia (Zhang et al., 2010; Sloothaak et al., 2014; Tournoy et al., 2008). Criteria for abnormal size differ by region and are chosen to give a reasonable sensitivity and specificity for LNMets. For example, 5mm is used for perirectal nodes, 6mm for internal mammary LNs, and 10mm is used for most thoracic and abdominal LNs. Values of 4, 5, and 6mm have been proposed for cervical LNs (Zhang et al., 2010).

Normal LNs have a hypointense (dark) cortex and hyperintense (bright) fatty hilum on T1 weighted imaging (T1WI). With dynamic contrast-enhanced T1WI, the cortex enhances rapidly and washes out on later images, features that are similar to those seen in malignant tumors and therefore not useful in detection of LNMets. As nodes are replaced with metastasis, altered enhancement may become apparent, but is accompanied by other changes in morphology and size. On T2-weighted imaging (T2WI), the normal cortex is hyperintense, corresponding to water content in the cortex, and the hilum is hypointense. When tumor metastases involve all or part of a node, water is displaced from that portion of the node and the T2 signal may become heterogeneous or hypointense. Thin section T2-weighted fast spin echo sequences are used to evaluate rectal carcinomas. In addition to size, nodes with irregular borders or heterogeneous T2 signal have been diagnosed as suspicious for metastasis with a sensitivity of 85% and specificity of 98% compared with a sensitivity of 48% and a specificity of 99% for nodes with mixed signal intensity alone (Brown et al., 2003).

Diffusion weighted imaging (DWI) is based on measuring the Brownian motion of water molecules, which is restricted in highly cellular tissue such as tumor tissue compared with water motion in normal and low cellular benign tissue. The greater the restriction of water movement, the higher the DWI signal. However, a high DWI signal can be mimicked by the high T2 signal from water in the target. Therefore, a second image, the apparent diffusion coefficient (ADC) map, is generated from the ADC, which measures the area in which a water molecule moves per unit of time. On an ADC map, regions with restricted diffusion have low intensity signal. Correlating the two images, DWI would show high signal and the ADC map would show low signal in areas with highly cellular tissue (Brandao et al., 2013). DWI has been used to evaluate metastases in axillary lymph nodes and in pelvic lymph nodes. In a study of metastasis to axillary lymph nodes in patients with breast cancer, DWI was 94.7% sensitive and 91.7% specific (Fornasa et al., 2012).

Negative contrast agents continue to be investigated. Ultrasmall superparamagnetic particles of iron oxide enhanced MRI looks at the uptake of iron atoms by macrophages in normal lymph node tissue. On MR sequences, the normal part of the node will show signal loss due to the presence of iron and abnormal portion will exhibit no signal loss. Nodes, which have heterogeneous or total loss of signal, contain metastasis. Although studies have shown promise, no intravenous iron agents have been approved by the Food and Drug Administration (FDA) at this time (Misselwitz, 2006).

### 4.2.4 *Computed Tomography*

On noncontrast CT scans, lymph nodes have the attenuation of other soft tissue structures and are best seen when surrounded by fat. The fatty hilum is often visible in normal LNs. Intravenous (IV) contrast administration helps better delineate the nodes. Diagnosis of LNMets is based on nodal enlargement, measured by maximal short axis diameter.

CT is useful in detecting enlargement of LNs and showing the distribution of abnormal nodes, both regional nodes and distant metastasis. Nodal size is the main criterion for detecting abnormal LNs. It is frequently used in conjunction with other imaging modalities (Hoang et al., 2013).

### 4.2.5 *Positron Emission Tomography With CT*

Positron emission tomography (PET)-CT evaluates the metabolism of glucose in tumors. 2-Deoxy-2-fluoro[18F]-D-glucose (FDG) is metabolized by active cells, which then accumulate the radionuclide. PET and CT images are obtained contemporaneously and the images are overlaid. The fused images provide more anatomically specific localization of the structures demonstrating glucose hypermetabolism than does PET alone. FDG PET is the most common tool for staging in the clinical setting. However, it has lower sensitivity and specificity for clinically node negative disease. PET is limited to tumor sizes of 7–8 mm in most settings, although some report detection of 5 mm tumor deposits. Sensitivity and specificity for LNMets are 61% and 80%, respectively (Wahl et al., 2004). Because of the insensitivity to small metastases, PET-CT is used for staging only after initial diagnosis and assessment of the likelihood of LNMets.

Standard uptake values are measurements that attempt to provide a reproducible quantitative measure of FDG update in tumors. This facilitates monitoring the response of tumors to treatment for an individual and making comparisons between patients (Kinahan and Fletcher, 2010).

### 4.2.6 *Lymphoscintigraphy*

Lymphoscintigraphy with Tc-99m sulfur colloid injection is used to identify the sentinel node draining a tumor bed, permitting surgical removal for pathological analysis. Dynamic image sequences help identify the presence of multiple sentinel nodes and nodes distal to the SLN (Intenzo et al., 2009). The old technique of lymphangiography, popularized in the 1950s and 1960s for identifying LNs in Hodgkin's disease, is no longer used for diagnostic purposes.

## 4.3 Treatment of Lymph Node Metastases

### 4.3.1 *Surgery*

The radical mastectomy was one of the first oncologic operations to consider locoregional lymphatics in the spread of tumor (Halsted, 1907). This rather morbid surgery involved removal of the breast, regional axillary lymph nodes, and chest wall musculature. Radical operations for rectal, uterine, pancreatic, colon, stomach, and lung cancers adopted the same principle of removing the primary tumor, the organ from which it arose and regional lymph nodes en bloc. Based on the importance of removing all lymphatics associated with the primary tumor and its nodal basin, early operations for melanoma described the removal of the primary tumor, the RLNs, and a long strip of contiguous skin and subcutaneous tissue thought to contain the lymphatics.

Removal of entire nodal basins subjected patients to devastating complications, including lymphedema, decreased limb function, sensory loss, and diminished quality of life. The morbidity of these resections forced surgeons to consider less invasive operations based on increasing knowledge of lymphatics and their role in cancer metastasis. The movement away from morbid oncologic surgeries was supported by scientific studies that demonstrated no differences in recurrence or survival between radical and "modified" resections (Robinson et al., 1976; Turner et al., 1981).

The goal of CLND is to maximize cure by minimizing locoregional failure. Until the landmark reports of no significant advantage to prophylactic CLND (Veronesi et al., 2003), this operation was performed in many patients with breast cancer or melanoma and a clinically negative LN basin. CALND in breast cancer involves dissection of level-I and -II and Rotter's nodes (lateral to and posterior to the pectoralis minor muscle and those nodes that reside in between pectoralis major and minor muscles) with anatomical borders being axillary vein cephalad, latissimus dorsi muscle laterally and medial edge of the pectoralis minor muscle medially, sparing level III nodes (medial to the pectoralis minor muscle) except in cases, where these nodes are palpably suspicious intraoperatively. Based on the lymphatic drainage patterns studied in CALND specimens, the distribution of breast cancer LNMets demonstrates a progressive spread from level I to level II to level III. Discontinuous or "skip" metastases that do not follow this graded lymphatic drainage pattern but drain cells directly from the breast to upper echelon nodes may occur in less than 2% of cases (Rosen et al., 1983). CALND involving level I, II, and III nodes remains the standard of care for melanoma of the upper extremity or trunk with a positive SLN. Complete removal of the draining LN basin is thought to provide locoregional control and potential cure.

Although the lymphatic metastasis of melanoma has also been shown to progress in an orderly, sequential manner (Reintgen et al., 1994), clinical practice mandated an anatomically complete dissection of all involved LN basins based on the variability of drainage patterns in second echelon nodes and risk of recurrence with a less aggressive nodal dissection (Reintgen et al., 1994; Norman et al., 1991; Muller et al., 2002; Essner, 2003; Uren, 2004; Namm et al., 2012).

In the last few decades, the purpose and extent of LN dissection has changed. Morton's landmark research validated the clinical use of intraoperative lymphatic mapping for early stage intermediate risk melanoma (Morton et al., 1992). SLN mapping, based on the orderly and predictable spread of cancer to the first RLN, became the validated standard of care and allowed many patients with a clinically negative axilla to avoid the morbidity associated with CLND. Accurate identification of occult RLN metastasis has helped surgeons avoid CALND in certain patients and has led to the upstaging and alteration of treatment algorithms in others. SLN biopsy (SLNB) has become the gold standard for patients with melanoma and breast cancer but its clinical application in other solid tumor types such as cancers of the esophagus, lung, prostate, kidney, bladder, stomach, colon and rectum, head and neck, penis, uterine cervix, and endometrium has been somewhat limited. SLNB is performed by the peritumoral injection of a tracer that is easily taken up into the lymphatics, allowing the SLN to be identified. The combined use of tracers, such as blue dye and radiolabeled colloid, yields a higher detection rate than one tracer alone (Cody et al., 2001).

The SLNB has become invaluable in the staging of breast cancer. The SLNB can also be offered to women with operable breast cancer who have multicentric tumors, ductal carcinoma in situ (DCIS) with simultaneous mastectomy, a prior history of breast or axillary surgery and to those who undergo preoperative systemic therapy. There is insufficient evidence to recommend SLNB in large or locally advanced breast cancer (T3/T4), inflammatory breast cancer, DCIS when breast conserving surgery is planned and in pregnancy (Lyman et al., 2014). Prior to the publication of the American College of Surgeons Oncology Group (ACSOG) Z0011 study that led to the paradigm shift in management of the axilla, an SLNB positive for metastasis mandated CALND. Giuliano and associates demonstrated no survival difference in patients with early stage breast cancer and limited axillary nodal burden who underwent SLNB alone versus CALND (Giuliano et al., 2011).

Indications for lymphatic mapping and subsequent SLNB in melanoma continue to evolve based on the recent results of landmark trials. The Multicenter Selective Lymphadenectomy Trial-I was the largest trial to address the role of lymphatic mapping with SLNB in determining prognosis and survival (Morton et al., 2014). Ten-year follow-up results confirmed the role of lymphatic mapping and SLNB-guided lymphadenectomy as a prognostic tool that improves disease specific survival compared to observation in intermediate thickness (1.20–3.50 mm) and thick (>3.50 mm) melanomas. Current National Comprehensive Cancer Network Guidelines (National Comprehensive Cancer Network) recommend SLNB for melanoma >1 mm thick or those ≤1 mm thick with high risk features, including ulceration or mitotic rate ≥1 per square millimeter. The American Joint Committee on Cancer guidelines emphasize lymphoscintigraphy followed by lymphatic mapping and SLNB as important components of melanoma staging that should be used to identify occult stage III regional nodal disease among patients with clinical stage IB or II melanoma (Gershenwald et al., 2010).

The decreased postoperative complications and psychological morbidity after SLNB when compared to CALND have been well studied (Schrenk et al., 2000; Schijven et al., 2003; Purushotham et al., 2005). SLNB, however, is occasionally complicated by extremity edema, paresthesias, decreased range of motion, and clinically significant seroma formation approaches 10% in certain studies (Wilke et al., 2006).

Radiation to the axilla has shown promising results in both locoregional control and decreased morbidity that is associated with surgical locoregional staging. The multicenter international EORTC AMAROS trial examined the use of axillary radiotherapy instead of axillary dissection after positive SLNB and found no difference in locoregional control or recurrence but did demonstrate a decreased incidence of axillary and extremity complications in the patients who received radiotherapy (5-year incidence of lymphedema was 14% in the radiotherapy group and 28% in the CALND group) (Rutgers et al., 2013).

New prospective randomized trials aim to determine if axillary US with fine needle biopsy is a reliable preoperative staging modality in the hopes that it will be a minimally invasive replacement of the SLNB. The SOUND study and other randomized trials are now examining the axillary recurrence, disease free and overall survival of patients with early stage breast cancer (T1–T2) with negative axillae who will be randomized to undergo SLNB versus no further axillary staging (Veronesi et al., 2010).

#### 4.3.2 *Other Treatments*

The trend toward less invasive locoregional staging in breast cancer has been so successful in part due to the advances in the fields of radiation and medical oncology. Much of the improvements in disease specific and overall survival in breast cancer can be attributed to the success of chemotherapeutic agents to treat the subclinical systemic disease and radiotherapy to control the axillary LN basin (Whelan et al., 2015).

Systemic chemotherapy has had less success in providing a reliable survival benefit in melanoma. A new era has evolved with the new approach to immunotherapy with "checkpoint control." Initially with CTL-4 antagonists and, more recently with remarkable results from anti-PD-1 and PDL-1 antibodies (Callahan and Wolchok, 2015; Hamid et al., 2013), there is, for the first time, a realistic possibility for systemic control of melanoma metastases and it is not unreasonable to speculate that these immune approaches to melanoma therapy may someday provide a significant benefit to those patients who harbor LNMets (Ribas, 2015; Ribas et al., 2011). New immunotherapeutic approaches to patients with lung, kidney, and pancreatic cancers also look promising. Along with these immune-based approaches are the exciting FDA-approved therapies that target aberrant genes, such as BRAF, mutated in half the melanoma population (Holderfield et al., 2014), and anti-HER2-neu antibodies and peptides in various tumors. In patients with LNMets neoadjuvant anti-HER2 therapy along with chemotherapy may result in a complete pathologic response in both the primary breast cancer and in the RLNs.

Radiation therapy following CLND in melanoma may improve locoregional control. Although some prospective trials of nodal radiotherapy in breast cancer improve locoregional control and overall survival, this has not been confirmed in other studies (Bastiaannet et al., 2005). Radiation is also valuable in the locoregional control of LNMets in anal, rectal, uterine, esophageal, gastric, pancreatic, and head and neck cancers.

## 5. PATHOLOGY

### 5.1 Gross Autopsy

Comprehensive postmortem examinations of all tissues in large-scale autopsy studies (diSibio and French, 2008), recently enhanced by radiologic imaging, yield valuable information regarding patterns of cancer metastasis. Tissues from many anatomic parts are also a great source for molecular genetic studies, allowing further dissection of the mechanisms of LN and systemic metastases.

Cancers arising in diverse anatomic sites often exhibit different patterns of metastasis. Primary breast cancer in patients treated by surgery alone accounted for 11.3% of all the cancers and contributed to 23.6% of all metastasis (diSibio and French, 2008). Breast cancer metastases were widespread and most often involved distant lymph nodes (12.4%), lungs (11.1%), regional lymph nodes (10.3%), liver (9.8%), and bone (9.5%). Prostate cancer metastasis differed in the frequency of anatomic metastatic sites compared to the breast and included regional lymph nodes (26.2%), bone (19.7%), distant lymph nodes (18.4%), lung (12.8%), and liver (7.8%). Even with modern systemic and locoregional therapies the patterns of metastasis learned from prior studies provide valuable insight in managing the cancer patients.

## 5.2 Histopathologic Techniques for Identifying Lymph Node Metastasis

### *5.2.1 Lymph Node Microanatomy (Willard-Mack, 2006)*

Understanding the normal LN anatomy familiarizes us with the pathophysiology of LNMets. A lymph node is an encapsulated discrete cluster of fibrovascular tissue enclosed within a dilated lymphatic sac/vessel where lymphocytes are transient migratory residents distributed in discrete lymphoid lobules, each divided into different anatomic and physiologic parts (Kelly, 1975; Belisle and Sainte-Marie, 1990; Gretz et al., 1997). The superficial cortex contains lymphoid follicles (predominantly B lymphocytes) that are surrounded and separated by interfollicular (or diffuse) cortex containing predominantly T lymphocytes. In the deep cortex (paracortex), T lymphocytes home and interact with dendritic antigen presenting cells. The medulla contains medullary cords, medullary sinuses, and the hilum. The medullary sinuses contain macrophages, lymphocytes, and erythrocytes.

Lymph flows through afferent lymphatics into a complex system of interconnecting spaces, including subcapsular, transverse, and medullary sinuses around the lymphoid lobule before exiting the LN through a single efferent LV at the hilum (Sainte-Marie et al., 1982). Since each lymphoid lobule gets an independent afferent lymphatic from a draining area, within the same LN there may be different levels of immunological activity leading to heterogenous morphologic appearance (Sainte-Marie et al., 1982). The anatomy (size, color) and histology including number/density of lymphoid lobules thus vary based on anatomical location, age, diet, and antigen exposure. LNs play an important part in the immune response and act as filter for particulate matter and cellular debris.

### *5.2.2 Pathologic Examination of Lymph Nodes*

#### 5.2.2.1 GROSS EVALUATION (LESTER ET AL., 2009; TANG ET AL., 2009)

Thorough gross dissection of LNs followed by histopathologic examination by review of the hematoxylin and eosin (H&E)-stained slides is an integral part of the pathologic examination of the surgical specimens (SLN biopsy, complete lymph node dissection, and/or excision biopsy of LNs) for appropriate staging of the disease.

Following dissection and harvesting of the LNs from the drainage basin, typically each of the grossly negative LNs are thinly sliced along the long axis of the node at 2 mm intervals, and all slices are submitted for microscopic examination. For grossly positive LNs (tumor recognized visually, either directly or with a lens), representative sections from tumor with LN capsule and surrounding tissue to demonstrate extracapsular extension are evaluated.

### 5.2.2.2 MICROSCOPIC EXAMINATION

Current best practice guidelines for the microscopic examination of an LN include a detailed examination of the entire node including capsule, cortex, paracortex, medulla, and surrounding attached soft tissues (Lester et al., 2009; Tang et al., 2009; Haley et al., 2005). At least one representative H&E level from all the LNs is examined. For SLNs, multiple levels with or without immunohistochemistry are often used to increase the sensitivity of morphologic detection of micrometastasis.

The different histologic patterns of metastasis and their implication in staging are as follows (Lester et al., 2009; Tang et al., 2009; Singletary et al., 2003):

1. Single or multiple clusters of tumor cells within different sinuses of a lymph node (Fig. 13.5):

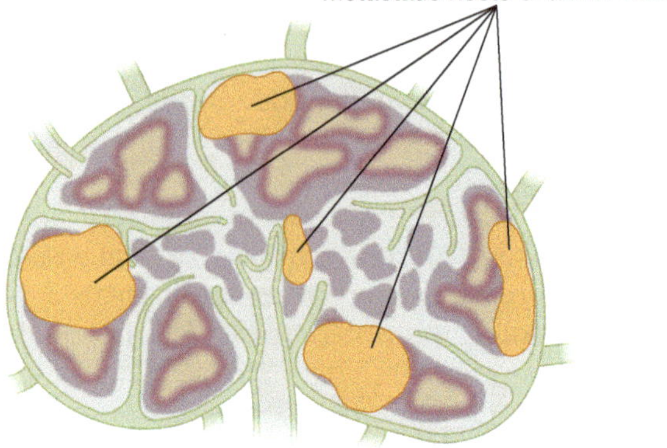


FIGURE 13.5 Single or multiple clusters of tumor cells within different sinuses of a lymph node.

Tumor emboli typically lodge first in the subcapsular sinuses in the early stage and may be challenging to identify under the microscope on H&E staining. Use of immunohistochemistry for antipancytokeratin enhances detection of small metastatic nests or isolated single cells (described below). The tumor emboli then travel through the transverse and medullary sinuses. The size of the metastatic deposit for N staging classification is based on the largest contiguous cluster of tumor cells, especially in the staging guidelines for breast cancer.

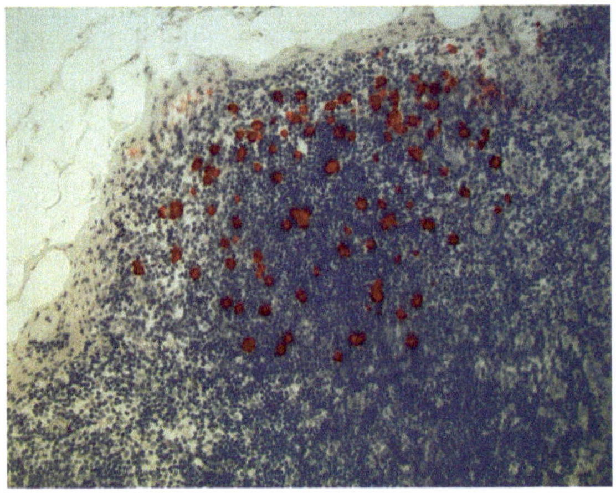

FIGURE 13.6 Dispersed pattern of lymph node metastasis in sentinel lymph node in a case of metastatic lobular carcinoma; cytokeratin immunostain (magnification: ×200).

2. Dispersed pattern of lymph node metastasis:

Some invasive carcinomas, particularly lobular carcinomas of the breast or signet ring cell carcinoma of stomach, colon or any other sites, may metastasize as individual tumor cells and not as cohesive nests or clusters. In such a scenario, the isolated tumor cells are often difficult to identify histologically as they frequently blend with the histiocytic cells within the LN. The estimation of the size of tumor present for N classification in such cases is difficult. Cytokeratin immunostain (Fig. 13.6) positively highlights the metastatic tumor cells to estimate the size as it affects the N stage in breast cancer staging. To avoid underclassification of such cases, an upper limit of 200 cells or size less than 0.2 mm in one node cross-section to classify as "isolated tumor cells" is recommended. The pathologist' judgment is required to determine the best N classification. The biologic significance of remains unknown and N0 staging is considered justified. In contrast, N staging for melanoma differs considerably as described below in detail.

**3.** Metastasis with extranodal or extracapsular tumor invasion (Fig. 13.7):

Metastatic carcinoma to the LN may invade through the capsule into adjacent adipose tissue, through the cortex or medulla. This finding usually correlates with the clinical impression of fixed or matted nodes upon palpation. These patients are at an increased risk of recurrence. Measurement of the extranodal extension including the surrounding desmoplastic response should be counted in determining the size of a LNMet. Tumor within the perinodal lymphatic spaces and in the axillary tissue without invasion of adipose tissue is not considered extranodal invasion. Direct invasion of an LN from a primary tumor is classified as LNMet.

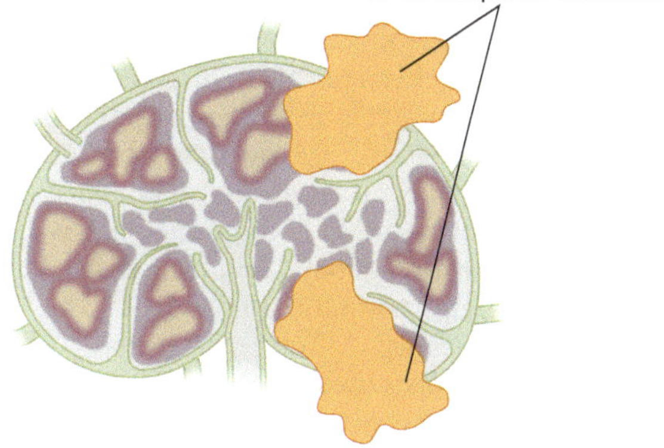


FIGURE 13.7 Metastasis with extranodal or extracapsular tumor invasion in case of breast carcinoma.

**4.** Tumor nodules in the perinodal adipose and soft tissues (Fig. 13.8):

Metastatic carcinoma can completely replace an LN. The presence of morphologically identifiable LN capsule or residual lymphoid tissue helps to confirm the findings as a metastatic LN. Sometimes foci of invasive carcinoma are seen in the perinodal adipose or soft tissues without evidence of metastasis in the adjacent LN. These tumor nodules without morphologic evidence of residual LN are counted as LNMets. These may occasionally represent direct stromal invasion from a primary carcinoma arising in axillary breast tissue or tumor emboli in afferent lymphatics.

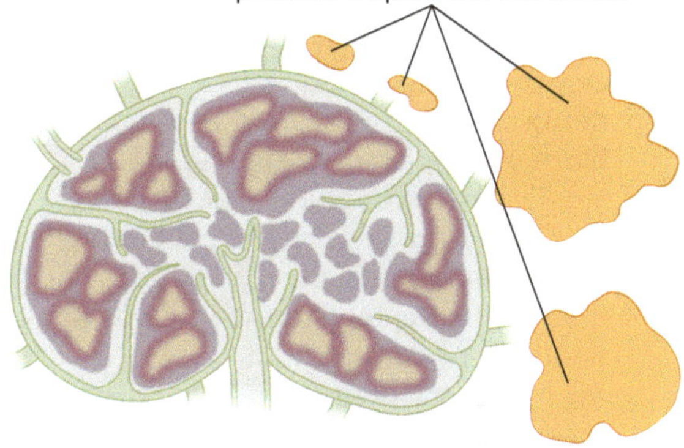


FIGURE 13.8 Tumor nodules in the perinodal adipose and soft tissues.

**5.** Examples of Specific patterns of LNMets

**a.** Tumors of endodermal origin.
Morphologic patterns of the metastatic LNs are similar to any other solid tumors based on morphologic subtypes. Like other primary sites, there still remains a major controversy regarding whether tumor cells arising in the lung metastasize first to the RLNs or invade and metastasize directly through tumor-associated BVs. Both routes are possible, as first described in 1947 (Ackerman and Regato, 1940; Wilson, 1961). The most commonly involved metastatic sites are mediastinal lymph nodes (84%), pleura (51%), liver (47%), bone (34%), brain (32%), pericardium (29%), and adrenals (29%) (Stenbygaard et al., 1999).
There are no unique patterns of LN spread from primary tumors of the stomach,

esophagus, colon, rectum, small bowel, or pancreas.

**b.** Tumors of ectodermal origin

A lot of attention has been directed at histologic analysis and reporting of the metastatic LNs for the purpose of staging of breast cancer (Cummings et al., 2014). The importance of LN micrometastasis in predicting outcome, and how that event compares to the presence of macrometastasis (>2 mm), was first defined and studied four decades ago (Huvos et al., 1971). The current approach (Edge et al., 2009) to LN evaluation is as follows.

Isolated tumor cell clusters (ITCs) are defined as single cells or small clusters of cells not larger than 0.2 mm and no more than 200 cells in a single cross-section, usually with no histologic evidence of malignant activity (such as desmoplastic proliferation or stromal reaction; Singletary et al., 2003). The RLNs are designated as pN0(i+) or pN0(i−) when ITCs are detected by H&E and/or immunohistochemistry. ITCs are not included in the total number of positive nodes for N classification.

Micrometastases measure more than 0.2 mm, but not more than 2 mm, and/or comprise more than 200 cells in a single cross-section. If only micrometastases are present, the N classification is pN1mi. If at least one macrometastasis is present, nodes with micrometastases are included in the total node count for N classification.

Macrometastases measure more than 2 mm (N1a and above based on the number of lymph nodes involved).

In the SLN era (Nathanson et al., 2015), pathologists have far fewer LNs to evaluate than in the previous era of CALND (Giuliano et al., 2011). If 30 nodes are presented for pathologic analysis at least 30 slides need to be processed. If one SLN is sampled, the pathologist can spend less time analyzing the LN and can easily do more than section. This suggests that the total volume of the LN evaluated is more than previously, and it is not surprising that a more thorough evaluation is possible and more metastases are found. Frozen section (FS) analysis of SLNs is indicated in patients undergoing mastectomy for breast cancer but may miss micrometastasis due to microscopic sampling errors, FS artifacts and tissue loss when tissues are trimmed twice during processing, once while sectioning frozen tissue and then again all through permanent sectioning. The grossing and microscopic examination protocol, size of metastatic carcinoma, and nodal staging for SLN is similar to that described above. The presence of lymphovascular invasion and metastasis to the SLN also predicts systemic metastasis (Nathanson et al., 2009b).

LNMets can invade through the capsule into perinodal soft tissue. The host response to this infiltration is wound healing in the form of excess fibrocollagenous tissue (desmoplastic reaction), which may mask tumor cells. However, the detection of extra-capsular extension does not upstage the N classification.

In melanoma, the gross handling of the LNs is no different to that described above but the microscopic N staging classification differs slightly (Edge et al., 2009). Melanoma metastases tend to populate the SLN in patterns that differ from those seen in breast cancer (Frishberg et al., 2009). In breast cancer, the metastases tend to congregate along the middle of the bisected LN, whereas melanoma cells tend to be distributed in other areas (Frishberg et al., 2009). Serial sectioning of SLNs in melanoma increases the likelihood of detecting micrometastasis compared to only one bisection of an LN, thus avoiding a false-negative rate as high as 10–15%. Similar to SLN in other epithelial malignancies, the sensitivity of detection of micrometastasis of melanoma is further enhanced by the use of immunohistochemical stains specific for melanoma (HMB-45 or Melan A).

The definitions of micrometastasis versus macrometastasis in melanoma LNs differ significantly compared to other organ systems where the general concept is based on the measured size of the metastatic tumor. In clinically and/or radiologically LN-negative melanoma patients, nodal metastasis diagnosed after pathologic examination is defined as micrometastasis. Macrometastases are defined as clinically detectable nodal metastases confirmed by therapeutic lymphadenectomy or when nodal metastasis exhibits gross extracapsular extension.

Regardless of quantity, size (ITC, micrometastasis, macrometastasis), or mode of detection such as by routine histologic examination or immunohistochemistry, melanoma LNMets are classified as N stage greater than N0 disease. Tumor burden or the patterns of metastasis in SLN, such as the S classification, do have clinical relevance as they have been shown to predict positive non-SLNs and such patients may benefit from CLND of the RLN basin (Cochran et al., 2000; Gershenwald et al., 2008; van Akkooi et al., 2008). By convention, the term RLN metastasis refers to disease confined to one draining nodal basin or two contiguous draining nodal basins. For example, in patients with nodal disease in combinations of femoral/iliac, axillary/supraclavicular, cervical/supraclavicular, axillary/femoral, or bilateral axillary or femoral metastases are considered RLNs. Metastasis to nondraining nodal basin(s) is considered M1 disease (Edge et al., 2009). The number of positive RLNs for metastatic melanoma rather than the size of lymph node metastasis more strongly predicts aggressive outcomes (Dewar et al., 2004; Balch et al., 2001a).

Satellitosis is the presence of a tumor nodule within 2 cm of the primary tumor, usually in the dermis or subcutaneous fat. In-transit metastasis is defined as intralymphatic tumor in skin or subcutaneous tissue more than 2 cm from the primary tumor but not beyond the nearest RLN basin. The presence of in-transit metastasis, clinically or pathologically, between the primary tumor and the RLNs indicates a poor prognosis and is used to define N2c disease (Edge et al., 2009).

## 6. CONCLUSIONS & SUMMARY

LNMets have been through five overlapping phases of discovery: clinical, radiologic, and pathologic observation of the concept; the development of surgical techniques to treat the disease; a phase of relating LNMets to survival (predicting the prognosis in patients with and without LNMets); an ongoing interest in learning about the mechanisms whereby tumors spread to LNs; and treatment of LNMets. The pace of discovery has increased with more scientists and clinicians contributing new and important basic and clinical ideas to lymphatics and the LNs, a classic example of how problems at the bedside can be solved by connecting clinicians and basic scientists with each other. We have learned a lot but there is a lot more to learn. With more knowledge about the subject, there will come a better understanding as to how best to manage patients with LNMets in the future.

## References


Avery, M., Nathanson, S.D., Hetzel, F.W., 1992. Lymph flow from murine footpad tumors before and after sublethal hyperthermia. Radiat. Res. 132, 50–53.

Alitalo, K., 2011. The lymphatic vasculature in disease. Nat. Med. 17, 1371–1380.

Adams, M.J., Jones, L., Walker, R.A., Pringle, J.H., Bell, S.C., 2002. Changes in tenascin-cisoform expression in invasive and pre-invasive breast disease. Cancer Res. 62, 3289–3297.

Alitalo, K., Carmeliet, P., 2002. Molecular mechanisms of lymphangiogenesis in health and disease. Cancer Cell 1, 219–227.

Achen, M.G., McColl, B.K., Stacker, S.A., 2005. Focus on lymphangiogenesis in tumor metastasis. Cancer Cell 7, 121–127.

Abe, H., Sch midt, R.A., Sennet, C.A., et al., 2007. US-guided core needle biopsy of axillary lymph nodes in patients with breast cancer: why and how to do it. Radiographics 27 (Suppl. 1), S91–S99.

Ackerman, L.V., Regato, U., 1940. Cancer: Diagnosis, Treatment and Prognosis. Mosby, St. Louis, p. 3084.
van Akkooi, A.C., Nowecki, Z.I., Voit, C., et al., 2008. Sentinel node tumor burden according to the Rotterdam criteria is the most important prognostic factor for survival in melanoma patients: a multicenter study in 388 patients with positive sentinel nodes. Ann. Surg. 248 (6), 949–955.
Butler, T.P., Grantham, F.H., Gullino, P.M., 1975. Bulk transfer of fluid in the interstitial compartment of mammary tumors. Cancer Res. 35, 3084–3088.
Boucher, Y., Jain, R.K., 1992. Microvascular pressure is the principal driving force for interstitial hypertension in solid tumors: implications for vascular collapse. Cancer Res. 52, 5110–5114.
Belisle, C., Sainte-Marie, G., 1990. Blood vascular network of the rat lymph node: tridimensional studies by light and scanning electron microscopy. Am. J. Anat. 189, 111–126.
Baluk, P., Fuxe, J., Hashizume, H., Romano, T., Lashnits, E., Butz, S., et al., 2007. Functionally specialized junctions between endothelial cells of lymphatic vessels. J. ExpMed 204, 2349–2362.
Baluk, P., Morikawa, S., Haskell, A., Mancuso, M., McDonald, D.M., 2003. Abnormalities of BM on BVs and endothelial sprouts in tumors. Amer. J. Pathol. 163, 1801–1825.
Benoit, J.N., Zasieja, D.C., Goodman, A.H., Granger, H.J., 1989. Characterization of intact mesenteric lymphatic pump and its responsiveness to acute edemagenic stress. Am. J. Physiol. 257, H2059–H2069.
Brown, G., Richards, C.J., Bourne, M.W., et al., 2003. Morphologic predictors of lymph node status in rectal cancer with use of high-spatial-resolution MR imaging with histopathologic comparison. Radiology 227 (2), 371–377.
Brandao, A.C., Lehman, C.D., Partridge, S.C., 2013. Breast magnetic resonance imaging: diffusion-weighted imaging. Magn. Reson. Imaging Clin. North America 21 (2), 321–336.
Bastiaannet, E., Beukema, J.C., Hoekstra, H.J., 2005. Radiation therapy following lymph node dissection in melanoma patients: treatment, outcome and complications. Canc. Treat. Rev. 31, 18–26.
Balch, C.M., Soong, S.J., Gershenwald, J.E., et al., 2001a. Prognostic factors analysis of 17,600 melanoma patients: validation of the American Joint Committee on Cancer melanoma staging system. J. Clin. Oncol. 19, 3622–3634.
Balch, C.M., Buzaid, A.C., Soong, S.J., et al., 2001b. Final version of the American Joint Committee on Cancer staging system for cutaneous melanoma. J. Clin. Oncol. 19 3635–3648-40.
Carr, I., 1983. Lymphatic metastasis. Cancer Met. Rev. 2, 307–317.
Cabanas, R., 1977. An approach for the treatment of penile carcinoma. Cancer 39, 456–466.
Carmeliet, P., Jain, R.K., 2000. Angiogenesis in cancer and other diseases. Nature 407, 249–257.
Condeelis, J., Segall, J.E., 2003. Intravital imaging of cell movement in tumours. Nat. Rev. Cancer 3, 921–930.
Carter, C.L., Allen, C., Henson, D.E., 1989. Relation of tumor size, lymph node status, and survival in 24,740 breast cancer cases. Cancer 63 (1), 181–187.
Chen, S.W., Bennett, G., Price, J., 1998. Axillary lymph node calcification due to metastatic papillary carcinoma. Australas Radiol. 42 (3), 241–243.
Cody III, H.S., Fey, J., Akhurst, T., Fazzari, M., Mazumdar, M., Yeung, H., Yeh, S.D.J., Borgen, P.I., 2001. Complementarity of blue dye and isotope in sentinel node localization for breast cancer: univariate and multivariate analysis of 966 procedures. Ann. Surg. Oncol. 8, 13–19.
Callahan, M.K., Wolchok, J.D., 2015. Clinical activity, toxicity, biomarkers, and future development of CTLA-4 checkpoint antagonists. Sem. Onc. 42, 573–586.
Cummings, M.C., Simpson, P.T., Reid, L.E., et al., 2014. Metastatic progression of breast cancer: insights from 50 years of autopsies. J. Pathol. 232, 23–31.
Cochran, A.J., Balda, B.-R., Starz, H., Bachter, D., Krag, D.N., Cruse, C.W., et al., 2000. The Augsburg consensus: techniques of lymphatic mapping, sentinel lymphadenectomy, and completion lymphadenectomy in cutaneous malignancies. Cancer 89, 236–241.
Das, S., Sarrou, E., Podgrabinska, S., Cassella, M., Mungamuri, S.K., Feirt, N., et al., 2013. Tumor cell entry into the lymph node is controlled by CCL1 chemokine expressed by lymph node lymphatic sinuses. J. Exp. Med. 210, 1509–1528.
Davies, P.F., Barbee, K.A., Violin, M.V., Robotewskyj, A., Chen, J., Joseph, L., et al., 1997. Spatial relationships in early signaling events of flow-mediated endothelial mechanotransduction. Ann. Rev. Physiol. 59, 527–549.
diSibio, G., French, S., 2008. Metastatic patterns of cancers. Arch. Pathol. Lab Med. 132, 931–939.
Dewar, D.J., Newell, B., Green, M.A., Topping, A.P., Powel, B., Cook, M.G., August 15, 2004. The microanatomic location of metastatic melanoma in sentinel lymph nodes predicts nonsentinel lymph node involvement. J. Clin. Oncol. 22, 3345–3349.
Essner, R., 2003. Surgical treatment of malignant melanoma. Surg. Clin. North America 83, 109–156.
Edge, S.B., Byrd, D.R., Carducci, M.A., Compton, C.A. (Eds.), 2009. AJCC Cancer Staging Manual, seventh ed. Springer, New York, NY.
Frishberg, D.P., Balch, C., Balzer, B.L., et al. Protocol for the Examination of Specimens from Patients with Melanoma of the Skin Protocol Applies to Melanoma of Cutaneous Surfaces Only. Arch. Pathol. Lab. Med. 133 (10), October 2009, 1560–1577.
Friedl, P., Alexander, S., 2011. Cancer invasion and the microenvironment: plasticity and reciprocity. Cell 147, 992–1009.
Fidler, I.J., 1973. Selection of successive tumour lines for metastasis. Nat. New Biol. 242, 148–149.

Fisher, B., 1980. Laboratory and clinical research in breast cancer—a personal adventure: the David A. Karnofsky memorial lecture. Cancer Res. 40 (11), 3863–3874.

Ferrara, N., Gerber, H.-P., LeCouter, J., 2003. The biology of VEGF and its receptors. Nat. Med. 9, 669–676.

Fleury, M.E., Boardman, K.C., Swartz, M.A., 2006. Autologous morphogen gradients by subtle interstitial flow and matrix interactions. Biophys. J. 91, 113–121.

Ferrone, C.R., Panageas, K.S., Busam, K., Brady, M.S., Coit, D.G., 2002. Multivariate prognostic model for patients with thick cutaneous melanoma: importance of sentinel lymph node status. Ann. Surg. Oncol. 9 (7), 637–645.

Fornasa, F., Nesoti, M.V., Bovo, C., et al., 2012. Diffusion-weighted magnetic resonance imaging in the characterization of axillary lymph nodes in patients with breast cancer. J. Magn. Reson. Imaging 36 (4), 858–864.

Gretz, J.E., Anderson, A.O., Shaw, S., 1997. Cords, channels, corridors and conduits: critical architectural elements facilitating cell interactions in the lymph node cortex. Immunol. Rev. 156, 11–24.

Gray, J.H., 1939. The relation of lymphatic vessels to the spread of cancer. Brit. J. Surg. 26, 462–495.

Giuliano, A.E., Hunt, K.K., Ballman, K.V., Beitsch, P.D., Whitworth, P.W., Blumencranz, P.W., et al., February 9, 2011. Axillary dissection vs no axillary dissection in women with invasive breast cancer and sentinel node metastasis: a randomized clinical trial. JAMA 305, 569–575.

Gentilini, O., Veronesi, U., 2012. Abandoning sentinel lymph node biopsy in early breast cancer? A new trial in progress at the European Institute of Oncology of Milan (SOUND: Sentinel node vs Observation after axillary UltraSouND). Breast 21, 678–681.

Gruber, M., Simon, M.C., 2006. Hypoxia-inducible factors, hypoxia and tumor angiogenesis. Curr. Opin. Hematol. 13, 169–174.

Gashev, A.A., 2008. Lymphatic vessels: pressure- and flow-dependent regulatory reactions. Ann. N.Y. Acad. Sci. 1131, 100–109.

Genis, L., Galvez, B.G., Gonzalo, P., Arroyo, A.G., 2006. MTI-MMP: universal or particular player in angiogenesis. Cancer Met. Rev. 25, 77–86.

Gershenwald, J.E., Thompson, W., Mansfield, P.F., Lee, J.E., Colome, M.I., Tseng, C., Lee, J.J., Balch, C.M., Reintgen, D.S., Ross, M.I., 1999. Multi-institutional melanoma lymphatic mapping experience: the prognostic value of sentinel lymph node status in 612 stage I or II melanoma patients. J. Clin. Oncol. 17 (3), 976–983.

Gershenwald, J.E., Soong, S.J., Balch, C.M., 2010. 2010 TNM staging system for cutaneous melanoma… and beyond. Ann. Surg. Oncol. 17, 1475–1477.

Gershenwald, J.E., Andtbacka, R.H., Prieto, V.G., et al., September 10, 2008. Microscopic tumor burden in sentinel lymph nodes predicts synchronous nonsentinel lymph node involvement in patients with melanoma. J. Clin. Oncol. 26 (26), 4296–4303.

Hoshida, T., Isaka, N., Hagendoorn, J., diTomaso, E., Chen, Y.L., Pytowski, B., et al., 2006. Imaging steps of lymphatic metastasis reveals that vascular endothelial growth factor-C increases metastasis by increasing delivery of cancer cells to lymph nodes: therapeutic implications. Cancer Res. 66, 8065–8075.

Hirakawa, S., Hong, Y.K., Harvey, N., Schacht, V., Matsuda, K., Libermann, T., et al., 2003. Identification of vascular lineage-specific genes by transcriptional profiling of isolated blood vascular and lymphatic endothelial cells. Amer. J. Pathol. 162, 575–586.

He, Y., Rajantie, I., Pajusola, K., Jeltsch, M., Holopainen, T., Yla-Herttuala, S., et al., 2005. Vascular endothelial growth factor receptor 3 mediated activation of lymphatic endothelium is crucial for tumor cell entry and spread via lymphatic vessels. Cancer Res. 65, 4739–4746.

Halsted, W., 1907. The results of radical operations for the cure of carcinoma of the breast. Ann. Surg. 46, 1–9.

Hirakawa, S., Brown, L.F., Kodama, S., Paavonen, K., Alitalo, K., Detmar, M., 2007. VEGF-C–induced lymphangiogenesis in sentinel lymph nodes promotes tumor metastasis to distant sites. Blood 109, 1010–1017.

Hanahan, D., Weinberg, R.A., 2011. Hallmarks of cancer; the next generation. Cell 144, 646–674.

Harrell, M.I., Iritani, B.M., Ruddell, A., 2007. Tumor-induced sentinel lymph node lymphangiogenesis and increased lymph flow precede melanoma metastasis. Am. J. Pathol. 170, 774–786.

Helvie, M.A., Rebner, N., Sickles, E.A., et al., 1988. Calcifications in metastatic breast carcinoma in axillary lymph nodes. Am. J. Roentgenol. 151 (5), 921–922.

Hoang, J.K., Vanka, J., Ludwig, B.J., et al., 2013. Evaluation of cervical lymph nodes in head and neck cancer with CT and MRI: tips, traps, and a systematic approach. Am. J. Roentgenol. 200 (1), W17–W25.

Hamid, O., Robert, C., Daud, A., Hodi, F.S., Hwu, W., Kefford, R., Wolchok, J.D., et al., 2013. Safety and tumor responses with lambrolizumab (anti–PD-1) in melanoma. N. Engl. J. Med. 369, 134–144.

Holderfield, M., Deuker, M.M., McCormick, F., McMahon, M., 2014. Targeting RAF kinases for cancer therapy: BRAF-mutated melanoma and beyond. Nat. Rev. Cancer 14, 455–467.

Haley, P., Perry, R., Ennulat, D., et al., 2005. STP position paper: best practice guideline for the routine pathology evaluation of the immune system. Toxicol. Pathol. 33, 404–407 discussion 408.

Huvos, A.G., Hutter, R.V.P., Berg, J.W., 1971. Significance of axillary macrometastases and micrometastases in mammary cancer. Ann. Surg. 173, 44–46.

Ingber, D.E., 1998. Cellular basis of mechano-transduction. Biol. Bull. 194, 323–327.

Intenzo, C.M., Truluck, C.A., Kushen, M.C., et al., 2009. Lymphoscintigraphy in cutaneous melanoma: an updated total body atlas of sentinel node mapping. Radiographics 29 (4), 1125–1135.

Jain, R.K., Martin, J.D., Stylianopoulos, T., 2014. The role of mechanical forces in tumor growth and therapy. Annu. Rev. Biomed. Eng. 16, 321–346.

Jonietz, E., 2012. The forces of cancer. Nature 491, S56–S57.

Joyce, J.A., Pollard, J.W., 2009. Microenvironmental regulation of metastasis. Nat. Rev. Cancer 9, 239–252.

Kelly, R.H., 1975. Functional anatomy of lymph nodes. I. The paracortical cords. Int. Arch. Allergy Appl. Immunol. 48, 836–849.

Kalluri, R., Zeisberg, M., 2006. Fibroblasts in cancer. Nat. Rev. Cancer 6, 392–401.

Kessenbrock, K., Plaks, V., Werb, Z., 2012. Matrix metalloproteinases: regulators of the tumor microenvironment. Cell 141, 52–67.

Kuhn, N.Z., Woodhouse, E.C., Mohla, S., 2015. The National Cancer Institute's efforts in promoting research in the tumor microenvironment. Cancer J. 21, 263–266.

Karaman, S., Detmar, M., 2014. Mechanisms of lymphatic metastasis. J. Clin. Invest. 124, 922–928.

Kohrt, H.E., Nouri, N., Nowels, K., Johnson, D., Holmes, S., Lee, P.P., 2005. Profile of immune cells in axillary lymph nodes predicts disease free survival in breast cancer. PLoS Med. 2, e284.

Kyzas, P.A., Evangelou, E., Denaxa-Kyza, D., et al., 2008. 18F-fluorodeoxyglucose positron emission tomography to evaluate cervical node metastases in patients with head and neck squamous cell carcinoma: a meta-analysis. J. Natl. Cancer Inst. 100 (10), 712–720.

Krasna, M.J., 2013. Radiographic and endosonographic staging in esophageal cancer. Thorac. Surg. Clin. 23 (4), 453–460.

Kinahan, P.E., Fletcher, J.W., 2010. Positron emission tomography-computed tomography standardized uptake values in clinical practice and assessing response to therapy. Semin. Ultrasound CT MR 31 (6), 496–505.

Liersch, R., Hirakawa, S., Berdel, W.E., Mesters, R.M., Detmar, M., 2012. Induced lymphatic sinus hyperplasia in sentinel lymph nodes by VEGF-C as the earliest premetastatic indicator. Int. J. Oncol. 41 (6), 2073–2078.

Liu, Y., Wu, B.Q., Geng, H., Xu, M.L., Zhong, H.H., 2015. Association of chemokine and chemokine receptor expression with the invasion and metastasis of lung carcinoma. Oncol. Lett. 10, 1315–1322.

Lee, J.H., Torisu-Itakara, H., Cochran, A.J., et al., 2005. Quantitative analysis of melanoma-induced cytokine-mediated immunosuppression in melanoma sentinel nodes. Clin. Cancer Res. 11, 107–112.

Leong, S.P., Peng, M., Zhou, Y.M., Vaquerano, J.E., Chang, J.W., 2002. Cytokine profiles of sentinel lymph nodes draining the primary melanoma. Ann. Surg. Oncol. 9, 82–87.

Li, Q., Teitz-Tennenbaum, S., Donald, E.J., Li, M., Chang, A.E., 2009. In vivo sensitized and in vitro activated B cells mediate tumor regression in cancer adoptive immunotherapy. J. Immunol. 183, 3195–3203.

Luke, G.P., Myers, J.N., Emelianov, S.Y., et al., 2014. Sentinel lymph node biopsy revisited: ultrasound-guided photoacoustic detection of micrometastases using molecularly targeted plasmonic nanosensors. Cancer Res. 74 (19), 5397–5408.

Lee, C.H., Giurescu, M.E., Philpotts, L.E., et al., 1997. Clinical importance of unilaterally enlarging lymph nodes on otherwise normal mammograms. Radiology 203 (2), 329–334.

Lyman, G.H., Temin, S., Edge, S.B., Newman, L.A., Turner, R.R., Weaver, D.L., Benson, A.B., Bosserman, L.D., Burstein, H.J., Cody, H., Hayman, J., Perkins, C.L., Podoloff, D.A., Giuliano, A.E., 2014. Sentinel lymph node biopsy for patients with early-stage breast cancer: American Society of Clinical Oncology clinical practice guideline update. J. Clin. Oncol. 32, 1365–1383.

Lester, S., Bose S., Cehn Y.Y., et al. Protocol for the Examination of Specimens from Patients with Invasive Carcinoma of the Breast Protocol Applies to All Invasive Carcinomas of the Breast Arch. Pathol. Lab. Med. 133, October 2009.

Mohammed, R.A.A., Martin, S.G., Gill, M.S., et al., 2007. Improved methods of detection of lymphovascular invasion demonstrate that it is the predominant method of vascular invasion in breast cancer and has important clinical consequences. Amer J. Surg. Pathol. 31, 1825–1833.

Mitchell, M.J., King, M.R., 2013. Computational and experimental models of cancer cell response to fluid shear stress. Front. Oncol. 3, 1–11.

McDonald, D.M., 2008. Angiogenesis and vascular remodeling in inflammation and cancer: biology and architecture of the vasculature. In: Figg, W.D., Folkman, J. (Eds.), Angiogenesis: An Integrative Approach From Science to Medicine. Springer, pp. 17–33 (Chapter 2).

Morton, D.L., Thompson, J.F., Cochran, A.J., Mozzillo, N., Nieweg, O.E., Roses, D.F., et al., 2014. Final trial report of sentinel-node biopsy versus nodal observation in melanoma. NEJM 370, 599–609.

Morton, D.L., Wen, D.R., Wong, J.H., Economou, J.S., Cagle, L.A., Storm, F.K., et al., 1992. Technical details of intraoperative lymphatic mapping for early stage melanoma. Arch. Surg. 127, 392–399.

Maeng, Y.S., Aguilar, B., Choi, S.I., Kim, E.K., 2016. Inhibition of TGFβIp expression reduces lymphangiogenesis and tumor metastasis. Oncogene 35, 196–205.

Muller, A., Homey, B., Soto, H., Ge, N., Catron, D., Buchanan, M.E., et al., 2001. Involvement of chemokine receptors in breast cancer metastasis. Nature 410, 50–56.

Munn, L.L., 2015. Mechanobiology of lymphatic contractions. Sem. Cell Develop. Biol. 38, 67–74.

Misselwitz, B., 2006. MR contrast agents in lymph node imaging. Eur. J. Radiol. 58 (3), 375–382.
Muller, M.G.S., Hennipman, F.A., van Leeuwen, P.A., Pijpers, R., Vuylsteke, R.J., Meijer, S., 2002. Unpredictability of lymphatic drainage patterns in melanoma patients. Eur. J. Nuc. Med. Mol. Imaging 29 (2), 255–261.
Nathanson, S.D., Shah, R., Rosso, K., 2015. Sentinel lymph node metastases in cancer: causes, detection and their role in disease progression. Sem. Cell Develop. Biol. 38, 106–116.
Nathanson, S.D., 2003. Insights into the mechanisms of lymph node metastasis. Cancer 98, 413–423.
Nathanson, S.D., Nelson, L.T., 1994. Interstitial fluid pressure in breast cancer, benign breast conditions and breast parenchyma. Ann. Surg. Oncol. 1, 333–338.
Nathanson, S.D., Avery, M., Anaya, P., Sarantou, T., Hetzel, F.W., 1997. Lymphatic diameters and lymph clearance rates in a murine melanoma model. Arch. Surg. 132, 311–315.
Nathanson, S.D., Miller, C.G., Paxton, J.H., 2009a. The role of lymphangiogenesis in regional lymph node metastasis: animal models. In: Leong, S.P.L. (Ed.), Current Clinical Oncology: From Local Invasion to Metastatic Cancer. Humana Press, pp. 211–226 (Chapter 19).
Nathanson, S.D., Kwon, D., Kapke, A., Alford, S.H., Chitale, D., 2009b. The role of lymph node metastasis in the systemic dissemination of breast cancer. Ann. Surg. Oncol. 16, 3396–3405.
Nathanson, S.D., Haas, G.P., Mead, M.J., Lee, M., 1986. Spontaneous regional lymph node metastases of three variants of the B16 melanoma: relationship to primary tumor size and pulmonary metastases. J. Surg. Oncol. 33, 41–45.
Nathanson, S.D., Mahan, M., 2011. Sentinel lymph node pressure in breast cancer. Ann. Surg. Oncol. 18, 3791–3796.
Nathanson, S.D., Shah, R., Chitale, D.A., Mahan, M., 2014. Intraoperative clinical assessment and pressure measurements of sentinel lymph nodes in breast cancer. Ann. Surg. Oncol. 21, 81–85.
Nathanson, S.D., Burke, M., Slater, R., et al., 2007. Preoperative identification of the sentinel lymph node in breast cancer. Ann. Surg. Oncol. 14 (11), 3102–3110.
Norman, J., Cruse, C.W., Espinosa, C., Cox, C., Berman, C., Clark, R., Saba, H., Wells, K., Reintgen, D., 1991. Redefinition of cutaneous lymphatic drainage with the use of lymphoscintigraphy for malignant melanoma. Am. J. Surg. 162, 432–437.
Ng, C.P., Hinz, B., Swartz, M.A., 2005. Interstitial fluid flow-induces myofibroblast differentiation and collagen alignmentin in vitro. J. Cell Sci. 118, 4731–4739.
Namm, J.P., Chang, A.E., Cimmino, V.M., Rees, R.S., Johnson, T.M., Sabel, M.S., 2012. Is a level III dissection necessary for a positive sentinel lymph node in melanoma? J. Surg. Onc. 105 (3), 225–228.
National Comprehensive Cancer Network. NCCN clinical practice guidelines in oncology for melanoma. Available from: URL: http://www.nccn.org.
Orimo, A., Gupta, P.B., Sgroi, F., Arenzana-Seisdedos, T., Delauney, T., Naeem, R., et al., 2002. Stromal fibroblasts present in invasive human breast carcinomas promote tumor growth and angiogenesis through elevated SDF-1/CXCL12 secretion. Cell 121, 335–348.
Olszewski, W.L., Engeset, A., 1980. Intrinsic contractility of prenodal lymph vessels and lymph flow in human leg. Am. J. Physiol. 239, H263–H269.
Pang, M.F., Georgoudaki, A.M., Lambut, L., Johansson, J., Tabor, V., Hagikura, K., Jin, Y., et al., 2016. TGF-β1-induced EMT promotes targeted migration of breast cancer cells through the lymphatic system by the activation of CCR7/CCL21-mediated chemotaxis. Oncogene 35, 748–760.
Pathak, A., Kumar, S., 2011. Biophysical regulation of tumor cell invasion: moving beyond matrix stiffness. Integr. Biol. 3, 267–278.
Padera, T.P., Stoll, B.R., Tooredman, J.B., Capen, D., di Tomaso, E., Jain, R.K., 2004. Cancer cells compress intratumor vessels. Nature 427, 695.
Pedersen, J.A., Lichter, S., Swartz, M.A., 2010. Cells in 3D matrices under interstitial flow: effects of extracellular matrix alignment on cell shear stress and drag forces. J. Biomech. 43, 900–905.
Patey, D.H., Dyson, W.H., 1948. The prognosis of carcinoma of the breast in relation to the type of operation performed. Br. J. Cancer 2, 7–13.
Podgrabinska, S., Braub, P., Velasco, P., Kloos, B., Pepper, M.S., Jackson, D.G., et al., 2002. Molecular characterization of lymphatic endothelial cells. Proc. Natl. Acad. Sci. U.S.A. 99, 16069–16074.
Padera, T.P., Kadambi, A., di Tomaso, E., Carreira, C.M., Brown, E.B., Boucher, Y., et al., 2002. Lymphatic metastasis in the absence of functional intratumor lymphatics. Science 296, 1883–1886.
Pereira, E.R., Jones, D., Jug, K., Padera, T.P., 2015. The lymph node micro-environment and its role in the progression of metastatic cancer. Sem. Cell Dev. Biol. 38, 98–105.
Purushotham, A.D., Upponi, S., Klevesath, M.B., Bobrow, L., Millar, K., Myles, J.P., Duffy, S.W., 2005. Morbidity after sentinel lymph node biopsy in primary breast cancer: results from a randomized controlled trial. J. Clin. Oncol. 23, 4312–4321.
Qian, C.N., Berghuis, B., Tsarfaty, G., Bruch, M., Kort, E.J., Ditlev, J., et al., 2006. Preparing the soil: the primary tumor induces vasculature reorganization in the sentinel lymph node before he arrival of metastatic cancer cells. Cancer Res. 66, 10365–10376.
Reintgen, D., Cruse, C.W., Wells, K., 1994. The orderly progression of melanoma nodal metastases. Ann. Surg. 220, 759–767.

Resnick, N., Yahav, H., Shay-Salit, A., Shushy, M., Schubert, S., Zilberman, L.C.M., et al., 2003. Fluid shear stress and the vascular endothelium: for better and for worse. Prog. Biophys. Mol. Biol. 81, 177–199.

Rutgers, E.J., Donker, M., Straver, M.E., Meijnen, P., Van De Velde, C.J.H., 2013. Radiotherapy or surgery of the axilla after a positive sentinel node in breast cancer patients: final analysis of the EORTC AMAROS trial (10981/22023). J. Clin. Oncol. 31 (Suppl.) abstr LBA1001.

Rosenberg, S.A., 2008. Principles of surgical oncology. In: DeVita, V.T., Lawrence, T.S., Rosenberg, S.A. (Eds.), Cancer: Principles and Practice of Oncology. Wolters Kluwer/Lippincott Williams & Wilkins, pp. 283–293.

Ribas, A., 2015. Releasing the breaks on cancer immunotherapy. NEJM 373, 1490–1492.

Robinson, G.N., Van Heerden, J.A., Payne, W.S., Taylor, W.F., Gaffey, T.A., 1976. The primary surgical treatment of carcinoma of the breast: a changing trend toward modified radical mastectomy. Mayo Clin. Proceed. 51, 433–442.

Rosen, P.P., Lesser, M.L., Kinne, D.W., Beattie, E.J., 1983. Discontinuous or "skip" metastases in breast carcinoma. Analysis of 1228 axillary dissections. Ann. Surg. 197, 276.

Ribas, A., Weber, J.S., Chmielowski, B., Comin-Anduix, B., Lu, D., Douek, M., Ragavendra, N., et al., 2011. Intra–lymph node prime-boost vaccination against melan A and tyrosinase for the treatment of metastatic melanoma: results of a phase 1 clinical trial. Clin. Cancer Res. 17, 2987–2996.

Sainte-Marie, G., Peng, F.S., Belisle, C., 1982. Overall architecture and pattern of lymph flow in the rat lymph node. Am. J. Anat. 164, 275–309.

Shieh, A.C., 2011. Biomechanical forces shape the tumor microenvironment. Ann. Biomed. Engin. 39, 1379–1389.

Sleeman, J.P., Nazarenko, I., Thiele, W., 2011. Do all roads lead to Rome? Routes to metastasis development. Int. J. Cancer 128, 2511–2526.

Swartz, M.A., Iida, N., Roberts, E.W., Sangaletti, S., Wong, M.H., Yull, F.E., et al., 2014. Tumor microenvironment complexity: emerging roles in cancer therapy. Cancer Res. 72, 2473–2480.

Stacker, S.A., Williams, S.P., Karnezis, T., Shayan, R., Fox, S.B., Achen, M.G., 2014. Lymphangiogenesis and lymphatic vessel remodeling in cancer. Nat. Rev. Cancer 14, 159–172.

Skobe, M., Hamberg, L.M., Hawighorst, T., Schirner, M., Wolf, G.L., Alitalo, K., et al., 2001. Concurrent induction of lymphangiogenesis, angiogenesis, and macrophage recruitment by vascular endothelial growth factor-C in melanoma. Amer. J. Pathol. 159, 893–903.

Stelzner, F., Steinau, H.U., Friedrichs, N., von Mallek, D., 2013. Der Weg regionerer Lymphbahn metastasen bei Weichgewebs sarkomen. Chirurg 84, 502–510.

Singer, C., Blankstein, E., Koenigsberg, T., et al., 2001. Mammographic appearance of axillary lymph node calcification in patients with metastatic ovarian carcinoma. Am. J. Roentgenol. 176 (6), 1437–1440.

Sloothaak, D.A., Grewal, S., Doornewaard, H., et al., 2014. Lymph node size as a predictor of lymphatic staging in colonic cancer. Br. J. Surg. 101 (6), 701–706.

Schrenk, P., Rieger, R., Shamiyeh, A., Wayand, W., 2000. Morbidity following sentinel lymph node biopsy versus axillary lymph node dissection for patients with breast carcinoma. Cancer 88, 608–614.

Schijven, M.P., Vingerhoets, A.J.J.M., Rutten, H.J.T., Nieuwenhuijzen, G.A.P., Roumen, R.M.H., Van Bussel, M.E., Voogd, A.C., 2003. Comparison of morbidity between axillary lymph node dissection and sentinel node biopsy. Europ J. Surg. Onc. 29 (4), 341–350.

Singletary, S.E., Greene, F.L., Sobin, L.H., 2003. Classification of isolated tumor cells: clarification of the 6th edition of the American Joint Committee on Cancer staging manual. Cancer 90, 2740–2741.

Stenbygaard, L.E., Sørensen, J.B., Larsen, H., Dombernowsky, P., 1999. Metastatic pattern in non-resectable non-small cell lung cancer. Acta Oncol. 38 (8), 993–998.

Tammela, T., Alitalo, K., 2010. Lymphangiogenesis: molecular mechanisms and future promise. Cell 140, 460–476.

Topper, N., Gimbrone Jr., M.A., 1999. Blood flow and vascular gene expression: fluidshear stress as a modulator of endothelial phenotype. Mol. Med. Today 5, 40–46.

Tutunea-Fatan, E., Mousumi, M., Xin, X., Lala, P.K., 2015. The role of CCL21/CCR7 chemokine axis in breast cancer-induced lymphangiogenesis. Mol. Cancer 14, 35.

Takeuchi, H., Fujimoto, A., Tanaka, M., Yamano, T., Hsueh, E., Hoon, D.S., 2004. CCL21chemokine regulates chemokine receptor CCR7 bearing malignant melanoma cells. Clin. Cancer Res. 10, 2351–2358.

Thoeny, H.C., Froehlich, J.M., Triantafyllou, M., et al., 2014. Metastases in normal-sized pelvic lymph nodes: detection with diffusion-weighted MR imaging. Radiology 273 (1), 125–135.

Tournoy, K.G., Ryck, F.D., Vanwalleghem, L., et al., 2008. The yield of endoscopic ultrasound in lung cancer staging: does lymph node size matter? J. Thorac. Oncol. 3 (3), 245–249.

Turner, L., Swindell, R., Bell, W.G., Hartley, R.C., Tasker, J.H., Wilson, W.W., Alderson, M.R., Leck, I.M., 1981. Radical versus modified radical mastectomy for breast cancer. Ann. Roy. Col. Surg. Eng. 63, 239.

Tang, L.J., Berlin, J., Branton, P. et al. Protocol for the Examination of Specimens from Patients with Primary Carcinoma of the Colon and Rectum Arch. Pathol. Lab. Med. 133 (10), October 2009, 1539–1551.

Urban, J.A., Baker, H.W., 1952. Radical mastectomy in continuity with en bloc resection of the internal mammary lymph node chain: a new procedure for primary operable cancer of the breast. Cancer 5, 992–1008.

Uren, R.F., 2004. Lymphatic drainage of the skin. Ann. Surg. Oncol. 11, 179S–185S.

Veronesi, U., Viale, G., Paganelli, G., Zurrida, S., Luini, A., Galimberti, V., et al., 2010. Sentinel lymph node biopsy in breast cancer: ten year results of a randomized controlled study. Ann. Surg. 251, 595–600.

Virchow, R.L.K., 1860. Cellular Pathology. John Churchill.

Veronesi, U., Paganelli, G., Viale, G., Luini, A., Zurrida, S., Galimberti, V., Intra, M., et al., 2003. A randomized comparison of sentinel-node biopsy with routine axillary dissection in breast cancer. N. Engl. J. Med. 349, 546–553.

Weil, P.H., Buchberger, R., 1999. From Billroth to PCV: a century of gastric surgery. World J. Surg. 23, 736–742.

Wiley, H.E., Gonzalez, E.B., Maki, W., Wu, M.-T., Hwang, S.T., 2001. Expression of CC chemokine receptor-7 and regional lymph node metastasis of B16 murine melanoma. J. Natl. Cancer Inst. 93, 1638–1643.

Witte, M.H., Jones, K., Wilting, J., Dictor, M., Seig, M., McHale, N., et al., 2006. Structure function relationships in the lymphatic system and implications for cancer biology. Cancer Met Rev. 25, 159–184.

Wang, W., Gosswami, S., Sahai, E., Wyckoff, J.B., Segall, J.E., Condeelis, J.S., 2005. Tumor cells caught in the act of invading: their strategy for enhanced cell motility. Trends Cell Biol. 15, 138–145.

Wyckoff, J., Wang, W., Lin, E.Y., Wang, Y., Pixley, F., Stanley, E.R., et al., 2004. A paracrine loop between tumor cells and macrophages is required for tumor cell migration in mammary tumors. Cancer Res. 64 (19), 7022–7029.

Weaver Am Page, J.M., Guelcher, S.A., Parekh, A., 2013. Synthetic and tissue derived models for studying rigidity effects on invadopodia activity. Methods Mol. Biol. 1046, 171–189.

Wang, W., Nepiyushchikh, Z.V., Zawieja, D.C., Chakraborty, S., Zawieja, S.D., Gaschev, A.A., et al., 2009. Inhibition of myosin light chain phosphorylation decreases rat mesenteric lymphatic contractile activity. Am. J. Physiol. Heart Circ. Physiol. 297, H726–H734.

Wahl, R.L., Siegel, B.A., Coleman, R.E., et al., 2004. Prospective multicenter study of axillary nodal staging by positron emission tomography in breast cancer: a report of the staging breast cancer with PET Study Group. J. Clin. Oncol. 22 (2), 277–285.

Wilke, L.G., McCall, L.M., Posther, K.E., Whitworth, P.W., Reintgen, D.S., Leitch, A.M., Gabram, S.G.A., et al., 2006. Surgical complications associated with sentinel lymph node biopsy: results from a prospective international cooperative group trial. Ann. Surg. Oncol. 13, 491–500.

Whelan, T.J., Olivotto, I.A., Parulekar, W.R., Ackerman, I., Chua, B.H., Nabid, A., Vallis, K.A., et al., 2015. Regional nodal irradiation in early-stage breast cancer. N. Engl. J. Med. 373, 307–316.

Willard-Mack, C.L., 2006. Normal structure, function, and histology of lymph nodes. Toxicol. Path. 34, 409–424.

Wilson, I., 1961. Patterns of metastasis in lung cancer-review. Cancer Res. 21.

Xia, X., Liu, H., Zhang, H., Shang, Z., 2015. Correlation between CCR7 expression and lymph node metastatic potential of human tongue carcinoma. Oral Dis. 21, 123–131.

Xu, Y., Yuan, L., Mak, J., Pardanaud, L., Caunt, M., Kasman, I., et al., 2010. Neuropilin-2mediates VEGF-C-induced lymphatic sprouting together with VEGFR3. J. Cell Biol. 188, 115–130.

Yamaguchi, H., Wyckoff, J., Condeelis, J., 2005. Cell migration in tumors. Curr. Opin. Cell Biol. 17 (5), 559–564.

Zlotnik, A., 2006. Involvement of chemokine receptors in organ specific metastasis. Contrib. Microbiol. 13, 113–120.

Zhang, G.-Y., Liu, L.-Z., Wei, W.-H., et al., 2010. Radiologic criteria of retropharyngeal lymph node metastasis in nasopharyngeal carcinoma treated with radiation therapy. Radiology 255 (2), 605–612.

CHAPTER

# 14

# Molecular Involvement of the Bone Marrow Microenvironment in Bone Metastasis

*S.F. Miler, C.Y. Thomas, Y. Shiozawa*

**Wake Forest School of Medicine, Winston-Salem, NC, United States**

## 1. INTRODUCTION

Cancer becomes much more difficult to treat, and can become incurable, once tumors metastasize to distant sites, such as bone. Indeed, only 20% of breast cancer patients survive longer than 5 years after the discovery of bone metastases (Roodman, 2004), whereas the 5-year survival rate of patients with only localized disease has exceeded 90%. In addition, in a study of 500 cases of women with breast cancer, the most common site of distant relapse, resulting in increased mortality, was the bone (Isales et al., 1987). Consistent with these findings, approximately 35% of the one million people who die from cancer each year in the United States die due to bone metastases (Mundy, 2002). Indeed, although successful treatments for primary malignancies have been achieved, metastatic disease, including bone metastasis, remains the major cause of cancer-related deaths (Raubenheimer and Noffke, 2006). There is, therefore, an increasing need to understand the biology of bone metastasis to better control this severe mortality rate.

Bone is the third most frequent site of metastasis, after lung and liver (Coleman, 2001). Traditionally, blood flow has been considered the sole determinant of metastatic deposit site (Raubenheimer and Noffke, 2006). This belief is supported because the most frequent sites of solid tumor metastases are organs that receive high volumes of blood, like the lungs or, in this context, vascular areas of the skeleton, particularly red bone marrow. However, site determinants other than only blood flow must be involved, because tumors rarely metastasize to some very vascular tissues like skeletal muscle, which receives over a fourth of cardiac output (Guise and Mundy, 1998). One theory is that the type of cancer helps determine the location of metastasis. Although over 70% of patients with advanced breast or prostate cancer develop bone metastases (Coleman and Rubens, 1987), it is only seen in 15–30% of patients with carcinoma of the lung, colon, rectum, thyroid, kidney, bladder, uterus, or stomach (Mundy, 2002). Perhaps breast and prostate cancers develop bone metastases at a higher rate than other cancer types

*Introduction to Cancer Metastasis*
http://dx.doi.org/10.1016/B978-0-12-804003-4.00014-1

because they have slowly progressing tumors, leading to longer survival and more time to metastasize (Selvaggi and Scagliotti, 2005). Additionally, breast and prostate cancers may have specific characteristics that cause them to find bone more suitable than other locations. Such physiological mechanisms of the bone metastatic process have not been fully uncovered due to the lack of appropriate tools and techniques (e.g., animal models) (Raubenheimer and Noffke, 2006). Short duration of patient survival after bone metastasis also makes it more difficult to investigate clinical indications of this disease (Selvaggi and Scagliotti, 2005).

Early detection is crucial for improving the prognosis of cancer patients with bone metastasis. Standard imaging techniques are radiography, 99mTc-MDP bone scintigraphy (Fig. 14.1) (Hage

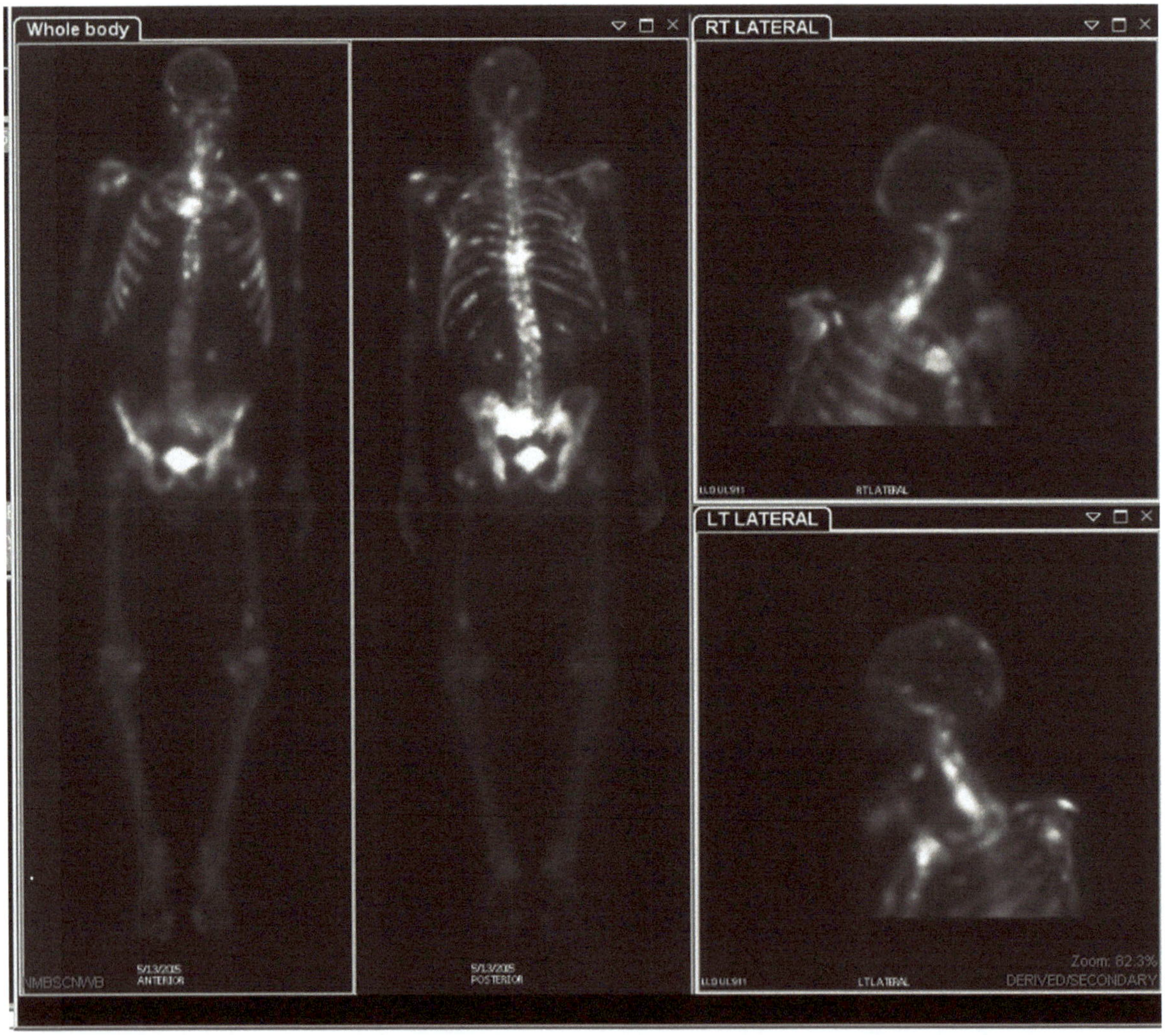


FIGURE 14.1 **Evidence of bone metastasis.** Bone scintigraphy shows abnormal, intense radiotracer uptake in the calvaria, scapulas, ribs, vertebral bodies, pelvis, and left femur of a patient with advanced prostate cancer.

et al., 2000), and, more recently, fluorine-18-fluorodeoxyglucose positron emission tomography/computed tomography (Chang et al., 2014). Significant progress has been made in the early diagnosis and treatment of primary tumors, but patients still die from metastatic disease, which can manifest decades later. Patients with bone metastases and their families play a tremendous price physically, emotionally, and financially, and the cost of providing health care to these patients is growing steadily. Therefore, new approaches to treatment and prevention of bone metastases are urgently needed.

It is unclear why some tumors selectively metastasize to and proliferate in bone. In fact, 17–57% of breast cancer patients with bone metastases develop metastasis only in bone when diagnosed (James et al., 2003; Koizumi et al., 2003; Scheid et al., 1986; Plunkett et al., 2000; Ahn et al., 2013; Lee et al., 2011). Although the dissemination process appears to involve multiple factors, growing evidence suggests that the bone marrow microenvironment plays a role in this complicated metastatic process. In this review, we highlight the phenotype of bone metastasis, the process of tumor growth in bone, and potential treatment options for this disease, by focusing on the crosstalk between the bone marrow microenvironment and the metastatic tumor.

## 2. TYPES OF BONE METASTASIS

Bone metastases are characterized as osteolytic (bone destructive) or osteoblastic (bone forming), but a mixture of osteolytic and osteoblastic behavior is usually present. In fact, most skeletal metastases fall somewhere in a range from mostly osteolytic (e.g., breast cancer) to mostly osteoblastic (e.g., prostate cancer) (Mundy, 2002), except in the case of multiple myeloma, which is purely osteolytic (Roodman, 2004). Bone growth and density depend on the balance between osteoclastogenesis and osteoblastogenesis. Osteoclasts are giant, multinucleated cells that destroy the bone, and develop from monocyte–macrophage lineage cells (Nijweide et al., 1986). Osteoblasts are bone marrow stromal cells that build bone and express factors that affect osteoclast differentiation and function (Takahashi et al., 1999). The activities of these two cell types are regulated not only by factors found in the bone marrow, but also by one another. Additionally, in bone metastasis, factors influencing bone remodeling are also produced by tumor cells. The combination of factors produced by tumor cells and bone cells creates crosstalk between the bone metastatic cancer cells and the bone marrow microenvironment, which results in a more complex phenotypic appearance of bone metastases.

### 2.1 Osteolytic Bone Metastasis

The prototypic osteolytic tumor is breast cancer, which metastasizes to bone and causes destruction, pain, and fracture. Multiple factors produced by breast cancer tumors are known to stimulate osteoclastic activities. These factors include parathyroid hormone-related protein (PTHrP), receptor activator of nuclear factor κB (RANK) ligand (RANKL), interleukin (IL)-11, IL-6, and IL-8 (Bendre et al., 2005; Kakonen et al., 2002; Yin et al., 1999; Kang et al., 2005). RANKL is also expressed by the cells of osteoblastic lineage and regulates osteoclast formation and the activities of osteoclasts, by binding to RANK (Yasuda et al., 1998; Simonet et al., 1997). IL-6, known to stimulate the production of RANKL in osteoblasts (Mundy, 2002) and expressed by MDA-MB-231 breast cancer cells, promotes the RANKL production of osteoblasts and induces osteolytic bone lesions in mouse models (Zheng et al., 2014). Blocking IL-6 with neutralizing antibody or blocking RANKL with its inhibitor osteoprotegerin in animal models significantly prevents tumor growth and osteolytic lesions in bone tissues (Zheng et al., 2014). Bone-derived transforming growth factor-β (TGF-β) stimulates PTHrP secretion by MDA-MB-231 breast cancer cells through both Smad and mitogen-activated protein kinase (MAPK) pathways, and this PTHrP from breast cancer cells mediates osteoclastogenesis, resulting in osteolytic bone metastasis in mouse models (Kakonen et al., 2002; Yin et al., 1999). These findings suggest that the interplay between bone metastatic breast cancer and osteoblast lineage cells plays an important role in the molecular regulation of osteolytic lesions of bone metastasis, leading to further dissemination of metastatic tumor cells (the "vicious cycle" mentioned frequently in cancer literature). Furthermore, the roles of microRNAs (miRNAs), short noncoding RNAs, in the development of osteolytic bone metastases have recently been revealed. The upregulation of miR-16 and miR-378 has been correlated with the increased secretion of soluble intercellular adhesion molecule 1 from breast cancer, leading to osteolytic bone metastasis (Ell et al., 2013). In addition, miR-141 and miR219 were seen to be significantly downregulated during osteoclastogenesis by targeting Mitf/Calcr and Mitf/Traf6, respectively (Ell et al., 2013). Systemic treatment of miR-141 and miR219 significantly reduced osteolytic metastatic lesions by inhibiting osteoclastic activities in an in vivo animal study (Ell et al., 2013). Additionally, it has been demonstrated that miR-34a blocks osteolytic bone metastasis by inhibiting osteoclastogenesis through directly targeting transforming growth factor β-induced factor 2 (Tgif2) (Krzeszinski et al., 2014). These data suggest that miRNAs may be useful as therapeutic targets or biomarkers for bone metastatic disease (Hayes et al., 2014).

## 2.2 Osteoblastic Bone Metastasis

Osteoblastic bone metastases occur mostly in prostate cancer, and have been associated with osteoblastogenesis stimulated by tumor production of growth factors, including bone morphogenetic protein (BMP), TGF-β, insulin-like growth factor, fibroblast growth factor, platelet-derived growth factor, vascular endothelial growth factor, and Wnt (Yi et al., 2002; Logothetis and Lin, 2005). Additionally, factors that regulate osteoclastic activities also contribute to the development of a bone microenvironment that is suitable and sustainable for osteoblastic metastases. Along with this notion, growing evidence suggests that bisphosphonate, an inhibitor of osteoclastogenesis, significantly reduces osteoblastic skeletal complications in patients with hormone-refractory metastatic prostate cancer compared to placebo-treated patients (Saad et al., 2004). Surprisingly, like breast cancer, osteoblastic prostate cancer highly expresses PTHrP, which is usually associated with osteolytic behavior, but in the case of prostate cancer PTHrP induces osteoblastic metastatic lesions (Liao et al., 2008). Osteoblastic metastasis is less understood than osteoclastic but a possible explanation for this conundrum is that fragments of PTHrP activate endothelin A receptor ($ET_AR$), which stimulates bone formation (Schluter et al., 2001). These fragments might be mimicking the behavior of a ligand for $ET_AR$, endothelin-1 (ET-1), a vasoactive peptide known to promote osteoblastogenesis. The high levels of ET-1 are observed in the plasma of metastatic prostate cancer patients and several prostate cancer cell lines (Nelson et al., 1995). In an animal study, ET-1 stimulated new bone formation associated with prostate cancer osteoblastic metastases, and astrasentan, an antagonist for the ET-1 receptor $ET_AR$, prevented osteoblastic metastases (Yin et al., 2003). Astrasentan delayed the increase of bone alkaline phosphatase expression in men with metastatic castration-resistant prostate cancer (mCRPC), however, it did not improve disease progression (Quinn et al., 2013; Carducci et al., 2007). ET-1 also activates osteoblastic formation indirectly in the marrow by downregulating dickkopf Wnt-signaling pathway inhibitor 1 (DKK1) (Clines et al., 2007). DKK1 is a secreted inhibitor of the Wnt-signaling pathway, which is known to be involved in osteoblastic differentiation (Baron and Kneissel, 2013). Indeed, the high level of DKK1 in the plasma of bone marrow and peripheral blood in multiple myeloma patients appears to correlate with osteolytic bone lesions (Tian et al., 2003).

# 3. PATHOPHYSIOLOGY OF BONE METASTASIS

There are several steps involved in the pathophysiological development of bone metastases. The first are similar to those of any cancer metastasis. Primary tumor cells produce proteolytic enzymes, which invade the surrounding tissue by traversing the walls of small blood vessels (Liotta and Kohn, 2001). These blood vessels may be preexisting or formed via angiogenesis (Weidner et al., 1991). Once in circulation, the cancer cells can travel to distant organ sites. Many cancer cells fail to survive, due to the normal, protective host-surveillance mechanisms during these initial stages (Zetter, 1990). If they do survive, these disseminated cancer cells may lie dormant in the metastatic site, such as bone, for long periods of time, retaining the potential for activation leading to metastatic growth (Pantel et al., 2009; Morgan et al., 2009; Nash et al., 2007). Specifically, for cancer cells to migrate toward the bone, they must be able to traverse the sinusoidal wall, invade the marrow stroma, activate angiogenesis, and locate the endosteal bone surface. The metastatic location depends on cancer type and the individual patient. Stephen Paget proposed this idea in 1989 as the "seed and soil" theory of metastasis (Paget, 1989), where "seed" refers to the cancer cell and "soil" refers to the microenvironment of the metastatic site. He pointed out that one reason cancers have high affinity for bone may be sheer probability, because most tumor cells circulate through the highly vascular bone marrow. Some other highly vascularized organs rarely host metastases, however, so the capillary bed in bone must provide a fertile environment for any arriving cells (Paget, 1989).

## 3.1 Dissemination Process of Bone Metastatic Tumor Cells

Full understanding of the more complex interactions between tumor cells and the host microenvironment resulting in active bone metastasis remains elusive. Although multiple factors are thought to be involved in the dissemination process, growing evidence suggests that several, specific molecules that play a role in the homing of hematopoietic stem cells (HSCs) to the bone marrow are also used by tumor cells to metastasize and establish hold in the marrow. There are two possible routes for tumor cells to spread to the distant sites: lymphatic and hematogenous dissemination (Banys et al., 2012; Wong and Hynes, 2006). Most prostate cancer patients with lymph node metastases develop bone metastases as well, whereas the presence of disseminated tumor cells in the bone marrow does not always correlate with lymph node metastases (Wong and Hynes, 2006). This suggests that bone metastatic cancer cells use hematogenous dissemination during their early diffusion, rather than lymphatic dissemination. During the homing process to the bone marrow, HSCs migrate toward the stromal cell-derived factor 1 (SDF-1) expressed by bone marrow stromal cells, using one of its receptors, C-X-C chemokine receptor type 4 (CXCR4) (Aiuti et al., 1997; Peled et al., 1999). This CXCR4/SDF-1 chemokine axis is also crucial for the early dissemination of breast cancer (Muller et al., 2001; Zhang et al., 2013) and prostate cancer (Taichman et al., 2002) to the bone. In fact, CXCR4 is highly expressed in bone metastatic samples from breast cancer (Kang et al., 2003) and prostate cancer (Sun et al., 2003). In in vivo studies, the pharmacological blockage of the interaction of CXCR4 and SDF-1 with CXCR4 inhibitor AMD3100 prevents prostate cancer bone metastasis (Gravina et al., 2015; Sun et al., 2005) and induces the susceptibility of prostate cancer cells to chemotherapeutic agent docetaxel (Domanska et al., 2012). In addition, Annexin II, an adhesion molecule that controls HSCs homing to the marrow (Jung et al., 2007), is expressed by bone marrow osteoblasts and also plays a central role in prostate cancer metastasis (Shiozawa et al., 2008). Blockage of Annexin II or its receptor in animal models prevents prostate cancer bone metastasis (Shiozawa et al., 2008). These findings illuminate the parallels between HSC homing to the marrow and the bone metastatic process; bone metastatic cancer cells utilize the same methods as HSC homing to gain access to the bone microenvironment, which the cancer exploits to promote bone metastases.

## 3.2 Bone Metastatic Tumor Cell Targets the Osteoblastic Niche in the Bone Marrow

Along with these findings, a recent study reveals direct evidence that bone metastatic prostate cancer cells directly target the microenvironment for HSCs, the HSC niche, and compete for occupancy of this niche; and that, once there, prostate cancer cells begin a process of either evicting HSCs into the peripheral blood or driving HSCs into progenitor cell pools (Shiozawa et al., 2011a). Additional evidence suggesting that prostate cancer targets the HSC niche during metastasis was obtained when the osteoblastic HSC niche was ablated using conditional osteoblast knockout tissues in metastatic assays (Shiozawa et al., 2011a). The results showed that fewer metastatic cells home to the marrow when the HSC niche is compromised. Conversely, increasing the HSC niche size with PTH, which enhances the engraftment of HSCs to the osteoblastic HSC niche (Calvi et al., 2003), was also found to promote bone metastasis (Shiozawa et al., 2011a). Importantly, bone metastatic prostate cancer cells are mobilized into peripheral blood with agents that mobilize HSCs from the marrow (Shiozawa et al., 2011a). This study offers a new paradigm for deeper understanding of which areas in the bone marrow are preferential targets for prostate cancer during dissemination. The osteoblastic niche is also involved in the early dissemination and colonization of bone metastatic breast cancer. Breast cancer cells colocalized with the cells of osteoblast lineage via E-cadherin (E-cad, cancer cells) and N-cadherin (N-cad, osteoblastic niche) (Wang et al., 2015). When either the expression of E-cad in breast cancer or of N-cad in osteoblastic niche was downregulated, the early dissemination and colonization of breast cancer cells was prevented (Wang et al., 2015). The functions of the osteoblastic niche affect not only the dissemination process but also the circulation of tumor cells. When animals are treated with a single dose of zoledronic acid (ZA) prior to inoculation with breast cancer cell line MDA-MB-231-NW1, the number of osteoblasts in mice rapidly decreased, along with the expected reduction of osteoclast numbers (Haider et al., 2014). ZA treatment also significantly decreased the number of circulating tumor cells, although there are no differences in the number of disseminated tumor cells in the bone (Haider et al., 2014).

## 3.3 Molecular Mechanisms of Survival, Tumor Colonization, Tumor Dormancy, and Metastatic Outgrowth

Once in the bone marrow, bone metastatic tumor cells also hijack specific functions of the bone marrow microenvironment to survive and develop into full-blown bone metastases. As previously stated, the early dissemination and colonization of bone metastasis depend on the functions of the osteoblastic niche (Shiozawa et al., 2011a; Wang et al., 2015). This niche was shown to support the survival of disseminated breast cancer cells by activating mammalian target of rapamycin (mTOR)/S6K pathway through direct binding (Wang et al., 2015). The inhibition of this pathway with mTOR inhibitors torin 1 and rapamycin prevented the survival of disseminated tumor cells in the marrow, but not in lung (Wang et al., 2015). The osteoblastic niche also plays a critical role in establishing metastatic tumor cell dormancy. The niche inhibited the proliferation of prostate cancer and simultaneously protected prostate cancer from chemotherapy-induced apoptosis through growth arrest-specific 6 (GAS6) expressed by the osteoblastic niche interacting with its receptor Axl on prostate cancer cells (Shiozawa et al., 2010; Taichman et al., 2013). This dormancy induction by the niche seems to be tissue dependent; the expression of GAS6 in skeletal tissues controls the growth of bone metastatic tumor cells. High levels of GAS6 are observed in the bones of the forelimb of mice, whereas the bones of the hindlimb express low levels of GAS6 (Jung et al., 2012). When osteoblasts have relatively lower GAS6 expression (in the hindlimb), bone metastatic prostate cancer cells are likely to be more proliferative (Jung et al., 2012). Bone marrow stromal cells, a precursor of osteoblasts, are also involved in the dormancy process of prostate cancer in the marrow. BMP7 produced by bone marrow stromal cells upregulated the cell cycle inhibitor p21, the metastasis suppressor gene N-myc downstream-regulated gene 1, and p38 MAPK phosphorylation, through BMP receptor 2 (BMPR2), but downregulated the ERK MAPK pathway in prostate cancer cells, leading to reversible growth inhibition of prostate cancer (Kobayashi et al., 2011). Once the tumor cells become dormant, they survive for years, even a decade or two, and can eventually outgrow. A recent study revealed the potential molecular mechanisms whereby occult bone metastatic breast cancer cells survive and regrow in late stage cancer. Bone metastatic breast cancer cells highly expressed a nonreceptor tyrosine kinase, Src, and the reduction of Src inhibited the bone metastatic growth in animal studies (Zhang et al., 2009). The activation of Src is necessary for the phosphorylation of the Akt survival signaling pathway mediated by the CXCR4/SDF-1 (Zhang et al., 2009). Furthermore, the expression of Src was associated with the inhibition of apoptosis signals by TNF-related apoptosis-inducing ligand (TRAIL) (Zhang et al., 2009).

# 4. TREATMENT OPTIONS FOR BONE METASTASIS

Generally, current treatment options for bone metastasis include antiresorptive agents (bisphosphonates, denosumab), hormone therapy, chemotherapy, local or systematic radiation therapy, surgery, and analgesics to control pain (Selvaggi and Scagliotti, 2005; Pinski and Dorff, 2005; Lipton, 2005). Unfortunately, bone metastasis is currently a progressive and incurable disease. Thus, treatments for bone metastasis have focused on preventing skeletal related events (SREs), reducing bone pain, and improving the quality of life (QOL), as much as on the continuous efforts to prolong survival.

## 4.1 Antiresorptive Agents

A key class of bone-targeting osteoclast inhibitors, bisphosphonates, are used to reduce SREs, alleviate bone pain, and treat hypercalcemia of malignancy, although they do not improve survival in women with metastatic breast cancer (Body, 2006; Stewart, 2005). Bisphosphonates can be classified into groups that contain or lack nitrogen. Nonnitrogen containing bisphosphonates (etidronate, clodronate, and tiludronate) are converted into methylene-containing analogs of adenosine trisphosphate, which causes apoptosis in macrophages and osteoclasts (Wong and Pavlakis, 2011). Nitrogen-containing bisphosphonates (pamidronate, alendronate, ibandronate, and ZA) inhibit a rate-limiting enzyme of the mevalonate pathway to prevent protein prenylation of small guanosine triphosphatases such as Ras, Rho, and Rac, which regulate osteoclast cell survival (Roelofs et al., 2006). Nitrogen-containing bisphosphonates also have potential antitumor properties at higher concentration in vitro (Roelofs et al., 2006). ZA strongly delayed the SREs in prostate cancer, lung cancer, renal cancer, bladder cancer, and colon cancer over placebo treatments (Saad et al., 2004; Rosen et al., 2003), whereas nonnitrogen containing bisphosphonates did not show overall clinical benefit in metastatic prostate cancer patients (Polascik and Mouraviev, 2008). Additionally, ZA has significantly lowered skeletal morbidity in breast cancer or multiple myeloma patients when combined with radiotherapy or with hormone therapy (Rosen et al., 2001). Nitrogen-containing bisphosphonates also have potential antitumor properties at higher concentration in vitro (Roelofs et al., 2006). Bisphosphonates are well tolerated, but sometimes induce osteonecrosis of the jaw (ONJ), which is one of the serious adverse events of antiresorptive agents.

Denosumab, a fully human monoclonal antibody to RANKL, has been tested for its effects on bone metastatic lesions in clinical trials. In a phase III trial of breast cancer patients with bone metastasis, denosumab treatments (120 mg every 4 weeks) were superior in delaying the onset of SREs and improving QOL compared to ZA treatments (4 mg every 4 weeks) (Stopeck et al., 2010). There are no significant differences in overall survival, disease progression, and adverse events, including ONJ (Stopeck et al., 2010). Likewise, denosumab (120 mg every 4 weeks) delayed the onset of SREs in a phase III trial of men with mCRPC compared to ZA (4 mg every 4 weeks), whereas no significant impacts were observed on overall survival, disease progression, and adverse events (Fizazi et al., 2011). Interestingly, this study also revealed the cost effectiveness of denosumab over ZA (Snedecor et al., 2013).

## 4.2 Pain Management

One of the principal symptoms of skeletal metastases is pain (Coleman, 2001). Up to 90% of patients with bone metastasis suffer from cancer-induced bone pain, which can be extremely debilitating (Smith and Barkin, 2014). Once cancer cells invade bone marrow, they interfere with the normal physiology of the microenvironment, resulting in bone pain that significantly impairs patients' QOL. Bone distraction initiated by metastatic tumor cells and the invasion of the metastatic tumor into the surrounding soft tissue appears to be a cause of bone pain (Reale et al., 2001). Cancer-related pain—both its causes and its management—poses a tremendous challenge to patients and their caregivers.

The standard treatment for cancer bone pain is to follow the World Health Organization's analgesic stepladder guidelines approach (McGrath, 1996; Burton and Cleeland, 2001). The first step is nonsteroidal anti-inflammatory analgesics, the second step is weak opioids, progressing on to strong opioids (McGrath, 1996; Burton and Cleeland, 2001). However, clinical management for cancer bone pain with these drugs remains challenging, due to side effects of these analgesics and/or rapidly increasing pain even with treatment (Kane et al., 2015). Alternative pain management for bone metastases includes antiresorptive therapy, surgery, radiotherapy (external beam radiotherapy or radiopharmaceuticals), hormone therapy, and chemotherapy. ZA significantly reduces bone pain in both breast cancer and prostate cancer patients with bone metastasis (Furlow, 2006; Fulfaro et al., 2005). Surgery can be performed to prevent or delay pathological fractures and/or spinal cord compression (Lipton, 2005). External beam radiotherapy has been shown to relieve pain within 4–6 weeks in 80% of bone metastatic patients, and serial treatment for recurrent pain is possible (Hoskin, 2003). Food and Drug Administration-approved radiopharmaceuticals, Striontium-89 (Paes and Serafini, 2010), and Samarium-153 (Sartor et al., 2007) have also been used for cancer bone pain. Although hormone therapy or chemotherapy itself is not sufficient to reduce SREs, hormone therapy or chemotherapy can be used as an analgesic for cancer bone pain (Lipton, 2005).

## 5. DISCUSSION/CONCLUSION

In this article, we discussed the molecular mechanisms whereby the microenvironment controls the process of bone metastasis (Fig. 14.2). The subsequent metastatic type, whether osteolytic or osteoblastic, depends on how metastatic cancer cells interact with the bone marrow microenvironment, especially the cells involved in bone remodeling (e.g., osteoclasts, osteoblasts) (Kingsley et al., 2007). We also outlined several therapeutic strategies that may help manage this painful disease. Although these treatments are still palliative, there are several novel agents that may be used to modulate bone metastatic disease. Currently, a monoclonal antibody against TGF-β and a neutralizing antibody against DKK1 (BHQ880) are in clinical trial (Li et al., 2014). These antibodies may be useful for targeting bone metastasis, since both TGF-β and DKK1 are involved in the progression of osteolytic bone metastasis. Other potential agents that have shown optimistic preclinical effects include (phosphoinositide 3-kinase) PI3K/Akt/mTOR pathway inhibitors, CXCR4 antagonist AMD3100, and inhibitors targeting Src. As stated above, these factors are associated with survival signals mediated by the bone marrow microenvironment. Therefore, disruption between bone metastatic tumor cells and the bone marrow microenvironment through these inhibitors or antagonists may have therapeutic potential for bone metastatic disease. One hypothesis worth pursuing is mobilization of metastatic tumor cells using HSC mobilizing agents, which may enhance their susceptibility to cytotoxic agents (Shiozawa et al., 2011b).

How the host microenvironment influences bone metastatic disease is a key unanswered question in cancer biology, however, studies such as the ones mentioned in this review are providing essential clues to the puzzle. The crosstalk between tumor cells and host cells is involved in each step of the metastatic process, offering many potential targets for therapy (Keller et al., 2001). Since metastatic progression of disseminated tumor cells in the bone may be environmentally directed by influences from within the bone marrow microenvironment, further studies to clarify the precise molecular roles of the bone marrow microenvironment in bone metastasis may provide a gateway to new treatment strategies. Critical questions remain unanswered regarding complete elucidation of the mechanisms of bone metastasis.

1. Are there specific genes that regulate bone metastasis?
2. Does the number of circulating tumor cells positively correlate with the progression of bone metastasis?
3. What are the differences in the roles of lymphatic and hematogenous dissemination in bone metastasis?
4. Why do some cancer cells metastasize to the bone? or What is unique about the bone marrow microenvironment that attracts cells from some primary tumors and not others?
5. Can targeting the bone marrow microenvironment serve as a potential therapy for bone metastatic disease?
6. Is it realistic to target a single cell type of the metastatic niche, especially in the marrow?
7. Do other cell types of the marrow (e.g., cells of hematopoietic origin, mesenchymal stem cells) also participate in the development of bone metastasis?
8. Does the bone marrow microenvironment induce the resistance of cancer cells to cytotoxic agents?
9. Does the marrow microenvironment influence cancer bone pain? If so, how?
10. Does cancer bone pain affect metastatic outgrowth?

Although the life expectancy of patients with bone metastases has increased in recent years due to early detection and diagnosis, this extended prognosis is not without severe QOL impairment, including increased pain,

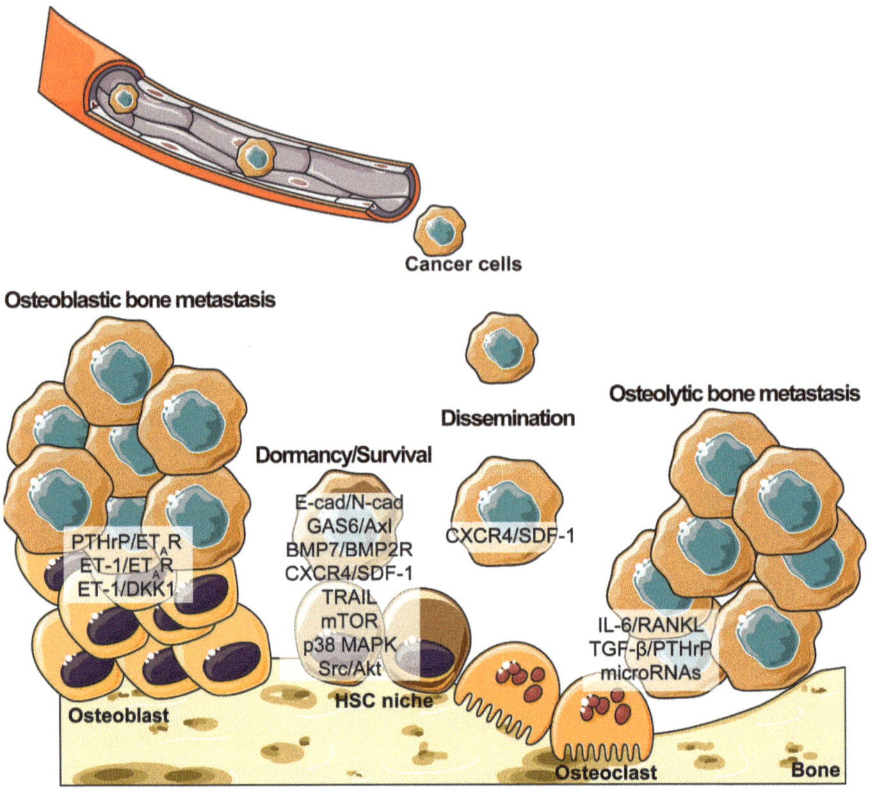


FIGURE 14.2 **Crosstalk between cancer cells and the bone marrow microenvironment in the process of bone metastasis.** Growing evidence supports the idea that tumor cells require aid from the surrounding microenvironment for their metastatic process. Bone marrow contains cells that control bone remodeling: osteoblasts and osteoclasts. Once in the bone marrow, disseminated tumor cells get involved in this diverse cellular community, resulting in osteoblastic (e.g., PTHrP/ET$_A$R, ET-1/ET$_A$R, ET-1/DKK1) or osteolytic (e.g., IL-6/RANKL, TGF-β/PTHrP, microRNAs) metastatic lesions. During their early dissemination, tumor cells may target a preexisting microenvironment for hematopoietic stem cells (HSCs), or the osteoblastic HSC niche, using similar mechanisms as the homing process of HSCs (e.g., CXCR4/SDF-1). Thereafter, disseminated tumor cells engage in and adapt to this metastatic microenvironment to survive for a long time as an occult tumor (e.g., E-cad/N-cad, GAS6/Axl, BMP7/BMPR2) and subsequently develop a metastatic progression (e.g., CXCR4/SDF-1, TRAIL). The activation of cellular signaling pathways (e.g., mTOR, p38 MAPK, Src/Akt) mediated by the bone marrow microenvironment influences the survival and metastatic outgrowth of bone metastatic tumor cells. *BMP7*, bone morphogenetic protein 7; *BMPR2*, BMP receptor 2; *CXCR4*, C-X-C chemokine receptor type 4; *DKK1*, dickkopf Wnt-signaling pathway inhibitor 1; *E-cad*, E-cadherin; *ET-1*, endothelin-1; *ET$_A$R*, endothelin A receptor; *GAS6*, growth arrest-specific 6; *IL-6*, interleukin-6; *MAPK*, mitogen-activated protein kinase; *mTOR*, mammalian target of rapamycin; *N-cad*, N-cadherin; *PTHrP*, parathyroid hormone-related protein; *RANKL*, receptor activator of nuclear factor κB ligand; *SDF-1*, stromal cell-derived factor 1; *TGF-β*, transforming growth factor-β; and *TRAIL*, TNF-related apoptosis-inducing ligand.

hypercalcemia, impaired mobility, bone marrow infiltration, fracture, spinal cord compression, and other nerve-compression syndromes. For patients with symptoms, the first and most distinguishing feature is bone pain, which is particularly challenging to treat. Therefore, it is also important to consider broadening the knowledge of consequences so that the QOL of cancer patients with bone metastases is improved. A deeper understanding of the cancer/microenvironment interaction will ultimately lead to the development of novel molecular targeted therapies for controlling currently incurable bone metastatic disease. Further studies are clearly warranted.

## Acknowledgments

We thank Ms. Janice E. Berry for critical reading of this manuscript. This work is directly supported by the National Cancer Institute (CA163124, Y. Shiozawa), the Department of Defense (W81XWH-14-1-0403, Y. Shiozawa), and the Prostate Cancer Foundation (Y. Shiozawa). Y Shiozawa research is supported by the National Center for Advancing Translational Sciences (NCATS), National Institutes of Health, through Grant Award Number UL1TR001420. Graphics adapted from Servier Medical Art (www.servier.com).

## References

Ahn, S.G., Lee, H.M., Cho, S.H., et al., 2013. Prognostic factors for patients with bone-only metastasis in breast cancer. Yonsei Med. J. 54 (5), 1168–1177.

Aiuti, A., Webb, I.J., Bleul, C., Springer, T., Gutierrez-Ramos, J.C., 1997. The chemokine SDF-1 is a chemoattractant for human CD34+ hematopoietic progenitor cells and provides a new mechanism to explain the mobilization of CD34+ progenitors to peripheral blood. J. Exp. Med. 185 (1), 111–120.

Bendre, M.S., Margulies, A.G., Walser, B., et al., 2005. Tumor-derived interleukin-8 stimulates osteolysis independent of the receptor activator of nuclear factor-kappaB ligand pathway. Cancer Res. 65 (23), 11001–11009.

Baron, R., Kneissel, M., 2013. WNT signaling in bone homeostasis and disease: from human mutations to treatments. Nat. Med. 19 (2), 179–192.

Banys, M., Gruber, I., Krawczyk, N., et al., 2012. Hematogenous and lymphatic tumor cell dissemination may be detected in patients diagnosed with ductal carcinoma in situ of the breast. Breast Cancer Res. Treat. 131 (3), 801–808.

Body, J.J., 2006. Breast cancer: bisphosphonate therapy for metastatic bone disease. Clin. Cancer Res. 12 (20 Pt 2), 6258s–6263s.

Burton, A.W., Cleeland, C.S., 2001. Cancer pain: progress since the WHO guidelines. Pain Pract. 1 (3), 236–242.

Coleman, R.E., 2001. Metastatic bone disease: clinical features, pathophysiology and treatment strategies. Cancer Treat. Rev. 27 (3), 165–176.

Coleman, R.E., Rubens, R.D., 1987. The clinical course of bone metastases from breast cancer. Br. J. Cancer 55 (1), 61–66.

Chang, C.Y., Gill, C.M., Joseph Simeone, F., et al., 2014. Comparison of the diagnostic accuracy of 99 m-Tc-MDP bone scintigraphy and 18 F-FDG PET/CT for the detection of skeletal metastases. Acta Radiol.

Carducci, M.A., Saad, F., Abrahamsson, P.A., et al., 2007. A phase 3 randomized controlled trial of the efficacy and safety of atrasentan in men with metastatic hormone-refractory prostate cancer. Cancer 110 (9), 1959–1966.

Clines, G.A., Mohammad, K.S., Bao, Y., et al., 2007. Dickkopf homolog 1 mediates endothelin-1-stimulated new bone formation. Mol. Endocrinol. 21 (2), 486–498.

Calvi, L.M., Adams, G.B., Weibrecht, K.W., et al., 2003. Osteoblastic cells regulate the haematopoietic stem cell niche. Nature 425 (6960), 841–846.

Domanska, U.M., Timmer-Bosscha, H., Nagengast, W.B., et al., 2012. CXCR4 inhibition with AMD3100 sensitizes prostate cancer to docetaxel chemotherapy. Neoplasia 14 (8), 709–718.

Ell, B., Mercatali, L., Ibrahim, T., et al., 2013. Tumor-induced osteoclast miRNA changes as regulators and biomarkers of osteolytic bone metastasis. Cancer Cell 24 (4), 542–556.

Fizazi, K., Carducci, M., Smith, M., et al., 2011. Denosumab versus zoledronic acid for treatment of bone metastases in men with castration-resistant prostate cancer: a randomised, double-blind study. Lancet 377 (9768), 813–822.

Furlow, B., 2006. Zoledronic acid palliation in bone-metastatic breast cancer. Lancet Oncol. 7 (11), 894.

Fulfaro, F., Leto, G., Badalamenti, G., et al., 2005. The use of zoledronic acid in patients with bone metastases from prostate carcinoma: effect on analgesic response and bone metabolism biomarkers. J. Chemother. 17 (5), 555–559.

Guise, T.A., Mundy, G.R., 1998. Cancer and bone. Endocr. Rev. 19 (1), 18–54.

Gravina, G.L., Mancini, A., Muzi, P., et al., 2015. CXCR4 pharmacological inhibition reduces bone and soft tissue metastatic burden by affecting tumor growth and tumorigenic potential in prostate cancer preclinical models. Prostate 75 (12), 1227–1246.

Hage, W.D., Aboulafia, A.J., Aboulafia, D.M., 2000. Incidence, location, and diagnostic evaluation of metastatic bone disease. Orthop. Clin. North Am. 31 (4), 515–528 vii.

Hayes, J., Peruzzi, P.P., Lawler, S., 2014. MicroRNAs in cancer: biomarkers, functions and therapy. Trends Mol. Med. 20 (8), 460–469.

Haider, M.T., Holen, I., Dear, T.N., Hunter, K., Brown, H.K., 2014. Modifying the osteoblastic niche with zoledronic acid in vivo-potential implications for breast cancer bone metastasis. Bone 66, 240–250.

Hoskin, P.J., 2003. Bisphosphonates and radiation therapy for palliation of metastatic bone disease. Cancer Treat. Rev. 29 (4), 321–327.

Isales, C., Carcangiu, M.L., Stewart, A.F., 1987. Hypercalcemia in breast cancer. Reassessment of the mechanism. Am. J. Med. 82 (6), 1143–1147.

James, J.J., Evans, A.J., Pinder, S.E., et al., 2003. Bone metastases from breast carcinoma: histopathological–radiological correlations and prognostic features. Br. J. Cancer 89 (4), 660–665.

Jung, Y., Wang, J., Song, J., et al., 2007. Annexin II expressed by osteoblasts and endothelial cells regulates stem cell adhesion, homing, and engraftment following transplantation. Blood 110 (1), 82–90.

Jung, Y., Shiozawa, Y., Wang, J., et al., 2012. Prevalence of prostate cancer metastases after intravenous inoculation provides clues into the molecular basis of dormancy in the bone marrow microenvironment. Neoplasia 14 (5), 429–439.

Koizumi, M., Yoshimoto, M., Kasumi, F., Ogata, E., 2003. Comparison between solitary and multiple skeletal metastatic lesions of breast cancer patients. Ann. Oncol. 14 (8), 1234–1240.

Kakonen, S.M., Selander, K.S., Chirgwin, J.M., et al., 2002. Transforming growth factor-beta stimulates parathyroid hormone-related protein and osteolytic metastases via Smad and mitogen-activated protein kinase signaling pathways. J. Biol. Chem. 277 (27), 24571–24578.

Kang, Y., He, W., Tulley, S., et al., 2005. Breast cancer bone metastasis mediated by the Smad tumor suppressor pathway. Proc. Natl. Acad. Sci. U. S. A. 102 (39), 13909–13914.

Krzeszinski, J.Y., Wei, W., Huynh, H., et al., 2014. miR-34a blocks osteoporosis and bone metastasis by inhibiting osteoclastogenesis and Tgif2. Nature 512 (7515), 431–435.

Kang, Y., Siegel, P.M., Shu, W., et al., 2003. A multigenic program mediating breast cancer metastasis to bone. Cancer Cell 3 (6), 537–549.

Kobayashi, A., Okuda, H., Xing, F., et al., 2011. Bone morphogenetic protein 7 in dormancy and metastasis of prostate cancer stem-like cells in bone. J. Exp. Med. 208 (13), 2641–2655.

Kane, C.M., Hoskin, P., Bennett, M.I., 2015. Cancer induced bone pain. BMJ 350, h315.

Kingsley, L.A., Fournier, P.G., Chirgwin, J.M., Guise, T.A., 2007. Molecular biology of bone metastasis. Mol. Cancer Ther. 6 (10), 2609–2617.

Keller, E.T., Zhang, J., Cooper, C.R., et al., 2001. Prostate carcinoma skeletal metastases: cross-talk between tumor and bone. Cancer Metastasis Rev. 20 (3–4), 333–349.

Lee, S.J., Park, S., Ahn, H.K., et al., 2011. Implications of bone-only metastases in breast cancer: favorable preference with excellent outcomes of hormone receptor positive breast cancer. Cancer Res. Treat. 43 (2), 89–95.

Logothetis, C.J., Lin, S.H., 2005. Osteoblasts in prostate cancer metastasis to bone. Nat. Rev. Cancer 5 (1), 21–28.

Liao, J., Li, X., Koh, A.J., et al., 2008. Tumor expressed PTHrP facilitates prostate cancer-induced osteoblastic lesions. Int. J. Cancer 123 (10), 2267–2278.

Liotta, L.A., Kohn, E.C., 2001. The microenvironment of the tumour-host interface. Nature 411 (6835), 375–379.

Lipton, A., 2005. Management of bone metastases in breast cancer. Curr. Treat. Options Oncol. 6 (2), 161–171.

Li, B.T., Wong, M.H., Pavlakis, N., 2014. Treatment and prevention of bone metastases from breast cancer: a comprehensive review of evidence for clinical practice. J. Clin. Med. 3 (1), 1–24.

Mundy, G.R., 2002. Metastasis to bone: causes, consequences and therapeutic opportunities. Nat. Rev. Cancer 2 (8), 584–593.

Morgan, T.M., Lange, P.H., Porter, M.P., et al., 2009. Disseminated tumor cells in prostate cancer patients after radical prostatectomy and without evidence of disease predicts biochemical recurrence. Clin. Cancer Res. 15 (2), 677–683.

Muller, A., Homey, B., Soto, H., et al., 2001. Involvement of chemokine receptors in breast cancer metastasis. Nature 410 (6824), 50–56.

McGrath, P.A., 1996. Development of the World Health Organization Guidelines on Cancer Pain Relief and Palliative Care in Children. J. Pain Symptom Manage. 12 (2), 87–92.

Nijweide, P.J., Burger, E.H., Feyen, J.H., 1986. Cells of bone: proliferation, differentiation, and hormonal regulation. Physiol. Rev. 66 (4), 855–886.

Nelson, J.B., Hedican, S.P., George, D.J., et al., 1995. Identification of endothelin-1 in the pathophysiology of metastatic adenocarcinoma of the prostate. Nat. Med. 1 (9), 944–949.

Nash, K.T., Phadke, P.A., Navenot, J.M., et al., 2007. Requirement of KISS1 secretion for multiple organ metastasis suppression and maintenance of tumor dormancy. J. Natl. Cancer Inst. 99 (4), 309–321.

Plunkett, T.A., Smith, P., Rubens, R.D., 2000. Risk of complications from bone metastases in breast cancer. implications for management. Eur. J. Cancer 36 (4), 476–482.

Pantel, K., Alix-Panabieres, C., Riethdorf, S., 2009. Cancer micrometastases. Nat. Rev. Clin. Oncol. 6 (6), 339–351.

Paget, S., 1989. The distribution of secondary growths in cancer of the breast. 1889. Cancer Metastasis Rev. 8 (2), 98–101.

Peled, A., Petit, I., Kollet, O., et al., 1999. Dependence of human stem cell engraftment and repopulation of NOD/SCID mice on CXCR4. Science 283 (5403), 845–848.

Pinski, J., Dorff, T.B., 2005. Prostate cancer metastases to bone: pathophysiology, pain management, and the promise of targeted therapy. Eur. J. Cancer 41 (6), 932–940.

Polascik, T.J., Mouraviev, V., 2008. Zoledronic acid in the management of metastatic bone disease. Ther. Clin. Risk Manag. 4 (1), 261–268.

Paes, F.M., Serafini, A.N., 2010. Systemic metabolic radiopharmaceutical therapy in the treatment of metastatic bone pain. Semin. Nucl. Med. 40 (2), 89–104.

Quinn, D.I., Tangen, C.M., Hussain, M., et al., 2013. Docetaxel and atrasentan versus docetaxel and placebo for men with advanced castration-resistant prostate cancer (SWOG S0421): a randomised phase 3 trial. Lancet Oncol. 14 (9), 893–900.

Roodman, G.D., 2004. Mechanisms of bone metastasis. N. Engl. J. Med. 350 (16), 1655–1664.

Raubenheimer, E.J., Noffke, C.E., 2006. Pathogenesis of bone metastasis: a review. J. Oral Pathol. Med. 35 (3), 129–135.

Roelofs, A.J., Thompson, K., Gordon, S., Rogers, M.J., 2006. Molecular mechanisms of action of bisphosphonates: current status. Clin. Cancer Res. 12 (20 Pt 2), 6222s–6230s.

Rosen, L.S., Gordon, D., Tchekmedyian, S., et al., 2003. Zoledronic acid versus placebo in the treatment of skeletal metastases in patients with lung cancer and other solid tumors: a phase III, double-blind, randomized trial—the Zoledronic Acid Lung Cancer and Other Solid Tumors Study Group. J. Clin. Oncol. 21 (16), 3150–3157.

Rosen, L.S., Gordon, D., Kaminski, M., et al., 2001. Zoledronic acid versus pamidronate in the treatment of skeletal metastases in patients with breast cancer or osteolytic lesions of multiple myeloma: a phase III, double-blind, comparative trial. Cancer J. 7 (5), 377–387.

Reale, C., Turkiewicz, A.M., Reale, C.A., 2001. Antalgic treatment of pain associated with bone metastases. Crit. Rev. Oncol. Hematol. 37 (1), 1–11.

Selvaggi, G., Scagliotti, G.V., 2005. Management of bone metastases in cancer: a review. Crit. Rev. Oncol. Hematol. 56 (3), 365–378.

Scheid, V., Buzdar, A.U., Smith, T.L., Hortobagyi, G.N., 1986. Clinical course of breast cancer patients with osseous metastasis treated with combination chemotherapy. Cancer 58 (12), 2589–2593.

Simonet, W.S., Lacey, D.L., Dunstan, C.R., et al., 1997. Osteoprotegerin: a novel secreted protein involved in the regulation of bone density. Cell 89 (2), 309–319.

Saad, F., Gleason, D.M., Murray, R., et al., 2004. Long-term efficacy of zoledronic acid for the prevention of skeletal complications in patients with metastatic hormone-refractory prostate cancer. J. Natl. Cancer Inst. 96 (11), 879–882.

Schluter, K.D., Katzer, C., Piper, H.M., 2001. A N-terminal PTHrP peptide fragment void of a PTH/PTHrP-receptor binding domain activates cardiac ET(A) receptors. Br. J. Pharmacol. 132 (2), 427–432.

Sun, Y.X., Wang, J., Shelburne, C.E., et al., 2003. Expression of CXCR4 and CXCL12 (SDF-1) in human prostate cancers (PCa) in vivo. J. Cell Biochem. 89 (3), 462–473.

Sun, Y.X., Schneider, A., Jung, Y., et al., 2005. Skeletal localization and neutralization of the SDF-1(CXCL12)/CXCR4 axis blocks prostate cancer metastasis and growth in osseous sites in vivo. J. Bone Miner. Res. 20 (2), 318–329.

Shiozawa, Y., Havens, A.M., Jung, Y., et al., 2008. Annexin II/annexin II receptor axis regulates adhesion, migration, homing, and growth of prostate cancer. J. Cell Biochem. 105 (2), 370–380.

Shiozawa, Y., Pedersen, E.A., Havens, A.M., et al., 2011a. Human prostate cancer metastases target the hematopoietic stem cell niche to establish footholds in mouse bone marrow. J. Clin. Invest. 121 (4), 1298–1312.

Shiozawa, Y., Pedersen, E.A., Patel, L.R., et al., 2010. GAS6/AXL axis regulates prostate cancer invasion, proliferation, and survival in the bone marrow niche. Neoplasia 12 (2), 116–127.

Stewart, A.F., 2005. Clinical practice. Hypercalcemia associated with cancer. N. Engl. J. Med. 352 (4), 373–379.

Stopeck, A.T., Lipton, A., Body, J.J., et al., 2010. Denosumab compared with zoledronic acid for the treatment of bone metastases in patients with advanced breast cancer: a randomized, double-blind study. J. Clin. Oncol. 28 (35), 5132–5139.

Snedecor, S.J., Carter, J.A., Kaura, S., Botteman, M.F., 2013. Denosumab versus zoledronic acid for treatment of bone metastases in men with castration-resistant prostate cancer: a cost-effectiveness analysis. J. Med. Econ. 16 (1), 19–29.

Smith, H.S., Barkin, R.L., 2014. Painful boney metastases. Am. J. Ther. 21 (2), 106–130.

Sartor, O., Reid, R.H., Bushnell, D.L., Quick, D.P., Ell, P.J., 2007. Safety and efficacy of repeat administration of samarium Sm-153 lexidronam to patients with metastatic bone pain. Cancer 109 (3), 637–643.

Shiozawa, Y., Pienta, K.J., Taichman, R.S., 2011b. Hematopoietic stem cell niche is a potential therapeutic target for bone metastatic tumors. Clin. Cancer Res. 17 (17), 5553–5558.

Takahashi, N., Udagawa, N., Suda, T., 1999. A new member of tumor necrosis factor ligand family, ODF/OPGL/TRANCE/RANKL, regulates osteoclast differentiation and function. Biochem. Biophys. Res. Commun. 256 (3), 449–455.

Tian, E., Zhan, F., Walker, R., et al., 2003. The role of the Wnt-signaling antagonist DKK1 in the development of osteolytic lesions in multiple myeloma. N. Engl. J. Med. 349 (26), 2483–2494.

Taichman, R.S., Cooper, C., Keller, E.T., Pienta, K.J., Taichman, N.S., McCauley, L.K., 2002. Use of the stromal cell-derived factor-1/CXCR4 pathway in prostate cancer metastasis to bone. Cancer Res. 62 (6), 1832–1837.

Taichman, R.S., Patel, L.R., Bedenis, R., et al., 2013. GAS6 receptor status is associated with dormancy and bone metastatic tumor formation. PLoS One 8 (4), e61873.

Weidner, N., Semple, J.P., Welch, W.R., Folkman, J., 1991. Tumor angiogenesis and metastasis—correlation in invasive breast carcinoma. N. Engl. J. Med. 324 (1), 1–8.

Wong, S.Y., Hynes, R.O., 2006. Lymphatic or hematogenous dissemination: how does a metastatic tumor cell decide? Cell Cycle 5 (8), 812–817.

Wang, H., Yu, C., Gao, X., et al., 2015. The osteogenic niche promotes early-stage bone colonization of disseminated breast cancer cells. Cancer Cell 27 (2), 193–210.

Wong, M., Pavlakis, N., 2011. Optimal management of bone metastases in breast cancer patients. Breast Cancer (Dove Med. Press) 3, 35–60.

Yin, J.J., Selander, K., Chirgwin, J.M., et al., 1999. TGF-beta signaling blockade inhibits PTHrP secretion by breast cancer cells and bone metastases development. J. Clin. Invest. 103 (2), 197–206.

Yasuda, H., Shima, N., Nakagawa, N., et al., 1998. Osteoclast differentiation factor is a ligand for osteoprotegerin/osteoclastogenesis-inhibitory factor and is identical to TRANCE/RANKL. Proc. Natl. Acad. Sci. U. S. A. 95 (7), 3597–3602.

Yi, B., Williams, P.J., Niewolna, M., Wang, Y., Yoneda, T., 2002. Tumor-derived platelet-derived growth factor-BB plays a critical role in osteosclerotic bone metastasis in an animal model of human breast cancer. Cancer Res. 62 (3), 917–923.

Yin, J.J., Mohammad, K.S., Kakonen, S.M., et al., 2003. A causal role for endothelin-1 in the pathogenesis of osteoblastic bone metastases. Proc. Natl. Acad. Sci. U. S. A. 100 (19), 10954–10959.

Zheng, Y., Chow, S.O., Boernert, K., et al., 2014. Direct crosstalk between cancer and osteoblast lineage cells fuels metastatic growth in bone via auto-amplification of IL-6 and RANKL signaling pathways. J. Bone Miner. Res. 29 (9), 1938–1949.

Zetter, B.R., 1990. The cellular basis of site-specific tumor metastasis. N. Engl. J. Med. 322 (9), 605–612.

Zhang, X.H., Jin, X., Malladi, S., et al., 2013. Selection of bone metastasis seeds by mesenchymal signals in the primary tumor stroma. Cell 154 (5), 1060–1073.

Zhang, X.H., Wang, Q., Gerald, W., et al., 2009. Latent bone metastasis in breast cancer tied to Src-dependent survival signals. Cancer Cell 16 (1), 67–78.

CHAPTER

# 15

# An Overview on Hepatic Metastasis

F. Uggeri[1], E. Pinotti[1], M. Garancini[2], F. Romano[1]

[1]University of Milano-Bicocca, Monza, Italy; [2]San Gerardo Hospital, Monza, Italy

## 1. INTRODUCTION

Metastases commonly affect the liver parenchyma, about 50–60% of liver cancer disease have secondary origin (Kasper et al., 2005). Due to its role as a filter in the portal circulation, the liver is often the first organ involved in metastases from the gastroenteric tract. In the course of the disease, more than 50% of patients with colorectal cancer (CRC) will be diagnosed with a synchronous or metachronous hepatic metastases (Alberts and Wagman, 2008). Also regarding neuroendocrine tumors (NETs), the liver is a preferential site of metastases and liver metastases occur in 40–93% of cases (Alagusundaramoorthy and Gedaly, 2014). Several studies have focused on the indications for surgery and outcomes of these two groups of patients (Ismaili, 2011; Jagannath et al., 2012; Lee et al., 2012). On the other hand, many other primary sites develop metastases in hepatic parenchyma, including stomach, pancreas, biliary, and genitourinary tract, soft tissue, skin, breast, melanoma, and gynecologic system (Groeschl et al., 2012; Page et al., 2014). Here are some reasons why liver is an important site of metastases:

- Liver is the largest gland of the body and its cellular composition is unique. Its morphological structure and the heterogeneity of hepatic cells make liver an excellent substrate for cell growth. The slow blood flow in the sinusoidal circulation, the capillary meandering, the fenestrations of the hepatic sinusoidal endothelium can make the engraftment of tumor cells easier.
- Hepatic parenchyma also plays a major role in the elimination of toxic substances from visceral and systemic circulation; in fact, the blood that drains from a visceral circulation accounts for 70% of the liver blood supply (Greenway, 1971).
- The anatomical location of the liver can promote a direct disease propagation from intraperitoneal organs.
- The ability of synthesis and liver regeneration with production of growth and biohumoral factors may favor the ability of neoplastic cell adhesion.

Until 20 years ago, the presence of metastatic hepatic disease, an unequivocal sign of systemic disease, in many cases contraindicated medical or surgical treatment. Encouraging data have been obtained in the last decades in the management of liver metastases from colorectal and NETs. The patients who undergo complete resections (R0) have a 5-year survival rate of approximately 40% and a 10-year overall survival rate of 25% (Malik et al., 2007; Osada et al.,

*Introduction to Cancer Metastasis*
http://dx.doi.org/10.1016/B978-0-12-804003-4.00015-3

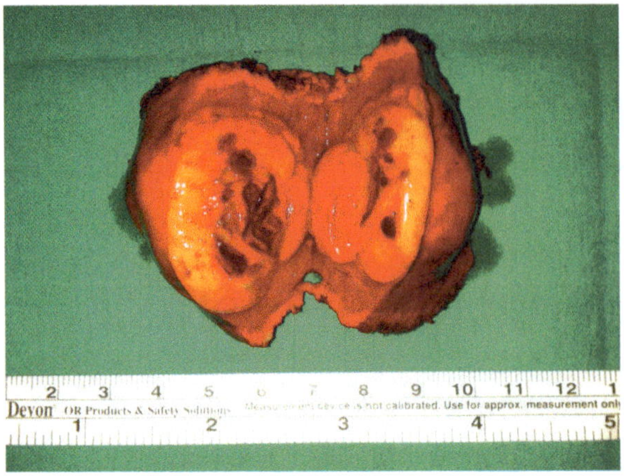


FIGURE 15.1 Wedge resection for colorectal liver metastases.

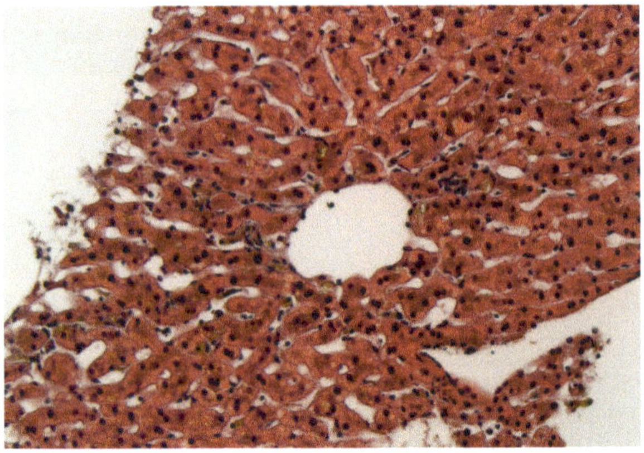

FIGURE 15.2 Centrilobular vein in healthy liver.

2012; Vladov et al., 2012; Quan et al., 2012) (Fig. 15.1) due to improved surgical device, imaging, and advances in oncologic therapy, allowing us to expand the curative treatment (Helling, 2005). Recently, several studies have reported an aggressive surgical treatment of liver metastases from other sites with a median 5- and 10-year overall survival of 32.3% and 24%, respectively (Uggeri et al., 2015). These results allowed the treatment of more patients, thereby helping to change the guidelines on the management of such diseases.

## 2. HEPATIC PHYSIOLOGY AND CANCER BIOLOGY

The liver is covered with a thin foil of collagen and elastic fibers containing sensitive nerve endings (Glisson's capsule): it extends from the parenchyma around vascular and bile structures that constitute the portal spaces. The parenchyma is composed by a mass of cancellous bone cells placed in close connection with a complex network formed by capillary sinusoids. The sinusoids are derived from the branch of portal vein and differ from classic capillaries for the presence on the their endothelium of specialized phagocytes (Kupffer cells) and for a greater permeability to molecules. Under the microscope, the liver is made up of functional units or "lobules"; these structures are pseudohexagonal, whose skeleton is represented by sheets of hepatocytes separated by sinusoids. At the center of the lobule there is a tributary of the venous hepatic system, which leads filtered blood to the right heart through the inferior vena cava. The smallest functional unit of the liver is the "acinus." A liver acinus is an oval-shaped section of the parenchyma, with the major axis connecting two centrilobular veins (Fig. 15.2) and the smaller approximately one side of the hexagon formed by the liver lobule. There are four major cell types that are found in the liver:

1. Hepatocytes are the most numerous cells and account for about 80% of the liver volume. They are often multinucleated and possess large amounts of endoplasmic reticulum, ribosomes, mitochondria due to the high metabolic needs and varied tasks they have to perform.
2. Stellate cells are of mesenchymal origin; they are star or irregular shaped. Their job is to secrete the principal substances forming the matrix, including type III collagen and reticulin. They are essential in the liver

regeneration as they secrete growth factors responsible for the regenerative capacity of the liver.

3. Sinusoidal endothelial cells constitute the fenestrated endothelium of venous sinusoids of the liver. The sinusoid fenestrations are very large (average diameter 100 μm) so blood can more easily spill in the Disse spaces and come into contact with the hepatocytes' microvilli.
4. Kupffer cells are macrophages of the liver; they are derived from monocytes and are placed in the lumen of venous sinusoids. Their function is to remove by phagocytosis any debris present in the blood before it comes into contact with the hepatocytes, but they can also stimulate the immune system by the secretion of numerous factors and cytokines.

These different cell populations collaborate in physiological functions of the liver. The liver is responsible for the synthesis and excretion of bile and many functions in the metabolism of glucose, fat, and protein; it is also responsible for the production of coagulation factors and vitamin metabolism. Like these important functions, the liver is responsible for the detoxifying action and together kidney is the main organ delegated for the removal of all metabolic wastes from the blood that accumulate in the passage in its tissues. The reticuloendothelial system of the liver also contains specialized cells of the immune system (Kupffer cells), which act as a filter against the antigens carried by the portal system.

The ability to organize into tissues requires a mechanism of recognition between cells and protein components of the extracellular matrix. Numerous studies have shown the existence, in the cell membrane, of receptor systems that regulate membership and recognition between cells of the same tissue. These interactions between cell and extracellular matrix regulate differentiation and cell proliferation by the production of soluble factors such as growth factors and cytokines (Adam and Watt, 1989). Contact with a substrate or cellular matrix is a fundamental factor for the growth of healthy cells; neoplastic cells possess the ability to disengage from this condition. This capability enables cancer cells to proliferate and move away from the primary tumor site resulting in metastasis (Comoglio and Nicolin, 1997). Metastasis may be displayed as the progression of a tumor cell through successive stages, starting with the detachment of the cell from the tissue and later crossing of the stroma and its injection into the bloodstream. The next steps include imprisonment in the capillary bed of the target organ and the beginning of tumor proliferation that is closely connected to the formation of new blood vessels (angiogenesis). The cell adhesion receptors play a key role in the early stages of neoplastic transformation. At the receptor level, an alteration of the mechanisms of cell adhesion or a failure signal in the cytoplasmic components that control these interactions is one of the essential processes for the establishment of tumor phenotype (Uleminckx et al., 1991). The alteration of the receptors is not an irreversible process but is due to changes of the tissue microenvironment that does not depend on genetic mutations. The disparate functions of the liver, the complex cellular structure of the liver, and the liver's role as a filter in the splanchnic circulation explain why the liver is a privileged branch for the neoplastic engraftment. The ability of cancer cells to grow in a secondary organ depends, as already mentioned, on the complementarity between the properties of cancer cells and the microenvironment of the target organ (Paget, 1989; Fidler, 2003). The functional aspects in the liver that regulate the tumor growth can be explained by several factors. The development of hepatic metastases is associated to physiopathologic processes such as aberrant regeneration, inflammation, and fibrosis that can change the microenvironment of the liver and facilitate the formation of liver metastases. Moreover, metastatic growth may be regulated by gene expression of neoplastic cells. Cancer cells reach the liver via portal circulation together with bacterial toxins; the cells that are in the liver circulation may stop in the sinusoidal epithelium. At this level, areas of infarction can be formed that damage the parenchyma. The tissue reoxygenation starts the inflammatory response with the release of substances such as nitric oxide and intermediate reactive oxygen (Wang et al., 2000; Anasagasti et al., 1996). These molecules are released by the sinusoids that also stimulate platelet aggregation and Kupffer cells contribute, with their phagocytic action, to the destruction of neoplastic cells too. Inflammatory active factors lead to the production of a cascade of cytokines and growth factors like tumor necrosis factor alpha and interferon alpha that activate natural killer cells. Neoplastic cells surviving the liver inflammatory response are able to interact with sinusoidal and Kupffer cells bringing to a production of metabolic factor modulating the immune defense, altering the receptor expression and ultimately leading to the production of angiogenesis stimulating factors (Kolios et al., 2006; Fig. 15.3).

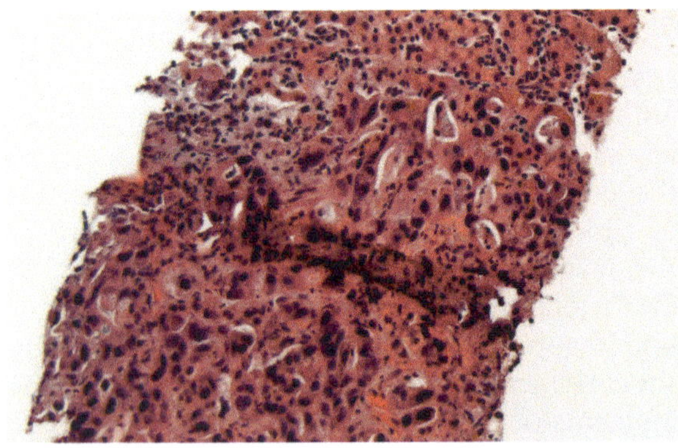

FIGURE 15.3 Interface between healthy liver and metastatic pancreatic adenoarcinoma.

## 3. HEPATIC FUNCTIONAL AND SURGICAL ANATOMY

Based on external appearance, four lobes are traditionally described in the liver: right, left, quadrate, and caudate. The liver is divided into right and left anatomic lobes by the attachment of the falciform ligament on the anterior–superior surface (portal–umbilical fissure). On the visceral surface of the liver, the fissures for the ligamentum venosum and ligamentum teres provide the demarcation. The quadrate lobe is demarcated in the visceral surface of the liver by the gall bladder fossa, porta hepatis, and the portal–umbilical fissure. The caudate lobe is demarcated by the groove of the inferior vena cava and the fissure of the venous ligament. The right portion of the caudate lobe is continuous with the right lobe by the caudate process, which form the superior boundary of the epiploic foramen. The quadrate lobe has been considered as a subdivision of the right anatomic lobe. In the literature and in anatomy treatises, the term lobes for quadrate and caudate are often used, but it is important to remember that these structures are not true lobes.

The functional anatomy of the liver is composed of eight segments, which was first described by Couinaud. Each of these eight segments is supplied by a single portal triad (also called pedicle) composed of a portal vein, hepatic artery, and bile duct. The eight segments are described: one on the caudate lobe (segment I), three on the left (segments II, III, and IV, which is further subdivided in two sections a–b), and four on the right (segments V,VI, VII, and VIII) (Fig. 15.4). Bismuth took into specific consideration the quadrate lobe (segment I) and he divided left lobe into three segments, II (left lateral superior subsegment), III (left lateral inferior subsegment), and IV (left medial subsegment), and the right lobe in four segments, V (right anterior inferior subsegment), VI (right anterior superior subsegment), VII (right posterior inferior subsegment), and VIII

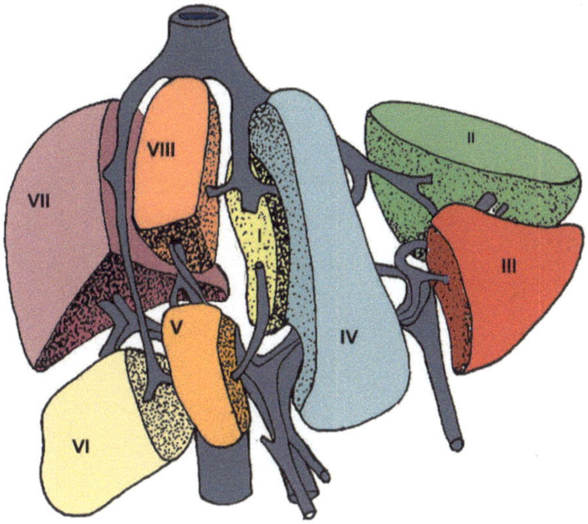


FIGURE 15.4 Liver segments.

(right posterior superior subsegment). These segments are further organized into four sectors that are separated by scissure containing the three main hepatic veins. The four sectors are even further organized into right and left liver (the phrase right and left liver is preferable to right and left lobe because there is no external mark that allows the identification of the right and left liver). The functional anatomy is more often seen as a cross-sectional imaging; Fig. 15.5 shows the anatomy in this context.

# 4. CLINICAL PRESENTATION AND DIAGNOSIS

## 4.1 Clinical Presentation

The majority of patients with solid liver lesions or hepatic metastases are asymptomatic for a long period and patients with minute or solitary liver metastases usually show a normal biochemical profile because of the vigorous compensatory ability of the liver. When symptoms occur they are usually nonspecific (loss of appetite, weight loss, nausea, mild pain in the right upper abdomen, and itching). Associated with the progression of metastatic disease, patients may experience cholestatic disease due to neoplastic tissue compression of portal–biliary pedicle. When the disease progresses and metastases increase in volume, with a consequent reduction of healthy parenchyma, symptoms of liver failure could be present (jaundice, ascites, bleeding diathesis). Patients with metastatic liver tumors may have findings related to the underlying malignancy (e.g., rectal bleeding in a patient colon cancer). When discussing the natural history of liver metastases it should be kept in mind that metastases reflect systemic disease and that the prognosis may be influenced by multiple factors, including percent of liver involvement, concomitant lymph node metastases, localization, and malignancy grade of the primary tumor and the general condition of the patient. Purkiss has documented the tumor doubling times varying between 52 and 199 days, also with the fastest growth rate in the smallest tumors (Purkiss and Williams, 1993). Although there is clearly a large variation in the growth pattern, the doubling times seem to be shorter than those found in primary hepatic tumors such as hepatocellular carcinoma (Barbara et al., 1992). The medial survival time for patients with untreated synchronous colorectal liver metastases ranges from 6.6 to 11 months after the diagnosis. Median survival time for patients with unresected solitary lesions is 15–20 months in different series. In a large series of 484 untreated patients, Stangl et al. found survival rates in patients with colorectal liver metastases of 31.3% at 1 year, 7.9% at 2 years, and 2.6% at 3 years (Stangl et al., 1994).

Laboratory tests are usually normal in patients with liver metastases. In the case of massive liver replacement by the tumor, biochemical signs may be present: increase in serum bilirubin, (prothrombin time) PT increased time, increased levels of $NH_3$, and hypoalbuminemia. There are no specific tumor markers for liver metastases. Tumor markers may be useful in the case of suspected liver metastases from unknown

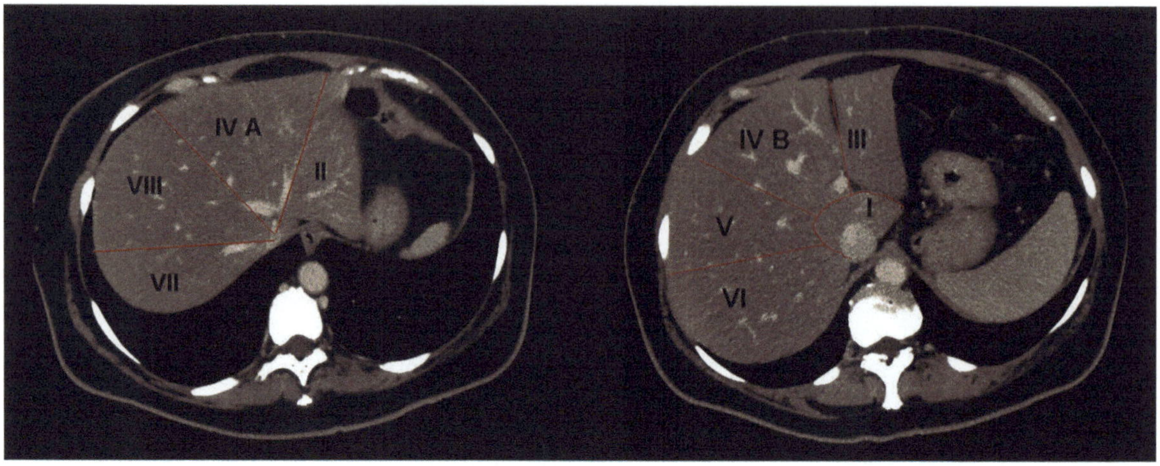


FIGURE 15.5 Liver functional CT anatomy.

TABLE 15.1 Tumor Markers in Patients With Liver Metastases

| Marker | Primary Tumor (s) | Normal Value |
|---|---|---|
| Carcinoembryonic antigen (CEA) | Colorectal, breast, lung, gastric, pancreatic, bladder, medullary thyroid, head and neck, cervical, and hepatic cancers, lymphoma, melanoma | <3–5ng/mL |
| Gastrointestinal cancer antigen (GICA) or CA19.9 | Pancreatic cancer, biliary tract, colon, esophageal, and hepatic cancers | <37U/mL |
| Alpha-fetoprotein (AFP) | Hepatocellular carcinoma, nonseminomatous germ cell tumors, gastric, biliary, and pancreatic cancers | <5.4ng/mL |
| CA125 | Ovarian, endometrial, fallopian tube, breast, lung, esophageal, gastric, hepatic, and pancreatic cancers | <35U/mL |

primitive but often are negative or unspecific. The tumor markers, if positive, can be helpful to place the suspect as a particular source of liver metastases. The most important tumor markers for liver metastases are described in Table 15.1 (Perkins et al., 2003).

## 4.2 Ultrasound

Ultrasound (US) is an economical noninvasive imaging modality for screening liver metastases. US findings in metastatic liver tumors are variable: most metastases are hypoechoic, but they can be hyperechoic and isoechoic too. The ability of US to detect a focal lesion depends on a number of factors: size, location, echogenicity, and mass effect. For lesions smaller than 1cm, the false-negative rate is as high as 80% (Wernecke et al., 1991). Metastases from adenocarcinoma are the most common, they usually are multiple and hypoechoic in comparison with the surrounding liver parenchyma (Sheiner and Brower, 1994).

Intraoperative US is important for liver surgeons; it can help to recognize metastatic lesions smaller than 1cm and can be useful in guiding the hepatic resection. In a study summarizing the results of a combined series for prospective evaluation of intraoperative US compared with preoperative US, arteriography, computerized tomography, and surgeon palpation, intraoperative US showed the best sensitivity, specificity, and diagnostic accuracy (Sheiner and Brower, 1994).

## 4.3 Computed Tomography Scan

Abdominal computed tomography (CT) enhanced with specific contrast agents is the recommended procedure of choice in evaluating liver metastases (Bennett and Bova, 1990). This technique is highly accurate in defining the total number and anatomical position of liver tumors. In addition, extrahepatic abdominal spread is also well evaluated and is fairly reproducible worldwide. Except for infiltrative tumors, such as lymphoma, most metastatic disease to the liver manifests on CT as multiple masses, usually with ill-defined margins and an irregular rim. On triphasic CT scan of the liver, metastatic liver lesions from the colon, stomach, and pancreas usually are hypodense and show lower attenuation in contrast to the brighter surrounding liver parenchyma (Ros and Davis, 1998). The lesions that enhance rapidly, such as metastatic disease from breast cancer, renal cell carcinoma, thyroid cancer, and neuroendocrine or carcinoid tumors, appear hyperdense compared with normal liver parenchyma, particularly on arterial phases of imaging and may appear isodense, or hypodense, to the normal liver parenchyma on the portal-venous phase of imaging (Figs. 15.6 and 15.7) (Oto et al., 2005; Pinotti et al., 2015).

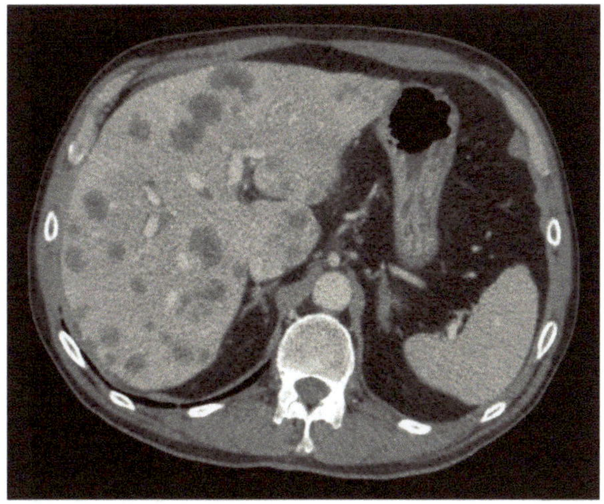

FIGURE 15.6 Hypodense liver metastases from colorectal cancer.

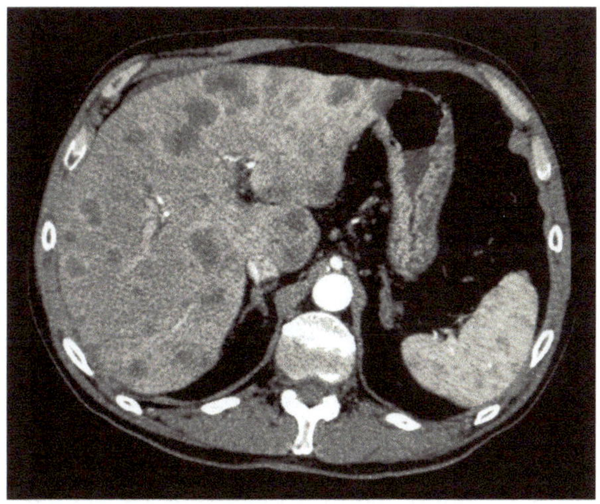

FIGURE 15.7 Hyperdense liver metastases from kidney cancer.

## 4.4 Magnetic Resonance Imaging

Magnetic resonance imaging (MRI) has a similar accuracy in the detection of liver metastases to CT (Stark et al., 1987) but MRI has a significantly greater accuracy than dynamic CT in demonstrating vascular involvement (Nelson et al., 1989). Specialized contrast agents, such as superparamagnetic iron oxide agents, improve the sensitivity of MRI for metastases. On MRI metastatic lesions appear as low-signal areas on T1-weighted images and moderately high-signal areas on T2-weighted images (Zech et al., 2007). Most common malignant hepatic tumors can be further classified as hypovascular or hypervascular based on imaging features. Hypovascular metastases are the most common and usually arise from colon adenocarcinoma. These tumors tend to have mixed arterial and portal venous supply and usually develop central necrosis presumably because of poor vascular perfusion to the center of enlarging tumors. Arterial phase perilesional enhancement may be observed and is typically seen with colon, pancreatic, or stomach adenocarcinoma metastases (Larson et al., 1994). Hypervascular metastases most commonly are related to carcinoid, or kidney origin, but include pancreatic islet cell tumors and melanoma. T2-weighted images show variable signal, usually moderately elevated, and low signal on T1-weighted images (Martin et al., 2005).

### 4.5 Positron Emission Tomography

Positron emission tomography (PET) provides a noninvasive determination of tumorous lesions and this technique has the ability to detect cancerous lesions by tracing increased accumulation of fluoro-2-deoxy-D-glucose (FDG) according to the mechanism that cancers have increased glucose uptake. Zhuang (Zhuang et al., 2000) compared PET and CT scans in patients with colorectal carcinomas and found a better sensitivity for PET. Similar results were also reported by Fröhlich (Fröhlich et al., 1999) for liver metastases of pancreatic carcinomas. A meta-analysis of the literature on detection of hepatic metastases from colorectal, gastric, and esophageal cancers by US, CT, MRI, and PET found that in studies with a specificity higher than 85%, the mean weighted sensitivity was 55% for US; 72% for CT; 76% for MRI; and 90% for PET. The conclusion was that at equivalent specificity, PET is the most sensitive noninvasive imaging modality for the diagnosis of hepatic metastases (Kinkel et al., 2002). Furthermore, a combined PET and CT system in a single imaging setting provides optimal coregistration of image data and may provide a highly accurate demonstration of liver metastases.

### 4.6 Fine Needle Aspiration Biopsy

The use of fine needle aspiration biopsy of a liver biopsy mass is much debated in the literature. The presence of a single nodule that does not show in other diagnostic characteristics in patients with no history of prior malignancies may be useful for the diagnosis. In patients with a previous history of malignancy in which are found single or multiple liver lesions during follow-up or during an occasional check, even if the imaging studies are not diagnostic, biopsy is not indicated because the risk that the lesions are metastatic is very high. The diagnostic accuracy of fine needle biopsy is low (Jones et al., 2005) and violation of the tumor capsule may promote needle track metastases. In 1983, Ryd has shown that a fine needle biopsy is possible to seed along the needle tract from 1000 to 100,000 cells (Ryd et al., 1983). Sadly, a significant number of patients with potentially resectable liver disease still undergo biopsy of the lesion prior to referral to a hepatobiliary surgical unit. Although unusual, tumor seeding along the biopsy needle tract can convert a potentially curable case to one with disseminated peritoneal disease, while adding virtually nothing to the diagnostic pathway. Preoperative biopsy is therefore not only unnecessary, but damages the long-term outcome for patients with resectable disease and must be strongly discouraged (Cresswell et al., 2009).

## 5. TREATMENT OF HEPATIC METASTASES

Over the past 30 years, there have been many advances in the treatment of liver metastases, due to the advancement of biological studies, of surgical knowledge, of tools, and of preoperative therapies; however, it is not always coincided to an increase in a statistically significant manner the survival of patients. Nevertheless, guidelines were created to direct the therapeutic approach and surgery, proving that hepatic metastases are frequent and that the surgical treatment is critical to even improve the prognosis of patients, both in terms of survival and both in terms of disease-free survival. Now it is believed that the best results are obtained in cases of metastatic cancer of the large intestine and NETs. Several studies have been published about it and consider liver metastases of these tumors, neoplasms with typically more favorable prognosis than metastases of other origins. Recent studies seem to show that surgery in the treatment of liver metastases from sarcoma is the right approach and with the best prognosis. On the contrary there is no clear and definite indication for resection of liver metastases from other primitiveness.

## 5.1 Surgery

The progress of medical oncology, which over time has developed increasingly effective adjuvant and neoadjuvant treatment, has emphasized the need of surgery in the management of neoplastic disease. It is well known that tumor stimulates the production of immunosuppressive cytokines (Kavanaught and Carbone, 1996); surgical resection can remove the main source of dissemination of the tumor with the effect of modifying the immune conflict between the host and the tumor. Surgery removes the "vascular sanctuaries," the central and poorly vascularized parts of the tumors where chemotherapy, immunotherapy, hormone therapy can not be effective. Also it is a very precise therapy in time and space; in this way, the tumor does not have the possibility of adaptation as happens in the resistances to chemotherapy drugs. These seemingly trivial and intuitive powers of the surgery have always been at the center of cancer therapies, in particular for what is regarded as solid tumors. Its role is currently increasingly complex and at the center of a multidisciplinary approach often extensive, which includes surgeons, radiologists, interventional radiologists, nuclear physicians, oncologists, immunologists, all employed in the management of cancer patients and determining the appropriate timing of surgery. Today surgical excision remains the best option in terms of survival and performance status in chemosensitive tumors such as melanoma, sarcomas, and pancreatic and kidney adenocarcinomas. From the point of view of the surgical techniques the last three decades have allowed us to develop the anatomical studies performed in the late fifties by Hepp and Couinaud preceded in the thirties by the Vietnamese Tung and Lin (Abdel-Misih and Bloomston, 2010; Helling and Azoulay, 2014; Vauthey et al., 2012). New medical devices have completed this technical–anatomic knowledge; examples are the development and dissemination using intraoperative US, tools for parenchymal dissection that use water jet ultrasonic, instruments for hemostasis of vessels of different sizes and the development of hemostatic materials. Moreover, the transplant surgery has allowed important clarifications, especially as regards the time of warm and cold ischemia applied to vascular clamp, the tolerance of the liver to total vascular exclusion, and the evaluation of the percentage of "remnant liver" to avoid postsurgical liver failure. All of which make liver elective surgery safer and more predictable than once. The fundamental problems that can make a surgical approach impossible in patients with liver metastases are essentially two: nonanatomical resection for the invasion of the tumor of the hilar structures and the volumetric unresectability. The onset of a postoperative hepatocellular insufficiency depends on the ability of the residue liver to ensure a sufficient metabolic function; the latter is due to the volume of remaining liver parenchyma, its preoperative status (healthy liver, liver cirrhosis, chemotreated liver), and postoperative status related to eventual ischemia during surgery. Postoperative liver failure is the leading cause of death in patients undergoing major liver resection. Mortality rates after this surgery vary between 0% and 5% in the case of healthy liver and may rise to 10% in the case of chronic disease with a morbidity rate higher than 30% (Kinoshita et al., 1986). The regenerative capacity of the liver parenchyma after resection has long been known; it was noted in patients with liver cancer associated with a thrombosis of a main branch of portal vein, as the portal vein thrombosis causes atrophy of the ipsilateral segments, and hypertrophy of the contralateral (Farges and Denys, 2001). Following this evidence, many procedures have been developed to induce a liver regeneration, thereby leading a greater number of patients to surgery by decreasing the risks of "postoperative liver failure" after major resections. Currently, in particular, portal embolization is a developed and tuning method. This procedure today is used to obtain a volumetric increase of the residual liver after resection; the gain in volumetric terms corresponds to a gain in functional terms, the latter is more rapid and evident compared with the first (Farges and Denys, 2001; Farges et al., 2003). CT with volumetric and vascular reconstruction allows to plan and evaluate surgical treatment with the following indices: total liver volume, resected liver volume, tumor volume, and future remnant liver; through these data, you may modify the surgical tactics and possibly set a presurgical embolization treatment. As regards the respective technique, thanks to anatomical features of the liver, precisely described by Couinaud and Tung (Helling and Azoulay, 2014; Vauthey et al., 2012), it is possible to perform a surgery also targeted to only one segment; this is because each segment, except the first, receives a segmental vascular portal, arterial, and biliary inflow. Instead, much attention should be made to the venous drainage through the hepatic veins, right, mean, and left, that drain more segments: from this point of view resection must be well planned because the sacrifice of a hepatic vein often forces to more extensive resections. The improvement of the surgical technique in recent decades permitted to perform an increasing number of laparoscopic liver resection. When this technique is applicable, the advantages are real and can be summarized in a greater exposure of the operative field through the use of optical systems that magnify the vision; laparoscopic exploration can avoid unnecessary laparotomies and finally leads to less trauma preserving the integrity of the abdominal wall that has a positive effect on the complication rate and on the drop in immune function postoperatively (Dionigi, 2011).

## 5.2 Liver Metastases From Colorectal Tumors

CRC accounts for over 9% of all cancer incidence; it is the third most common cancer worldwide and the fourth most common cause of death (Torre et al., 2015). The CRC has been widely studied and on it there are guidelines on the treatment of both primary lesion and the metastases. The liver is the first site of metastatic spread of CRC, we find them in 50–60% of patients, and it is the only site of metastases in 30–40% of cases of advanced colorectal tumor. The 20–25% of patients with initially diagnosed CRC have contemporaneous (synchronous) liver metastases, whereas approximately 40–50% will develop hepatic metastases after resection of the primary colorectal tumor (metachronous) most frequently within the first 3 years of follow-up (Konopke et al., 2009; Nordlinger et al., 1996).

Several studies have been published about it, proving the specific need of surgery and emphasizing the usefulness of adjuvant therapies such as chemotherapy, both in patients with resectable metastases and in patients with unresectable lesions. It is used for either cytoreductive or palliation intent. The natural history of colorectal liver metastases is variable. The median survival without treatment is around 8 months from the presentation, and the prognosis of these patients is improved in the case of single lesion or metastases confined to a single liver lobe. In only 20–30% of patients with colorectal liver metastases the disease is confined to the liver and potentially resectable. Patients who undergo curative resection have a 5-year survival ranging from 28% to 40% than patients who do not undergo resection, the 5-year survival being 0%. Patients who underwent curative resection have an average survival at 1 year of 90%, at 3 years of 59%, and at 5 years of 37%. The survival rate increases to 47% at 5 years when the surgical resection margins are performed with microscopic disease free more than 10 mm. Patients with disease-free surgical margin of less than 10 mm have a lower survival (29% at 5 years; Imamura et al., 2001).

## 5.3 Hepatic Metastases from Neuroendocrine Tumors

NETs are a group of neoplasms characterized by a relatively slow growth, and they can lead various clinical syndromes for their ability to produce functioning hormones. All these tumors arise from neuroectodermal tissues and are characterized by the skill to secrete specific substances that are metabolically active. They are relatively rare (one to five cases/100,000 people for year), but their incidence is increasing between 3% and 10% of annual cases, probably the increases in incidence of this injury are related to improved diagnostics and screening technologies for cancer (Hallet et al., 2015). About 85% of NETs have gastrointestinal origin and at the time of diagnosis liver metastases are found in 80% of patients, and in view of these data about 10% of all hepatic metastases are of neuroendocrine origin. In most of the cases metastases are synchronous and completely resectable only in 7–15% of patients. Although at one time NETs were considered "indolent" tumors, today it has shown a significant higher survival in patients who are able to get a curative surgical treatment (75–99% at 5 years against 13–54% of nonresected) (Jagannath et al., 2012). The prognosis also depends on the neuroendocrine primitive lesion with the literature supporting a lower survival for pancreatic NETs than those of intestinal origin (30–60% vs. 60–90% at 5 years). Several articles confirm that aggressive surgical approach against NETs provides good long-term survival, particularly if the resection is accompanied by adjuvant therapies. The usefulness of liver resection has been demonstrated in terms of survival and improving quality of life, today we get survivals between 45% and 85% at 5 years and between 35% and 79% at 10 years (Sarmiento et al., 2003; Norton et al., 2003). The data lead us to affirm that the resection of hepatic metastases and primary tumor is not indicated only in those cases where the primary tumor is small and asymptomatic and if there are the presence of unresectable liver lesions (Lee et al., 2012).

## 5.4 Liver Metastases from Sarcoma

Liver is often a site of metastases for primary tumors other than those of gastrointestinal origin. Regarding liver metastases from sarcoma, the treatment of choice has always been chemotherapy, many patients undergo chemoembolization with regression of mass only in 50% of cases. However, both of these methods give short survival posttreatment. Today surgery is considered a safe treatment for this type of pathology with a low rate of mortality (less than 5%). Studies demonstrate the high rate of postoperative recurrence (≥35%) that afflict these patients, but despite this, surgery improves the long-term outcome with more satisfactory results than chemoembolization and chemotherapy. In selected cases, the presence of recurrence only in the liver does not contraindicate repeated resections. Furthermore, the presence of extrahepatic disease has not appeared as a contraindication to surgery when it is possible for the complete removal of all the masses. After surgery, the median survival amounted to 32 months and 20% of patients were alive at 5 years after complete resection (R0). The worst prognosis, of course, is not curative resection (R1–R2) or in the presence of unresectable extrahepatic metastases. In these cases, the debulking surgery seems to have a better prognosis compared to medical therapy alone (Stavrou et al., 2006; Lang et al., 2000). DeMatteo et al. (2001) reported the experience of 56 hepatectomies for sarcomas metastatic to the liver, most of these were gastrointestinal stromal tumor and gastrointestinal leiomyosarcoma. The authors reported a complete resection rate of 15% with a median survival of 39 months and at 5-year survival of 30%. Data analysis has shown that the interval between the resection of the primitive and hepatic resection is of prognostic importance when metastases appears in a time of over 24 months. This emphasizes, as with other histological types, the importance of the tumor biology of primary lesion as a predictor of long-term outcome. These facts are very similar to the conclusions reached in various studies regarding liver metastases from colorectal or neuroendocrine cancer, and for this reason even today hepatic metastases from sarcomas are treated surgically.

## 5.5 Hepatic Metastases From Other Primary Sites

The utility of a correct surgical approach in the treatment of liver metastases from CRC, NETs, and sarcoma was observed and demonstrated, it seems appropriate to analyze the data relating to the treatment of liver metastases from other primary tumors, no colorectal, no neuroendocrine, no sarcoma, and to consider whether surgery would be in any case the best treatment possible in terms of survival, outcome, quality of life, and disease-free survival. In these patients, more meaningful factors that affect prognosis are given below:

- The interval between the treatment of the primary tumor and the diagnosis of liver metastases (metachronous): The best prognosis was associated with a delay of this occurrence greater than or equal to 24 months; the prognosis is intermediate between 12 and 24 months but obviously better than when the appearance of the lesion is less than 1 year after resection of the primary tumor (Earle et al., 2006; Takada et al., 2001; Lendoire et al., 2007).
- Size and number of metastases: The prognosis worsens especially with the number of metastases exceeding three to four and if the size of one of it is greater than 5 cm (Duan et al., 2012; O'Rourke et al., 2007; Groeschl et al., 2012).
- Primary site and histological subtype: Liver metastases from breast, kidney, uterus, ovary, testicle, ampulla of Vater, and adrenal gland cancer had a 5-year overall survival greater than 30%, whereas metastases from gastric and duodenum cancer had a survival between 30% and 15%; liver metastases from pancreas, anus, esophagus, cardia, and lung cancer had a 5-year overall survival less than 15% (Bresadola et al., 2011; Ercolani et al., 2009; Slotta et al., 2014). Surely tumors of gastrointestinal origin have a worse prognosis than those of other primary sites.
- Surgical margin status: R0 resection remains the rule and R2 resection is outlawed, whereas R1 resection does not have catastrophic results if you can associate it with an effective chemotherapy (Adam et al., 2006; Harrison et al., 1997; Marudanayagam et al., 2011). The safety surgical margin is indicated among 0, 5, and 1 cm, in particular, a margin less than 5 mm is associated with greater rates of recurrence and with a less hepatic disease-free survival. However, this argument has always been debated, and today it seems reasonable and prognostically significant performing resection to get a margin less than a centimeter, rather than not do it.
- Type of intervention performed: The importance of hepatectomy is also an independent prognostic factor, which

indirectly reflects the extent of liver disease (Harrison et al., 1997; Hemming et al., 2000).

Nowadays hepatic resection also for noncolorectal, nonneuroendocrine, and nonsarcoma metastases is safe and feasible with an outcome that depends on certain clinical and pathological features such as the type of primary tumor and its histology. A statistical model for the patient based on correct prognostic factors that can predict the long-term survival. The contraindications for surgery seem to be represented in particular by metastases rapidly evolving beyond the control of systemic treatments, but in some chemoresistant tumors, like kidney and melanoma, surgery should be considered immediately as the only curative intent therapy and should not be delayed (Slotta et al., 2014; Adam et al., 2006). Currently, the acceptable risk of elective liver surgery, the existence of more than a third of patients alive at 5-year demonstrates that surgical treatment deserves to be discussed for many patients. The discussion should be multidisciplinary in function of the patients, of the primary disease, of the characteristics of the metastatic disease, of the type of intervention planned, and of the adjuvant treatments available.

## 5.6 Besides Surgery: Alternative Therapies and Adjuvants

In an alternative to surgical approach, particularly for those patients in whom surgery is contraindicated, the "minimally invasive" therapies, many of these are percutaneous methods, play an increasingly important role. The use of these therapies in alternative or combined to surgery allows a multinodal treatment of patients with primary or secondary liver cancer.

Generally interventional radiology techniques that have earned a place in the treatment of liver tumors are directed to the cure of nonsurgically resectable lesions. At an early stage of metastatic liver disease the use of locoregional therapies, combined with systemic chemotherapy, can increase rates of clinical response, allowing you to submit more cases to radical surgery. The percutaneous local therapy, represented above all by radiofrequency, has therefore grown in association with surgery and respect to which is not an alternative, but a complementary tool in the treatment of metastases (Treska et al., 2012). Instead transarterial techniques are closest to the medical treatment, acting mostly palliative, through the targeted delivery of chemotherapeutic. The techniques listed in Table 15.2 can be used in a sequential manner and not to force in an exclusive way.

## 5.7 Ablation: Radiofrequency and Chemical Ablation

Necrosis of the tumor can be obtained by means of chemical ablation (percutaneous ethanol injection or PEI), which includes the infusion of denaturants such as ethanol or acetic acid in the tumor mass or by thermal ablation, which causes the death of neoplastic tissues by cold (cryoablation) or by infusion of heat (radiofrequency ablation [RFA], microwave, or laser). These methods can be performed via laparoscopy, percutaneous, or open surgery. The RFA of liver metastases is entered rightfully into the locoregional treatments as percutaneous methodical guided radiologically, both as intraoperative procedure to complete a liver resection (Fig. 15.8). A review of the American Society of Clinical Oncology analyzed the effectiveness of RFA in patients with liver metastases, considering three aspects:

1. The efficacy of RFA liver resection for resectable tumors
2. The utility of RFA for unresectable tumors
3. The different approaches to RFA (open, percutaneous, laparoscopic)

The RFA correlates with a particularly wide range of 5-year survival (14–55%) and local recurrence (36–60%), low mortality rate (0–2%), and a low risk of complications (6–9%; Wong

TABLE 15.2 Nonsurgical Therapies

| | |
|---|---|
| Percutaneous techniques | • Radiofrequency ablation (RFA)<br>• Percutaneous ethanol injection (PEI)<br>• Injection of sailing boiling (PSIT)<br>• Injection of acetic acid<br>• Laser therapy (YAG)<br>• Microwave ablation<br>• Cryotherapy |
| Transarterial techniques | • Chemoembolization (TACE)<br>• Arterial embolization (TAE)<br>• Radioembolization (SIRT)<br>• TAE with radioactive lipiodol<br>• Intra-arterial chemotherapy (HAI) |
| Others | • Radiotherapy<br>• Accelerated protons<br>• Therapy with somatostatin analogues<br>• PRRT (peptide receptor radionuclide therapy)<br>• Chemotherapy (adjuvant) |

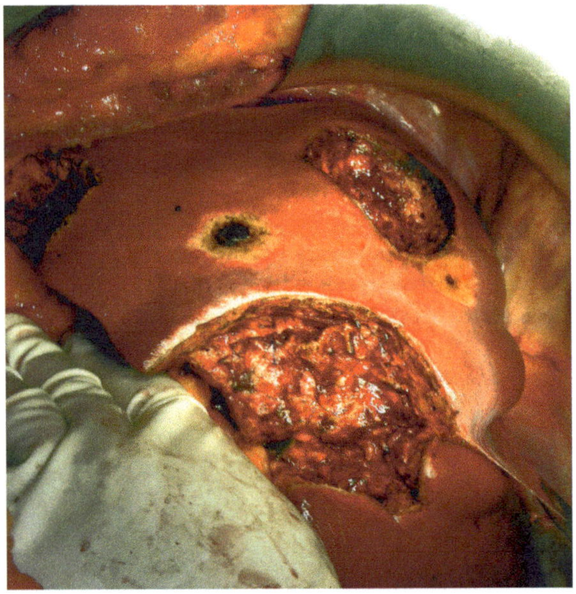

FIGURE 15.8 Atypical liver resections associated with intraoperative radiofrequency.

et al., 2010; Curley, 2003). The range of survival after RFA suggests that the selection criteria of patients was not unique for this procedure to the detriment of the proof of its healing potential. In terms of survival, the evidence suggested that liver resection increases the overall survival, particularly in those patients in whom a curative resection (R0) is obtained and without extrahepatic disease. In unresectable tumors, the RFA has shown a significant efficacy in the palliative treatment of patients extensively pretreated with multiple lines of systemic chemotherapy, delaying the local progression of the disease. However, it can allow a sterilization of a liver lobe as being a prelude to a portal embolization or hepatic resection in two stages for metastases initially unresectable or as a first approach in cases of resectable metastases, but surgically difficult to remove in a short time. Summing, the RFA can be used as follows:

- In combination with major liver resection for contralateral injury of size equal to or slightly greater than 3 cm;
- In anticipation of surgical treatment two or three times, such a treatment of downstaging;
- In patients and not candidates for chemotherapeutic or surgical treatments.

In view of a treatment in more steps especially for liver metastases from CRC, the RFA plays an important role in tumor cytoreduction of a liver lobe before a possible portal embolization. This technique is also applicable in case of relapse after hepatectomy (Elias et al., 2002; Livraghi et al., 2003).

## 5.8 Arterial Embolization

The general criterion for selecting patients eligible for embolization procedures is the presence of unresectable disease not susceptible to ablative therapy alone. Embolization in its forms: trans-arterial embolization (TAE), trans-arterial chemoembolization (TACE), and selective hepatic radiotherapy or hepatic radioembolization (SIRT) are based on the selective infusion of particles in the branch of hepatic artery that supplies the tumor lesions. The ration of TAE and TACE is to reduce the blood flow to the tumor resulting in ischemia and necrosis. The TACE is different from TAE. The administration of a concentrated dose of chemotherapy combined with an emulsifying agent is through the arterial catheter, before the embolization particle enters into the bloodstream. The TACE is used in the palliative treatment of liver metastases often as a rescue in patients extensively pretreated. SIRT is a locoregional treatment that across systemic circulation reach the lesion and deposit in its context cytotoxic substances. The technique involves the arterial catheterization and the release of resin or glass microspheres that contain yttrium-90 (beta issuer). The latter, trapped in the vascular bed of metastases, destroys cancer cells by emitting radiation (Murthy et al., 2002). In the last years, the SIRT has been used in patients with secondary lesions from different histologies (colorectal, breast, NETs, etc.), but certainly the great majority of the cases treated refers to colorectal adenocarcinomas (Kennedy et al., 2001). Undoubtedly, the reason why metastases from CRC were treated with this technique is due to the fact that the microspheres for their size (~30 μm) can easily reach the peripheral region of the lesion where the neoplastic growth is more exuberant (Ackerman et al., 1969; Gray et al., 1992). In the context of this pathology many clinical trials of phase I and II have long been validated for clinical safety also in patients treated with multiple lines of systemic chemotherapy (Coldwell et al., 2001; Van Hazel et al., 2004; Douglas et al., 2007).

## 5.9 Portal Embolization

The portal embolization is a technique recently introduced in clinical practice; the aim is to increase the possibility to perform a liver major resection, increasing artificially the volume of the part of the liver that must not be detached. The contralateral liver should go against hypertrophy that allows the reduction of postoperative complications related to liver failure caused by the reduced amounts of the remaining liver after resection. The reduction of blood flow from the portal vein to half of the liver causes the increase of the factors that stimulate hepatic growth of the other half, thus leading to an increase in the volume. The procedure can be performed under local anesthesia; in the angiography room after a proper study of hepatic vascularization inserting a thin catheter that allows the portal branch of interest (Fig. 15.9). Portal embolization can be programmed as follows:

- Portal vein embolization can be considered before major hepatic resection and in all cases in which the future liver remnant is estimated lower than 40%.
- In major resections planned after neoadjuvant chemotherapy.
- In two-stage strategy in patients affected by bilobar hepatic metastases (Jaeck et al., 2004; Abulkhir et al., 2008; De Baere et al., 1996).

The relative hypertrophy of the liver lobe embolized will be in this case an index of the degree of impairment of the parenchyma due to drugs. The valuation of the patient undergoing the procedure is carried out 2–4 weeks after repeating a CT with hepatic volume and possibly with a functional test with green indocyanine.

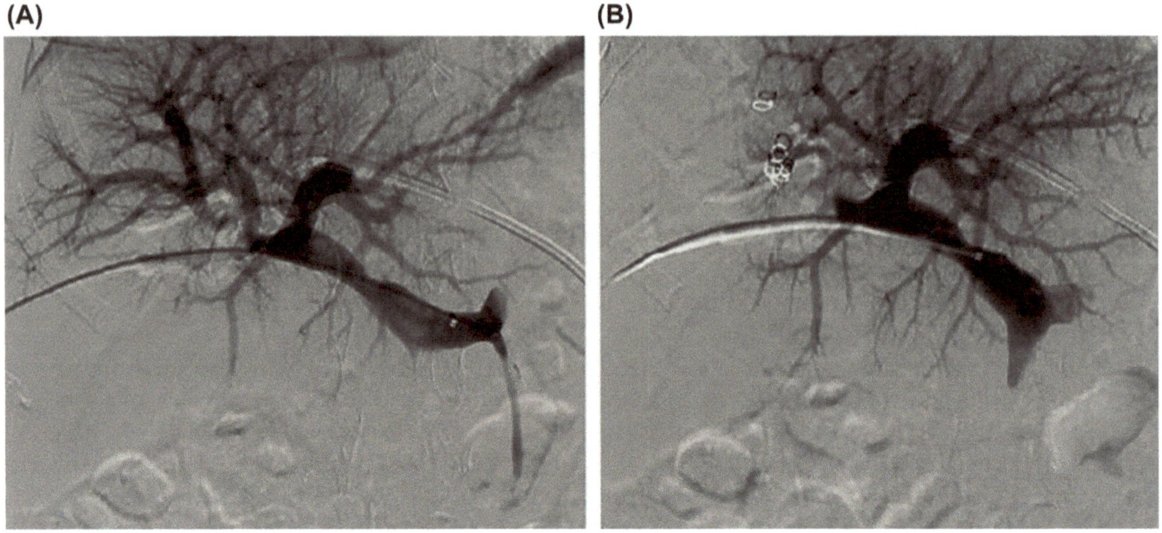


FIGURE 15.9 Portal embolization A: pre B: post the procedure.

## 5.10 Chemotherapy: Neoadjuvant Preoperative and Adjuvant Postoperative

In the case of colorectal tumors resection of primary tumor associated with chemotherapy and radiotherapy proves effective: nevertheless about 60% of patients develop local, regional, or distant metastases; in 85% of cases within two and half years after colonic resection and in the remaining 15% in the subsequent years. In over 50% of cases the cause of death is the presence of liver metastases (Gerunda, 2012). In the last decades, an improvement of the prognosis of patients with liver metastases due to drafting of multidisciplinary treatment protocols that combine surgical resection to more effective chemotherapy has led to an effective increase of survival of these patients (Adam et al., 2004). The appearance of ablative techniques, alternatives to surgery, in combination with chemotherapy appears to improve survival also in those patients in whom surgery resection is contraindicated (Ruers et al., 2015). Chemotherapy can be performed preoperative (neoadjuvant) in the case of nonresectable lesions at the diagnosis and it is aimed at trying to reduce the tumor mass or classically postliver resection as an aid to surgery. Recent protocols in the treatment of metastatic colorectal tumors based on neoadjuvant therapy have been shown to achieve an objective response in 40–69% of patients and a recovery to radical surgery in 16–30%. These patients have an overall and disease-free survival not dissimilar to whom resected in first instance (Giacchetti et al., 1999). Normally, if you adhere to neoadjuvant chemotherapy, reassessment of the patient has to be programmed after an interval of at least 4 weeks of treatment. The interval can be changed depending on therapeutic response obtained. A recent analysis of the experimental and retrospective studies published that have analyzed the resection of liver metastases in patients initially inoperable, after neoadjuvant chemotherapy, was able to demonstrate a resection rate ranging from 24% to 54% (Mentha et al., 2006; Folprecht et al., 2005). This confirms that the chemotherapy can have a decisive influence on patient survival, allowing a good percentage to achieve a curative surgery.

## 6. CONCLUSIONS

Liver metastases are the most common liver cancer. The liver is often the site of metastasis from CRC and NETs. However, liver metastases are also frequent in the case of renal, gastric, ovarian, breast, lung cancer, and melanoma. In the last decades the prognosis of liver metastases is significantly improved but this disease still remains a major challenge for clinicians and requires a multidisciplinary approach. Nowadays, many weapons have been added to deal with this disease: new chemotherapy drugs, new angiographic techniques (portal embolization, TACE, TAE, SIRT), improved technologies, and extensive surgical approach (two-staged hepatectomy, (associating *liver* partition and portal vein ligation for staged hepatectomy) ALPPS). In future, the knowledge and technology improvement will hopefully allow us to increase the number of potentially treatable patients, significantly impacting their prognosis and quality of life.

## References


Abdel-Misih, S.R., Bloomston, M., August 2010. Liver anatomy. Surg. Clin. North Am. 90 (4), 643–653.

Abulkhir, A., Limongelli, P., Healey, A.J., Damrah, O., Tait, P., Jackson, J., et al., 2008. Pre-operative portal vein embolization for major liver resection: a meta-analysis. Ann. Surg. 247 (1), 49–57.

Ackerman, N.B., Lien, W.M., Kondi, E.S., Silverman, N.A., 1969. The blood supply of experimental liver metastases. I. The distribution of hepatic artery and portal vein blood to small and large tumors. Surgery 66 (6), 1067–1072.

Adam, J.C., Watt, F.M., 1989. Fibronectin inhibits the terminal differentiation of human keratinocytes. Nature 340, 307–309.

Adam, R., Delvart, V., Pascal, G., Valeanu, A., Castaing, D., Azoulay, D., et al., October 2004. Rescue surgery for unresectable colorectal liver metastases downstaged by chemotherapy. A model to predict long-term survival. Ann. Surg. 240 (84)), 644–658.

Adam, R., Chiche, L., Aloia, T., Elias, D., Salmon, R., Rivoire, M., et al., 2006. Hepatic resection from noncolorectal nonendocrine liver metastases: analysis of 1452 patients and development of a prognostic model. Ann. Surg. 244, 524–535.

Alagusundaramoorthy, S.S., Gedaly, R., October 21, 2014. Role of surgery and transplantation in the treatment of hepatic metastases from neuroendocrine tumor. World J. Gastroenterol. 20 (39), 14348–14358.

Alberts, S.R., Wagman, L.D., 2008. Chemotherapy for colorectal cancer liver metastases. Oncologist 13, 1063–1073.

Anasagasti, M.J., Alvarez, A., Avivi, C., Vidal-Vanaclocha, F., 1996. Interleukin-1-mediated $H_2O_2$ production by hepatic sinusoidal endothelium in response to B16 melanoma cell adhesion. J. Cell. Physiol. 167 (2), 314–323.

Barbara, L., Benzi, G., Gaiani, S., Fusconi, F., Zironi, G., Siringo, S., et al., 1992. Natural history of small untreated hepatocellular carcinoma in cirrhosis: a multivariate analysis of prognostic factors of tumor growth rate and patients survival. Hepatology 16 (1), 132–137.

Bennett, W.F., Bova, J.G., October 1990. Review of hepatic imaging and a problem-oriented approach to liver masses. Hepatology 12 (4 Pt 1), 761–775.

Bresadola, V., Rossetto, A., Adani, G.L., Baccarani, U., Lorenzin, D., Favero, D., et al., 2011. Liver resection for noncolorectal and nonneuroendocrine metastases: results of a study on 56 patients at a single institution. Tumori 97 (3), 316–322.

Coldwell, D., Kennedy, A.S., Echo, D.A., 2001. Feasibility of treatment of hepatic tumors utilizing Yttrium-90 microspheres. J. Vasc. Interv. Radiol. 12 (S1), S113.

Comoglio, P.M., Nicolin, A., 1997. The neoplastic cell. UTET 23, 286–297.

Cresswell, A.B., Welsh, F.K., Rees, M., November 2009. A diagnostic paradigm for resectable liver lesions: to biopsy or not to biopsy? HPB (Oxford) 11 (7), 533–540.

Curley, S.A., 2003. Radiofrequency ablation of malignant liver tumors. Ann. Surg. Oncol. 10 (4), 338–347.

De Baere, T., Roche, A., Elias, D., Lasser, P., Lagrange, C., Bousson, V., 1996. Preopertaive portal vein embolization for extension of hepatectomy indications. Hepatology 24 (6), 1386–1391.

DeMatteo, R.P., Shah, A., Fong, Y., Jarnagin, W.R., Blumgart, L.H., Brennan, M.F., 2001. Results of hepatic resection fro sarcoma metastatic to the liver. Ann. Surg. 234 (4), 540–547.

Dionigi, R., 2011. Surgery Theoretical Basis and General Surgery, fifth ed. Elsevier.

Douglas, M., Codwell, D., Nutting, D.O., 2007. Use of Yttrium-90 microspheres in the treatment of unresectable hepatic metastases from breast cancer. Int. J. Radiat. Oncol. Biol. Phys. 69 (3), 800–804.

Duan, X.F., Dong, N.N., Zhang, T., Li, Q., 2012. Comparison of surgical outcomes in patients with colorectal liver metastases versus noncolorectal liver metastases: a Chinese experience. Hepatol. Res. 42 (3), 296–303.

Earle, S.A., Perez, E.A., Gutierrez, J.C., Sleeman, D., Livingstone, A.S., Franceschi, D., et al., 2006. Hepatectomy enables prolonged survival in select patients with isolated non-colorectal liver metastases. J. Am. Coll. Surg. 203, 436–446.

Elias, d, De Beare, T., Smayra, T., Ouellet, J.F., Roche, A., Lasser, P., 2002. Percutaneous radiofrequency thermoablation as an alternative to surgery for treatment of liver tumor recurrence after hepatectomy. Br. J. Surg. 89 (6), 752–756.

Ercolani, G., Vetrone, G., Grazi, G.L., Cescon, M., Di Gioia, P., Ravaioli, M., et al., 2009. The role of liver surgery in the treatment of non-colorectal non-neuroendocrine metastases (NCRNNE). Analysis of 134 resected patients. Minerva Chir 64 (6), 551–558.

Farges, O., Denys, A., November 2001. Portal vein embolization prior to hepatectomy. Techniques, indications and results. Ann. Chir 126 (9), 836–844.

Farges, O., Belghiti, J., Kianmanesh, R., Regimbeau, J.M., Santoro, R., Vilgrain, V., et al., February 2003. Portal vein embolization before right hepatectomy, prospective clinical trial. Ann. Surg. 237 (2), 208–217.

Fidler, I.J., 2003. The pathogenesis of cancer metastasis: the "seed and soil" hypothesis revisited. Nat. Rev. Cancer 3, 453–458.

Folprecht, G., Grothey, A., Alberts, S., Raab, H.R., Kohne, C.H., August 2005. Neoadjuvant treatment of unresectable colorectal liver metastases: correlation between tumour response and resection rates. Ann. Oncol. 16 (8), 1311–1319.

Fröhlich, A., Diederichs, C.G., Staib, L., Vogel, J., Beger, H.G., Reske, S.N., February 1999. Detection of liver metastases from pancreatic cancer using FDG PET. J. Nucl. Med. 40 (2), 250–255.

Gerunda, G.E., 2012. Guidelines in Liver Metastases. Liver and Multivisceral Transplant Center, Modena.

Giacchetti, S., Itzhaki, M., Gruia, G., Adam, R., Zidani, R., Kunstlinger, F., et al., 1999. Long-term survival of patients with unresectable colorectal cancer liver metastases following infusional chemotherapy with %-fluorouracil, leucovorin, oxaliplatin and surgery. Ann. Oncol. 10, 663–669.

Gray, B.N., Anderson, J.E., Burton, M.A., Van Hazel, G., Codde, J., Morgan, C., et al., 1992. Regression of liver metastases following treatment with Yttrium 90 microspheres. Aust. N. Z. Surg. 62, 105–110.

Greenway, C.V., Stark, R.D., 1971. Hepatic vascular bed. Physiol. Rev. 51, 23–65.

Groeschl, R.T., Nachmany, I., Steel, J.L., Reddy, S.K., Glazer, E.S., de Jong, M.C., Pawlik, T.M., Geller, D.A., Tsung, A., Marsh, J.W., Clary, B.M., Curley, S.A., Gamblin, T.C., May 2012. Hepatectomy for noncolorectal non-neuroendocrine metastatic cancer: a multi-institutional analysis. J. Am. Coll. Surg. 214 (5), 769–777.

Hallet, J., Law, C.H., Cukier, M., Saskin, R., Liu, N., Singh, S., February 15, 2015. Exploring the rising incidence of neuroendocrine tumors: a population-based analysis of epidemiology, metastatic presentation, and outcomes. Cancer 121 (4), 589–597.

Harrison, L.E., Brennan, M.F., Newman, E., Fortner, J.G., Picardo, A., Blumgart, L.H., et al., 1997. Hepatic resection for noncolorectal non neuroendocrine metastases: a fifteen-year experience with ninety-six patients. Surgery 121 (6), 625–632.

Helling, T.S., Azoulay, D., June 2014. Ton that Tung's livers. Ann. Surg. 259 (6), 1245–1252.

Helling, T.S., November-December 2005. Advances in liver surgery. Mo. Med. 102 (6), 560–564.

Hemming, A.W., Sielaff, T.D., Gallinger, S., Cattral, M.S., Taylor, B.R., Greig, P.D., et al., 2000. Hepatic resection of noncolorectal nonneuroendocrine metastases. Liver Transpl. 6 (1), 97–101.

Imamura, I., Matsuyama, Y., Shimada, R., Kubota, M., Nakayama, A., Kobayashi, A., et al., 2001. A study of factors influencing prognosis after resection of hepatic metastases from colorectal and gastric carcinoma. Am. J. Gastroenterol. 96, 3178–3184.

Ismaili, N., 2011. Treatment of colorectal liver metastases. World J. Surg. Oncol. 9, 154.

Jaeck, D., Oussoultzoglou, E., Rosso, E., Greget, M., Weber, J.C., Bchellier, P., et al., 2004. A two-stage hepatectomy procedure with portal vein embolization to achieve curative resection for initially unresectable multiple and bilobar colorectal liver metastases. Ann. Surg. 240 (6), 1037–1051.

Jagannath, P., Chhabra, D., Shrikhande, S., Shah, R., 2012. Surgical treatment of liver metastases in neuroendocrine neoplasms. Int. J. Hepatol. 2012 ID 782672.

Jones, O.M., Rees, M., John, T.G., Bygrave, S., Plant, G., September 2005. Biopsy of resectable colorectal liver metastases causes tumour dissemination and adversely affects survival after liver resection. Br. J. Surg. 92 (9), 1165–1168.

Kasper, H.U., Drebber, U., Dries, V., Dienes, H.P., October 2005. Liver metastases: incidence and histogenesis. Z. Gastroenterol. 43 (10), 1149–1157.

Kavanaught, D.Y., Carbone, D.P., 1996. Immunologic dysfunction in cancer. Hematol. Oncol. Clin. North Am. 10, 927–951.

Kennedy, A.S., Murthy, R., Van Echo, D., 2001. Preliminary results of outpatient hepatic artery brachytherapy for colorectal hepatic metastases. Eur. J. Cancer 37, S289–S290.

Kinkel, K., Lu, Y., Both, M., Warren, R.S., Thoeni, R.F., September 2002. Detection of hepatic metastases from cancers of the gastrointestinal tract by using noninvasive imaging methods (US, CT, MR imaging, PET): a meta-analysis. Radiology 224 (3), 748–756.

Kinoshita, H., Sakai, K., Hirohashi, K., Igawa, S., Yamasaki, O., Kubo, S., 1986. Peroperative portal vein embolization for hepatocellular carcinoma. World J. Surg. 10, 803–808.

Kolios, G., Valatas, V., Kouroumalis, E., 2006. Role of Kupffer cells in the pathogenesis of liver disease. World J. Gastroenterol. 12 (46), 7412–7420.

Konopke, R., Kersting, S., Distler, M., Dietrich, J., Gastmeier, J., Heller, A., et al., 2009. Prognostic factors and evaluation of a clinical score for predicting survival after resection of colorectal liver metastases. Liver Int. 29, 89–102.

Lang, H., Nussbaum, K.T., Kaudel, P., Fruhauf, N., Flemming, P., Raab, R., 2000. Hepatic metastases from leiomyosarcoma: a single-center experience with 34 liver resections during a 15-year period. Ann. Surg. 231 (4), 500–505.

Larson, R.E., Semelka, R.C., Bagley, A.S., Molina, P.L., Brown, E.D., Lee, J.K., August 1994. Hypervascular malignant liver lesions: comparison of various MR imaging pulse sequences and dynamic CT. Radiology 192 (2), 393–399.

Lee, S.Y., Cheow, P.C., Teo, J.Y., Ooi, L.L.P.J., 2012. Surgical treatment of neuroendocrine liver metastases. Int. J. Hepatol. 2012 ID 146590PJ.

Lendoire, J., Moro, M., Andriani, O., Grondona, J., Gil, O., Raffin, G., et al., 2007. Liver resection for noncolorectal, non-neuroendocrine metastases: analysis of a multicenter study from Argentina. HPB (Oxford) 9 (6), 435–439.

Livraghi, T., Solbiati, L., Meloni, F., Ierace, T., Goldberg, S.N., Gazelle, G.S., 2003. Percutaneous radiofrequency ablation of liver metastases in potential candidates for resection. Cancer 97 (12), 3027–3035.

Malik, H.Z., Prasad, K.R., Halazun, K.J., Aldoori, A., Al-Mukhtar, A., Gomez, D., et al., 2007. Preoperative prognostic score for predicting survival after hepatic resection for colorectal liver metastases. Ann. Surg. 246 (5), 806–814.

Martin, D.R., Danrad, R., Hussain, S.M., September 2005. MR imaging of the liver. Radiol. Clin. North Am. 43 (5), 861–886.

Marudanayagam, R., Sandhu, B., Perera, M.T.P.R., Taniere, P., Coldham, C., Bramhall, S., et al., 2011. Hepatic resection for non-colorectal, non-neuroendocrine, non-sarcoma metastasis: a single-centre experience. HPB (Oxford) 13 (4), 286–292.

Mentha, G., Majno, P.E., Andres, A., Rubbia-Brandt, L., Morel, P., Roth, A.D., 2006. Neoadjuvant chemotherapy and resection of advanced synchronous liver metastases before treatment of the colorectal primary. Br. J. Surg. 93 (7), 872–878.

Murthy, R., Kennedy, A.S., Coldwell, D., 2002. Technical aspects of theraspheres (TS) infusion. J. Vasc. Interv. Radiol. 13 (2), S2.

Nelson, R.C., Chezmar, J.L., Sugarbaker, P.H., Bernardino, M.E., July 1989. Hepatic tumors: comparison of CT during arterial portography, delayed CT, and MR imaging for preoperative evaluation. Radiology 172 (1), 27–34.

Nordlinger, B., Guiguet, M., Valliant, J.C., Balladur, P., Boudjema, K., Bachellier, P., et al., 1996. Surgical resection of colorectal carcinoma metastases to the liver: a prognostic scoring system to improve case selection, based on 1568 patients. Cancer 77 (7), 1254–1262.

Norton, A.J., Warren, R.S., Kelly, M.G., Zuraek, M.B., Jensen, R.T., 2003. Aggressive surgery for metastatic liver neuroendocrine tumors. Surgery 134 (6), 1057–1063.

O'Rourke, T.R., Tekkis, P., Yeung, S., Fawcett, J., Linch, S., Strong, R., et al., 2007. Long-term results of liver resection for non-colorectal, non-neuroendocrine metastases. Ann. Surg. Oncol. 15 (1), 207–218.

Osada, S., Imai, H., Sasaki, Y., Tanaka, Y., Tokuyama, Y., Okumura, N., et al., 2012. Strategy for synchronous and multiple liver metastases. Hepatogastroenterology 59 (113), 198–203.

Oto, A., Tamm, E.P., Szklaruk, J., September 2005. Multidetector row CT of the liver. Radiol. Clin. North Am. 43 (5), 827–848.

Page, A.J., Weiss, M.J., Pawlik, T.M., October 15, 2014. Surgical management of noncolorectal cancer liver metastases. Cancer 120 (20), 3111–3121.

Paget, S., 1989. The distribution of secondary growths in cancer of the breast. Lancet 133, 571–573.

Perkins, G.L., Slater, E.D., Sanders, G.K., Prichard, J.G., September 15, 2003. Serum tumor markers. Am. Fam. Physician 68 (6), 1075–1082.

Pinotti, E., Ratti, F., Cipriani, F., Aldrighetti, L., 2015. Liver metastases from kidney Cancer. In: Noncolorectal, Nonneuroen-docrine Liver Metastases: Diagnosis and Current Therapies. Springer, pp. 33–48.

Purkiss, S.F., Williams, N.S., 1993. Growth rate and percentage of hepatic replacement of colorectal liver metastases. Br. J. Surg. 80, 1036–1038.

Quan, D., Gallinger, S., Nhan, C., Auer, R.A., Biagi, J.J., Fletcher, G.G., et al., 2012. The role of liver resection for colorectal cancer metastases in an era of multimodality treatment: a systematic review. Surgery 151 (6), 860–870.

Ros, P.R., Davis, G.L., 1998. The incidental focal liver lesion: photon, proton, or needle? Hepatology 27 (5), 1183–1190.

Ruers, T., Put, C.J.A., Van Coevorden, F., Pierie, J.P., Rinkes, I.B., Ledermann, J.A., et al., 2015. Radiofrequency ablation (RFA) combined with chemotherapy for unresectable colorectal liver metastases (CRC LM): long-term survival results of a randomized phase II study of the EORTC-NCRI CCSG-ALM intergroup 40004 (CLOCC). J. Clin. Oncol. 33 Suppl. abstr 3501.

Ryd, W., Hagmar, B., Eriksson, O., January 1983. Local tumour cell seeding by fine-needle aspiration biopsy. A semiquantitative study. Acta Pathol. Microbiol. Immunol. Scand. A 91 (1), 17–21.

Sarmiento, J.M., Heywood, G., Rubin, J., Ilstrup, D.M., Nagorney, D.M., Que, F.G., 2003. Surgical treatment of neuroendocrine metastases to the liver: a plea for resection to increase survival. J. Am. Coll. Surg. 197 (1), 29–37.

Sheiner, P.A., Brower, S.T., 1994. Treatment of metastatic cancer to the liver. Semin. Liver Dis. 14, 169–177.

Slotta, J.E., Schuld, J., Distler, S., Richter, S., Schilling, M.K., Kollmar, O., 2014. Hepatic resection of noncolorectal and non-neuroendocrine liver metastases-survival benefit for patients with gastrointestinal primary cancers-a case-controlled study. Int. J. Surg. 12, 163–168.

Stangl, R., Altendorf-Hofmann, A., Charnley, R.M., Scheele, J., 1994. Factors influencing the natural history of colorectal liver metastases. Lancet 343, 1405–1410.

Stark, D.D., Wittenberg, J., Butch, R.J., Ferrucci Jr., J.T., November 1987. Hepatic metastases: randomized, controlled comparison of detection with MR imaging and CT. Radiology 165 (2), 399–406.

Stavrou, G.A., Flemming, P., Oldhafer, K.J., 2006. Liver resection for metastasis due to malignant mesenchymal tumours. HPB 8 (2), 110–113.

Takada, Y., Otsuka, M., Seino, K., Taniguchi, H., Koike, N., Kawamoto, T., et al., 2001. Hepatic resection for metastastic tumors from noncolorectal carcinoma. Hepatogastroenterology 48 (37), 83–86.

Torre, L.A., Bray, F., Siegel, R.L., Ferlay, J., Lortet-Tieulent, J., Jemal, A., March 2015. Global cancer statistics, 2012. CA Cancer J. Clin. 65 (2), 87–108.

Treska, V., Liska, V., Skalicky, T., Sutnar, A., Treskova, I., Narsanska, A., et al., 2012. Non colorectal liver metastases: surgical treatment options. Hepatogastroenterology 59 (113), 245–248.

Uggeri, F., Ronchi, P.A., Goffredo, P., Garancini, M., Degrate, L., Nespoli, L., et al., 2015. Metastatic liver disease from non-colorectal, non-neuroendocrine, non-sarcoma cancers: a systematic review. World J. Surg. Oncol. 13, 191.

Uleminckx, K., Vackat, L., Marrel, M., Fiers, W., Van Roy, F.V., 1991. Genetic manipulation of E-cadherin expression by epithelial tumors cells reveals an invasion suppressor role. Cell 66, 107–119.

Van Hazel, G., Blavkell, A., Anderson, J., Price, D., Moroz, P., Bower, G., et al., 2004. Randomized phase II trial of SIR-spheres plus Fluorouracil/Leucovorin alone in advanced colorectal cancer. J. Surg. Oncol. 88, 78–85.

Vauthey, J.N., Zimmitti, G., Shindoh, J., August 2012. From Couinaud to molecular biology: the seven virtues of hepato-pancreato-biliary surgery. HPB (Oxford) 14 (8), 493–499.

Vladov, N., Vasilevski, I., Takorov, I., Mutafchiyski, V., Sergeev, S., Odisseva, E., et al., 2012. Rational surgical aggression in multimodal treatment of liver colorectal metastases. Hepatogastroenterology 59 (113), 241–244.

Wang, H., Mcintosh, A., Hasinoff, B.B., Rector, E.S., Ahmed, N., Nance, D.M., et al., 2000. B16 melanoma cell arrest in the mouse liver induces nitric oxide release and sinusoidal cytotoxicity: a natural hepatic defense against metastasis. Cancer Res. 60, 5862–5869.

Wernecke, K., Rummeny, E., Bongartz, G., Vassallo, P., Kivelitz, D., Wiesmann, W., et al., October 1991. Detection of hepatic masses in patients with carcinoma: comparative sensitivities of sonography, CT, and MR imaging. AJR Am. J. Roentgenol 157 (4), 731–739.

Wong, S., Mangu, P.B., Choti, M.A., Crocenzi, T.S., Dodd, G.D., Dorfman, G.S., et al., 2010. American Society of Clinical Oncology evidence review on radiofrequency ablation of hepatic metastases from colorectal cancer. J. Clin. Oncol 28 (3), 493–508.

Zech, C.J., Herrmann, K.A., Reiser, M.F., Schoenberg, S.O., 2007. MR imaging in patients with suspected liver metastases. Value of liver-specific contrast agent Gd-EOB-DTPA. Magn. Reson. Med. Sci. 6 (1), 43–52.

Zhuang, H., Sinha, P., Pourdehnad, M., Duarte, P.S., Yamamoto, A.J., Alavi, A., September 2000. The role of positron emission tomography with fluorine-18-deoxyglucose in identifying colorectal cancer metastases to liver. Nucl. Med. Commun. 21 (9), 793–798.

CHAPTER

# 16

# Pulmonary Metastasis

M. Gonzalez, T. Krueger, J.Y. Perentes
**Centre Hospitalier Universitaire Vaudois, Lausanne, Switzerland**

## 1. DEVELOPMENT OF PULMONARY METASTASES

Pulmonary metastases may potentially develop along the course of any malignant tumor (30% of prevalence) in patients diagnosed with cancer (Downey, 1999; Putnam, 2002; Harvey et al., 1994). Pulmonary metastases are thought to result from the dissemination of tumor cells from the primary tumor site to the lungs. The metastatic spread is a complex process, involving invasion of tumor cells through the basement membrane and stroma, intravasation into either the lymphatic system or blood circulation and survival in the circulation until they reach the metastatic site (Fidler, 2003). The cells may then form a premetastatic niche and will eventually form macrometastases. There are currently two major metastasis concepts that could influence the surgical approach of metastasis (Sleeman, 2012). The first theory suggests a late dissemination of metastasis from the primary tumor after the primary tumor and surrounding environment have acquired sufficient genetic changes for migration and distant implementation function acquisition (Fidler, 2003). A second theory suggests that dissemination occurs very early in the cancer process, sometimes at the premalignant stage, and that single dormant cells can but will not necessarily develop into proper metastasis following genetic and environmental alterations (Morgan et al., 2009). These theories could occur separately in different tumor types/individuals or may occur together in a same patient.

## 2. RATIONALE FOR LOCAL TREATMENT OF PULMONARY METASTASES

The management of patients who develop lung metastases is generally palliative as other organs are also invaded in general. Treatments include chemotherapy and life quality supportive treatments. Pulmonary metastasectomy refers to the surgical excision of metastasis to the lung originating from an extrathoracic location. The resection of pulmonary metastases may be indicated for palliation in highly selected patients presenting, for example, uncontrollable pain associated chest wall invasion by the metastasis, massive hemoptysis, or retention pneumonia due to centrally located metastases. However, there is a group of patients with pulmonary metastases who may benefit from lung metastasis resection with a so-called curative intent.

*Introduction to Cancer Metastasis*
http://dx.doi.org/10.1016/B978-0-12-804003-4.00016-5

Surgical resection of pulmonary metastases is nowadays widely performed in selected metastatic patients, although there are no prospective randomized studies comparing pulmonary metastasectomy with chemotherapy or observation alone (Treasure, 2007). Several retrospective studies have suggested an increased survival for patients who underwent complete resection of lung metastases in comparison to retrospective series of patients who did not benefit from surgery. The International Registry of Lung Metastases reviewed retrospectively 5206 patients from 18 centers (United States, Canada, and Europe; The International Registry of Lung Metastases, 1997). Overall operative mortality was only 1.3% demonstrating that pulmonary metastasectomy was safe. Complete resection could be achieved in 88% of patients and was considered as an important prognostic factor for survival: the 5-year overall survival was of 36% in patients with complete resection compared to 13% in patients with incomplete resection. These results supported that pulmonary metastasectomy offered a survival advantage when complete resection was achievable. In addition, other prognoses of survival factors were identified in this study: (1) a tumor recurrence free interval of >36 months, (2) single pulmonary metastasis, and (3) the histological type of the primary tumor. In addition, several retrospective single-center studies have supported the role of pulmonary metastasectomy in different tumors types with a 5-year survival rate ranging from 20% to 80% depending on the primary tumor type (Tables 16.1 and 16.2; Casiraghi et al., 2011; Robert et al., 1997; Younes et al., 2012). Nonetheless, it is currently not possible to identify patients who may benefit mostly from this surgical strategy.

The improvement of surgical techniques [video assisted thoracoscopic surgery (VATS)], radiological imaging [thin slices helical computed tomography (CT) scan], more frequent use of positron emission tomography (PET) CT, and advances with new chemotherapeutic agents have contributed to increase the number of pulmonary metastasectomy procedures and enhance the survival of metastatic patients (Erhunmwunsee & D'Amico, 2009; Pastorino et al., 2003; Perentes et al., 2015). The following are the established criteria for lung metastasectomy (Erhunmwunsee and D'Amico, 2009; Pastorino, 1997): (1) a controlled or controllable primary tumor with no evidence of active disease; (2) the absence of extrathoracic metastatic disease. However, involvement of other extrathoracic sites may not be a contraindication to the resection of the pulmonary disease if all metastatic sites can be resected completely before the lung resection; (3) complete resection of lung metastasis must be achievable; (4) the patient has sufficient pulmonary and cardiovascular functional reserves to tolerate pulmonary resection;

TABLE 16.1 Five-Year Survival After Pulmonary Metastasectomy According to Histological Type

| Primary Tumor | Five-Year Survival (%) |
|---|---|
| Colorectal carcinoma | 24–68 |
| Soft tissue sarcoma | 21–52 |
| Osteosarcoma | 7–65 |
| Renal cell carcinoma | 42–74 |
| Nonseminomatous germ-cell tumor | 79–94 |
| Melanoma | 21–35 |
| Mammary carcinoma | 36–51 |
| Head and neck | 26–59 |

TABLE 16.2 Indication for a Pulmonary Metastasectomy

1. The primary site of cancer is controlled or controllable
2. Absence of other extrathoracic metastases (with exception for metastases that are potentially resectable)
3. Resection of pulmonary metastases can be complete
4. Sufficient pulmonary reserves to tolerate pulmonary resection
5. No valid alternative therapy

TABLE 16.3 Summary of the Different Approaches

| Approach | Advantages | Disadvantages |
|---|---|---|
| Thoracotomy | Good unilateral exposure | Painful, second intervention |
| | Bimanual palpation | Bilateral lesion |
| Sternotomy | Bilateral exposure | Poor exposure to lower lobe |
| | Less painful | |
| Clamshell incision | Bilateral exposure | Painful, sacrifice of both internal mammary arteries |
| Video-assisted thoracoscopic surgery | Less painful | No bimanual palpation, detection of centrally located lesion |
| | Low morbidity | |
| | Faster recovery | |
| | Less immune response | |
| | Repetitive surgery | |
| Radiofrequency | Less invasive | General anesthesia |
| | | Morbidity |
| | Good results for lesion less than 3cm | Local control |
| | | Peripheral lesions |
| Stereotaxic radiotherapy | Less invasive | Local control |
| | No general anesthesia | Peripheral lesions |
| | | Repeat procedure |

(5) there is no valid alternative therapy. Despite this aggressive strategy, a series report up to 50% of patients with local or distant metastatic relapse (The International Registry of Lung Metastases, 1997).

The surgical metastasectomy approach is compatible with a late dissemination process if the five criteria described above are fulfilled. However, in case of an early dissemination process theory, surgical metastasectomy may not be able to achieve complete metastatic resection as premetastatic niches are too small to be palpated. Therefore, a resection approach with surveillance and possibility to redo resection should be favored in this case (Perentes et al., 2015).

## 3. DIAGNOSIS OF PULMONARY METASTASES

The probability of a new lung lesion being a metastatic is directly related to the original primary tumor. In situations of sarcoma or melanoma primary tumors, the likelihood of a new lung nodule being metastasis is approximately 80% (Downey, 1999; Putnam, 2002; Harvey et al., 1994). If the primary tumor is urothelial or colorectal, the likelihood of a new lung nodule being metastatic is 50%. Interestingly, in situations of head and neck cancer, a new pulmonary nodule is twice as likely to being a primary tumor rather than a metastatic tumor (Rusch, 1995). The latter is mostly explained by the exposure to tobacco. For these reasons, the histological analysis of pulmonary nodules should be performed preoperatively or intraoperatively to determine the surgical strategy of resection. Fine needle aspiration biopsy can sometimes be a good alternative; however, this approach does not always yield enough material to differentiate primary lung cancer from lung metastasis of colorectal or breast cancer. In centrally located lesion, bronchoscopy may be useful to obtain biopsy to confirm metastatic process.

## 4. PREOPERATIVE EXAMS

### 4.1 Control of the Primary Tumor

Surgical resection of lung metastases may be considered only if the site of the primary disease is controlled or controllable. Surgery for lung metastases should not be the treatment of choice if the primary tumor is considered as unresectable. Staging of the primary site depends on the underlying tumor and the disease-free interval. Local tumor recurrence must always be excluded within the 6 weeks preceeding pulmonary metastasectomy. Radiological investigations should include magnetic resonance imaging (MRI) or CT scan of the primary tumor region. In colorectal cancer patients, a recent colonoscopy will be indicated to exclude local recurrence. In cases of primary tumor and metastasis diagnosed synchronously, a combined approach can be applied (Erhunmwunsee and D'Amico, 2009).

### 4.2 Exclusion of Extrathoracic Metastases

Extrathoracic metastatic disease should be ruled out before the pulmonary resection, depending on the primary tumor. In colorectal cancer patients, abdominal CT scan should be performed to exclude hepatic metastases or peritoneal carcinomatosis. PET scan is nowadays currently performed to assess the extrathoracic metastatic disease in patients with sarcoma, epithelial tumors (breast, colon, lung), and melanomas offering a better sensitivity than the CT scan alone (Pastorino et al., 2003; Dalrymple-Hay et al., 2002; Fortes et al., 2008). PET scan presents lower sensitivity than chest CT scan for the detection of pulmonary nodules smaller than 10 mm in diameter and should not be recommended for the detection of additional pulmonary lesions (Detterbeck et al., 2010). However, PET scan may be useful to determine hilar or mediastinal lymph node involvement. Lymph node invasion, which is found in approximately 15–20% of cases, represents a factor of poor prognosis (Garcia-Yuste et al., 2010b). If PET scan is positive for mediastinal lymph node involvement, some authors recommend preoperative mediastinal staging with endoscopic bronchial ultrasonography (EBUS) or mediastinoscopy to histologically assess the mediastinum and potentially exclude patients from subsequent pulmonary metastasectomy (Garcia-Yuste et al., 2010b). Finally, in cases of neurological symptoms, and in some tumor types with cerebral dissemination (e.g., melanoma), a cerebral MRI is indicated to exclude the presence of brain metastases (Erhunmwunsee and D'Amico, 2009).

### 4.3 Complete Resection is Feasible

Preoperative pulmonary radiologic exams are crucial to determine whether the patient can benefit from a complete resection of lung metastases and to plan the type of procedure (Detterbeck et al., 2010). Metastases may have various radiological appearances, but generally are round, well delineated, and peripherally located in the lower lobes due to a better vascularization of this part of the lung. The single breath thin-slice (1 mm) chest CT scan with injection of contrast product is the standard for the evaluation of pulmonary nodules and invasion of adjacent structures (great vessels, chest wall, spinal column, and esophagus). The examination is recommended to be performed within 4 weeks of the pulmonary metastasectomy (Detterbeck et al., 2010). Nevertheless, despite the progress of radiological imaging with thin-slice multidetector row CT scans, the ability to detect all metastatic nodules remains controversial and many surgeons still recommend bimanual palpation of the lung to detect metastases that are not visible on the preoperative CT scan (Margaritora et al., 2002; Kayton et al., 2006). Various studies with old generation CT scans (>5 mm thin-slice) have shown that 16–46% of pulmonary lesions discovered during the bimanual palpation of the lung were not identified on preoperative chest CT scan (McCormack et al., 1996; Cerfolio et al., 2009; Nakas et al., 2009). Nowadays, spiral CT scan allows an analysis of the entire lung volume during a single breath-hold thus improving the accuracy of lung nodule detection by eliminating both the respiratory motion artifact and volume averaging. This technique has shown a nodule detection sensitivity of 82% versus 75% for high-resolution CT scan (Margaritora et al., 2002). More recent thin-slice 1 mm CT scans were shown to have an even better spatial resolution and are predicted to achieve metastasis detection comparable to a bimanual palpation. In a recent series of 27 patients, thin sliced (1 mm) chest CT scan detection of lung metastasis was challenged by bimanual palpation of the lung by thoracotomy (Kang et al., 2008). The sensitivity and the negative predictive value were 97% and 96%, respectively, in patients with nonsarcomatous tumors. CT scan is also able to detect small nodules of less than 5 mm, which, in turn, increase the rate of false positive intrapulmonary nodules. Interestingly, Nakajima et al. (2007) reported that more than 40% of pulmonary nodules smaller than 5 mm in diameter (detected by chest CT scan) in metastatic patients were not pulmonary metastases. Eckardt and Licht (2012) reported similar results in 2012 in a prospective sequentially controlled study enrolling 37 patients eligible for pulmonary metastasectomy by VATS undergoing first VATS resection by one team, immediately followed by thoracotomy by another team. They found 29 additional nodules during thoracotomy, but 76% of them were benign. In addition, Ludwig et al. in 2008, compared the number of pulmonary metastases in 276 patients suspected of pulmonary metastases detected in the preoperative CT scan and during the thoracotomy. The number of lesions found and removed during the operation was higher in 39% than visualized on the preoperative CT scan. For patients with a single nodule on preoperative CT, only 7% presented more lesions during thoracotomy of which 84% were malignant. Overall, this suggests that 95% of patients with solitary metastasis on chest CT scan undergo a useless bimanual palpation by thoracotomy according to the current guidelines (Perentes et al., 2015). Altogether, this suggests that the radiologic assessment of the lung with modern CT scans has a good sensitivity in detecting metastatic lesions. This probably explains a shift in the surgical approach from thoracotomy or more minimal invasive techniques such as thoracoscopy.

## 4.4 Can the Patient Tolerate Lung Resection?

Complete pulmonary function and a cardiac evaluation should be performed prior to any kind of management of pulmonary metastases (Brunelli et al., 2009). Postoperative pulmonary reserve of the patient must be estimated before pulmonary resection considering the number and the location of lung metastases. Peripheral lesions are recommended to be resected by atypical resection (wedge resection; Erhunmwunsee and D'Amico, 2009). But, centrally located metastases may sometimes require more extended resection with anatomical resections (segmentectomy, lobectomy, or pneumonectomy; The International Registry of Lung Metastases, 1997). In case of bilateral involvement with multiple metastases, the estimation of postoperative pulmonary function may be particularly difficult, if lesions are centrally located and multiple anatomical resections are required. In this situation, bilateral access such as sternotomy may be necessary to palpate both lungs to assess the resectability of bilateral pulmonary lesions.

## 5. SURGICAL RESECTION

### 5.1 Extent of Surgical Resection

The main goal of pulmonary metastasectomy is the complete resection of the metastases while sparing pulmonary parenchyma. Most of the pulmonary metastatic lesions are generally located at the periphery of the lung and easily accessible to atypical resection. In contrast to primary lung cancer that requires anatomical resection, atypical resection (wedge resection) is as effective as anatomical resection to manage metastasis to the lung parenchyma (Erhunmwunsee and D'Amico, 2009). In this situation, wedge resection is generally performed using standard stapler, which achieves clear surgical margins and efficient aerostasis. Others techniques are proposed as an alternative to staplers for multiple lesions to spare lung parenchyma (Venuta et al., 2010). Cautery resection (precision resection) with subsequent individual ligation of small vessels and bronchi is ideal for deeply located or peripheral lesion in patients who cannot tolerate anatomical resection (Cooper et al., 1986). Precise resection can also be achieved with the use of lasers (Rolle et al., 2006a,b). This is usually performed by thoracotomy and allows a superficial excision limited without deformation of the pulmonary parenchyma. This technique has been reported to achieve complete resection for lesions located close to major bronchi and/or vessels. The 1318nm (neodymium-doped yttrium aluminium garnet) Nd:YAG laser is the only laser device capable of dissecting lung parenchyma due to its high absorption of water. Some authors recommend that parenchymal defect after laser resection should not be sutured avoiding pulmonary distortion without increase of air leak (Kirschbaum et al., 2014). Other surgical ablative techniques such as ligasure vessel sealing system (Shigemura et al., 2004) or ultracision (Eichfeld et al., 2000) have been proposed for atypical resection, but data are lacking in thoracic surgery on lung parenchyma with these techniques.

For large or central lesions, segmentectomy, lobectomy, or occasionally pneumonectomy may be required (Migliore et al., 2010; Putnam et al., 1993). Within the International Registry of Lung Metastases, pulmonary metastasectomy was achieved with wedge resection in 67%, segmentectomy in 9%, lobectomy in 21%, and pneumonectomy in 3% (The International Registry of Lung Metastases, 1997). Some authors consider pneumonectomy as a relative contraindication to pulmonary metastasectomy and should be reserved for solitary centrally located tumor with a long tumor-free interval without previous pulmonary resection. Recently, Casiraghi et al. (2015) reported their experience with extended pulmonary metastasectomy in 29 patients: three resections of the chest wall, one azygos, one diaphragm, four vascular resections/reconstructions, six sleeve resections, and 14 pneumonectomies. Complete resection was obtained in all patients. Thirty-day operative morbidity and mortality rates were 38% and 0%, respectively. Overall survival after a complete extended metastasectomy was 66% at 2years, 42% at 5years suggesting that resection extended to other structures (chest wall, great vessels) and may be justified in selected patients if complete resection can be achieved.

### 5.2 Surgical Approach

Surgical approach is determined by the number of metastases, their location, the functional reserves of the patient, and whether or not the disease is limited to one side of the thorax. Surgical approach should obtain complete resection while saving as much pulmonary parenchyma as possible (Kaifi et al., 2010). The optimal approach remains controversial and different approaches are proposed each with their advantages and disadvantages: thoracotomy (anterior or posterolateral), median sternotomy, bilateral trans-sternal thoracotomy (Clamshell incision), and thoracoscopy (Table 16.3; Erhunmwunsee and D'Amico, 2009; Kaifi et al., 2010; Davidson et al., 2001).

The most commonly used approach is the thoracotomy (anterolateral or posterolateral), which allows the opportunity to perform bimanual palpation of the entire lung. Postoperative pain and the inability to reach contralateral lung are the major disadvantages. Systematic one-stage bilateral thoracotomies provide no survival advantage when unilateral disease is evident on radiological findings (Younes et al., 2002). Staged thoracotomies are recommended in patients with bilateral disease. Median sternotomy may expose both lungs allowing bilateral resection (Roth et al., 1986). Postoperative pain is generally less important than bilateral thoracotomy. However, access to the lower lobes, lymph node dissection, and anatomical resections can be more difficult than with a standard thoracotomy. Clamshell incision (bilateral anterior thoracotomy with transverse sternotomy) allows excellent visualization and palpation of the lung, but involves a sacrifice of two mammary arteries and important postoperative pain (Erhunmwunsee and D'Amico, 2009).

VATS has gained progressive interest for pulmonary metastasectomy (Dong et al., 2014). VATS may present some important advantages for the management of isolated pulmonary lesions: smaller incisions, better visualization of the pleural cavity, less postoperative pain, less surgical morbidity, shorter length of hospitalization, fewer adhesion in cases of redo metastasectomy, shorter interval between surgery and adjuvant therapy with better treatment compliance (Sonett, 1999). In addition, most pulmonary metastases are generally located in the periphery of the lung and can be easily resected by VATS. However, these advantages stay in conflict with the inability to practice a bimanual palpation of the entire lung to detect nodules that are not visible on preoperative radiological examinations. Another potential concern was whether or not a safe resection margin could be adequately obtained by VATS and the risk of local recurrence at the resection or port site (Downey et al., 1996; Mutsaerts et al., 2001). Even if there is no prospective randomized studies comparing thoracoscopic approach to open bimanual thoracotomy, several surgical series have demonstrated comparable survivals between the two approaches for patients with less than three lesions (Nakas et al., 2009; Mutsaerts et al., 2002; Nakajima et al., 2008; Gossot et al., 2009; Carballo et al., 2009; Chao et al., 2012). Undetectable lesions by VATS cannot be assimilated to incomplete resection, which was determined as a strong factor of poor prognosis (Perentes et al., 2015; Naunheim, 2007). These missed nodules may be resected later once they become detectable on a follow-up CT scan. Multiple repetitive resections of metachronous pulmonary metastases do not seem to adversely affect the overall outcome because recurrence is frequent (>50%; The International Registry of Lung Metastases, 1997) and many curable patients benefit from repeated resection, and repeated thoracoscopies are better tolerated than repeated thoracotomies.

Finally, VATS resection is less traumatic and significantly decreases the immune response with a possible effect on the progression of the disease. Thus, the aim of this minimally invasive approach is to obtain complete resection, fast recover and early return to adjuvant therapy (Perentes et al., 2015). Local recurrence has been reported during the initial experience of VATS approach both in the original resection and thoracoscopic port sites (Downey et al., 1996). These complications have been mainly published before the years 2000 and multiple improvements for tumor removal (development of extraction bags) have since been developed. In general, it is important to secure free surgical margin by macroscopic examination of the stapler line in relation to the metastasis and, in case of doubt, by confirming the completeness of resection by frozen section.

Several adjunctive procedures have been suggested to help in the localization of nodules during VATS approach. Needle localization, methylene blue injection, and sonographic evaluation have been used to identify nodules not easily palpable on the visceral pleural surface (Zaman et al., 2012). However, these maneuvers may help for the resection of solitary radiological detectable lesion and will not allow for detection of lesion less than 5mm. Nowadays, the VATS approach has gained acceptance in the surgical community.

In conclusion, the optimal approach for pulmonary metastasectomy continues to evolve and the patient must be informed of the different surgical options with their advantages and limitations. Given continued advancement in both imaging and operative technology, the role of VATS is expected to grow (Perentes et al., 2015).

## 6. RADIOLOGICAL ABLATION

Alternative therapies for patients that cannot undergo surgery are currently under investigation. Radiological ablation under guidance therapies has been proposed using three different techniques: radiofrequency (King et al., 2004), microwave (Simon et al., 2005), and cryoablation (Beland et al., 2005). These different techniques that are typically performed using computed tomography guidance. Radiofrequency ablation (RFA) is most frequently used causing focal coagulation necrosis in tissue (Hiraki et al., 2014). RFA is now commonly used in both primary and metastatic lung cancer in nonoperable patients. RFA delivers high frequency alternating electric current through a needle electrode. This causes local ionic oscillation and tissue heating. This thermal injury causes cell death. The electrode is introduced percutaneously into the tumor similar to percutaneous lung biopsy (Hiraki et al., 2014). The most common complication is pneumothorax, which occurs in up to 50% of procedures requiring chest tube placement for in approximately 25% of patients. Other severe complications, such as pleural effusion, hemoptysis, hemorrhage, or bronchopleural fistula, have been described (King et al., 2004). The local efficacy depends on tumor size with better results reported for tumors less than 3cm (de Baere et al., 2015). Local progression after RFA is not rare, occurring in 10% or more of patients. Repeat RFA may be proposed in case of local recurrence. Short- to midterm survival after RFA appears promising and is approximately 85–95% at 1year and 45–55% at 3-years for metastatic colorectal cancer patients (Hiraki et al., 2014; de Baere et al., 2015; Petre et al., 2013).

Schlijper et al. (2014) performed a systematic review for ablative treatment of pulmonary metastases in colorectal cancer patients. They could include only four RFA series of more than 50 patients with a follow-up period of more than 24months. In comparison with surgical series, RFA ablation presented similar mortality, but morbidity rates were more important in RFA series (42% vs. 1–14% for surgery). For RFA the 5-year survival rate ranged from 34.9% to 45%, whereas the 5-year survival rate reported for surgery ranged from 29% to 71.2%. Survival outcomes were influenced by the size of the intrapulmonary lesion, especially in patients treated with RFA. Tumors larger than 3cm give significantly shorter long-term survival and local control rates than lesions smaller than 3cm. Recently, de Baere et al. (2015) reported a large experience with different type of tumor from two centers with 566 patients over an 8-year period. Median overall survival was 62months. Four-year local efficacy was 89%. In patients treated for colorectal cancer metastases, size of >2cm (hazard ratio (HR) 2.1) and number of metastases >3 (HR 1.86) were significantly associated with poorer survival.

## 7. RADIOTHERAPY

Hypofractionated stereotactic body radiotherapy (SBRT) is an emerging noninvasive technique for the treatment of pulmonary metastases (Siva et al., 2010). There is currently no consensus on a standard approach for the stereotactic radiotherapy of pulmonary metastases regarding treatment techniques and dose fractionation schemes. However, the application of small numbers of large doses of radiation should, theoretically, obtain better local control. The biologically effective dose at the isocenter of the pulmonary metastasis is recommended to be more than 100 Gy to achieve local control (Siva et al., 2010).

Stereotactic ablative radiation therapy has, however, been shown to achieve satisfactory local control and promising survival rates for lung metastases, with 2-year local control reaching up to 89% (Takeda et al., 2011; Ricardi et al., 2012; Norihisa et al., 2008). Recently, Filippi et al. (2015) reported 40 colorectal cancer patients who received stereotactic ablative radiation therapy. The median delivered biological effective dose was 96 Gy, in one to eight daily fractions. Median follow-up time was 20 months (range, 3–72 months). Overall survival rates at 1, 2, and 5 years were, respectively, 84%, 73%, and 39%, and median overall survival was 46 months. Progression occurred in 25 patients (62.5%), at a median interval of 8 months; failure at stereotactic ablative body radiotherapy (SABR) site was observed in three patients (7.5%). Progression-free survival rates were 49% and 27% at 1 and 2 years, respectively.

Toxicity rates are low after SABR for pulmonary metastases, with published series showing consistently less than a 5% incidence of grade 3+ toxicities (or those requiring hospitalization; Siva et al., 2010).

Efficacy of SBRT may be difficult to assess in terms of tumor response. Large doses per fraction used in SBRT may cause segmental atelectasis or focal fibrosis, which can confound response assessment with conventional CT.

The safety and effectiveness of SBRT is promising with good compliance and fewer toxicity. However, the optimal dose and number of fraction are not standardized. Long-term results are still lacking and this technique should be proposed under strict clinical studies or in nonoperable patients.

## 8. LYMPH NODE INVOLVEMENT

Mediastinal lymph node dissection or sampling is not routinely performed during lung metastasectomy in comparison with primary lung cancer. The probability of survival seems considerably lower if lymph node involvement is present. An (European Society of Thoracic Surgery) ESTS survey reported that 56% of surgeons performed mediastinal sampling during lung metastasectomy, 13% performed complete dissection, and 32% did not perform lymph node biopsy (Internullo et al., 2008). Lymph node may be involved in two distinct scenarios: (1) lymph node involvement discovered on the preoperative imaging or invasive mediastinal staging, and (2) incidental lymph node involvement discovered during surgery after mediastinal sampling or complete dissection. The sensitivity of PET scan for detecting mediastinal lymph node metastases has been reported to be between 35% and 100% (Pastorino et al., 2003; Hamaji et al., 2012). Given this unclear sensitivity, neither CT nor PET seems to be an appropriate screening tool for mediastinal lymph nodes (LNs), and authors have suggested that at the time of pulmonary metastasectomy, thoracic and mediastinal lymph nodes dissection (LND) are the best diagnostic procedures to evaluate mediastinal LN metastasis (Hamaji et al., 2012; Seebacher et al., 2015). The incidence of mediastinal LN metastases with pulmonary metastases ranges between 12% and 32% (Hamaji et al., 2012; Seebacher et al., 2015; Pfannschmidt et al., 2006a; Ercan et al., 2004; Veronesi et al., 2007; Szoke et al., 2010). Systematic mediastinal lymph node dissection has been reported to have a significative difference in survival between patients with positive lymph node in comparison with negative lymph node. Pfannschmidt reported 245 patients who underwent systematic lymph node dissection during pulmonary metastasectomy and founded hilar or mediastinal lymph node involvement in 32% (Pfannschmidt et al., 2006a). Median survival was 64 months for patients without nodal metastases and 33 and 21 months for patients with hilar and mediastinal metastases, respectively.

Recently, Hamaji et al. (2012) reported retrospectively 518 patients operated for lung metastasectomy of colorectal cancer. One hundred ninety-nine patients had no lymphadenectomy, 279 patients underwent mediastinal LND that was negative, and 40 patients underwent LND that contained one or more positive LNs for metastases. Median survivals for the no LND group, negative LND group, and positive LND group were 52, 58.5, and 34 months, respectively.

The incidence of positive lymph node may be different according to different primary tumors. Seebacher et al. (2015) reported recently 256 procedures of pulmonary metastasectomy with lymph node sampling or complete dissection. Interestingly, unexpected LN involvement was found in 35% with breast cancer, in 9.2% with colorectal cancer, and in 20.8% with renal cell carcinoma. In addition, they did not find statistical difference on a 5-year survival between LN sampling and radical removal. Central localization of pulmonary metastases may also be associated with increased rate of lymph node involvement.

The complete mediastinal lymphadenectomy may prolong operative time, but associated morbidity is very rare (Allen et al., 2006). Although survival benefit after lymph node dissection is unclear, most authors consider that complete mediastinal lymphadenectomy or sampling should be recommended during pulmonary metastasectomy to achieve accurate staging and guide additional chemotherapy. When chemotherapy is less effective, a radical removal of lymph node may be of importance to achieve maximal local control. For example, metastatic lymph node involvement in renal cancer has been reported up to 46% of patients (Renaud et al., 2013). Recently, Kudelin et al. (2013) reported a median overall survival of 72 months for pN0 patients, 51 months for pN1, and 36 months for pN2 patients without significative difference ($p = .2$). When mediastinal lymph node involvement is suspected preoperatively by chest CT scan or PET-CT, mediastinal investigation should be realized including mediastinoscopy, EBUS, or EUS (endoscopic ultrasound). Pulmonary metastasectomy remains controversial for preoperative histological proven mediastinal involvement and some authors propose to exclude patients with positive mediastinal lymph node from pulmonary metastasectomy (Gonzalez and Gervaz, 2015).

# 9. RESECTION OF RECURRENT METASTASES

Recurrence of pulmonary metastasis is frequent and may occur in up to 50% of patients (The International Registry of Lung Metastases, 1997). The International Registry of Lung Metastases reported recurrence in 53% and patients who underwent a second metastasectomy had a 44% survival at 5 years. A longer disease-free interval between the first metastasectomy and recurrent pulmonary metastases is associated with favorable prognosis. Surgical resection of recurrent pulmonary metastases has been studied particularly colorectal cancer with a 5-year survival ranging from 29% to 85% (Kim et al., 2008; Welter et al., 2007; Watanabe et al., 2009). Recently, Salah et al. (2013) reported in a pooled analysis 148 colorectal cancer patients operating from repeat pulmonary metastasectomy. The 5-year survival rate was 52% for patients who first had metastasectomy and comparable to 57.9% from the second metastasectomy for patients who had repeated resection.

More than two metastatic pulmonary nodules and maximum diameter of largest pulmonary nodule ≥3cm were the only independent factors associated with inferior survival following repeated pulmonary resection. These results suggest that lesions that are not detected by preoperative CT scan may be resected later without significant impact on overall survival. Even if palpation at the time of metastasectomy was performed, regular follow-up is recommended to detect rapid recurrence and aggressive strategy. Closed clinical or radiological follow-up is recommended with chest CT scan in the postoperative period, repeated every 6 months for the first 2 years, then yearly. If the lung has not been palpated or the tumor doubling time is short, a more frequent radiological monitoring should be discussed (Detterbeck et al., 2010).

## 10. LUNG METASTASIS IN GIVEN TUMOR TYPES

### 10.1 Colorectal Cancer

Colorectal cancer is the third most common malignancy with an estimated incidence of 1.2 million new cases each year. At least 50% of colorectal cancer (CRC) patients will develop metastatic disease, which is localized in the lung in 5–15% of them (Gonzalez et al., 2013). Twenty-five percent of patients present initially with synchronous metastases, and 25% patients subsequently develop metachronous metastases. Pulmonary metastasectomy is therefore a treatment option in 1–2% of all patients with colorectal carcinoma. Pulmonary metastases may potentially develop along different scenarios: (1) synchronous pulmonary metastases; (2) after previous hepatic metastases; (3) metachronous development of isolated lung metastases; and (4) after previous lung metastasectomy (Gonzalez et al., 2012). Multiple retrospective studies have investigated the outcome of colorectal cancer patients who underwent resection of lung metastasis with a 5-year survival varying from 24% to 68%, and with a median survival of 18–67 months (Gonzalez et al., 2012, 2013; Pfannschmidt et al., 2010a, 2007). Despite many surgical series reports, it is difficult to determine which patient will clearly benefit from lung metastasectomy. In 2007, Pfannschmidt et al. published an extensive systematic review on pulmonary metastases of CRC. Twenty series were collected over a 10-year period for a total of >1500 patients (Pfannschmidt et al., 2007). Operative mortality after lung metastasectomy was <3%. Complete resection was a factor of better survival with anecdotic survival after incomplete resection. A recent meta-analysis including 25 series for a total of 2925 patients permitted to identify four important prognostic of poorer survival (Gonzalez et al., 2013): (1) short disease-free interval between colonic tumor and lung metastasis occurrence, (2) elevated prethoracotomy carcinoembryonic antigen (CEA) level, (3) metastatic of hilar or mediastinal lymph node, and (4) multiple pulmonary metastases. In addition, a pooled analysis of individual data of patients of eight previously published studies reported a 5-year survival rate of 54% (Salah et al., 2012). Three independent factors of poor survival were identified: (1) prethoracotomy CEA level >5 ng/mL; (2) number of lung metastases >1; and (3) disease-free interval <36 months between primary tumor resection and the appearance of lung metastases. The prognosis was correlated with the presence of these preoperative risk factors. Five-year survival rate was >45% for intermediate and low-risk patients, and thoracic surgery should be restricted to this group of patients.

Patients presenting with both hepatic and pulmonary metastases have a 5-year survival rate after lung metastasectomy ranging from 11% to 61% (Gonzalez et al., 2012). A recent pooled analysis of patients operated for both hepatic and pulmonary metastases reported a 5-year survival rate of 54%, well comparable with patients who underwent only pulmonary metastasectomy (Salah et al., 2015). None of these factors can be considered as an absolute contraindication to pulmonary resection. Recurrence rate is frequent after pulmonary metastasectomy in approximately 50% of the patients. Repeated pulmonary resection has also been largely investigated with promising results with survival at 5 years ranging between 29% and 85% (Gonzalez et al., 2012). Salah et al. (2013) observed that 20% of patients will benefit of repeated resection, with a 5-year survival rate of 58% comparable after a first metastasectomy. These results highlight the importance of follow-up and aggressive management of recurrence, when it is possible.

The recent development of new chemotherapeutic agents in colorectal patient allowed a prolonged progression-free survival and overall survival reaching up 25% of survival at 5 years. The exact role of pulmonary metastasectomy is currently questioned in colorectal patients. A randomized controlled trial comparing surgery with best medical treatment (Pulmonary Metastasectomy in Colorectal Cancer) is currently conducted in the United Kingdom to determine the real utility of surgery (Treasure et al., 2010).

## 10.2 Soft Tissue Sarcoma

Sarcomas of the bone and soft tissue are heterogeneous malignancies. Twenty to fifty percent of patients with soft tissue sarcoma will develop pulmonary metastases. The lung is generally the unique metastatic site of distant sarcoma and remains the primary cause of death. This is why its management is crucial in these patients. As systemic chemotherapy remains limited with poor response rate of less than 25%, pulmonary metastasectomy has been proposed as long as complete surgical resection is feasible despite the lack of randomized studies to support this aggressive management. For patients with disease limited to the lungs, pulmonary metastasectomy can result in significant long-term survival in comparison to historical controls with survival rate at 5 years varying from 29% to 52% (Kim et al., 2011; Porter et al., 2004; Pfannschmidt et al., 2006b; Canter et al., 2007; Predina et al., 2011; Schur et al., 2014; Mizuno et al., 2013; Dossett et al., 2015; Okiror et al., 2016). The complete resection is the principal prognostic factor of better survival. Other risk factors of poorer prognosis have also been described: (1) high grade tumor, (2) tumor of >5 cm, (3) multiple metastases, (4) a bilateral disease, and (5) a short disease-free interval. Due to retrospective study designs and the fact that chemotherapy regimens as well as surgical approaches are highly individualized, it is difficult to establish the exact role of perioperative chemotherapy. Repeat surgery in case of recurrence is also considered to be an important treatment to obtain longer survival rate. Several studies have suggested that re-recurrent patients who undergo redo pulmonary metastasectomy have a survival advantage compared with those who undergo single pulmonary metastasectomy (Kim et al., 2011; Chen et al., 2009; Blackmon et al., 2009).

## 10.3 Osteosarcoma

Metastatic disease is frequent during osteosarcoma with 10–20% at initial diagnosis and 30–40% within the 18 months after diagnosis (Salah and Toubasi, 2015). Pulmonary metastasectomy may be attractive for these patients because metastases of osteosarcoma get preferentially to the lung. Treatment of metastatic osteosarcoma generally includes the combination of chemotherapy and surgery. If lung metastases are synchronous, chemotherapy is started and lung metastasectomy is performed after resection of the primary tumor. Postoperatively, chemotherapy may be added to increase the control rate. However, some studies do not show relevant survival with adjuvant chemotherapy after complete metastatic resection especially with patients presenting less than three metastases with long disease-free survival (Ferrari et al., 2003; Bacci et al., 2005). The 5-year survival rate varies from 7% to 65% (Salah and Toubasi, 2015; Briccoli et al., 2010;

Pfannschmidt et al., 2006c; Harting and Blakely, 2006). Completeness of resection, long disease-free survival interval, and the number of metastases have been described as prognostic factors for better survival.

The effect of histological subtype on survival remains unclear following PM (pulmonary metastases) in osteosarcoma. Recently, Salah and Toubasi (2015) reported worse survival with chondroblastic subtypes. Repeat pulmonary metastasectomy for osteosarcoma should be considered as a valid treatment and remissions or possible cure may be obtained even in patients who experience two or more lung relapses (Bacci et al., 2005; Briccoli et al., 2005).

## 10.4 Renal Cell Carcinoma

Approximately 30% of renal cancer patients may present metastases at the time of diagnosis, and another 25% patients develop metachronous metastases (Renaud et al., 2014). Several studies have shown benefit on survival after pulmonary metastasectomies in case of metastatic renal carcinoma. Five-year survival ranges between 42% and 74% after complete resection in comparison with 8% and 22% in the case of incomplete resection (Kudelin et al., 2013; Renaud et al., 2014; Pfannschmidt et al., 2002; Alt et al., 2011; Assouad et al., 2007). Hilar or mediastinal lymph nodes invasion was found as a poor prognosis with survival ranging from 0% to 38%. Chemotherapy and radiotherapy are not efficient in metastatic renal cancer with a 5-year survival rate of less than 11% in nonoperated patients. Surgery with complete resection and systematic mediastinal lymph node dissection seems to offer better locoregional control in these patients.

## 10.5 Melanoma

After primary tumor treatment, more than 30% of patients with a malignant melanoma will develop lung metastases usually associated with a poor prognosis. Despite recent progresses with systemic therapies, such as cytokines, antibodies, or vaccination, 5-year survival remains below 20% in metastatic melanoma patients. When confined to the lung, pulmonary metastasectomy seems to offer a benefit on survival with a 5-year survival ranging from 21% to 35%, as opposed to less than 5% for nonsurgical patients (Petersen et al., 2007; Younes et al., 2013; Oliaro et al., 2010). These results suggest that pulmonary resection of metastases may offer benefit in selected patients especially when complete resection is achieved. Indeed, complete resection was reported in a large series as an important prognostic factor of survival with a 5-year survival of 21% and median survival of 19 months in comparison with 13% and 11 months, respectively, for incomplete resection (Petersen et al., 2007). Disease-free interval of less than 12 months and invasion of thoracic lymph node have been described as poor prognosis of survival (Oliaro et al., 2010).

## 10.6 Nonseminomatous Testicular Germ Tumor

Pulmonary metastases are encountered in 30% of patients with nonseminomatous testicular germ tumors (NSGCT; Pfannschmidt et al., 2010b). Chemotherapy is generally the primary option treatment in case of metastatic NSGCT (Pfannschmidt et al., 2006d). Approximately 20% of these patients present residual pulmonary lesions after initial chemotherapy. Resection of residual pulmonary lesion is generally indicated after chemotherapy in case normalization of tumor markers (beta-hCG (human chorionic gonadotropin), alpha-fetoprotein) and may result in three distinct lesions, which cannot be clearly differentiated by radiological exams: (1) necrotic lesions, (2) viable tumor, and (3) mature teratoma (Pfannschmidt et al., 2010b). PET-CT has been evaluated to correlate uptake of fluorodeoxyglucose (FDG) with histological necrosis.

The sensitivity and specificity of low uptake of FDG-PET were only 70% and 48% to diagnose necrosis (Oechsle et al., 2008). Although surgical resection does not prolong survival in patients with solely necrosis, complete resection of mature teratoma or viable cancer may be associated with prolonged disease-free interval and overall survival with 5-year survival ranging from 79% to 94% in these situations (Pfannschmidt et al., 2010b; Schirren et al., 2012). Viable tumor and an absence of tumor markers normalization after chemotherapy have been reported of poorer prognosis of survival. Pulmonary metastasectomy may also be indicated in case of lack of response or partial response to chemotherapy when complete resection is feasible.

### 10.7 Breast Cancer

Metastatic disease in breast cancer is often widespread and pulmonary metastasectomy is rarely indicated. In a large series of 13,502 breast cancer patients, 60 patients (0.4%) with isolated pulmonary metastases were reported and only 40 of them could benefit from lung resection (McDonald et al., 1994). However, some patients will develop solitary isolated lung metastases amenable to lung metastasectomy. Five-year survival after complete resection varies between 35% and 59% (Garcia-Yuste et al., 2010a; Meimarakis et al., 2013). A recent meta-analysis over 16 studies including a total of 1937 patients reported a 5-year survival rate of 46% (Fan et al., 2015). They could identify four prognostic factors of poorer survival: incomplete resection, multiple metastases, disease-free interval of less than 3 years, and negative hormonal status receptor of the metastases.

### 10.8 Head and Neck Tumor

Distant metastases from head and neck squamous cell carcinoma represent 5% to 47% of patients following primary treatment (Florescu and Thariat, 2014). Pulmonary metastases account for two-thirds of distant metastases. Pulmonary metastases are difficult to distinguish from a new primary pulmonary tumor on imaging and even after histological analysis, especially in case of solitary lesion. When the analysis of loss of heterozygosity is used to differentiate metastases from second primary tumors, up to 50% of lesions clinically interpreted as metastases were finally primary lung cancer (Geurts et al., 2005). Five-year survival rates after pulmonary metastasectomy from head and neck cancer range from 26% to 59% (Florescu and Thariat, 2014; Shiono et al., 2009). Risk factors for poorer survival after pulmonary metastasectomy are oral cavity cancers, lymph nodes metastases, glandular cancer, incomplete resection, and a short disease-free interval.

### 10.9 Other Tumors

Pulmonary metastasectomy may also be considered for patients with other histological types such as uterine (Yamamoto et al., 2004), esophageal (Kobayashi et al., 2014), pancreatic (Arnaoutakis et al., 2011), or hepatocellular carcinoma (Kitano et al., 2012). Criteria of pulmonary resection are rarely indicated for these tumors.

## 11. CONCLUSION

Pulmonary metastases are frequent during the history of cancer and are found with different types of primary tumors. Various retrospective studies suggest that pulmonary metastasectomy has an impact on overall survival and should be proposed if the primary tumor is controlled, if there are no extrathoracic metastasis and if complete resection can be performed with adequate pulmonary and cardiac functional reserves. Several prognostic factors may influence medical decision (single metastasis, unilateral, long interval free-disease, type of primary cancer). Completeness of resection is the cornerstone for better prognosis and survival. Pulmonary resection should be as economical on lung parenchyma as possible with atypical wedge resections as preferred approach. VATS combined with regular high resolution CT scans seems to offer equivalent oncologic results to the classical thoracotomy with bimanual palpation of the lung and allows easy reoperation, which is frequent along the course of these diseases. Radiofrequency and stereotaxic radiotherapy are currently under evaluation with promising results.

## References


Allen, M.S., Darling, G.E., Pechet, T.T., et al., 2006. Morbidity and mortality of major pulmonary resections in patients with early-stage lung cancer: initial results of the randomized, prospective ACOSOG Z0030 trial. Ann. Thorac. Surg. 81 (3), 1013–1019 discussion 1019–1020.

Alt, A.L., Boorjian, S.A., Lohse, C.M., Costello, B.A., Leibovich, B.C., Blute, M.L., 2011. Survival after complete surgical resection of multiple metastases from renal cell carcinoma. Cancer 117 (13), 2873–2882.

Arnaoutakis, G.J., Rangachari, D., Laheru, D.A., et al., 2011. Pulmonary resection for isolated pancreatic adenocarcinoma metastasis: an analysis of outcomes and survival. J. Gastrointest. Surg. 15 (9), 1611–1617.

Assouad, J., Petkova, B., Berna, P., Dujon, A., Foucault, C., Riquet, M., 2007. Renal cell carcinoma lung metastases surgery: pathologic findings and prognostic factors. Ann. Thorac. Surg. 84 (4), 1114–1120.

Bacci, G., Briccoli, A., Longhi, A., et al., 2005. Treatment and outcome of recurrent osteosarcoma: experience at Rizzoli in 235 patients initially treated with neoadjuvant chemotherapy. Acta Oncol. 44 (7), 748–755.

de Baere, T., Auperin, A., Deschamps, F., et al., 2015. Radiofrequency ablation is a valid treatment option for lung metastases: experience in 566 patients with 1037 metastases. Ann. Oncol. 26 (5), 987–991.

Beland, M.D., Dupuy, D.E., Mayo-Smith, W.W., 2005. Percutaneous cryoablation of symptomatic extraabdominal metastatic disease: preliminary results. Am. J. Roentgenol. 184 (3), 926–930.

Blackmon, S.H., Shah, N., Roth, J.A., et al., 2009. Resection of pulmonary and extrapulmonary sarcomatous metastases is associated with long-term survival. Ann. Thorac. Surg. 88 (3), 877–884 discussion 884–875.

Briccoli, A., Rocca, M., Salone, M., et al., 2005. Resection of recurrent pulmonary metastases in patients with osteosarcoma. Cancer 104 (8), 1721–1725.

Briccoli, A., Rocca, M., Salone, M., Guzzardella, G.A., Balladelli, A., Bacci, G., 2010. High grade osteosarcoma of the extremities metastatic to the lung: long-term results in 323 patients treated combining surgery and chemotherapy, 1985–2005. Surg. Oncol. 19 (4), 193–199.

Brunelli, A., Charloux, A., Bolliger, C.T., et al., 2009. ERS/ESTS clinical guidelines on fitness for radical therapy in lung cancer patients (surgery and chemo-radiotherapy). Eur. Respir. J. 34 (1), 17–41.

Canter, R.J., Qin, L.X., Downey, R.J., Brennan, M.F., Singer, S., Maki, R.G., 2007. Perioperative chemotherapy in patients undergoing pulmonary resection for metastatic soft-tissue sarcoma of the extremity: a retrospective analysis. Cancer 110 (9), 2050–2060.

Carballo, M., Maish, M.S., Jaroszewski, D.E., Holmes, C.E., 2009. Video-assisted thoracic surgery (VATS) as a safe alternative for the resection of pulmonary metastases: a retrospective cohort study. J. Cardiothorac. Surg. 4, 13.

Casiraghi, M., De Pas, T., Maisonneuve, P., et al., 2011. A 10-year single-center experience on 708 lung metastasectomies: the evidence of the "international registry of lung metastases". J. Thorac. Oncol. 6 (8), 1373–1378.

Casiraghi, M., Maisonneuve, P., Brambilla, D., et al., 2015. The role of extended pulmonary metastasectomy. J. Thorac. Oncol. 10 (6), 924–929.

Cerfolio, R.J., McCarty, T., Bryant, A.S., 2009. Non-imaged pulmonary nodules discovered during thoracotomy for metastasectomy by lung palpation. Eur. J. Cardiothorac. Surg. 35 (5), 786–791 discussion 791.

Chao, Y.K., Chang, H.C., Wu, Y.C., et al., 2012. Management of lung metastases from colorectal cancer: video-assisted thoracoscopic surgery versus thoracotomy—a case-matched study. Thorac. Cardiovasc. Surg. 60 (6), 398–404.

Chen, F., Fujinaga, T., Sato, K., et al., 2009. Significance of tumor recurrence before pulmonary metastasis in pulmonary metastasectomy for soft tissue sarcoma. Eur. J. Surg. Oncol. 35 (6), 660–665.

Cooper, J.D., Perelman, M., Todd, T.R., Ginsberg, R.J., Patterson, G.A., Pearson, F.G., 1986. Precision cautery excision of pulmonary lesions. Ann. Thorac. Surg. 41 (1), 51–53.

Dalrymple-Hay, M.J., Rome, P.D., Kennedy, C., Fulham, M., McCaughan, B.C., 2002. Pulmonary metastatic melanoma – the survival benefit associated with positron emission tomography scanning. Eur. J. Cardiothorac. Surg. 21 (4), 611–614 discussion 614–615.

Davidson, R.S., Nwogu, C.E., Brentjens, M.J., Anderson, T.M., 2001. The surgical management of pulmonary metastasis: current concepts. Surg. Oncol. 10 (1–2), 35–42.

Detterbeck, F.C., Grodzki, T., Gleeson, F., Robert, J.H., 2010. Imaging requirements in the practice of pulmonary metastasectomy. J. Thorac. Oncol. 5 (6 Suppl. 2), S134–S139.

Dong, S., Zhang, L., Li, W., Du, J., Liu, X., Chen, X., 2014. Evaluation of video-assisted thoracoscopic surgery for pulmonary metastases: a meta-analysis. PLoS One 9 (1), e85329.

Dossett, L.A., Toloza, E.M., Fontaine, J., et al., 2015. Outcomes and clinical predictors of improved survival in a patients undergoing pulmonary metastasectomy for sarcoma. J. Surg. Oncol. 112 (1), 103–106.

Downey, R.J., McCormack, P., LoCicero 3rd, J., 1996. Dissemination of malignant tumors after video-assisted thoracic surgery: a report of twenty-one cases. The Video-Assisted Thoracic Surgery Study Group. J. Thorac. Cardiovasc. Surg. 111 (5), 954–960.

Downey, R.J., 1999. Surgical treatment of pulmonary metastases. Surg. Oncol. Clin. N. Am. 8 (2), 341.

Eckardt, J., Licht, P.B., 2012. Thoracoscopic versus open pulmonary metastasectomy: a prospective, sequentially controlled study. Chest 142 (6), 1598–1602.

Eichfeld, U., Tannapfel, A., Steinert, M., Friedrich, T., 2000. Evaluation of ultracision in lung metastatic surgery. Ann. Thorac. Surg. 70 (4), 1181–1184.

Ercan, S., Nichols 3rd, F.C., Trastek, V.F., et al., 2004. Prognostic significance of lymph node metastasis found during pulmonary metastasectomy for extrapulmonary carcinoma. Ann. Thorac. Surg. 77 (5), 1786–1791.

Erhunmwunsee, L., D'Amico, T.A., 2009. Surgical management of pulmonary metastases. Ann. Thorac. Surg. 88 (6), 2052–2060.

Fan, J., Chen, D., Du, H., Shen, C., Che, G., 2015. Prognostic factors for resection of isolated pulmonary metastases in breast cancer patients: a systematic review and meta-analysis. J. Thorac. Dis. 7 (8), 1441–1451.

Ferrari, S., Briccoli, A., Mercuri, M., et al., 2003. Postrelapse survival in osteosarcoma of the extremities: prognostic factors for long-term survival. J. Clin. Oncol. 21 (4), 710–715.

Fidler, I.J., 2003. The pathogenesis of cancer metastasis: the 'seed and soil' hypothesis revisited. Nat. Rev. Cancer 3 (6), 453–458.

Filippi, A.R., Badellino, S., Ceccarelli, M., et al., 2015. Stereotactic ablative radiation therapy as first local therapy for lung oligometastases from colorectal cancer: a single-institution cohort study. Int. J. Radiat. Oncol. Biol. Phys. 91 (3), 524–529.

Florescu, C., Thariat, J., 2014. Local ablative treatments of oligometastases from head and neck carcinomas. Crit. Rev. Oncol. Hematol. 91 (1), 47–63.

Fortes, D.L., Allen, M.S., Lowe, V.J., et al., 2008. The sensitivity of 18F-fluorodeoxyglucose positron emission tomography in the evaluation of metastatic pulmonary nodules. Eur. J. Cardiothorac. Surg. 34 (6), 1223–1227.

Garcia-Yuste, M., Cassivi, S., Paleru, C., 2010a. Pulmonary metastasectomy in breast cancer. J. Thorac. Oncol. 5 (6 Suppl. 2), S170–S171.

Garcia-Yuste, M., Cassivi, S., Paleru, C., 2010b. Thoracic lymphatic involvement in patients having pulmonary metastasectomy: incidence and the effect on prognosis. J. Thorac. Oncol. 5 (6 Suppl. 2), S166–S169.

Geurts, T.W., Nederlof, P.M., van den Brekel, M.W., et al., 2005. Pulmonary squamous cell carcinoma following head and neck squamous cell carcinoma: metastasis or second primary? Clin. Cancer Res. 11 (18), 6608–6614.

Gonzalez, M., Gervaz, P., 2015. Risk factors for survival after lung metastasectomy in colorectal cancer patients: systematic review and meta-analysis. Future Oncol. 11 (2 Suppl.), 31–33.

Gonzalez, M., Ris, H.B., Krueger, T., Gervaz, P., 2012. Colorectal cancer and thoracic surgeons: close encounters of the third kind. Expert Rev. Anticancer Ther. 12 (4), 495–503.

Gonzalez, M., Poncet, A., Combescure, C., Robert, J., Ris, H.B., Gervaz, P., 2013. Risk factors for survival after lung metastasectomy in colorectal cancer patients: a systematic review and meta-analysis. Ann. Surg. Oncol. 20 (2), 572–579.

Gossot, D., Radu, C., Girard, P., et al., 2009. Resection of pulmonary metastases from sarcoma: can some patients benefit from a less invasive approach? Ann. Thorac. Surg. 87 (1), 238–243.

Hamaji, M., Cassivi, S.D., Shen, K.R., et al., 2012. Is lymph node dissection required in pulmonary metastasectomy for colorectal adenocarcinoma? Ann. Thorac. Surg. 94 (6), 1796–1800.

Harting, M.T., Blakely, M.L., 2006. Management of osteosarcoma pulmonary metastases. Semin. Pediatr. Surg. 15 (1), 25–29.

Harvey, J.C., Lee, K., Beattie, E.J., 1994. Surgical management of pulmonary metastases. Chest Surg. Clin. N. Am. 4 (1), 55–66.

Hiraki, T., Gobara, H., Iguchi, T., Fujiwara, H., Matsui, Y., Kanazawa, S., 2014. Radiofrequency ablation as treatment for pulmonary metastasis of colorectal cancer. World J. Gastroenterol. 20 (4), 988–996.

Internullo, E., Cassivi, S.D., Van Raemdonck, D., Friedel, G., Treasure, T., Group, E.P.M.W., 2008. Pulmonary metastasectomy: a survey of current practice amongst members of the European Society of Thoracic Surgeons. J. Thorac. Oncol. 3 (11), 1257–1266.

Kaifi, J.T., Gusani, N.J., Deshaies, I., et al., 2010. Indications and approach to surgical resection of lung metastases. J. Surg. Oncol. 102 (2), 187–195.

Kang, M.C., Kang, C.H., Lee, H.J., Goo, J.M., Kim, Y.T., Kim, J.H., 2008. Accuracy of 16-channel multi-detector row chest computed tomography with thin sections in the detection of metastatic pulmonary nodules. Eur. J. Cardiothorac. Surg. 33 (3), 473–479.

Kayton, M.L., Huvos, A.G., Casher, J., et al., 2006. Computed tomographic scan of the chest underestimates the number of metastatic lesions in osteosarcoma. J. Pediatr. Surg. 41 (1), 200–206 discussion 200–206.

Kim, A.W., Faber, L.P., Warren, W.H., et al., 2008. Repeat pulmonary resection for metachronous colorectal carcinoma is beneficial. Surgery 144 (4), 712–717 discussion 717–718.

Kim, S., Ott, H.C., Wright, C.D., et al., 2011. Pulmonary resection of metastatic sarcoma: prognostic factors associated with improved outcomes. Ann. Thorac. Surg. 92 (5), 1780–1786 discussion 1786–1787.

King, J., Glenn, D., Clark, W., et al., 2004. Percutaneous radiofrequency ablation of pulmonary metastases in patients with colorectal cancer. Br. J. Surg. 91 (2), 217–223.

Kirschbaum, A., Steinfeldt, T., Gockel, A., Di Fazio, P., Quint, K., Bartsch, D.K., 2014. Airtightness of lung parenchyma without a closing suture after atypical resection using the Nd:YAG Laser LIMAX 120. Interact. Cardiovasc. Thorac. Surg. 18 (1), 92–95.

Kitano, K., Murayama, T., Sakamoto, M., et al., 2012. Outcome and survival analysis of pulmonary metastasectomy for hepatocellular carcinoma. Eur. J. Cardiothorac. Surg. 41 (2), 376–382.

Kobayashi, N., Kohno, T., Haruta, S., et al., 2014. Pulmonary metastasectomy secondary to esophageal carcinoma: long-term survival and prognostic factors. Ann. Surg. Oncol. 21 (Suppl. 3), S365–S369.

Kudelin, N., Bolukbas, S., Eberlein, M., Schirren, J., 2013. Metastasectomy with standardized lymph node dissection for metastatic renal cell carcinoma: an 11-year single-center experience. Ann. Thorac. Surg. 96 (1), 265–270 discussion 261–270.

Ludwig, C., Cerinza, J., Passlick, B., Stoelben, E., 2008. Comparison of the number of pre-, intra- and postoperative lung metastases. Eur. J. Cardiothorac. Surg. 33 (3), 470–472.

Margaritora, S., Porziella, V., D'Andrilli, A., et al., 2002. Pulmonary metastases: can accurate radiological evaluation avoid thoracotomic approach? Eur. J. Cardiothorac. Surg. 21 (6), 1111–1114.

McCormack, P.M., Bains, M.S., Begg, C.B., et al., 1996. Role of video-assisted thoracic surgery in the treatment of pulmonary metastases: results of a prospective trial. Ann. Thorac. Surg. 62 (1), 213–216 discussion 216–217.

McDonald, M.L., Deschamps, C., Ilstrup, D.M., Allen, M.S., Trastek, V.F., Pairolero, P.C., 1994. Pulmonary resection for metastatic breast cancer. Ann. Thorac. Surg. 58 (6), 1599–1602.

Meimarakis, G., Ruttinger, D., Stemmler, J., et al., 2013. Prolonged overall survival after pulmonary metastasectomy in patients with breast cancer. Ann. Thorac. Surg. 95 (4), 1170–1180.

Migliore, M., Jakovic, R., Hensens, A., Klepetko, W., 2010. Extending surgery for pulmonary metastasectomy: what are the limits? J. Thorac. Oncol. 5 (6 Suppl. 2), S155–S160.

Mizuno, T., Taniguchi, T., Ishikawa, Y., et al., 2013. Pulmonary metastasectomy for osteogenic and soft tissue sarcoma: who really benefits from surgical treatment? Eur. J. Cardiothorac. Surg. 43 (4), 795–799.

Morgan, T.M., Lange, P.H., Porter, M.P., et al., 2009. Disseminated tumor cells in prostate cancer patients after radical prostatectomy and without evidence of disease predicts biochemical recurrence. Clin. Cancer Res. 15 (2), 677–683.

Mutsaerts, E.L., Zoetmulder, F.A., Rutgers, E.J., 2001. Port site metastasis as a complication of thoracoscopic metastatectomy. Eur. J. Surg. Oncol. 27 (3), 327–328.

Mutsaerts, E.L., Zoetmulder, F.A., Meijer, S., Baas, P., Hart, A.A., Rutgers, E.J., 2002. Long term survival of thoracoscopic metastasectomy vs metastasectomy by thoracotomy in patients with a solitary pulmonary lesion. Eur. J. Surg. Oncol. 28 (8), 864–868.

Nakajima, J., Murakawa, T., Fukami, T., Sano, A., Sugiura, M., Takamoto, S., 2007. Is finger palpation at operation indispensable for pulmonary metastasectomy in colorectal cancer? Ann. Thorac. Surg. 84 (5), 1680–1684.

Nakajima, J., Murakawa, T., Fukami, T., Takamoto, S., 2008. Is thoracoscopic surgery justified to treat pulmonary metastasis from colorectal cancer? Interact. Cardiovasc. Thorac. Surg. 7 (2), 212–216 discussion 216–217.

Nakas, A., Klimatsidas, M.N., Entwisle, J., Martin-Ucar, A.E., Waller, D.A., 2009. Video-assisted versus open pulmonary metastasectomy: the surgeon's finger or the radiologist's eye? Eur. J. Cardiothorac. Surg. 36 (3), 469–474.

Naunheim, K., 2007. Thoracoscopy versus the open approach for resection of solitary pulmonary metastases. In: Ferguson, M. (Ed.), Difficult Decisions in Thoracic Surgery: An Evidence Based Approach. Springer-Verlag, London, pp. 151–157.

Norihisa, Y., Nagata, Y., Takayama, K., et al., 2008. Stereotactic body radiotherapy for oligometastatic lung tumors. Int. J. Radiat. Oncol. Biol. Phys. 72 (2), 398–403.

Oechsle, K., Hartmann, M., Brenner, W., et al., 2008. [18F] Fluorodeoxyglucose positron emission tomography in nonseminomatous germ cell tumors after chemotherapy: the German multicenter positron emission tomography study group. J. Clin. Oncol. 26 (36), 5930–5935.

Okiror, L., Peleki, A., Moffat, D., et al., 2016. Survival following pulmonary metastasectomy for sarcoma. Thorac. Cardiovasc. Surg. 64 (2), 146–149.

Oliaro, A., Filosso, P.L., Bruna, M.C., Mossetti, C., Ruffini, E., 2010. Pulmonary metastasectomy for melanoma. J. Thorac. Oncol. 5 (6 Suppl. 2), S187–S191.

Pastorino, U., Veronesi, G., Landoni, C., et al., 2003. Fluorodeoxyglucose positron emission tomography improves preoperative staging of resectable lung metastasis. J. Thorac. Cardiovasc. Surg. 126 (6), 1906–1910.

Pastorino, U., 1997. Lung metastasectomy: why, when, how. Criti. Rev. Oncol. Hematol. 26 (3), 137–145.

Perentes, J.Y., Krueger, T., Lovis, A., Ris, H.B., Gonzalez, M., 2015. Thoracoscopic resection of pulmonary metastasis: current practice and results. Crit. Rev. Oncol. Hematol. 95 (1), 105–113.

Petersen, R.P., Hanish, S.I., Haney, J.C., et al., 2007. Improved survival with pulmonary metastasectomy: an analysis of 1720 patients with pulmonary metastatic melanoma. J. Thorac. Cardiovasc. Surg. 133 (1), 104–110.

Petre, E.N., Jia, X., Thornton, R.H., et al., 2013. Treatment of pulmonary colorectal metastases by radiofrequency ablation. Clin. Colorectal Cancer 12 (1), 37–44.

Pfannschmidt, J., Hoffmann, H., Muley, T., Krysa, S., Trainer, C., Dienemann, H., 2002. Prognostic factors for survival after pulmonary resection of metastatic renal cell carcinoma. Ann. Thorac. Surg. 74 (5), 1653–1657.

Pfannschmidt, J., Klode, J., Muley, T., Dienemann, H., Hoffmann, H., 2006a. Nodal involvement at the time of pulmonary metastasectomy: experiences in 245 patients. Ann. Thorac. Surg. 81 (2), 448–454.

Pfannschmidt, J., Klode, J., Muley, T., Dienemann, H., Hoffmann, H., 2006b. Pulmonary metastasectomy in patients with soft tissue sarcomas: experiences in 50 patients. Thorac. Cardiovasc. Surg. 54 (7), 489–492.

Pfannschmidt, J., Klode, J., Muley, T., Hoffmann, H., Dienemann, H., 2006c. Pulmonary resection for metastatic osteosarcomas: a retrospective analysis of 21 patients. Thorac. Cardiovasc. Surg. 54 (2), 120–123.

Pfannschmidt, J., Zabeck, H., Muley, T., Dienemann, H., Hoffmann, H., 2006d. Pulmonary metastasectomy following chemotherapy in patients with testicular tumors: experience in 52 patients. Thorac. Cardiovasc. Surg. 54 (7), 484–488.

Pfannschmidt, J., Dienemann, H., Hoffmann, H., 2007. Surgical resection of pulmonary metastases from colorectal cancer: a systematic review of published series. Ann. Thorac. Surg. 84 (1), 324–338.

Pfannschmidt, J., Hoffmann, H., Dienemann, H., 2010a. Reported outcome factors for pulmonary resection in metastatic colorectal cancer. J. Thorac. Oncol. 5 (6 Suppl. 2), S172–S178.

Pfannschmidt, J., Hoffmann, H., Dienemann, H., 2010b. Thoracic metastasectomy for nonseminomatous germ cell tumors. J. Thorac. Oncol. 5 (6 Suppl. 2), S182–S186.

Porter, G.A., Cantor, S.B., Walsh, G.L., et al., 2004. Cost-effectiveness of pulmonary resection and systemic chemotherapy in the management of metastatic soft tissue sarcoma: a combined analysis from the University of Texas M.D. Anderson and Memorial Sloan-Kettering Cancer Centers. J. Thorac. Cardiovasc. Surg. 127 (5), 1366–1372.

Predina, J.D., Puc, M.M., Bergey, M.R., et al., 2011. Improved survival after pulmonary metastasectomy for soft tissue sarcoma. J. Thorac. Oncol. 6 (5), 913–919.

Putnam Jr., J.B., 2002. New and evolving treatment methods for pulmonary metastases. Semin. Thorac. Cardiovasc. Surg. 14 (1), 49–56.

Putnam Jr., J.B., Suell, D.M., Natarajan, G., Roth, J.A., 1993. Extended resection of pulmonary metastases: is the risk justified? Ann. Thorac. Surg. 55 (6), 1440–1446.

Renaud, S., Falcoz, P.E., Olland, A., Massard, G., 2013. Should mediastinal lymphadenectomy be performed during lung metastasectomy of renal cell carcinoma? Interact. Cardiovasc. Thorac. Surg. 16 (4), 525–528.

Renaud, S., Falcoz, P.E., Alifano, M., et al., 2014. Systematic lymph node dissection in lung metastasectomy of renal cell carcinoma: an 18 years of experience. J. Surg. Oncol. 109 (8), 823–829.

Ricardi, U., Filippi, A.R., Guarneri, A., et al., 2012. Stereotactic body radiation therapy for lung metastases. Lung Cancer 75 (1), 77–81.

Robert, J.H., Ambrogi, V., Mermillod, B., Dahabreh, D., Goldstraw, P., 1997. Factors influencing long-term survival after lung metastasectomy. Ann. Thorac. Surg. 63 (3), 777–784.

Rolle, A., Pereszlenyi, A., Koch, R., Bis, B., Baier, B., 2006a. Laser resection technique and results of multiple lung metastasectomies using a new 1318 nm Nd:YAG laser system. Lasers Surg. Med. 38 (1), 26–32.

Rolle, A., Pereszlenyi, A., Koch, R., Richard, M., Baier, B., 2006b. Is surgery for multiple lung metastases reasonable? A total of 328 consecutive patients with multiple-laser metastasectomies with a new 1318-nm Nd:YAG laser. J. Thorac. Cardiovasc. Surg. 131 (6), 1236–1242.

Roth, J.A., Pass, H.I., Wesley, M.N., White, D., Putnam, J.B., Seipp, C., 1986. Comparison of median sternotomy and thoracotomy for resection of pulmonary metastases in patients with adult soft-tissue sarcomas. Ann. Thorac. Surg. 42 (2), 134–138.

Rusch, V.W., 1995. Pulmonary metastasectomy. Current indications. Chest 107 (6 Suppl.), 322S–331S.

Salah, S., Toubasi, S., 2015. Factors predicting survival following complete surgical remission of pulmonary metastasis in osteosarcoma. Mol. Clin. Oncol. 3 (1), 157–162.

Salah, S., Watanabe, K., Welter, S., et al., 2012. Colorectal cancer pulmonary oligometastases: pooled analysis and construction of a clinical lung metastasectomy prognostic model. Ann. Oncol. 23 (10), 2649–2655.

Salah, S., Watanabe, K., Park, J.S., et al., 2013. Repeated resection of colorectal cancer pulmonary oligometastases: pooled analysis and prognostic assessment. Ann. Surg. Oncol. 20 (6), 1955–1961.

Salah, S., Ardissone, F., Gonzalez, M., et al., 2015. Pulmonary metastasectomy in colorectal cancer patients with previously resected liver metastasis: pooled analysis. Ann. Surg. Oncol. 22 (6), 1844–1850.

Schirren, J., Trainer, S., Eberlein, M., Lorch, A., Beyer, J., Bolukbas, S., 2012. The role of residual tumor resection in the management of nonseminomatous germ cell cancer of testicular origin. Thorac. Cardiovasc. Surg. 60 (6), 405–412.

Schlijper, R.C., Grutters, J.P., Houben, R., et al., 2014. What to choose as radical local treatment for lung metastases from colo-rectal cancer: surgery or radiofrequency ablation? Cancer Treat. Rev. 40 (1), 60–67.

Schur, S., Hoetzenecker, K., Lamm, W., et al., 2014. Pulmonary metastasectomy for soft tissue sarcoma—report from a dual institution experience at the Medical University of Vienna. Eur. J. Cancer 50 (13), 2289–2297.

Seebacher, G., Decker, S., Fischer, J.R., Held, M., Schafers, H.J., Graeter, T.P., 2015. Unexpected lymph node disease in resections for pulmonary metastases. Ann. Thorac. Surg. 99 (1), 231–236.

Shigemura, N., Akashi, A., Nakagiri, T., Ohta, M., Matsuda, H., 2004. A new tissue-sealing technique using the Ligasure system for nonanatomical pulmonary resection: preliminary results of sutureless and stapleless thoracoscopic surgery. Ann. Thorac. Surg. 77 (4), 1415–1418 discussion 1419.

Shiono, S., Kawamura, M., Sato, T., et al., 2009. Pulmonary metastasectomy for pulmonary metastases of head and neck squamous cell carcinomas. Ann. Thorac. Surg. 88 (3), 856–860.

Simon, C.J., Dupuy, D.E., Mayo-Smith, W.W., 2005. Microwave ablation: principles and applications. Radiographics 25 (Suppl. 1), S69–S83.

Siva, S., MacManus, M., Ball, D., 2010. Stereotactic radiotherapy for pulmonary oligometastases: a systematic review. J. Thorac. Oncol. 5 (7), 1091–1099.

Sleeman, J.P., 2012. Metastasis: understanding is the beginning of order in chaos. Semin. Cancer Biol. 22 (3), 173.

Sonett, J.R., 1999. Pulmonary metastases: biologic and historical justification for VATS. Video assisted thoracic surgery. Eur. J. Cardiothorac. Surg. 16 (Suppl. 1), S13–S15 discussion S15–S16.

Szoke, T., Kortner, A., Neu, R., et al., 2010. Is the mediastinal lymphadenectomy during pulmonary metastasectomy of colorectal cancer necessary? Interact. Cardiovasc. Thorac. Surg. 10 (5), 694–698.

Takeda, A., Kunieda, E., Ohashi, T., Aoki, Y., Koike, N., Takeda, T., 2011. Stereotactic body radiotherapy (SBRT) for oligometastatic lung tumors from colorectal cancer and other primary cancers in comparison with primary lung cancer. Radiother. Oncol. 101 (2), 255–259.

The International Registry of Lung Metastases, 1997. Long-term results of lung metastasectomy: prognostic analyses based on 5206 cases. J. Thorac. Cardiovasc. Surg. 113 (1), 37–49.

Treasure, T., Fallowfield, L., Lees, B., 2010. Pulmonary metastasectomy in colorectal cancer: the PulMiCC trial. J. Thorac. Oncol. 5 (6 Suppl. 2), S203–S206.

Treasure, T., 2007. Pulmonary metastasectomy: a common practice based on weak evidence. Ann. R. Coll. Surg. Engl. 89 (8), 744–748.

Venuta, F., Rolle, A., Anile, M., Martucci, N., Bis, B., Rocco, G., 2010. Techniques used in lung metastasectomy. J. Thorac. Oncol. 5 (6 Suppl. 2), S145–S150.

Veronesi, G., Petrella, F., Leo, F., et al., 2007. Prognostic role of lymph node involvement in lung metastasectomy. J. Thorac. Cardiovasc. Surg. 133 (4), 967–972.

Watanabe, K., Nagai, K., Kobayashi, A., Sugito, M., Saito, N., 2009. Factors influencing survival after complete resection of pulmonary metastases from colorectal cancer. Br. J. Surg. 96 (9), 1058–1065.

Welter, S., Jacobs, J., Krbek, T., Krebs, B., Stamatis, G., 2007. Long-term survival after repeated resection of pulmonary metastases from colorectal cancer. Ann. Thorac. Surg. 84 (1), 203–210.

Yamamoto, K., Yoshikawa, H., Shiromizu, K., et al., 2004. Pulmonary metastasectomy for uterine cervical cancer: a multivariate analysis. Ann. Thorac. Surg. 77 (4), 1179–1182.

Younes, R.N., Gross, J.L., Deheinzelin, D., 2002. Surgical resection of unilateral lung metastases: is bilateral thoracotomy necessary? World J. Surg. 26 (9), 1112–1116.

Younes, R.N., Fares, A.L., Gross, J.L., 2012. Pulmonary metastasectomy: a multivariate analysis of 440 patients undergoing complete resection. Interact. Cardiovasc. Thorac. Surg. 14 (2), 156–161.

Younes, R., Abrao, F.C., Gross, J., 2013. Pulmonary metastasectomy for malignant melanoma: prognostic factors for long-term survival. Melanoma Res. 23 (4), 307–311.

Zaman, M., Bilal, H., Woo, C.Y., Tang, A., 2012. In patients undergoing video-assisted thoracoscopic surgery excision, what is the best way to locate a subcentimetre solitary pulmonary nodule in order to achieve successful excision? Interact. Cardiovasc. Thorac. Surg. 15 (2), 266–272.

CHAPTER

17

# Brain Metastasis: Basic Biology, Clinical Management, and Insight From Experimental Model Systems

*D.H. Murrell*[1,2], *F. Perera*[3], *A.F. Chambers*[2,3], *P.J. Foster*[1,2]

[1]Robarts Research Institute, London, ON, Canada; [2]Western University, London, ON, Canada; [3]London Health Sciences Centre, London, ON, Canada

## 1. INTRODUCTION

A brain metastasis is a tumor that develops in the brain as a result of a primary tumor elsewhere spreading from its original location. It may also be known as a secondary brain tumor, or a metastatic brain tumor. Primary tumors that commonly metastasize to the brain include lung, breast, melanoma, renal, and colorectal. Other cancers such as testicular, lymphoma, and leukemia can also metastasize to the brain, but do so at a much lower frequency.

Brain metastases are the most commonly diagnosed neoplasms in the brain and they outnumber primary malignant brain tumors by a factor of 10 (Davis et al., 2012). Nearly 10% of metastatic cancer patients will develop a symptomatic brain metastasis (Schouten et al., 2002; Barnholtz-Sloan et al., 2004). Jemal et al. (2010) estimate that there will be more than 1.5 million new cases of cancer annually in the United States, suggesting brain metastasis will affect approximately 150,000 people every year. Still, this is likely an underestimation of true brain metastasis incidence; some metastases may be overlooked because patients are asymptomatic, or already too sick and a diagnosis would be irrelevant to their care. Historically, autopsy studies have reported substantially higher proportions of patients with brain metastasis than what was diagnosed clinically (Lin et al., 2004).

The incidence of brain metastasis is increasing in the clinic. This is partially because the ability to detect small tumors has improved as better neuroimaging techniques have become available. Furthermore, brain metastases often manifest later in cancer progression; recent advances in treatment options for primary cancers have resulted in longer patient survival and better control of extracranial disease thereby allowing the brain and central nervous system (CNS) to declare as a sanctuary site for metastases. Thus, the prevalence of brain metastasis is expected to rise further as systemic therapy improves. For example, human epidermal growth factor receptor 2 (HER2+) breast cancer patients treated

*Introduction to Cancer Metastasis*
http://dx.doi.org/10.1016/B978-0-12-804003-4.00017-7

TABLE 17.1 Brain Metastasis Incidence for Populations in Detroit, USA, Sweden, and Maastricht, the Netherlands

| | Brain Metastasis Incidence (%) | | |
|---|---|---|---|
| **Primary Cancer** | **Metropolitan Detroit Cancer Surveillance System (1973–2001), *n* = 16,210** | **Sweden (1987–2006), *n* = 15,517** | **Maastricht Cancer Registry (1986–1995), *n* = 2724** |
| Lung | 19.9 | 44.1 (M), 33.5 (F) | 16.3 |
| Breast | 5.1 | 32.6 (F) | 5.0 |
| Melanoma | 6.9 | 12.3 (M), 5.8 (F) | 7.4 |
| Renal | 6.5 | 7.8 (M), 4.7 (F) | 9.8 |
| Colorectal | 1.8 | 9.0 (M), 7.4 (F) | 1.2 |

*F*, female; *M*, male.

with trastuzumab develop CNS metastases in 31–48% of cases and 33–50% of those women are responding to therapy or have stable disease at other sites when diagnosed with CNS involvement (Stemmler et al., 2006; Shmueli et al., 2004; Bendell et al., 2003; Lai et al., 2004). Three large studies have estimated the incidence of brain metastases from population-based cohorts (Schouten et al., 2002; Barnholtz-Sloan et al., 2004; Smedby et al., 2009). The reported incidences of brain metastasis in these studies, due to the five most common primary cancers, are listed in Table 17.1.

Brain metastases that develop in the parenchyma follow patterns of blood circulation. They frequently develop at the border between gray and white matter where vessel size decreases and traps cancer cells. Moreover, the general distribution of metastases in the brain parallels blood flow; thus 80% grow in the cerebral hemispheres, 15% in the cerebellum, and 5% in the brain stem (Gil-Gil et al., 2014; Klos and O'Neill, 2004). Perera et al. (2012) used magnetic resonance imaging (MRI) to compare the spatial distribution of brain metastases due to breast cancer in two different mouse models (HER2±) and showed that more than 50% of metastases grew in the cortex and 25–30% of metastases were identified in the central brain for either cell line. The proportion of tumors that grew in the posterior brain was higher for the HER2+ cell line (Perera et al., 2012). In addition to parenchymal lesions, metastases can form in the meninges of the brain. This is common for brain metastases due to melanoma. There are three clinically distinct manifestations of brain metastasis: solitary brain metastasis, multiple brain metastasis, and leptomeningeal metastasis. Presentation with multiple brain metastases is most common in breast cancer and melanoma patients (Pestalozzi, 2009; Fife et al., 2004). In breast cancer, multiple brain metastases occur in 78% of cases, whereas solitary and leptomeningeal metastasis are less common (14% and 8%, respectively; Pestalozzi, 2009).

## 2. BIOLOGY

### 2.1 Brain Metastasis Formation

A series of single steps with varying efficiency are involved when a primary cancer metastasizes to the brain. The first steps of metastasis are inefficient; a cell must escape the primary tumor, invade local tissue, and intravasate into the vasculature (Chambers et al., 2000). It must then survive in the blood and disseminate via the circulation. Many cancer cells will reach the brain in this way and some will arrest in the microcirculation. Initially, cells arrest at

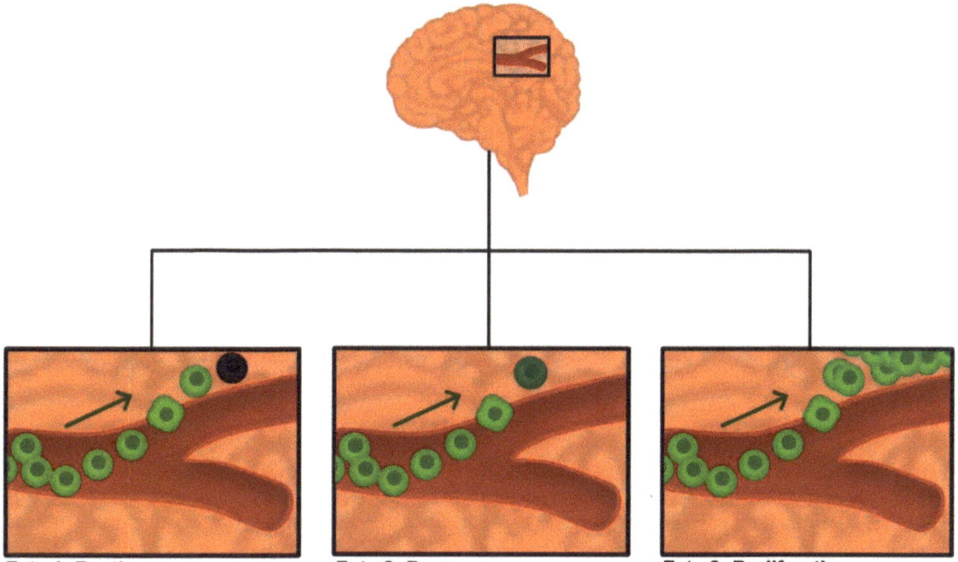


FIGURE 17.1 Brain colonization by metastatic cancer cells requires that cells survive in the vasculature and arrest in the microcirculation. Single cancer cells may then extravasate out of the vasculature and into the brain where they will experience one of three fates: (1) death, (2) dormancy, where the cell remains viable but nonproliferative, or (3) proliferation to form brain metastases.

blood vessel branches where the cell to vessel diameter ratio approaches one (Kienast et al., 2010). Single cancer cells may then extravasate via holes in the vascular wall, and into the surrounding tissue (Kienast et al., 2010). Cancer cells can execute these steps with relatively high efficiency (Chambers et al., 2000). Once the metastatic cancer cell arrives in the brain, it may experience one of three fates (Fig. 17.1): (1) it may die; (2) it may remain viable but dormant; or (3) it may proliferate and form a brain metastasis. Many cells do not survive the required steps for brain colonization (Chambers et al., 2000). This relies on cancer cell survival outside of the vasculature and invasion of the surrounding tissue, both of which are inefficient steps in the metastatic cascade (Chambers et al., 2000). Subsequent development or recruitment of tumor-associated vasculature is required for macrometastatic growth; until this vasculature develops, surviving extravasated cancer cells adhere to the outer vessel wall to migrate and access oxygen and nutrients (Kienast et al., 2010; Steeg et al., 2011). The mechanisms of brain metastasis vascularization are discussed below.

## 2.2 Brain Microenvironment

The brain microenvironment offers a unique metastatic site with vasculature that differs from most other organs (Fig. 17.2). Extracranial capillaries are thin and composed of a single endothelial cell layer with various combinations of fenestra, clefts, and pores (Deeken and Löscher, 2007). Structurally, this allows for uncomplicated exchange of oxygen, carbon dioxide, nutrients, and waste between the blood and surrounding tissue. In contrast, a highly selective barrier surrounds capillaries in the brain. It is comprised of endothelial cells connected by tight junctions, a basement membrane, pericytes, and astrocytes (Deeken and Löscher, 2007). Together, these

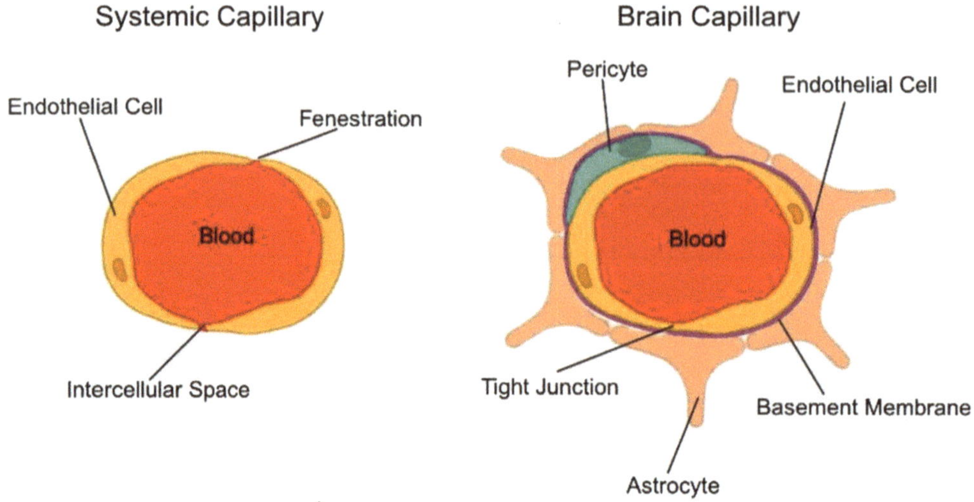


FIGURE 17.2 Cross-sections of a systemic and brain capillary illustrate differences in their general structure. In contrast to systemic capillaries, brain capillaries are surrounded by many protective components. Brain capillaries lack fenestrations and intercellular spaces; instead, tight junctions between endothelial cells create a diffusion barrier. Furthermore, brain capillaries are surrounded by astrocytes, pericytes, and an intact basement membrane, which provide additional protection for the brain.

components form the blood–brain barrier (BBB), which serves to protect the brain by tightly regulating the diffusion of substrates from the blood into the brain parenchyma. The BBB has very low passive permeability and high levels of efflux transporters, which maintain the internal brain microenvironment and prevent neurotoxicity (Deeken and Löscher, 2007). Under normal conditions, the BBB is extremely exclusive and prevents traditional systemically administered chemo- and molecular therapeutics from entering the brain.

As a cancer cell extravasates from a brain capillary, it penetrates the BBB and the protective cells surrounding the vasculature are displaced. This compromises BBB integrity and a new barrier, known as the blood–tumor barrier (BTB), is formed in its place around the growing tumor. Relatively little is known about the BTB in patients, but its patency appears to vary substantially. For example, tumor detection by clinical imaging relies on extravasation of contrast agents across the BBB/BTB; yet, breast cancer therapeutics such as paclitaxel, doxorubicin, and trastuzumab, which have been highly successful for treating extracranial metastases, do not cross the BBB/BTB well and are ineffective against brain metastases (Stemmler et al., 2006; Connell et al., 2013). It is presumed that exclusion from the brain causes these and other chemotherapies to fail, thus rendering the brain a "sanctuary site" for metastatic growth.

## 2.3 Alternative Vascularization Methods

The development of a brain metastasis creates an increasingly high demand for oxygen and nutrients. For most tumors, this need is typically met by the promotion of tumor-associated neovasculature in a process called angiogenesis. The brain is a unique metastatic site because it is the most highly vascularized organ in the human body and, for this reason, cerebral metastases may grow without inducing angiogenesis. Instead, tumors can exploit the rich preexisting vasculature of the brain through alternative vascularization mechanisms. Tumors may co-opt the innate dense capillary bed in the brain by growing alongside preexisting microvessels and invading the surrounding brain tissue (Leenders et al., 2004; Döme et al., 2007; Donnem et al., 2013). Co-opted vessel density is lower within brain metastases compared to normal brain; however, dilated vessels within tumors compensate to meet increased metabolic demands (Bugyik et al., 2011). The role of angiogenesis in brain metastasis development was studied by Leenders et al. (2004); they investigated the outcome of antiangiogenic drugs on the development of brain metastasis and found that although angiogenesis was blocked, tumors continued to progress via co-option of the preexisting vasculature. Thus, brain tumors may develop without angiogenesis.

### 2.4 Molecular Mechanisms of Brain Metastasis

Relatively little is known about the molecular mechanisms that are associated with brain metastases. Several recent studies have taken steps to address this question but much remains to be learned. Signal transducer and activator of transcription 3 have been suggested to play a role in brain metastasis from melanoma (Xie et al., 2006; Huang et al., 2008). In a literature review, Huang and Ouyang (2013) identified a number of potential biomarkers associated with brain metastases from lung cancer and noted that these associations would need to be assessed in prospective clinical trials.

Studies of primary tumor samples have led to the identification of potential biomarkers of brain metastasis. For example, expression profiling comparing 23 brain metastases and matched primary breast tumor samples identified numerous genes involved in DNA repair pathways, and functionally confirmed two genes, BARD1 and RAD51, in vitro and in vivo (Woditschka et al., 2014). Similarly, McMullin et al. (2014) identified a "BRCA1 Deficient-Like" gene signature that was associated with brain metastasis in HER2+ positive patients, compared to unmatched primary tumors. Duchnowska studied cohorts of HER2+ breast cancer patients and identified 13- and 3-gene signatures associated with early brain metastasis formation (Duchnowska et al., 2015). However, these signatures were not confirmed in a validation cohort, although the genes identified may provide clues to molecular mechanisms for further study (Duchnowska et al., 2015). Recently, Saunus et al. (2015) carried out multiple analyses (copy number, RNA, and exome sequencing) of 36 surgical samples of brain metastases from patients with primary breast cancer, lung, melanoma or esophageal cancers, and identified multiple genomic alterations in these tumors. This study identified HER3 as being overexpressed and activated in brain metastases. Overall, while many molecular pathways have been identified that may contribute to brain metastases, much remains to be learned both to better predict which patients are at higher risk for brain metastasis development and to develop strategies for targeted therapies that will be useful, given the unique challenges of treatment of metastases in the brain.

## 3. CLINICAL MANAGEMENT

### 3.1 Diagnosis

In patients, brain metastasis development often results in headaches, memory loss, and general weakness. Depending on tumor location, neurologic dysfunction may include ataxia, seizures, loss of sensory or motor function, cranial neuropathy, visual or speech changes, and/or cognitive decline (Steeg et al., 2011; Son et al., 2012).

Suspicion of brain metastasis is evaluated by imaging—usually computed tomography (CT) or MRI—where lesions of several millimeters in size are generally radiographically detectable. Contrast-enhanced cranial CT can detect many

TABLE 17.2 Magnetic Resonance Imaging (MRI) Protocols for Brain Metastasis Detection and Evaluation

| MRI Protocol | Tumor Appearance |
|---|---|
| T1 weighted | |
| Pregadolinium (pre-Gd) intravenously (IV) | Isointense or mildly hypointense; melanin may cause hyperintensities in some brain metastases due to melanoma; hemorrhagic metastases may also be hyperintense |
| Post-Gd IV | Intense enhancement throughout, in nodes, or ring enhancing |
| T2 weighted | Hyperintense; melanin may cause some brain metastases due to melanoma to appear hypointense |
| Diffusion-weighted imaging | Hyperintense on apparent diffusion coefficient map |
| MR perfusion | Increased cerebral blood volume and blood flow |
| MR spectroscopy | Increased metabolic activity; choline and N-acetyl aspartate are common metabolites that are compared with creatine, which is used as an internal control |

brain metastases; however, diagnosis by MRI is considered superior due to better soft tissue contrast, ability to detect smaller tumors, fewer artifacts, and ability to acquire images in all three planes and without the risk of ionizing radiation dose to the patient (Schellinger et al., 1999). The radiological appearance of brain metastasis is diverse and varies substantially with primary tumor origin. Generally, the lesion is well demarcated from the brain parenchyma though it may also be surrounded by cerebral edema.

Contrast-enhanced T1-weighted MRI is widely regarded as the most accurate method for clinical brain tumor detection (Brem et al., 2011). Images are acquired before and after the intravenous administration of a gadolinium (Gd)-based contrast agent. In normal brain, the BBB prevents Gd from diffusing out of the circulation. By contrast, the BBB is often disrupted in the local area of a tumor due to tumor infiltration, increased vascular permeability, or poorly constructed angiogenic vessels. Tumors are identified by diffusion of Gd across the impaired BBB/BTB, which results in signal enhancement (brightness) in the post-Gd image.

An MRI session may include many image acquisitions to create different contrasts and gain information about the structure of the brain and tumor. Employing multiple contrast mechanisms aids radiologists to differentiate a brain metastasis from a primary brain tumor, cerebra abscess, or stroke. Furthermore, these protocols may be used to evaluate response after treatment. MR sequences that are used for brain metastasis visualization and evaluation, as well as tumor appearance, are described in Table 17.2 (Fink and Fink, 2013).

If imaging suggests clinical presentation of metastatic lesions in the brain, stereotactic or open biopsy may be indicated to confirm diagnosis and characterize the metastasis (Brem et al., 2011). Though invasive, biopsy is useful to rule out inflammation or infection, which may appear similar to a metastatic lesion by radiology or MRI.

## 3.2 Prognosis

The prognosis of patients diagnosed with brain metastasis is poor. The median survival and 1-year survival found in a study by Smedby et al. (2009) for the five most common primary cancers that metastasize to the brain are listed in Table 17.3.

The management of patients with brain metastases is complicated and considers several prognostic factors to maximize survival and neurological function. Important prognostic

TABLE 17.3 Median Survival and 1-Year Survival for Cancer Patients After a Brain Metastasis Diagnosis

| Primary Cancer | Median Survival (Months) | 1-Year Survival (%) |
|---|---|---|
| Lung | 2.5 | 10.4 |
| Breast | 3.2 | 19 |
| Melanoma | 2.7 | 9.8 |
| Renal | 4.7 | 21.1 |
| Colorectal | 2.6 | 6.7 |

factors include Karnofsky performance status (KPS), number of brain metastases, tumor histology, pathological grading, extracranial disease status, hormone receptor status (if applicable), and age less than 65 years (Gil-Gil et al., 2014; Eichler and Loeffler, 2007). The availability of potentially effective systemic treatment options postdiagnosis of brain metastases also affects the outlook.

KPS is a scaled index that ranks patients based on their functional abilities. The scale ranges from 0 to 100 in increments of 10, where 0 is dead and 100 represents normal functioning with no complaints and no outward evidence of disease. A KPS score of 70 is routinely used as a threshold for clinical decision making. A KPS score greater than or equal to 70 means the patient is able to care for him or herself, whereas a KPS score below 70 indicates the patient requires at least occasional assistance in his or her daily life.

### 3.2.1 *Prognostic Scores*

Outcomes for patients with brain metastasis are heterogeneous and vary by diagnosis as well as several prognostic factors that can be disease specific. Prognostic indexes are useful to estimate survival for patients with brain metastasis and to tailor treatment appropriately. Several scoring systems exist, including the recursive partitioning analysis (RPA) of prognostic factors class system, graded prognostic assessment (GPA), and disease-specific GPA, among others.

In 1997, the Radiation Therapy Oncology Group (RTOG) developed an index to aid in estimating the prognosis for individual patients with brain metastasis. The prognostic index consisted of three classes derived by RPA using prognostic factors their database from RTOG brain metastasis trials (7916, 8528, and 8905; Gaspar et al., 1997). RPA class I included patients with KPS ≥70, age <70, with controlled primary cancer and no extracranial metastasis; these patients had the most favorable survival of approximately 7 months. RPA class III was all patients with KPS <70 and had poor survival of about 2 months. RPA class II included all other patients and the expected survival was 4 months (Gaspar et al., 2000).

In 2007, the GPA prognostic index was introduced to incorporate new data (RTOG 9508) and address limitations of the RPA system, such as the subjective nature of estimating systemic disease extent, and inclusion of number of brain metastasis (Sperduto et al., 2008). This system forms four statistically different groups based on a sum of scores (0, 0.5, or 1.0) of four factors: age, KPS, presence or absence of extracranial metastasis, and number of brain metastases. The median survival was best, 11 months, for a GPA index of 3.5–4. A GPA score of 3 came with survival of 6.9 months, whereas a GPA of 1.5–2.5 meant survival of 2.8 months, and a GPA of 0–1 meant 2.6 months. This method was found to be as prognostic as the RPA index and more quantitative (Sperduto et al., 2008).

A multi-institutional analysis of over 4000 patients with brain metastasis found that the importance of prognostic factors was diagnosis specific; yet, the RPA and GPA prognostic scoring systems were not disease specific because they did not consider primary tumor origin, or histological tumor type (Sperduto et al., 2010). The GPA was updated to the diagnosis-specific GPA (DS-GPA) in 2010 to address this concern (Sperduto et al., 2010). For patients with lung cancer, prognosis depended on age, KPS, number of brain metastases, and the presence of extracranial metastases, whereas for melanoma and renal primary cancers, prognosis depended only on KPS and number of brain metastasis; finally, for breast and gastrointestinal cancers KPS was the only significant prognostic factor (Sperduto et al., 2010). The DS-GPA was further refined for breast cancer to include considerations for breast cancer subtype (based upon estrogen receptor, progesterone receptor, and HER2 status). The updated DS-GPA for breast cancer found that, in addition to KPS, age and subtype were significant prognostic factors (Sperduto et al., 2012). In this model, the median survival for patients with breast cancer brain metastases ranged from 3.35 months for patients with a DS-GPA score of 1.0 or less to 25.3 months for patients with a score of 3.5–4 (Sperduto et al., 2012). This likely underscores the importance of breast cancer subtype and the availability of effective systemic therapy and, in particular, anti-HER2/neu targeted therapy in affecting the patient outcome.

## 3.3 Treatment

In general, systemic chemo- and molecular anticancer therapies, which are highly successful at treating extracranial metastases, are ineffective for brain metastases because they are unable to cross the intact BBB/BTB in sufficient concentrations (Stemmler et al., 2006; Connell et al., 2013). This leaves few treatment options for brain metastases. The National Comprehensive Cancer Network published their consensus on clinical practice guidelines in oncology for CNS cancers, which included management strategies for brain metastases (Brem et al., 2011). Symptoms can be managed with steroids and antiepileptic drugs. Corticosteroids, such as dexamethasone, are used to reduce swelling in the brain and to help alleviate symptoms. This treatment alone offers survival of only 1–2 months (Weissman, 1988). Primary anticancer treatment possibilities to extend survival include whole brain radiotherapy (WBRT), surgical resection, stereotactic radiosurgery (SRS), or some combination of these (Fig. 17.3).

WBRT is a mainstay in the treatment for brain metastasis; in this method, radiation dose is delivered to the entire brain volume, regardless of tumor location. It is often used as an adjuvant therapy, though it can be used alone for palliation. WBRT alone improves neurologic symptoms and may extend survival to 4–6 months (Viani et al., 2007).

Surgical resection involves craniotomy followed by neurosurgery to remove or debulk the tumor mass. In 1990, Patchell et al. published results indicating that surgical resection followed by WBRT was superior to WBRT alone in a randomized trial including 48 patients with a solitary brain metastasis. In this study, the surgical group outlived the radiation group by more than 2.5 times and remained functionally independent nearly five times longer (Patchell et al., 1990).

In 1998, Patchell et al. published another randomized clinical trial comparing surgical resection alone versus surgical resection followed by WBRT and determined that adjuvant WBRT was beneficial for patients with a single metastasis, even when complete surgical resection of the tumor was achieved. Recurrence in the brain was reduced in the radiotherapy group and patients were less likely to die from neurologic causes than patients who underwent surgical resection alone (Patchell et al., 1998).

SRS offers an alternative and less invasive treatment option compared to surgery. This is beneficial for tumors that develop in eloquent cortical areas, which were previously considered untreatable due to potentially devastating neurological side effects from surgery. Furthermore, SRS is advantageous for treatment of deep-seated and otherwise unresectable tumors. This technique uses multiple convergent radiation beams to deliver very high radiation doses to a precise tumor volume.

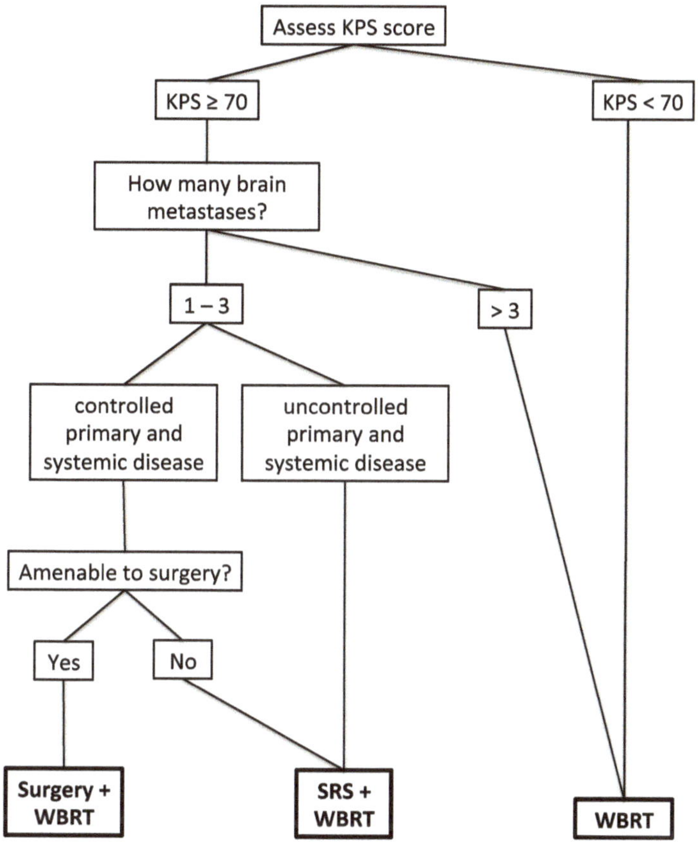


FIGURE 17.3 A traditional treatment decision tree for patients diagnosed with brain metastasis. Recently, disease-specific prognostic indexes have expanded this simple overview to include considerations for age and primary cancer type or subtype; moreover, each prognostic factor (age, Karnofsky performance status, number of brain metastases, presence or absence of systemic disease) is scored and the sum of these scores classifies patients into DS-GPA group with different expected survival rates. Aggressive [surgery or stereotactic radiosurgery + whole brain radiotherapy (WBRT)] or conservative treatment (WBRT or supportive care) is then based on the average DS-GPA group survival.

### 3.3.1 *Oligometastatic Disease (1–3 Metastases)*

Patients presenting with few metastatic lesions on MRI (typically 1–3) and controlled extracranial disease may undergo surgical resection (Brem et al., 2011). Surgical excision of brain metastasis may also be indicated, regardless of systemic disease status, if the mass effect by the tumor causes significant symptoms for the patient. Tumor location and accessibility are critical factors for determining if surgical intervention is the most feasible and advantageous approach.

Following resection, WBRT and/or SRS are used as adjuvant therapies to control micrometastatic disease in the surgical bed and surrounding brain tissue (Brem et al., 2011). WBRT is delivered in radiation doses ranging from 20 to 40 Gy in 5–20 fractions, though 30 Gy/10 or 37.5/15 are common treatment regimens (Son et al., 2012; Brem et al., 2011). SRS typically uses single doses of 15–24 Gy (Brem et al., 2011). In some cases, WBRT may be delivered in addition to an SRS "boost" to the surgical bed.

The primary treatment for patients with oligometastatic disease that is not amenable to surgery is WBRT and/or SRS. In RTOG 9508 clinical trial of patients with one to three brain metastases, Andrews et al. (2004) found that SRS in combination with WBRT improved survival when compared to WBRT alone but only in the subset of patients with single metastases or favorable features (RPA class I). In addition, patients who underwent SRS had better neurologic function 6 months posttreatment than those who had WBRT alone (Andrews et al., 2004). There is concern that WBRT leads to neurocognitive decline compared to SRS or surgical resection. Nevertheless, in the comparison of SRS or surgical resection of one to three brain metastases versus the same treatments followed by WBRT, the (European Organization for Research and Treatment of Cancer) EORTC 22,952–26,001 study found that adding WBRT reduced recurrence of initial brain metastases and the development of new metastases but did not improve overall survival or functional decline (Kocher et al., 2011).

If oligometastatic disease is presented with disseminated systemic disease, then WBRT followed by appropriate systemic therapy are used as primary treatment (Brem et al., 2011).

### 3.3.2 *Multiple Metastatic Lesions (>3 Metastases)*

WBRT is the clinical standard for treating multiple (e.g., >3) brain metastases and diffuse disease, though this treatment alone is generally considered palliative. Surgical resection may be considered to relieve mass effect (Brem et al., 2011). In patients with unfavorable RPA or GPA scores, supportive care only may be an option.

### 3.3.3 *Preventative Treatment*

Prophylactic cranial irradiation (PCI) is a preventative treatment for patients at high risk of developing clinically relevant brain metastasis. It is a WBRT-like regimen that is employed if developing micrometastatic foci are suspected in the brain, but are not yet detectable by radiology. PCI decreased the subsequent incidence of brain metastasis in a study of 286 patients with extensive small-cell lung cancer (SCLC) responding to chemotherapy; moreover, PCI prolonged both disease-free and overall survival (Slotman et al., 2007). Despite the significant survival benefit in SCLC patients, PCI does not guarantee prevention of brain metastasis and may cause unwanted neurocognitive side effects. For this reason, the risk to benefit ratio for this method is not favorable in many other cancers. The utility of PCI is therefore mainly limited to SCLC patients, though it has been suggested for some patients with advanced metastatic breast cancer (Huang et al., 2009).

### 3.3.4 *Recent Advances in Systemic Therapy*

Systemic therapies that are able to cross the BBB/BTB, or increase BBB/BTB permeability, and exert a therapeutic effect are in development. Promising new brain-permeable compounds and chemotherapeutic approaches for the prevention or treatment of brain metastases were discussed in a review by Dr. Steeg (Steeg et al., 2011). Several of these new approaches are disease specific. In a subset of patients with HER2 positive breast cancer brain metastases, the use of capecitabine and lapatinib as initial therapy for brain metastases led to a 65% objective (at least 50% or more volumetric improvement) response in 44 of 45 patients, a median time to radiation of 8.3 months, and a median survival of 17 months in the phase II LANDSCAPE clinical trial (Bachelot et al., 2013). Niwińska et al. (2010) performed a review of 420 breast cancer patients with brain metastases to access the significance of molecular subtype and the effect of systemic treatment after WBRT. Other than the triple-negative subtypes, follow-up systemic treatment significantly improved the median survival compared to no systemic treatment (Niwińska et al., 2010).

# 4. EXPERIMENTAL MODEL SYSTEMS

Laboratory mouse models are most commonly used to study brain metastasis in vivo. Several experimental model systems have been described, including specific models for brain metastases due to lung cancer, breast cancer, and melanoma (Mathieu et al., 2004; Yoneda et al., 2001; Palmieri et al., 2007; Tanner et al., 2004; Tao et al., 2008; Alterman and Stackpole, 1989; Henry et al., 2015). These models provide insight into brain metastatic development and enable investigation that is not possible in the clinic (Fig. 17.4).

Both syngeneic and xenograft transplant models exist. Syngeneic transplant models introduce murine cancer cells into mouse recipients of the same genetic background. Some examples of this type of model include the murine 4T1-BR5 brain metastatic breast cancer cell line injected into BALB/c mice, or K-1735 melanoma cell line injected into C3H$^{-}$ mice (Tao et al., 2008; Lockman et al., 2010; Fidler et al., 1981). This type of model offers the opportunity to investigate tumor development in the presence of a fully functioning immune system; however, it lacks specificity to human cancer and may not reflect human disease heterogeneity. Xenograft transplant models inoculate immune-compromised hosts with human cancer cell lines that are derived from patient tissue samples. Several mouse strains are available for this research with varying degrees of immune deficiency. This approach lacks functioning immune response, but allows for the direct metastasis modeling that is relevant to human cancer development.

Syngeneic and xenograft transplant models can be initiated in similar ways. Some experimental models involve brain metastases that can form spontaneously; alternatively, cancer cells or tumor fragments can be delivered orthotopically

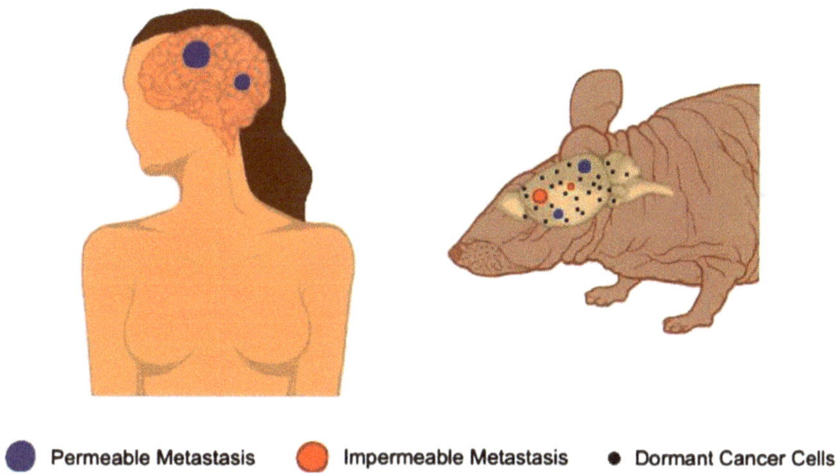


FIGURE 17.4 A comparison of radiological findings in human (left) and animal models (right). Current clinical magnetic resonance imaging (MRI) strategies detect brain metastases on the order of several millimeters in size, or larger. Contrast-enhanced MRI is routinely used, though it only detects metastases that are permeable to gadolinium-based contrast agents, which generally means the blood–brain barrier (BBB)/blood–tumor barrier (BTB) has been disrupted. Experimental microanatomical and cellular MRI techniques can codetect brain metastases and discrete regions of signal void, which represent nonproliferative cancer cells in animal models. Metastases on the order of a couple hundred cubic microns are easily detectable; in addition, multiple contrast mechanisms can be exploited to detect gadolinium-impermeable brain metastases where the BBB/BTB remains intact.

or hematogenously for brain metastasis development. Spontaneous brain metastasis models are advantageous because the metastatic cascade develops naturally. These models generally involve subcutaneous or orthotopic implantation of a metastatic primary cancer cell line, such as the B16 melanoma clone 3.5, and brain metastases naturally develop after a latency period (Alterman and Stackpole, 1989). Spontaneous models are difficult to initiate and are not frequently reported. Furthermore, the presence of a primary tumor can be challenging and may require surgical excision due to its detrimental effect on host health. In orthotopic injection models, tumor cells or tumor fragments are implanted directly into the brain parenchyma or cerebrospinal fluid space for metastasis development. This model is the simplest to initiate and can be useful because it develops only a single tumor in a known location. Moreover, it does not require the use of specific metastatic or brain-tropic cell lines. Though simple, implantation models are limiting because they bypass natural brain colonization and therefore may not adequately model all aspects of metastatic growth. Hematogenous delivery of cancer cells mimics parts of natural metastatic spread by transport of cancer cells via the circulation and therefore creates a more physiologically relevant model. A cancer cell suspension can be injected into the beating left ventricle of the mouse heart (intracardiac injection) or into the internal carotid artery (intra-arterial injection) for delivery to the brain. Intracardiac injection results in 3.5–9.5% of cells delivered to the mouse brain by cardiac output (Quintana et al., 1979; Basse et al., 1988). This model is relatively straightforward, but cancer cells also disseminate to other sites and may cause unwanted tumor growth. Intra-arterial injection guarantees that all injected cells are delivered to the brain and permanent ligation of the internal carotid artery ensures that only cerebral metastases develop. This model is difficult to initiate and advanced microsurgical skill is required to successfully execute inoculation procedures.

Several cell lines with biased organ tropism have been developed to use in breast cancer brain metastasis models, such as MDA-MB-231-BR, MDA-MB-231-BR-HER2, JIMT-1-BR3, SUM-190-BR3, and 4T1-BR5 (Table 17.4; Yoneda et al., 2001; Palmieri et al., 2007; Tanner et al., 2004; Lockman et al., 2010; Murrell et al., 2015). These brain metastatic cell lines are derived from primary breast cancer cell lines by repeated rounds of selection. A parental primary cancer cell line that is known to be metastatic is grown in culture and introduced hematogenously into mice. Metastases will develop into various organs over time, including in the brain. Brain tumors are isolated and harvested; these cells are then regrown in culture. This is referred to as a "brain derivative" of the parental cell line. Biased tropism for the brain is improved by several sequential rounds of reinjection of the "brain derivative" cell line, isolation of brain tumors, and regrowth in culture. These cells lines then inherently prefer the brain microenvironment, which is advantageous in hematogenous delivery models to mitigate metastatic growth in other organs.

Complete research models that accurately reflect clinical disease are rare. The

TABLE 17.4 Brain-Tropic (BR) Breast Cancer Cell Lines Derived From Parental Cell Lines by Successive Selection Rounds for Use in Preclinical Model Systems

| Cell Line | Origin |
|---|---|
| MDA-MB-231-BR | Human breast carcinoma; ER (estrogen receptor)-/PR (progesterone receptor)-/HER2 (human epidermal growth factor receptor 2)- |
| MDA-MB-231-BR-HER2 | Above, transfected with HER2 |
| JIMT-1-BR3 | Human breast carcinoma isolated from a Herceptin-resistant patient; ER-/PR-/HER2+ |
| SUM-190-BR3 | Human inflammatory breast cancer; ER-/PR-/HER2+ |
| 4T1-BR5 | Murine breast cancer |

MDA-MB-231-BR cell line delivered by intracardiac injection has been well characterized and produces a model that is similar to resected human brain metastases from breast cancer in many ways, including proliferation and apoptosis rates (Fitzgerald et al., 2008). For this reason, the "231BR" model has been used extensively to study brain metastases from triple-negative breast cancer (Perera et al., 2012; Yoneda et al., 2001; Fitzgerald et al., 2008; Heyn et al., 2006; Ribot et al., 2011; Kil et al., 2008; McGowan et al., 2011; Taskar et al., 2012; Gril et al., 2008). Moreover, the 231BR cell line was transduced with HER2 cDNA (MDA-MB-231-BR-HER2) to create a brain metastasis model for the HER2 subtype of breast cancer. This resulted in the formation of greater numbers of large metastases (>50 μm) compared with the non-HER2 version, but was irrelevant to the development of micrometastases (Perera et al., 2012; Palmieri et al., 2007). Further in vivo investigation compared this model to two other HER2+ brain metastasis models (JIMT-1-BR3, SUM-190-BR3) and found substantial disease heterogeneity between models with respect to tumor incidence, volume, morphology, and radiographic appearance despite that they were all meant to represent metastases from HER2+ breast cancer (Murrell et al., 2015). This may reflect the natural heterogeneity in human cancer and suggests that a panel of different models should be employed when evaluating new diagnostics and therapeutics.

# 5. PRECLINICAL FINDINGS

## 5.1 Blood–Tumor Barrier Patency

The BTB has been studied in preclinical models to determine its patency in brain metastasis and to investigate its role in systemic drug and contrast agent delivery. Lockman et al. (2010) found significant heterogeneity in the degree of BTB compromise associated with 231BR-HER2 and 4T1-BR5 brain metastases and that it was not correlated with tumor size. Moreover, they demonstrated that BTB permeability was generally greater than in normal brain but still remained a substantial impediment for drug delivery. They discovered partial BTB compromise in more than 89% of brain metastases, but drug uptake (paclitaxel, doxorubicin) only reached cytotoxic levels in the top 10% of permeable tumors (Lockman et al., 2010). Percy et al. (2012) also found heterogeneous BTB permeability within the 231BR-HER2 model. Their study showed that BTB patency prevented detection of small impermeable brain metastases by contrast-enhanced MRI, especially at early time points in disease progression; similar results suggested that BTB permeability affected tumor detection by MRI and these were also demonstrated in other brain metastasis models due to breast cancer or melanoma (Henry et al., 2015; Murrell et al., 2015; Percy et al., 2011). An example of contrast-enhanced and noncontrast-enhanced MRI in a 231-BR-HER2 model illustrates the effect of BTB permeability on brain metastasis detection (Fig. 17.5).

## 5.2 Dormancy

Cellular MRI is a technique that enables tracking of single cancer cells over time in model systems. First, cells are labeled in culture with an iron oxide contrast agent. After implantation into the host, cells appear as discrete regions of signal void within the brain. This technology was employed to follow the fate of brain metastatic cancer cells in the 231BR model over 1 month. Iron-labeled solitary cells, and small clusters of cells, were identified as regions of signal void in the brains of mice early after their injection. At the experimental endpoint MRI revealed three distinct cell fates: (1) a population of cells that were called "transient" (94%); these signal voids disappeared over time, (2) a population of cells that were called "nonproliferative" (4.5%); these remained as discrete

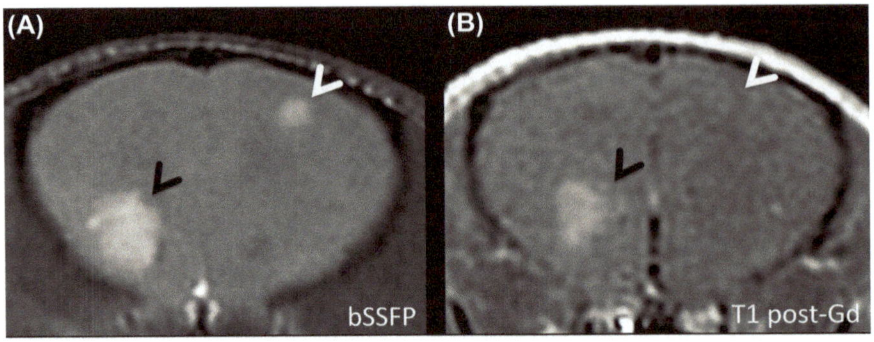


FIGURE 17.5 Experimental magnetic resonance imaging (MRI) of brain metastases in the 231-BR-HER2 (human epidermal growth factor receptor 2) model illustrates the effect of blood–brain barrier (BBB)/blood–tumor barrier (BTB) permeability on radiological tumor detection. Balanced steady-state free precession (bSSFP) imaging utilizes endogenous contrast for tumor detection; in this image, two metastases are detectable (A). T1-weighted contrast-enhanced [T1 postgadolinium (post-Gd)] MRI relies on tumor detection by intense enhancement postadministration of Gd intravenously, which occurs in lesions where the BBB/BTB is compromised; in this image, only one tumor is visible (B). A comparison of bSSFP and T1 post-Gd MRI suggests that the BBB/BTB associated with one metastasis has become damaged (*black arrow*), whereas the second tumor is able to grow with an intact BBB/BTB (*white arrow*).

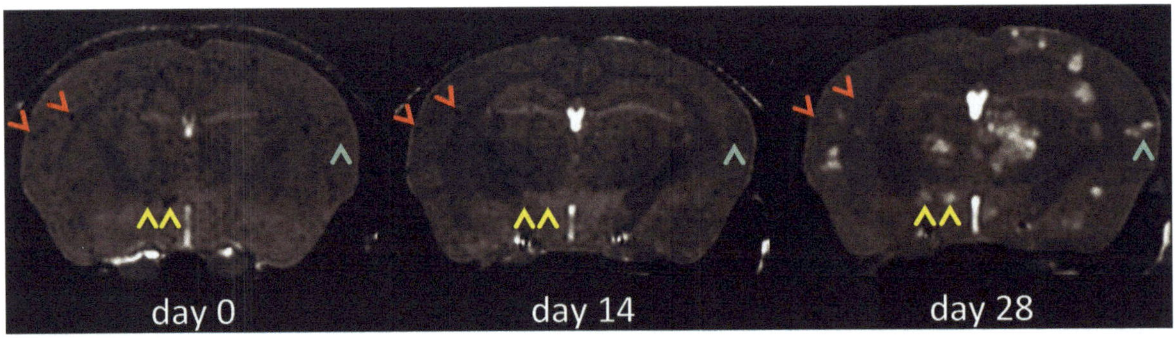


FIGURE 17.6 Cellular magnetic resonance imaging (MRI) illustrates the evolution of brain metastatic growth in the 231-BR-HER2 (human epidermal growth factor receptor 2) model. This technique enables the codetection of cancer cells with varying latency periods. After intracardiac injection (day 0), iron-labeled cancer cells are visualized as discrete regions of signal void. Proliferative cancer cells (*yellow arrows*) become undetectable over time (day 14) because the iron label is diluted between daughter cells with each cell division. By day 28 postcell injection, brain metastases are detectable in this region; they appear hyperintense on MRI due to tumor-associated edema. A subset of nonproliferating cancer cells retain their iron label and remain detectable as discrete signal voids over time (*red arrows*) or begin to proliferate after an extended latency period (*blue arrows*).

signal voids over time, and (3) a population of cells that we called "proliferative" (1.5%); these were signal voids in a region where a tumor developed (Heyn et al., 2006). Fig. 17.6 illustrates the evolution of these cell fates over time in a 231-BR-HER2 model. The subpopulation of nonproliferative, "quiescent" or "dormant" cancer cells that persist in the brain are especially interesting for their potential modulation of treatment response. Anticancer therapies traditionally target actively proliferating cancer cells, which may allow nonproliferating cancer cells to evade treatment (Townson et al., 2009). If these cells remain viable and are able to reinitiate growth, they could drive cancer recurrence by proliferating to form clinically relevant metastases in the future.

## 6. CONCLUSION

Brain metastases occur in 10% and 50% of cancer patients, depending on the type of primary tumor. They are much more common than primary brain tumors. The brain microenvironment supports tumor growth by providing a rich blood supply, and limits the effectiveness of therapeutics, by preventing access to the brain via the BBB. A diagnosis of brain metastasis is terrifying since treatment options are limited, mostly palliative, and survival is typically less than 1 year. Experimental models of brain metastasis are a powerful way to study mechanisms underlying metastasis, dormancy, and the role of different subpopulations of cancer cells. Advances in imaging technologies, such as MRI, positron emission tomography, and bioluminescence imaging have provided insight into brain metastasis evolution by allowing experimental models to be studied in vivo, over long periods of time and at ultra-high resolution. Advances in the field of cellular and molecular MRI are emerging as methods to improve diagnosis, predict prognosis, and treatment of brain metastasis. For example, the development of novel MRI contrast agents that can identify molecular targets will improve specificity. Recognition of the increasing incidence of brain metastasis points to a critical need for novel approaches to investigating this complex and feared consequence of systemic cancer.

## References


Andrews, D.W., Scott, C.B., Sperduto, P.W., et al., 2004. Whole brain radiation therapy with or without stereotactic radiosurgery boost for patients with one to three brain metastases: phase III results of the RTOG 9508 randomised trial. Lancet 363, 1665–1672.

Alterman, A.L., Stackpole, C.W., 1989. B16 melanoma spontaneous brain metastasis: occurrence and development within leptomeninges blood vessels. Clin. Exp. Metastasis 7, 15–23.

Barnholtz-Sloan, J.S., Sloan, A.E., Davis, F.G., Vigneau, F.D., Lai, P., Sawaya, R.E., 2004. Incidence proportions of brain metastases in patients diagnosed (1973 to 2001) in the Metropolitan Detroit Cancer Surveillance System. J. Clin. Oncol. 22, 2865–2872.

Bendell, J.C., Domchek, S.M., Burstein, H.J., et al., 2003. Central nervous system metastases in women who receive trastuzumab-based therapy for metastatic breast carcinoma. Cancer 97, 2972–2977.

Bugyik, E., Dezso, K., Reiniger, L., et al., 2011. Lack of angiogenesis in experimental brain metastases. J. Neuropathol. Exp. Neurol. 70, 979–991.

Brem, S.S., Bierman, P.J., Black, P., et al., 2011. Central nervous system cancers: Clinical Practice Guidelines in Oncology. J. Natl. Compr. Cancer Netw. 9, 352–400.

Bachelot, T., Romieu, G., Campone, M., et al., 2013. Lapatinib plus capecitabine in patients with previously untreated brain metastases from HER2-positive metastatic breast cancer (LANDSCAPE): a single-group phase 2 study. Lancet Oncol. 14, 64–71.

Basse, P., Hockland, P., Heron, I., Hokland, M., 1988. Fate of tumor cells injected into left ventricle of heart in BALB/c mice: role of natural killer cells. J. Natl. Cancer Inst. 80, 657–665.

Chambers, A.F., Naumov, G.N., Vantyghem, S.A., Tuck, A.B., 2000. Molecular biology of breast cancer metastasis. Clinical implications of experimental studies on metastatic inefficiency. Breast Cancer Res. 2, 400–407.

Connell, J.J., Chatain, G., Cornelissen, B., et al., 2013. Selective permeabilization of the blood–brain barrier at sites of metastasis. J. Natl. Cancer Inst. 105, 1634–1643.

Davis, F.G., Dolecek, T.A., McCarthy, B.J., Villano, J.L., 2012. Toward determining the lifetime occurrence of metastatic brain tumors estimated from 2007 United States cancer incidence data. Neuro Oncol. 14, 1–7.

Deeken, J.F., Löscher, W., 2007. The blood–brain barrier and cancer: transporters, treatment, and Trojan horses. Clin. Cancer Res. 13, 1663–1674.

Döme, B., Hendrix, M.J.C., Paku, S., Tóvári, J., Tímár, J., 2007. Alternative vascularization mechanisms in cancer: pathology and therapeutic implications. Am. J. Pathol. 170, 1–15.

Donnem, T., Hu, J., Ferguson, M., et al., 2013. Vessel co-option in primary human tumors and metastases: an obstacle to effective anti-angiogenic treatment? Cancer Med. 2, 427–436.

Duchnowska, R., Jassem, J., Goswami, C.P., et al., 2015. Predicting early brain metastases based on clinicopathological factors and gene expression analysis in advanced HER2-positive breast cancer patients. J. Neuro Oncol. 122, 205–216.

Eichler, A.F., Loeffler, J.S., 2007. Multidisciplinary management of brain metastases. Oncologist 12, 884–898.

Fife, K.M., Colman, M.H., Stevens, G.N., et al., 2004. Determinants of outcome in melanoma patients with cerebral metastases. J. Clin. Oncol. 22, 1293–1300.

Fink, K.R., Fink, J.R., 2013. Imaging of brain metastases. Surg. Neurol. Int. 4, S209–S219.

Fidler, I.J., Gruys, E., Cifone, M.A., Barnes, Z., Bucana, C., 1981. Demonstration of multiple phenotypic diversity in a murine melanoma of recent origin. J. Natl. Cancer Inst. 67, 947–956.

Fitzgerald, D.P., Palmieri, D., Hua, E., et al., 2008. Reactive glia are recruited by highly proliferative brain metastases of breast cancer and promote tumor cell colonization. Clin. Exp. Metastasis 25, 799–810.

Gil-Gil, M.J., Martinez-Garcia, M., Sierra, A., et al., 2014. Breast cancer brain metastases: a review of the literature and a current multidisciplinary management guideline. Clin. Transl. Oncol. 16, 436–446.

Gaspar, L., Scott, C., Rotman, M., et al., 1997. Recursive partitioning analysis (RPA) of prognostic factors in three Radiation Therapy Oncology Group (RTOG) brain metastases trials. Int. J. Radiat. Oncol. Biol. Phys. 37, 745–751.

Gaspar, L.E., Scott, C., Murray, K., Curran, W., 2000. Validation of the RTOG recursive partitioning analysis (RPA) classification for brain metastases. Int. J. Radiat. Oncol. Biol. Phys. 47, 1001–1006.

Gril, B., Palmieri, D., Bronder, J.L., et al., 2008. Effect of lapatinib on the outgrowth of metastatic breast cancer cells to the brain. J. Natl. Cancer Inst. 100, 1092–1103.

Huang, F.J., Steeg, P.S., Price, J.E., et al., 2008. Molecular basis for the critical role of suppressor of cytokine signaling-1 in melanoma brain metastasis. Cancer Res. 68, 9634–9642.

Huang, Q., Ouyang, X., 2013. Predictive biochemical-markers for the development of brain metastases from lung cancer: clinical evidence and future directions. Cancer Epidemiol. 37, 703–707.

Huang, F., Alrefae, M., Langleben, A., Roberge, D., 2009. Prophylactic cranial irradiation in advanced breast cancer: a case for caution. Int. J. Radiat. Oncol. 73, 752–758.

Henry, M.N., Chen, Y., McFadden, C.D., Simedrea, F.C., Foster, P.J., 2015. In-vivo longitudinal MRI study: an assessment of melanoma brain metastases in a clinically relevant mouse model. Melanoma Res. 25, 127–137.

Heyn, C., Ronald, J.A., Ramadan, S.S., et al., 2006. In vivo MRI of cancer cell fate at the single-cell level in a mouse model of breast cancer metastasis to the brain. Magn. Reson. Med. 56, 1001–1010.

Jemal, A., Siegel, R., Xu, J., Ward, E., 2010. Cancer statistics, 2010. CA Cancer J. Clin. 60, 277–300.

Klos, K.J., O'Neill, B.P., 2004. Brain metastases. Neurologist 10, 31–46.

Kienast, Y., von Baumgarten, L., Fuhrmann, M., et al., 2010. Real-time imaging reveals the single steps of brain metastasis formation. Nat. Med. 16, 116–122.

Kocher, M., Soffietti, R., Abacioglu, U., et al., 2011. Adjuvant whole-brain radiotherapy versus observation after radiosurgery or surgical resection of one to three cerebral metastases: results of the EORTC 22952-26001 study. J. Clin. Oncol. 29, 134–141.

Kil, W.J., Cerna, D., Burgan, W.E., et al., 2008. In vitro and in vivo radiosensitization induced by the DNA methylating agent temozolomide. Clin. Cancer Res. 14, 931–938.

Lin, N.U., Bellon, J.R., Winer, E.P., 2004. CNS metastases in breast cancer. J. Clin. Oncol. 22, 3608–3617.

Lai, R., Dang, C.T., Malkin, M.G., Abrey, L.E., 2004. The risk of central nervous system metastases after trastuzumab therapy in patients with breast carcinoma. Cancer 101, 810–816.

Leenders, W.P.J., Küsters, B., Verrijp, K., et al., 2004. Antiangiogenic therapy of cerebral melanoma metastases results in sustained tumor progression via vessel co-option. Clin. Cancer Res. 10, 6222–6230.

Lockman, P.R., Mittapalli, R.K., Taskar, K.S., et al., 2010. Heterogeneous blood-tumor barrier permeability determines drug efficacy in experimental brain metastases of breast cancer. Clin. Cancer Res. 16, 5664–5678.

McMullin, R.P., Wittner, B.S., Yang, C., et al., 2014. A BRCA1 deficient-like signature is enriched in breast cancer brain metastases and predicts DNA damage-induced poly (ADP-ribose) polymerase inhibitor sensitivity. Breast Cancer Res. 16, R25.

Mathieu, A., Remmelink, M., D'Haene, N., et al., 2004. Development of a chemoresistant orthotopic human nonsmall cell lung carcinoma model in nude mice. Cancer 101, 1908–1918.

Murrell, D.H., Hamilton, A.M., Mallett, C.L., van Gorkum, R., Chambers, A.F., Foster, P.J., 2015. Understanding heterogeneity and permeability of brain metastases in murine models of HER2-positive breast cancer through magnetic resonance imaging: implications for detection and therapy. Transl. Oncol. 8, 176–184.

McGowan, P.M., Simedrea, C., Ribot, E.J., et al., 2011. Notch1 inhibition alters the CD44hi/CD24lo population and reduces the formation of brain metastases from breast cancer. Mol. Cancer Res. 9, 834–844.

Niwińska, A., Murawska, M., Pogoda, K., 2010. Breast cancer brain metastases: differences in survival depending on biological subtype, RPA RTOG prognostic class and systemic treatment after whole-brain radiotherapy (WBRT). Ann. Oncol. 21, 942–948.

Perera, M., Ribot, E.J., Percy, D.B., et al., 2012. In vivo magnetic resonance imaging for investigating the development and distribution of experimental brain metastases due to breast cancer. Transl. Oncol. 5, 217–225.

Pestalozzi, B.C., 2009. Brain metastases and subtypes of breast cancer. Ann. Oncol. 20, 803–805.
Patchell, R.A., Tibbs, P.A., Walsh, J.W., et al., 1990. A randomized trial of surgery in the treatment of single metastases to the brain. N. Engl. J. Med. 322, 494–500.
Patchell, R.A., Tibbs, P.A., Regine, W.F., et al., 1998. Postoperative radiotherapy in the treatment of single metastases to the brain. J. Am. Med. Assoc. 280, 1485–1489.
Palmieri, D., Bronder, J.L., Herring, J.M., et al., 2007. Her-2 overexpression increases the metastatic outgrowth of breast cancer cells in the brain. Cancer Res. 67, 4190–4198.
Percy, D.B., Ribot, E.J., Chen, Y., et al., 2011. In vivo characterization of changing blood-tumor barrier permeability in a mouse model of breast cancer metastasis: a complementary magnetic resonance imaging approach. Invest. Radiol. 46, 718–725.
Quintana, A., Raczka, E., Bonaccorsi, A., 1979. Cardiac output distribution measured with radioactive microspheres in the mouse. Pharmacol. Res. Commun. 11, 245–252.
Ribot, E.J., Martinez-Santiesteban, F.M., Simedrea, C., et al., 2011. In vivo single scan detection of both iron-labeled cells and breast cancer metastases in the mouse brain using balanced steady-state free precession imaging at 1.5 T. J. Magn. Reson. Imaging 34, 231–238.
Schouten, L.J., Rutten, J., Huveneers, H.A.M., Twijnstra, A., 2002. Incidence of brain metastases in a cohort of patients with carcinoma of the breast, colon, kidney, and lung and melanoma. Cancer 94, 2698–2705.
Stemmler, H.J., Kahlert, S., Siekiera, W., Untch, M., Heinrich, B., Heinemann, V., 2006. Characteristics of patients with brain metastases receiving trastuzumab for HER2 overexpressing metastatic breast cancer. Breast 15, 219–225.
Shmueli, E., Wigler, N., Inbar, M., 2004. Central nervous system progression among patients with metastatic breast cancer responding to trastuzumab treatment. Eur. J. Cancer 40, 379–382.
Smedby, K.E., Brandt, L., Bäcklund, M.L., Blomqvist, P., 2009. Brain metastases admissions in Sweden between 1987 and 2006. Br. J. Cancer 101, 1919–1924.
Steeg, P.S., Camphausen, K.A., Smith, Q.R., 2011. Brain metastases as preventive and therapeutic targets. Nat. Rev. Cancer 11, 352–363.
Saunus, J.M., Quinn, M.C., Patch, A.-M., et al., 2015. Integrated genomic and transcriptomic analysis of human brain metastases identifies alterations of potential clinical significance. J. Pathol. 237, 363–378.
Son, C.H., Jimenez, R., Niemierko, A., Loeffler, J.S., Oh, K.S., Shih, H.A., 2012. Outcomes after whole brain reirradiation in patients with brain metastases. Int. J. Radiat. Oncol. Biol. Phys. 82, e167–e172.
Schellinger, P.D., Meinck, H.M., Thron, A., 1999. Diagnostic accuracy of MRI compared to CCT in patients with brain metastases. J. Neurooncol. 44, 275–281.
Sperduto, P.W., Berkey, B., Gaspar, L.E., Mehta, M., Curran, W., 2008. A new prognostic index and comparison to three other indices for patients with brain metastases: an analysis of 1,960 patients in the RTOG database. Int. J. Radiat. Oncol. 70, 510–514.
Sperduto, P.W., Chao, S.T., Sneed, P.K., et al., 2010. Diagnosis-specific prognostic factors, indexes, and treatment outcomes for patients with newly diagnosed brain metastases: a multi-institutional analysis of 4,259 patients. Int. J. Radiat. Oncol. Biol. Phys. 77, 655–661.
Sperduto, P.W., Kased, N., Roberge, D., et al., 2012. Summary report on the graded prognostic assessment: an accurate and facile diagnosis-specific tool to estimate survival for patients with brain metastases. J. Clin. Oncol. 30, 419–425.
Slotman, B., Faivre-Finn, C., Kramer, G., et al., 2007. Prophylactic cranial irradiation in extensive small-cell lung cancer. N. Engl. J. Med. 357, 664–672.
Tanner, M., Kapanen, A.I., Junttila, T., et al., 2004. Characterization of a novel cell line established from a patient with Herceptin-resistant breast cancer. Mol. Cancer Ther. 3, 1585–1592.
Tao, K., Fang, M., Alroy, J., Sahagian, G.G., 2008. Imagable 4T1 model for the study of late stage breast cancer. BMC Cancer 8, 228–246.
Taskar, K.S., Rudraraju, V., Mittapalli, R.K., et al., 2012. Lapatinib distribution in HER2 overexpressing experimental brain metastases of breast cancer. Pharm. Res. 29, 770–781.
Townson, J.L., Ramadan, S.S., Simedrea, C., et al., 2009. Three-dimensional imaging and quantification of both solitary cells and metastases in whole mouse liver by magnetic resonance imaging. Cancer Res. 69, 8326–8331.
Viani, G.A., Castilho, M.S., Salvajoli, J.V., et al., 2007. Whole brain radiotherapy for brain metastases from breast cancer: estimation of survival using two stratification systems. BMC Cancer 7, 53–61.
Woditschka, S., Evans, L., Duchnowska, R., et al., 2014. DNA double-strand break repair genes and oxidative damage in brain metastasis of breast cancer. J. Natl. Cancer Inst. 106, 1–13.
Weissman, D.E., 1988. Glucocorticoid treatment for brain metastases and epidural spinal cord compression: a review. J. Clin. Oncol. 6, 543–551.
Xie, T.X., Huang, F.J., Aldape, K.D., et al., 2006. Activation of stat3 in human melanoma promotes brain metastasis. Cancer Res. 66, 3188–3196.
Yoneda, T., Williams, P.J., Hiraga, T., Niewolna, M., Nishimura, R., 2001. A bone-seeking clone exhibits different biological properties from the MDA-MB-231 parental human breast cancer cells and a brain-seeking clone in vivo and in vitro. J. Bone Miner. Res. 16, 1486–1495.

PART III

# EMERGING TARGETS FOR CANCER METASTASIS

CHAPTER

# 18

# Developmental Pathways: Emerging Therapeutic Targets for Cancer Metastasis

A.H. Nwabo Kamdje[1], L. Vecchio[2], P. Takam Kamga[3], P.F. Seke Etet[2], J.M. Muller[4], G. Bassi[3], M. Krampera[3]

[1]University of Ngaoundere, Ngaoundere, Cameroon; [2]Qassim University, Buraydah, Saudi Arabia; [3]University of Verona, Verona, Italy; [4]University of Poitiers, Poitiers, France

## 1. INTRODUCTION

The development of metastasis from a primary tumor site is a multistep phenomenon including epithelial–mesenchymal transition (EMT), tumor neoangiogenesis, and spread of malignancy to target tissues and organs. Such spreading results from malignant cell transport through blood vessels to target tissues and organs, the invasion of the latter by infiltrating malignant cells, and finally, the development of secondary tumors (Geiger and Peeper, 2009; van Zijl et al., 2011). Master developmental pathways, such as Notch, Wnt, and Hedgehog, are signaling pathways playing pivotal roles all along embryonic development. The role of Notch in solid tumor and hematological malignancy initiation and development has been extensively documented. Overall, although it may also act as a tumor suppressor, in most cases Notch is a major oncogene associated with tumor progression to metastasis, anoikis resistance, EMT, neoangiogenesis, malignant cell proliferation, and changes in tissue microenvironment promoting the homing of metastasis promoting cells (Espinoza and Miele, 2013; Ntziachristos et al., 2014; Nwabo Kamdje et al., 2014; Seke Etet et al., 2012).

The overexpression of Wnt signaling is also common in many hematological malignancies and solids tumors. Clinical and experimental evidence suggests that Wnt/β-catenin activation is critical for cancer development, angiogenesis, migration, and invasion (Clevers, 2006; Nwabo Kamdje et al., 2014; Seke Etet et al., 2012). Antagonists of Wnt pathway, such as Wnt inhibitory factor 1, Dickkopf proteins, the secreted frizzled-related proteins, and disheveled-axin domain containing 1, mitigated tumorigenic and metastatic processes of various

*Introduction to Cancer Metastasis*
http://dx.doi.org/10.1016/B978-0-12-804003-4.00018-9

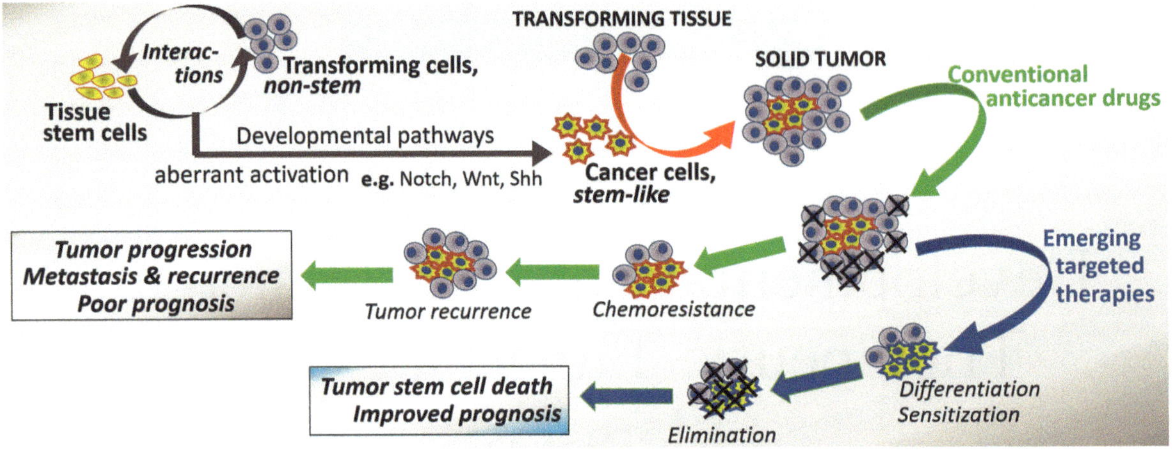


FIGURE 18.1 **Effects of emerging combined therapies.** Effects of conventional anticancer drugs and their combined therapies with emerging therapies targeting developmental pathways of cancer stem-like cells and the treatment outcome.

cancer types in vitro and in vivo (Chung et al., 2009; Lin et al., 2013; Xu et al., 2014). As Notch and Wnt pathways, hedgehog is an evolutionary and developmental pathway involved in cellular differentiation, but also in physiological and tumorigenic control of postnatal cellular events such as proliferation, cell death, motility, migration, and invasion (Nwabo Kamdje et al., 2014; Seke Etet et al., 2012).

Emerging therapies targeting master developmental pathways represent promising approaches for metastasis prevention and cancer stem cell elimination (see Fig. 18.1). We herein summarize data supporting a role for developmental pathways in prometastatic processes, with an emphasis on the master developmental pathways Notch, Wnt, and Shh, considering their therapeutic potential. Emerging antimetastatic strategies targeting these developmental pathways are also discussed.

## 2. PROMETASTATIC CHANGES AND DEVELOPMENTAL PATHWAYS

### 2.1 Resistance to Anoikis

As other anchorage-dependent cells, survival of tumor cells following detachment from surrounding epithelial tissue or extracellular matrix requires the ability to resist against the form of programmed cell death resulting from such detachment termed as anoikis. Indeed, unlike most normal anchorage-dependent cells that undergo cell death when detached from the extracellular matrix, tumor cells can implement several molecular mechanisms promoting their survival in suspension and the spread of metastases. Experimental and clinical evidence strongly suggests that subversion of the repertoire of integrin developmental pathways is the main molecular mechanism favoring anoikis resistance in tumor cells.

Cell–cell and cell–extracellular matrix contacts support anchorage-dependent cell survival in an integrin-dependent fashion, partly due to integrin-mediated activation of antiapoptotic signaling pathways such as phosphoinositide 3-kinase (PI3K)/protein kinase B (Akt), mitogen-activated protein kinase (MAPK), and focal adhesion kinase pathways (Chiarugi and Giannoni, 2008; Paoli et al., 2013). Aberrant constitutive activation of these antiapoptotic signaling pathways results in posttranslational modifications and changes in levels of Bcl-2 antiapoptotic protein family members (i.e., multidomain and Bcl-2 homologous 3 (BH3) domain-only proteins). In turn, overexpression of the latter proteins in tumor cells prevents the mitochondrial function disruptions that normally lead to programmed cell death (Martin and Vuori, 2004), as well as related proapoptotic mechanisms such as metabolism-induced autophagy (Chiarugi and Giannoni, 2008; Paoli et al., 2013).

In addition to these intrinsic apoptotic mechanisms, metastatic competent cells also achieve resistance to anoikis and anticancer therapy by interfering with extrinsic apoptosis pathways, such as death receptors' families like Fas receptors (Su et al., 2015). Notably, the prevention of Fas receptor activity results in increased levels of cellular fas-associated protein with death domain (FADD)-like interleukin-1 beta (IL-1ß)-converting enzyme inhibitory protein, a master antiapoptotic regulator, in these tumor cells. This increase results in a crosstalk between extrinsic and intrinsic caspase pathways that impair the ability to use common apoptotic caspases, in particular effector caspases, favoring tumor cell survival (Su et al., 2015). Furthermore, integrin signaling has also reported the ability to induce ligand-independent activation of other developmental signaling molecules that are able to trigger resistance to anoikis. Examples include loss in cell adhesion factors like E-cadherin and overactivation of other EMT signaling molecules like mammalian target of rapamycin (mTORC1 and mTORC2) (Barnes et al., 2010; Gulhati et al., 2011; Kumar et al., 2011), growth factors like epidermal growth factor (EGF) receptor and vascular endothelial growth factor (VEGF) receptor signaling pathways (Paoli et al., 2013; Sher et al., 2009), and oncogenes like Wnt, β-catenin, Notch, and Shh (Kandala and Srivastava, 2012; Kwon et al., 2014; Yu et al., 2012).

The potential of the pharmacological targeting of developmental signaling pathways favoring resistance to anoikis is also raised by the fact that they support the survival of metastatic cells throughout the metastatic journey, i.e., from the primary site to target tissue invasion and secondary tumor development (Kumar et al., 2011; Paoli et al., 2013).

## 2.2 Epithelial-to-Mesenchymal Transition

EMT is a reversible specialized process taking place during the normal embryonic development, which is pivotal for cell specialization and tissue patterning. During development, early embryonic epithelial cells detachment from the initial tissue occurs without activation of anoikis. Then, these cells acquire the motile shape that will allow them to migrate to new sites where they will undergo mesenchymal-to-epithelial transition (MET) providing them with the ability to build new tissues (for review see Kalluri, 2009; Wang et al., 2015). This process, which also occurs during wound healing, is also used by transforming cells during tumorigenesis and tumor progression. Seemingly, primary tumor cells undergo EMT during cancer progression, which confers them a mesenchymal-like phenotype associated with high motility, but also resistance to anoikis and anticancer drugs (Diaz and de Herreros, 2016; Kalluri, 2009; Wang et al., 2015; Wei and Yang, 2016).

Although normal development EMT is controlled and highly organized, this event appears to be more stochastic in cancer (Arwert et al., 2012). Physiological EMT results from the coordination of molecular events associated with new cell capacity. More specifically, EMT starts by the loss of cell junction proteins associated with epithelial organization such as E-cadherin, claudins, occludins, and catenins, followed closely by the expression of mesenchymal markers such as N-cadherin and vimentin (Garg, 2013; Lamouille et al., 2014). A large number of oncogenic factors known to induce and/or sustain tumor cell motility and metastasis, such as transforming growth factor beta (TGF-β) and insulin-like growth factor 1, induce the loss of the epithelial marker E-cadherin (Margadant and

Sonnenberg, 2010; Vazquez-Martin et al., 2013). E-cadherin is at the center of EMT and anoikis signaling networks, as its loss is common along cancer progression where it associates with aberrant EMT, tumor cell migration and invasion, and resistance to anoikis. Besides, the loss of E-cadherin and other adherens junction molecules disrupts epithelial tissue, leading cells to the adoption of a less polarized phenotype with higher invasive properties. The switch in cell morphologies and properties is mediated by key transcription factors termed as EMT-activating transcription factors. Examples include the transcription factors of Snail family (e.g., Snail, Slug, Smuc), zinc-finger E-box-binding (ZEB) factors [e.g., the two-handed zinc-finger factors of d-crystallin/E2 box factor (dEF1) family proteins dEF1/ZEB1 and Smad-interacting protein (SIP)1/ZEB2], and basic helix–loop–helix factors (e.g., Twist1, Twist2, E12, E47, and Tbx3) (Nieto, 2002; Ozawa and Kobayashi, 2015; Perez-Moreno et al., 2001).

## 2.3 Motility and Migration

Metastatic competent cells detached from primary tumors need to migrate to invade others sites. Migration and other motility events result from specific cytoskeletal changes mediated and regulated mainly by small GTPase proteins of the Rho family (Pajic et al., 2015; Rathinam et al., 2011). Members of the rho GTPase family process extracellular signal into information that will lead to the reorganization of the cytoskeleton. Notably, as discussed above, the molecular events that trigger EMT and confer anoikis can activate the migratory machine in tumor cells through members of the Rho GTPase family (Martin-Villar et al., 2006).

Mounting evidence supports Rho GTPase involvement in all tumorigenic events, from primary tumor initiation to metastasis. These small GTPases direct actin dynamic and protrusive patterns in migrating cells, including the projection of the cytoskeletal protein actin on the leading edge of the cell (lamellipodia), the extension of slender cytoplasmic projections beyond the leading edge of lamellipodia (filopodia), and extracellular matrix degradation-associated protrusions of the plasma membrane (invadopodia) (Rathinam et al., 2011). Actin dynamics result from its polymerization induced by actin-related protein 2 (Arp2) and Arp3. Invadopodia is characteristic of invasive tumor cells, whereas the other patterns can also occur in normal processes (Stoletov and Lewis, 2015). RhoA, Cdc42, and Rac1 are the most documented Rho GTPases involved in cell migration and metastasis. Besides these common properties, RhoA is also linked to stress fiber, whereas Cdc42 and rac1 are involved in filopodia and lamellipodia, respectively (Citi et al., 2014; Fortin Ensign et al., 2013).

Moreover, in addition to EMT-induced protrusive processes, migrating tumor cells can also undergo collective ameboid transition and mesenchymal to ameboid transition, with a characteristic invadopodia (Alizadeh et al., 2014; Stoletov and Lewis, 2015). Notably, such invadopodia requires the enzymatic digestion of the extracellular matrix by the leading edge, mediated by matrix metalloproteinases. This type of migration is observed in many solid tumors, including breast carcinoma (Chen et al., 1994), for instance.

## 2.4 Neoangiogenesis

The angiogenic processes triggered during tumorigenesis (termed as neoangiogenesis) provide the neoplastic tissue with new blood vessels that will supply the high needs of this tissue in nutrients and oxygen. Such high needs are triggered by paracrine growth signaling aberrantly activated directly or indirectly in transforming cells in response to extracellular matrix digestion (Villanueva, 2015). The new blood vessels are also used as highways by invading tumor cells, which, thanks to the motility provided by EMT, are able to travel great distances in the body to start metastases in other organs.

Neoangiogenic dynamics mainly rely on angiogenic signaling molecules of VEGF family (VEGF-A/B/C/D) and their receptors (VEGFR-1/2/3/4) (Przybylski, 2009; Pugh and Ratcliffe, 2003). The main activator of the angiogenic cascade in the transforming and neoplastic tissue is the hypoxic microenvironment. The low oxygen tension observed in the microenvironment induces the stabilization of hypoxia-inducible factors (HIFs) preventing the von Hippel–Lindau binding protein-mediated degradation of these transcription factors. Then, HIFs enter tumor cell nuclei, where they dimerize and associate with p/300 to form a transactivation complex. In turn, this transcriptional complex will stimulate the expression of hypoxia target genes by binding to their hypoxia response elements. Hypoxia targets genes, including genes involved in the angiogenesis cascade, such as VEGF system genes, as well as other ligand–receptor systems with high angiogenic potential such as angiopoetin/Tie2, platelet-derived growth factor (PDGF)/PDGF receptor, Delta-like ligand 4 (DLL4)/Notch, and fibroblast growth factor (FGF)/FGF receptor (Johnson et al., 2015; Schodel et al., 2016).

# 3. NOTCH SIGNALING AND METASTASIS

## 3.1 Notch Family of Signaling Molecules

Notch family of transmembrane proteins encompasses four receptors (Notch1–4) and five ligands (Jagged1/2, Dll-1/3/4). Notch receptors are activated when they bind to a ligand expressed on the membrane of an adjacent cell. Then, the receptors undergo two proteolytic cleavages at S2 and S3 sites performed by a disintegrin and metalloproteinase (ADAM) 10/17 metalloprotease and gamma secretase complex, respectively. These two proteolytic events result in the release of an intracellular active form of the receptors called Notch intracellular domain (NICD). Subsequently, NICD domains enter the nucleus where they form a transcriptional activation complex with maltose operon periplasmic protein 1 (MalM1) and recombination signal binding protein for immunoglobulin Kappa J (RBP-jk) proteins. This complex will promote the expression of various genes involved in cell fate determination, including for instance *myc*, *cyclinD*, and genes of the helix basic family such as *hes1* and *hey1*. Functionally, products of this transcription will control cell proliferation, cell death, adhesion, invasion, and migration.

## 3.2 Notch in the Metastatic Cascade

Notch signaling controls survival pathways important for anoikis resistance such as nuclear factor-κB (NF-κB), Akt, Sat3, mTOR (Guo et al., 2011). It was reported that the ability to modulate the expression of EMT transcription factors such as Snail, Slug, Zinc finger E-box-binding homeobox 1/2 (ZEB1/2), transforming growth factor beta (TGF-β), fibroblast growth factor (FGF), and platelet-derived growth factor ( PDGF) to increase the expression of mesenchymal markers (e.g., Slug) and to downregulate epithelial markers and activate Rho GTPases (Espinoza and Miele, 2013; Kwon et al., 2014; Shao et al., 2015). For instance, Kwon et al. (2014) assessed Notch signaling prometastatic role in a mouse model of prostate cancer. Constitutive expression of Notch1 intracellular domain in mouse prostate luminal cells: (1) impaired secretory functions; (2) suppressed anoikis via Hes1 nondependent NF-κB activity; and (3) stimulated luminal cell proliferation by potentiating PI3K/Akt signaling (Kwon et al., 2014). In a more recent report by Shao et al. (2015) Jagged1-induced Notch signaling triggered, migration, invasion, and a Slug-dependent EMT transition in breast cancer cells. All these effects were abrogated by Notch silencing (Shao et al., 2015), underlining the potential of Notch targeting for the modulation of tumor cell motility. Furthermore, Rho-GTPase upregulation is common in patients diagnosed with T-cell acute lymphoblastic leukemia, a hematological malignancy where Notch activating mutations are observed in more than 50% of cases (Bhavsar et al., 2013).

## 3.3 Notch Targeting as Antimetastatic Strategy

### 3.3.1 *Notch-Based Therapeutic Strategies*

Notch-based therapeutic strategies were developed for various steps of the signaling cascade including ligand–receptor interactions, enzymatic cleavages of the receptors, cytoplasmic interactions, and the transcriptional activation complex. Notably, the requirement of S2 and S3 cleavage for the activation of Notch receptors made these events particularly attractive for the modulation of Notch-induced repression of E-cadherin, which associates with a number of protumorigenic events and poor prognosis in solid cancers such as breast cancer (Leong et al., 2007). As already mentioned, S2 and S3 are due, respectively, to ADAM10/17 metalloprotease and secretase complex, respectively. Unlike S2 cleavage inhibitors, gamma secretase inhibitors (GSIs) have been widely used in anticancer preclinical research, partly because they were developed first and tested in Alzheimer's disease and experimental models where they proved relatively safe (Shih and Wang, 2007). Interestingly, GSIs achieving total blockade of Notch signaling display strong antineoplastic responses, whereas even those achieving partial blockade can display decent antineoplastic activities (Nwabo Kamdje et al., 2012; Schwarzer et al., 2014).

GSI treatment mitigated the development of tumor cell invasion and metastasis in vivo and in vitro by preventing EMT, migration, invasion, and neoangiogenesis (Lopez-Guerra et al., 2015). A number of GSIs have recently entered clinical trials, including BMS906024, MK0752, PF03084014, and R04929097 (Andersson and Lendahl, 2014). PF03084014 is currently in phase 1 trial in metastatic pancreatic adenocarcinoma patients not previously treated with anticancer therapies. R04929097 is in phase 2 trial in metastatic melanoma (Lee et al., 2015). MK0752 is in phases 1 and 2 trials for metastatic and advanced primary tumors, respectively, in breast cancer (Andersson and Lendahl, 2014). Moreover, blocking antibodies with GSI activity were also developed. These agents target components of GSI complex, such as nicastrin, the largest member of the complex. Anti-nicastrin monoclonal antibodies clone 2H6 elicited pleiotropic antimetastatic activities on invasive cancer cell lines, including an attenuation of invadopodia degradation of the extracellular matrix and delayed cancer cell extravasation through endothelial cells in the in vitro Boyden chamber invasion assay (Filipovic et al., 2014).

Enthusiasm for anticancer and antimetastatic application of GSIs is mitigated by the limitation of gamma secretase targeting, considering that this Notch signaling component is not always critical in the pathological phenotype of malignancies. In addition, a number of tissues physiologically require Notch activity for tissue plasticity. Gastric epithelium is the most relevant example of such tissue. Its functional alterations partly explain the gastrointestinal toxicity observed following GSI treatment (Andersson and Lendahl, 2014; Lee et al., 2015; Schwarzer et al., 2014). Interestingly, in vivo experiments suggest that anti-nicastrin monoclonal antibodies are more potent than GSIs currently in clinical trials, with less marked gastrointestinal signs (Filipovic et al., 2014). Altogether, these observations suggested the possibility to improve the therapeutic outcome of Notch signaling targeting by changing the specific targets and the approaches used. These findings raised the question whether the development of inhibitors for specific Notch receptors or ligand would also improve therapeutic outcome.

### 3.3.2 *Targeting Notch Ligands and Receptors*

Notch ligand and receptor targeting is a particularly interesting approach since it allows specific targeting of Notch receptors or ligands critical in the pathological phenotype of malignancies, including for the metastatic phenotype. For instance, Notch1 was reported to control metastatic processes in small lung cancer cell lines and to initiate EMT transition and invasion of breast cancer cells (Hassan et al., 2014; Shao et al., 2015). Blocking monoclonal antibodies targeting specifically Notch receptor subtypes have been developed. These molecules mainly act on the EGF-like repeats of Notch receptors. Various Notch receptor blocking antibodies have entered clinical trials for metastatic and advanced solid cancers. For instance, tarextumab (OMP-59R5), anti-Notch2/3 receptors currently in phase 1 trials in patients with untreated metastatic pancreatic cancer, is showing promising therapeutic effects in association with the antineoplastic (chemotherapy) drugs Nab-Paclitaxel and Gemcitabine (O'Reilly et al., 2014). Examples of other Notch receptor or ligand blocking antibodies currently in clinical trials include Notch1 monoclonal antibody OMP-52M5 and the anti-DLL4 demcizumab (OMP-21M18) (Andersson and Lendahl, 2014).

Besides, small molecules acting as receptor or ligand decoys were also developed. A recent report by Kangsamaksin et al. (2015) provided experimental evidence for the potential of decoy molecules for silencing Notch receptors and ligands specifically. These authors developed Notch decoys N1–13 and N1–24, interfering with Dll–Notch and Jagged–Notch interactions, respectively. The decoy molecules potently disrupted tumor growth and promoted normal endothelial sprouting in tumor microenvironments partly by reducing angiogenic sprouting, vessel perfusion, and pericyte coverage, among other proangiogenic processes (Kangsamaksin et al., 2015). Besides the possibility to improve the therapeutic outcome, Notch receptor or ligand specific targeting may also decrease the severity or abrogate adverse gastrointestinal effects observed with unspecific targeting of Notch signaling.

# 4. ANTIMETASTATIC POTENTIAL OF WNT AND HEDGEHOG TARGETING

## 4.1 Wnt Targeting

### *4.1.1 Wnt Signaling and Cancer*

Wnt family of signaling molecules encompass a number of cysteine-rich glycosylated secreted ligands that bind to the extracellular domain of frizzled family of receptors. Wnt binding triggers a signaling cascade resulting in the activation of genes involved in stem cell maintenance, cell survival, proliferation, motility, migration, and fate determination during the development. Wnt aberrant overexpression is used by tumorigenic processes to activate pathogenic developmental-like Wnt signaling activity in transforming cells, favoring stemness and chemotherapy resistance (Logan and Nusse, 2004; Nwabo Kamdje et al., 2014; Seke Etet et al., 2012).

Canonical Wnt/β-catenin pathway is the most studied Wnt signaling pathway. In the canonical pathway, Wnt binding to frizzled-7 leads to the disaggregation of β-catenin destruction complex made of the tumor suppressor adenomatous polyposis coli (APC), serine/threonine protein kinase glycogen synthase kinase 3, and casein kinase. Canonical Wnt/β-catenin pathway requires the coreceptors LRP5 and LRP6. The disaggregation of β-catenin destruction complex results in the inhibition of β-catenin destruction, and conversely, in the cytoplasmic accumulation of β-catenin. Once in sufficiently high levels, a fraction of cytoplasmic β-catenin migrates to the nucleus where it interacts with member of transcription factor (TCF)/lymphoid enhancer-binding factor 1 (LEF-1) family of transcription factors. In turn, these transcription factors activate genes encoding for proteins involved in EMT, survival, angiogenesis motility, and invasion such as c-Myc, Jagged1, vascular endothelial growth factor (VEGF), C-C motif chemokine ligand 2 (CCL2), Snail, Slug, vimentin, and metalloproteinases (Brabletz et al., 1999; Jiang et al., 2007; Marchenko et al., 2002; Parmalee and Kitajewski, 2008; Qi et al., 2016; Yeo et al., 2014).

Two other Wnt signaling pathways termed as noncanonical Wnt pathways were also described, namely the planar cell polarity pathway and Wnt/calcium pathway. Wnt noncanonical pathway does not require the coreceptors low-density lipoprotein receptor-related protein (LRP)5/6. The noncanonical planar polarity pathway activates small Rho-GTPases like Ras-related C3 botulinum toxin substrate 1 (RAC1),

Rho, and Cdc42, which control actin dynamics through Rho kinase, the actin-binding protein cofilin, and the MAPK kinase c-Jun N-terminal kinase (JNK) (Gomez-Orte et al., 2013; Komiya and Habas, 2008; Wang, 2009).

Wnt/calcium pathway, which is initiated by noncanonical Wnt ligands like Wnt5a, activates phosphoinositide phospholipase C (PLC) and phosphodiesterase 6. PLC-mediated cleavage of the plasma membrane phospholipid phosphatidylinositol 4,5-bisphosphate releases inositol trisphosphate and diacyl-glycerol. The resulting signaling cascades induce the release of mitochondrial $Ca^{2+}$ in the cytoplasm, with subsequent activation of $Ca^{2+}$-dependent enzymes such as $Ca^{2+}$/calmodulin-dependent kinase II, protein kinase C (PKC), and the serine/threonine protein phosphatase calcineurin (Gomez-Orte et al., 2013; Komiya and Habas, 2008).

### 4.1.2 Role of Wnt Signaling in Metastasis and Therapeutic Potential

Evidence sustaining a role for aberrant Wnt pathway activation in metastasis was provided by reports of the ability of Wnt antagonism to suppress tumorigenesis and metastasis, partly (1) by enhancing the expression of epithelial marker like E-cadherin, and keratin 8/18 (preventing EMT) and (2) by decreasing the expression of prometastatic factors such as Slug, Twist, Snail, and metalloproteinases (Chung et al., 2009; Lin et al., 2013; Xu et al., 2014). The aberrant accumulation of β-catenin in the cytoplasm observed in colorectal tumors and many other solid cancers results from tumor-promoting epigenetic events such as DNA demethylation on *cadherin associated protein beta 1 (CTNNB1)* gene promoter that activates β-catenin synthesis and DNA hypermethylation on *Apc* gene promoter that silence this gene, resulting in the decrease in cellular levels of β-catenin destruction complex member APC (Vecchio et al., 2013; Wang et al., 2015). Similarly, the absence of degradation of β-catenin observed in many cancers may also result from silencing mutations of *Apc* gene, whereas activating mutations of *CTNNB1* gene may also increase β-catenin cellular levels (Kumamoto et al., 2015; Morin et al., 1997). Genetic and epigenetic suppression of the other members of destruction complex were also reported (Hayes et al., 2015; Tejeda-Munoz and Robles-Flores, 2015). Furthermore, a fraction of β-catenin is bound to E-cadherin in the cytoskeleton, thus, EMT-associated loss of E-cadherin results in the release β-catenin (Tian et al., 2011). Failure of misshaped β-catenin (product of mutations) to bind to E-cadherin also increases β-catenin cytoplasmic levels.

Canonical Wnt3 signaling consistently promoted a partial EMT-like transition with increased N-cadherin, Twist, Slug; decreased E-cadherin in trastuzumab-resistant breast tumor cells; and Wnt3 knockdown by small interfering RNA (siRNA) decreased the expression of EMT triggering factor epidermal growth factor receptor (EGFR) in these malignant cells (Wu et al., 2012). In addition, activation of Wnt/β-catenin signaling correlates with EMT and the potency of proliferation and invasiveness in prostate cancer cells (Jiang et al., 2007; Zhao et al., 2011). Nuclear localization of β-catenin is a molecular marker of EMT in colon cancer (Lee et al., 2006). The inhibition of Wnt signaling with β-catenin short hairpin RNA (shRNA) reversed HIF-1α-induced EMT in human prostate cancer (Jiang et al., 2007; Zhao et al., 2011). Moreover, Wnt/β-catenin may promote EMT via direct transactivation of EMT transcription factors such as ZEB1 (Sanchez-Tillo et al., 2011).

Noncanonical Wnt signaling pathways were also reported to play pivotal roles in cancer metastasis. For instance, the well-known activator of noncanonical Wnt pathway Wnt5a increased metastasis in a PKC- and Stat3-dependent fashion in melanoma and also the expression of melanoma immunogens and severity markers GP100, melan-A protein (MART-1), and tyrosinase (de Vries et al., 2001; Dissanayake et al., 2008).

Wnt5a is upregulated in epithelial tumors, where it promotes EMT and migration in a β-catenin-independent fashion (Ford et al., 2014;

Zhu et al., 2014). Treatment of these tumor cells with Wnt5a resulted in an upregulation of EMT stimulating factors, including vimentin, Snail1, and Slug (Ford et al., 2014). In addition, Gujral et al. (2014) reported an upregulation of Wnt5 receptor Frizzled2 in metastatic liver, lung, colon, and breast cancer cell lines and in high-grade tumors. Such expression correlated with the overexpression of EMT markers (Gujral et al., 2014). Still in the same study, pharmacological and genetic perturbation analyses suggested that Wnt5a/Frizzled2 signaling induces EMT and cell migration via a previously unrecognized Fyn/Stat3-dependent noncanonical pathway.

Besides, considering the involvement of planar polarity pathway in the regulation of actin cytoskeleton mediated via its ability to activate Rho-GTPases mentioned above (Section 4.1.1) (Gomez-Orte et al., 2013; Komiya and Habas, 2008; Wang, 2009), a role for this noncanonical Wnt pathway in the cytoskeleton reorganization supporting motility, migration, and invasion can be guessed. Not surprisingly, experimental evidence suggests the involvement of this pathway in cellular motility and invasive capacity of metastasis promoting cells in various cancers, including in pancreatic ductal adenocarcinoma and breast cancer (Cetin et al., 2015; Hatakeyama et al., 2014). Moreover, Frizzled receptors that can trigger both canonical and noncanonical signaling such as Frizzled2 and Frizzled7 reported the ability to interact with Rho-A and other small GTPases during metastatic cascade. Consequently and for instance, Wnt3a/Frizzled2 signaling stimulates cell migration and invasion in a Rho-A-dependent fashion in multiple myeloma (Qiang et al., 2003, 2005), and Wnt5a was reported to promote cell migration in breast cancer through the Disheveled2/Daam1/Rho-A axis (Zhu et al., 2012).

Canonical Wnt signaling pathway inhibitors such as the CREB binding protein (CBP)/β-catenin antagonist PRI-724 have been developed. Despite the potential for Wnt signaling targeting for cancer stem cell elimination (Krause et al., 2014; Lenz and Kahn, 2014) and metastasis suppression, the pivotal roles of this signaling pathway in the maintenance of tissue stem cells' stemness, and in the regulation of tissue homeostasis and cell fate all along lifespan raise concerns for safety of Wnt-based anticancer strategies (Kahn, 2014), discouraging their development. Surprisingly, preclinical studies (Lenz and Kahn, 2014) and phase 1 clinical trials (Nagaraj et al., 2015) suggested that PRI-724 is relatively well tolerated and the drug is currently in phase 2 trials in combination with chemotherapy drugs and the angiogenesis inhibitor bevacizumab (National Institute of Health (NIH) trial identification codes 3C-13-3, NCI-2015-00436, and NCT02413853). Experimental evidence suggests differences in Wnt signaling mediating stemness in malignant and nonmalignant cells (Khan et al., 2016). The characterization of such differences may open new avenues for safer targeting of Wnt signaling.

## 4.2 Hedgehog, Metastatic Cascade and Emerging Therapies

### 4.2.1 *Hedgehog Family*

Vertebrate Hedgehog family of lipid modified proteins encompasses Desert Hedgehog, Indian Hedgehog, and Sonic Hedgehog (Shh). Shh is the most studied Hedgehog family member. Hedgehog proteins are ligand of 12-pass transmembrane protein receptors termed as Patched. In the absence of interactions with ligands, Patched receptors catalytically maintain the 7-pass transmembrane protein Smoothened (Smo) in an inactive state. Upon ligand binding, the inhibitory activity of Patched is abrogated and Smo activates transcription factors of Gli zinc-finger protein family in the cytoplasm. Then, Gli proteins migrate to the nucleus where they transactivate or repress various genes (Echelard et al., 1993; Varjosalo and Taipale, 2008).

Hedgehog has been reported to play pivotal roles in cancer development (for review see Takebe et al., 2015) and in metastatic cancer. For instance, loss and gain of function experiments and also pharmacological inhibitions suggested that Hedgehog signaling is required all along the metastatic cascade (Lee et al., 2015; Varnat et al., 2009; Xu et al., 2014; Yoo et al., 2008; Zheng et al., 2012). Examples included the following studies: (1) high Hedgehog-Gli signatures coincided with the development of metastases in colon cancer (Varnat et al., 2009); also (2) Gli1 transcription factor activity is associated with tumor aggressiveness in papillary thyroid cancer (Lee et al., 2015). In addition, Hedgehog promotes EMT in various solid tumors, including pancreatic, colon, and breast cancers (Xu et al., 2014). Gli proteins are involved in TGF-β mediated EMT in hepatocellular carcinoma, marked by an increased expression of Snail1 (Zheng et al., 2012). Interestingly, Gli1 knockdown resulted in decreased hepatocellular carcinoma migration and invasion (Zheng et al., 2012). Shh enhanced gastric cancer cell motility and invasiveness, whereas no increase was observed in cells treated with Shh signaling inhibitor 3-Keto-N-aminoethylaminocaproyldihydrocinnamoyl cyclopamine (cyclopamine-KAAD) or anti-Shh monoclonal antibodies (Yoo et al., 2008). Moreover, Yoo et al. (2011) elegantly demonstrated that Shh signaling promotes gastric cancer metastasis at least partly through PI3K/Akt signaling-mediated matrix metallopeptidase 9 activation and EMT.

#### 4.2.2 *Hedgehog-Based Therapeutic Strategies*

Smo inhibitors are the largest group of Hedgehog signaling inhibitors. The naturally occurring alkaloid cyclopamine is a typical example. Cyclopamine inhibits tumor cell growth and invasiveness both in vitro and in vivo (Fan et al., 2011; Ma et al., 2013; Warzecha et al., 2012). However, the translation to cancer treatment has been challenging due to the poor stability and solubility of cyclopamine and also due to its severe adverse effects resulting from the activity of the products of its metabolism (Ma et al., 2013; Ng and Curran, 2011). More soluble and stable derivatives of cyclopamine were developed, which included the following: (1) the emerging prodrug cyclopamine glucuronide tested in glioblastoma multiform models (Bensalma et al., 2015); and (2) vismodegib (GDC-0449) currently in phase 2 clinical trials in patients with metastatic pancreatic adenocarcinoma (Kim et al., 2014), advanced primary tumors and metastatic basal cell carcinoma (Sekulic et al., 2015), and pediatric brain tumors (Robinson et al., 2015). Other Smo inhibitors currently in clinical trials for refractory and metastatic cancer treatment include IPI-926 (Ko et al., 2016), LDE225 (Migden et al., 2015), BMS-833923 (Justilien and Fields, 2015), PF-0449913 (Fei et al., 2015). Most of these prodrugs and drugs have an acceptable safety profile with only low grade or relatively minor adverse effects, together with promising anticancer and antimetastatic effects. Moreover, Gli inhibitors, another class of Hedgehog inhibitors, include prodrugs like Gant-58 and Gant-61, which displayed a strong antimetastatic potential in preclinical studies (Fu et al., 2013; Lauth et al., 2007; Li et al., 2015; Xu et al., 2014).

A number of blocking antibodies targeting Hedgehog signaling are available, whereas only few inhibitors of Hedgehog ligands were developed. The monoclonal antibody 5E1, one of the most studied Hedgehog-targeting antibodies, antagonizes all three Hedgehog ligands. Studies in nude mice revealed 5E1 ability to inhibit metastasis of pancreatic cancer (Bailey et al., 2009). The Hedgehog ligand inhibitor robotnikinin was discovered by Stanton et al. (2009) using small-molecule microarrays and diversity-oriented synthesis approaches. This small molecule mediates extracellular Shh inhibition by binding to this protein, preventing interactions with patched receptors. Data on the effects of robotnikinin on metastatic processes are lacking.

## 5. CONCLUDING REMARKS

Emerging master developmental pathways-based targeted therapies represent promising approaches for metastasis prevention and cancer stem cell elimination. Such therapeutic potential is possible because the tumorigenesis-mediated hijacking of EMT, cell motility, migration, stemness, angiogenesis, and other key phenomena determining cell fate controlled by developmental pathways are pivotal for all steps of solid tumor initiation, maintenance, and development. Various classes of Notch signaling inhibitors are currently in clinical trials. The development of therapeutics targeting specifically Notch signaling components critical for malignancy pathological phenotype is expected to improve the therapeutic outcome and decrease adverse effects. Wnt signaling targeting, the safety of which being a serious concern, is surprisingly well-tolerated in clinical trials and is expected to be a game changer for cancer stem cell elimination and metastasis prevention. Unlike Notch and Wnt pathways, translation of Shh signaling blockade to cancer treatment has been challenging due to severe adverse effects and poor therapeutic potential of various monoclonal antibodies and naturally occurring Smo inhibitors. Further mechanistic studies of developmental pathway involvement in refractory and metastatic cancers may further improve targeting efficiency and the therapeutic outcome.

## References


Alizadeh, A.M., Shiri, S., Farsinejad, S., 2014. Metastasis review: from bench to bedside. Tumour Biol. 35, 8483–8523.

Andersson, E.R., Lendahl, U., 2014. Therapeutic modulation of Notch signalling—are we there yet? Nat. Rev. Drug Discov. 13, 357–378.

Arwert, E.N., Hoste, E., Watt, F.M., 2012. Epithelial stem cells, wound healing and cancer. Nat. Rev. Cancer 12, 170–180.

Bailey, J.M., Mohr, A.M., Hollingsworth, M.A., 2009. Sonic hedgehog paracrine signaling regulates metastasis and lymphangiogenesis in pancreatic cancer. Oncogene 28, 3513–3525.

Barnes, E.A., Kenerson, H.L., Jiang, X., Yeung, R.S., 2010. Tuberin regulates E-cadherin localization: implications in epithelial-mesenchymal transition. Am. J. Pathol. 177, 1765–1778.

Bensalma, S., Chadeneau, C., Legigan, T., Renoux, B., Gaillard, A., de Boisvilliers, M., Pinet-Charvet, C., Papot, S., Muller, J.M., 2015. Evaluation of cytotoxic properties of a cyclopamine glucuronide prodrug in rat glioblastoma cells and tumors. J. Mol. Neurosci. 55, 51–61.

Bhavsar, P.J., Infante, E., Khwaja, A., Ridley, A.J., 2013. Analysis of Rho GTPase expression in T-ALL identifies RhoU as a target for Notch involved in T-ALL cell migration. Oncogene 32, 198–208.

Brabletz, T., Jung, A., Dag, S., Hlubek, F., Kirchner, T., 1999. beta-catenin regulates the expression of the matrix metalloproteinase-7 in human colorectal cancer. Am. J. Pathol. 155, 1033–1038.

Cetin, G.O., Toylu, A., Atabey, N., Sercan, Z., Sakizli, M., 2015. Downregulation of VANGL1 inhibits cellular invasion rather than cell motility in hepatocellular carcinoma cells without stimulation. Genet. Test. Mol. Biomarkers 19, 283–287.

Chen, W.T., Lee, C.C., Goldstein, L., Bernier, S., Liu, C.H., Lin, C.Y., Yeh, Y., Monsky, W.L., Kelly, T., Dai, M., 1994. Membrane proteases as potential diagnostic and therapeutic targets for breast malignancy. Breast Cancer Res. Treat. 31, 217–226.

Chiarugi, P., Giannoni, E., 2008. Anoikis: a necessary death program for anchorage-dependent cells. Biochem. Pharmacol. 76, 1352–1364.

Chung, M.T., Lai, H.C., Sytwu, H.K., Yan, M.D., Shih, Y.L., Chang, C.C., Yu, M.H., Liu, H.S., Chu, D.W., Lin, Y.W., 2009. SFRP1 and SFRP2 suppress the transformation and invasion abilities of cervical cancer cells through Wnt signal pathway. Gynecol. Oncol. 112, 646–653.

Citi, S., Guerrera, D., Spadaro, D., Shah, J., 2014. Epithelial junctions and Rho family GTPases: the zonular signalosome. Small GTPases 5, 1–15.

Clevers, H., 2006. Wnt/beta-catenin signaling in development and disease. Cell 127, 469–480.

de Vries, T.J., Smeets, M., de Graaf, R., Hou-Jensen, K., Brocker, E.B., Renard, N., Eggermont, A.M., van Muijen, G.N., Ruiter, D.J., 2001. Expression of gp100, MART-1, tyrosinase, and S100 in paraffin-embedded primary melanomas and locoregional, lymph node, and visceral metastases: implications for diagnosis and immunotherapy. A study conducted by the EORTC Melanoma Cooperative Group. J. Pathol. 193, 13–20.

Diaz, V.M., de Herreros, A.G., 2016. F-box proteins: keeping the epithelial-to-mesenchymal transition (EMT) in check. Semin. Cancer Biol. 36, 71–79.

Dissanayake, S.K., Olkhanud, P.B., O'Connell, M.P., Carter, A., French, A.D., Camilli, T.C., Emeche, C.D., Hewitt, K.J., Rosenthal, D.T., Leotlela, P.D., Wade, M.S., Yang, S.W., Brant, L., Nickoloff, B.J., Messina, J.L., Biragyn, A., Hoek, K.S., Taub, D.D., Longo, D.L., Sondak, V.K., Hewitt, S.M., Weeraratna, A.T., 2008. Wnt5A regulates expression of tumor-associated antigens in melanoma via changes in signal transducers and activators of transcription 3 phosphorylation. Cancer Res. 68, 10205–10214.

Echelard, Y., Epstein, D.J., St-Jacques, B., Shen, L., Mohler, J., McMahon, J.A., McMahon, A.P., 1993. Sonic hedgehog, a member of a family of putative signaling molecules, is implicated in the regulation of CNS polarity. Cell 75, 1417–1430.

Espinoza, I., Miele, L., 2013. Deadly crosstalk: notch signaling at the intersection of EMT and cancer stem cells. Cancer Lett. 341, 41–45.

Fan, Q., Gu, D., He, M., Liu, H., Sheng, T., Xie, G., Li, C.X., Zhang, X., Wainwright, B., Garrossian, A., Garrossian, M., Gardner, D., Xie, J., 2011. Tumor shrinkage by cyclopamine tartrate through inhibiting hedgehog signaling. Chin. J. Cancer 30, 472–481.

Fei, H., Ke, P., Wang, N., Shen, H., Huang, J., Tan, J., Liang, L., Song, X., 2015. An evaluation comparing Californium252 neutron brachytherapy with neoadjuvant intra-arterial embolism chemotherapy assisted surgery effect for treating advanced cervical carcinoma patients. Eur. J. Gynaecol. Oncol. 36, 442–446.

Filipovic, A., Lombardo, Y., Faronato, M., Abrahams, J., Aboagye, E., Nguyen, Q.D., d'Aqua, B.B., Ridley, A., Green, A., Rahka, E., Ellis, I., Recchi, C., Przulj, N., Sarajlic, A., Alattia, J.R., Fraering, P., Deonarain, M., Coombes, R.C., 2014. Anti-nicastrin monoclonal antibodies elicit pleiotropic anti-tumour pharmacological effects in invasive breast cancer cells. Breast Cancer Res. Treat. 148, 455–462.

Ford, C.E., Punnia-Moorthy, G., Henry, C.E., Llamosas, E., Nixdorf, S., Olivier, J., Caduff, R., Ward, R.L., Heinzelmann-Schwarz, V., 2014. The non-canonical Wnt ligand, Wnt5a, is upregulated and associated with epithelial to mesenchymal transition in epithelial ovarian cancer. Gynecol. Oncol. 134, 338–345.

Fortin Ensign, S.P., Mathews, I.T., Symons, M.H., Berens, M.E., Tran, N.L., 2013. Implications of rho GTPase signaling in glioma cell invasion and tumor progression. Front. Oncol. 3, 241.

Fu, J., Rodova, M., Roy, S.K., Sharma, J., Singh, K.P., Srivastava, R.K., Shankar, S., 2013. GANT-61 inhibits pancreatic cancer stem cell growth in vitro and in NOD/SCID/IL2R gamma null mice xenograft. Cancer Lett. 330, 22–32.

Garg, M., 2013. Epithelial-mesenchymal transition – activating transcription factors – multifunctional regulators in cancer. World J. Stem Cells 5, 188–195.

Geiger, T.R., Peeper, D.S., 2009. Metastasis mechanisms. Biochim. Biophys. Acta 1796, 293–308.

Gomez-Orte, E., Saenz-Narciso, B., Moreno, S., Cabello, J., 2013. Multiple functions of the noncanonical Wnt pathway. Trends Genet. 29, 545–553.

Gujral, T.S., Chan, M., Peshkin, L., Sorger, P.K., Kirschner, M.W., MacBeath, G., 2014. A noncanonical Frizzled2 pathway regulates epithelial-mesenchymal transition and metastasis. Cell 159, 844–856.

Gulhati, P., Bowen, K.A., Liu, J., Stevens, P.D., Rychahou, P.G., Chen, M., Lee, E.Y., Weiss, H.L., O'Connor, K.L., Gao, T., Evers, B.M., 2011. mTORC1 and mTORC2 regulate EMT, motility, and metastasis of colorectal cancer via RhoA and Rac1 signaling pathways. Cancer Res. 71, 3246–3256.

Guo, S., Liu, M., Gonzalez-Perez, R.R., 2011. Role of Notch and its oncogenic signaling crosstalk in breast cancer. Biochim. Biophys. Acta 1815, 197–213.

Hassan, W.A., Yoshida, R., Kudoh, S., Hasegawa, K., Niimori-Kita, K., Ito, T., 2014. Notch1 controls cell invasion and metastasis in small cell lung carcinoma cell lines. Lung Cancer 86, 304–310.

Hatakeyama, J., Wald, J.H., Printsev, I., Ho, H.Y., Carraway III, K.L., 2014. Vangl1 and Vangl2: planar cell polarity components with a developing role in cancer. Endocr. Relat. Cancer 21, R345–R356.

Hayes, J.D., Chowdhry, S., Dinkova-Kostova, A.T., Sutherland, C., 2015. Dual regulation of transcription factor Nrf2 by Keap1 and by the combined actions of beta-TrCP and GSK-3. Biochem. Soc. Trans. 43, 611–620.

Jiang, Y.G., Luo, Y., He, D.L., Li, X., Zhang, L.L., Peng, T., Li, M.C., Lin, Y.H., 2007. Role of Wnt/beta-catenin signaling pathway in epithelial-mesenchymal transition of human prostate cancer induced by hypoxia-inducible factor-1alpha. Int. J. Urol. 14, 1034–1039.

Johnson, R.W., Schipani, E., Giaccia, A.J., 2015. HIF targets in bone remodeling and metastatic disease. Pharmacol. Ther. 150, 169–177.

Justilien, V., Fields, A.P., 2015. Molecular pathways: novel approaches for improved therapeutic targeting of Hedgehog signaling in cancer stem cells. Clin. Cancer Res. 21, 505–513.

Kahn, M., 2014. Can we safely target the WNT pathway? Nat. Rev. Drug Discov. 13, 513–532.

Kalluri, R., 2009. EMT: when epithelial cells decide to become mesenchymal-like cells. J. Clin. Invest. 119, 1417–1419.

Kandala, P.K., Srivastava, S.K., 2012. Diindolylmethane-mediated Gli1 protein suppression induces anoikis in ovarian cancer cells in vitro and blocks tumor formation ability in vivo. J. Biol. Chem. 287, 28745–28754.

Kangsamaksin, T., Murtomaki, A., Kofler, N.M., Cuervo, H., Chaudhri, R.A., Tattersall, I.W., Rosenstiel, P.E., Shawber, C.J., Kitajewski, J., 2015. NOTCH decoys that selectively block DLL/NOTCH or JAG/NOTCH disrupt angiogenesis by unique mechanisms to inhibit tumor growth. Cancer Discov. 5, 182–197.

Khan, A.H., Bloom, J.S., Faridmoayer, E., Smith, D.J., 2016. Genetic screening reveals a link between Wnt signaling and antitubulin drugs. Pharmacogenomics. J. 16 (Suppl. 2), 164–172.

Kim, S., Kang, M., Lee, S., Bae, S., Han, S., Jang, J.Y., Park, T., 2014. Identifying molecular subtypes related to clinicopathologic factors in pancreatic cancer. Biomed. Eng. Online 13 (Suppl. 2), S5.

Ko, A.H., LoConte, N., Tempero, M.A., Walker, E.J., Kate, K.R., Lewis, S., Chang, W.C., Kantoff, E., Vannier, M.W., Catenacci, D.V., Venook, A.P., Kindler, H.L., 2016. A phase I study of FOLFIRINOX Plus IPI-926, a hedgehog pathway inhibitor, for advanced pancreatic adenocarcinoma. Pancreas. 45 (Suppl. 3), 370–375.

Komiya, Y., Habas, R., 2008. Wnt signal transduction pathways. Organogenesis 4, 68–75.

Krause, U., Ryan, D.M., Clough, B.H., Gregory, C.A., 2014. An unexpected role for a Wnt-inhibitor: Dickkopf-1 triggers a novel cancer survival mechanism through modulation of aldehyde-dehydrogenase-1 activity. Cell Death Dis. 5, e1093.

Kumamoto, K., Ishida, H., Ohsawa, T., Ishibashi, K., Ushiama, M., Yoshida, T., Iwama, T., 2015. Germline and somatic mutations of the gene in papillary thyroid carcinoma associated with familial adenomatous polyposis: analysis of three cases and a review of the literature. Oncol. Lett. 10, 2239–2243.

Kumar, S., Park, S.H., Cieply, B., Schupp, J., Killiam, E., Zhang, F., Rimm, D.L., Frisch, S.M., 2011. A pathway for the control of anoikis sensitivity by E-cadherin and epithelial-to-mesenchymal transition. Mol. Cell Biol. 31, 4036–4051.

Kwon, O.J., Valdez, J.M., Zhang, L., Zhang, B., Wei, X., Su, Q., Ittmann, M.M., Creighton, C.J., Xin, L., 2014. Increased Notch signalling inhibits anoikis and stimulates proliferation of prostate luminal epithelial cells. Nat. Commun. 5, 4416.

Lamouille, S., Xu, J., Derynck, R., 2014. Molecular mechanisms of epithelial-mesenchymal transition. Nat. Rev. Mol. Cell Biol. 15, 178–196.

Lauth, M., Bergstrom, A., Shimokawa, T., Toftgard, R., 2007. Inhibition of GLI-mediated transcription and tumor cell growth by small-molecule antagonists. Proc. Natl. Acad. Sci. U.S.A. 104, 8455–8460.

Lee, J.M., Dedhar, S., Kalluri, R., Thompson, E.W., 2006. The epithelial-mesenchymal transition: new insights in signaling, development, and disease. J. Cell Biol. 172, 973–981.

Lee, J., Jeong, S., Lee, C.R., Ku, C.R., Kang, S.W., Jeong, J.J., Nam, K.H., Shin, D.Y., Chung, W.Y., Lee, E.J., Jo, Y.S., 2015. GLI1 transcription factor affects tumor aggressiveness in patients with papillary thyroid cancers. Medicine (Baltimore) 94, e998.

Lee, S.M., Moon, J., Redman, B.G., Chidiac, T., Flaherty, L.E., Zha, Y., Othus, M., Ribas, A., Sondak, V.K., Gajewski, T.F., Margolin, K.A., 2015. Phase 2 study of RO4929097, a gamma-secretase inhibitor, in metastatic melanoma: SWOG 0933. Cancer 121, 432–440.

Lenz, H.J., Kahn, M., 2014. Safely targeting cancer stem cells via selective catenin coactivator antagonism. Cancer Sci. 105, 1087–1092.

Leong, K.G., Niessen, K., Kulic, I., Raouf, A., Eaves, C., Pollet, I., Karsan, A., 2007. Jagged1-mediated Notch activation induces epithelial-to-mesenchymal transition through Slug-induced repression of E-cadherin. J. Exp. Med. 204, 2935–2948.

Li, H., Da, L.J., Fan, W.D., Long, X.H., Zhang, X.Q., 2015. Transcription factor glioma-associated oncogene homolog 1 is required for transforming growth factor-beta1-induced epithelial-mesenchymal transition of non-small cell lung cancer cells. Mol. Med. Rep. 11, 3259–3268.

Lin, C.H., Guo, Y., Ghaffar, S., McQueen, P., Pourmorady, J., Christ, A., Rooney, K., Ji, T., Eskander, R., Zi, X., Hoang, B.H., 2013. Dkk-3, a secreted wnt antagonist, suppresses tumorigenic potential and pulmonary metastasis in osteosarcoma. Sarcoma 2013, 147541.

Logan, C.Y., Nusse, R., 2004. The Wnt signaling pathway in development and disease. Annu. Rev. Cell Dev. Biol. 20, 781–810.

Lopez-Guerra, M., Xargay-Torrent, S., Rosich, L., Montraveta, A., Roldan, J., Matas-Cespedes, A., Villamor, N., Aymerich, M., Lopez-Otin, C., Perez-Galan, P., Roue, G., Campo, E., Colomer, D., 2015. The gamma-secretase inhibitor PF-03084014 combined with fludarabine antagonizes migration, invasion and angiogenesis in NOTCH1-mutated CLL cells. Leukemia 29, 96–106.

Ma, H., Li, H.Q., Zhang, X., 2013. Cyclopamine, a naturally occurring alkaloid, and its analogues may find wide applications in cancer therapy. Curr. Top. Med. Chem. 13, 2208–2215.

Marchenko, G.N., Marchenko, N.D., Leng, J., Strongin, A.Y., 2002. Promoter characterization of the novel human matrix metalloproteinase-26 gene: regulation by the T-cell factor-4 implies specific expression of the gene in cancer cells of epithelial origin. Biochem. J. 363, 253–262.

Margadant, C., Sonnenberg, A., 2010. Integrin-TGF-beta crosstalk in fibrosis, cancer and wound healing. EMBO Rep. 11, 97–105.

Martin, S.S., Vuori, K., 2004. Regulation of Bcl-2 proteins during anoikis and amorphosis. Biochim. Biophys. Acta 1692, 145–157.

Martin-Villar, E., Megias, D., Castel, S., Yurrita, M.M., Vilaro, S., Quintanilla, M., 2006. Podoplanin binds ERM proteins to activate RhoA and promote epithelial-mesenchymal transition. J. Cell Sci. 119, 4541–4553.

Migden, M.R., Guminski, A., Gutzmer, R., Dirix, L., Lewis, K.D., Combemale, P., Herd, R.M., Kudchadkar, R., Trefzer, U., Gogov, S., Pallaud, C., Yi, T., Mone, M., Kaatz, M., Loquai, C., Stratigos, A.J., Schulze, H.J., Plummer, R., Chang, A.L., Cornelis, F., Lear, J.T., Sellami, D., Dummer, R., 2015. Treatment with two different doses of sonidegib in patients with locally advanced or metastatic basal cell carcinoma (BOLT): a multicentre, randomised, double-blind phase 2 trial. Lancet Oncol. 16, 716–728.

Morin, P.J., Sparks, A.B., Korinek, V., Barker, N., Clevers, H., Vogelstein, B., Kinzler, K.W., 1997. Activation of beta-catenin-Tcf signaling in colon cancer by mutations in beta-catenin or APC. Science 275, 1787–1790.

Nagaraj, A.B., Joseph, P., Kovalenko, O., Singh, S., Armstrong, A., Redline, R., Resnick, K., Zanotti, K., Waggoner, S., DiFeo, A., 2015. Critical role of Wnt/beta-catenin signaling in driving epithelial ovarian cancer platinum resistance. Oncotarget 6, 23720–23734.

Ng, J.M., Curran, T., 2011. The Hedgehog's tale: developing strategies for targeting cancer. Nat. Rev. Cancer 11, 493–501.

Nieto, M.A., 2002. The snail superfamily of zinc-finger transcription factors. Nat. Rev. Mol. Cell Biol. 3, 155–166.

Ntziachristos, P., Lim, J.S., Sage, J., Aifantis, I., 2014. From fly wings to targeted cancer therapies: a centennial for notch signaling. Cancer Cell 25, 318–334.

Nwabo Kamdje, A.H., Bassi, G., Pacelli, L., Malpeli, G., Amati, E., Nichele, I., Pizzolo, G., Krampera, M., 2012. Role of stromal cell-mediated Notch signaling in CLL resistance to chemotherapy. Blood Cancer J. 2, e73.

Nwabo Kamdje, A.H., Seke Etet, P.F., Vecchio, L., Muller, J.M., Krampera, M., Lukong, K.E., 2014. Signaling pathways in breast cancer: therapeutic targeting of the microenvironment. Cell Signal. 26, 2843–2856.

O'Reilly, E., Smith, L., Bendell, J., Strickler, J., Zalupski, M., Gluck, W., Kapoun, A., Yen, W., Xu, L., Hill, D., Zhou, L., Dupont, J., Cohn, A., 2014. In: Final Results of Phase Ib of Anticancer Stem Cell Antibody Tarextumab (OMP-59R5, TRXT, Anti-notch 2/3) in Combination With Nab-Paclitaxel and Gemcitabine (Nab-P+Gem) in Patients (Pts) With Untreated Metastatic Pancreatic Cancer (MPC), pp. 233–234.

Ozawa, M., Kobayashi, W., 2015. Reversibility of the snail-induced epithelial-mesenchymal transition revealed by the Cre-loxP system. Biochem. Biophys. Res. Commun. 458, 608–613.

Pajic, M., Herrmann, D., Vennin, C., Conway, J.R., Chin, V.T., Johnsson, A.K., Welch, H.C., Timpson, P., 2015. The dynamics of Rho GTPase signaling and implications for targeting cancer and the tumor microenvironment. Small GTPases 6, 123–133.

Paoli, P., Giannoni, E., Chiarugi, P., 2013. Anoikis molecular pathways and its role in cancer progression. Biochim. Biophys. Acta 1833, 3481–3498.

Parmalee, N.L., Kitajewski, J., 2008. Wnt signaling in angiogenesis. Curr. Drug Targets 9, 558–564.

Perez-Moreno, M.A., Locascio, A., Rodrigo, I., Dhondt, G., Portillo, F., Nieto, M.A., Cano, A., 2001. A new role for E12/E47 in the repression of E-cadherin expression and epithelial-mesenchymal transitions. J. Biol. Chem. 276, 27424–27431.

Przybylski, M., 2009. A review of the current research on the role of bFGF and VEGF in angiogenesis. J. Wound Care 18, 516–519.

Pugh, C.W., Ratcliffe, P.J., 2003. Regulation of angiogenesis by hypoxia: role of the HIF system. Nat. Med. 9, 677–684.

Qi, J., Yu, Y., Akilli, O.O., Holland, J.D., Besser, D., Fritzmann, J., Wulf-Goldenberg, A., Eckert, K., Fichtner, I., Birchmeier, W., 2016. New Wnt/beta-catenin target genes promote experimental metastasis and migration of colorectal cancer cells through different signals. Gut 65 (Suppl. 10), 1690–1701.

Qiang, Y.W., Endo, Y., Rubin, J.S., Rudikoff, S., 2003. Wnt signaling in B-cell neoplasia. Oncogene 22, 1536–1545.

Qiang, Y.W., Walsh, K., Yao, L., Kedei, N., Blumberg, P.M., Rubin, J.S., Shaughnessy Jr., J., Rudikoff, S., 2005. Wnts induce migration and invasion of myeloma plasma cells. Blood 106, 1786–1793.

Rathinam, R., Berrier, A., Alahari, S.K., 2011. Role of Rho GTPases and their regulators in cancer progression. Front. Biosci. (Landmark. Ed.) 16, 2561–2571.

Robinson, G.W., Orr, B.A., Wu, G., Gururangan, S., Lin, T., Qaddoumi, I., Packer, R.J., Goldman, S., Prados, M.D., Desjardins, A., Chintagumpala, M., Takebe, N., Kaste, S.C., Rusch, M., Allen, S.J., Onar-Thomas, A., Stewart, C.F., Fouladi, M., Boyett, J.M., Gilbertson, R.J., Curran, T., Ellison, D.W., Gajjar, A., 2015. Vismodegib exerts targeted efficacy against recurrent sonic hedgehog-subgroup medulloblastoma: results from phase II pediatric brain tumor consortium studies PBTC-025B and PBTC-032. J. Clin. Oncol. 33, 2646–2654.

Sanchez-Tillo, E., de, B.O., Siles, L., Cuatrecasas, M., Castells, A., Postigo, A., 2011. beta-catenin/TCF4 complex induces the epithelial-to-mesenchymal transition (EMT)-activator ZEB1 to regulate tumor invasiveness. Proc. Natl. Acad. Sci. U.S.A. 108, 19204–19209.

Schodel, J., Grampp, S., Maher, E.R., Moch, H., Ratcliffe, P.J., Russo, P., Mole, D.R., 2016. Hypoxia, hypoxia-inducible transcription factors, and renal cancer. Eur. Urol. 69 (Suppl. 4), 646–657.

Schwarzer, R., Nickel, N., Godau, J., Willie, B.M., Duda, G.N., Schwarzer, R., Cirovic, B., Leutz, A., Manz, R., Bogen, B., Dorken, B., Jundt, F., 2014. Notch pathway inhibition controls myeloma bone disease in the murine MOPC315.BM model. Blood Cancer J. 4, e217.

Seke Etet, P.F., Vecchio, L., Nwabo Kamdje, A.H., 2012. Interactions between bone marrow stromal microenvironment and B-chronic lymphocytic leukemia cells: any role for Notch, Wnt and Hh signaling pathways? Cell Signal. 24, 1433–1443.

Sekulic, A., Migden, M.R., Lewis, K., Hainsworth, J.D., Solomon, J.A., Yoo, S., Arron, S.T., Friedlander, P.A., Marmur, E., Rudin, C.M., Chang, A.L., Dirix, L., Hou, J., Yue, H., Hauschild, A., 2015. Pivotal ERIVANCE basal cell carcinoma (BCC) study: 12-month update of efficacy and safety of vismodegib in advanced BCC. J. Am. Acad. Dermatol. 72, 1021–1026.

Shao, S., Zhao, X., Zhang, X., Luo, M., Zuo, X., Huang, S., Wang, Y., Gu, S., Zhao, X., 2015. Notch1 signaling regulates the epithelial-mesenchymal transition and invasion of breast cancer in a Slug-dependent manner. Mol. Cancer 14, 28.

Sher, I., Adham, S.A., Petrik, J., Coomber, B.L., 2009. Autocrine VEGF-A/KDR loop protects epithelial ovarian carcinoma cells from anoikis. Int. J. Cancer 124, 553–561.

Shih, I., Wang, T.L., 2007. Notch signaling, gamma-secretase inhibitors, and cancer therapy. Cancer Res. 67, 1879–1882.

Stanton, B.Z., Peng, L.F., Maloof, N., Nakai, K., Wang, X., Duffner, J.L., Taveras, K.M., Hyman, J.M., Lee, S.W., Koehler, A.N., Chen, J.K., Fox, J.L., Mandinova, A., Schreiber, S.L., 2009. A small molecule that binds Hedgehog and blocks its signaling in human cells. Nat. Chem. Biol. 5, 154–156.

Stoletov, K., Lewis, J.D., 2015. Invadopodia: a new therapeutic target to block cancer metastasis. Expert Rev. Anticancer Ther. 15, 733–735.

Su, Z., Yang, Z., Xu, Y., Chen, Y., Yu, Q., 2015. Apoptosis, autophagy, necroptosis, and cancer metastasis. Mol. Cancer 14, 48.

Takebe, N., Miele, L., Harris, P.J., Jeong, W., Bando, H., Kahn, M., Yang, S.X., Ivy, S.P., 2015. Targeting Notch, Hedgehog, and Wnt pathways in cancer stem cells: clinical update. Nat. Rev. Clin. Oncol. 12, 445–464.

Tejeda-Munoz, N., Robles-Flores, M., 2015. Glycogen synthase kinase 3 in Wnt signaling pathway and cancer. IUBMB Life. 67 (Suppl. 12), 914–922.

Tian, X., Liu, Z., Niu, B., Zhang, J., Tan, T.K., Lee, S.R., Zhao, Y., Harris, D.C., Zheng, G., 2011. E-cadherin/beta-catenin complex and the epithelial barrier. J. Biomed. Biotechnol. 2011, 567305.

van Zijl, F., Krupitza, G., Mikulits, W., 2011. Initial steps of metastasis: cell invasion and endothelial transmigration. Mutat. Res. 728, 23–34.

Varjosalo, M., Taipale, J., 2008. Hedgehog: functions and mechanisms. Genes Dev. 22, 2454–2472.

Varnat, F., Duquet, A., Malerba, M., Zbinden, M., Mas, C., Gervaz, P., Altaba, A., 2009. Human colon cancer epithelial cells harbour active HEDGEHOG-GLI signalling that is essential for tumour growth, recurrence, metastasis and stem cell survival and expansion. EMBO Mol. Med. 1, 338–351.

Vazquez-Martin, A., Cufi, S., Oliveras-Ferraros, C., Torres-Garcia, V.Z., Corominas-Faja, B., Cuyas, E., Bonavia, R., Visa, J., Martin-Castillo, B., Barrajon-Catalan, E., Micol, V., Bosch-Barrera, J., Menendez, J.A., 2013. IGF-1R/epithelial-to-mesenchymal transition (EMT) crosstalk suppresses the erlotinib-sensitizing effect of EGFR exon 19 deletion mutations. Sci. Rep. 3, 2560.

Vecchio, L., Seke Etet, P.F., Kipanyula, M.J., Krampera, M., Nwabo Kamdje, A.H., 2013. Importance of epigenetic changes in cancer etiology, pathogenesis, clinical profiling, and treatment: what can be learned from hematologic malignancies? Biochim. Biophys. Acta 1836, 90–104.

Villanueva, M.T., 2015. Angiogenesis: a sudden rush of blood to the tumour. Nat. Rev. Cancer 15, 135.

Wang, N., Wang, Z., Wang, Y., Xie, X., Shen, J., Peng, C., You, J., Peng, F., Tang, H., Guan, X., Chen, J., 2015. Dietary compound isoliquiritigenin prevents mammary carcinogenesis by inhibiting breast cancer stem cells through WIF1 demethylation. Oncotarget 6, 9854–9876.

Wang, S.S., Jiang, J., Liang, X.H., Tang, Y.L., 2015. Links between cancer stem cells and epithelial-mesenchymal transition. Onco. Targets Ther. 8, 2973–2980.

Wang, Y., 2009. Wnt/Planar cell polarity signaling: a new paradigm for cancer therapy. Mol. Cancer Ther. 8, 2103–2109.

Warzecha, J., Dinges, D., Kaszap, B., Henrich, D., Marzi, I., Seebach, C., 2012. Effect of the Hedgehog-inhibitor cyclopamine on mice with osteosarcoma pulmonary metastases. Int. J. Mol. Med. 29, 423–427.

Wei, S.C., Yang, J., 2016. Forcing through tumor metastasis: the interplay between tissue rigidity and epithelial-mesenchymal transition. Trends Cell Biol. 26 (Suppl. 2), 111–120.

Wu, Y., Ginther, C., Kim, J., Mosher, N., Chung, S., Slamon, D., Vadgama, J.V., 2012. Expression of Wnt3 activates Wnt/beta-catenin pathway and promotes EMT-like phenotype in trastuzumab-resistant HER2-overexpressing breast cancer cells. Mol. Cancer Res. 10, 1597–1606.

Xu, X., Su, B., Xie, C., Wei, S., Zhou, Y., Liu, H., Dai, W., Cheng, P., Wang, F., Xu, X., Guo, C., 2014. Sonic hedgehog-Gli1 signaling pathway regulates the epithelial mesenchymal transition (EMT) by mediating a new target gene, S100A4, in pancreatic cancer cells. PLoS One 9, e96441.

Xu, Y., An, Y., Wang, X., Zha, W., Li, X., 2014. Inhibition of the Hedgehog pathway induces autophagy in pancreatic ductal adenocarcinoma cells. Oncol. Rep. 31, 707–712.

Xu, Z., Liu, D., Fan, C., Luan, L., Zhang, X., Wang, E., 2014. DIXDC1 increases the invasion and migration ability of non-small-cell lung cancer cells via the PI3K-AKT/AP-1 pathway. Mol. Carcinog. 53, 917–925.

Yeo, E.J., Cassetta, L., Qian, B.Z., Lewkowich, I., Li, J.F., Stefater III, J.A., Smith, A.N., Wiechmann, L.S., Wang, Y., Pollard, J.W., Lang, R.A., 2014. Myeloid WNT7b mediates the angiogenic switch and metastasis in breast cancer. Cancer Res. 74, 2962–2973.

Yoo, Y.A., Kang, M.H., Kim, J.S., Oh, S.C., 2008. Sonic hedgehog signaling promotes motility and invasiveness of gastric cancer cells through TGF-beta-mediated activation of the ALK5-Smad 3 pathway. Carcinogenesis 29, 480–490.

Yoo, Y.A., Kang, M.H., Lee, H.J., Kim, B.H., Park, J.K., Kim, H.K., Kim, J.S., Oh, S.C., 2011. Sonic hedgehog pathway promotes metastasis and lymphangiogenesis via activation of Akt, EMT, and MMP-9 pathway in gastric cancer. Cancer Res. 71, 7061–7070.

Yu, M., Ting, D.T., Stott, S.L., Wittner, B.S., Ozsolak, F., Paul, S., Ciciliano, J.C., Smas, M.E., Winokur, D., Gilman, A.J., Ulman, M.J., Xega, K., Contino, G., Alagesan, B., Brannigan, B.W., Milos, P.M., Ryan, D.P., Sequist, L.V., Bardeesy, N., Ramaswamy, S., Toner, M., Maheswaran, S., Haber, D.A., 2012. RNA sequencing of pancreatic circulating tumour cells implicates WNT signalling in metastasis. Nature 487, 510–513.

Zhao, J.H., Luo, Y., Jiang, Y.G., He, D.L., Wu, C.T., 2011. Knockdown of beta-Catenin through shRNA cause a reversal of EMT and metastatic phenotypes induced by HIF-1alpha. Cancer Invest. 29, 377–382.

Zheng, X., Vittar, N.B., Gai, X., Fernandez-Barrena, M.G., Moser, C.D., Hu, C., Almada, L.L., McCleary-Wheeler, A.L., Elsawa, S.F., Vrabel, A.M., Shire, A.M., Comba, A., Thorgeirsson, S.S., Kim, Y., Liu, Q., Fernandez-Zapico, M.E., Roberts, L.R., 2012. The transcription factor GLI1 mediates TGFbeta1 driven EMT in hepatocellular carcinoma via a SNAI1-dependent mechanism. PLoS One 7, e49581.

Zhu, Y., Tian, Y., Du, J., Hu, Z., Yang, L., Liu, J., Gu, L., 2012. Dvl2-dependent activation of Daam1 and RhoA regulates Wnt5a-induced breast cancer cell migration. PLoS One 7, e37823.

Zhu, H.H., Zhu, X.Y., Zhou, M.H., Cheng, G.Y., Lou, W.H., 2014. Effect of WNT5A on epithelial-mesenchymal transition and its correlation with tumor invasion and metastasis in nasopharyngeal carcinoma. Asian Pac. J. Trop. Med. 7, 488–491.

CHAPTER

# 19

# Emerging Therapeutic Targets for Cancer Metastasis: From the Perspective of Embryo Implantation

*D.-Q. Li, Z.-M. Shao*

**Fudan University, Shanghai, China**

## 1. INTRODUCTION

Distant metastasis to specific target organs represents a fundamental challenge for effective management of human cancer, which accounts for over 90% of cancer-related deaths (Hanahan and Weinberg, 2000; Valastyan and Weinberg, 2011; Gupta and Massague, 2006; Sethi and Kang, 2011). Substantial evidence has demonstrated that tumor metastasis is a complex multistep process, in which a subpopulation of cancer cells with highly invasive and metastatic potential detach from their original locations, invade, and degrade the surrounding extracellular matrix (ECM), intravasate into the blood or the lymphatic vessels, travel with and survive in the circulation, and extravasate to and colonize new terrain in the secondary organs (Hanahan and Weinberg, 2000; Valastyan and Weinberg, 2011; Gupta and Massague, 2006; Sethi and Kang, 2011). Accumulating evidence shows that the location of tumor metastases is not random. It is rather highly selective and specific; different types of human cancer tend to metastasize to specific target organs (Nguyen and Massague, 2007; Nguyen et al., 2009). For example, breast cancer preferentially metastasizes to the lung, liver, bone, and brain in the context of its molecular characteristics, whereas about 65% prostate cancer metastasizes to the bones (Weigelt et al., 2005; Schroeder et al., 2012). The development of organ-specific metastases is governed by a combinatorial code of anatomical, genetic, and biochemical determinants (Valastyan and Weinberg, 2011; Nguyen and Massague, 2007; Fidler, 2003; Paget, 1989). Among them, the well-recognized seed-soil hypothesis, originally proposed by Stephen Paget one century ago, underlines the importance of the delicate interplay between the metastatic cancer cells (the "seed") and their target organ microenvironments (the "soil") in the successful establishment of metastases (Chambers et al., 2002; Strickland and Richards, 1992).

Although remarkable advances have been made in the treatment of localized malignancies

*Introduction to Cancer Metastasis*
http://dx.doi.org/10.1016/B978-0-12-804003-4.00019-0

during the past several decades, successful therapeutic strategies have yet to be developed to manage metastatic disease (Chambers et al., 2002). It is much appreciated that an incomplete understanding of the mechanistic underpinnings for cancer metastasis is a major impediment to designing effective interventions for prevention or treatment of this fatal disease. Unfortunately, progress in addressing this intractable issue is largely limited due to the lack of suitable model systems. Thus, new perspectives to understand the molecular events driving tumor metastasis and to discover new therapeutic targets for metastatic disease are urgently needed.

Embryo implantation is a key physiological process for mammalian reproduction, which involves attachment of the trophoblasts to the ECM components, degradation of the ECM, and migration through the connective tissue, establishment of uteroplacental blood flow by remodeling the maternal vasculature, and growth of fetus in the maternal endometrium (Strickland and Richards, 1992; Menkhorst et al., 2016; Morgan et al., 1998). A growing body of evidence has shown that trophoblast invasion of the maternal endometrium during embryo implantation and cancer invasion of host tissues share some common biological features, such as cell migration, invasion, angiogenesis, and immune escape (Strickland and Richards, 1992; Even-Ram et al., 1998; Murray and Lessey, 1999; Yagel et al., 1988) (Fig. 19.1). Moreover, recent studies by us and others demonstrated that mouse trophoblastic cells even exhibit a dominant invasiveness phenotype over cancer cells in an in vitro coculture system (Fang et al., 2010; Ding et al., 2012). In contrast to unlimited tumor invasion and metastasis, however, embryo implantation is considered as a natural model of successfully controlled tissue invasion (Arck et al., 2000; Knoeller et al., 2003), in which trophoblast invasion into the maternal endometrium is strictly regulated through a delicate dialogue between the trophoblasts and the maternal endometrial microenvironment (Strickland and Richards, 1992; Perry et al., 2010; Staun-Ram and Shalev, 2005; Cohen et al., 2010). In this instance, trophoblast invasion is limited in time, only during first and early second trimester of pregnancy, and in space, limited to the proximal third of myometrium of the maternal endometrium (Cohen et al., 2010; Cohen and Bischof, 2007; Bischof et al., 2000a). Consequently, dysregulation of the invasive behaviors of the trophoblasts is intimately linked with various clinical pathological conditions (Soundararajan and Rao, 2004). For instance, deficient or inadequate trophoblast invasion is a common feature of pregnancy pathologies such as preeclampsia, intrauterine growth retardation, and spontaneous abortion (Lewis et al., 1994). Conversely, excessive invasion is implicated in the pathogenesis of choriocarcinoma, a highly metastatic and lethal trophoblastic cancer (Strickland and Richards, 1992; Soundararajan and Rao, 2004). Thus, unraveling trophoblast invasion and its control mechanisms during embryo implantation are key to understand the metastatic behaviors of malignant cancer cells and to define new therapeutic strategies for the prevention and treatment of metastatic disease (Murray and Lessey, 1999; Perry et al., 2010; Serman et al., 2012).

Collectively, trophoblast invasion into the maternal endometrium during embryo implantation is a coordinated process, which is governed by an intricate balance between the intrinsic invasive properties of the trophoblasts and the inhibitory factors derived from the maternal endometrium (Strickland and Richards, 1992; Cohen et al., 2010; Bischof et al., 2000a). In this chapter, we describe the molecular constituents controlling trophoblast invasion with a specific focus on trophoblast invasion inhibitory factors derived from the maternal endometrial microenvironment during embryo implantation (Table 19.1). In addition, we also discuss what we have learned from embryo implantation about tumor metastasis and the emerging potential therapeutic targets to block

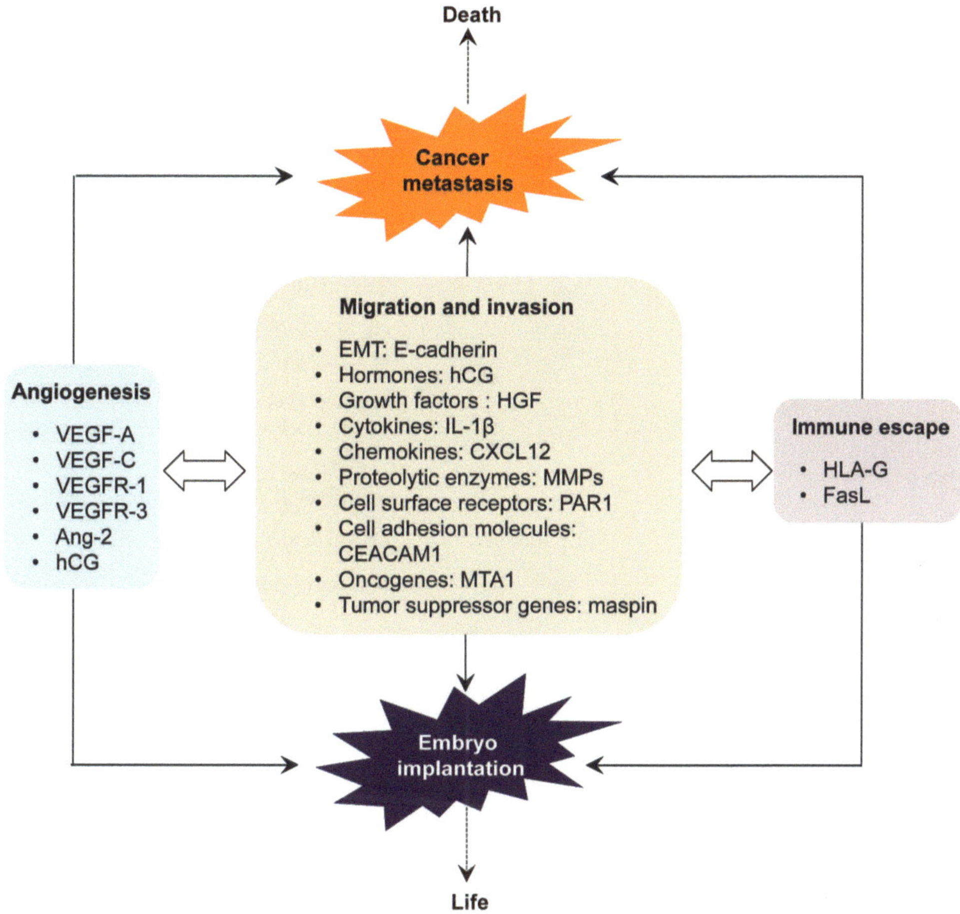


FIGURE 19.1 The similarity between embryo implantation and tumor metastasis in angiogenesis, cell migration and invasion, and immune escape.

tumor invasion and metastasis from the point of view of embryo implantation.

## 2. THE SIMILARITY BETWEEN EMBRYO IMPLANTATION AND TUMOR METASTASIS

### 2.1 Angiogenesis

It has been well documented that angiogenesis and vascular remodeling are crucial processes in tumor invasion and metastasis as well as in embryo implantation (Murray and Lessey, 1999; Zygmunt et al., 2002). Numerous studies have demonstrated that initiation of the angiogenic process is regulated by numerous proangiogenic factors, such as the vascular endothelial growth factor (VEGF) family of growth factors consisting of VEGF A–E and placental growth factor (PIGF) (Heath and Bicknell, 2009). This family of growth factors induces angiogenesis through binding to and activating the tyrosine kinase VEGF receptors 1–3 (VEGFR1–3; Heath and Bicknell, 2009). To establish uteroplacental circulation, the invasive human trophoblasts

TABLE 19.1 Trophoblast Invasion Inhibitory Factors Derived From the Maternal Endometrium and Their Implications in Cancer Metastasis

| Classification | Genes | Implication in Cancer Metastasis | References |
|---|---|---|---|
| ECM-degrading enzyme inhibitors | TIMP-1 | Yes | Waterhouse et al. (1993) |
| | TIMP-2 | Yes | Waterhouse et al. (1993) |
| | TIMP-3 | Yes | Higuchi et al. (1995) and Leco et al. (1996) |
| | TIMP-4 | Yes | Yang et al. (2006) |
| | α2-MG | Yes | Gu et al. (1992), Hannan et al. (2010) and Esadeg et al. (2003) |
| | PAI-1 | Yes | Lockwood et al. (1999), Lockwood and Schatz (1996), Lockwood et al. (1997) and Schatz et al. (1999) |
| | Cystatin C | Yes | Afonso et al. (1997) and Afonso et al. (2002) |
| | CTLA-2β | Unknown | Cheon et al. (2004) |
| Cytokines | TGF-β | Yes | Lala and Graham (1990) and Graham and Lala (1992) |
| | TNFα | Yes | Huber et al. (2006) and Otun et al. (2011) |
| | IFNγ | Yes | Lash et al. (2006) |
| | Uteroglobin | Yes | Kundu et al. (1996) |
| Metastasis suppressor genes | NME1 | Yes | Xie et al. (2010) |
| | CD82 (KAI1) | Yes | Gellersen et al. (2007), Zhang et al. (2012), Koo et al. (2013) and Li et al. (2010) |
| | KiSS-1 | Yes | Zhang et al. (2014) |
| | GPR54 | Yes | Zhang et al. (2014) |
| HtrA family of serine proteases | HtrA1 | Yes | Chen et al. (2014) |
| | HtrA3 | Yes | Singh et al. (2011, 2010) and Chen et al. (2014) |
| Glycoproteins | Mucin 15 | Yes | Shyu et al. (2007) |
| | Chondromodulin-I | Yes | Miura et al. (2011) |

*CTLA*, cytotoxic T-lymphocyte antigen; *ECM*, extracellular matrix; *GPR54*, G protein-coupled receptor 54; *HtrA*, high-temperature requirement factor A; *KAI1*, kangai 1; *KiSS*, Kisspeptin; *NME*, nonmetastatic cells; *PAI*, plasminogen activator inhibitor; *TGF-β*, transforming growth factor beta; *TIMP*, tissue inhibitors of metalloproteinase; *TNF*, tumor necrosis factor; *α2-MG*, α2-macroglobulin.

stimulate conventional vasculogenesis and angiogenesis through expressing VEGF-A, VEGF-C, PlGF, and their receptors VEGFR-1, VEGFR-3, and angiopoietin-2 (Zhou et al., 2003). In the same vein, the VEGF–VEGFR signaling axis-mediated angiogenesis significantly contributes to the growth and metastasis of different types of cancer cells (Su et al., 2006). Consequently, targeting the VEGF pathway using various antiangiogenic drugs suppresses tumor progression and metastasis (Ebos and Kerbel, 2011). In addition, human chorionic gonadotropin (hCG), a hormone produced by the trophoblasts during embryo implantation, is also expressed in approximate 20–40% of human epithelial cancers (Fukuda et al., 2008). Interestingly, this hormonal factor derived from both the trophoblasts and cancer cells promotes angiogenesis by stimulating the migration and capillary sprout formation (Zygmunt et al., 2002). Together, these accumulating evidence suggests that cancer cells and trophoblastic cells exploit the same angiogenesis mechanisms to regulate both tumor progression and placental development, respectively (Murray and Lessey, 1999) (Fig. 19.1).

## 2.2 Migration and Invasion

A growing body of evidence has demonstrated that the trophoblasts and metastatic cancer cells share the same migratory and invasive properties and related signaling pathways (Strickland and Richards, 1992; Murray and Lessey, 1999; Yagel et al., 1988, 1989; Ferretti et al., 2007) (Fig. 19.1). For instance, an important mechanism for governing trophoblast invasion is transition of the trophoblasts from epithelial to mesenchymal (EMT) phenotype (Kalluri and Weinberg, 2009). In this process, expression of the cell adhesion molecule E-cadherin, a hallmark of EMT, is high in the syncytiotrophoblasts and is downregulated when cytotrophoblasts invade the maternal decidual tissue (Batistatou et al., 2007; Shih Ie et al., 2002). Similarly, EMT is considered as a hallmark of cancer, and loss or downregulation of functional E-cadherin expression plays a dominant role in facilitating cancer invasion and metastasis (Kalluri and Weinberg, 2009; Hanahan and Weinberg, 2011). In addition, the trophoblasts and cancer cells also use other same molecular mediators for migration and invasion, including hormones (such as hCG; Zygmunt et al., 1998), growth factors (such as hepatocyte growth factor (HGF); Cartwright et al., 1999), cytokines (such as interleukin-1β, IL-1β; Librach et al., 1994), chemokines (such as C-X-C motif chemokine 12, CXCL12; Ren et al., 2012), proteolytic enzymes (such as matrix metalloproteinases, MMPs; Lala and Graham, 1990; Staun-Ram et al., 2004), cell surface receptors (such as proteinase-activated receptor 1, PAR1; Even-Ram et al., 1998), cell adhesion molecules (such as carcinoembryonic antigen-related cell adhesion molecule 1, CEACAM1; Bamberger et al., 2000), oncogenes (such as metastasis-associated protein 1, MTA1; Bruning et al., 2009), and tumor suppressor genes (such as maspin; Dokras et al., 2002). Thus, the highly migrative and invasive trophoblast has been termed "pseudomalignant" and embryo implantation has been defined as a "physiological metastasis" process (Strickland and Richards, 1992; Even-Ram et al., 1998; Ferretti et al., 2007).

## 2.3 Immune Escape

The semiallogeneic trophoblast cells of fetal origin and cancer cells are all recognized as nonself by the immune system and are vulnerable to immunological attack (Robertson and Moldenhauer, 2014). To invade neighboring tissues, both the trophoblasts and cancer cells use a number of similar mechanisms to prevent rejection by the host through escaping the immune response (Mullen, 1998) (Fig. 19.1). For instance, trophoblast cells and approximately 40–90% of human cancer fail to express the major histocompatibility complex (MHC) class I molecules, which is an important mechanism in immune escape from T-cell-mediated immune responses (Ferretti et al., 2007). In contrast, the trophoblasts strongly express the nonclassical MHC class I molecule human leucocyte antigen-G (HLA-G), which protects the fetus from destruction by the maternal immune system and thus contributes to maternal tolerance of the fetus during pregnancy (Ferretti et al., 2007; Carosella et al., 2015; Hamai et al., 1999). In support of this notion, absent or reduced expression of HLA-G expression in the trophoblasts is associated with preeclampsia (Goldman-Wohl et al., 2000; Colbern et al., 1994). In human cancer, high expression of HLA-G is a common feature and is intimately implicated in immune escape, tumor metastasis, and poor prognosis (Amodio et al., 2014). In addition, abundant expression of the Fas ligand (FasL or CD95 L) in the trophoblasts and cancer cells represent another common mechanism for the development of maternal immune tolerance during embryo implantation (Uckan et al., 1997) and tumor immune escape (Hahne et al., 1996). In this instance, FasL enables to eliminate Fas-positive lymphocytes by inducing apoptosis, thus evading host immune surveillance and destruction (Green and Ferguson, 2001). Thus, a comprehensive understanding of the immune escape mechanisms during embryo implantation may lead to new strategies for developing cancer immunotherapy against metastatic disease.

# 3. TROPHOBLAST INVASION INHIBITORY FACTORS DERIVED FROM THE MATERNAL ENDOMETRIUM DURING EMBRYO IMPLANTATION AND THEIR IMPLICATIONS IN CANCER METASTASIS

As discussed above, trophoblast invasion into the maternal endometrium during embryo implantation shares a lot of biological similarities with tumor invasion and metastasis (Strickland and Richards, 1992; Even-Ram et al., 1998; Murray and Lessey, 1999; Fang et al., 2010; Kalluri and Weinberg, 2009) (Fig. 19.1). The key question is why tumor invasion and metastasis is an uncontrolled process, whereas trophoblastic invasion is a highly controlled event. One of the important reasons is that the intrinsic invasiveness of trophoblast cells is strictly regulated by the microenvironment of the uterus (Menkhorst et al., 2016; Graham and Lala, 1992). Uterine control of trophoblast invasion is due, at least in part, to the decidualizing stroma, which produce a lot of cell migration and invasion inhibitory factors participating in the control of trophoblast invasion in a temporal and spatial pattern (Afonso et al., 1997). Thereafter, we will discuss the trophoblast invasion of inhibitory factors derived from the maternal endometrium during embryo implantation and their connections with cancer invasion and metastasis.

## 3.1 Inhibitors of the ECM-Degrading Enzymes

Degradation of ECM components by various ECM-degrading enzymes, including matrix metalloproteinases (MMPs), plasmin, and cathepsins, is critical for trophoblast invasion and tumor invasion and metastasis (Egeblad and Werb, 2002) (Fig. 19.2). Among them, the MMPs have been the focus of numerous studies, in particular MMP-2 and MMP-9, which specifically degrade type IV collagen and gelatins. The trophoblasts produce both MMP-2 and MMP-9 (Morgan et al., 1998; Isaka et al., 2003; Librach et al., 1991). In early trophoblasts, MMP-2 is the main gelatinase and the key enzyme in trophoblast invasion, whereas both MMP-2 and -9 participate in trophoblast invasion in late first trimester trophoblasts (Staun-Ram et al., 2004). Consequently, dysregulation of MMP-2 and -9 contributes to preeclamptic pregnancies (Campbell et al., 2004). Additionally, the trophoblasts also express the putative MMP-2 activators such as MMP-14 (also known as membrane-type 1 MMP, MT1-MMP; Nawrocki et al., 1996; Tanaka et al., 1998) and MMP-15 (MT2-MMP; Bjorn et al., 2000) for invasion of the maternal endometrium. Similarly, involvement of various MMPs including the above-mentioned MMP-2, -9, -14, and -15 in cancer invasion and metastasis has been well documented in a vast body of literature (Egeblad and Werb, 2002; Zhai et al., 2005). Thus, MMPs have been considered as promising targets in the control of metastatic disease (Overall and Lopez-Otin, 2002).

Unlike the uncontrolled tumor invasion and metastasis, however, the balance between

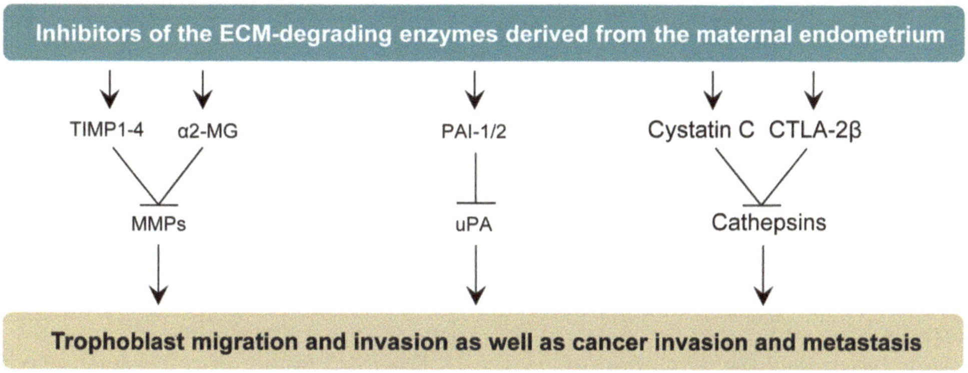


FIGURE 19.2 Involvement of inhibitors of the extracellular matrix-degrading enzymes derived from the maternal endometrium in embryo implantation and cancer metastasis.

MMPs and tissue inhibitors of metalloproteinases (TIMPs) is critical for coordinated invasion of the trophoblasts into the maternal endometrium. To date, four TIMPs have been characterized in human and designated as TIMP-1 to -4, which reversibly inhibit MMPs in a 1:1 stoichiometric fashion (Egeblad and Werb, 2002). Emerging evidence shows that both TIMP-1 and -2 are expressed in maternal decidual cells with a dramatic increase of TIMP-1 at the term of pregnancy (Polette et al., 1994), and the invasive ability of early trophoblasts is inhibited by the TIMP antibodies in a dose-dependent manner (Isaka et al., 2003). A comparative study of the expression levels of TIMP-1 and -2 between clearly invasive carcinomas and in situ trophoblast invasion revealed that TIMP-1 and -2 are expressed more often and intense in decidual cells than in cancer stromal cells, respectively (Dimo et al., 2012). Thus, reduced expression of TIMP-1 and -2 in cancer stromal cells as compared with the maternal endometrium during embryo implantation highlights its importance in uncontrolled tumor invasion and metastasis (Dimo et al., 2012). In addition to TIMP-1 and -2, both TIMP-3 and TIMP-4 have been reported to be expressed in the maternal decidual cells during early pregnancy (Higuchi et al., 1995; Yang et al., 2006; Whiteside et al., 2001), indicating that TIMP-3 and -4 are involved in controlling trophoblast invasion by regulating MMPs in the maternal endometrium during early pregnancy (Higuchi et al., 1995; Yang et al., 2006). In addition, MMP activity is also controlled by other endogenous inhibitors, such as α2-macroglobulin (α2-MG), an abundant plasma protein (Egeblad and Werb, 2002). Evidence shows that α2-MG is expressed in the maternal decidua (Gu et al., 1992; Hannan et al., 2010), which exerts significant decidual regulation on trophoblast invasion through modulating MMP activities (Esadeg et al., 2003). In contrast to tightly regulated trophoblast invasion, loss or inactivation of TIMPs through, at least in part, promoter hypermethylation, thus resulting in an imbalance between MMPs and TIMPs, is a common feature of human cancer progression and metastasis (Galm et al., 2005; Anania et al., 2011; Lu et al., 2014; Bachman et al., 1999). In line with these observations, induced expression of TIMPs shows an inhibitory effect on tumor invasion and metastasis (DeClerck et al., 1992). Thus, restoring the physiological balance between MMPs and TIMPs could be one of the therapeutic strategies against tumor metastasis. To this point, great efforts have been made to develop selective MMP inhibitors (such as Batimastat) as anticancer therapeutics (Overall and Lopez-Otin, 2002), but results from the clinical trials are disappointing. Alternatively, whether enhanced expression of TIMPs or α2-MG alone or in combination with selective MMP inhibitors has a better therapeutic effect on tumor metastasis needs to be further investigated in the near future.

In addition to MMPs, another important proteolytic enzyme is the serine proteinase plasmin, which is generated by the cleavage of the proenzyme plasminogen through the action of the urokinase-type plasminogen activator (uPA) and its receptor (uPAR; Reuning et al., 1998). The uPA system exerts its action on matrix degradation, in part indirectly, by proteolytic activation of pro-MMPs (Koizumi et al., 2010). Naturally, the uPA activity is negatively regulated by the plasminogen activator inhibitors (PAI-1 and -2) (Reuning et al., 1998). A number of experimental and clinical studies have well defined the important and causal roles for the uPA and its receptor in facilitating tumor invasion and metastasis (Fazioli and Blasi, 1994; Duffy et al., 1999). In support of this notion, selective synthetic uPA inhibitors effectively inhibit tumor invasive and metastatic potential in multiple cancer model systems (Xing et al., 1997; Zhu et al., 2007). Similarly, during embryo implantation, uPA is expressed by human preimplantation embryos as well as invasive trophoblasts and facilitates trophoblast invasion of the uterus (Khamsi et al., 1996; Feng et al., 2001, 2000). In contrast to tumor metastasis, however, decidual cells of pregnant endometrium express PAI-1, thus providing a mechanism by which decidual cells control local hemostasis during trophoblast invasion (Feng et al., 2001; Lockwood et al., 1999; Lockwood and Schatz, 1996). In support of these notions, trophoblast invasion-stimulatory factors, such as polypeptide adrenomedullin (Wong et al., 2013) and IL-1β (Karmakar and Das, 2002), enhance trophoblast migration and invasion through inducing uPA expression. In contrast, trophoblast invasion-inhibitory factors, such as transforming growth factor-β (TGF-β) and tumor necrosis factor α (TNFα), exert their anti-invasive effect by suppressing secretion of uPA and enhancing PAI-1 production and secretion (Karmakar and Das, 2002; Graham, 1997; Huber et al., 2006; Chakraborty et al., 2002). These discoveries from embryo implantation further validate PAI-1 as a potential therapeutic target for the development of antimetastatic therapeutic agents. As expected, accumulating experimental evidence has shown that transfer of the PAI-1 gene inhibits glioma cell invasion and motility (Hjortland et al., 2003) and liver metastasis of pancreatic cancer (Inoue et al., 2005). Clearly, these encouraging results in experimental tumors need to be further verified in clinic.

The third family of proteolytic enzymes is the cysteine cathepsins, which belong to the papain subfamily of cysteine proteases and have biological roles in degrading ECM (Mohamed and Sloane, 2006). Extensive studies have documented that cysteine cathepsins are highly upregulated in a wide variety of human cancers and play causal roles in tumor migration, invasion, angiogenesis, and metastasis (Mohamed and Sloane, 2006). Not surprisingly, the peri-hatching blastocysts and invasive trophoblasts express cathepsins to invade the maternal endometrium (Afonso et al., 1997, 1999; Sireesha et al., 2008; Amarante-Paffaro et al., 2011). Meanwhile, cystatin C, the main endogenous inhibitor of cathepsins, is synthesized by the uterine decidua and localized to the cells in close contact with the trophoblast during implantation, thus highlighting that the balance between trophoblast cathepsins and decidual cystatin C expression is pivotal in physiological trophoblast development (Afonso et al., 1997, 2002; Quinn et al., 2006; Thilaganathan et al., 2009). Consequently, enhanced maternal serum cystatin C is an early pregnancy marker for preeclampsia characterized by defective trophoblast invasion (Thilaganathan et al., 2009). In addition, cystatin C is also expressed in the blastocysts, indicating that the invading embryo regulates the depth of its own invasion through the expression of the cathepsin inhibitors (Baston-Buest et al., 2010). In contrast, cystatin C has been reported to be down-regulated in prostate cancer, thus enhancing prostate cancer invasion via mitogen-activated protein kinase/extracellular signal-regulated kinase

(MAPK/ERK) and androgen receptor pathways (Wegiel et al., 2009). Conversely, induced expression of cystatin C inhibits lung metastasis of human fibrosarcoma (Kopitz et al., 2005) and impairs the invasiveness of human glioblastoma (Konduri et al., 2002). These results highlight that cystatin C is a potential therapeutic target against cancer metastasis. In addition, cytotoxic T-lymphocyte antigen-2β (CTLA-2β), another cathepsin L protease inhibitor (Delaria et al., 1994), is also expressed in the maternal decidua during embryo implantation, which has a role in the regulation of implantation of the embryo by neutralizing the activities of one or more proteases generated by the proliferating trophoblast (Cheon et al., 2004). To date, there is no report about the role of CTLA-2β in cancer metastasis. These results further support the notion that trophoblast invasion is controlled in part by a balance of trophoblast-derived proteinases and uterine decidual proteinase inhibitors (Afonso et al., 1997, 2002).

## 3.2 Cytokines

Extensive evidence has suggested that endometrial-derived cytokines play key roles in tightly controlling trophoblast migration and invasion within the maternal uterus during embryo implantation (Bischof et al., 2000b; Chard, 1995) (Fig. 19.3). One of such decidua-derived cytokines is the TGF-β (Graham and Lala, 1991; Graham et al., 1994). Several lines of evidence have shown that the latent form of TGF-β is activated by trophoblast-derived proteinases, and then activated TGF-β in turn suppresses trophoblast migration and invasion through multiple distinct mechanisms. First, TGF-β induces expression of the protease inhibitors, such as TIMP-1, -2, PAI-1 and -2, and downregulates uPA, in both the decidua and the trophoblasts (Karmakar and Das, 2002; Chakraborty et al., 2002). Second, TGF-β promotes differentiation of invasive trophoblast cells into noninvasive syncytiotrophoblasts (Lala and Graham, 1990; Graham and Lala, 1992, 1991; Karmakar and Das, 2002; Rama et al., 2003). Third, TGF-β decreases HGF-induced trophoblast motility and invasion (Tse et al., 2002). Fourth, TGF-β mediates upregulation of cell-to-cell adhesion along with an increased ezrin and E-cadherin expression (Karmakar and Das, 2004). Thus, TGF-β provides the key control of trophoblast

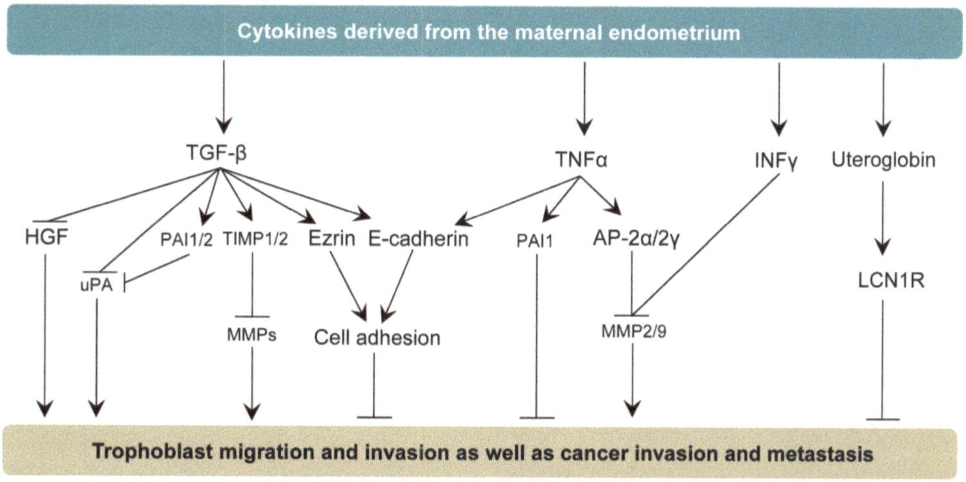


FIGURE 19.3 Signaling pathways mediated by the maternal endometrium-derived cytokines in embryo implantation and tumor metastasis.

invasiveness in situ (Graham and Lala, 1992). Similarly, a TGF-β binding proteoglycan, termed decorin, has been shown to be colocalized with TGF-β in the decidual ECM and to suppress proliferation, migration, and invasiveness of human invasive trophoblasts (Xu et al., 2002). Unlike normal trophoblast cells, however, trophoblast-derived choriocarcinoma cells are refractory to the antiproliferative and anti-invasive effects of both TGF-β and decorin (Graham et al., 1994; Xu et al., 2002), which are highly relevant to choriocarcinoma progression and metastasis. The underlying mechanism for resistance of choriocarcinoma cells to the anti-invasive action of TGF-β has been found to be associated with the failure of TGF-β to upregulate TIMPs and PAIs, in contrast to the noted effect of TGF-β on normal trophoblast cells (Khoo et al., 1998). Evasion of the TGF-β-mediated growth- and invasion-inhibitory effects was also found in other types of human cancer (Hanahan and Weinberg, 2011). Interestingly, induced expression of decorin has been shown to effectively block invasion and metastasis in several cancer model systems, including breast (Yang et al., 2015; Reed et al., 2005), prostate (Xu et al., 2015), and colorectal cancer (Bi et al., 2012) as well as murine osteosarcoma (Shintani et al., 2008). These results demonstrate a functional role for decorin in reduction or prevention of tumor metastases and could eventually lead to improved therapeutics for metastatic cancer (Reed et al., 2005).

Another decidua-derived cytokine with the capacity to inhibit trophoblast invasion is the TNFα (Vince et al., 1992; Chen et al., 1991; Otun et al., 2011), which restricts trophoblast invasion partially by increasing the expression of PAI-1 through the transcription factor nuclear factor-kappa B (Huber et al., 2006). Consistently, over-induction of PAI-1 by TNFα has been linked to restricted trophoblast invasion in preeclampsia (Bauer et al., 2004). In addition, TNFα has been shown to induce expression of activator protein-2α (AP-2α) and AP-2γ in trophoblasts, which in turn suppress the invasion of trophoblast cells by repression of MMP-2 and MMP-9 and upregulation of E-cadherin (Kotani et al., 2009). Similarly, transcription factor AP-2α has been reported to inhibit malignant phenotype of human melanoma through downregulation of PAR1 (Tellez et al., 2003). Regarding cancer, TNFα has a dual role in tumor promotion and suppression in a context-dependent manner (Wang and Lin, 2008; Qin et al., 1998).

In addition, interferon γ (IFNγ) is produced in the maternal decidual tissue and uterine natural killer cells and inhibits trophoblast invasion via a mechanism involving increased apoptosis and decreased MMP-2 activity (Lash et al., 2006). In human cancer, it has been reported that IFNγ attenuates astroglioma cell invasiveness properties through transcriptional suppression of MMP-2 expression (Qin et al., 1998) and potentiates the therapeutic effect of immunotherapy on melanoma invasion and metastasis through the integrin αVβ3 signaling pathway (Gong et al., 2008).

During early pregnancy, IL-11 is maximally expressed in the decidua and IL-11 inhibits human trophoblast invasion via transcription factor signal transducer and activator of transcription 3 (STAT3), indicating a potential role for IL-11 in the decidual restraint of trophoblast invasion during normal pregnancy (Paiva et al., 2009). However, it appears that IL-11 is a metastasis promoter in some types of human cancer (Li et al., 2012), reflecting the complexity of cancer biology.

Uteroglobin (also known as blastokinin) is a steroid-dependent, multifunctional, cytokine-like protein, which was originally isolated from the uterus of rabbits during early pregnancy (Beier, 1968). Recent studies suggest that uteroglobin suppresses trophoblast invasiveness via a novel class of high-affinity cell surface binding site (Kundu et al., 1996). It has been proposed that uteroglobin exerts the antimotility and anti-invasive effects through interacting with the lipocalin-1 receptor (Zhang et al., 2006). Consistently, multiple lines of evidence have defined uteroglobin as a potential novel tumor suppressor and molecular therapeutic for prostate cancer (Weeraratna et al., 1997; Patierno et al., 2002).

## 3.3 Metastasis Suppressor Genes

Metastasis suppressor genes have been defined as a group of genes that suppress the metastatic potential of cancer cells without affecting tumorigenicity through targeting one of the key steps of the invasion–metastasis cascade (Smith and Theodorescu, 2009; Steeg, 2003). To date, only few metastasis suppressor genes have been characterized (Smith and Theodorescu, 2009; Steeg, 2003) (Fig. 19.4).

The first identified metastasis suppressor is the NME1 (nonmetastatic cells 1; also known as NM23 or NM23-H1), which was discovered in 1988 by analyzing differentially expressed genes between high- and low-metastatic murine melanoma cell lines using differential hybridization technologies (Steeg et al., 1988). Since then, a vast body of literature has established NME1 as a putative metastasis suppressor gene with predominant roles in suppressing the invasion and metastasis of wide types of human cancer (Smith and Theodorescu, 2009; Steeg, 2003; Shoushtari et al., 2011). Interestingly, it was demonstrated that NME1 is expressed in both human trophoblast cells and decidual stromal cells (DSCs) in early pregnancy (Xie et al., 2010), and the DSC-expressed NME1 suppresses invasiveness of human first-trimester trophoblast cells via the MAPK/ERK pathway (Xie et al., 2010). Moreover, attenuation of NME1 expression is involved in the thymic stromal lymphopoietin-stimulated trophoblast invasion (Wang et al., 2012).

Another well-recognized tumor metastasis suppressor is the CD82 (also known as kangai 1 (KAI1)), which was originally identified as a metastasis suppressor gene for prostate cancer (Dong et al., 1995). Subsequent studies pointed out a wide spectrum metastasis suppressor function for CD82 (Smith and Theodorescu, 2009; Steeg, 2003). During embryo implantation, CD82 is specifically expressed on decidualized endometrial stromal cells and inhibits trophoblast invasion through, at least in part, upregulation of the expression of TIMP-1 in an autocrine manner (Gellersen et al., 2007; Zhang et al., 2012; Koo et al., 2013; Li et al., 2010). Consistently, the expression level of CD82 in decidua of the miscarriage is significantly higher than that of the normal early pregnancy, indicating that the abnormal higher CD82 expression in decidua

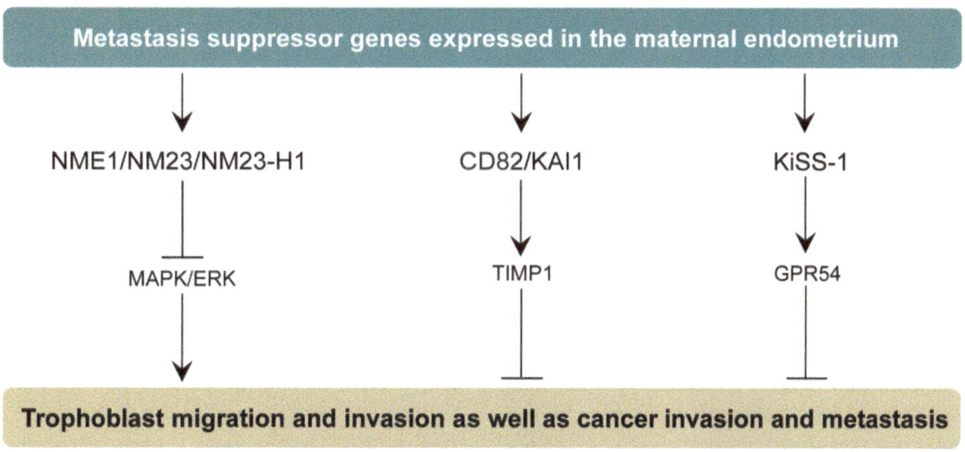


FIGURE 19.4 Metastasis suppressor genes expressed in the maternal endometrium during embryo implantation.

restricts appropriate invasion of trophoblasts that leads to early pregnancy wastage (Li et al., 2010).

Kisspeptin (KiSS-1) gene, initially identified as a melanoma metastasis suppressor gene (Lee et al., 1996), encodes a number of peptides (KiSS -10, -13, -,14, and -54), which are endogenous ligands to a G protein–coupled receptor (GPR54; Ohtaki et al., 2001). Substantial evidence supports a suppressor role for KiSS-1 in across a range of human cancers (Lee and Welch, 1997; Ikeguchi et al., 2004; Dhar et al., 2004; Navenot et al., 2005). Recent studies reveal that expression levels of KiSS-1 and GPR54 are dynamically increased in decidualizing stromal cells in intact pregnant females as well as in pseudopregnant mice undergoing artificially induced decidualization (Zhang et al., 2014). Given the key inhibitory roles for the KiSS-1/GPR54 signaling axis in cell migration and invasion, these results indicate that the KiSS-1/GPR54 system is involved in promoting uterine decidualization and tight regulation of trophoblast invasion during early pregnancy (Zhang et al., 2014).

Taken together, multiple metastasis suppressor genes including NME1, CD82, and KiSS-1, are spatiotemporally expressed in the maternal endometrium during embryo implantation, which play key roles in tightly regulating trophoblast invasion (Fig. 19.4). However, these genes are upregulated during early cancer but are generally downregulated, inactivated, or lost during cancer progression, which enable cancer cells to metastasize to distant target organs (Smith and Theodorescu, 2009; Dong et al., 1996; Guo et al., 1996). Thus, a better understanding of the spatio-temporal expression patterns and functions of metastasis-suppressor genes in the maternal endometrium during embryo implantation are likely to lead to discover new metastasis suppressor genes and develop new treatment strategies using metastasis suppressor genes against metastatic disease. Given the key roles for metastasis suppressor genes in cancer progression, restoring metastasis suppressor gene function by means of gene transfer, induction of previously suppressed gene expression and exogenous administration would be novel therapeutic approaches to block metastatic disease in the future (Smith and Theodorescu, 2009; Shoushtari et al., 2011).

## 3.4 High-Temperature Requirement Factor A Family of Serine Proteases

The high-temperature requirement factor A (HtrA) family of serine proteases was initially identified in *Escherichia coli* by the phenotype of null mutants that were unable to grow at elevated temperatures (Chien et al., 2009; Lipinska et al., 1988). To date, four human homologues of *E. coli* HtrA have been identified, including HtrA 1–4 (Chien et al., 2009). Accumulating evidence shows that the HtrA family of serine protease is involved in multiple pathophysiological processes, including cancer progression, mammalian reproduction, and neurodegenerative disease (Chien et al., 2009; Zurawa-Janicka et al., 2010). In this instance, HtrA1 has been documented to be downregulated in human melanoma (Baldi et al., 2002) and endometrial cancer (Bowden et al., 2006) and can be used to predict response to platinum-based combination therapies in gastric cancer (Catalano et al., 2011). Similarly, HtrA3 is downregulated in endometrial (Bowden et al., 2006) and ovarian (Bowden et al., 2010; Singh et al., 2013) cancer. These results highlight a potential tumor suppressive functions for HtrA proteins. In contrast, HtrA1 is upregulated in both endometrial glands and decidual cells during endometrial preparation for embryo implantation and during first-trimester pregnancy at placentation and inhibits trophoblast invasion (Nie et al., 2006). Similarly, HtrA3 is also highly expressed in the decidual cells in the late secretory phase of the menstrual cycle and throughout pregnancy, and negatively regulates trophoblast invasion (Singh et al., 2011, 2010). However,

HtrA4 is highly expressed in the invasive trophoblasts and facilitates trophoblast invasion (Chen et al., 2014). More interestingly, it has been found that decidua-secreted HtrA1 and HtrA3 antagonize HtrA4-mediated trophoblast invasion through interacting with and degrading HtrA4 (Chen et al., 2014). These results highlight an interesting regulatory mechanism of trophoblast invasion through a crosstalk between the HtrA family members (Chen et al., 2014).

### 3.5 Glycoproteins

Mucins belong to the family of high-molecular-weight glycoproteins characterized by the presence of a heavily O-glycosylated tandem repeat region and consist of 21 members (Kaur et al., 2013). Accumulating evidence shows that mucins are involved in diverse biological functions, including tumor invasion, metastasis, and drug resistance (Kaur et al., 2013; Kufe, 2009). Among them, mucin 15 was found to be present in the glandular epithelium of the decidua and inhibit trophoblast invasion through upregulation of TIMP-1 and -2 (Shyu et al., 2007). Similarly, expression of mucin 15 in hepatocellular carcinoma cells reduces their aggressive behavior in vitro and in vivo by inducing dimerization of epidermal growth factor receptor and decreasing phosphoinositide 3-kinase-AKT signaling (Wang et al., 2013). Thus, mucin 15 may represent an attractive therapeutic target for cancer metastasis.

Chondromodulin-I (ChM-I) was previously defined as an antiangiogenic glycoprotein (Hiraki et al., 1999). In situ hybridization and immunohistochemical analysis revealed that ChM-I is localized to the mature decidua surrounding the MMP-9-expressing trophoblasts and inhibits trophoblast migration (Miura et al., 2011). Moreover, it was found that ChM-I is specifically lost in chondrosarcomas, and administration of recombinant human ChM-I protein suppresses angiogenesis and growth of chondrosarcomas and colon adenocarcinoma, implying its therapeutic potential for solid tumors (Hayami et al., 1999).

## 4. CONCLUSIONS AND PERSPECTIVES

A considerable body of evidence has demonstrated that there are striking similarities between embryo implantation and tumor metastasis including migration, invasion, angiogenesis, and immune escape (Strickland and Richards, 1992; Even-Ram et al., 1998; Murray and Lessey, 1999; Kalluri and Weinberg, 2009) (Fig. 19.1), but distinct differences do exist. In this instance, unlike uncontrolled tumor invasion and metastasis, trophoblast invasion into the maternal endometrium is essential for successful implantation and is tightly regulated by molecular and cellular interactions between the trophoblasts and the maternal microenvironment (Staun-Ram and Shalev, 2005). At this point, the maternal decidual cells temporally and spatially express numerous cell migration- and invasion-inhibitory factors, such as the ECM-degrading enzyme inhibitors (Fig. 19.2), cytokines (Fig. 19.3), metastasis suppressor genes (Fig. 19.4), serine proteases, and glycoproteins (Table 19.1), to counteract trophoblast invasion. As discussed above, however, all of these endometrium-derived regulatory factors are dysregulated during cancer progression and metastasis. Thus, deciphering the regulatory mechanisms of embryo implantation that has been defined as a natural model of successful controlled tissue invasion (Arck et al., 2000) and a "physiological metastasis" process (Even-Ram et al., 1998) is particularly important for, not only the biology of mammalian reproduction but also the biology of cancer metastasis (Karmakar et al., 2004). We hope that a comprehensive investigation of the similarities and differences between embryo implantation and tumor metastasis will provide new perspectives for understanding the molecular mechanisms of cancer metastasis and for developing novel anti-metastatic therapy strategies, ultimately leading to better treatments for patients with metastatic disease. It is believed that it will come true in the near future.

## Acknowledgments

We are in debt to our colleagues in this field whose original work may have not been cited here due to space limitations. The work in the Li lab is supported by the National Natural Science Foundation of China (No. 81372847 and 81572584), the Program for Professor of Special Appointment (Eastern Scholar) at Shanghai Institutions of Higher Learning (No. 2013-06), and the Innovation Program of Shanghai Municipal Education Commission (No. 2015ZZ007).

## References

Arck, P.C., Hertwig, K., Hagen, E., Hildebrandt, M., Klapp, B.F., 2000. Pregnancy as a model of controlled invasion might be attributed to the ratio of CD3/CD8 to CD56. Am. J. Reprod. Immunol. 44 (1), 1–8.

Amodio, G., Sales de Albuquerque, R., Gregori, S., 2014. New insights into HLA-G mediated tolerance. Tissue Antigens 84 (3), 255–263.

Afonso, S., Romagnano, L., Babiarz, B., 1997. The expression and function of cystatin C and cathepsin B and cathepsin L during mouse embryo implantation and placentation. Development 124 (17), 3415–3425.

Anania, M.C., Sensi, M., Radaelli, E., Miranda, C., Vizioli, M.G., Pagliardini, S., et al., 2011. TIMP3 regulates migration, invasion and in vivo tumorigenicity of thyroid tumor cells. Oncogene 30 (27), 3011–3023.

Amarante-Paffaro, A.M., Hoshida, M.S., Yokota, S., Goncalves, C.R., Joazeiro, P.P., Bevilacqua, E., et al., 2011. Localization of cathepsins D and B at the maternal-fetal interface and the invasiveness of the trophoblast during the postimplantation period in the mouse. Cells Tissues Organs 193 (6), 417–425.

Afonso, S., Romagnano, L., Babiarz, B., 1999. Expression of cathepsin proteinases by mouse trophoblast in vivo and in vitro. Dev. Dyn. 216 (4–5), 374–384.

Afonso, S., Tovar, C., Romagnano, L., Babiarz, B., 2002. Control and expression of cystatin C by mouse decidual cultures. Mol. Reprod. Dev. 61 (2), 155–163.

Bischof, P., Meisser, A., Campana, A., 2000a. Paracrine and autocrine regulators of trophoblast invasion—a review. Placenta 21 (Suppl. A), S55–S60.

Batistatou, A., Makrydimas, G., Zagorianakou, N., Zagorianakou, P., Nakanishi, Y., Agnantis, N.J., et al., 2007. Expression of dysadherin and E-cadherin in trophoblastic tissue in normal and abnormal pregnancies. Placenta 28 (5–6), 590–592.

Bamberger, A.M., Sudahl, S., Loning, T., Wagener, C., Bamberger, C.M., Drakakis, P., et al., 2000. The adhesion molecule CEACAM1 (CD66a, C-CAM, BGP) is specifically expressed by the extravillous intermediate trophoblast. Am. J. Pathol. 156 (4), 1165–1170.

Bruning, A., Makovitzky, J., Gingelmaier, A., Friese, K., Mylonas, I., 2009. The metastasis-associated genes MTA1 and MTA3 are abundantly expressed in human placenta and chorionic carcinoma cells. Histochem. Cell Biol. 132 (1), 33–38.

Bjorn, S.F., Hastrup, N., Larsen, J.F., Lund, L.R., Pyke, C., 2000. Messenger RNA for membrane-type 2 matrix metalloproteinase, MT2-MMP, is expressed in human placenta of first trimester. Placenta 21 (2–3), 170–176.

Bachman, K.E., Herman, J.G., Corn, P.G., Merlo, A., Costello, J.F., Cavenee, W.K., et al., 1999. Methylation-associated silencing of the tissue inhibitor of metalloproteinase-3 gene suggest a suppressor role in kidney, brain, and other human cancers. Cancer Res. 59 (4), 798–802.

Baston-Buest, D.M., Schanz, A., Buest, S., Fischer, J.C., Kruessel, J.S., Hess, A.P., 2010. The embryo's cystatin C and F expression functions as a protective mechanism against the maternal proteinase cathepsin S in mice. Reproduction 139 (4), 741–748.

Bischof, P., Meisser, A., Campana, A., 2000b. Mechanisms of endometrial control of trophoblast invasion. J. Reprod. Fertil. Suppl. 55, 65–71.

Bi, X., Pohl, N.M., Qian, Z., Yang, G.R., Gou, Y., Guzman, G., et al., 2012. Decorin-mediated inhibition of colorectal cancer growth and migration is associated with E-cadherin in vitro and in mice. Carcinogenesis 33 (2), 326–330.

Bauer, S., Pollheimer, J., Hartmann, J., Husslein, P., Aplin, J.D., Knofler, M., 2004. Tumor necrosis factor-alpha inhibits trophoblast migration through elevation of plasminogen activator inhibitor-1 in first-trimester villous explant cultures. J. Clin. Endocrinol. Metab. 89 (2), 812–822.

Beier, H.M., 1968. Uteroglobin: a hormone-sensitive endometrial protein involved in blastocyst development. Biochim. Biophys. Acta 160 (2), 289–291.

Baldi, A., De Luca, A., Morini, M., Battista, T., Felsani, A., Baldi, F., et al., 2002. The HtrA1 serine protease is down-regulated during human melanoma progression and represses growth of metastatic melanoma cells. Oncogene 21 (43), 6684–6688.

Bowden, M.A., Di Nezza-Cossens, L.A., Jobling, T., Salamonsen, L.A., Nie, G., 2006. Serine proteases HTRA1 and HTRA3 are down-regulated with increasing grades of human endometrial cancer. Gynecol. Oncol. 103 (1), 253–260.

Bowden, M.A., Drummond, A.E., Fuller, P.J., Salamonsen, L.A., Findlay, J.K., Nie, G., 2010. High-temperature requirement factor A3 (Htra3): a novel serine protease and its potential role in ovarian function and ovarian cancers. Mol. Cell. Endocrinol. 327 (1–2), 13–18.

Chambers, A.F., Groom, A.C., MacDonald, I.C., 2002. Dissemination and growth of cancer cells in metastatic sites. Nat. Rev. Cancer 2 (8), 563–572.

Cohen, M., Wuillemin, C., Irion, O., Bischof, P., 2010. Role of decidua in trophoblastic invasion. Neuro Endocrinol. Lett. 31 (2), 193–197.

Cohen, M., Bischof, P., 2007. Factors regulating trophoblast invasion. Gynecol. Obstet. Invest. 64 (3), 126–130.

Cartwright, J.E., Holden, D.P., Whitley, G.S., 1999. Hepatocyte growth factor regulates human trophoblast motility and invasion: a role for nitric oxide. Br. J. Pharmacol. 128 (1), 181–189.

Carosella, E.D., Rouas-Freiss, N., Roux, D.T., Moreau, P., LeMaoult, J.H.L.A.-G., 2015. An immune checkpoint molecule. Adv. Immunol. 127, 33–144.

Colbern, G.T., Chiang, M.H., Main, E.K., 1994. Expression of the nonclassic histocompatibility antigen HLA-G by preeclamptic placenta. Am. J. Obstet. Gynecol. 170 (5 Pt 1), 1244–1250.

Campbell, S., Rowe, J., Jackson, C.J., Gallery, E.D., 2004. Interaction of cocultured decidual endothelial cells and cytotrophoblasts in preeclampsia. Biol. Reprod. 71 (1), 244–252.

Chakraborty, C., Gleeson, L.M., McKinnon, T., Lala, P.K., 2002. Regulation of human trophoblast migration and invasiveness. Can. J. Physiol. Pharmacol. 80 (2), 116–124.

Cheon, Y.P., DeMayo, F.J., Bagchi, M.K., Bagchi, I.C., 2004. Induction of cytotoxic T-lymphocyte antigen-2beta, a cysteine protease inhibitor in decidua: a potential regulator of embryo implantation. J. Biol. Chem. 279 (11), 10357–10363.

Chard, T., 1995. Cytokines in implantation. Hum. Reprod. Update 1 (4), 385–396.

Chen, H.L., Yang, Y.P., Hu, X.L., Yelavarthi, K.K., Fishback, J.L., Hunt, J.S., 1991. Tumor necrosis factor alpha mRNA and protein are present in human placental and uterine cells at early and late stages of gestation. Am. J. Pathol. 139 (2), 327–335.

Chien, J., Campioni, M., Shridhar, V., Baldi, A., 2009. HtrA serine proteases as potential therapeutic targets in cancer. Curr. Cancer Drug Targets 9 (4), 451–468.

Catalano, V., Mellone, P., d'Avino, A., Shridhar, V., Staccioli, M.P., Graziano, F., et al., 2011. HtrA1, a potential predictor of response to cisplatin-based combination chemotherapy in gastric cancer. Histopathology 58 (5), 669–678.

Chen, Y.Y., Chuang, P.Y., Chen, C.P., Chiu, Y.H., Lo, H.F., Cheong, M.L., et al., 2014. Functional antagonism between high temperature requirement protein A (HtrA) family members regulates trophoblast invasion. J. Biol. Chem. 289 (33), 22958–22968.

Ding, X., Fang, L., Zhang, H., Qiao, H., Wang, Z.B., 2012. Invasiveness of mouse embryos to human ovarian cancer cells HO8910PM and the role of MMP-9. Cancer Cell Int. 12 (1), 23.

Dokras, A., Gardner, L.M., Kirschmann, D.A., Seftor, E.A., Hendrix, M.J., 2002. The tumour suppressor gene maspin is differentially regulated in cytotrophoblasts during human placental development. Placenta 23 (4), 274–280.

Dimo, B., Ioannidis, I., Karameris, A., Vilaras, G., Tzoumakari, P., Nonni, A., et al., 2012. Comparative study of the immunohistochemical expression of tissue inhibitors of metalloproteinases 1 and 2 between clearly invasive carcinomas and "in situ" trophoblast invasion. Med. Oncol. 29 (3), 2270–2275.

DeClerck, Y.A., Perez, N., Shimada, H., Boone, T.C., Langley, K.E., Taylor, S.M., 1992. Inhibition of invasion and metastasis in cells transfected with an inhibitor of metalloproteinases. Cancer Res. 52 (3), 701–708.

Duffy, M.J., Maguire, T.M., McDermott, E.W., O'Higgins, N., 1999. Urokinase plasminogen activator: a prognostic marker in multiple types of cancer. J. Surg. Oncol. 71 (2), 130–135.

Delaria, K., Fiorentino, L., Wallace, L., Tamburini, P., Brownell, E., Muller, D., 1994. Inhibition of cathepsin L-like cysteine proteases by cytotoxic T-lymphocyte antigen-2 beta. J. Biol. Chem. 269 (40), 25172–25177.

Dong, J.T., Lamb, P.W., Rinker-Schaeffer, C.W., Vukanovic, J., Ichikawa, T., Isaacs, J.T., et al., 1995. KAI1, a metastasis suppressor gene for prostate cancer on human chromosome 11p11.2. Science 268 (5212), 884–886.

Dhar, D.K., Naora, H., Kubota, H., Maruyama, R., Yoshimura, H., Tonomoto, Y., et al., 2004. Downregulation of KiSS-1 expression is responsible for tumor invasion and worse prognosis in gastric carcinoma. Int. J. Cancer 111 (6), 868–872.

Dong, J.T., Suzuki, H., Pin, S.S., Bova, G.S., Schalken, J.A., Isaacs, W.B., et al., 1996. Down-regulation of the KAI1 metastasis suppressor gene during the progression of human prostatic cancer infrequently involves gene mutation or allelic loss. Cancer Res. 56 (19), 4387–4390.

Even-Ram, S., Uziely, B., Cohen, P., Grisaru-Granovsky, S., Maoz, M., Ginzburg, Y., et al., 1998. Thrombin receptor overexpression in malignant and physiological invasion processes. Nat. Med. 4 (8), 909–914.

Ebos, J.M., Kerbel, R.S., 2011. Antiangiogenic therapy: impact on invasion, disease progression, and metastasis. Nat. Rev. Clin. Oncol. 8 (4), 210–221.

Egeblad, M., Werb, Z., 2002. New functions for the matrix metalloproteinases in cancer progression. Nat. Rev. Cancer 2 (3), 161–174.

Esadeg, S., He, H., Pijnenborg, R., Van Leuven, F., Croy, B.A., 2003. Alpha-2 macroglobulin controls trophoblast positioning in mouse implantation sites. Placenta 24 (10), 912–921.

Fidler, I.J., 2003. The pathogenesis of cancer metastasis: the 'seed and soil' hypothesis revisited. Nat. Rev. Cancer 3 (6), 453–458.

Fang, L.Q., Zhang, H., Ding, X.Y., Li, D.Q., Hou, X.L., Qiao, H., et al., 2010. Mouse trophoblastic cells exhibit a dominant invasiveness phenotype over cancer cells. Cancer Lett. 299 (2), 111–118.

Fukuda, M.N., Sugihara, K., Nakayama, J., 2008. Trophinin: what embryo implantation teaches us about human cancer. Cancer Biol. Ther. 7 (8), 1165.

Ferretti, C., Bruni, L., Dangles-Marie, V., Pecking, A.P., Bellet, D., 2007. Molecular circuits shared by placental and cancer cells, and their implications in the proliferative, invasive and migratory capacities of trophoblasts. Hum. Reprod. Update 13 (2), 121–141.

Fazioli, F., Blasi, F., 1994. Urokinase-type plasminogen activator and its receptor: new targets for anti-metastatic therapy? Trends Pharmacol. Sci. 15 (1), 25–29.

Feng, Q., Liu, K., Liu, Y.X., Byrne, S., Ockleford, C.D., 2001. Plasminogen activators and inhibitors are transcribed during early macaque implantation. Placenta 22 (2–3), 186–199.

Feng, Q., Liu, Y., Liu, K., Byrne, S., Liu, G., Wang, X., et al., 2000. Expression of urokinase, plasminogen activator inhibitors and urokinase receptor in pregnant rhesus monkey uterus during early placentation. Placenta 21 (2–3), 184–193.

Gupta, G.P., Massague, J., 2006. Cancer metastasis: building a framework. Cell 127 (4), 679–695.

Goldman-Wohl, D.S., Ariel, I., Greenfield, C., Hochner-Celnikier, D., Cross, J., Fisher, S., et al., 2000. Lack of human leukocyte antigen-G expression in extravillous trophoblasts is associated with pre-eclampsia. Mol. Hum. Reprod. 6 (1), 88–95.

Green, D.R., Ferguson, T.A., 2001. The role of Fas ligand in immune privilege. Nat. Rev. Mol. Cell Biol. 2 (12), 917–924.

Graham, C.H., Lala, P.K., 1992. Mechanisms of placental invasion of the uterus and their control. Biochem. Cell Biol. 70 (10–11), 867–874.

Gu, Y., Jayatilak, P.G., Parmer, T.G., Gauldie, J., Fey, G.H., Gibori, G., 1992. Alpha 2-macroglobulin expression in the mesometrial decidua and its regulation by decidual luteotropin and prolactin. Endocrinology 131 (3), 1321–1328.

Galm, O., Suzuki, H., Akiyama, Y., Esteller, M., Brock, M.V., Osieka, R., et al., 2005. Inactivation of the tissue inhibitor of metalloproteinases-2 gene by promoter hypermethylation in lymphoid malignancies. Oncogene 24 (30), 4799–4805.

Graham, C.H., 1997. Effect of transforming growth factor-beta on the plasminogen activator system in cultured first trimester human cytotrophoblasts. Placenta 18 (2–3), 137–143.

Graham, C.H., Lala, P.K., 1991. Mechanism of control of trophoblast invasion in situ. J. Cell. Physiol. 148 (2), 228–234.

Graham, C.H., Connelly, I., MacDougall, J.R., Kerbel, R.S., Stetler-Stevenson, W.G., Lala, P.K., 1994. Resistance of malignant trophoblast cells to both the anti-proliferative and anti-invasive effects of transforming growth factor-beta. Exp. Cell Res. 214 (1), 93–99.

Gong, W., Zhang, G.M., Liu, Y., Lei, Z., Li, D., Yuan, Y., et al., 2008. IFN-gamma withdrawal after immunotherapy potentiates B16 melanoma invasion and metastasis by intensifying tumor integrin alphavbeta3 signaling. Int. J. Cancer 123 (3), 702–708.

Gellersen, B., Briese, J., Oberndorfer, M., Redlin, K., Samalecos, A., Richter, D.U., et al., 2007. Expression of the metastasis suppressor KAI1 in decidual cells at the human maternal-fetal interface: regulation and functional implications. Am. J. Pathol. 170 (1), 126–139.

Guo, X., Friess, H., Graber, H.U., Kashiwagi, M., Zimmermann, A., Korc, M., et al., 1996. KAI1 expression is up-regulated in early pancreatic cancer and decreased in the presence of metastases. Cancer Res. 56 (21), 4876–4880.

Hanahan, D., Weinberg, R.A., 2000. The hallmarks of cancer. Cell 100 (1), 57–70.

Heath, V.L., Bicknell, R., 2009. Anticancer strategies involving the vasculature. Nat. Rev. Clin. Oncol. 6 (7), 395–404.

Hanahan, D., Weinberg, R.A., 2011. Hallmarks of cancer: the next generation. Cell 144 (5), 646–674.

Hamai, Y., Fujii, T., Yamashita, T., Miki, A., Hyodo, H., Kozuma, S., et al., 1999. The expression of human leukocyte antigen-G on trophoblasts abolishes the growth-suppressing effect of interleukin-2 towards them. Am. J. Reprod. Immunol. 41 (2), 153–158.

Hahne, M., Rimoldi, D., Schroter, M., Romero, P., Schreier, M., French, L.E., et al., 1996. Melanoma cell expression of Fas(Apo-1/CD95) ligand: implications for tumor immune escape. Science 274 (5291), 1363–1366.

Higuchi, T., Kanzaki, H., Nakayama, H., Fujimoto, M., Hatayama, H., Kojima, K., et al., 1995. Induction of tissue inhibitor of metalloproteinase 3 gene expression during in vitro decidualization of human endometrial stromal cells. Endocrinology 136 (11), 4973–4981.

Hannan, N.J., Stephens, A.N., Rainczuk, A., Hincks, C., Rombauts, L.J., Salamonsen, L.A., 2010. 2D-DiGE analysis of the human endometrial secretome reveals differences between receptive and nonreceptive states in fertile and infertile women. J. Proteom. Res. 9 (12), 6256–6264.

Huber, A.V., Saleh, L., Bauer, S., Husslein, P., Knofler, M., 2006. TNFalpha-mediated induction of PAI-1 restricts invasion of HTR-8/SVneo trophoblast cells. Placenta 27 (2–3), 127–136.

Hjortland, G.O., Bjornland, K., Pettersen, S., Garman-Vik, S.S., Emilsen, E., Nesland, J.M., et al., 2003. Modulation of glioma cell invasion and motility by adenoviral gene transfer of PAI-1. Clin. Exp. Metastasis 20 (4), 301–309.

Hiraki, Y., Mitsui, K., Endo, N., Takahashi, K., Hayami, T., Inoue, H., et al., 1999. Molecular cloning of human chondromodulin-I, a cartilage-derived growth modulating factor, and its expression in Chinese hamster ovary cells. Eur. J. Biochem. 260 (3), 869–878.

Hayami, T., Shukunami, C., Mitsui, K., Endo, N., Tokunaga, K., Kondo, J., et al., 1999. Specific loss of chondromodulin-I gene expression in chondrosarcoma and the suppression of tumor angiogenesis and growth by its recombinant protein in vivo. FEBS Lett. 458 (3), 436–440.

Isaka, K., Usuda, S., Ito, H., Sagawa, Y., Nakamura, H., Nishi, H., et al., 2003. Expression and activity of matrix metalloproteinase 2 and 9 in human trophoblasts. Placenta 24 (1), 53–64.

Inoue, M., Sawada, T., Uchima, Y., Kimura, K., Nishihara, T., Tanaka, H., et al., 2005. Plasminogen activator inhibitor-1 (PAI-1) gene transfection inhibits the liver metastasis of pancreatic cancer by preventing angiogenesis. Oncol. Rep. 14 (6), 1445–1451.

Ikeguchi, M., Yamaguchi, K., Kaibara, N., 2004. Clinical significance of the loss of KiSS-1 and orphan G-protein-coupled receptor (hOT7T175) gene expression in esophageal squamous cell carcinoma. Clin. Cancer Res. 10 (4), 1379–1383.

Knoeller, S., Lim, E., Aleta, L., Hertwig, K., Dudenhausen, J.W., Arck, P.C., 2003. Distribution of immunocompetent cells in decidua of controlled and uncontrolled (choriocarcinoma/hydatidiform mole) trophoblast invasion. Am. J. Reprod. Immunol. 50 (1), 41–47.

Kalluri, R., Weinberg, R.A., 2009. The basics of epithelial-mesenchymal transition. J. Clin. Invest. 119 (6), 1420–1428.

Koizumi, M., Momoeda, M., Hiroi, H., Nakazawa, F., Nakae, H., Ohno, T., et al., 2010. Inhibition of proteases involved in embryo implantation by cholesterol sulfate. Hum. Reprod. 25 (1), 192–197.

Khamsi, F., Armstrong, D.T., Zhang, X., 1996. Expression of urokinase-type plasminogen activator in human preimplantation embryos. Mol. Hum. Reprod. 2 (4), 273–276.

Karmakar, S., Das, C., 2002. Regulation of trophoblast invasion by IL-1beta and TGF-beta1. Am. J. Reprod. Immunol. 48 (4), 210–219.

Kopitz, C., Anton, M., Gansbacher, B., Kruger, A., 2005. Reduction of experimental human fibrosarcoma lung metastasis in mice by adenovirus-mediated cystatin C overexpression in the host. Cancer Res. 65 (19), 8608–8612.

Konduri, S.D., Yanamandra, N., Siddique, K., Joseph, A., Dinh, D.H., Olivero, W.C., et al., 2002. Modulation of cystatin C expression impairs the invasive and tumorigenic potential of human glioblastoma cells. Oncogene 21 (57), 8705–8712.

Karmakar, S., Das, C., 2004. Modulation of ezrin and E-cadherin expression by IL-1beta and TGF-beta1 in human trophoblasts. J. Reprod. Immunol. 64 (1–2), 9–29.

Khoo, N.K., Bechberger, J.F., Shepherd, T., Bond, S.L., McCrae, K.R., Hamilton, G.S., et al., 1998. SV40 tag transformation of the normal invasive trophoblast results in a premalignant phenotype. I. Mechanisms responsible for hyperinvasiveness and resistance to anti-invasive action of TGFbeta. Int. J. Cancer 77 (3), 429–439.

Kotani, T., Iwase, A., Ino, K., Sumigama, S., Yamamoto, E., Hayakawa, H., et al., 2009. Activator protein-2 impairs the invasion of a human extravillous trophoblast cell line. Endocrinology 150 (9), 4376–4385.

Kundu, G.C., Mantile, G., Miele, L., Cordella-Miele, E., Mukherjee, A.B., 1996. Recombinant human uteroglobin suppresses cellular invasiveness via a novel class of high-affinity cell surface binding site. Proc. Natl. Acad. Sci. U.S.A. 93 (7), 2915–2919.

Koo, T.B., Han, M.S., Tadashi, Y., Seong, W.J., Choi, J.Y., 2013. Differential expression of the metastasis suppressor KAI1 in decidual cells and trophoblast giant cells at the feto-maternal interface. BMB Rep. 46 (10), 507–512.

Kaur, S., Kumar, S., Momi, N., Sasson, A.R., Batra, S.K., 2013. Mucins in pancreatic cancer and its microenvironment. Nat. Rev. Gastroenterol. Hepatol. 10 (10), 607–620.

Kufe, D.W., 2009. Mucins in cancer: function, prognosis and therapy. Nat. Rev. Cancer 9 (12), 874–885.

Karmakar, S., Dhar, R., Das, C., 2004. Inhibition of cytotrophoblastic (JEG-3) cell invasion by interleukin 12 involves an interferon gamma-mediated pathway. J. Biol. Chem. 279 (53), 55297–55307.

Lewis, M.P., Sullivan, M.H., Elder, M.G., 1994. Regulation by interleukin-1 beta of growth and collagenase production by choriocarcinoma cells. Placenta 15 (1), 13–20.

Librach, C.L., Feigenbaum, S.L., Bass, K.E., Cui, T.Y., Verastas, N., Sadovsky, Y., et al., 1994. Interleukin-1 beta regulates human cytotrophoblast metalloproteinase activity and invasion in vitro. J. Biol. Chem. 269 (25), 17125–17131.

Lala, P.K., Graham, C.H., 1990. Mechanisms of trophoblast invasiveness and their control: the role of proteases and protease inhibitors. Cancer Metastasis Rev. 9 (4), 369–379.

Librach, C.L., Werb, Z., Fitzgerald, M.L., Chiu, K., Corwin, N.M., Esteves, R.A., et al., 1991. 92-kD type IV collagenase mediates invasion of human cytotrophoblasts. J. Cell Biol. 113 (2), 437–449.

Lu, H., Cao, X., Zhang, H., Sun, G., Fan, G., Chen, L., et al., 2014. Imbalance between MMP-2, 9 and TIMP-1 promote the invasion and metastasis of renal cell carcinoma via SKP2 signaling pathways. Tumour Biol. 35 (10), 9807–9813.

Lockwood, C.J., Krikun, G., Schatz, F., 1999. The decidua regulates hemostasis in human endometrium. Semin. Reprod. Endocrinol. 17 (1), 45–51.

Lockwood, C.J., Schatz, F., 1996. A biological model for the regulation of peri-implantational hemostasis and menstruation. J. Soc. Gynecol. Invest. 3 (4), 159–165.

Lash, G.E., Otun, H.A., Innes, B.A., Kirkley, M., De Oliveira, L., Searle, R.F., et al., 2006. Interferon-gamma inhibits extravillous trophoblast cell invasion by a mechanism that involves both changes in apoptosis and protease levels. FASEB J. 20 (14), 2512–2518.

Li, T.M., Wu, C.M., Huang, H.C., Chou, P.C., Fong, Y.C., Tang, C.H., 2012. Interleukin-11 increases cell motility and up-regulates intercellular adhesion molecule-1 expression in human chondrosarcoma cells. J. Cell. Biochem. 113 (11), 3353–3362.

Li, M.Q., Hou, X.F., Shao, J., Tang, C.L., Li, D.J., 2010. The DSCs-expressed CD82 controls the invasiveness of trophoblast cells via integrinbeta1/MAPK/MAPK3/1 signaling pathway in human first-trimester pregnancy. Biol. Reprod. 82 (5), 968–979.

Lee, J.H., Miele, M.E., Hicks, D.J., Phillips, K.K., Trent, J.M., Weissman, B.E., et al., 1996. KiSS-1, a novel human malignant melanoma metastasis-suppressor gene. J. Natl. Cancer Inst. 88 (23), 1731–1737.

Lee, J.H., Welch, D.R., 1997. Suppression of metastasis in human breast carcinoma MDA-MB-435 cells after transfection with the metastasis suppressor gene, KiSS-1. Cancer Res. 57 (12), 2384–2387.

Lipinska, B., Sharma, S., Georgopoulos, C., 1988. Sequence analysis and regulation of the htrA gene of *Escherichia coli*: a sigma 32-independent mechanism of heat-inducible transcription. Nucleic Acids Res. 16 (21), 10053–10067.

Leco, K.J., Edwards, D.R., Schultz, G.A., 1996. Tissue inhibitor of metalloproteinases-3 is the major metalloproteinase inhibitor in the decidualizing murine uterus. Mol. Reprod. Dev. 45 (4), 458–465.

Lockwood, C.J., Krikun, G., Hausknecht, V., Wang, E.Y., Schatz, F., 1997. Decidual cell regulation of hemostasis during implantation and menstruation. Ann. N.Y. Acad. Sci. 828, 188–193.

Menkhorst, E., Winship, A., Van Sinderen, M., Dimitriadis, E., 2016. Human extravillous trophoblast invasion: intrinsic and extrinsic regulation. Reprod. Fertil. Dev. 28, 406–415.

Morgan, M., Kniss, D., McDonnell, S., 1998. Expression of metalloproteinases and their inhibitors in human trophoblast continuous cell lines. Exp. Cell Res. 242 (1), 18–26.

Murray, M.J., Lessey, B.A., 1999. Embryo implantation and tumor metastasis: common pathways of invasion and angiogenesis. Semin. Reprod. Endocrinol. 17 (3), 275–290.

Mullen, C.A., 1998. Review: analogies between trophoblastic and malignant cells. Am. J. Reprod. Immunol. 39 (1), 41–49.

Mohamed, M.M., Sloane, B.F., 2006. Cysteine cathepsins: multifunctional enzymes in cancer. Nat. Rev. Cancer 6 (10), 764–775.

Miura, S., Shukunami, C., Mitsui, K., Kondo, J., Hiraki, Y., 2011. Localization of chondromodulin-I at the feto-maternal interface and its inhibitory actions on trophoblast invasion in vitro. BMC Cell Biol. 12, 34.

Nguyen, D.X., Massague, J., 2007. Genetic determinants of cancer metastasis. Nat. Rev. Genet. 8 (5), 341–352.

Nguyen, D.X., Bos, P.D., Massague, J., 2009. Metastasis: from dissemination to organ-specific colonization. Nat. Rev. Cancer 9 (4), 274–284.

Nawrocki, B., Polette, M., Marchand, V., Maquoi, E., Beorchia, A., Tournier, J.M., et al., 1996. Membrane-type matrix metalloproteinase-1 expression at the site of human placentation. Placenta 17 (8), 565–572.

Navenot, J.M., Wang, Z., Chopin, M., Fujii, N., Peiper, S.C., 2005. Kisspeptin-10-induced signaling of GPR54 negatively regulates chemotactic responses mediated by CXCR4: a potential mechanism for the metastasis suppressor activity of kisspeptins. Cancer Res. 65 (22), 10450–10456.

Nie, G., Hale, K., Li, Y., Manuelpillai, U., Wallace, E.M., Salamonsen, L.A., 2006. Distinct expression and localization of serine protease HtrA1 in human endometrium and first-trimester placenta. Dev. Dyn. 235 (12), 3448–3455.

Overall, C.M., Lopez-Otin, C., 2002. Strategies for MMP inhibition in cancer: innovations for the post-trial era. Nat. Rev. Cancer 2 (9), 657–672.

Otun, H.A., Lash, G.E., Innes, B.A., Bulmer, J.N., Naruse, K., Hannon, T., et al., 2011. Effect of tumour necrosis factor-alpha in combination with interferon-gamma on first trimester extravillous trophoblast invasion. J. Reprod. Immunol. 88 (1), 1–11.

Ohtaki, T., Shintani, Y., Honda, S., Matsumoto, H., Hori, A., Kanehashi, K., et al., 2001. Metastasis suppressor gene KiSS-1 encodes peptide ligand of a G-protein-coupled receptor. Nature 411 (6837), 613–617.

Paget, S., 1989. The distribution of secondary growths in cancer of the breast. 1889. Cancer Metastasis Rev. 8 (2), 98–101.

Perry, J.K., Lins, R.J., Lobie, P.E., Mitchell, M.D., 2010. Regulation of invasive growth: similar epigenetic mechanisms underpin tumour progression and implantation in human pregnancy. Clin. Sci. (Lond.) 118 (7), 451–457.

Polette, M., Nawrocki, B., Pintiaux, A., Massenat, C., Maquoi, E., Volders, L., et al., 1994. Expression of gelatinases A and B and their tissue inhibitors by cells of early and term human placenta and gestational endometrium. Lab. Invest. 71 (6), 838–846.

Paiva, P., Salamonsen, L.A., Manuelpillai, U., Dimitriadis, E., 2009. Interleukin 11 inhibits human trophoblast invasion indicating a likely role in the decidual restraint of trophoblast invasion during placentation. Biol. Reprod. 80 (2), 302–310.

Patierno, S.R., Manyak, M.J., Fernandez, P.M., Baker, A., Weeraratna, A.T., Chou, D.S., et al., 2002. Uteroglobin: a potential novel tumor suppressor and molecular therapeutic for prostate cancer. Clin. Prostate Cancer 1 (2), 118–124.

Quinn, C.E., Simmons, D.G., Kennedy, T.G., 2006. Expression of Cystatin C in the rat endometrium during the peri-implantation period. Biochem. Biophys. Res. Commun. 349 (1), 236–244.

Qin, H., Moellinger, J.D., Wells, A., Windsor, L.J., Sun, Y., Benveniste, E.N., 1998. Transcriptional suppression of matrix metalloproteinase-2 gene expression in human astroglioma cells by TNF-alpha and IFN-gamma. J. Immunol. 161 (12), 6664–6673.

Ren, L., Liu, Y.Q., Zhou, W.H., Zhang, Y.Z., 2012. Trophoblast-derived chemokine CXCL12 promotes CXCR4 expression and invasion of human first-trimester decidual stromal cells. Hum. Reprod. 27 (2), 366–374.

Robertson, S.A., Moldenhauer, L.M., 2014. Immunological determinants of implantation success. Int. J. Dev. Biol. 58 (2–4), 205–217.

Reuning, U., Magdolen, V., Wilhelm, O., Fischer, K., Lutz, V., Graeff, H., et al., 1998. Multifunctional potential of the plasminogen activation system in tumor invasion and metastasis (review). Int. J. Oncol. 13 (5), 893–906.

Rama, S., Suresh, Y., Rao, A.J., 2003. TGF beta1 induces multiple independent signals to regulate human trophoblastic differentiation: mechanistic insights. Mol. Cell. Endocrinol. 206 (1–2), 123–136.

Reed, C.C., Waterhouse, A., Kirby, S., Kay, P., Owens, R.T., McQuillan, D.J., et al., 2005. Decorin prevents metastatic spreading of breast cancer. Oncogene 24 (6), 1104–1110.

Sethi, N., Kang, Y., 2011. Unravelling the complexity of metastasis – molecular understanding and targeted therapies. Nat. Rev. Cancer 11 (10), 735–748.

Schroeder, A., Heller, D.A., Winslow, M.M., Dahlman, J.E., Pratt, G.W., Langer, R., et al., 2012. Treating metastatic cancer with nanotechnology. Nat. Rev. Cancer 12 (1), 39–50.

Strickland, S., Richards, W.G., 1992. Invasion of the trophoblasts. Cell 71 (3), 355–357.

Staun-Ram, E., Shalev, E., 2005. Human trophoblast function during the implantation process. Reprod. Biol. Endocrinol. 3, 56.

Soundararajan, R., Rao, A.J., 2004. Trophoblast 'pseudo-tumorigenesis': significance and contributory factors. Reprod. Biol. Endocrinol. 2, 15.

Serman, A., Simon, F., Fabijanovic, D., Serman, L., 2012. Epigenetic control of cell invasion – the trophoblast model. Biomol. Concepts 3 (6), 487–494.

Su, J.L., Yang, P.C., Shih, J.Y., Yang, C.Y., Wei, L.H., Hsieh, C.Y., et al., 2006. The VEGF-C/Flt-4 axis promotes invasion and metastasis of cancer cells. Cancer Cell 9 (3), 209–223.

Shih Ie, M., Hsu, M.Y., Oldt 3rd, R.J., Herlyn, M., Gearhart, J.D., Kurman, R.J., 2002. The role of E-cadherin in the motility and invasion of implantation site intermediate trophoblast. Placenta 23 (10), 706–715.

Staun-Ram, E., Goldman, S., Gabarin, D., Shalev, E., 2004. Expression and importance of matrix metalloproteinase 2 and 9 (MMP-2 and -9) in human trophoblast invasion. Reprod. Biol. Endocrinol. 2, 59.

Sireesha, G.V., Mason, R.W., Hassanein, M., Tonack, S., Navarrete Santos, A., Fischer, B., et al., 2008. Role of cathepsins in blastocyst hatching in the golden hamster. Mol. Hum. Reprod. 14 (6), 337–346.

Shintani, K., Matsumine, A., Kusuzaki, K., Morikawa, J., Matsubara, T., Wakabayashi, T., et al., 2008. Decorin suppresses lung metastases of murine osteosarcoma. Oncol. Rep. 19 (6), 1533–1539.

Smith, S.C., Theodorescu, D., 2009. Learning therapeutic lessons from metastasis suppressor proteins. Nat. Rev. Cancer 9 (4), 253–264.

Steeg, P.S., 2003. Metastasis suppressors alter the signal transduction of cancer cells. Nat. Rev. Cancer 3 (1), 55–63.

Steeg, P.S., Bevilacqua, G., Kopper, L., Thorgeirsson, U.P., Talmadge, J.E., Liotta, L.A., et al., 1988. Evidence for a novel gene associated with low tumor metastatic potential. J. Natl. Cancer Inst. 80 (3), 200–204.

Shoushtari, A.N., Szmulewitz, R.Z., Rinker-Schaeffer, C.W., 2011. Metastasis-suppressor genes in clinical practice: lost in translation? Nat. Rev. Clin. Oncol. 8 (6), 333–342.

Singh, H., Li, Y., Fuller, P.J., Harrison, C., Rao, J., Stephens, A.N., et al., 2013. HtrA3 is downregulated in cancer cell lines and significantly reduced in primary serous and granulosa cell ovarian tumors. J. Cancer 4 (2), 152–164.

Singh, H., Endo, Y., Nie, G., 2011. Decidual HtrA3 negatively regulates trophoblast invasion during human placentation. Hum. Reprod. 26 (4), 748–757.

Singh, H., Makino, S.I., Endo, Y., Nie, G., 2010. Inhibition of HTRA3 stimulates trophoblast invasion during human placental development. Placenta 31 (12), 1085–1092.

Shyu, M.K., Lin, M.C., Shih, J.C., Lee, C.N., Huang, J., Liao, C.H., et al., 2007. Mucin 15 is expressed in human placenta and suppresses invasion of trophoblast-like cells in vitro. Hum. Reprod. 22 (10), 2723–2732.

Schatz, F., Krikun, G., Runic, R., Wang, E.Y., Hausknecht, V., Lockwood, C.J., 1999. Implications of decidualization-associated protease expression in implantation and menstruation. Semin. Reprod. Endocrinol. 17 (1), 3–12.

Tanaka, S.S., Togooka, Y., Sato, H., Seiki, M., Tojo, H., Tachi, C., 1998. Expression and localization of membrane type matrix metalloproteinase-1 (MT1-MMP) in trophoblast cells of cultured mouse blastocysts and ectoplacental cones. Placenta 19 (1), 41–48.

Thilaganathan, B., Ralph, E., Papageorghiou, A.T., Melchiorre, K., Sheldon, J., 2009. Raised maternal serum cystatin C: an early pregnancy marker for preeclampsia. Reprod. Sci. 16 (8), 788–793.

Tse, W.K., Whitley, G.S., Cartwright, J.E., 2002. Transforming growth factor-beta1 regulates hepatocyte growth factor-induced trophoblast motility and invasion. Placenta 23 (10), 699–705.

Tellez, C., McCarty, M., Ruiz, M., Bar-Eli, M., 2003. Loss of activator protein-2alpha results in overexpression of protease-activated receptor-1 and correlates with the malignant phenotype of human melanoma. J. Biol. Chem. 278 (47), 46632–46642.

Uckan, D., Steele, A., Cherry, W.B.Y., Chamizo, W., Koutsonikolis, A., et al., 1997. Trophoblasts express Fas ligand: a proposed mechanism for immune privilege in placenta and maternal invasion. Mol. Hum. Reprod. 3 (8), 655–662.

Valastyan, S., Weinberg, R.A., 2011. Tumor metastasis: molecular insights and evolving paradigms. Cell 147 (2), 275–292.

Vince, G., Shorter, S., Starkey, P., Humphreys, J., Clover, L., Wilkins, T., et al., 1992. Localization of tumour necrosis factor production in cells at the materno/fetal interface in human pregnancy. Clin. Exp. Immunol. 88 (1), 174–180.

Weigelt, B., Peterse, J.L., van't Veer, L.J., 2005. Breast cancer metastasis: markers and models. Nat. Rev. Cancer 5 (8), 591–602.

Whiteside, E.J., Jackson, M.M., Herington, A.C., Edwards, D.R., Harvey, M.B., 2001. Matrix metalloproteinase-9 and tissue inhibitor of metalloproteinase-3 are key regulators of extracellular matrix degradation by mouse embryos. Biol. Reprod. 64 (5), 1331–1337.

Wong, B.S., Lam, K.K., Lee, C.L., Wong, V.H., Lam, M.P., Chu, I.K., et al., 2013. Adrenomedullin enhances invasion of human extravillous cytotrophoblast-derived cell lines by regulation of urokinase plasminogen activator expression and s-nitrosylation. Biol. Reprod. 88 (2), 34.

Wegiel, B., Jiborn, T., Abrahamson, M., Helczynski, L., Otterbein, L., Persson, J.L., et al., 2009. Cystatin C is downregulated in prostate cancer and modulates invasion of prostate cancer cells via MAPK/Erk and androgen receptor pathways. PLoS One 4 (11), e7953.

Wang, X., Lin, Y., 2008. Tumor necrosis factor and cancer, buddies or foes? Acta Pharmacol. Sin. 29 (11), 1275–1288.

Weeraratna, A.T., Cajigas, J.A., Schwartz, A., Enquist, E.G., Manyak, M.J., Patierno, S.R., 1997. Loss of uteroglobin expression in prostate cancer: relationship to advancing grade. Clin. Cancer Res. 3 (12 Pt 1), 2295–2300.

Wang, Y., Fan, D.X., Duan, J., Li, M.Q., Zhu, X.Y., Jin, L.P., 2012. Thymic stromal lymphopoietin downregulates NME1 expression and promotes invasion in human trophoblasts via the activation of STAT3 signaling pathway. Clin. Immunol. 143 (1), 88–95.

Wang, R.Y., Chen, L., Chen, H.Y., Hu, L., Li, L., Sun, H.Y., et al., 2013. MUC15 inhibits dimerization of EGFR and PI3K-AKT signaling and is associated with aggressive hepatocellular carcinomas in patients. Gastroenterology 145 (6), 1436–1448 e1–e12.

Waterhouse, P., Denhardt, D.T., Khokha, R., 1993. Temporal expression of tissue inhibitors of metalloproteinases in mouse reproductive tissues during gestation. Mol. Reprod. Dev. 35 (3), 219–226.

Xing, R.H., Mazar, A., Henkin, J., Rabbani, S.A., 1997. Prevention of breast cancer growth, invasion, and metastasis by antiestrogen tamoxifen alone or in combination with urokinase inhibitor B-428. Cancer Res. 57 (16), 3585–3593.

Xu, G., Guimond, M.J., Chakraborty, C., Lala, P.K., 2002. Control of proliferation, migration, and invasiveness of human extravillous trophoblast by decorin, a decidual product. Biol. Reprod. 67 (2), 681–689.

Xu, W., Neill, T., Yang, Y., Hu, Z., Cleveland, E., Wu, Y., et al., 2015. The systemic delivery of an oncolytic adenovirus expressing decorin inhibits bone metastasis in a mouse model of human prostate cancer. Gene Ther. 22 (3), 247–256.

Xie, K.M., Hou, X.F., Li, M.Q., Li, D.J., 2010. NME1 at the human maternal-fetal interface downregulates titin expression and invasiveness of trophoblast cells via MAPK pathway in early pregnancy. Reproduction 139 (4), 799–808.

Yagel, S., Parhar, R.S., Jeffrey, J.J., Lala, P.K., 1988. Normal nonmetastatic human trophoblast cells share in vitro invasive properties of malignant cells. J. Cell. Physiol. 136 (3), 455–462.

Yagel, S., Khokha, R., Denhardt, D.T., Kerbel, R.S., Parhar, R.S., Lala, P.K., 1989. Mechanisms of cellular invasiveness: a comparison of amnion invasion in vitro and metastatic behavior in vivo. J. Natl. Cancer Inst. 81 (10), 768–775.

Yang, Q., Wang, H.X., Zhao, Y.G., Lin, H.Y., Zhang, H., Wang, H.M., et al., 2006. Expression of tissue inhibitor of metalloproteinase-4 (TIMP-4) in endometrium and placenta of rhesus monkey (*Macaca mulatta*) during early pregnancy. Life Sci. 78 (24), 2804–2811.

Yang, Y., Xu, W.W., Neill, T., Hu, Z., Wang, C.H., Xiao, X., et al., 2015. Systemic delivery of an oncolytic adenovirus expressing decorin for the treatment of breast cancer bone metastases. Hum. Gene Ther. 26, 813–825.

Zygmunt, M., Herr, F., Keller-Schoenwetter, S., Kunzi-Rapp, K., Munstedt, K., Rao, C.V., et al., 2002. Characterization of human chorionic gonadotropin as a novel angiogenic factor. J. Clin. Endocrinol. Metab. 87 (11), 5290–5296.

Zhou, Y., Bellingard, V., Feng, K.T., McMaster, M., Fisher, S.J., 2003. Human cytotrophoblasts promote endothelial survival and vascular remodeling through secretion of Ang2, PlGF, and VEGF-C. Dev. Biol. 263 (1), 114–125.

Zygmunt, M., Hahn, D., Munstedt, K., Bischof, P., Lang, U., 1998. Invasion of cytotrophoblastic JEG-3 cells is stimulated by hCG in vitro. Placenta 19 (8), 587–593.

Zhai, Y., Hotary, K.B., Nan, B., Bosch, F.X., Munoz, N., Weiss, S.J., et al., 2005. Expression of membrane type 1 matrix metalloproteinase is associated with cervical carcinoma progression and invasion. Cancer Res. 65 (15), 6543–6550.

Zhu, M., Gokhale, V.M., Szabo, L., Munoz, R.M., Baek, H., Bashyam, S., et al., 2007. Identification of a novel inhibitor of urokinase-type plasminogen activator. Mol. Cancer Ther. 6 (4), 1348–1356.

Zhang, Z., Kim, S.J., Chowdhury, B., Wang, J., Lee, Y.C., Tsai, P.C., et al., 2006. Interaction of uteroglobin with lipocalin-1 receptor suppresses cancer cell motility and invasion. Gene 369, 66–71.

Zhang, Q., Tan, D., Luo, W., Lu, J., Tan, Y., 2012. Expression of CD82 in human trophoblast and its role in trophoblast invasion. PLoS One 7 (6), e38487.

Zhang, P., Tang, M., Zhong, T., Lin, Y., Zong, T., Zhong, C., et al., 2014. Expression and function of kisspeptin during mouse decidualization. PLoS One 9 (5), e97647.

Zurawa-Janicka, D., Skorko-Glonek, J., Lipinska, B., 2010. HtrA proteins as targets in therapy of cancer and other diseases. Expert Opin. Ther. Targets 14 (7), 665–679.

CHAPTER

# 20

# Wnt Signaling as a Therapeutic Target in Cancer and Metastasis

*R. Morgan*[1], *R. Ankrah*[1], *S. El-Tanani*[1], *P.M. Loadman*[1], *L. Pattterson*[1], *P.S. Rudland*[2], *M. El-Tanani*[1]

[1]University of Bradford, Bradford, United Kingdom; [2]University of Liverpool, Liverpool, United Kingdom

## 1. INTRODUCTION

For 20 years, the Wnt signaling pathway has known to be associated with tumor formation. The first evidence for this was the recognition of a common site of integration of the mouse mammary tumor virus in the promoter of a gene thereafter named *Int-1*, the forced expression of which results in mammary tumors in mice. Sequencing of the *Int-1* gene showed it to be homologous to the *Drosophila Wingless* gene and the combination of the two names produced the term *Wnt* (Clevers, 2006; Ilyas, 2005). The Wnt family of genes is now known to contain 19 members and regulates three pathways: the canonical pathway, the planar cell polarity pathway, and the $Ca^{2+}$ pathway (Fig. 20.1) (El-Tanani et al., 2006; Katoh and Katoh, 2007). It is primarily aberrant behavior of the canonical pathway that leads to tumor formation.

Normally the canonical pathway helps determine cell fate and regulate cell proliferation (Katoh and Katoh, 2007). The Wnt signal is transduced when Wnt proteins bind to the membrane-bound frizzled receptors and coreceptors low-density lipoprotein receptor-related protein (LRP) 5 and 6. In the absence of this signal (Fig. 20.1A, right), β-catenin is bound to a protein complex consisting of axin, adenomatous polyposis coli (APC) proteins, casein-kinase 1α (CK1), and glycogen synthase kinase 3β (GSK3β). This complex is sometimes known as the destruction complex and phosphorylates β-catenin, allowing it to be targeted for ubiquitinylation and subsequently destroyed by proteasomes (Ilyas, 2005; Katoh and Katoh, 2007). This destruction keeps the level of β-catenin low and stops its relocation to the nucleus, where it binds to the T-cell factor/lymphoid enhancer factor (TCF/LEF) proteins. TCF/LEFs are a subfamily of the high mobility group of transcription factors and can act either as transcriptional repressors or transcriptional activators depending on the complex they form (Hoppler and Kavanagh, 2007). If β-catenin is absent from the nucleus, they interact with corepressors groucho or C-terminal binding protein (CtBP) (Hoppler and Kavanagh, 2007). These complexes then bind to the Wnt response element of target genes to suppress them.

*Introduction to Cancer Metastasis*
http://dx.doi.org/10.1016/B978-0-12-804003-4.00020-7

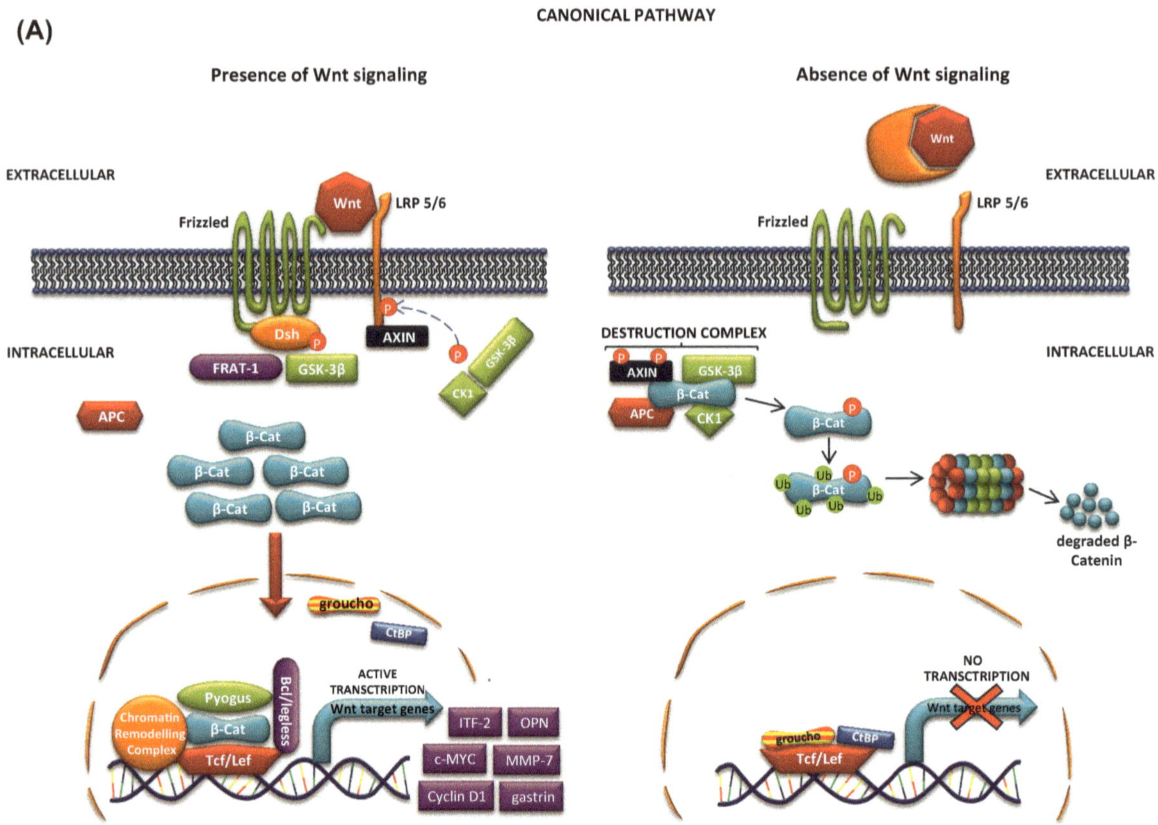


FIGURE 20.1 **Wnt signaling in the canonical and noncanonical pathways.** Canonical pathway: (A, left) Canonical Wnt signal transduction begins with binding of Wnt proteins to the membrane-bound frizzled receptors and coreceptors, (low-density lipoprotein receptor-related protein) LRP5 and LRP6. This halts β-catenin (β-cat) degradation via two ways. First, binding to LRP5/6 causes phosphorylation of the intracellular domain of LRP5/6 by casein-kinase 1α (CK1) and glycogen synthase kinase (GSK-3β). This increases their affinity for axin, removing it from the destruction complex, and axin is subsequently dephosphorylated and degraded by an unknown mechanism. Concurrently, the cytoplasmic protein Disheveled (Dsh) is recruited to frizzled at the cell membrane where it is phosphorylated, allowing it to react with the proto-oncogene product, Frat1, and GSK-3β to disrupt GSK-3β activity in the destruction complex. These combined events halt β-catenin degradation and result in increased cytoplasmic β-catenin, and its ensuing translocation into the nucleus. In the nucleus, β-catenin competes with transcription corepressors groucho and C-terminal Binding Protein (CtBP) to bind with the T-cell factor/lymphoid enhancer factors (TCF/LEFs). When bound to TCF/LEFs, β-catenin acts as a scaffold for the association of TCF/LEFs with chromatin-remodeling complexes and the transcriptional coactivators Bcl-9/legless and Pygopus. This complex activates transcription of Wnt target genes including c-Myc, cyclin D1, matrix metalloproteinase 7, gastrin, immunoglobulin transcription factor 2, and osteopontin. (A, right) In the absence of Wnt signaling, β-catenin is bound to a protein complex consisting of axin, adenomatous polyposis coli, CK1 and GSK3β. This complex, known as the "destruction complex," phosphorylates and targets β-catenin for ubiquitinylation and subsequent destruction by proteasomes. This keeps the levels of cytoplasmic β-catenin low, preventing its translocation into the nucleus. Absence of β-catenin from the nucleus permits the interaction of groucho or C-terminal binding protein with Tcf/Lef to maintain a transcriptionally repressed state. Noncanonical pathways: (B, left) The $Ca^{2+}$ pathway begins with activation of protein kinase C via Dsh and possibly upstream heterotrimeric G proteins. This activation results in an increase in intracellular calcium. The increased intracellular calcium affects multiple functions, which include gastrulation of vertebrates, polarization of cochlear hair cells, cardiogenesis, and myogenesis. (B, right) In the planar cell polarity pathway, phosphorylation of Dsh at the cell membrane mediates activation of Rho and/or Rac, which in turn activate Rho kinases and c-Jun N-terminal kinases, respectively. This results in cell polarization alterations and actin cytoskeleton remodeling. The noncanonical pathways are not known to have any direct role in cancer formation, although they may act as antagonists to the canonical pathway.

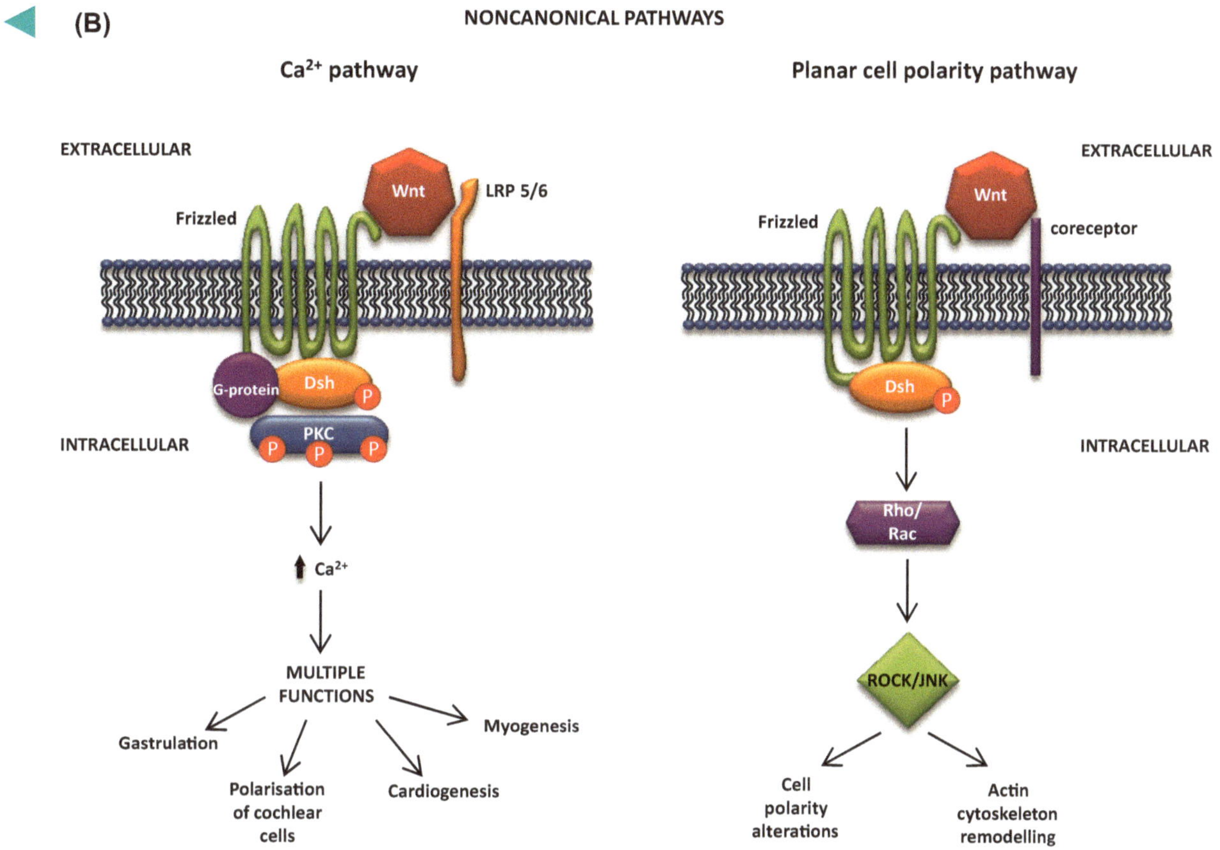


FIGURE 20.1 (cont'd)

When the Wnt signal is transduced from the frizzled receptors, two events occur that stop the degradation of β-catenin (Fig. 20.1A, left) (Clevers, 2006). First, binding of Wnt proteins to LRP5/6 causes the intracellular domain to become phosphorylated by CK1 and GSK3β. This increases their affinity for axin leading to its removal from the destruction complex. Axin is then dephosphorylated and degraded via a mechanism that is as yet not understood. The low levels of axin in the cytoplasm make this a very effective way of stopping β-catenin degradation. Second, the cytoplasmic protein disheveled (Dsh) is recruited to the cell membrane where it is phosphorylated. Dsh can then react with the proto-oncogene product, Frat1 and GSK3β to disrupt GSK3β activity in the destruction complex (Katoh and Katoh, 2007). The combined effect is an increase in cytoplasmic β-catenin and its subsequent translocation into the nucleus. The mechanism of this transportation has not been identified, although the protein contains no nuclear location signal.

Once in the nucleus, β-catenin competes with groucho and CtBP to bind with the TCF/LEFs and hence activate transcription. In this activation complex, β-catenin acts as a scaffold for the TCF/LEFs to associate with chromatin-remodeling complexes and the transcriptional coactivators Bcl-9/legless and Pygopus (Ilyas, 2005; Katoh and Katoh, 2007; Hoppler and Kavanagh, 2007). This complex activates transcription of target genes including *c-myc*, cyclin D1, matrix metalloproteinase 7 (*MMP7*), gastrin, immunoglobulin transcription factor 2, and osteopontin (*OPN*) (El-Tanani et al., 2006).

The two noncanonical Wnt signaling pathways are the planar cell polarity (PCP) pathway and the $Ca^{2+}$ pathway (Fig. 20.1B). The PCP pathway (Fig. 20.1B, right) is mediated through c-Jun N-terminal kinase and changes the cell polarity by reconstructing the actin cytoskeleton (Ilyas, 2005; El-Tanani et al., 2006). The $Ca^{2+}$ pathway (Fig. 20.1B, left) activates protein kinase C via Dsh and possibly upstream heterotrimeric G proteins. This activation results in an increase in intracellular calcium. Evidence suggests that the noncanonical pathways are involved in gastrulation of vertebrates, polarization of cochlear hair cells, cardiogenesis, and myogenesis (Ilyas, 2005; Katoh and Katoh, 2007). The noncanonical pathways are not known to have any direct role in cancer formation, although they may act as antagonists to the canonical pathway and thus have a tumor suppressor function.

## 2. WNT SIGNALING AND CANCER

Aberrant Wnt signaling is associated with various types of cancer and usually leads to increased levels of β-catenin in the nucleus. The malignancy most strongly linked to aberrant Wnt signaling is colorectal cancer, which accounts for 55,000 deaths within the United States every year or 10% of all cancer-related deaths (Barker and Clevers, 2006). Wnt signaling is thus a prime area for research into new anticancer therapies.

Within the colon there are four levels of epithelium that invaginate to form what are known as crypts. At the bottom of the crypts lie stem cells that produce proliferating precursor epithelial cells. These populate the bottom two-thirds of the crypt and are capable of migrating up to the top third of the crypt (Schneikert and Behrens, 2007). This process is controlled by Wnt proteins that are secreted by the stromal cells under the stem cell compartment or by a subset of the epithelial cells. Once the precursor cells reach the top third of the crypt, they stop proliferating, probably because they are no longer exposed to Wnt, but continue to migrate up the crypt to replace the gut lining (Schneikert and Behrens, 2007). After about a week they undergo apoptosis and are replaced by new cells. In colorectal cancer, mutations in the canonical Wnt pathway distort this process and the cells divide uncontrollably and do not undergo programmed death (Schneikert and Behrens, 2007).

Most Wnt signaling pathway mutations in colorectal cancer occur in the *APC* gene (80%; Barker and Clevers, 2006), leading to the initial discovery of the affected allele in familial adenomatous polyposis (FAP), a disease that causes polyps to form in the colon and often leads to carcinoma (Barker and Clevers, 2006). These *APC* mutations result in a protein that lacks the C-terminal functions required for microtubule binding, polarity, and chromosome segregation. Most importantly, loss of the C-terminal Serine-Alanine-Methionine-Proline (SAMP) repeats, which mediate axin binding, affects the formation of the β-catenin destruction complex. β-Catenin then stabilizes and locates to the nucleus, where it can function as a transcription factor (Segditsas and Tomlinson, 2006). In a smaller portion of colorectal cancers (10%), a mutation is found in the β-catenin gene itself. Mutations in exon 3 usually lead to the phosphorylation and hence stabilization of β-catenin (Barker and Clevers, 2006; Segditsas and Tomlinson, 2006). Mutations or transcriptional silencing of frizzled receptors can also lead to a change in the level of Wnt signaling in cells. In addition, some mutations in Axin I, Axin II, β-catenin, and TCF4 tend to be found in microsatellite-unstable colonic neoplasms (Segditsas and Tomlinson, 2006).

Abnormal Wnt signaling is also found in other cancers such as melanoma, hepatoblastomas, endometrial carcinomas, and prostate, lung, hepatocellular, ovarian, and breast carcinomas. In addition, a subset of germline *APC* mutations causes Turcot's syndrome that is characterized by the formation of multiple tumors including those of the colon and brain (Morin, 1999). Leukemia has also been shown to involve aberrant Wnt signaling, which is perhaps unsurprising given its role in stem cell self-renewal and proliferation. It is particularly noteworthy that in acute myeloid leukemia, mutations of receptor tyrosine kinases play an important role in activating events downstream of Wnt (Mikesch et al., 2007). Upregulated β-catenin also has a role in the formation of around 75% of pilomatricomas, a common skin cancer (Morin, 1999).

## 3. THE ROLE OF WNT SIGNALING IN EPITHELIAL–MESENCHYMAL TRANSITION AND CANCER STEM CELLS

Aberrant Wnt signaling is a crucial factor in the epithelial–mesenchymal transition (EMT) of cancer cells, which in turn promotes cancer invasion and metastasis. Several proteases regulated by Wnt signaling decrease cell–cell adhesion and favor EMT and metastasis. MMP7, upregulated by Wnt canonical pathway activation, and MMP3 show increased expression in colorectal cancers and have been shown to promote metastasis through degradation of the extracellular matrix (Wagenaar-Miller et al., 2004; Lochter et al., 1997) and E-cadherin (Lochter et al., 1997; Noe et al., 2001). The cell adhesion molecule L1 (L1CAM) is also known to mediate E-cadherin degradation (Shtutman et al., 2006). Proteolysis of cadherins by caspase 3 or a disintegrin and metalloproteinase domain-containing protein 10 (ADAM10) releases β-catenin from its complex with E-cadherin and α-catenin leading to its translocation from cell–cell junctions to the cytoplasm and nucleus and the resulting increase in β-catenin signaling promotes metastasis (Onder et al., 2008). Furthermore, repression of E-cadherin can be mediated by several transcription factors including members of the Slug/Snail family, Smad interacting protein 1 (SIP-1), E12/E47, zinc finger E-Box binding homeobox 1 (ZEB1) and Twist, through interaction with E box sequences in the promoter of E-cadherin (Batlle et al., 2000; Cano et al., 2000; Perez-Moreno et al., 2001; Grooteclaes and Frisch, 2000; Nieto, 2002; Bolos et al., 2003) and these can also act as molecular switches of EMT (Nieto, 2002; Hartwell et al., 2006; Yang et al., 2004). Snail is abnormally produced in many cell lines and primary tumors and is associated with E-cadherin loss, altered cancer cell morphology, and increased cancer aggression and metastasis. Stabilization of Snail when phosphorylated by GSK3β in the Wnt-signaling pathway enhances its role as an inducer of EMT (Barrallo-Gimeno and Nieto, 2005). Snail expression was shown to be necessary for lymph node metastasis of human breast carcinoma MDA-MB-231 cells (Olmeda et al., 2007), and was found to be elevated in ovarian cancer metastatic lesions (Jin et al., 2010). Snail-induced EMT is also reported to affect immunosuppression and results in accelerated metastasis (Kudo-Saito et al., 2009). Direct and indirect targets of Snail—inhibitor of DNA-binding protein 1, secreted protein acidic and rich in cysteine, and MMP2—have been identified as upregulated in lung metastasis (Minn et al., 2005; Peinado et al., 2007). Furthermore, following the knockdown of Snail in mouse melanoma (B16-F10) models, tumor growth and metastasis were significantly inhibited as a result of increased systemic immune responses and elevated numbers of tumor-infiltrating lymphocytes (Kudo-Saito et al., 2009).

There is also evidence that Twist plays a critical role in the early stages of metastasis such as intravasation, including the finding that *Twist* small interfering ribonucleic acid (siRNA) treatment of cell lines with a high metastatic potential reduces the number of circulating tumor cells in animal models. Furthermore, increased Twist expression has been shown to promote EMT progression in metastatic human mammary epithelial cells through decreased E-cadherin, β- and γ-catenin expression and increased vimentin and fibronectin expression. Additionally, cells display a spindle like morphology and exhibit increased migratory abilities. Slug expression is also correlated with distant metastasis (Bolos et al., 2003;

Cheung et al., 2005) and is reported to be necessary for the metastasis of transformed melanoma cells (Gupta et al., 2005). It plays a role as a vital mediator of Twist-induced EMT and metastasis, in which it suppresses the epithelial phase and then functions together with Twist to promote EMT and metastasis (Casas et al., 2011). Elevated Twist and Snail expression in lung cancer (Shi et al., 2013) and the expression of Snail in prostate cancer (Neal et al., 2012, 2011) and breast cancer (Dhasarathy et al., 2007, 2011), Slug in gastric cancer (Uchikado et al., 2011) and Snail and Twist in adenocarcinomas (Shi et al., 2013) are all associated with a poor prognosis. In breast cancer cells, the activity of Snail and Slug has been shown to result in TGF-β signaling (Noe et al., 2001).

OPN, another target gene of the Wnt/TCF signaling pathway, is known to contribute to the malignant transformation and metastasis of cells through its interaction with αv integrins. Overexpression of OPN in tumors is involved in the progression of human gastric (Ue et al., 1998; Rudland et al., 2002) and breast cancers (Rudland et al., 2002). High serum levels of OPN have also been linked with metastasis in breast cancers (Singhal et al., 1997). The binding of β-catenin to TCF1/LEF has been found to initiate OPN transcription activation (El-Tanani et al., 2006). Additionally in the presence of β-catenin, TCF4 stimulates the activation and protein expression of OPN (El-Tanani et al., 2006). In the absence of β-catenin, however, binding of TCF4 to A/TA/TCAAAG regions within the OPN promoter impedes the transcription and translation of OPN (El-Tanani et al., 2006). In interactions with the estrogen receptor and Ets transcription factors, TCF4 acts to suppress OPN activation, whereas TCF1 acts synergistically to activate the OPN promoter (El-Tanani et al., 2004). Transient transfection of *TCF4* cDNA has been shown to block the promoter and protein expression of OPN in rat mammary carcinomas (El-Tanani et al., 2001). Additionally, TCF4 transcripts have been found to be downregulated in breast cancers (Nosho et al., 2006). An inverse association of OPN and TCF4 expression has also been demonstrated in human breast carcinomas. This resulted in overexpressed OPN, which induces tumorigenesis and metastasis through the activation of Ras (Behrend et al., 1994).

T-cell lymphoma invasion and metastasis 1 is a Wnt target gene that selectively activates Rac1 and may consequently promote the invasion and metastasis of colorectal tumors (Malliri et al., 2006). Recently, the overexpression of three other genes, block of proliferation 1 (BOP1), cyclin-dependent kinases regulatory subunit 2 (CKS2), and nuclear factor, interleukin 3 regulated (NFIL3), has been reported to promote EMT and cell motility in vitro and metastasis in mice with colorectal cancer (Qi et al., 2015). MicroRNAs are also implicated in EMT and metastasis. miR-374a has been found to be markedly overexpressed in metastatic breast cancer and has been identified as a metastasis-promoting miRNA through activation of the β-catenin pathway and suppression of Phosphatase and tensin homolog (PTEN), Wnt family member 5A (WNT5A) and Wnt inhibitory factor 1 (WIF1) (Cai et al., 2013).

The program of EMT in cancer cells has been associated with CSCs (Mani et al., 2008; Blick et al., 2010; Polyak and Weinberg, 2009). Studies have demonstrated that either exposure to TGF-β1 or the overexpression of Twist or Snail in a differentiated human mammary epithelial cell line (HMLE cells) results in the development of the CD44 high/CD24 low stem cell profile (Mani et al., 2008). In another study, an increase in the number of stem cells was demonstrated in mammary epithelial cells following TGF-β treatment (Morel et al., 2008). In ovarian cancer cells transfected with Snail1 and Snail2 there was a significant increase in the population of CD44 high/CD117 high CSCs and reexpression of *Nanog* and *KLF4*, known stem-cell-associated genes (Kurrey et al., 2009). It is believed that CSCs having undergone EMT possess migratory and the stem cell abilities that facilitate the formation of distant metastases (Brabletz et al., 2005). In support of this, disseminated breast cancer cells in the bone marrow and circulation have been shown to have enriched CD44 high/CD24 low antigen phenotype populations (Riethdorf and Pantel, 2008; Riethdorf et al.,

2008; Slade et al., 2009). Furthermore, the finding that CSCs can express chemokine receptors such as chemokine receptor type 4 (CXCR4) and migrate in response to the chemokine CXCL12 produced in hematopoietic regions and other sites suggests that CSCs are crucial in the distant metastasis of cancer cells from the primary site (Hermann et al., 2007; Gelmini et al., 2009; Krohn et al., 2009). Thus, specific chemokine receptor and ligand expression regimes or profiles may ultimately determine the specific metastatic sites of different cancers. At distant metastatic sites, cells undergo a mesenchymal–epithelial transition and exhibit an epithelial phenotype that facilitates their adhesion and the formation of secondary tumors. It has been suggested that inhibition of nuclear β-catenin signaling by overexpression of dickkopf-related protein 1 (DKK1) or secreted frizzled-related protein 1 (SFRP1) reduces the expression of Slug and Twist and results in the upregulation of breast epithelial markers in breast cancer (DiMeo et al., 2009).

Based on the above report, potential disruption of the Wnt signaling pathway could be a very important therapeutic approach in many cancers. A number of techniques are already under development with this goal in view and are achieving varying degrees of success.

## 4. THERAPEUTIC STRATEGIES TARGETING WNT SIGNALING

The therapeutic techniques being tried include both the use of existing drugs and newly developed ones. Genetic strategies may also have a role in disrupting Wnt signaling pathway. These are divulged below and summarized in Table 20.1.

### 4.1 Nonsteroidal Anti-inflammatory Drugs

Nonsteroidal anti-inflammatory drugs (NSAIDs) are traditionally used in the treatment of pain, inflammation, and fever. These drugs work by inhibiting cyclooxygenase (COX), a key component in the synthesis of arachidonic acid, and therefore stop the production of thromboxane and prostaglandins (Takahashi-Yanaga and Sasaguri, 2007). However, recent evidence suggests that they may have an additional role as antineoplastic agents. This antineoplastic role has been shown in a number of epidemiological trials where aspirin and other NSAIDs have shown chemoprotective effects against cancers including those of the breast and lung (Hernandez-Diaz and Garcia Rodriguez, 2007; Harris et al., 2006). This anticancer role of NSAIDs seems particularly evident in colon cancer where aberrant Wnt signaling behavior is most often observed.

The action of NSAIDs in chemoprotection is linked to the suppression of COX activity, which becomes elevated in some cancers. COX elevation contributes to tumorigenesis by inhibiting apoptosis, stimulating angiogenesis, activating carcinogens, promoting a more invasive phenotype, and modulating inflammation/immune suppression (Watson, 2006). However, NSAIDs that do not reduce COX activity also have an antineoplastic effect (Smith et al., 2000). This shows that the chemoprotective effect of NSAIDs is dependent on multiple mechanisms.

NSAIDs inhibit cancer by acting at the initial stages of the adenoma to carcinoma conversion, suggesting a link with the Wnt signaling pathway. This link was first recognized when NSAIDs were found to reverse FAP polyp growth in humans and in mouse models (Barker and Clevers, 2006). Recently, Boon et al. tested five FAP patients who were treated using the NSAID Sulindac for 6 months. They found a significant decrease in nuclear β-catenin within the polyps, providing evidence for modified Wnt signaling that results from a COX-independent, antineoplastic effect of NSAIDs (Boon et al., 2004). It was assumed that the COX and Wnt signaling chemoprotective effects of NSAIDs acted separately, although subsequent studies on colon cancer indicated that they may be linked (Barker and Clevers, 2006). In colon cancer the increase in COX activity upregulates formation of prostaglandin, which, in turn, may block β-catenin degradation, providing a link between the two systems (Barker and Clevers, 2006).

Different NSAIDs seem to inhibit the Wnt signaling cycle in different ways. Aspirin and indomethacin were both found to decrease aberrant Wnt activity through enhanced phosphorylation, whereas Sulindac seems to promote the proteasomal degradation of β-catenin (Barker and Clevers, 2006; Dihlmann et al., 2003). Sulindac sulfate, an oxidative metabolite of Sulindac, induces apoptosis in cancer cell lines by targeting cyclic guanosine monophosphate (cGMP) phosphodiesterase, activating protein kinase G, and hence causing an increase in cGMP levels (Li et al., 2002). However, the doses of NSAIDs required for chemoprotection are often toxic, with major side effects including gastric bleeding, abdominal pain, and liver abnormalities, thus limiting their use in humans.

The COX enzyme has two main isoforms, COX-1 and COX-2. The former is involved in biological homeostasis including cytoprotection of the gastric mucosa and platelet aggregation and is expressed in most adult cells. This contrasts with COX-2, which is recruited to areas in which there is an ongoing inflammatory response, and is also expressed by many cell lines including those derived from head and neck, colon, and skin cancer (Takahashi-Yanaga and Sasaguri, 2007). Both COX-1 and COX-2 are usually inhibited by most NSAIDs and the inhibition of COX-1 is generally responsible for toxicity. Therefore, COX-2 selective NSAIDs should be safer and more effective cancer drugs, and thus agents such as celecoxib and rofecoxib were developed. Celecoxib has been shown to decrease polyp formation in FAP by 28% and decreases nuclear β-catenin levels in vitro (Barker and Clevers, 2006; Maier et al., 2005). These results suggest that part of celecoxib's antineoplastic effect is due to the suppression of aberrant Wnt signaling. However, randomized controlled trials in humans have shown an increased risk of cardiovascular events with COX inhibitors (coxibs), thus limiting their use and leading to the withdrawal of rofecoxib from the market. Celecoxib is now the only US Food and Drug Administration that approved NSAID for the treatment of FAP (Barker and Clevers, 2006). Another type of NSAID that shows promise in reducing cancer growth is nitric oxide (NO)-releasing aspirin. In APC mice models it has proved to be more effective than normal aspirin and importantly shows no intestinal toxicity (Barker and Clevers, 2006).

In summary, NSAIDs show great promise in the prevention of cancer involving the dysregulation of Wnt signaling. However, the mechanistic basis of their anticancer effect remains unclear, and their associated toxicity may prevent them from being adopted as first-line chemoprotective drugs.

## 4.2 Vitamin A and D

Vitamin A is converted in the body to a number of different metabolites, which are collectively called retinoids. Retinoids act through two types of receptor: retinoic acid receptors and retinoid X receptors. Both types contain several subtypes and are ligand-dependent transcription factors (Clarke et al., 2004). These retinoid receptors have been shown to reduce the function of both AP-1 and the Wnt signaling pathway and also act to stabilize cell to cell adhesions (Takahashi-Yanaga and Sasaguri, 2007).

Vitamin D regulates calcium and potassium within the body to maintain the skeleton. One of its active forms, 1α,25-dihydroxyvitamin D3, inhibits Wnt signaling, and in mouse models it has been shown to suppress both colorectal and breast cancer. This effect is mediated by an increase in E-cadherin expression, which in turn promotes the relocation of β-catenin from the nucleus to the cell membrane (Palmer et al., 2001). It also activates vitamin D receptors; these

TABLE 20.1 Summary of Potential Therapeutic Strategies Targeting Aberrant Wnt Signaling in Cancer

| Therapeutic Strategy | Therapeutic Agent | Target | Mode of Action | Effect |
|---|---|---|---|---|
| Nonsteroidal anti-inflammatory drugs | Celecoxib | Cyclooxygenase 2 (COX-2) | Suppression of COX activity | Decreased nuclear β-catenin levels in vitro, decreased polyp formation in familial adenomatous polyposis coli |
| | Aspirin, indomethacin | β-Catenin | Phosphorylation of β-catenin | Decreased aberrant Wnt activity |
| | Sulindac | β-Catenin | Promotes proteasomal degradation of β-catenin | Inhibition of Wnt signaling |
| | Sulindac sulfate | Cyclic guanosine monophosphate (cGMP) phosphodiesterase | Increase in cGMP levels | Chemoprotection |
| Vitamin A | Retinoids | Retinoic acid, retinoid X receptors | Signaling via retinoic acid and retinoid X receptors | Decreased AP-1<br>Decreased Wnt signaling<br>Stabilization of cell–cell adhesions |
| Vitamin D | 1α,25–dihydroxyvitamin D3 | E-cadherin<br><br><br>Vitamin D receptor | Increased E-cadherin expression promotes β-catenin cell membrane relocalization<br>Vitamin D receptor activation | Decreased nuclear β-catenin<br>Inhibition of Wnt signaling<br>Decreased β-catenin activation |
| Antibody-based | Anti-Wnt-1<br>Anti-Wnt-2 | Wnt-1<br>Wnt-2 | Binding and inhibition of Wnt-1 and Wnt-2 proteins | Decreased Tcf/Lef transcription activity<br>Induction of apoptosis and tumor suppression |
| Small-molecule inhibitors | ICAT (inhibitor of β-catenin and TCF-4) (Mikesch et al., 2007) | Tcf/Lef complexes | Disrupts Tcf/lef complexes | |
| | Three small-molecule inhibitors (PKF115-584, PKF118-310, CGP049090) | Tcf/β-catenin | Disrupt Tcf/β-catenin complexes and adenomatous polyposis coli (APC)/β-catenin binding | |
| | Three small-molecule inhibitors | Tcf/β-catenin | Inhibit β-catenin–Tcf interactions | |
| | ICG-001 | CBP | Inhibition of CBP/β-catenin complex | Induces apoptosis in colon cancer cell lines<br>Decreased tumor formation in vivo |
| | FJ9 | Frizzled-7/Dsh | Disrupts interaction between Frizzled-7 and Disheveled | Decreased survivin protein levels, Increased apoptosis<br>Cessation of tumor growth in mouse models |

*Continued*

TABLE 20.1 Summary of Potential Therapeutic Strategies Targeting Aberrant Wnt Signaling in Cancer—cont'd

| Therapeutic Strategy | Therapeutic Agent | Target | Mode of Action | Effect |
|---|---|---|---|---|
| Protein knockdown (chimeric F-box proteins) | F3APCbc4 | β-Catenin | Target pathogenic nuclear β-catenin<br>Deactivation of canonical Wnt pathway | Decreased β-catenin<br>Arrest of colorectal cancer cells in vitro and tumor growth in vivo |
| | F-TrCP-Ecad | β-Catenin | Decreased cytoplasmic β-catenin | |
| Viral-based (replicating oncolytic viruses) | Adenovirus containing fas-associated via death domain placed under Tcf/Lef response element control | | Direct cell killing | Increased selectivity for cells with intrinsic Wnt activity |
| | Adenovirus containing diphtheria toxin A under β-catenin/TCF-dependent promoter 4 (CTP4) control | | | |
| | Adenovirus expressing herpes simplex virus (HSV) thymidine kinase | | | |
| (Prodrugs) | | | Conversion of prodrug ganciclovir into active form | Increased cell death<br>Decrease in colon cancer growth |
| Genetic | OPN antisense cDNA<br>c-MYC antisense treatment<br>Wnt-1 small interfering Ribonucleic Acid (siRNA) | OPN<br>c-MYC<br>Wnt-1 | Inactivation of OPN mRNA, effectively switching off OPN gene<br>Inactivation of c-MYC mRNA<br>Wnt-1 gene silencing | Inhibition of tumor formation and growth<br>Inhibition of proliferation<br>Induction of differentiation<br>Induction of apoptosis<br>Increased apoptosis |

receptors actively compete with Tcf for binding to β-catenin. The overall effect is to suppress β-catenin activation. However, the prophylactic potential of this approach is limited to cancers that express vitamin D receptors, and this is demonstrated by the inefficacy of vitamin D treatment in a highly tumorigenic version of colonic cancer deficient in vitamin D receptors (Palmer et al., 2001).

## 4.3 Antibody-Based Strategies

Most cancers that exhibit deregulated Wnt signaling do so as a result of mutations in the *APC*, *axin*, or *β-catenin* genes. However, Wnt signaling can also be inappropriately activated by overexpression of Wnt protein or frizzled receptors, and this in turn can facilitate tumor development. In these cases, antibodies that bind to Wnt ligands or frizzled receptors may act to block Wnt signaling (Barker and Clevers, 2006). In head and neck squamous carcinoma cell lines, Rhee et al. (2002) tested the function of five *Wnt* (*Wnt-1*, *5a*, *7a*, *10b*, and *13*) and two *frizzled* (*Fz2* and *5*) genes. Of these, they found that *Wnt-1*, *7a*, *10b*, and *13* and *Fz-2* were overexpressed at the RNA level and that Wnt-1 and 10b and Fz-2 were also highly expressed at the protein level. They produced anti-Wnt-1 antibodies and found their application reduced transcriptional activity of Tcf/Lef and induced apoptosis (Rhee et al., 2002). Wnt-1 protein is also overexpressed in non-small-cell lung cancer (NSCLC) cell lines and treatment with the same anti-Wnt-1 antibodies also induced apoptosis in these cells (He et al., 2005a, 2004). Monoclonal anti-Wnt-1 antibodies have also been shown to suppress tumors in mouse models of these cancers (He et al., 2004).

Wnt-2 proteins are also expressed at abnormally high levels in various cancers including NSCLC, cell lines derived from which can be induced to undergo apoptosis using a monoclonal anti-Wnt-2 antibody (You et al., 2004). This result again demonstrates the potential of anti-Wnt antibodies as therapeutic agents. Furthermore, as this would only affect cells that are dependent on elevated Wnt signaling, toxicity to normal adult cells could be minimized. Upregulation of both frizzled 1 and 2 proteins was found in human breast cancer and was associated with tumor invasiveness in colon carcinoma cell lines, suggesting that they are also potential targets for antibody therapy (Milovanovic et al., 2004; Holcombe et al., 2002). However, blocking these receptors may induce changes to other signaling pathways that use them and hence cause further side effects.

Of particular note is the recent findings that colon cancer cells induced by downstream events, such as *APC* mutations, also express high levels of Wnt ligands, which, in turn, enhance overall Wnt signal transduction, sometimes to such an extent that reactivation of SFRP or the use of anti-Wnt-1 antibodies induces apoptosis in these colon cancer cell lines (Suzuki et al., 2004; He et al., 2005b). More importantly though, it suggests that this increased Wnt-Frizzled coupling leading to aberrant Wnt signaling may be an early event in carcinogenesis, and underscores the potential value of a therapeutic monoclonal antibody strategy (Suzuki et al., 2004; Taketo, 2004).

## 4.4 Small-Molecule Inhibitors

Aberrant Wnt activity can arise through a number of different mechanisms, although translocation of β-catenin into the nucleus and assembly of the Tcf/β-catenin complex is common to all of these (Barker and Clevers, 2006). Development of an effective inhibitor has been facilitated by the availability of structural data for the binding site between the β-catenin and Tcf. Crystal structures show that β-catenin has a 12 armadillo repeat that forms a nicked right-handed superhelix, which has a high positive charge. This superhelix contains a groove 20 Å wide and 95 Å long with which it binds the amino terminal domain of Tcf, using its three distinct binding modules (Knapp et al., 2001). Like most protein–protein interactions, the binding region occurs over a large surface area and has a very high binding affinity (~8nM) (Poy et al., 2001). This high affinity and delocalized site makes it very difficult for a small inhibitor to effectively block the overall binding site. However, there are frequently hot spots within these interaction domains that may act as docking sites for small-molecule inhibitors (Fasolini et al., 2003). Tcf has Asp residues and β-catenin contains hydrophobic pockets that show potential in this regard (Fasolini et al., 2003).

β-Catenin has a number of functions within the cell. It forms complexes with E-cadherin that act in cell adhesion and also it binds with APC and Axin to regulate normal β-catenin levels in healthy tissue (Barker and Clevers, 2006). It is, therefore, essential that small-molecule inhibitors interact with the Tcf/β-catenin complex without affecting the other roles carried out by β-catenin, which could result in significant toxicity, especially if cell adhesions were reduced due to a lack of E-cadherin (Barker and Clevers, 2006). Another difficulty in the development of small-molecule inhibitors is the overlap between Tcf, cadherin, and APC structural binding surfaces, disruption of which could also lead to toxicity and/or be oncogenic (Poy et al., 2001). The Tcf antagonist, inhibitor of beta-catenin and TCF (ICAT), reportedly disrupts Tcf complexes without affecting cadherin, although the mechanism by which this is achieved remains unclear. Unfortunately, due to the high electrostatic properties and the high degree of cooperativity associated with the structure of Tcf, effective Tcf antagonists are difficult to make (Fasolini et al., 2003).

Even though there are many hurdles to developing effective small-molecule inhibitors, their antineoplastic potential is great. Three natural small molecule inhibitors have been identified by Lepourcelet et al. (2004). These were shown to disrupt the Tcf/β-catenin complex, although again the mechanism through which they do this is unclear. Subsequent analysis of the inhibitors' docking sites to either β-catenin or Tcf has proved difficult. However, all the inhibitors were observed to interfere with APC binding to β-catenin, suggesting that their actions are mediated through β-catenin (Lepourcelet et al., 2004). This inability to determine their specific docking site or interaction with either Tcf or β-catenin could reduce their therapeutic usefulness. Furthermore, the toxicity associated with these compounds still needs to be fully tested.

By using virtual screening and medium-throughput biophysical assays, Trosset et al. (2006) were able to identify three small-molecule inhibitors that bind to the K435/R469 hot spot to inhibit the interaction between β-catenin and Tcf. This appears to be a very useful approach, since high-throughput screening has not revealed any small-molecule inhibitors (Barker and Clevers, 2006; Trosset et al., 2006).

The Tcf/β-catenin complex requires other transcriptional coactivators to activate the transcription of target genes. This provides an opportunity to produce small-molecule inhibitors that target these additional coactivators; these coactivators include cAMP response element-binding protein (CREB) binding protein (CBP) and B-cell lymphoma 9 (BCL9)/pygopus (Barker and Clevers, 2006). Selective inhibition of these coactivators of the Tcf/β-catenin complex may reduce toxicity, although it is possible that they may also function in other signal transduction pathways, which could lead to other unwanted off-target effects.

CBP binds to the C-terminal of β-catenin through its CREB domain, allowing the Tcf/β-catenin complex to activate transcription, probably by recruitment of basal transcriptional machinery. A lead compound in the inhibition of this CBP–β-catenin complex is ICG-001, which prevents binding to the N-terminal of CBP but does not interact with the homologous region of p300. Thus, ICG-001 selectively disrupts CBP/β-catenin interaction while not interfering with p300/β-catenin binding. ICG-001-induced apoptosis in colon cancer cell lines in a dose-dependent manner (Emami et al., 2004). In vivo treatment with ICG-001 decreased tumor formation in mouse$^{min}$ and SW620 nude mouse xenograft models without causing mortality, toxicity, or weight loss, highlighting its therapeutic potential. This antineoplastic effect is believed to be mediated through the downregulation of the survivin protein that inhibits caspase activation (Emami et al., 2004). Furthermore, CBP possesses intrinsic histone acetyl transferase activity that may act to remodel chromatin structure and remove Groucho repressor factors (Takemaru and Moon, 2000).

Another potential target of small-molecule inhibition is the BCL9/β-catenin interaction. Studies have shown that this binding site is distinct from most other β-catenin binding sites, making it a good target for therapy (Sampietro et al., 2006). It is also smaller than the Tcf/β-catenin binding site and has a medium binding affinity ($K_d = 0.5\,\mu M$), making it a far easier target (Sampietro et al., 2006). Importantly, it has been shown that this region is essential for BCL9 interaction but not for cadherin binding, so possibly avoiding any unwanted effects on cell adhesion (Sampietro et al., 2006). A small-molecule inhibitor for this binding site has been reported to show efficacy in vivo, although a lack of data makes it hard to validate this claim (Barker and Clevers, 2006).

It seems clear then that small-molecule inhibitors of the Tcf/β-catenin complex have a great potential in cancer therapy. However, there have also been considerable efforts to develop inhibitors of other steps in the Wnt signaling pathway. For example, it has been shown that the PDZ domain of Dsh mediates binding to the frizzled receptors and so provides an attractive target for small-molecule inhibition. Dapper acts as an endogenous inhibitor of the Wnt signaling pathway by interacting with Dsh via the dishevelled, Egl-10 and pleckstrin domain (DEP) interface (Barker and Clevers, 2006; Zhang et al., 2006). This interaction causes degradation of Dsh through a mechanism that is insensitive to lysosome and proteasome inhibitors. Shan et al. used the crystal structure of the *Xenopus* Dsh PDZ domain bound to Dapper to discover small-molecule inhibitors of this interaction. Using the UNITY software they identified a small organic molecule, NSC668036, which bound to the PDZ domain of Dsh with a high degree of specificity (Barker and Clevers, 2006; Shan et al., 2005). In *Xenopus*, NSC668036 was able to block axis duplication without affecting β-catenin, a result that suggested that it inhibits Wnt signaling at the Frizzled receptor level. However, NSC668036 has only a limited ability to reduce transcription of Tcf target genes stimulated by Wnt signaling, and does not seem to be a viable candidate for cancer therapy. An inhibitor designed by Fujii et al. (2007), named FJ9, designed specifically to disrupt the interaction between Frizzled-7 and the PDZ domain of Dsh, shows more promise. FJ9 has a high degree of specificity, little or no effect on global gene expression, appears not to act on the noncanonical Wnt pathway, causes apoptosis in human cancer cell lines, and stops tumor growth in xenograft mouse models.

Overall, small-molecule inhibitors show great potential as a next generation of cancer therapeutics, although more research is needed to identify more effective and specific drugs.

## 5. PROTEIN KNOCKDOWN STRATEGIES

When Wnt signaling is in its "off" state, β-cadherin is phosphorylated near the destruction complex and is degraded by the ubiquitin-conjugation machinery. This ubiquitin-conjugation machinery is an elaborate mechanism responsible for the destruction of many proteins that are misfolded or oxidized, or are involved in the cell cycle or cell signaling (Dihlmann and von Knebel Doeberitz, 2005; Su et al., 2003). The process is mediated by a multimeric E3 ubiquitin ligase and its associated F-box protein may be responsible for targeting phosphorylated β-catenin (Hart et al., 1999). When the destruction complex fails to form, as is the case in many cancers, β-catenin is not phosphorylated and therefore cannot be targeted for degradation; this failure results in its stabilization and nuclear relocation. A potential antioncogenic therapy would be to redirect this ubiquitin-conjugation machinery to target and degrade β-catenin regardless of its phosphorylated status (Dihlmann and von Knebel Doeberitz, 2005). To accomplish this degradation of β-catenin, researchers have tried to engineer chimeric F-box proteins (CFPs) that only recognize pathogenic β-catenin (Dihlmann and von Knebel Doeberitz, 2005). In the Su et al. study, a CFP (F3APCbc4) was designed to target β-catenin in colorectal cancer. This CFP had achieved a marked reduction in β-catenin after 24h and no detectable nuclear β-catenin after 48h. The CFP was significantly more effective than the endogenous tumor suppressor APC, perhaps through suppression of the Skp, Cullin, F-box containing (SCF) component. This arrested both colorectal cancer cells in vitro and tumor growth in nude mouse models in vivo through the deactivation of the canonical Wnt pathway (Su et al., 2003).

Cong et al. produced a CFP (F-TrCP-Ecad) with the binding domain of E-cadherin attached to the beta-transducin repeat-containing protein (βTrCP) ubiquitin-protein ligase. They showed a decrease in cytoplasmic β-catenin that led to impaired growth and clonogenic ability of colonic cancer cells and their loss of carcinogenesis when introduced into nude mice (Cong et al., 2003). Remarkably in both these tests, membrane-associated β-catenin binding to E-cadherin was unaffected, whereas F3APCbc4 specifically targeted nuclear β-catenin and F-TrCP-Ecad was shown to have a short half-life, possibly providing any E-cadherin-bound β-catenin protection from degradation (Su et al., 2003; Cong et al., 2003). This modification reduces the chance of side effects due to failure of cell–cell adhesion and highlights its potential therapeutic usefulness.

As an alternative strategy, Liu et al. produced a CFP by replacing the WD40-repeat of BTrCP with the β-catenin binding domains of E-cadherin and Tcf. They showed that there was a marked reduction of β-catenin in the absence of APC protein, again validating the potential usefulness of this approach (Liu et al., 2004). Another advantage of the CFP knockdown approaches is the ability to reveal gene function, since, unlike genomic knockout or gene interference approaches, they do not silence the expression of the entire gene. This ability could be particularly useful in the study of multifunctional proteins. However, some cancer cell lines show resistance to pathogenic β-catenin deletion and currently there is no viable delivery system to patients; both of these factors remain obstacles to be overcome in CFP development (Dihlmann and von Knebel Doeberitz, 2005).

## 6. VIRAL-BASED THERAPIES

Adenovirus-based vectors have been used in experiments in cancer gene therapy for around a decade (Zhang et al., 2006). Two main strategies have been investigated, the first of which involves the direct killing of the cancer cell through replicating oncolytic viruses, or the use of nonreplicating viruses encoding tumor repressor genes, suicide genes, or antiangiogenic genes. The second is the use of immunotherapeutic approaches leading to host antitumor immune responses that kill the cells at both local and metastatic locations. Although clinical trials have been carried out using both methods and results have looked promising, robust evidence of efficacy has remained elusive (Zhang et al., 2006). The high levels of β-catenin present in cells with aberrant Wnt signaling have been used to direct the adenovirus-mediated killing of target colon cancer cells. In one such experiment, an adenovirus-containing fas-associated via death domain was placed under the control of the Tcf/LEF responsive element and showed a high level of selectivity for cells with intrinsic Wnt activity, although its safety and efficacy in vivo are as yet unknown (Chen and McCormick, 2001). Another study by Lipinski et al. used an adenovirus-containing diphtheria toxin A under the control of the CTP4 promoter. CTP4 is expressed at nearly undetectable levels in normal tissue but at high levels in cells with deregulated overexpressing β-catenin, suggesting that it may offer a high degree of selectivity in vivo (Lipinski et al., 2004). Although this shows promise as a cancer therapy, high levels of specificity are important due to the destructive nature of the genes that these adenoviruses contain (Barker and Clevers, 2006).

Other approaches to adenovirus therapy include the use of prodrugs that are converted by enzymes encoded in the adenovirus genome and of replicating adenoviruses. Kwong et al. used an adenovirus that expressed the herpes simplex virus thymidine kinase that converted the prodrug ganciclovir into its active form to kill the cell. This suppressed tumors produced by the colon cancer line DLD-1 in nude mice, indicating that it could function as a therapeutic (Kwong et al., 2002). Brunori et al. developed replicating adenoviruses, which expressed the viral genes E2 and E1B with their promoters controlled by the Tcf4 transcription factor. Replicating adenoviruses would seem to be more effective than their nonreplicating counterparts, since the initial therapeutic dose is supplemented by virus made within the tumor. However, some aberrant Wnt signaling cell lines from colon cancer did not activate the E2 and E1B promoters and replication decreased 50- to 100-fold in lung cancer cell lines (Brunori et al., 2001). This approach could be useful in the treatment of liver metastases from colon cancers, but changes will need to be introduced to produce a virus that can treat all colon cancers. Some researchers have also used genes rather than promoters to enhance the spread of their virus, thereby increasing therapeutic efficacy, an example being adenovirus death protein (ADP), an adenovirus (Ad) cytolytic protein (Toth et al., 2004).

Recent examples of such adenoviruses include genes encoding cytosine deaminase that converts 5-fluorocytosine into the highly toxic 5-fluorouracil and *Escherichia coli* nitroreductase that activates prodrug CB1954. Both of these approaches can kill colonic cancer cell lines and the latter has suppressed tumor formation in xenograft mouse models (Fuerer and Iggo, 2004; Lukashev et al., 2005). From the above, we can see that this rapidly expanding area of research shows potential in cancer treatment. However, a number of obstacles remain, namely the efficacy of tumor cell transduction and the inhibition of efficacy by anti-Ad host immune responses (Zhang et al., 2006).

## 6.1 Genetic Strategies

It has been observed that the deregulation and overexpression of β-catenin in the cell nucleus can promote the expression of genes that lead to cancer formation. This observation provides an additional therapeutic target by first identifying and then disrupting these oncogenes. Furthermore, the ability of interfering RNA to block the expression of a gene provides an interesting opportunity not only to disrupt the target genes of Wnt signaling but also to inhibit genes that encode Wnt ligands.

Microarray analysis suggests that there are around 300–400 Wnt signaling cascade target genes that encode proteins with diverse functions (van de Wetering et al., 2002; Van der Flier et al., 2007). Several of these genes such as *OPN*, *c-MYC*, and *cyclin D1* have been linked directly to the formation of various cancers. OPN is a secreted extracellular glycophosphoprotein that, when overexpressed in a primary tumor, promotes the metastatic phenotype and poor outcome in many cancers, particularly of the breast (El-Tanani et al., 2006). OPN expression is regulated in part by the Wnt signaling pathway. It has been shown that antisense cDNA targeting OPN can actively inhibit tumor formation and growth in experimental systems, highlighting the potential of a genomic strategy as a therapeutic (El-Tanani et al., 2006).

*c-MYC* was discovered as a cellular homologue to the *v-MYC* retroviral oncogene and has a major role in development. *c-MYC* expression has been found to be activated in various human cancers and is associated with cancers that cause as many as 70,000 cancer deaths per year in the United States (Dang, 1999). This gene was also found to be upregulated by the Wnt signaling pathway and there is obvious interest in developing therapeutics to control its activity. Chen et al. showed that antisense treatment of *c-MYC* inhibited proliferation and induced differentiation and apoptosis in acute myeloblastic leukemia cell lines, whereas Iverson et al. showed that treating prostate cancer cell lines with an antisense phosphorodiamidate morpholino oligomer caused growth inhibition and apoptosis (Chen et al., 2006; Iversen et al., 2003). Although this approach is promising, there is as yet no evidence to suggest that it could have efficacy in Wnt driven cancers (Barker and Clevers, 2006). Moreover, trials to assess whether *c-MYC* antisense treatments can be used in conjunction with chemotherapy are currently being performed. There are early indications that the efficacy of the chemotherapy can be affected by reduced *c-MYC* expression, for example, enhancing the effect of vinblastine and lowering that of 5-fluorouracil (Barker and Clevers, 2006). Cyclin D1 is important in the regulation of the cell cycle and has been shown to be highly expressed within some cancers (Sansom et al., 2005). Evidence from colon cancer mouse models suggests that it is not involved in the formation of early adenomas, but acts to enhance the growth of tumors in vivo (Barker and Clevers, 2006).

Wnt ligand genes could also be a novel target for cancer therapeutics. He et al. (2004) showed that silencing the *Wnt-1* gene using siRNA produced apoptosis in human cancer cell lines. However, removal of an entire Wnt protein may cause toxicity and reduce the therapeutic index using this approach.

# 7. CONCLUSION

In the last 2 years increased knowledge of the Wnt signaling pathway has led to a great interest in the development of drugs that can block its carcinogenic effects. This complex pathway lends itself to potential inhibition at many levels. Drugs that work at the level of the Tcf/β-catenin complex, perhaps, hold the greatest therapeutic potential in combating Wnt driven cancers, since they target the convergence point of the pathway—nuclear β-catenin. We still have a long way to go in designing a perfectly selective drug that does not disrupt membrane-located β-catenin and the possibility of instead targeting its transcriptional coactivators should be further explored. Furthermore, the prolonged use of drugs that disrupt Wnt signaling may produce side effects in the skin and intestine where it functions in cell renewal and may therefore represent another barrier to the development of these drugs. The development of the Wnt gene program is also discovering new targets for treatment and may prove invaluable in long-term cancer therapy.

For the most part, these new drugs are still in the early stages of development. However, existing drugs such as the NSAIDs and vitamins A and D derivatives have shown moderate efficacy in combating Wnt-derived cancer. New generations of these drugs may provide improved treatments in the near future, until more specific and effective agents become available in the longer term. Overall, Wnt signaling is a rich environment for therapeutic targeting and drug discovery.

## References

Barker, N., Clevers, H., 2006. Mining the Wnt pathway for cancer therapeutics. Nat. Rev. Drug Discov. 5, 997–1014.

Barrallo-Gimeno, A., Nieto, M.A., 2005. The snail genes as inducers of cell movement and survival: implications in development and cancer. Development 132, 3151–3161.

Batlle, E., et al., 2000. The transcription factor snail is a repressor of E-cadherin gene expression in epithelial tumour cells. Nat. Cell Biol. 2, 84–89.

Behrend, E.I., Craig, A.M., Wilson, S.M., Denhardt, D.T., Chambers, A.F., 1994. Reduced malignancy of ras-transformed NIH 3T3 cells expressing antisense osteopontin RNA. Cancer Res. 54, 832–837.

Blick, T., et al., 2010. Epithelial mesenchymal transition traits in human breast cancer cell lines parallel the CD44(hi/)CD24 (lo/-) stem cell phenotype in human breast cancer. J. Mammary Gland Biol. Neoplasia 15, 235–252.

Bolos, V., et al., 2003. The transcription factor slug represses E-cadherin expression and induces epithelial to mesenchymal transitions: a comparison with snail and E47 repressors. J. Cell Sci. 116, 499–511.

Boon, E.M., et al., 2004. Sulindac targets nuclear beta-catenin accumulation and Wnt signalling in adenomas of patients with familial adenomatous polyposis and in human colorectal cancer cell lines. Br. J. Cancer 90, 224–229.

Brabletz, T., Jung, A., Spaderna, S., Hlubek, F., Kirchner, T., 2005. Opinion: migrating cancer stem cells – an integrated concept of malignant tumour progression. Nat. Rev. Cancer 5, 744–749.

Brunori, M., Malerba, M., Kashiwazaki, H., Iggo, R., 2001. Replicating adenoviruses that target tumors with constitutive activation of the wnt signaling pathway. J. Virol. 75, 2857–2865.

Cai, J., et al., 2013. MicroRNA-374a activates Wnt/beta-catenin signaling to promote breast cancer metastasis. J. Clin. Invest. 123, 566–579.

Cano, A., et al., 2000. The transcription factor snail controls epithelial-mesenchymal transitions by repressing E-cadherin expression. Nat. Cell Biol. 2, 76–83.

Casas, E., et al., 2011. Snail2 is an essential mediator of Twist1-induced epithelial mesenchymal transition and metastasis. Cancer Res. 71, 245–254.

Chen, R.H., McCormick, F., 2001. Selective targeting to the hyperactive beta-catenin/T-cell factor pathway in colon cancer cells. Cancer Res. 61, 4445–4449.

Chen, J.P., Chen, L., Leek, J., Lin, C., 2006. Antisense c-myc fragments induce normal differentiation cycles in HL-60 cells. Eur. J. Clin. Invest. 36, 49–57.

Cheung, M., et al., 2005. The transcriptional control of trunk neural crest induction, survival, and delamination. Dev. Cell 8, 179–192.

Clarke, N., Germain, P., Altucci, L., Gronemeyer, H., 2004. Retinoids: potential in cancer prevention and therapy. Expert Rev. Mol. Med. 6, 1–23.

Clevers, H., 2006. Wnt/beta-catenin signaling in development and disease. Cell 127, 469–480.

Cong, F., Zhang, J., Pao, W., Zhou, P., Varmus, H., 2003. A protein knockdown strategy to study the function of beta-catenin in tumorigenesis. BMC Mol. Biol. 4, 10.

Dang, C.V., 1999. c-Myc target genes involved in cell growth, apoptosis, and metabolism. Mol. Cell. Biol. 19, 1–11.

Dhasarathy, A., Kajita, M., Wade, P.A., 2007. The transcription factor snail mediates epithelial to mesenchymal transitions by repression of estrogen receptor-alpha. Mol. Endocrinol. 21, 2907–2918.

Dhasarathy, A., Phadke, D., Mav, D., Shah, R.R., Wade, P.A., 2011. The transcription factors snail and slug activate the transforming growth factor-beta signaling pathway in breast cancer. PLoS One 6, e26514.

Dihlmann, S., von Knebel Doeberitz, M., 2005. Wnt/beta-catenin-pathway as a molecular target for future anti-cancer therapeutics. Int. J. Cancer 113, 515–524.

Dihlmann, S., Klein, S., Doeberitz Mv, M., 2003. Reduction of beta-catenin/T-cell transcription factor signaling by aspirin and indomethacin is caused by an increased stabilization of phosphorylated beta-catenin. Mol. Cancer Ther. 2, 509–516.

DiMeo, T.A., et al., 2009. A novel lung metastasis signature links Wnt signaling with cancer cell self-renewal and epithelial-mesenchymal transition in basal-like breast cancer. Cancer Res. 69, 5364–5373.

El-Tanani, M.K., et al., 2006. The regulation and role of osteopontin in malignant transformation and cancer. Cytokine Growth Factor Rev. 17, 463–474.

El-Tanani, M., Fernig, D.G., Barraclough, R., Green, C., Rudland, P., 2001. Differential modulation of transcriptional activity of estrogen receptors by direct protein-protein interactions with the T cell factor family of transcription factors. J. Biol. Chem. 276, 41675–41682.

El-Tanani, M., Platt-Higgins, A., Rudland, P.S., Campbell, F.C., 2004. Ets gene PEA3 cooperates with beta-catenin-Lef-1 and c-Jun in regulation of osteopontin transcription. J. Biol. Chem. 279, 20794–20806.

Emami, K.H., et al., 2004. A small molecule inhibitor of beta-catenin/CREB-binding protein transcription [corrected]. Proc. Natl. Acad. Sci. U.S.A. 101, 12682–12687.

Fasolini, M., et al., 2003. Hot spots in Tcf4 for the interaction with beta-catenin. J. Biol. Chem. 278, 21092–21098.

Fuerer, C., Iggo, R., 2004. 5-fluorocytosine increases the toxicity of Wnt-targeting replicating adenoviruses that express cytosine deaminase as a late gene. Gene Ther. 11, 142–151.

Fujii, N., et al., 2007. An antagonist of dishevelled protein-protein interaction suppresses beta-catenin-dependent tumor cell growth. Cancer Res. 67, 573–579.

Gelmini, S., et al., 2009. The CXCR4/CXCL12 axis in endometrial cancer. Clin. Exp. Metastasis 26, 261–268.

Grooteclaes, M.L., Frisch, S.M., 2000. Evidence for a function of CtBP in epithelial gene regulation and anoikis. Oncogene 19, 3823–3828.

Gupta, P.B., et al., 2005. The melanocyte differentiation program predisposes to metastasis after neoplastic transformation. Nat. Genet. 37, 1047–1054.

Harris, R.E., Beebe-Donk, J., Alshafie, G.A., 2006. Reduction in the risk of human breast cancer by selective cyclooxygenase-2 (COX-2) inhibitors. BMC Cancer 6, 27.

Hart, M., et al., 1999. The F-box protein beta-TrCP associates with phosphorylated beta-catenin and regulates its activity in the cell. Curr. Biol. 9, 207–210.

Hartwell, K.A., et al., 2006. The Spemann organizer gene, Goosecoid, promotes tumor metastasis. Proc. Natl. Acad. Sci. U.S.A. 103, 18969–18974.

He, B., et al., 2004. A monoclonal antibody against Wnt-1 induces apoptosis in human cancer cells. Neoplasia 6, 7–14.

He, B., et al., 2005a. Wnt signaling in stem cells and non-small-cell lung cancer. Clin. Lung Cancer 7, 54–60.

He, B., et al., 2005b. Blockade of Wnt-1 signaling induces apoptosis in human colorectal cancer cells containing downstream mutations. Oncogene 24, 3054–3058.

Hermann, P.C., et al., 2007. Distinct populations of cancer stem cells determine tumor growth and metastatic activity in human pancreatic cancer. Cell Stem Cell 1, 313–323.

Hernandez-Diaz, S., Garcia Rodriguez, L.A., 2007. Nonsteroidal anti-inflammatory drugs and risk of lung cancer. Int. J. Cancer 120, 1565–1572.

Holcombe, R.F., et al., 2002. Expression of Wnt ligands and frizzled receptors in colonic mucosa and in colon carcinoma. Mol. Pathol. 55, 220–226.

Hoppler, S., Kavanagh, C.L., 2007. Wnt signalling: variety at the core. J. Cell Sci. 120, 385–393.

Ilyas, M., 2005. Wnt signalling and the mechanistic basis of tumour development. J. Pathol. 205, 130–144.

Iversen, P.L., Arora, V., Acker, A.J., Mason, D.H., Devi, G.R., 2003. Efficacy of antisense morpholino oligomer targeted to c-myc in prostate cancer xenograft murine model and a phase I safety study in humans. Clin. Cancer Res. 9, 2510–2519.

Jin, H., et al., 2010. Snail is critical for tumor growth and metastasis of ovarian carcinoma. Int. J. Cancer 126, 2102–2111.

Katoh, M., Katoh, M., 2007. WNT signaling pathway and stem cell signaling network. Clin. Cancer Res. 13, 4042–4045.

Knapp, S., et al., 2001. Thermodynamics of the high-affinity interaction of TCF4 with beta-catenin. J. Mol. Biol. 306, 1179–1189.

Krohn, A., et al., 2009. CXCR4 receptor positive spheroid forming cells are responsible for tumor invasion in vitro. Cancer Lett. 280, 65–71.

Kudo-Saito, C., Shirako, H., Takeuchi, T., Kawakami, Y., 2009. Cancer metastasis is accelerated through immunosuppression during snail-induced EMT of cancer cells. Cancer Cell 15, 195–206.

Kurrey, N.K., et al., 2009. Snail and slug mediate radioresistance and chemoresistance by antagonizing p53-mediated apoptosis and acquiring a stem-like phenotype in ovarian cancer cells. Stem Cells 27, 2059–2068.

Kwong, K.Y., Zou, Y., Day, C.P., Hung, M.C., 2002. The suppression of colon cancer cell growth in nude mice by targeting beta-catenin/TCF pathway. Oncogene 21, 8340–8346.

Lepourcelet, M., et al., 2004. Small-molecule antagonists of the oncogenic Tcf/beta-catenin protein complex. Cancer Cell 5, 91–102.

Li, H., et al., 2002. Pro-apoptotic actions of exisulind and CP461 in SW480 colon tumor cells involve beta-catenin and cyclin D1 down-regulation. Biochem. Pharmacol. 64, 1325–1336.

Lipinski, K.S., et al., 2004. Optimization of a synthetic beta-catenin-dependent promoter for tumor-specific cancer gene therapy. Mol. Ther. 10, 150–161.

Liu, J., Stevens, J., Matsunami, N., White, R.L., 2004. Targeted degradation of beta-catenin by chimeric F-box fusion proteins. Biochem. Biophys. Res. Commun. 313, 1023–1029.

Lochter, A., et al., 1997. Matrix metalloproteinase stromelysin-1 triggers a cascade of molecular alterations that leads to stable epithelial-to-mesenchymal conversion and a premalignant phenotype in mammary epithelial cells. J. Cell Biol. 139, 1861–1872.

Lukashev, A.N., Fuerer, C., Chen, M.J., Searle, P., Iggo, R., 2005. Late expression of nitroreductase in an oncolytic adenovirus sensitizes colon cancer cells to the prodrug CB1954. Hum. Gene Ther. 16, 1473–1483.

Maier, T.J., Janssen, A., Schmidt, R., Geisslinger, G., Grosch, S., 2005. Targeting the beta-catenin/APC pathway: a novel mechanism to explain the cyclooxygenase-2-independent anticarcinogenic effects of celecoxib in human colon carcinoma cells. FASEB J. 19, 1353–1355.

Malliri, A., et al., 2006. The rac activator Tiam1 is a Wnt-responsive gene that modifies intestinal tumor development. J. Biol. Chem. 281, 543–548.

Mani, S.A., et al., 2008. The epithelial-mesenchymal transition generates cells with properties of stem cells. Cell 133, 704–715.

Mikesch, J.H., Steffen, B., Berdel, W.E., Serve, H., Muller-Tidow, C., 2007. The emerging role of Wnt signaling in the pathogenesis of acute myeloid leukemia. Leukemia 21, 1638–1647.

Milovanovic, T., et al., 2004. Expression of Wnt genes and frizzled 1 and 2 receptors in normal breast epithelium and infiltrating breast carcinoma. Int. J. Oncol. 25, 1337–1342.

Minn, A.J., et al., 2005. Genes that mediate breast cancer metastasis to lung. Nature 436, 518–524.

Morel, A.P., et al., 2008. Generation of breast cancer stem cells through epithelial-mesenchymal transition. PLoS One 3, e2888.

Morin, P.J., 1999. Beta-catenin signaling and cancer. BioEssays 21, 1021–1030.

Neal, C.L., et al., 2012. Snail transcription factor negatively regulates maspin tumor suppressor in human prostate cancer cells. BMC Cancer 12, 336.

Neal, C.L., McKeithen, D., Odero-Marah, V.A., 2011. Snail negatively regulates cell adhesion to extracellular matrix and integrin expression via the MAPK pathway in prostate cancer cells. Cell Adh. Mig. 5, 249–257.

Nieto, M.A., 2002. The snail superfamily of zinc-finger transcription factors. Nat. Rev. Mol. Cell Biol. 3, 155–166.

Noe, V., et al., 2001. Release of an invasion promoter E-cadherin fragment by matrilysin and stromelysin-1. J. Cell Sci. 114, 111–118.

Nosho, K., et al., 2006. Overexpression of poly(ADP-ribose) polymerase-1 (PARP-1) in the early stage of colorectal carcinogenesis. Eur. J. Cancer 42, 2374–2381.

Olmeda, D., et al., 2007. SNAI1 is required for tumor growth and lymph node metastasis of human breast carcinoma MDA-MB-231 cells. Cancer Res. 67, 11721–11731.

Onder, T.T., et al., 2008. Loss of E-cadherin promotes metastasis via multiple downstream transcriptional pathways. Cancer Res. 68, 3645–3654.

Palmer, H.G., et al., 2001. Vitamin D(3) promotes the differentiation of colon carcinoma cells by the induction of E-cadherin and the inhibition of beta-catenin signaling. J. Cell Biol. 154, 369–387.

Peinado, H., Olmeda, D., Cano, A., 2007. Snail, Zeb and bHLH factors in tumour progression: an alliance against the epithelial phenotype? Nat. Rev. Cancer 7, 415–428.

Perez-Moreno, M.A., et al., 2001. A new role for E12/E47 in the repression of E-cadherin expression and epithelial-mesenchymal transitions. J. Biol. Chem. 276, 27424–27431.

Polyak, K., Weinberg, R.A., 2009. Transitions between epithelial and mesenchymal states: acquisition of malignant and stem cell traits. Nat. Rev. Cancer 9, 265–273.

Poy, F., Lepourcelet, M., Shivdasani, R.A., Eck, M.J., 2001. Structure of a human Tcf4-beta-catenin complex. Nat. Struct. Biol. 8, 1053–1057.

Qi, J., Yu, Y., Özlem Akilli Öztürk, Holland, J.D., Besser, D., Fritzmann, J., Wulf-Goldenberg, A., Eckert, K., Fichtner, I., Birchmeier, W., 2016. New Wnt/β-catenin target genes promote experimental metastasis and migration of colorectal cancer cells through different signals. Gut. 65(10) 1690–1701 http://dx.doi.org/10.1136/gutjnl-2014-307900.

Rhee, C.S., et al., 2002. Wnt and frizzled receptors as potential targets for immunotherapy in head and neck squamous cell carcinomas. Oncogene 21, 6598–6605.

Riethdorf, S., Pantel, K., 2008. Disseminated tumor cells in bone marrow and circulating tumor cells in blood of breast cancer patients: current state of detection and characterization. Pathobiology 75, 140–148.

Riethdorf, S., Wikman, H., Pantel, K., 2008. Review: biological relevance of disseminated tumor cells in cancer patients. Int. J. Cancer 123, 1991–2006.

Rudland, P.S., et al., 2002. Prognostic significance of the metastasis-associated protein osteopontin in human breast cancer. Cancer Res. 62, 3417–3427.

Sampietro, J., et al., 2006. Crystal structure of a beta-catenin/BCL9/Tcf4 complex. Mol. Cell 24, 293–300.

Sansom, O.J., et al., 2005. Cyclin D1 is not an immediate target of beta-catenin following Apc loss in the intestine. J. Biol. Chem. 280, 28463–28467.

Schneikert, J., Behrens, J., 2007. The canonical Wnt signalling pathway and its APC partner in colon cancer development. Gut 56, 417–425.

Segditsas, S., Tomlinson, I., 2006. Colorectal cancer and genetic alterations in the Wnt pathway. Oncogene 25, 7531–7537.

Shan, J., Shi, D.L., Wang, J., Zheng, J., 2005. Identification of a specific inhibitor of the dishevelled PDZ domain. Biochemistry 44, 15495–15503.

Shi, Y., et al., 2013. Expression of the epithelial-mesenchymal transition-related proteins and their clinical significance in lung adenocarcinoma. Diagn. Pathol. 8, 89.

Shtutman, M., Levina, E., Ohouo, P., Baig, M., Roninson, I.B., 2006. Cell adhesion molecule L1 disrupts E-cadherin-containing adherens junctions and increases scattering and motility of MCF7 breast carcinoma cells. Cancer Res. 66, 11370–11380.

Singhal, H., et al., 1997. Elevated plasma osteopontin in metastatic breast cancer associated with increased tumor burden and decreased survival. Clin. Cancer Res. 3, 605–611.

Slade, M.J., et al., 2009. Comparison of bone marrow, disseminated tumour cells and blood-circulating tumour cells in breast cancer patients after primary treatment. Br. J. Cancer 100, 160–166.

Smith, M.L., Hawcroft, G., Hull, M.A., 2000. The effect of non-steroidal anti-inflammatory drugs on human colorectal cancer cells: evidence of different mechanisms of action. Eur. J. Cancer 36, 664–674.

Su, Y., Ishikawa, S., Kojima, M., Liu, B., 2003. Eradication of pathogenic beta-catenin by Skp1/Cullin/F box ubiquitination machinery. Proc. Natl. Acad. Sci. U.S.A. 100, 12729–12734.

Suzuki, H., et al., 2004. Epigenetic inactivation of SFRP genes allows constitutive WNT signaling in colorectal cancer. Nat. Genet. 36, 417–422.

Takahashi-Yanaga, F., Sasaguri, T., 2007. The Wnt/beta-catenin signaling pathway as a target in drug discovery. J. Pharmacol. Sci. 104, 293–302.

Takemaru, K.I., Moon, R.T., 2000. The transcriptional coactivator CBP interacts with beta-catenin to activate gene expression. J. Cell Biol. 149, 249–254.

Taketo, M.M., 2004. Shutting down Wnt signal-activated cancer. Nat. Genet. 36, 320–322.

Toth, K., et al., 2004. An oncolytic adenovirus vector combining enhanced cell-to-cell spreading, mediated by the ADP cytolytic protein, with selective replication in cancer cells with deregulated wnt signaling. Cancer Res. 64, 3638–3644.

Trosset, J.Y., et al., 2006. Inhibition of protein-protein interactions: the discovery of druglike beta-catenin inhibitors by combining virtual and biophysical screening. Proteins 64, 60–67.

Uchikado, Y., et al., 2011. Increased slug and decreased E-cadherin expression is related to poor prognosis in patients with gastric cancer. Gastric Cancer 14, 41–49.

Ue, T., et al., 1998. Co-expression of osteopontin and CD44v9 in gastric cancer. Int. J. Cancer 79, 127–132.

van de Wetering, M., et al., 2002. The beta-catenin/TCF-4 complex imposes a crypt progenitor phenotype on colorectal cancer cells. Cell 111, 241–250.

Van der Flier, L.G., et al., 2007. The intestinal Wnt/TCF signature. Gastroenterology 132, 628–632.

Wagenaar-Miller, R.A., Gorden, L., Matrisian, L.M., 2004. Matrix metalloproteinases in colorectal cancer: is it worth talking about? Cancer Metastasis Rev. 23, 119–135.

Watson, A.J., 2006. An overview of apoptosis and the prevention of colorectal cancer. Crit. Rev. Oncol. Hematol. 57, 107–121.

Yang, J., et al., 2004. Twist, a master regulator of morphogenesis, plays an essential role in tumor metastasis. Cell 117, 927–939.

You, L., et al., 2004. Inhibition of Wnt-2-mediated signaling induces programmed cell death in non-small-cell lung cancer cells. Oncogene 23, 6170–6174.

Zhang, L., Gao, X., Wen, J., Ning, Y., Chen, Y.G., 2006. Dapper 1 antagonizes Wnt signaling by promoting dishevelled degradation. J. Biol. Chem. 281, 8607–8612.

# Index

'*Note*: Page numbers followed by "f" indicate figures and "t" indicate tables.'


## A

Acinar-ductal metaplasia (ADM), 118
Actinic keratosis, 199
Adenocarcinoma, 81
Adherens junctions (AJs), 4–5
Adjuvant chemotherapy, 186
Advanced breast cancer (ABC), 20–21
Androgen deprivation therapy (ADT), 36
Androgen receptor (AR) signaling pathway
  AR-V7, 37
  CRPC, 36
  DBD, 35–36
  functional domains, 35–36
  hinge region, 35–36
  LBD, 35–36
  NTD, 35–36
  RB, 36
  testosterone, 35
Angiogenesis, 239–240, 355–357
Animal models, 327f
Anthracyclines, 22
Anticytotoxic T lymphocyte associated antigen (Anti-CTLA-4), 192
Apparent diffusion coefficient (ADC), 184, 247
Aromatase inhibitors (AIs)
  classes of, 21
  nonsteroidal, 21
  steroidal, 21
Astrocyte elevated gene 1 (AEG-1), 85

## B

Barcelona clinic liver cancer (BCLC) staging system, 170
BCAAs. *See* Branched-chain amino acids (BCAAs)
Bevacizumab, 25, 109
Biochemical recurrence (BCR), 34
  diagnosis of, 48
Biomarkers, 99t–105t
  diagnostic, 106
  predictive, 106
  prognostic, 106
Bisphosphonates, 26, 269–270
Bladder cancer
  biomarkers, 184–185
  cell cycle regulators, 191
  clinical presentations, 186
  clinical prognosticators
    postchemotherapy prognosticators, 188–189
    prechemotherapy prognosticators, 188
  epidemiology
    ASR, 177–178, 178t
    global age adjusted death rates, 178–179, 179t–180t
    MIBC, 178–179
    nonmuscle invasive disease, 178–179
    progression, risk of, 178–179, 178t
  grading, 182, 182t
  immunotherapy
    anti-CTLA-4, 192
    PD-L1 inhibitors, 191–192
  management of
    alternative chemotherapeutic agents, 190
    GC, 189–190
    MVAC, 189
    neoadjuvant chemotherapy, 190
    radiosensitizing chemotherapy, 190
    standard chemotherapy regimes, 189
  positive pelvic nodes and metastatic disease
    lymphadenectomy levels, 186
    MIBC, 186
    neoadjuvant chemotherapy, 187
    perioperative chemotherapy, radical cystectomy, 187
    postchemotherapy lymphadenectomy, 187
    trimodal bladder-sparing therapy, 188
  prognostic factors, 185–186
  staging
    AJCC staging of, 181, 181t
    CIS, 181, 182f
    clinical stage, 179
    pathological staging, 179, 181f
    TNM classification system, 181
  staging tools
    CT scan, 182–183
    mpMRI, 184
    MRI, 182–183
    PET, 183–184
    TURBT, 183
    USPIO, 184
  surgery, role of
    extended lymph node dissection, 192
    palliative chemotherapy, 192
    palliative radiation, 192
    palliative surgery, 192
    patients, supportive measures in, 193
  tyrosine kinase receptor inhibitors, 191
Blastokinin. *See* Uteroglobin
Blood–brain barrier (BBB), 319–320
Blood–tumor barrier (BTB), 320
Bone metastasis, 8, 19
  bone formation, 44–45
  early detection, 264–265
  evidence of, 264f
  osteoblasts, 45–46
  osteolysis, 45
  pathophysiology of
    bone marrow stromal cells, 268–269
    dissemination process of, 267–268
    GAS6, 268–269
    mTOR inhibitors, 268–269

Bone metastasis *(Continued)*
survival, molecular mechanisms of, 268–269
tumor cell targets, osteblastic niche, 268
tumor colonization, 268–269
tumor dormancy, 268–269
RANKL, 44–45
treatment options
antiresorptive agents, 269–270
pain management, 270
types of
osteoblastic, 266–267
osteolytic, 265–266
uPA, 45–46
Bowen's disease, 199
Brain metastasis
biology
alternative vascularization methods, 320–321
brain metastasis formation, 318–319
brain microenvironment, 319–320
molecular mechanisms, 321
CNS, 317–318
diagnosis, 321–322
experimental model systems
brain derivative, 328
laboratory mouse models, 327
orthotopic injection models, 327–328
spontaneous brain metastasis models, 327–328
syngenic and xenograft transplant models, 327–328
incidence of, 317–318
preclinical findings
blood–tumor barrier patency, 329
dormancy, 329–331
prognostic scores, 323–324
treatment
multiple metastatic lesions (>3 metastases), 326
oligometastatic disease (1–3 metastases), 325–326
preventative treatment, 326
SRS, 324
surgical resection, 324
systemic therapy, recent advances in, 326
WBRT, 324
Branched-chain amino acids (BCAAs), 120
Breast cancer cells
bone metastatic, 268–269
early dissemination and colonization, 268
Breast cancer metastasis
CSCs, 17
dormancy, 16–17
DTCs and CTCs, 16
intrinsic subtypes, 17–18
metastatic dissemination
bone metastasis, 19
brain metastasis, 19–20
clinical symptoms, 18
liver metastasis, 19
locoregional recurrence, 18–19
lung metastasis, 19
peritoneal dissemination, 20
pleural metastasis, 19
microenvironment,
adipocytes and adipose progenitor cells, 16
postmenopausal breast cancer patients, 15–16
stromal fibroblasts, 16
prognostic factors, recurrence, 20
proliferation signals
ER, 14, 14f
HER family, 14–15, 15f
treatment for
bisphosphonate, 26
chemotherapy, 22–23
endocrine therapy, 21–22
metronomic chemotherapy, 23–24
molecular targeted therapy, 24–26
palliative care, 27–28
radiotherapy, 26–27
RANKL, 26
surgery, 27

## C

Cancer bone pain, 270
Cancer cells, 62
anoikis of, 63–64
Cancer metastasis
defined, 3
EMT
Snail and Zeb transcription factors, 6–7
TGF-β pathway, 7
TNF-α, 7
Twist1 induces, 6–7
metastatic cascade
AJs, 4–5
initiation of, 3–4
overview of, 3, 4f
TJ, 5
microRNAs and CSCs, 8
MMPs, 5–6
organ-specific metastasis, 7–8
perspectives, 8–9
Cancer stem cells (CSCs), 8, 17, 63, 69–71
concept, 125
Capecitabine, 23–24
Carboplatin, 23
Carcinoma in situ (CIS), 181
Castrate-resistant prostate cancer (CRPC), 34
β-Catenin, 40
11C-choline, 183–184
Cell adhesion molecules (CAMs), 41–42
Cell cycle pathways, 38–39
Central nervous system (CNS), 317–318
Cervical cancer metastasis
dissemination, routes of
direct local invasion, 81–82
hematogenous dissemination, 82–83
lymphatic dissemination, 82
high morbidity and mortality, 78
histopathology, 81
HPV vaccination, 77–78
major molecular factors
EGFR, 84
HOX genes, 83
miRNAs, 86–87
PDGFR, 84–85
PI3K/AKT/mTOR signaling pathway, 83
VEGF, 85
National data, 78
Pap screening programs, 77–78
pathogenesis, 80–81
risk factors, 79f
age, first birth and multiparity, 80
cigarette smoking, 80
family history, 80
high-risk sexual partner, 79
HPV infections, 78
immunosuppression, 79
low socioeconomic status, 80
multiple sexual partners, 79
oral contraceptives, 80

sexual activity, 78
sexually transmitted diseases, 79
vulvar/vaginal squamous intraepithelial neoplasia, 79
Chemotherapy
anthracyclines, 22
capecitabine, 23
eribulin, 23
platinum salts, 23
taxanes, 22–23
vinorelbine, 23
Chimeric F-box proteins (CFPs), 387–388
Chondromodulin-I (ChM-I), 365
Circulating endothelial progenitor cells (CEPs), 23–24
Circulating tumor cells (CTCs), 16
Cirrhosis, 165, 173
Cisplatin, 23
Claudin, 5
Colorectal cancer metastasis
challenges and probable solutions
immunotherapy, 111–112
personalized medicine, 111
main sites of
liver, 107–108
lung, 109
peritoneum, 108
mechanistic pathways, 95–96
molecular aspects of
biomarkers for, 98–106, 99t–105t
gene expression level, 97
signaling pathways, involvement of, 97–98
Vogelstein's model, 96–97
multistep process, 96
prevention strategies, 95
pulmonary metastases, 307
survival rate of, 95
therapeutic options for
chemotherapy, 110
CT scan, 109
5-FU monotherapy, 109
surgical intervention, 109–110
Colorectal liver metastasis (CLM), 107, 286–287
natural history of, 286–287
wedge resection for, 278f
Compared chemotherapy (CHT), 154–155
Complete axillary lymph node dissection (CALND), 236–237
Complete resection (CR), 227
Computed tomography (CT) scan, 183, 247
Cooperative Osteosarcoma Study Group (COSS), 224–225
CRPC. *See* Castrate-resistant prostate cancer (CRPC)
CSCs. *See* Cancer stem cells (CSCs)
Cutaneous squamous cell carcinoma (cSCC)
chemotherapy agents, 214
13-cis-retinoic acid, 214
clinical features, 200t
genetic disorders, 202–203
immunosuppression, 200–202
lesion site, 203–204
lesion size, 203
preexisting lesion, 203
recurrence, 204
electrochemotherapy, 213–214
high risk, histologic features of
histologic differentiation, 208
histologic subtype, 208
lymphovascular invasion, 206–208
PNI, 205–206
tumor thickness, 205
HPV, 204–205
lymph node metastasis, 213
molecular high-risk markers
cytogenetic alterations, 209
EGFR, 209
EMT, 209–210
IKKalpha, 210–211
radiotherapy, 213
sentinel node biopsy, 213
staging systems, 211–212, 211t
treating high-risk, 212–213
Cyclin-dependent kinases (CDK), 36
Cyclooxygenase (COX), 381–382
Cyclooxygenase 2 (COX-2), 86
Cytotoxic T-lymphocyte antigen-2β (CTLA-2β), 360–361

D

Decidual stromal cells (DSCs), 363
2-Deoxy-2-fluoro[18F]-d-glucose (FDG), 247–248
Denosumab, 26, 270
Developmental pathways
motility and migration, 340
neoangiogenesis, 340–341
notch signaling and metastasis
metastatic cascade, 341
notch-based therapeutic strategies, 342
signaling molecules, notch family of, 341
targeting notch ligands and receptors, 342–343
prometastatic changes
EMT, 339–340
resistance to anoikis, 338–339
Wnt and Hedgehog targeting, 343–346
Diffusion weighted imaging (DWI), 247
Disease-free interval (DFI), 20
Disseminated tumor cells (DTCs), 16
Distant metastases, 310
DNA repair pathways, 39
Docetaxel, 22
Doxorubicin, 22
Ductal carcinoma in situ (DCIS), 249

E

EBRT. *See* External beam radiation therapy (EBRT)
E-cadherin, 4–5, 41–42, 210
ECM-degrading enzymes, inhibitors of
CTLA-2β, 360–361
cystatin C, 360–361
MMP-2, 358
PAI-1, 360
TIMPs, 358–360
uPA, 360
EHS. *See* Extrahepatic spread (EHS)
Embryo implantation
and tumor metastasis
angiogenesis, 355–357
immune escape, 357–358
migration and invasion, 357
ECM components, 354
trophoblast invasion, 354–355
cytokines, 361–363
ECM-degrading enzymes, inhibitors of, 358–361
glycoproteins, 365
metastasis suppressor genes, 363–364
serine proteases, HtrA family of, 364–365
EMT. *See* Epithelial–mesenchymal transition (EMT)

Endocrine therapy
AIs, 21
SERDs, 21–22
SERMs, 21
Endoscopic bronchial ultrasonography (EBUS), 300
Endoscopic ultrasonography (EUS), 119–120
Endothelin-1 (ET-1), 266–267
Epidermal growth factor (EGF), 43–44
Epidermal growth factor receptor (EGFR), 84, 209
Epidermolysis bullosa (EB), 202
Epirubicin, 22
Epithelial to mesenchymal transition (EMT), 4–5, 42, 63, 67–69, 209–210, 339–340
Snail and Zeb transcription factors, 6–7
TGF-β pathway, 7
TNF-α, 7
Twist1 induces, 6–7
Eribulin, 23
EUS. *See* Endoscopic ultrasonography (EUS)
Everolimus, 26
External beam radiation therapy (EBRT), 171–172
Extracellular matrix (ECM), 5–6, 62, 354
Extrahepatic spread (EHS), 165

F
Fibroblast growth factor receptor (FGFR3), 184
Fibronectin (FN), 64
5-fluorouracil (5-FU) therapy, 23, 98–106
Focal adhesion kinase (FAK), 85
Food and Drug Administration (FDA), 247
Fulvestrant, 22

G
Gamma secretase inhibitors (GSIs), 342
Gap junctions (GJs), 3–5
Gastric cancer metastasis
epidemiology and screening programs
*H. pylori*, 138
radiographic and endoscopic screening, 138
serologic testing, 138
survival rates, 137–138
symptoms, 139
patterns, 139–141
hematogenous spread, 140
hepatic metastases, 140
KT, 140
Lauren's criteria, 140
peritoneal spread, 139–140
diagnostics, importance of, 141
18-F-FDG PET/CT, GC diagnostics, 151
current role of, 141
distant metastasis (M1) and prognostic role, 144–145, 147t–148t
improvement and development of, 151–152
laparoscopy, 149
limitations of, 146–149
MRI, 151
preoperative staging of, 142–144
radical treatment, 150–151
prognostic and predictive factors, 152–153
biomarkers
CA, 153
CEA, 153
lncRNAs, 154
metabolomisc, 154
miRNAs, 153–154
numerous genetic and epigenetic alterations, 153
treatment methods of, 154–155
Gemcitabine and cisplatin (GC), 189–190
Genetically engineered mouse models (GEMMs), 128–129
Glassy cell tumors, 82–83
Graded prognostic assessment (GPA), 26–27
Growth arrest-specific 6 (GAS6), 268–269

H
Heat shock proteins (HSP), 191
Hedgehog signaling
Gli proteins, 345–346
Hedgehog-based therapeutic strategies, 346
Hematologic malignancies, 201
Hematopoietic stem cell transplant (HSCT), 201–202
Hematopoietic stem cells (HSCs), 267–268
Hepatic metastasis
adjuvant postoperative, 292–293
alternative therapies and adjuvants, 289
arterial embolization, 291
clinical presentation, 281–282
CT scan, 283
fine needle aspiration biopsy, 284
hepatic functional and surgical anatomy
caudate lobe, 280
liver segments, 280–281, 281f
quadrate lobe, 280
hepatic physiology and cancer biology
cancer cells, 279–280
cell adhesion receptors, 279–280
healthy liver, centrilobular vein in, 278, 278f
hepatocytes, 278
inflammatory active factors, 279–280
Kupffer cells, 279–280
neoplastic cells, 279–280
sinusoidal endothelial cells, 279
stellate cells, 278–279
liver metastases
colorectal tumors, 286–287
sarcoma, 287–288
metastases, important site of, 277
MRI, 283–284
neoadjuvant preoperative, 292–293
NETs, 287
PET, 284
portal embolization, 291, 292f
primary sites, 288–289
radiofrequency and chemical ablation, 289–291
surgery, 285–286
US, 282–283
Hepatitis B virus (HBV), 164–165
Hepatitis C virus (HCV), 163–165
Hepatocellular carcinoma (HCC)
diagnosis
biopsy, 169
CT, 167–168
MRI imaging, 167–168
PET, 168
US, 167

epidemiology
EHS, 165–167
incidence, 163–164
TNM staging for, 165, 166t
underlying disease/risk factors, 164–165
prognosis and staging, 169f, 170–171
treatment
EBRT, 171–172
IFNa functions, 172
RFA, 171
sorafenib, 172–173
sunitinib, 173
surgical options, 171
systemic chemotherapy, 172
Highly active antiretroviral therapy (HAART), 202
High-resolution computed tomography (HRCT), 224
Homeobox (HOX) genes, 83
HOX Antisense Intergenic RNA (HOTAIR), 83
Human immunodeficiency virus (HIV) infection, 202
Human papillomavirus (HPV), 77–78, 204–205
Hypoxia-inducible factors (HIFs), 341

## I

Immunosuppression
hematologic malignancies, 201
HIV infection, 202
HSCT, 201–202
solid organ transplant, 201
Interferon alfa (IFNa), 172
Invasion, 62
Isolated tumor cell clusters (ITCs), 254

## K

Karnofsky performance status (KPS), 322–323
Kisspeptin (KiSS-1), 364

## L

Laparoscopy, 149
Lapatinib, 24
Liver cancer
colorectal tumors, 286–287
hepatic metastasis. *See* Hepatic metastasis
sarcoma, 287–288
Liver resection, 170
Liver transplantation, 168
Lung cancer metastasis
CSCs, 69–71
EMT, 67–69
information, 61
mechanism of, 62–63
regulation, molecular mechanism in
αvβ3 integrin, 64–65
anoikis, 63–64
Cav-1, 66
chemokines, 67
CXCR4, 67
ECM, 66–67
EGFR, 63–65
FAK activity, 65–66
FN receptor, 64
HER2, 64
integrin α5β1, 64–65
migration and invasion, 64–65
MMPs, 66–67
SFKs, 63–64
TGF-β, 65
Lung metastasis. *See* Pulmonary metastasis
Lung, 8
Luteinizing hormone releasing hormone (LHRH), 21
Lymph node metastasis (LNMets), 46, 152, 227
anatomic and mechanical logic, 235
clinical aspects of
(PET)-CT, 247–248
clinical diagnosis, 244
CT scans, 247
lymphoscintigraphy, 248
mammography, 245
MRI, 246–247
US, 245–246, 246f
clinical importance of, 236
history
MRM, 237
radical operations, 237
sentinel node concept, 237–238
mechanical (anatomic) aspects
angiogenesis, 239–240
lymphangiogenesis, 240
lymphatic pathways, 241–242
mechanobiology and tumor progression, 242
SLN microenvironment, 242–244
molecular aspects
CCR7-CCL21, 238–239
chemokines, 239
TGFβIp, 238–239
VEGF-C, 238
pathology
gross autopsy, 250–251
gross evaluation, 251
lymph node microanatomy, 251
microscopic examination, 252–255
prevention of, 236
surgery
CALND, 248–249
radiation, 250
radical mastectomy, 248
SLN biopsy, 249
systemic chemotherapy, 250
surgical management of, 236–237
Lymphangion systole, 241
Lymphatic vessel density (LVD), 82
Lymphovascular invasion
lymphatic vessels, new markers for, 206
podoplanin, 207–208
Prox1, 207
VEGF-C, 206–207
Lysosome-associated protein transmembrane 4 beta (LAPTM4B), 86

## M

α2-Macroglobulin (α2-MG), 358–360
Magnetic resonance imaging (MRI), 119–120, 246–247
Major histocompatibility complex (MHC) class I molecules, 357–358
Mammalian target of rapamycin (mTOR), 268–269
Mammography, 245
Matrix metalloproteinases (MMPs), 5–6, 66–67
Maximum tolerable dose (MTD), 23–24
Melanoma, 244
Metastatic bladder cancer
biomarkers, 184–185
cell cycle regulators, 191
clinical presentations, 186
clinical prognosticators
postchemotherapy prognosticators, 188–189
prechemotherapy prognosticators, 188

Metastatic bladder cancer *(Continued)*
epidemiology
ASR, 177–178, 178t
global age adjusted death rates, 178–179, 179t–180t
MIBC, 178–179
nonmuscle invasive disease, 178–179
progression, risk of, 178–179, 178t
grading, 182, 182t
immunotherapy
anti-CTLA-4, 192
PD-L1 inhibitors, 191–192
management of
alternative chemotherapeutic agents, 190
GC, 189–190
MVAC, 189
neoadjuvant chemotherapy, 190
radiosensitizing chemotherapy, 190
standard chemotherapy regimes, 189
positive pelvic nodes and metastatic disease
lymphadenectomy levels, 186
MIBC, 186
neoadjuvant chemotherapy, 187
perioperative chemotherapy, radical cystectomy, 187
postchemotherapy lymphadenectomy, 187
trimodal bladder-sparing therapy, 188
prognostic factors, 185–186
staging
AJCC staging of, 181, 181t
CIS, 181, 182f
clinical stage, 179
pathological staging, 179, 181f
TNM classification system, 181
staging tools
CT scan, 182–183
mpMRI, 184
MRI, 182–183
PET, 183–184
TURBT, 183
USPIO, 184
surgery, role of
extended lymph node dissection, 192
palliative chemotherapy, 192
palliative radiation, 192
palliative surgery, 192
patients, supportive measures in, 193
tyrosine kinase receptor inhibitors, 191
Metastatic breast cancer (MBC), 13–14
bisphosphonate, 26
chemotherapy
anthracyclines, 22
capecitabine, 23
eribulin, 23
platinum salts, 23
taxanes, 22–23
vinorelbine, 23
CSCs, 17
dormancy, 16–17
DTCs and CTCs, 16
endocrine therapy
AIs, 21
SERDs, 21–22
SERMs, 21
intrinsic subtypes, 17–18
metastatic dissemination
bone metastasis, 19
brain metastasis, 19–20
clinical symptoms, 18
liver metastasis, 19
locoregional recurrence, 18–19
lung metastasis, 19
peritoneal dissemination, 20
pleural metastasis, 19
metronomic chemotherapy, 23–24
microenvironment
adipocytes and adipose progenitor cells, 16
postmenopausal breast cancer patients, 15–16
stromal fibroblasts, 16
molecular targeted therapy
bevacizumab, 25
everolimus, 26
lapatinib, 24
pertuzumab, 25
T-DM1, 24–25
trastuzumab, 24
palliative care, 27–28
prognostic factors, recurrence, 20
proliferation signals
ER, 14, 14f
HER family, 14–15, 15f
radiotherapy, 26–27
RANKL, 26
surgery, 27
Metastatic colorectal cancer (mCRC). *See* Colorectal cancer metastasis
Metastatic cutaneous squamous cell carcinomas (cSCC)
chemotherapy agents, 214
13-cis-retinoic acid, 214
clinical features, 200t
genetic disorders, 202–203
immunosuppression, 200–202
lesion site, 203–204
lesion size, 203
preexisting lesion, 203
recurrence, 204
electrochemotherapy, 213–214
high risk, histologic features of
histologic differentiation, 208
histologic subtype, 208
lymphovascular invasion, 206–208
PNI, 205–206
tumor thickness, 205
HPV, 204–205
lymph node metastasis, 213
molecular high-risk markers
cytogenetic alterations, 209
EGFR, 209
EMT, 209–210
IKKalpha, 210–211
radiotherapy, 213
sentinel node biopsy, 213
staging systems, 211–212, 211t
treating high-risk, 212–213
Metastatic prostate cancer. *See* Prostate cancer metastasis
Methotrexate, vinblastine, doxorubicin, and cisplatin (MVAC), 189
Micropapillary variant bladder cancer, 185
MicroRNA (miRNAs), 69, 86–87
Microsatellite stable (MSS), 106
Microscopic examination, lymph node( LN)
extranodal/extracapsular tumor invasion, metastasis with, 253, 253f
LNMets, specific patterns of, 253–254
lymph node metastasis, dispersed pattern of, 252, 252f
micrometastasis vs. macrometastasis, 255

perinodal adipose and soft tissues, tumor nodules in, 253, 253f
RLNs, 255
SLN, 254
tumor cells, single or multiple clusters of, 252, 252f
tumor emboli, 252
Migration, 62
Mitogen-activated protein kinase (MAPK), 265–266
MMPs. *See* Matrix metalloproteinases (MMPs)
Modified radical mastectomy (MRM), 237
Molecular targeted therapy
bevacizumab, 25
everolimus, 26
lapatinib, 24
pertuzumab, 25
T-DM1, 24–25
trastuzumab, 24
MTD. *See* Maximum tolerable dose (MTD)
Multiparametric MRI (mpMRI), 184
Muscle invasive bladder cancer (MIBC), 178–179

## N
Nab-paclitaxel therapy, 22–23
National Comprehensive Cancer Network (NCCN), 141, 154–155
Nested variant urothelial cancer, 185
Netrin-1, 106
Neuroendocrine cell carcinomas, 81–83
Neuroendocrine nonurothelial variant, 185
Neuroendocrine tumors (NETs), 287
Nonalcoholic fatty liver disease (NAFLD), 165
Nonalcoholic steatohepatitis (NASH), 165
Non-small-cell lung cancer (NSCLC), 61, 385
CXCR4, 67
EMT, 69
integrin α5β1, 64–65
Notch ligand and receptor, 69
Nonsteroidal anti-inflammatory drugs (NSAIDs), 381
Nonurothelial bladder cancer, 185–186
Notch intracellular domain (NICD), 341
Notch-based therapeutic strategies, 342
Nuclear factor kappa B (NF-κB), 85

## O
Organ transplant recipients (OTRs), 201, 201f
Osteoblastic bone metastasis, 266–267
Osteolytic bone metastasis, 265–266
Osteosarcoma (OS) metastasis
clinical features
incidence, 223–224
detection, 224
prognostic factors, 224–227, 226t
surgical treatment, 227–228
VATS, 228
radiotherapy, 228
emerging treatment option, 228–229
Overall survival (OS), 16, 27, 46

## P
p53 protein, 184
Paclitaxel, 22
Paget disease, 224
Palliative care, 27–28
Pancreatic ductal adenocarcinoma (PDAC)
ADM, 118
causes of, 117–118
computational model, 130
genetics and biology of, 118–119
in vivo experimental models
GEMMs, 128–129
organoid models, 129
PDX, 127–128
KRAS mutations, 118
management of
drugs targeting tumor metabolism, 121–122
immunotherapy and cancer vaccines, 122
laboratory tests and radiological imaging, 120
next-generation sequencing, 121
presenting symptoms, 120
screening, 119–120
staging, 120–121
stromal compartment, 122
treatment and clinical trials, 121
metastatic cells, 129–130
molecular mechanisms driving metastasis
seed, 123–125
soil, 125–126
TGF-β pathway, 126–127
tumor-derived exosomes and metastatic niche, 127
preneoplastic lesions, 118
Panitumumab, 109
Papanicolaou (Pap) test screening, 77–78
Parathyroid hormone-related protein (PTHrP), 265–266
Patient-derived xenografts (PDX), 127–128
Perineural invasion (PNI), 205–206
Peritoneal metastasis (PM), 108
Pertuzumab, 25
PET. *See* Positron emission tomography (PET)
PFS. *See* Progression free survival (PFS)
Phospholipase C (PLC), 344
Plasminogen activator inhibitors (PAI-1 and -2), 360
Platelet-derived growth factor receptors (PDGFR), 84–85
Platinum salts, 23
Platinum-based chemotherapy, 189
PM. *See* Peritoneal metastasis (PM)
Poly ADP-ribose polymerase (PARP), 39
Portal vein invasion (PVI), 170
Positron emission tomography (PET), 168, 183–184
Positron emission tomography (PET)-CT, 247–248
Prognostic indices (PIs), 152–153
Progression free survival (PFS), 16, 26, 145
Prophylactic cranial irradiation (PCI), 326
Prostate cancer metastasis
cellular biology of
AR signaling pathway, 35–37
cell cycle pathways, 38–39
coregulator factors, 40–41
DNA repair pathways, 39
PI3K/Akt-signaling pathway, 37–38
Wnt pathways, 40
epidemiology and overview of
CRPC, 34
natural history of, 33
osteoblastic bony metastases, 33–34, 34f
therapeutic options, 35

Prostate cancer metastasis *(Continued)*
metastatic cascade, 41–53
abiraterone and enzalutamide, 50
ADT, 49
bone metastases, 44–46
cabazitaxel, 50
cell adhesion, loss of, 41–42
chemokines, 43–44
clinical considerations in, 47–48
CTCs, 42
cysteine proteinases, 42
developments, 52
docetaxel, 50–51
EGF, 43–44
GnRH agonist, 49
imaging modalities, 52, 52f
immune system, 42–43
integrins, 42
investigational therapies, 52–53
liquid biopsy and precision medicine, 52
lymph node metastases, 46
lymphangiogenesis, 41
lymphatic system, 43–44
medical castration, 49
MMPs, 42
neovasculature, 41
PAR-1, 42–43
serine proteinases, 42
surgical castration, 49
symptomatic metastatic disease, treatment of, 51–52
visceral metastases, 46–47
Prostate-specific antigen (PSA), 33, 48
Protease-activated receptor 1 (PAR-1), 42–43
Protein kinase C epsilon (PKCε), 64–65
Protein tyrosine phosphatase receptor type D (PTPRD), 209
PTEN mutation, 38
Pulmonary metastasis, 225–227
development of, 297
diagnosis of, 299–300
established criteria for, 298–299
indication for, 298, 298t
lymph node involvement, 305–306
preoperative exams
complete resection of, 300–301
extrathoracic metastases, exclusion of, 300
peripheral lesions, 301–302
primary tumor, control of, 300
resection of, 297
radiological ablation, 304
radiotherapy, 305
recurrent, resection of, 306–307
surgical approach
clamshell incision, 303
postoperative pain, 303
VATS, 303–304
surgical metastasectomy approach, 299
surgical resection, extent of, 302
surgical techniques, improvement of, 298–299
survival factors, prognoses of, 298
tumor types
breast cancer, 310
colorectal cancer, 307–308
head and neck tumor, 310
melanoma, 309
nonseminomatous testicular germ tumor, 309–310
osteosarcoma, 308–309
renal cell carcinoma, 309
soft tissue sarcoma, 308
PVI. *See* Portal vein invasion (PVI)

## Q

Quality of life (QOL), 20–21, 27

## R

Rab coupling protein (RCP), 64–65
Radiation Therapy Oncology Group (RTOG), 323
Radical prostatectomy (RP), 46
Radiofrequency ablation (RFA), 109, 171, 228–229
liver metastases, aspects, 289
Radiotherapy, 26–27
Ramucirumab, 154–155
Receptor activator of nuclear factor kappa B-Ligand (RANKL), 26, 265–266
Regional lymph nodes (RLNs), 235, 255
Retinoblastoma (RB), 36
Rhabdomyosarcoma, 81

## S

Samarium-153 ethylenediaminetetra-methylene phosphonic acid (153Sm-EDTMP), 229
SBRT. *See* Stereotactic body radiotherapy (SBRT)
Seed and soil theory
CSCs concept, 125
genetic events
CDKN2A, 123
KRAS activating mutations, 123
SMAD4, 123–124
TP53, inactivating mutations in, 123–124
loss of differentiation and EMT, 124–125
tumor microenvironment, 125–126
Selective estrogen receptor modulators, 21
Semaphorin 4D (Sema4D), 85
Sentinel lymph node (SLN), 235, 254
Sentinel lymph node biopsy (SLNB), 237–238
Sipuleucel-T, 50
Skin cancer. *See* Cutaneous squamous cell carcinoma (cSCC)
Skip metastasis, 227
Small cell carcinomas, 81
Small molecule inhibitors
CBP–β-catenin complex, 386–387
Tcf/β-catenin complex, 385–387
Small-cell lung cancer (SCLC), 61, 326
Solid organ transplant, 201
Sonic hedgehog (SHH), 122
Sorafenib, 172–173
Squamous cell carcinoma, 81
Squamous variant, 185–186
Src family kinases (SFKs), 63–64
Stereotactic body radiotherapy (SBRT), 228, 305
Stereotactic radiosurgery (SRS), 26–27, 324
Stereotactic radiotherapy (SRT), 26–27
Steroid receptor coactivator (SRC), 40–41
Stromal cell-derived factor 1 (SDF-1), 267–268
Stromal cells, 5–6
Stromal-interaction molecule 1 (STIM1), 86
Syngenic transplant models, 327–328

## T

T2-weighted imaging (T2WI), 247
TACE. *See* Transarterial chemoembolization (TACE)
Tamoxifen, 21, 28
TAMs. *See* Tumor associated macrophages (TAMs)

Taxanes, 22–23
TGF-β. *See* Transforming growth factor-β (TGF-β)
Thoracoscopy, 300–303
Tight junctions (TJs), 5
Tissue inhibitors of metalloproteinases (TIMPs), 358–360
Transarterial chemoembolization (TACE), 170
Transforming acidic coiled-coil protein family (TACC3), 84
Transforming growth factor-β (TGF-β), 5–6, 65
Transforming growth factor-β induced protein (TGFβIp), 238–239
Trastuzumab, 23–24, 154–155
Trastuzumab emtansine (T-DM1), 24–25
Trimodal bladder-sparing therapy, 188
Tumor associated macrophages (TAMs), 126
Tumor necrosis factor alpha (TNF-α), 5–6
Tumor-node-metastasis (TNM), 181
Turcot's syndrome, 378–379

U
Ultrasmall superparamagnetic particles of iron oxide (USPIO), 184
Ultrasound (US), 245–246, 246f, 282–283
Ultraviolet radiation (UVR), 199
Urokinase-type plasminogen activator (uPA), 45–46, 360
Urothelial bladder cancer, 185–186
Urothelial cancer, 177
  bladder cancer. *See* Bladder cancer
  incidence of, 177–178
Uteroglobin, 362–363

V
Vascular endothelial growth factor (VEGF), 25
  AEG-1 expression, 85
  CHI3L1, 85
  COX-2, 86
  FAK, 85
  Fibulin-4, 85
  Gal-3, 85
  LAPTM4B, 86
  Sema4D expression, 85
  STIM1, 86
  WNT1, 85–86
Vascular endothelial growth factor C (VEGF-C), 82
Video-assisted thoracoscopic surgery (VATS), 228
Vinflunine, 190
Vinorelbine, 23
Virchow's node, 244
Visceral metastases
  intracranial metastases, 47
  liver metastases, 47
  OS, 46
  pulmonary metastases, 47
Vogelstein's model, 96–97

W
Whole brain radiotherapy (WBRT), 26–27, 324
Wnt inhibitory factor 1 (WIF1), 85–86
Wnt signaling pathway
  APC gene, 378
  Ca2+ pathway, 378
  canonical Wnt/β-catenin pathway, 343, 375
  colorectal cancer, 378
  E-cadherin, 344
  EMT
    CSCs, 380–381
    E-cadherin, 379
    miRNA, 380
    OPN, 380
    Slug expression, 379–380
    Snail expression, 379–380
    TCF4, 380
    Twist expression, 379–380
  noncanonical Wnt signaling pathways, 344–345
  PCP pathway, 378
  protein knockdown strategies, 387–388
  stromal cells, 378
  role for, 344
  therapeutic strategies
    antibody-based strategies, 385
    NSAIDs, 381
    small molecule inhibitors, 385–387
    vitamin A and D, 382–385
  viral-based therapies
    adenovirus therapy, 389
    genetic strategies, 389–390
    immunotherapeutic approaches, 388–389
    oncolytic viruses, 388–389
    Tcf-4 transcription factor, 389
  Wnt/calcium pathway, 344

X
Xenograft transplant models, 327–328

Kolibri 360 GmbH
Ferdinand-Jühlke-Straße 7
99095 Erfurt, Deutschland
produktsicherheit@kolibri360.de